Nonlinear Finite Element Analysis in Structural Mechanics

Proceedings of the Europe-U.S. Workshop
Ruhr-Universität Bochum, Germany,
July 28-31, 1980

Editors
W. Wunderlich, E. Stein, and K.-J. Bathe

With 272 Figures

Springer-Verlag Berlin Heidelberg GmbH

Prof. Dr.-Ing. W. WUNDERLICH
Institut für Konstruktiven Ingenieurbau
Ruhr-Universität Bochum
Universitätsstraße 150, 4630 Bochum, Germany

Prof. Dr.-Ing. E. STEIN
Fachbereich Bauingenieur- und Vermessungswesen
Universität Hannover
Callinstraße 32, 3000 Hannover, Germany

Prof. K.-J. BATHE, Ph. D.
Department of Mechanical Engineering, MIT
Cambridge, MA 02139, USA

ISBN 978-3-642-81591-1 ISBN 978-3-642-81589-8 (eBook)
DOI 10.1007/978-3-642-81589-8

Originally published by Springer-Verlag Berlin Heidelberg New York in 1981
Softcover reprint of the hardcover 1st edition 1981

2061/3020 - 543210

Preface

With the rapid development of computational capabilities, nonlinear finite element analysis in structural mechanics has become an important field of research. Its objective is the realistic assessment of the actual behavior of structures by numerical methods. This requires that all nonlinear effects, such as the nonlinear characteristics of the material and large deformations be taken into account. The activities in this field being worldwide, direct interaction between the various research groups is necessary to coordinate future research and to overcome the time gap between the generation of new results and their appearance in the literature.

The first U.S.-Germany Symposium was held in 1976 at the Massachusetts Institute of Technology. Under the general topic "Formulations and Computational Algorithms in Finite Element Analysis" it provided an opportunity for about 20 researchers from each country to present lectures, hold discussions, and establish mutual contacts. The success of this first symposium was so encouraging that it seemed natural to organize a second bilateral meeting, this time in Germany, and to invite researchers from other European countries as well.

The purpose of the Europe-U.S. Workshop held in July 1980 in Bochum was to exchange new results of research in the field of nonlinear finite element analysis with emphasis on structural problems. It was felt that giving the meeting a workshop character would encourage thorough and active discussions and establish new or tighten old contacts between individuals and research organizations. The organizers also hoped that the advantages of a workshop with a relatively small group of researchers - representing most universities and organizations working in the field - would outweigh some shortcomings lying in the limitation of the number of participants. On the

European side the organizing committee consisted of W. Wunderlich of the Ruhr-University, Bochum, Chairman, E. Stein of the Technical University, Hannover and O.C. Zienkiewicz of University College of Wales, Swansea. On the U.S. side the organizing committee consisted of K.J. Bathe of MIT, Cambridge, Chairman, J.T. Oden of the University of Texas, Austin and S. Key of Sandia Laboratories, Albuquerque. Financial support was provided by the Deutsche Forschungsgemeinschaft and the U.S. National Science Foundation. The workshop was held at the Ruhr-University, Bochum during the week of July 28 - 31, 1980 with 15 participants from the U.S. and 30 participants from various European countries. In all, 38 lectures were given and supplemented by numerous discussions. This volume contains the papers presented at the symposium.

The finite element method has been created as a numerical tool for the analysis of engineering problems. For its full development and application to nonlinear problems knowledge from a number of disciplines is important: continuum mechanics, materials science, numerical analysis, functional analysis and computer science. Use of this knowledge and an understanding of the various, sometimes diverse, viewpoints prevalent in these fields is indispensible if continued success in the development of nonlinear finite element methods is to be achieved. It is in this spirit that the participants gathered at the workshop to exchange their ideas and experiences and to report on their most recent research results. We believe that the papers presented there, and compiled in this book, give a true representation of the state-of-the-art of nonlinear finite element analysis in structural mechanics.

The editors wish to thank all participants for their contributions and Dr. Obrecht for his help in organizing the meeting. The financial assistance of the Deutsche Forschungsgemeinschaft and the U.S. National Science Foundation is most gratefully acknowledged, and thanks are extended to Springer-Verlag for a speedy publication of these proceedings.

October 1980

W. Wunderlich
E. Stein
K.J. Bathe

Contributors

S.N. Atluri, H. Murakawa, School of Civil Engineering, Georgia Institute of Technology, Atlanta, Georgia 30332, USA.

I. Babuska, Institute for Physical Science and Technology, University of Maryland, College Park, MD 20742, USA.

J. Banovec, University Edvard Kardelj of Ljubljana, Faculty for Architecture, Civil Engineering and Survey, Yugoslavia.

K.J. Bathe, L.W. Ho, Department of Mechanical Engineering, Massachusetts Institute of Technology, Cambridge, Massachusetts 02139, USA.

T. Belytschko, Department of Civil Engineering, Northwestern University, Evanston, Illinois 60201, USA.

P.G. Bergan, Division of Structural Mechanics, The Norwegian Institute of Technology, Trondheim, Norway.

J.F. Besseling, Laboratory for Engineering Mechanics, University of Technology Delft, The Netherlands.

F. Brezzi, Istituto di Analisi Numerica del C.N.R. and Istituto di Matematica Applicata dell'Università di Pavia 27100-Pavia, Italy.

K. Brink, W.B. Krätzig, Ruhr-Universität Bochum, Institut für Konstruktiven Ingenieurbau III, 4630 Bochum, West-Germany.

G.F. Carey, Texas Institute for Computational Mechanics, University of Texas, Austin, Texas 78712, USA.

L. Corradi, G. Maier, Department of Structural Engineering, Technical University (Politecnico), Milano, Italy.

M. A. Crisfield, Transport and Road Research Laboratory, Crowthorne, Berkshire, U.K..

O.M. Eidsheim, P.K. Larsen, Department of Civil Engineering, The Norwegian Institute of Technology, Trondheim, Norway.

R. Glowinski, P. Le Tallec, V. Ruas de Barros, Paris VI University, L.A. 189, Tour 55.65, 75230 Paris, France.

T.J.R. Hughes, W.K. Liu, I. Levit, Division of Engineering and Applied Science, California Institute of Technology, Pasadena, California 91125, USA.

C. Johnson, R. Scott, Chalmers University of Technology, Department of Computer Sciences, 402 20 Göteborg 5, Sweden.

J.J. Kalker, H.J.C. Allaert, J. de Mul, Delft University of Technology, Delft, The Netherlands.

H.A. Kamel, J.M. Tan, Aerospace and Mechanical Engineering Department, University of Arizona, Tucson, Arizona 85721, USA.

D. Karamanlidis, A. Honecker, K. Knothe, Technische Universität Berlin, Institut für Mechanik, Straße des 17. Juni 135, 1000 Berlin 12, West Germany.

S.W. Key, C.M. Stone, R.D. Krieg, Applied Mechanics Division, Sandia Laboratories, Albuquerque, N.M. 87115, USA.

B.H. Kröplin, H. Duddeck, Technische Universität Braunschweig, Institut für Statik, Beethovenstr. 51, 3300 Braunschweig, West Germany.

D.S. Malkus, Department of Mathematics, Illinois Institute of Technology, Chicago, Illinois 60616, USA.

G. Mehlhorn, D. Klein, Technische Hochschule Darmstadt, Institut für Massivbau, Alexanderstr. 5, 6100 Darmstadt, West Germany

M.J. Mikkola, H.S. Sinisalo, The Academy of Finland, Department of Civil Engineering, Helsinki University of Technology, 02150 Espoo 15, Finland.

H.D. Mittelmann, Universität Dortmund, Abteilung Mathematik, August-Schmidt-Str. 1, 4600 Dortmund, West Germany.

A.K. Noor, J.M. Peters, George Washington University Center at NASA Langley Research Center, Hampton, Virginia 23665, USA.

J.T. Oden, Texas Institute for Computational Mechanics, The University of Texas, Austin, Texas 78712, USA.

P.J. Pahl, Technische Universität Berlin, Institut für Allgemeine Bauingenieurmethoden, Straße des 17. Juni 135, 1000 Berlin 12, West Germany.

J. Paulun, E. Stein, Technische Universität Hannover, Lehrstuhl für Baumechanik, Callinstr. 15, 3000 Hannover, West Germany.

T.H.H. Pian, K. Kubomura, Massachusetts Institute of Technology, Cambridge, Massachusetts 02139.

E. Ramm, Universitat Stuttgart, Institut für Baustatik, Pfaffenwaldring 7, 7000 Stuttgart, West Germany.

F.G. Rammerstorfer, D.F. Fischer, Voest-Alpine AG, Hauptverwaltung Linz, A-4010 Linz, Austria.

A. Samuelsson, M. Fröier, Department of Structural Mechanics, Chalmers University of Technology, 41296 Göteborg, Sweden.

R.L. Taylor, E.L. Wilson, S. J. Sackett, Department of Civil Engineering, University of California, Berkeley, CA 94720, USA.

J.C.F. Telles, C.A. Brebbia, Civil Engineering Department, University of California, Irvine, CA 92717, USA.

L.C. Wellford, Jr., S.M. Hamdan, Department of Civil Engineering, University of Southern California, Los Angeles, California, USA.

W. Wunderlich, H. Obrecht, Ruhr-Universität Bochum, Institut für Konstruktiven Ingenieurbau IV, 4630 Bochum West Germany.

O.C. Zienkiewicz, Institute of Numerical Methods in Engineering, University College of Swansea, U.K..

Contents

Part I: General Formulations 1

A-posteriori error estimation for the finite element method
I. Babuska 3

Another look at the application of the principle of virtual work with particular reference to finite plate and shell elements
J.F. Besseling 11

New general and complementary energy theorems, finite strain, rate sensitive inelasticity and finite elements: some computational studies
S.N. Atluri and H. Murakawa 28

Formulation of contact problems by assumed stress hybrid elements
T.H.H. Pian and K. Kubomura 49

Part II: Geometrically Nonlinear Problems - Rods and Shells 61

Strategies for tracing the nonlinear response near limit points
E. Ramm 63

Parameter sensitivity of nonlinear structures concerning stability limit
F.G. Rammerstorfer and D.F. Fischer 90

Behaviour of finite element solutions near a bifurcation point
F. Brezzi 109

Some results in the analysis of thin shell structures
K.J. Bathe and L.W. Ho 122

Nonlinear dynamic finite element analysis of shells
T.J.R. Hughes, W.K. Liu and I. Levit 151

Elastic collapse analysis of shells via global-local approach
A.K. Noor and J.M. Peters 169

Large spatial deformations of rods using generalized variational principles
W. Wunderlich and H. Obrecht 185

Large deflection finite element analysis of pre- and postcritical response of thin elastic frames
D. Karamanlidis, A. Honecker and K. Knothe........... 217

Geometrically correct formulations for curved finite bar elements under large deformations
K. Brink and W.B. Krätzig 236

Part III: Physically Nonlinear Problems 257

Some nonlinear problems of soil statics and dynamics
O.C. Zienkiewicz 259

Numerical methods in elasto-plasticity -
A comparative study
A. Samuelsson and M. Fröier 274

Finite element elastoplastic and limit analysis:
Some consistency criteria and their implications
L. Corradi and G. Maier 290

A finite element method for problems in perfect
plasticity using discontinuous trial functions
C. Johnson and R. Scott 307

Finite element analysis for combined material
and geometric nonlinearities
M.A. Crisfield 325

Incremental elastic-plastic deformations of
stiffened plates in compression and bending
J. Paulun and E. Stein 339

A study of some generalized constitutive models
for elasto-plastic shells
O.M. Eidsheim and P.K. Larsen 364

An efficient finite element method for elastic
plastic analysis of plane frames
J. Banovec 385

Elasto-plastic boundary element analysis
J.C.F. Telles and C.A. Brebbia 403

Simplified calculation models applied to
postbuckling analysis of thin plates
B.H. Kröplin and H. Duddeck 435

Finite element analysis of reinforced concrete
slabs and panels
G. Mehlhorn and D. Klein 452

Part IV: Nonlinear Dynamics 479

Finite element analysis of transient nonlinear
response of reinforced concrete structures
M.J. Mikkola and H.S. Sinisalo 481

An analysis of the stability and convergence properties of a Crank-Nicholson algorithm for
nonlinear elasto-dynamics problems
L.C. Wellford, Jr. and S.M. Hamdan 502

Part V: Solution Methods 519

Direct solution of equations by frontal and variable band, active column methods
R.L. Taylor, E.L. Wilson and S.J. Sackett 521

Solution by iteration in displacement and load spaces
P.G. Bergan 553

Partitioned and adaptive algorithms for explicit time integration
T. Belytschko 572

Dynamic relaxation applied to the quasi-static, large deformation, inelastic response of axisymmetric solids
S.W. Key, C.M. Stone and R.D. Krieg 585

On the efficient solution of nonlinear finite element systems
H.D. Mittelmann 621

The numerical calculation of the contact problem in the theory of elasticity
J.J. Kalker, H.J.C. Allaert and J. de Mul............ 637

Exterior penalty methods for contact problems in elasticity
J.T. Oden 655

Approximate solution of nonlinear problems in incompressible finite elasticity
R. Glowinski, P. Le Tallec and V. Ruas de Barros..... 666

Incompressible finite elements: The LBB condition and the discrete eigenstructure
D.S. Malkus 696

Part VI: Computational Algorithms 713

Data management in finite element analysis
P.J. Pahl 715

Performance of finite element algorithms on an array processor-minicomputer based system
H.A. Kamel and J.M. Tan 742

High-speed processors and implication for algorithms and methods
G.F. Carey 758

Part I:
General Formulations

A-posteriori Error Estimation for the Finite Element Method

I. BABUSKA
University of Maryland, College Park, USA

Summary

The paper shows how a finite element solver for linear elliptic equations with an a-posteriori estimator can be extended to the case of eigenvalue problems and nonlinear elliptic problems.

Introduction

Recently in the project FEARS we developed a finite element solver for linear elliptic equations which has adaptive features and provides a-posteriori error estimation of the error. In this presentation we show how this kind of solver with a-posteriori estimator can be extended to the eigenvalue problems and nonlinear elliptic problems. Finally typical results of FEARS are illustrated in two examples.

The Linear Problem

Let us consider an elliptic linear boundary value problem on $\Omega \subset R^n$:

$$Lu = f \quad , \tag{1a}$$

$$u \equiv (u_1, \ldots u_m) \quad , \quad f \equiv (f_1, \ldots f_m)$$

with boundary conditions

$$u = 0 \quad \text{on} \quad \partial\Omega \quad . \tag{1b}$$

We will assume that there exists $0 < c_1, c_2 < \infty$ such that for any $u,v \in H_0^1(\Omega)$

$$c_1 ||u||^2_{H^1(\Omega)} \leq (Lu,u), \quad |(Lu,v)| \leq c_2 ||u||_{H^1(\Omega)} ||v||_{H^1(\Omega)} \tag{2}$$

We denoted by $H^1(\Omega)$ (respective to $H_0^1(\Omega)$) the usual Sobolev space of functions with square integrable first derivatives (respective with zero traces on $\partial\Omega$). By $L_p(\Omega)$, $1 \leq p \leq \infty$ we denote the usual space of functions with p^{th} power integrable. The respective norms are denoted by $||\cdot||_{H^1(\Omega)}$, $||\cdot||_{L_p(\Omega)}$, etc.

Further we denote $||u||_E = (Lu,u)^{1/2}$. Because of (2) $||\cdot||_E$ and $||\cdot||_{H^1(\Omega)}$ are equivalent.

In addition we assume that $f \in R$, where R is a cone of admissible right hand sides and we assume that for every $f \in R$ there exists exactly one (weak) solution $u \in H_0^1(\Omega)$ of the problem (1).

Let S be a family of finite element meshes μ and by S^* we denote the totality of finite element functions associated to the meshes $\mu \in S$. For a fixed $\mu \in S$ we define $S^*(\mu)$ as the linear space of all finite element functions (in H_0^1) associated with the mesh μ. Further we associate to every $\mu \in S$ a cone $R^*(\mu) \subset R$. For given $\mu \in S$ $f \in R(\mu)$ we denote $u(\mu) \in S^*(\mu)$ the finite element solution of (1) (using the same test and trials functions) associated to the mesh μ and the right hand side f. We shall assume that $u(\mu)$ exists and is uniquely determined.

We will assume also that there is an error estimator $E(u(\mu))$ which is computable only by use of the finite element solution $u(\mu)$ and the input data such that

1) There exist constants D_i, $i = 1,2$ dependent only on S, R^*, L but are independent of particular mesh μ, f,u such that

$$D_1 E(u(\mu)) \leq ||u-u(\mu)||_E \leq D_2 E(u(\mu)) \tag{3}$$

2) For any $1 < p < \infty$

$$\frac{1}{\mathcal{E}} ||u-u(\mu)||_{L_p(\Omega)} \leq K(\mathcal{E},p) \quad (4)$$

where $K(\mathcal{E},p)$ is independent of u,μ etc. and $K(\mathcal{E},p) \to 0$ as $\mathcal{E} \to 0$.

3) There exists a sequence of meshes $\mu_i \in \mathcal{S}$ $i = 1,2,...$ such that for $i \to \infty$

$$||u(\mu_i) - u||_E \to 0 \quad (5a)$$

$$\theta = \frac{\mathcal{E}(u(\mu_i))}{||u(\mu_1)-u||_E} \to 1 \quad . \quad (5b)$$

The Eigenvalue Problem

Let us assume in this section that in addition the problem (1) is selfadjoint, i.e.

$$(Lu,v) = (u,Lv) \quad (6)$$

for any $u,v \in H_0^1(\Omega)$.

Consider now the eigenvalue problem

$$Lu = \lambda u \quad (7a)$$

$$u = 0 \quad \text{on} \quad \partial\Omega \quad (7b)$$

and assume that for any $v(\mu) \in \mathcal{S}^*(\mu), v(\mu) \in \mathcal{R}^*(\mu) \subset \mathcal{R}$. By $u(\mu) \in \mathcal{S}^*(\mu)$, $||u(\mu)||_{L_2(\Omega)} = 1$ we denote the finite element eigenfunction associated to the approximate eigenfunction $\lambda(\mu)$. Assume further that there exists sequence μ_i such that $\lambda(\mu_i) \to \lambda$ and $u(\mu_i) \to u$ where u and λ are eigenfunction and eigenvalue of (7). Obviously $u(\mu)$ is the finite element solution of (1) with $f(\mu) = \lambda(\mu)u(\mu)$.

Denote by $V(\mu) = u$ the exact solution of this problem. The error estimator $\mathcal{E}(u(\mu))$ estimates the norm $||V-u(\mu)||_E$ as explained in the previous section. Now we have

THEOREM 1. There exists eigenvalue λ_j such that

$$|\lambda(\mu_i)-\lambda_j| \leq \lambda^{1/2}(\mu_i)E(\mu_i)D_2 \qquad (8)$$

and in addition

$$|\lambda(\mu_i)-\lambda| \leq \lambda^{1/2}(\mu_i)E(\mu_i)\ (1+\sigma(1)) \quad . \qquad (9)$$

Proof. Denote $\phi_j \in H_0^1(\Omega)$ the complete system of normalized eigenfunctions. Let

$$u(\mu) = \Sigma c_i\phi_i \quad .$$

Then

$$\Sigma c_i^2 = 1 \ , \quad \Sigma c_i^2\lambda_i = \lambda(\mu) \quad .$$

Denoting $\xi = V-u(\mu)$ we have

$$(LV,v) = (Lu(\mu),v) + (L\xi,v) = \lambda(\mu)(u(\mu),v), \ v \in H_0^1(\Omega)$$

and of course

$$L(\phi_j,v) = \lambda_j(\phi_j,v) \quad \forall v \in H_0^1(\Omega) \quad .$$

Denoting $e_j = u(\lambda) - \phi_j$ we get

$$(Le_j,v) - \lambda_j(e_j,v) = (\lambda(\mu)-\lambda_j)(u(\mu),v) - (L\xi,u) \quad .$$

Set now $v = (\text{sign}(\lambda(\mu)-\lambda_j))c_j\phi_j$. Then we get

$$|(\lambda(\mu)-\lambda_j)|(u(\mu),\phi_j)c_j) = (L(\xi),\text{sign}(\lambda(\mu)-\lambda_j)c_j\phi_j)$$

and therefore we get

$$\min_j|\lambda(\mu)-\lambda_j| \leq ||u(\mu)||_E||\xi||_E \leq \lambda^{1/2}(\mu)\cdot E(\mu)D_2 \qquad (10)$$

where D_2 is given in (3). When μ_i is used for μ then the right side of (8) can be replaced by $\lambda^{1/2}(\mu_j)E(\mu_j)(1+\sigma(1))$. Let us remark that (8) is independent of the spectrum, but the $\sigma(1)$ term in (8) is dependent on it. The theorem 1 can be generalized in various ways. Let us mention one.

THEOREM 2. Let (5b) hold. Then $E^2(\mu_j)(1+\sigma(1)) \leq |\lambda(\mu_j)-\lambda| \leq E^2(\mu_j)(1+\sigma(1))$. The proof goes by analogous way of [1].

The Nonlinear Problem

Let us be interested in the nonlinear problem

$$Lu + g(u) = f \tag{11a}$$

$$u = 0 \quad \text{on} \quad \partial\Omega \tag{11b}$$

About g we will assume the following. For any $u,v \in H_0^1(\Omega)$ and $p \geq \frac{2m}{m+2}$ we have

$$||g(u)||_{L_p(\Omega)} \leq C||u||_E \tag{12}$$

$$(g(u+v) - g(u),v) \geq 0 \tag{13}$$

$$||g(u+v) - g(u)||_{L_p} \leq C||v||_E \tag{14}$$

In addition we will assume that for any $f \in R$ there exists unique solution (in H_0^1) of (11) and if $u \in S^*(\mu)$ then, $g(u) \in R^*(\mu)$, and $f \in R^*(\mu)$ for any $\mu \in S$. We assume also that the finite element solution $u(\mu)$ exists and is unique.

Let now V be the exact solution of (1) with the right hand side $f - g(u(\mu))$ and $v(\mu)$ be the finite element solution. Obviously $u(\mu) = v(\mu)$. Denote by $E(u(\mu))$ the estimator for $||V-u(\mu)||_E$. Now we have

THEOREM 3. Let μ_i be a sequence of meshes such that $E(\mu_i) \to 0$ as $i \to \infty$ and (5) is satisfied. Then

$$(1+\sigma(1))E(u(\mu)) \leq ||u-u(\mu)||_E \leq (1+\sigma(1))E(u(\mu)) \tag{15}$$

Proof. Write $V = \xi+u(\mu)$. Then for $\forall v \in H_0^1(\Omega)$

$$(LV,v)+(g(V),v) = (f,v) + ((g(V)-g(V-\xi)),v) = (f,v) + (R,v) \tag{16}$$

We have also

$$(Lu,v) + (g(u),v) = (f,u) \tag{17}$$

and therefore denoting $z = V-u$ we get $(Lz,v) + (g(u+z)-g(u),v) = (R,v)$ which yields

$$||z||_E^2 \leq ||z||_{L_q}||R||_{L_p} \leq CR(p,\varepsilon)||\xi||_E||z||_E$$

and therefore

$$||z||_E \leq \mathcal{E}(u(\mu))\sigma(1) \quad \text{as} \quad \mathcal{E} \to 0 .$$

Now

$$||u-u(\mu)||_E \leq ||u-V||_E + ||V-u(\mu)||_E \leq$$

$$\leq (1+\sigma(1))\mathcal{E}(u(\mu))$$

and because

$$||V-u||_E - ||u(\mu)-V||_E \leq ||u-u(\mu)||_E$$

we get the other side of (15).

This theorem can be generalized in various ways, but we will not elaborate more.

The estimators for the elliptic problem can be also used in other finite element problems as in linear and nonlinear evolution equations.

The FEARS

During recent years the solver FEARS (finite element adaptive research solver) was developed which has various special features which will not be discussed here. For more information, see [2], [3], [4], [5], [6].

Here we will illustrate the behaviour of the effectivity index $\theta = \mathcal{E}/||u-u(\mu)||_E$ introduced in (5b) on two examples.

Example 1. Elasticity problem for homogeneous isotropic medium ($\nu = .25$) on the ring

$$\Omega = [(x_1,x_2)|.25 < (x_1^2+x_2^2)^{1/2} < 1]$$

with displacement u_1,u_2 prescribed on internal boundary $S_1 = [(x_1,x_2)| (x_1^2+x_2^2)^{1/2} = .25]$ namely $u_1 = 1$, $u_2 = 0$. On the external boundary $S_2 = [x_1,x_2| (x_1^2+x_2^2)^{1/2} = 1]$ we take $u_1 = u_2 = 0$. The exact solution is known. The computation was performed on a quarter ring in the adaptive mode by (mapped) piecewise bilinear elements. The (energy) error is asymptotically of order $N^{-1/2}$ where N is the number of degrees of freedom. Table 1 shows essential results.

DOF	$100 \frac{\|\|e\|\|}{\|\|u\|\|}$	$\frac{\mathcal{E}}{\|\|e\|\|}$
20	14.697	1.015
40	11.664	1.095
52	10.168	.981
76	8.136	1.0002
130	6.332	.9895
173	5.744	.97107

Table 1. Elasticity problem on the ring Ω .

We see higher effectivity of the estimator when $||e|| \to 0$.

Example 2. Elasticity problem for homogeneous isotropic medium with $\nu = 0.0$ on the rectangle

$$R = [(x_1,x_2)| \; 0 < x_1 < 1, \quad -4 < x_2 < 0]$$

On the sides

$$R_1 = [(x_1,x_2)| \; 0 < x_1 < 1, \quad x_2 = -4]$$

$$R_2 = [(x_1,x_2)| \; 1/2 < x_1 < 1, \quad x_2 = 0]$$

we assume zero traction. On

$$R_3 = [(x_1,x_2)| \; 0 < x_1 < 1/2, \quad x_2 = 0]$$

we prescribe $u_1 = 0$, $u_2 = 1$. On

$$R_4 = [(x_1,x_2)| \; x_1 = 1, \quad -4 < x_2 < 0]$$

we assume $u_2 = 0$ and zero traction in x_1 direction. On $R_J = [(x_1,x_2) | \; x_1 = 0, \quad -4 < x_2 < 0]$ we assume $u_1 = 0$ and zero traction in x_2 direction.

Computation has been made in the adaptive mode. The essential data are presented in Table 2.

DOF	$100 \frac{\|\|e\|\|}{\|\|u\|\|}$	$\frac{\mathcal{E}}{\|\|e\|\|}$
32	41.76	.672
50	35.04	.672
80	25.25	.764
106	21.82	.764
168	17.29	.773
229	14.10	.827

Table 2. Elasticity problem on the rectangle 2.

The solution has oscillatory singularity in the point $x_1 = 0$, $x_2 = 1/2$ of the type $r^{1/2}$. For a uniform mesh the assymptotical order of convergence is $N^{-1/4}$, for adaptive meshes $N^{-1/2}$. The table shows that very clearly the effectivity index $\theta = \frac{\mathcal{E}}{||e||}$ converges to 1 and experience shows that about for 5% accuracy $.85 < \theta < 1.15$. In most cases much better effectivity index occurs.

References

[1] Babuška, I.; Rheinboldt, W.C.: A-posteriori error estimates for the finite element method. Int. J. Num. Mech. Eng. 12 (1978) 1597-1615.

[2] Babuška, I.; Rheinboldt, W.C.: Analysis of optimal finite element meshes in R^1. Math. of Comp. 33 (1979) 431-463.

[3] Babuška, I.; Rheinboldt, W.C.: Error estimates for adaptive finite element computations. SIAM J. Numer. Anal. 15 (1978) 736-719.

[4] Babuška, I.; Miller, A.: A-posteriori estimates and adaptive techniques for a finite element method. to appear.

[5] Babuška, I.; Rheinboldt, W.C.: Reliable error estimation and mesh adaptation for the finite element method. Comp. Meth. Nonlinear Mechanica, J.T. Oden, editor, North Holland Publ. Co. (1980) 61-108.

[6] Rheinboldt, W.C.: On a data structure for adaptive fielement mesh refinements. ACM transaction on Math. Software 6 (1980) 166-187.

Another Look at the Application of the Principle of Virtual Work with Particular Reference to Finite Plate and Shell Elements

J F BESSELING
University of Technology, Delft, The Netherlands

Summary

In the finite element model each element may be looked upon as a deformable body, however with a limited number of deformation modes, determined by the number of nodal displacement and rotation components on its boundary. In terms of a corresponding number of suitably defined strain parameters finite element properties are derived for plate and shell elements, valid for arbitrarily large displacements and rotations. The membrane properties are derived from a strain- or a elastic potential distribution in the domain of the element. The bending properties follow from an equilibrium field of internal moments and are restricted to small deformations of the element. It is shown that for triangular elements all properties can be given in closed form, but for large rotations the equations of equilibrium can only be solved as rate equations if bending is taken into account.

1. Introduction

The variational statement of the dynamics of solid bodies in terms of the principle of virtual work is a suitable starting point for the discrete modelling of the mechanical behaviour of such bodies, in particular if they represent complex structures. We shall employ this principle in the version which in [1] is attributed to Piola. The principle of virtual work according to Piola requires the virtual work of the external forces and of the inertial forces to be zero for all infinitesimal rigid body motions. Infinitesimal rigid body displacements and rotations can be specified as subsidiary conditions. Introduced into the virtual work condition with the aid of multipliers the resulting contribution may be interpreted as the virtual work of deformation, where the multipliers are components of a stress field.

We may conclude that for a discrete model it is sufficient to define for each finite element the necessary and sufficient conditions for rigid body motion in terms of the nodal velocities, or equivalently to define the deformation modes by properly chosen generalized strain measures. If these strains by constitutive equations are related to the dual stress multipliers, we arrive at the system of equations by which we can solve a stress- and deformation problem. This approach, that was indicated in [2], will be discussed here in some more detail.

2. Principle of virtual work

Let us consider a body B of massdensity ρ, subjected to surface tractions $\underline{t}$, a volume force density $\rho\underline{f}$ and an acceleration field $\ddot{\underline{u}}$. Written with rectangular cartesian components of the vector quantities involved, the principle of virtual work states as a necessary condition for any motion of the body B

$$\int_{\partial B} t_i \delta u_i dA + \int_B \rho f_i \delta u_i dV = \int_B \rho \ddot{u}_i \delta u_i dV \tag{2.1}$$

for all rigid virtual displacement fields δu_i. A rigid virtual displacement field is a field of infinitesimal, kinematically admissible displacements satisfying the conditions

$$\left(\frac{\partial \delta u_i}{\partial x_j} + \frac{\partial \delta u_j}{\partial x_i} \right) = \delta\varepsilon_{ij} = 0. \tag{2.2}$$

We observe that $\delta\varepsilon_{ij}$ is a symmetric tensor and hence the conditions (2.2) can be introduced into the variational statement of the principle of virtual work by means of a symmetric tensor field of multipliers t_{ij}. Thus (2.1) is transformed into

$$\int_{\partial B} t_i \delta u_i dA + \int_B \rho f_i \delta u_i dV = \int_B (\rho \ddot{u}_i \delta u_i + t_{ij} \delta\varepsilon_{ij})\, dV, \tag{2.3}$$

which now must hold for all kinematically admissible virtual displacements δu_i.

By applying Green's transformation we derive from (2.3)

$$t_i = t_{ij} n_j \tag{2.4}$$

and

$$\frac{\partial t_{ij}}{\partial x_j} + \rho f_i = \rho \ddot{u}_i, \tag{2.5}$$

which identifies t_{ij} as the stress tensor of Cauchy. Hence $t_{ij}\delta\varepsilon_i$ in the right-hand side of (2.3) may rightly be called the virtual work of deformation per unit volume.

In the case of elastic material behaviour, or if the inelastic deformations remain sufficiently small, deformations with respect to an initial geometry of the body are physically relevant variables. The deformation tensor may then be considered as the sum of an elastic and an inelastic part, such that for the elastic components hold unaltered the elasticity relations that are valid in the initial geometry. It should be stressed however that this is no longer true after large inelastic deformations [3].

If we restrict ourselves to elastic and small inelastic deformations it is appropriate to express the volume integrals in (2.3) in terms of the initial geometry B_o of the body. We define a_i as the cartesian coordinates of the material points in this initial geometry. They identify as material coordinates the material points during the deformation process.

The difference of the square of a line-element in the deformed and in the undeformed geometry is defined by

$$ds^2-ds_o^2 = 2E_{\alpha\beta}da_\alpha da_\beta, \tag{2.6}$$

where $E_{\alpha\beta}$ is the lagrangian strain tensor

$$E_{\alpha\beta} = \frac{1}{2}\left(\frac{\partial x_k}{\partial a_\alpha}\frac{\partial x_k}{\partial a_\beta} - \delta_{\alpha\beta}\right) = \frac{1}{2}\left(\frac{\partial u_\alpha}{\partial a_\beta}+\frac{\partial u_\beta}{\partial a_\alpha}+\frac{\partial u_k}{\partial a_\alpha}\frac{\partial u_k}{\partial a_\beta}\right). \tag{2.7}$$

A rigid virtual displacement field can now be defined by the condition that the first variation of the strain tensor (2.7) vanishes in all points of the body

$$\delta E_{\alpha\beta} = \frac{1}{2}\left(\frac{\partial \delta u_\alpha}{\partial a_\beta}+\frac{\partial \delta u_\beta}{\partial a_\alpha}+\frac{\partial u_k}{\partial a_\alpha}\frac{\partial \delta u_k}{\partial a_\beta}+\frac{\partial \delta u_k}{\partial a_\alpha}\frac{\partial u_k}{\partial a_\beta}\right) = 0. \tag{2.8}$$

Using the equality $\rho dV = \rho_o dV_o$ and again with the aid of a symmetric tensor field of multipliers, $\sigma_{\alpha\beta}$, the principle of virtual work is expressed with volume integrals in terms of the initial geometry B_o by

$$\int_{\partial B} t_i\delta u_i dA + \int_{B_o} \rho_o f_i \delta u_i dV_o = \int_{B_o} (\rho_o \ddot{u}_i \delta u_i + \sigma_{\alpha\beta}\delta E_{\alpha\beta})\, dV_o, \tag{2.9}$$

which must hold for all kinematically admissible virtual displacement fields.

The tensor $\sigma_{\alpha\beta}$ is the so-called pseudo stress tensor of Kirchhoff, related to the stress tensor of Cauchy by

$$\sigma_{\alpha\beta} = \frac{\rho_o}{\rho} \frac{\partial a_\alpha}{\partial x_i} t_{ij} \frac{\partial a_\beta}{\partial x_j} . \tag{2.10}$$

This relation follows from the identities

$$\int_B t_{ij} \delta\varepsilon_{ij} dV = \int_{B_o} \frac{\rho_o}{\rho} \frac{\partial a_\alpha}{\partial x_i} t_{ij} \frac{\partial a_\alpha}{\partial x_j} \frac{\partial x_k}{\partial a_\alpha} \frac{\partial \delta u_k}{\partial a_\beta} dV_o =$$
$$= \int_{B_o} \frac{\rho_o}{\rho} \frac{\partial a_\alpha}{\partial x_i} t_{ij} \frac{\partial a_\beta}{\partial x_j} \delta E_{\alpha\beta} dV_o . \tag{2.11}$$

The principal difference between $\delta\varepsilon_{ij}$ and $\delta E_{\alpha\beta}$ lies in the fact that $\delta\varepsilon_{ij}$ cannot, and $\delta E_{\alpha\beta}$ can be derived from a measure of finite strain as a first variation. The strains $E_{\alpha\beta}$ are by their elastic components $E'_{\alpha\beta} = E_{\alpha\beta} - E^o_{\alpha\beta}$ related to the stresses $\sigma_{\alpha\beta}$, that are the proper duals of the virtual deformations $\delta E_{\alpha\beta}$.

To allow for a maximum freedom in approximation methods we now write the virtual work condition (2.9) for an assembly of N finite elements with a number of multiplier functions, $\sigma^*_{\alpha\beta}$, t^*_i, and t^{**}_i.

$$- \int_{\partial B^t} \bar{t}_i \delta u^*_i dA + \sum_{e=1}^{N} \int_{B^e_o} \Big[-\rho_o (f_i - \ddot{u}^e_i) \delta u^e_i +$$
$$+ \sigma^e_{\alpha\beta} \delta E^e_{\alpha\beta} + \sigma^*_{\alpha\beta} \Big\{ \Big(\delta_{k\alpha} + \frac{\partial u^e_k}{\partial a_\alpha} \Big) \frac{\partial \delta u^e_k}{\partial a_\beta} - \delta E^e_{\alpha\beta} \Big\} +$$
$$+ \delta\sigma^*_{\alpha\beta} \Big\{ \frac{1}{2} \Big(\frac{\partial u^e_\alpha}{\partial a_\beta} + \frac{\partial u^e_\beta}{\partial a_\alpha} + \frac{\partial u^e_k}{\partial a_\alpha} \frac{\partial u^e_k}{\partial a_\beta} \Big) - E^e_{\alpha\beta} \Big\} \Big] dV_o +$$
$$+ \sum_{e=1}^{N} \int_{\partial B^e_o} \Big[(\delta u^*_i - \delta u^e_i) t^*_i + (u^*_i - u^e_i) \delta t^*_i \Big] dA_o +$$
$$+ \int_{\partial B^u} \Big[-\delta u^*_i t^{**}_i + (\bar{u}_i - u^*_i) \delta t^{**}_i \Big] dA = 0. \tag{2.12}$$

The displacement fields u^*_i are only defined on the external surface and on the element interfaces, while $\sigma^e_{\alpha\beta}$ is related to $E^e_{\alpha\beta}$, for instance through the linear isotropic constitutive relations

$$\sigma^e_{\alpha\beta} = C\left\{ E^e_{kk} - \alpha(T-T_o) \right\} \delta_{\alpha\beta} + 2G(E^e_{\alpha\beta} - E''_{\alpha\beta} - \frac{1}{3} E^e_{kk}\delta_{\alpha\beta}) . \quad (2.13)$$

Here $E''_{\alpha\beta}$ represents the small inelastic strains and $\alpha(T-T_o)$ is the specific thermal expansion. The prescribed quantities $\bar{u}_i$ and $\bar{t}_i$ will usually be given on the external boundary in the deformed state. By considering the variations $\delta E^e_{\alpha\beta}$, δu^e_k, and δu^*_i we derive the following equations

$$\sigma^*_{\alpha\beta} = \sigma^e_{\alpha\beta} , \quad (2.14a)$$

$$\frac{\partial}{\partial a_\beta}\left\{ \sigma^*_{\alpha\beta} \left(\delta_{k\alpha} + \frac{\partial u^e_k}{\partial a_\alpha}\right)\right\} + \rho_o f_k = \rho_o \ddot{u}^e_k \text{ in } B^e_o , \quad (2.14b)$$

$$\sigma^*_{\alpha\beta} \left(\delta_{k\alpha} + \frac{\partial u^e_k}{\partial a_\alpha}\right) n_\beta = t^*_k \qquad \text{on } \partial B^e_o , \quad (2.14c)$$

$$t^*_{i(\underline{n})} = - t^*_{i(-\underline{n})} \text{ on element interfaces,} \quad (2.14d)$$

$$t^{**}_i \, dA = t^*_i \, dA_o \qquad \text{on } \partial B^u , \quad (2.14e)$$

$$t^*_i \, dA_o = \bar{t}_i \, dA \qquad \text{on } \partial B^t . \quad (2.14f)$$

From the variation of the multipliers we have an additional set of equations

$$E^e_{\alpha\beta} = \frac{1}{2}\left\{ \frac{\partial u^e_\alpha}{\partial a_\beta} + \frac{\partial u^e_\beta}{\partial a_\alpha} + \frac{\partial u^e_k}{\partial a_\alpha}\frac{\partial u^e_k}{\partial a_\beta}\right\} \qquad \text{in } B^e_o , \quad (2.15a)$$

$$u^e_i = u^*_i \qquad \text{on } \partial B^e_o , \quad (2.15b)$$

$$u^*_i = \bar{u}_i \qquad \text{on } \partial B^u . \quad (2.15c)$$

The most common type of finite elements is the so-called kinematically conforming type. The equations (2.15) are satisfied apriori. With the aid of proper interpolation polynomials in (2.9) from the principle of virtual work a set of equations is obtained, which after differentiation of the appropriate constitutive equations are of the form

$$\left[\underset{\sim}{K}^o + \underset{\sim}{K}^u - \underset{\sim}{P}^o - \underset{\sim}{P}^u + \underset{\sim}{G}\right] \dot{\underline{u}} + \underset{\sim}{M}\dddot{\underline{u}} = \dot{\underline{f}} + \underline{g} . \quad (2.16)$$

The matrices are symmetric and the geometrical non-linearities present themselves in the matrices $\underset{\sim}{K}^u$, $\underset{\sim}{P}^u$ and $\underset{\sim}{G}$.

For the analysis of shells and of large bending deformations of plates this approach meets with fundamental difficulties. For geometrically nonlinear problems of initially flat plates, in addition to the rigid body motions, simple independent states of either constant membrane deformation or bending deformation must be represented in each finite element. For shells even in the linear case it is difficult to ensure that bending deformation of a finite element is not accompanied by spurious membrane deformations. If inextensional bending is permitted by the shell geometry, the finite shell element should reproduce exactly simple independent states of membrane and bending deformation. Otherwise, the relatively large stiffness against membrane deformation will obscure the inextensional bending modes of deformation in the finite element model of the shell.

It is not possible to meet the above requirements by a straightforward displacement approach to the finite element modelling on the basis of the variational statement (2.9). In the next paragraph we shall have another look at the application of the principle of virtual work.

3. The consequences of the representation of the displacement field by a finite dimensional vector space

In the finite element method at interelement boundary points, -lines, and -surfaces the adjoining elements must have common displacements in order to ensure continuity of the structure. If these displacements are to be approximated by a finite set of functions, the displacements at the interelement boundaries and at the natural boundary of the structure can be taken as interpolations between nodal displacements at the interface between any two elements or between an element and the surroundings of the structure. Continuity of the structure is then ensured for all values of the nodal displacements and the displacement field is represented by a finite dimensional vectorspace U^e with elements in R^n.

We recall that in the formulation of the principle of virtual work rigid virtual displacements had to be characterized. The interpolations between nodal displacements must be such that all possible rigid motions of the finite element are represented. Even for elements with curved boundaries this requirement is met by the isoparametric concept [4]. For the transformation to natural coordinates ζ_α of the finite element and for the interpolation between nodal displacements the same interpolation functions are used. In the case of material finite elements the coordinates of the material points of a finite element are then given by

$$x_i^e = \psi_N(\zeta_\alpha)(a_{Ni}^e + u_{Ni}^e), \quad \sum_{N=1}^{n} \psi_N(\zeta_\alpha) = 1. \tag{3.1}$$

Infinitesimal rigid body motions of a finite element are now defined by a subspace of $\mathcal{U}^e$, which will be called the vectorspace of degrees of freedom, $\mathcal{V}^e$. It is appropriate to call the complementary subspace, $\mathcal{E}^e$, the deformationspace of the finite element.

$$\mathcal{U}^e = \mathcal{V}^e \oplus \mathcal{E}^e. \tag{3.2}$$

The conditions for a rigid body motion are expressed by $\delta E_{\alpha\beta} = 0$. However, in the derivation of the finite element equations by application of the principle of virtual work it suffices to require locally $\delta E_{\alpha\beta} = 0$, such that for each finite element a number of linearly independent conditions is obtained that is equal to the dimension of its deformation space. In other words, for the finite element model we may derive the contribution of the subsidiary conditions in the variational statement (2.9),

$$\int_{B_o^e} \sigma_{\alpha\beta} \delta E_{\alpha\beta} dV_o, \tag{3.3}$$

from a strain distribution with parameters obtained by local application of (2.7). The number of these strain parameters is determined by the number of nodal displacements u_{Ni}^e minus the number of degrees of freedom of the finite element as a rigid body. The strain parameters are by application of (2.7) expressed as nonlinear functions of the nodal displacements with the aid of the interpolations (3.1):

$$\varepsilon_i = D_i(u_k). \tag{3.4}$$

The strain parameters thus defined are elements of a vector of generalized strains of the finite element, $\underline{\varepsilon}^e$, for which holds $\delta\varepsilon^e \in \mathcal{E}^e$.

Instead of (3.3) we now may add $\sigma_i \delta\varepsilon_i$ to the virtual work condition in order to characterize the rigid body motions. However, in order to end up with a physically defined set of equations we must be able to derive constitutive equations for the generalized stresses, $\underline{\sigma}$, which are the duals of the virtual strains, $\delta\underline{\varepsilon}$. It is then in the first place necessary to describe the deformed state of each finite element over its whole domain. For this we can introduce a set of functions of the natural coordinates, such that

$$E^e_{\alpha\beta} = \varepsilon_i \phi_{i\alpha\beta}(\zeta_k). \tag{3.5}$$

For the linear constitutive equations (2.13), provided also the inelastic and thermal strains are given as functions of the natural coordinates with ε^o_i as parameters, the virtual work condition (2.9) can be evaluated and we obtain ($D_{i,k} \equiv \frac{\partial D_i}{\partial u_k}$)

$$\delta u_k \left[-f_k + M_{k\ell}\ddot{u}_\ell + \sigma_i D_{i,k}\right] = 0 \;\forall\; \delta u_k, \tag{3.6}$$

where

$$\sigma_i = S_{ij}(\varepsilon_j - \varepsilon^o_j), \quad S_{ij} = S_{ji}. \tag{3.7}$$

In the case of nonlinear elasticity we have

$$\sigma_{\alpha\beta}\delta E_{\alpha\beta} = \rho_o \frac{\partial e}{\partial E_{\alpha\beta}} \delta E_{\alpha\beta}. \tag{3.8}$$

The elastic potential per unit undeformed volume, $\rho_o e$, is a complicated function of the components of the strain tensor, and hence for the finite element a complicated function of the generalized strains ε_i. But rather than resort to numerical integration for the evaluation of (3.3) in each step of the deformation process, we can determine first the elastic potential of the finite element as a whole, as a function of the generalized strains ε_i. With the aid of interpolation functions we obtain by integration over the domain of the finite element from local functional relationships,

$$\rho_o e = \rho_o e(E_{\alpha\beta}) = \rho_o e(\varepsilon_i, \zeta_\alpha), \tag{3.9}$$

the finite element expression for the elastic potential.

$$E(\varepsilon_i) = \int_{B^e_o} \rho_o e(\varepsilon_i, \zeta_a)\, dV_o . \tag{3.10}$$

The virtual work condition again assumes the form (3.6), but the generalized stresses σ_i are now determined by

$$\sigma_i = \frac{\partial E}{\partial \varepsilon_i} . \tag{3.11}$$

The approach outlined above proves to be particularly fruitful in dealing with the geometrical nonlinearities of the membrane deformations of plates and shells. It avoids the appearance of spurious membrane deformations, which are most detrimental to the accuracy of the solution, in particular if the plate or shell would permit inextensional deformation. The success of reduced integration techniques for the improvement of the conditioning of the matrix and of the accuracy of the solution can also be understood in terms of the local enforcement of the conditions $\delta E_{\alpha\beta}$, in number not exceeding the number of linearly independent virtual deformation parameters, as determined by the number of nodal displacements.
Since local application of the expressions (2.7) give the components of strain as simple quadratic functions of the nodal displacements all properties of triangular (and rectangular) elements can be determined in closed form.

4. Bending of plates and shells

The description of the bending of plates and shells by finite elements meets with the wellknown difficulty, that not only the deflections but also the rotations must be continuous at element interfaces. For a kinematically conforming approach this implies continuity conditions for the normal derivatives of the deflections, which are not easily satisfied. Here a derivation of the element properties on the basis of a bending and twisting moment distribution provides a simpler formulation than a derivation on the basis of displacement interpolation functions.
We observe that for the static case ($\ddot{u}_i^e = 0$) in the absence of volume forces ($f_i = 0$) the displacement fields u_i^e disappear from expression (2.12), if the equations (2.14b) and (2.14c) are satisfied. By a rigid rotation of the finite element, provided the element is chosen sufficiently small in relation with its deformations, these equations can always be reduced to the linear form with any desired degree of accuracy.

$$\frac{\partial \sigma^*_{\alpha\beta}}{\partial a_\beta} = 0, \quad \sigma^*_{\alpha\beta} n_\beta = t^*_\alpha . \tag{4.1}$$

When in addition we satisfy the relations (2.14a), $\sigma^e_{\alpha\beta} = \sigma^*_{\alpha\beta}$, expression (2.12) reduces to

$$-\int_{\partial B^t} \bar{t}_i \delta u_i^* dA + \sum_{e=1}^{N} \left[-\int_{B_o^e} \delta\sigma_{\alpha\beta}^* E_{\alpha\beta}^* dV_o + \int_{\partial B_o^e} (\delta u_i^* t_i^* + u_i \delta t_i^*) \, dA_o \right] +$$

$$+ \int_{\partial B^u} \left[-\delta u_i^* t_i^{**} + (\bar{u}_i - u_i^*) \delta t_i^{**} \right] dA = 0. \qquad (4.2)$$

We should realize that t_i^* in (4.2) and t_α^* in (4.1) differ by the rigid rotation, that had to be applied to each of the finite elements in order to free the equilibrium equations from the displacements. Since the orthogonal transformation, that represents this rigid rotation, has to be expressed in terms of displacements, the geometrical nonlinearity enters condition (4.2) through this transformation.

We shall now discuss the implications of the above reasoning for the analysis of the bending of plates by triangular finite elements. The deformation of each single element is assumed to remain small, such that with respect to a plane through the vertices of a triangular element we may apply linear bending theory. Let w^e denote the displacement component normal to the basic triangle. With moments $M_{\alpha\beta}$ and curvatures $k_{\alpha\beta}$ we obtain for each finite element, analogues to the linearized contributions of the elements in (2.12),

$$\int_{A_o^e} \left[M_{\alpha\beta}^e \delta k_{\alpha\beta}^e + M_{\alpha\beta}^* \left(\frac{-\partial^2 \delta w^e}{\partial a_\alpha \partial a_\beta} - \delta k_{\alpha\beta}^e \right) + \delta M_{\alpha\beta}^* \left(\frac{-\partial^2 w^e}{\partial a_\alpha \partial a_\beta} - k_{\alpha\beta}^e \right) \right] dA_o$$

$$+ \sum_{i=1}^{3} \int_{s_i} \left[(\delta w^* - \delta w^e) \, V_n^* + (w^* - w^e) \, \delta V_n^* + \left(\delta\phi^* + \frac{\partial \delta w^e}{\partial n} \right) M_n^* + \right.$$

$$\left. + \left(\phi^* + \frac{\partial w^*}{\partial n} \right) \delta M_n^* \right] ds + \sum_{i=1}^{3} \left\{ (\delta w^* - \delta w^e) \, H_{ci}^* + (w^* - w^e) \, \delta H_{ci}^* \right\}$$

$$(4.3)$$

If we let $M_{\alpha\beta}^* = M_{\alpha\beta}^e$, V_n^*, M_n^*, and H_{ci}^* satisfy the equations

$$\frac{\partial^2 M_{\alpha\beta}^*}{\partial a_\alpha \partial a_\beta} = 0, \qquad (4.4a)$$

$$V_n^* = \left(\frac{\partial M_{\alpha\beta}^*}{\partial a_\alpha} + \frac{\partial M_{\alpha\gamma}^*}{\partial a_\beta} t_\beta t_\gamma \right) n_\alpha, \quad M_n^* = M_{\alpha\beta}^* n_\alpha n_\beta, \qquad (4.4b)$$

$$H^*_{ci} = (M^*_{\alpha\beta} n_\alpha t_\beta)^{s_i^-} - (M^*_{\alpha\beta} n_\alpha t_\beta)^{s_i^+} , \qquad (4.4c)$$

then the displacement w^e disappears from (4.3) and we have for each element

$$- \int_{A_o^e} \delta M^*_{\alpha\beta} k^*_{\alpha\beta} dA_o + \sum_{i=1}^{3} \int_{s_i} \left[\delta w^* V^*_n + \delta\phi^* M^*_n + w^* \delta V^*_n + \phi^* \delta M^*_n \right] ds$$

$$+ \sum_{i=1}^{3} (\delta w^* H^*_{ci} + w^* \delta H^*_{ci}). \qquad (4.5)$$

On the basis of linear moment distributions, with the values of $M^*_{\alpha\beta}$ at the vertices as generalized stresses, in [5] element properties were derived. By a quadratic interpolation for w^*, and a linear interpolation for ϕ^* along each of the three sides of the basic triangle, nine linearly independent deformation modes were defined, corresponding to the nine generalized stresses. With the aid of the linear elasticity relations expression (4.5) can then be evaluated and we find:

$$-\delta\sigma^e_i (S^{e^{-1}}_{ij} \sigma^e_j + D^w_{ik} w^*_k + D^\phi_{i\ell} \phi^*_\ell) + \sigma^e_i (D^w_{ik} \delta w^*_k + D^\phi_{i\ell} \delta\phi^*_\ell). \qquad (4.6)$$

The matrices $[S^e_{ij}]$, $[D^w_{ik}]$, and $[D^\phi_{i\ell}]$ have been given in [5] 1).
The displacements w^*_k and the rotations ϕ^*_ℓ at the boundary are defined with respect to the basis triangle of the element under consideration. Continuity of the structure requires that they are expressed in terms of the displacement and rotation functions defined on the interfaces of the finite elements. Apart from a rigid translation, which has no consequences, this implies for the displacements an orthogonal transformation. The elements of the complete transformation matrix, as determined by the displacements of the vertices of the basic triangle in the case of small membrane deformations, are given in the appendix. However, for the rotations it is a different matter. Infinitesimal rotations can be decomposed into orthogonal components. The infinitesimal rotation $\delta\phi^*$ for a particular element is then equal to the difference between the tangential component of the

1) The submatrix C_{33}, as given in [5], must be multiplied by a factor 2.

infinitesimal rotation of the plate surface at the interelement boundary and the tangential component of the infinitesimal rotation of the basic triangle. The latter is determined by the displacements of the vertices and their variations, while the tangential component of the infinitesimal rotation of the plate surface at the interelement boundary must be the same for adjacent elements in order to ensure continuity of the structure. Thus the infinitesimal rotations in (4.6) can, just as the infinitesimal and finite displacements, be expressed in terms of functions defined on the interfaces of the finite elements, such that continuity between elements is ensured. This is not possible however for the finite rotations ϕ^*, if these rotations as well as the rotations of the basic triangle do not remain sufficiently small.

If the rotations do remain small, we may follow the approach indicated in [6]. Otherwise we meet with the difficulty that finite rotations can only be characterized by orthogonal transformations, defined in terms of angular coordinates, which obey complicated nonlinear transformation rules. Then we are compelled to define ϕ^* for each finite element separately as the normal slope of the plate at the element boundary with respect to the basic triangle of the element under consideration, $\phi^* = -\frac{\partial w^e}{\partial n}$. Continuity of the structure can now be ensured by deriving the rate of change of ϕ^* at any moment from the difference between the tangential component of the angular velocity of the plate surface at the interelement boundary and the tangential component of the rate of rotation of the respective triangle. As a consequence for arbitrarily large rotations for the finite element model under consideration only rate equations can be formulated. In view of the numerical procedures, required to solve the highly nonlinear equations in the case of large rotations, this is not felt as a serious disadvantage of the model.

The fact however that ϕ^* cannot be expressed in terms of nodal coordinates implies that the generalized strains from (4.6),

$$\varepsilon_i = D^w_{ik} w^*_k + D^\phi_{i\ell} \phi^*_\ell , \tag{4.7}$$

cannot be written as functions of the type (3.4). The rate equations on the basis of (3.6) are of the form

$$(D_{i,k} S_{ij} D_{j,\ell} + \sigma_i D_{i,k\ell})\, \dot{u}_\ell + \dot{M}_{k\ell}\, \dddot{u}_\ell = \dot{f}_k . \tag{4.8}$$

The symmetry of the geometrical stiffness matrix, $[\sigma_i D_{i,k\ell}]$, is clearly due to the analytic relation between generalized strains and nodal coordinates. It turns out that this symmetry is lost if bending is taken into account by (4.7), because the generalized bending strains (4.7) are determined by the nodal coordinates as follows

$$\varepsilon_i = D^w_{ik} A_{k\ell} u_\ell + D^\phi_{im}\phi^*_m = D^u_i(u_k) + D^\phi_{im}\phi^*_m,$$
$$\dot{\phi}^*_m = \dot{\phi}_m - B_{mk}\dot{u}_k. \qquad (4.9)$$

Here $\dot{\phi}$ is the angular velocity at the interelement boundary. The transformation coefficients $A_{k\ell}$ and B_{mk} are functions of the nodal displacements at the element vertices. Some details are given in the appendix.

The equilibrium equations according to the principle of virtual work now read

$$\sigma_i\,(D^u_{i,k} - D^\phi_{i\ell}B_{\ell k}) = f^u_k,$$
$$\sigma_i D^\phi_{i\ell} = f^\phi_\ell, \qquad (4.10)$$

from which the following rate equations are derived

$$[(D^u_{i,k} - D^\phi_{im}B_{mk})\,S_{ij}\,(D^u_{j,\ell} - D^\phi_{jn}B_{n\ell}) + \sigma_i(D^u_{i,k\ell} - D^\phi_{im}\,B_{mk,\ell})]\,\dot{u}_\ell$$
$$+ (D^u_{i,k} - D^\phi_{im}B_{mk})\,S_{ij}\,D^\phi_{j\ell}\,\dot{\phi}_\ell = \dot{f}^u_k,$$
$$D^\phi_{i\ell}\,S_{ij}\,(D^u_{j,m} - D^\phi_{jn}B_{nm})\,\dot{u}_m + D^\phi_{i\ell}S_{ij}D^\phi_{jm}\dot{\phi}_m = \dot{f}^\phi_\ell. \qquad (4.11)$$

Since $B_{mk,\ell}$ will in general be unsymmetric in the indices k and ℓ, the symmetry of the geometrical stiffness matrix is lost. It should be observed however that the geometrical nonlinearity of the bending deformations, embodied in the transformation $\underset{\sim}{B}$, need only be taken into account in the case of rotations for which no longer holds $\phi^2 \ll 1$. This in contrast with the geometrical nonlinearity of the membrane deformations, which for instance determines the buckling behaviour [7].

It has been shown in [8] that for doubly curved shell elements, provided they are taken sufficiently small, the bending behaviour can be approximated to any desired degree of accuracy by the bending behaviour of the flat triangular element. For the membrane behaviour a more accurate description for equal size of elements is obtained by going from the flat plate approximation to the flat plate with initial curvature, determined by the midside initial displacements w^{*o}.

Finally it can be observed that in case of distributed loading normal to the surface no improvement of accuracy may be expected going from equivalent nodal forces corresponding to a linear deflection interpolation to equivalent nodal forces found by the quadratic interpolation which is possible for the TRIM-6 element. This is due to the fact that the bending properties of the element are derived from a moment distribution, that satisfies the homogeneous equilibrium equations (4.4a). However, as it was mentioned in [6], for pressure loaded shells, that permit a membrane solution of the equilibrium equations, we can enter the finite element into the structure in a state of initial membrane deformation with the corresponding membrane forces acting on the surrounding structure. Thus spurious bending stresses are avoided.

References

1. C. Truesdell and R. Toupin, 'The classical field theories', p. 596, Encyclopedia of Physics, Vol. III/1, Springer Verlag, Berlin, Göttingen, Heidelberg (1960).
2. J.F. Besseling, 'Finite element methods', Trends in Solid Mechanics, Delft University Press, Sijthoff and Noordhoff Int. Publ. (1979), pp. 53-78.
3. J.F. Besseling, 'A thermodynamic approach to rheology', Proc. IUTAM Symp. on irreversible aspects of continuum mechanics, Springer Verlag, Wien (1968) pp. 16-53.
4. O.C. Zienkiewicz, The finite element method, third edition, McGraw-Hill Book Co., London (1977).
5. J.F. Besseling, Postbuckling and nonlinear analysis by the finite element method as a supplement to a linear analysis, ZAMM 55, (1975), pp. T3-T16.
6. J.F. Besseling, L.J. Ernst, A.U. de Koning, E. Riks, K. van der Werff, 'Geometrical and physical nonlinearities, some developments in the Netherlands', Proc. Fenomech 1978, North Holland Publ. Co., Amsterdam; Comp. Meths. Appl. Mech. Eng. 17/18 (1979), pp. 131-157.
7. L.J. Ernst, A geometrically nonlinear finite element shell theory; applications to the postbuckling behaviour of shells, Dept. Mech. Eng. T.H. Delft, WTHD-126 (1980).
8. L.J. Ernst, A finite element approach to shell problems, Dept. Mech. Eng. T.H. Delft, WTHD-114 (1979).

Appendix

The coordinates in the x-y plane of the three cornerpoints of a triangular element in the undeformed state define the following geometrical quantities

$$a_1 = x_3 - x_2 , \quad b_1 = y_2 - y_3 , \quad \ell_1^2 = a_1^2 + b_1^2 ,$$
$$a_2 = x_1 - x_3 , \quad b_2 = y_3 - y_1 , \quad \ell_2^2 = a_2^2 + b_2^2 ,$$
$$a_3 = x_2 - x_1 , \quad b_3 = y_1 - y_2 , \quad \ell_3^2 = a_3^2 + b_3^2 ,$$

and

$$2A = a_3 b_2 - a_2 b_3 = a_1 b_3 - a_3 b_1 = a_2 b_1 - a_1 b_2 .$$

For each element the following vectors are introduced where u, v, w represent displacement components of a cornerpoint.

$$d_1 = \frac{1}{2A}\begin{vmatrix} b_1 \\ b_2 \\ b_3 \end{vmatrix}, \quad d_2 = \frac{1}{2A}\begin{vmatrix} a_1 \\ a_2 \\ a_3 \end{vmatrix}, \quad u^e = \begin{vmatrix} u_1 \\ u_2 \\ u_3 \end{vmatrix}, \quad v^e = \begin{vmatrix} v_1 \\ v_2 \\ v_3 \end{vmatrix}, \quad w^e = \begin{vmatrix} w_1 \\ w_2 \\ w_3 \end{vmatrix} .$$

If the deformations of the basic triangle may be neglected its rotation can be characterized by the orthogonal transformation

$$R^T = \begin{bmatrix} 1+b^T u^e & a^T u^e & -b^T w^e + v^{eT}(ba^T - ab^T)\, w^e \\ b^T v^e & 1+a^T v^e & -a^T w^e - u^{eT}(ba^T - ab^T)\, w^e \\ b^T w^e & a^T w^e & 1 + b^T u^e + a^T v^e + u^{eT}(ba^T - ab^T)\, v^e \end{bmatrix}$$

Now the transformation coefficients $A_{k\ell}$ in (4.9) follow from

$$w^* = R_{3k} u_k , \quad k = 1, 2, 3.$$

The angular velocity of the basic triangle is determined by the spinmatrix

$$\Omega = \dot{R}^T R = \begin{bmatrix} 0 & -\omega_z & \omega_y \\ \omega_z & 0 & -\omega_x \\ -\omega_y & \omega_x & 0 \end{bmatrix} .$$

The tangential unit vector along the side pq of the basic triangle is given by

$$\underline{t}_i = \frac{x_q - x_p + u_q - u_p}{\ell_i}\, \underline{e}_x + \frac{y_q - y_p + v_q - v_p}{\ell_1}\, \underline{e}_y \,\frac{\omega_q - \omega_p}{\ell_i}\, \underline{e}_z .$$

The tangential component of the angular velocity of the basic triangle about the side pq, expressed in terms of u_k and $\dot{u}_k$, follows as the scalar product of the vectors $\underline{\omega}$ and $\underline{t}_i$. Thus the transformation coefficients $B_{k\ell}$ in (4.9) are determined.

New General and Complementary Energy Theorems, Finite Strain, Rate Sensitive Inelasticity and Finite Elements: Some Computational Studies

S. N. ATLURI, H. MURAKAWA
Georgia Institute of Technology, Atlanta, USA

Abstract:

General variational theorems for the rate problems of rate-dependent finite strain inelasticity, in terms of the appropriate rates of the first and second Piola-Kirchhoff stress tensors, the symmetrized Biot-Lure' stress tensor, and their conjugate measures of strain-rate, are discussed. Certain new rate-complementary-energy principles, involving the rate of spin and the rate of the symmetrized Biot-Lure' stress tensor as variables, are stated for finite strain analysis of rate-sensitive materials, such as those exhibiting elasto-visco-plastic and creep behavior. Uniqueness and stability criteria for those inelastic solids, using the finite element counterparts of the new complementary energy rate principles, are discussed. Computational studies, using the complementary energy methods, discussed herein include: (*i*) bifurcation necking and post-buckling analyses of initially perfect elasto-plastic bars, and (*ii*) post-buckling and large deformation analyses of thin elastic plates under inplane compression and transverse bending loads.

Introduction:

The topic of rate (incremental), multi-field, variational principles, in general, and the rate complementary energy principles, in particular, and the corresponding finite element methods, for finite strain analysis of compressible non-linear-elastic solids were discussed in detail by Atluri and Murakawa [1]. Also discussed in [1] were the contributions of Koiter, Zubov, and Fraeijs de Veubeke, dealing with the subject of complementary energy principles, governing the *total* deformations of semi-linear and/or nonlinear compressible isotropic elastic materials. It was shown in [1], that the concept of treating the angular momentum balance condition as an a posteriori constraint through a complementary energy principle involving the symmetrized Biot-stress (or what is also referred to as the symmetrized Lure'-stress or the Jaumann-stress) as well as the orthogonal tensor of rigid rotation, as variables, as first introduced by F. de Veubeke, has certain fundamentally novel features that makes it attractive for practical application.

The ideas of discretizing the angular momentum balance conditions through a complementary energy principle has been extended by the authors in (*i*) the incremental (rate) analysis of finite strains in compressible as well as incompressible nonlinear elastic materials [2-4], (*ii*) the rate problems of classical (rate-independent) finite strain, elasto-plasticity [5-9], and (*iii*) nonlinear stability and post-bifurcation analysis of semilinear isotropic elastic beams [10].

In the present paper the authors' earlier work, [2-10], is extended to the cases of finite strain analyses of materials with rate-sensitive behavior such as elasto-viscoplasticity and creep, and post-buckling and large-deformation behavior of structural members such as plates and shells, undergoing large rotations and large stretches.

The summary of the topics presented in the following is: (*i*) a discussion of general (multi-field) variational principles, with emphasis on complementary energy, in terms of alternate stress-rates and conjugate measures of strain-rate, for rate-sensitive inelastic materials, (*ii*) rate complementary energy potentials, for the chosen stress-rates, for rate-dependent as well as rate-independent materials, (*iii*) criteria for uniqueness and stability of solutions, (*iv*) numerical study of necking of an initially perfect elasto-plastic bar, and (*v*) numerical study of post-buckling of an axially compressed plate of semilinear isotropic elastic material, undergoing large rotations, as well that of a thin plate undergoing large displacements due to transverse loading.

PRELIMINARIES:

We use a fixed rectangular cartesian coordinate system, and employ the notation: ($\tilde{}$) denotes a second-order tensor; ($\approx$) denotes a fourth-order tensor; (-) implies a vector; $\underline{a} = \tilde{A}.\underline{b}$ implies $a_i = A_{ij}b_j$; $\tilde{A}.\tilde{B}$ implies a product such that $(\tilde{A}.\tilde{B})_{ij} = A_{ij}B_{jk}$; $\tilde{A}:\tilde{B} = \text{trace}\,(\tilde{A}^T.\tilde{B}) = A_{ij}B_{ij}$; and $\underline{u}.\underline{t} = u_i t_i$. A particle in the undeformed body has a position vector $\underline{x} = x_\alpha \underline{e}_\alpha (\alpha=1..3)$ where $\underline{e}_\alpha$ are unit cartesian bases. The gradient operator $\underline{\nabla}^o$ in the undeformed configuration C_o is $\underline{\nabla}^o = (\underline{e}_\alpha \partial/\partial x_\alpha)$. The position vector of the particle in the deformed configuration, say C_N, is $\underline{y} = y_i \underline{e}_i$. The gradient operator in C_N is $\underline{\nabla}^N = (\underline{e}_i \partial/\partial y_i)$. The deformation gradient tensor is $\tilde{F} \equiv (\underline{\nabla}^o \underline{y})^T$, such that $F_{i\alpha} = y_{i,\alpha} \equiv (\partial y_i/\partial x_\alpha)$. The nonsingular $\tilde{F}$ has the polar-decomposition, $\tilde{F} = \tilde{\alpha}.(\tilde{I}+\tilde{h})$ where the rotation $\tilde{\alpha}$ is orthogonal and the stretch $\tilde{h}$ is symmetric and +ve definite. The Green-Lagrange strain is $\tilde{g} = 1/2(\tilde{F}^T.\tilde{F}-\tilde{I}) = 1/2(\tilde{e}+\tilde{e}^T+\tilde{e}^T.\tilde{e})$ where $\tilde{e} = (\underline{\nabla}^o \underline{u})^T$, $\underline{u} = \underline{y}-\underline{x}$.

For the present purposes, we introduce the stress measures (*i*) the "true" Cauchy stress $\tilde{\tau}$; (*ii*) a weighted tensor, the Kirchhoff stress tensor $\tilde{\sigma} = J\tilde{\tau}$ where J is determinant of matrix $[y_{i,\alpha}]$; (*iii*) the first Piola-Kirchhoff stress tensor $\tilde{t}$; (*iv*) the second Piola-Kirchhoff stress tensor $\tilde{s}$, and (*v*) the symmetrized Biot stress tensor (or what is also often referred to as the symmetrized Lure', or the Jaumann stress tensor) $\tilde{r}$. As discussed in [1,5], and elsewhere, the above stress measures are related as:

$$\tilde{\tau} = \frac{1}{J}\tilde{F}.\tilde{t} = \frac{1}{J}\tilde{F}.\tilde{s}.\tilde{F}^T = \frac{1}{J}\tilde{\sigma} \tag{1}$$

$$\tilde{t} = \tilde{s}.\tilde{F}^T = J(\tilde{F}^{-1}.\tilde{\tau}); \quad \tilde{s} = J(\tilde{F}^{-1}.\tilde{\tau}.\tilde{F}^{-T}) \tag{2}$$

$$\tilde{r} = \frac{1}{2}(\tilde{t}.\tilde{\alpha}+\tilde{\alpha}^T.\tilde{t}^T) = \frac{1}{2}[\tilde{s}.(\tilde{I}+\tilde{h}) + (\tilde{I}+\tilde{h}).\tilde{s}] \tag{3}$$

The tensors $\tilde{\tau}$, $\tilde{\sigma}$, $\tilde{s}$, and $\tilde{r}$ are symmetric, while $\tilde{t}$ is unsymmetric. In the above, $\tilde{F}^{-T} = (\tilde{F}^{-1})^T$ and the superscript T denotes a transpose.

RATE FORMULATIONS:

Now, we consider the (incremental) rate analysis of finite strain problems of an inelastic solid with a rate-sensitive constitutive law. In doing so, one can choose an arbitrary reference frame. In practice, however, two choices, one the so-called total-Lagrangean (TL) and the other, the so-called updated-Lagrangean (UL) reference frames are appealing. Eventhough the choice of a reference frame does not, per se, affect the theoretical or computational approaches, we discuss the details of a UL formulation, since the rate consitutive relations of an inelastic solid depend, naturally, on the current state of true stress.

In the UL formulation, the solution variables in the generic state C_{N+1} are referred to the configuration of the body in the immediately preceding state, C_N, which is known. In the UL formulation, one is essentially concerned with an initial stress problem: the initial "true" stress in C_N is the Cauchy stress $\underset{\sim}{\tau}^N$, while the initial displacements in C_N as referred to C_N are, obviously, zero. Let y_i^N be the current spatial coordinates of a particle in C_N. Let $\underline{\nabla}^N$ be the gradient operator in C_N (ie., $\underline{\nabla}^N = \underline{e}_i \partial/\partial y_i^N$) and let $\underline{\dot{u}}$ be rate of deformation (velocities) from C_N. We define the rate of displacement gradient $\underset{\sim}{\dot{e}} = (\underline{\nabla}^N \underline{\dot{u}})^T$ and write $\underset{\sim}{\dot{e}} = \underset{\sim}{\dot{\varepsilon}} + \underset{\sim}{\dot{\omega}}$ where $\underset{\sim}{\dot{\varepsilon}}[\dot{\varepsilon}_{ij} = \frac{1}{2}(\partial \dot{u}_i/\partial y_j^N + \partial \dot{u}_j/\partial y_i^N)]$ is the symmetric UL strain-rate and $\underset{\sim}{\dot{\omega}}[\dot{\omega}_{ij} = \frac{1}{2}(\partial \dot{u}_i/\partial y_j^N - \partial \dot{u}_j/\partial y_i^N)$ is the skew-symmetric spin-rate. Let $\underset{\sim}{\dot{\tau}}$, and $\underset{\sim}{\dot{\sigma}}(\equiv J^N[\underset{\sim}{\dot{\tau}} + (\underset{\sim}{\dot{\varepsilon}}:\underset{\sim}{I})\underset{\sim}{\tau}^N])$ (where $J^N = \rho_o/\rho^N$, ρ_o and ρ_N being the mass-densities in C_o and C_N, respectively), be the substantial derivatives of the Cauchy and Kirchhoff stresses respectively. As is well-known, these stress-rates are not objective. Let $\underset{\sim}{\dot{t}}$, $\underset{\sim}{\dot{s}}$, and $\underset{\sim}{\dot{r}}$ represent the appropriate stress rates referred to C_N; ie., for instance, $\underset{\sim}{\dot{s}}\Delta t = \underset{\sim}{s}_N^{N+1} - \underset{\sim}{\tau}^N$ where $\underset{\sim}{s}_N^{N+1}$ is the second Piola-Kirchhoff stress in C_{N+1} referred to (and measured per unit area in) C_N. It is shown in [5] that:

$$\underset{\sim}{\dot{s}} = (\underset{\sim}{\dot{\sigma}} - \underset{\sim}{\dot{e}}.\underset{\sim}{\sigma}^N - \underset{\sim}{\sigma}^N.\underset{\sim}{\dot{e}}^T)/J^N; \quad \underset{\sim}{\dot{t}} = (\underset{\sim}{\dot{\sigma}} - \underset{\sim}{\dot{e}}.\underset{\sim}{\sigma}^N)/J^N \tag{4a,b}$$

$$\underset{\sim}{\dot{r}} = \frac{1}{2}(\underset{\sim}{\dot{t}} + \underset{\sim}{\dot{t}}^T + \underset{\sim}{\tau}^N.\underset{\sim}{\dot{\omega}} + \underset{\sim}{\dot{\omega}}^T.\underset{\sim}{\tau}^N) = \underset{\sim}{\dot{s}} + \frac{1}{2}(\underset{\sim}{\tau}^N.\underset{\sim}{\dot{\varepsilon}} + \underset{\sim}{\dot{\varepsilon}}.\underset{\sim}{\tau}^N) \tag{5a,b}$$

Unless large elastic deformations of a dilatational nature have preceded the inelastic straining, one may, without significant error, assume that $J^N \simeq 1.0$.

The equations of linear momentum balance (LMB), angular momentum balance (AMB), compatibility, and traction and displacement boundary conditions (TBC and DBC) in the UL rate formulation can be written as:

$$\text{LMB:} \quad \underline{\nabla}^N.[\underset{\sim}{\dot{s}} + \underset{\sim}{\tau}^N.(\underline{\nabla}^N \underline{\dot{u}})] + \rho^N \underline{\dot{B}} = \underline{0} \;\text{(or)}\; \underline{\nabla}^N.\underset{\sim}{\dot{t}} + \rho^N \underline{\dot{B}} = \underline{0} \tag{6a,b}$$

AMB: $\dot{\underset{\sim}{s}} = \dot{\underset{\sim}{s}}^T$; (or) $(\underline{\nabla}^N\dot{\underline{u}})^T.\underset{\sim}{\tau}^N + \dot{\underset{\sim}{t}} = \dot{\underset{\sim}{t}}^T + \underset{\sim}{\tau}^N.(\underline{\nabla}^N\dot{\underline{u}})$ (7a,b)

or, equivalently,

$$\dot{\underset{\sim}{\omega}}.\underset{\sim}{\tau}^N + \dot{\underset{\sim}{\varepsilon}}.\underset{\sim}{\tau}^N + \dot{\underset{\sim}{t}} = \dot{\underset{\sim}{t}}^T + \underset{\sim}{\tau}^N.\dot{\underset{\sim}{\varepsilon}} + \underset{\sim}{\tau}^N.\dot{\underset{\sim}{\omega}}^T \qquad (7c)$$

compatibility:

$$\dot{\underset{\sim}{e}} \equiv \dot{\underset{\sim}{\varepsilon}} + \dot{\underset{\sim}{\omega}} = (\underline{\nabla}^N\dot{\underline{u}})^T;\ \text{(or)}\ \dot{\underset{\sim}{\varepsilon}} = (\underline{\nabla}^N\dot{\underline{u}}) + (\underline{\nabla}^N\dot{\underline{u}})^T \qquad (8a,b)$$

TBC: $\underline{n}^*.[\dot{\underset{\sim}{s}} + \underset{\sim}{\tau}^N.(\underline{\nabla}^N\dot{\underline{u}})] \equiv \underline{n}^*.[\dot{\underset{\sim}{t}}] = \dot{\bar{\underline{t}}}$ at $S_{\sigma N}$ (9)

DBC: $\dot{\underline{u}} = \dot{\bar{\underline{u}}}$ at S_{uN} (10)

Let us suppose, for the moment, that the consitutive law for a rate-dependent material can be expressed (as shown later in this paper) in terms of certain rate-potentials, as:

$$\dot{\underset{\sim}{s}} = \partial\dot{W}/\partial\dot{\underset{\sim}{\varepsilon}};\ \dot{\underset{\sim}{t}} = \partial\dot{U}/\partial\dot{\underset{\sim}{e}}^T;\ \dot{\underset{\sim}{r}} = \partial\dot{Q}/\partial\dot{\underset{\sim}{\varepsilon}} \qquad (11a\text{-}c)$$

We consider the Legendre (contact) transformations of the type:

$$\dot{\underset{\sim}{s}}:\dot{\underset{\sim}{\varepsilon}} - \dot{W}(\dot{\underset{\sim}{\varepsilon}}) = \dot{S}^*(\dot{\underset{\sim}{s}});\ \dot{\underset{\sim}{t}}^T:\dot{\underset{\sim}{e}} - \dot{U}(\dot{\underset{\sim}{e}}) = \dot{E}^*(\dot{\underset{\sim}{t}});$$
$$\dot{\underset{\sim}{r}}:\dot{\underset{\sim}{\varepsilon}} - \dot{Q}(\dot{\underset{\sim}{\varepsilon}}) = \dot{R}^*(\dot{\underset{\sim}{r}}) \qquad (12a\text{-}c)$$

such that

$$\partial\dot{S}^*/\partial\dot{\underset{\sim}{s}} = \dot{\underset{\sim}{\varepsilon}};\ \partial\dot{E}^*/\partial\dot{\underset{\sim}{t}} = \dot{\underset{\sim}{e}}^T;\ \partial\dot{R}^*/\partial\dot{\underset{\sim}{r}} = \dot{\underset{\sim}{\varepsilon}}$$

As discussed in [1-6], the AMB conditions are embedded in the structure of $\dot{W}$ and $\dot{U}$. As shown in [5], the complementary energy principles, and the Hellinger-Reissner type principles, involving the stress-rates $\dot{\underset{\sim}{s}}$, $\dot{\underset{\sim}{t}}$, and $\dot{\underset{\sim}{r}}$ are as below. In each case the functional whose stationary condition is the principle in question is given. The respective functionals are, denoted by π with the subscript C and HR denoting complementary and Hellinger-Reissner type functionals, respectively.

$$\pi_C(\dot{\underline{u}},\dot{\underset{\sim}{s}}) = \int_{V_N}\{-\dot{S}^*(\dot{\underset{\sim}{s}}) - \tfrac{1}{2}\underset{\sim}{\tau}^N:[(\underline{\nabla}^N\dot{\underline{u}}).(\underline{\nabla}^N\dot{\underline{u}})^T]\}dv + \int_{S_{UN}}\dot{\underline{t}}.\dot{\bar{\underline{u}}}ds \qquad (13)$$

$$\pi_{HR}(\dot{\underline{u}},\dot{\underset{\sim}{s}}) = \int_{V_N}\{-\dot{S}^*(\underset{\sim}{s}) + \rho^N\dot{\underline{B}}.\dot{\underline{u}} + \tfrac{1}{2}\underset{\sim}{\tau}^N:[(\underline{\nabla}^N\dot{\underline{u}}).(\underline{\nabla}^N\dot{\underline{u}})^T]$$
$$+ \tfrac{1}{2}\dot{\underset{\sim}{s}}:[(\underline{\nabla}^N\dot{\underline{u}}) + (\underline{\nabla}^N\dot{\underline{u}})^T]\}dv - \int_{S_{\sigma N}}\dot{\bar{\underline{t}}}.\dot{\underline{u}}ds - \int_{S_{uN}}\dot{\underline{t}}.(\dot{\underline{u}}-\dot{\bar{\underline{u}}})ds \qquad (14)$$

$$\pi_C(\dot{\underset{\sim}{t}}) = \int_{V_N}-\dot{E}^*(\dot{\underset{\sim}{t}})dv + \int_{S_{uN}}\dot{\underline{t}}.\dot{\bar{\underline{u}}}\ ds \qquad (15)$$

$$\pi_{HR}(\dot{\underset{\sim}{t}},\dot{\underline{u}}) = \int_{V_N}\{-\dot{E}^*(\dot{\underset{\sim}{t}})dv - \rho^N\dot{\underline{B}}.\dot{\underline{u}} + \dot{\underset{\sim}{t}}^T:[(\underline{\nabla}^N\dot{\underline{u}})^T]\}dv \quad + \text{contd.}$$

$$- \int_{S_{\sigma N}} \dot{\bar{t}}.\dot{\underline{u}}ds - \int_{S_{uN}} \dot{\underline{t}}.(\dot{\underline{u}}-\dot{\bar{\underline{u}}})ds \qquad (16)$$

$$\pi_C(\dot{\underset{\sim}{t}},\dot{\underset{\sim}{\omega}}) = \int_{V_N} \{-\dot{R}^*(\dot{\underset{\sim}{r}}) + \frac{1}{2}\underset{\sim}{\tau}^N:(\dot{\underset{\sim}{\omega}}^T.\dot{\underset{\sim}{\omega}}) - \dot{\underset{\sim}{t}}^T:\dot{\underset{\sim}{\omega}}\}dv + \int_{S_{uN}} \dot{\underline{t}}.\dot{\bar{\underline{u}}}ds \qquad (17)$$

$$\pi_{HR}(\dot{\underset{\sim}{t}};\dot{\underset{\sim}{\omega}};\dot{\underline{u}}) = \int_{V_N} \{-\dot{R}^*(\dot{\underset{\sim}{r}}) + \frac{1}{2}\underset{\sim}{\tau}^N:(\dot{\underset{\sim}{\omega}}^T.\dot{\underset{\sim}{\omega}}) - \rho^N\dot{\underline{B}}.\dot{\underline{u}} + \dot{\underset{\sim}{t}}^T:[\underline{\nabla}^N\dot{\underline{u}}]^T - \dot{\underset{\sim}{t}}^T:\dot{\underset{\sim}{\omega}}\}dv$$

$$- \int_{S_{\sigma N}} \dot{\bar{\underline{t}}}.\dot{\underline{u}}ds - \int_{S_{uN}} \dot{\underline{t}}.(\dot{\underline{u}}-\dot{\bar{\underline{u}}})ds \qquad (18)$$

In the above V_N, $S_{\sigma N}$, and S_{uN} are the volume, prescribed-traction boundary, and prescribed displacement boundary, respectively, of the solid in C_N; and ρ^N and $\dot{\underline{B}}$, are respectively, the mass-density and rate-of-body-force in C_N. The above functionals are valid, in general, for non-conservative (deformation-dependent) surface tractions. If A is the set of conditions that are satisfied a priori, and B is the set of those conditions that are satisfied a posteriori in the variational principle, for each of the above functionals the sets A and B are as follows: (*i*) Eq. (13): Set A(Eqs. 6a, 7a, and 9), set B(Eqs. 8b, and 10) (*ii*) Eq. (14): Set A (the existence of $\dot{S}^*$ such that $\partial\dot{S}^*/\partial\dot{\underset{\sim}{s}} = \dot{\underset{\sim}{\varepsilon}}$, and Eq. 7a), set B(Eqs. 6a, 8b, 9, and 10). (*iii*) Eq. (17): Set A(Eqs. 6b, 9, the definition of $\dot{\underset{\sim}{r}}$ as in Eq. 5a and that $\dot{\underset{\sim}{\omega}}$ is skew-symmetric), set B(Eqs. 7c, 8a, and 10); (*iv*) Eq. 18: Set A(the definition of $\dot{\underset{\sim}{r}}$ as in Eq. (5a), and that $\dot{\underset{\sim}{\omega}}$ is skew symmetric), set B(6b, 7c, 8a, 9, and 10).

The complementary, and Hellinger-Reissner type principles as through Eqs. (17) and (18) were first stated by Atluri [5]. The general invalidity of the principles through Eq. (15) which was alluded to by Hill [11], and Eq. (16), were discussed in [5]. Eventhough, Eqs. (13) and (14), and the attendant variational principles, may be viewed as being consistent, the limitations of practical applicability of these are discussed in [5]. Especially, in the application of Eq. (13), the need to select a symmetric $\dot{\underset{\sim}{s}}$, such that Eq. (6a) (which involved coupling with $\dot{\underline{u}}$) is satisfied, a priori, is not an altogether easy proposition. Several interesting ways of satisfying Eq. (6a), and of application of Eq. (13), were discussed by Atluri [6,7,12].

However, the complementary principle of Eq. (17), introduced in [5], has several attractive features for practical application: (*i*) the LMB, Eq. (6b) can be easily satisfied by setting: $\dot{\underset{\sim}{t}} = \underline{\nabla}^N x \underset{\sim}{\psi} + \dot{\underset{\sim}{t}}^p$ where ψ are first-order (once differentiable) stress functions (*ii*) $\dot{\underset{\sim}{\omega}}$ can be selected to be skew symmetric, by setting $\dot{\omega}_{ij} = e_{kij}\Omega_k$ where e_{kij} is an alternating tensor). In general, even in a TL formulation, the constraint $\underset{\sim}{\alpha}.\underset{\sim}{\alpha}^T = I$ is easily met by taking $\underset{\sim}{\alpha}$ to be a function of the 3 Euler-angles of rigid rotation [4]. In the case of plates and shells, the concepts of a finite-rotation vector, as discussed later, may be employed.

The application of the complementary energy principle as through Eq. (17), and its TL rate counterpart, is illustrated later in this paper.

RATE POTENTIALS FOR RATE-SENSITIVE MATERIALS:

As discussed in [5], and elsewhere, the principle of objectivity is met, in writing the rate constitutive law of the material, by postulating the constitutive relation between the objective strain-rate $\dot{\underset{\sim}{\varepsilon}}$ and the objective co-rotational (or also at times referred to as the Zaremba, or the rigid-body or the Jaumann) rate of Kirchhoff stress, denoted here by $\overset{\circ}{\underset{\sim}{\sigma}}{}^*$. It is well-known that,

$$\overset{\circ}{\underset{\sim}{\sigma}}{}^* = \dot{\underset{\sim}{\sigma}} - \dot{\underset{\sim}{\omega}}.\underset{\sim}{\sigma}^N - \underset{\sim}{\sigma}^N.\dot{\underset{\sim}{\omega}}^T \tag{19}$$

Thus, in view of Eqs. (4-5),

$$\dot{\underset{\sim}{s}} = (\overset{\circ}{\underset{\sim}{\sigma}}{}^* - \dot{\underset{\sim}{\varepsilon}}.\underset{\sim}{\sigma}^N - \underset{\sim}{\sigma}^N.\dot{\underset{\sim}{\varepsilon}})/J^N; \quad \dot{\underset{\sim}{t}} = (\overset{\circ}{\underset{\sim}{\sigma}}{}^* - \dot{\underset{\sim}{\varepsilon}}.\underset{\sim}{\sigma}^N - \underset{\sim}{\sigma}^N.\dot{\underset{\sim}{\omega}})/J^N$$

$$\dot{\underset{\sim}{r}} = [\overset{\circ}{\underset{\sim}{\sigma}}{}^* - \tfrac{1}{2}(\dot{\underset{\sim}{\varepsilon}}.\underset{\sim}{\sigma}^N + \underset{\sim}{\sigma}^N.\dot{\underset{\sim}{\varepsilon}})]/J^N \tag{21}$$

Thus, if $\dot{V}$ is the postulated rate potential for $\overset{\circ}{\underset{\sim}{\sigma}}{}^*$ such that $\partial\dot{V}/\partial\dot{\underset{\sim}{\varepsilon}} = \overset{\circ}{\underset{\sim}{\sigma}}{}^*$, we can define:

$$J^N\dot{W} = \dot{V} - \underset{\sim}{\sigma}^N:(\dot{\underset{\sim}{\varepsilon}}.\dot{\underset{\sim}{\varepsilon}}); \quad \partial\dot{W}/\partial\dot{\underset{\sim}{\varepsilon}} = \dot{\underset{\sim}{s}} \tag{22}$$

$$J^N\dot{U} = \dot{W} + \tfrac{1}{2}\underset{\sim}{\sigma}^N:(\dot{\underset{\sim}{e}}^T.\dot{\underset{\sim}{e}}); \quad \partial\dot{U}/\partial\dot{\underset{\sim}{e}}^T = \dot{\underset{\sim}{t}} \tag{23}$$

$$J^N\dot{Q} = \dot{V} - \tfrac{1}{2}\underset{\sim}{\sigma}^N:(\dot{\underset{\sim}{\varepsilon}}.\dot{\underset{\sim}{\varepsilon}}); \quad \partial\dot{Q}/\partial\dot{\underset{\sim}{\varepsilon}} = \dot{\underset{\sim}{r}} \tag{24}$$

From these, one can establish $\dot{S}^*(\dot{\underset{\sim}{s}})$, $\dot{E}^*(\dot{\underset{\sim}{t}})$ and $\dot{R}^*(\dot{\underset{\sim}{r}})$ as defined in Eqs. (12a-c). Thus, we focus attention on the potential $\dot{V}$.

It is worth noting that for materials with rate-independent constitutive laws, such as classical elastic-plastic materials the derivative ($\dot{}$) is considered to be with respect to a fictition time. However, for rate-sensitive materials, such as elasto-viscoplastic and creeping materials, the derivative ($\dot{}$) is w.r.t. to natural time.

For rate-independent classical elastic-plastic materials, Hill [13] presented the postulation:

$$\dot{V} = \frac{1}{2} L_{ijkl}\dot{\varepsilon}_{ij}\dot{\varepsilon}_{kl} - \frac{\alpha}{g}(\lambda_{kl}\dot{\varepsilon}_{kl})^2 \tag{25}$$

where L_{ijkl} is a +ve definite symmetric (under $ij \leftrightarrow kl$ interchange) tensor of instantaneous elastic modulii, $\alpha = 1$ or 0 according to whether $\lambda_{kl}\,\dot{\varepsilon}_{kl}$ is positive or negative, g is a scalar related to the measure of hardening, and λ_{ij} is a tensor normal to the interface between elastic and plastic domain in the $\dot{\varepsilon}_{kl}$ space. Prandtl-Reuss type rate equations of type (25) for classical isotropically hardening materials can easily be derived, formally, to be [5]:

$$\dot{\underset{\sim}{\sigma}}^* = 2\mu\dot{\underset{\sim}{\varepsilon}} + \lambda(\dot{\underset{\sim}{\varepsilon}}:\underset{\sim}{I})\underset{\sim}{I} - 12\alpha\mu^2 \frac{(\dot{\underset{\sim}{\varepsilon}}:\underset{\sim}{\sigma}')\underset{\sim}{\sigma}'}{(\underset{\sim}{\sigma}':\underset{\sim}{\sigma}')[6\mu+2(\partial F_o/\partial W^P)]} \tag{26}$$

where, λ and μ are Lame' constants, $\underset{\sim}{\sigma}' \equiv \underset{\sim}{\sigma} - 1/3(\underset{\sim}{\sigma}:\underset{\sim}{I})\underset{\sim}{I}$ is deviatoric Kirchhoff stress, and the yield-surface is represented by $F = [3J_2(\underset{\sim}{\sigma}')]^{\frac{1}{2}} - F_o = 0$.
Here $F_o = F_o(W^p)$ where $W^p = \int \underset{\sim}{\sigma}:\dot{\underset{\sim}{\varepsilon}}^p dt$; and $J_2(\underset{\sim}{\sigma}') = (1/2)(\underset{\sim}{\sigma}':\underset{\sim}{\sigma}')$.
A rate-sensitive constitutive law of a considerable generality, as given by Perzyna [14], can be easily written for finite strains, when an associative flow-rule is used, as:

$$\dot{\underset{\sim}{\varepsilon}}^a = \gamma < \phi(F) > \frac{\partial F}{\partial \underset{\sim}{\sigma}} \tag{27}$$

where $<\phi>$ denotes a specific function, such that $<\phi> = \phi(F)$ for $F>0$, and $\phi=0$ for $F\leq 0$. The parameter γ is called the fluidity parameter and $\dot{\underset{\sim}{\varepsilon}}^a$ is the general anelastic strain. It has been shown by Zienkiewicz and Coworkers [15], and Argyris and Coworkers [16] that classical, rate-independent elasto-plastic solutions can be obtained from the above theory, when (*i*) either $\gamma \to \infty$ or (*ii*) a stationary solution of the visco-plastic flow is sought. Various forms of ϕ were reviewed by Perzyna [14]. For the Hencky-Mises-Huber yield criterion, one can define F to be:

$$F = [3J_2(\underset{\sim}{\sigma}')]^{\frac{1}{2}} - F_o \equiv \sigma_{eq} - F_o \tag{28}$$

where σ_{eq} is the equivalent Kirchhoff stress. A simple choice for $\phi(F)$ can be:

$$\phi(F) = F^n \tag{29}$$

With Eq. (29), the viscoplastic strain rate as in Eq. (27) can be seen to correspond to the well-known Norton's power law for steady-state creep when $F_o \to 0$. We now derive rate potentials $\Delta V(=\dot{V}\Delta t), \Delta W$, etc. for the viscoplastic constitutive laws given by Eqs. (27-29).

Let at times t_N and $t_N+\Delta t$, the Kirchhoff stresses be $\underset{\sim}{\sigma}^N + \overset{\circ}{\underset{\sim}{\sigma}}\Delta t \equiv \underset{\sim}{\sigma}^N + \Delta\underset{\sim}{\sigma}$, respectively, where $\overset{\circ}{\underset{\sim}{\sigma}}$ is the substantial derivative. The inelastic strain-rates corresponding to Eqs. (27-29) at times t_N and $t_N+\Delta t$, are given, respectively, by:

$$\dot{\underset{\sim}{\varepsilon}}^a_N \equiv (\dot{\underset{\sim}{\varepsilon}}^a)_N = \gamma F^N (\partial F/\partial \underset{\sim}{\sigma}) = \gamma (3/2\sigma_{eq})(\sigma_{eq}-F_o)^n \underset{\sim}{\sigma}' \tag{30}$$

In the above $\underset{\sim}{\sigma}' \equiv (\underset{\sim}{\sigma}^N)'$ (ie., superscript N dropped for convenience) and $\sigma^2_{eq} = (3/2)\underset{\sim}{\sigma}':\underset{\sim}{\sigma}'$. Likewise at a time Δt later,

$$\dot{\underset{\sim}{\varepsilon}}^a_{N+1} = \gamma(3/2)(\sigma_{eq}+\Delta\sigma_{eq})^{-1}(\sigma_{eq}+\Delta\sigma_{eq}-F_o)^n(\underset{\sim}{\sigma}'+\Delta\underset{\sim}{\sigma}') \tag{31}$$

By straight forward algebra, it can be shown that,

$$\dot{\underset{\sim}{\varepsilon}}^a_{N+1} = \dot{\underset{\sim}{\varepsilon}}^a_N + \gamma(3/2\sigma_{eq})\{-(\Delta\sigma_{eq}/\sigma_{eq})(\sigma_{eq}-F_o)^n\underset{\sim}{\sigma}' + (\sigma_{eq}-F_o)^n\Delta\underset{\sim}{\sigma}'$$
$$+ n(\sigma_{eq}-F_o)^{n-1}\underset{\sim}{\sigma}'\Delta\sigma_{eq}\} \tag{32}$$

However, it can be easily shown that

$$\Delta\sigma_{eq} = (3/2)[(\underset{\sim}{\sigma}':\Delta\underset{\sim}{\sigma}')/\sigma_{eq}] \tag{33}$$

Using (33) in (32) we obtain:

$$\dot{\underset{\sim}{\varepsilon}}^a_{N+1} = \dot{\underset{\sim}{\varepsilon}}^a_N + \underset{\approx}{V}:\Delta\underset{\sim}{\sigma} \tag{34}$$

wherein the definition of $\underset{\approx}{V}$ is apparent. Since $\Delta\underset{\sim}{\sigma}$ can include the effects of pure spin between t_N and t_{N+1}, one can replace Eq. (34) by:

$$\dot{\underset{\sim}{\varepsilon}}^a_{N+1} = \dot{\underset{\sim}{\varepsilon}}^a_N + \underset{\approx}{V}:\Delta\underset{\sim}{\sigma}^* \tag{35}$$

where $\Delta\underset{\sim}{\sigma}^* = \overset{\circ}{\underset{\sim}{\sigma}}{}^*\Delta t$ is the corotational increment of Kirchhoff stress. Now, the corotational rate $\overset{\circ}{\underset{\sim}{\sigma}}{}^*$ can be written as:

$$\overset{\circ}{\underset{\sim}{\sigma}}{}^* = \underset{\approx}{L}_e:(\dot{\underset{\sim}{\varepsilon}} - \dot{\underset{\sim}{\varepsilon}}^a) \tag{36}$$

Where $\underset{\approx}{L}_e$ is the tensor of instantaneous elastic moduli.

$$\text{Thus,}\quad \Delta\underset{\sim}{\sigma}^* = \underset{\approx}{L}_e:\Delta\underset{\sim}{\varepsilon} - L_e:\int_{t_N}^{t_N+\Delta t} \dot{\underset{\sim}{\varepsilon}}^a dt \tag{37}$$

one may use the approximation,

$$\dot{\underset{\sim}{\varepsilon}}^a = (1-\beta)\dot{\underset{\sim}{\varepsilon}}^a_N + \beta\dot{\underset{\sim}{\varepsilon}}^a_{N+1} \qquad t_N \le t \le t_N + \Delta t \tag{38}$$

when Eqs. (35) and (38) are used, Eq. (37) becomes,

$$\Delta\underset{\sim}{\sigma}^* = \underset{\approx}{L}_e : \Delta\underset{\sim}{\varepsilon} - \Delta t \underset{\approx}{L}_e : (\dot{\underset{\sim}{\varepsilon}}^a_N + \beta \underset{\approx}{V} \Delta\underset{\sim}{\sigma}^*) \tag{39}$$

From which, upon rearranging terms,

$$\Delta\underset{\sim}{\sigma}^* = \underset{\approx}{M} : (\Delta\underset{\sim}{\varepsilon} - \Delta\bar{\underset{\sim}{\varepsilon}}_a) \tag{40}$$

wherein the definition of $\underset{\approx}{M}$ is apparent, and $\Delta\bar{\underset{\sim}{\varepsilon}}_a$ is known and is given by: $\Delta\bar{\varepsilon}_a = \Delta t \dot{\varepsilon}^a_N$. From Eq. (40) one can immediately write

$$\Delta V = \frac{1}{2} M_{ijk\ell} \Delta\varepsilon_{ij} \Delta\varepsilon_{k\ell} - M_{ijk\ell} \Delta\bar{\varepsilon}_{aij} \Delta\varepsilon_{k\ell} \tag{41}$$

From Eq. (41), the potentials ΔW, ΔU, and ΔQ can easily be obtained through Eqs. (22-24).

It is noted that Wang [17] attempted to derive a relation between $\Delta\underset{\sim}{s}$, $\Delta\underset{\sim}{\varepsilon}$, $\Delta\bar{\underset{\sim}{\varepsilon}}_a$, and hence ΔW directly. However, this derivation appears to be in error, since, among other reasons, the transformation between the deviatoric part of $\underset{\sim}{s}$ and the deviatoric part of $\underset{\sim}{\tau}$ was assumed to be the same as that between $\underset{\sim}{s}$ and $\underset{\sim}{\tau}$ themselves.

All the above developments for the UL rate formulation can be converted to a TL rate formulation by noting the relations [5]:

$$\underset{\sim}{E}' = (\underset{\sim}{F}^N)^T . \dot{\underset{\sim}{\varepsilon}} . \underset{\sim}{F}^N ; \underset{\sim}{e}' = \dot{\underset{\sim}{e}} . \underset{\sim}{F}^N ; \ \underset{\sim}{s}' = J^N (\underset{\sim}{F}^N)^{-1} . \dot{\underset{\sim}{s}} . (\underset{\sim}{F}^N)^{-T}$$

$$\underset{\sim}{t}' = J^N (F^N)^{-1} . \dot{\underset{\sim}{t}} ; \ \underset{\sim}{r}' = (1/2)[\underset{\sim}{t}^N . \underset{\sim}{\alpha}' + \underset{\sim}{\alpha}'^T . \underset{\sim}{t}^{NT} + \underset{\sim}{t}' . \underset{\sim}{\alpha}^N + \underset{\sim}{\alpha}^{NT} . \underset{\sim}{t}'^T] \tag{42}$$

where $\underset{\sim}{E}'$ is TL rate of Gree-strain, and $\underset{\sim}{e}'$, $\underset{\sim}{s}'$, $\underset{\sim}{t}'$, and $\underset{\sim}{r}'$ are TL rates of $\underset{\sim}{e}$, $\underset{\sim}{s}$, $\underset{\sim}{t}$, and $\underset{\sim}{r}$ respectively. Now $\underset{\sim}{\alpha}'$ is subject to the constraint that $\underset{\sim}{\alpha}^{NT} . \underset{\sim}{\alpha}'$ is skew-symmetric.

UNIQUENESS & STABILITY CRITERION:

In the present paper, the application of the rate complementary energy principle as embodied in Eq. (17), or its TL counterpart, will be used in some computational studies. In the direct application of Eq. (17), the assumed stress field $\dot{\underset{\sim}{t}}$ must not only satisfy the LMB condition within each element, but also the traction reciprocity condition at the interelement boundary, viz., $(\underline{n}^* . \dot{\underset{\sim}{t}})^+ + (\underline{n}^* . \dot{\underset{\sim}{t}})^- = 0$ at ρ_{mN} (where + and -, respectively, indicate the two sides of ρ_{mN}, the interface between mth and (m+1)th elements in C_N). In the present work this interelement condition is introduced as a posteriori constraint, through a Lagrange multiplier $\dot{\underline{u}}_\rho$ at ρ_{mN}, in the functional of Eq. (17), thus leading to a 'hybrid' finite element method. The thus modified functional is:

$$\pi_{HS}(\dot{\underset{\sim}{\omega}},\dot{\underset{\sim}{t}},\dot{\underline{u}}_\rho)=\sum_m\{\int_{V_{mN}}[-\dot{R}^*(\dot{\underset{\sim}{r}})+(1/2)\underset{\sim}{\tau}^N:(\dot{\underset{\sim}{\omega}}^T.\dot{\underset{\sim}{\omega}})-\dot{\underset{\sim}{t}}^T.\dot{\underset{\sim}{\omega}}]dv$$

$$+\int_{S_{umN}}(\underline{n}^*.\dot{\underset{\sim}{t}}).\dot{\underline{u}}ds+\int_{\rho_{mN}}(\underline{n}^*.\dot{\underset{\sim}{t}}).\dot{\underline{u}}_\rho ds \qquad (43)$$

At the point of bifurcation, or instability, from the concept of adjacent compatible states, the following criterion can be shown to hold:

$$\pi_{HS}=0 \text{ and } \delta\pi_{HS}=0 \qquad (44)$$

with the constraints:

$$\underline{\nabla}^N.\dot{\underset{\sim}{t}}=0;\ \dot{\underset{\sim}{\omega}}=-\dot{\underset{\sim}{\omega}}^T;\ \underline{n}^*.\dot{\underset{\sim}{t}}=0 \text{ at } S_{\sigma mN};$$

$$\text{and}\quad \dot{\underline{u}}=0 \text{ at } S_{umN} \qquad (45)$$

In the case of linear pre-buckling states, the above criterion reduces to an eigen-value problem, with the eigen-value depending on $\underset{\sim}{\tau}^N$.

EXAMPLE PROBLEMS:

Necking of an Initially Perfect, Plane-Strain, Elastic-Plastic Bar.

Cowper and Onat [18] examined the above bifurcation necking problem, for a bar of rigid-plastic work-hardening material, under uniform tension applied at the ends of the bar. Mises' yield, and isotropic hardening criteria were used [18]. In [18] only the eigen-value problem for the applied tension at which necking would initiate in the bar was treated, but the phenomenon of post-bifurcation necking was not treated in [18]. $L_o(L)$ and $B_o(B)$ are the initial (current) length and width of the bar respectively. If y_1^N and y_2^N are the current cartesian coordinates of a material particle, the boundary conditions are: (*i*) at $y_1^N=0$: $\dot{u}_1=-V$; $\dot{t}_2=0$; (*ii*) at $y_1^N=L$: $\dot{u}_1=+V$; $\dot{t}_2=0$; (*iii*) at $y_2^N=\pm(B/2)$: $\dot{t}_1=\dot{t}_2=0$.

In the present analysis, the complementary energy formulation based on the TL rate equivalent of Eq. (17) was used. The problem parameters used are: $(B_o/L_o)=(1/3)$; τ_y (yield stress) $=4\times10^4$ psi; the true stress versus logarithmic strain $(\ln(l/l_o))$ curve was assumed to be bilinear, with the two slopes, $E=10^7$ psi, and $h=5\times10^4$ psi. The notation

$\eta = (L-L_o)L_o$; is used. The eigen-value solution for η at bifurcation-necking for the perfect bar, as obtained in [18] for the present linear-hardening but rigid-plastic material, is $\eta^c = 0.48$.

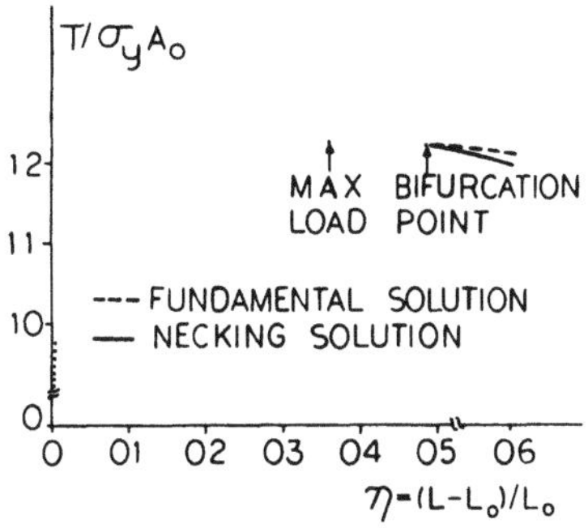

Figure 1

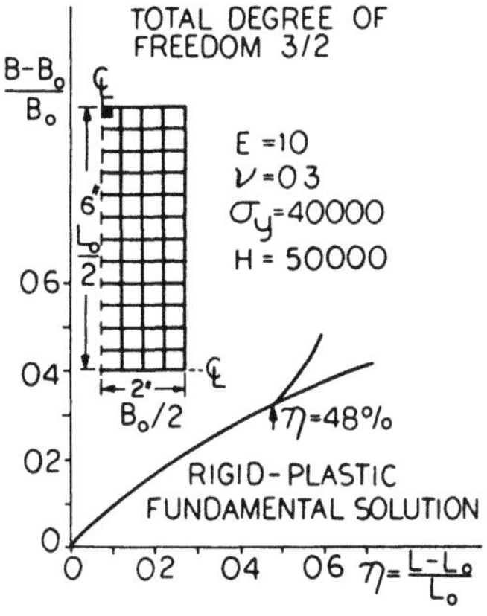

Figure 2

The variation of total applied load with η is shown in Fig. 1, from which it is seen that necking starts at $\eta^c = 0.482$ (which is in excellent agreement with the value of $\eta^c = 0.48$ of [18]). It is also seen that necking starts after the maximum load is attained. The convergence of η^c with the finite element mesh has been reported elsewhere [8,9], with the mesh as shown in Fig. 2, which is used to obtain the remainder of the reported results, being the finest mesh reported in [8]. The variation of the width reduction ratio, $(B-B_o)/B_o$ is shown in Fig. 2, from which it is seen that at $\eta = \eta^c$ the width reduction becomes much more pronounced as compared to the rigid-plastic fundamental solution.

The variation of δ/L_o (with δ being defined as the difference of widths at loading edge and the necking sections, respectively) with η is shown in Fig. 3. The slope of this curve at the beginning of necking, viz, $\eta = \eta^c = 0.48$, was obtained in an asymptotic analysis in [18]. The present result for this initial slope is in excellent agreement with [18]. However as necking develops, the slope $(\partial\eta/\partial\delta)$ decreases from the initial value at $\eta = \eta^c$, which appears to be in contradiction with the result of McMeeking and Rice [19]. The necked profile of the bar for $\eta \geq \eta^c$, are shown in Fig. 4.

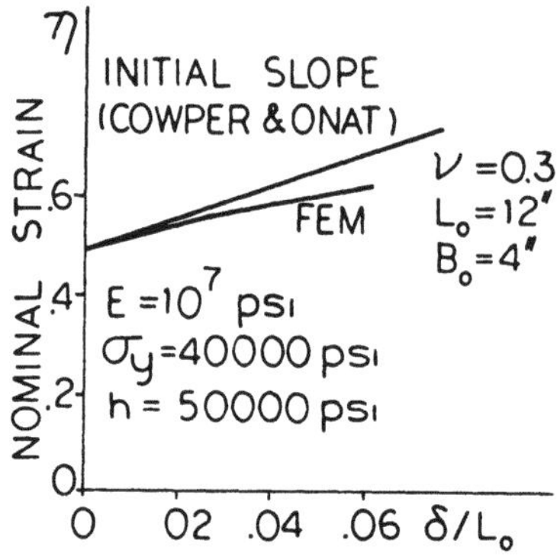

Figure 3

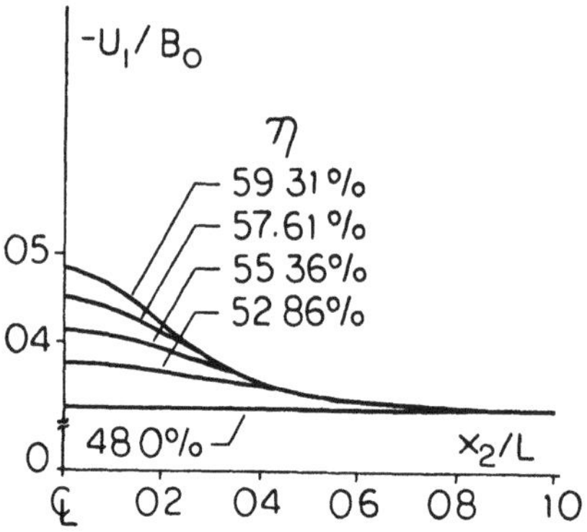

Figure 4

The necked profiles of the bar, and the progressive development of unloaded regions (shaded) are shown in Fig. 5 at various values of η. Note that unloading begins at the center of the loaded face of the bar at $\eta = \eta^c = 0.482$. Finally the distribution of Cauchy stresses, τ_{11} (in the direction of loading) τ_{22} (in the width loading), and τ_{33} (in the thickness direction of this plane-strain specimen), at the neck ($y_1^N = L/2$) are shown in Fig. 6. These results are in excellent qualitative agreement with those of Needleman [20] who also analyses the necking and post-necking problem of an initially perfect cylindrical bar. It is noted that

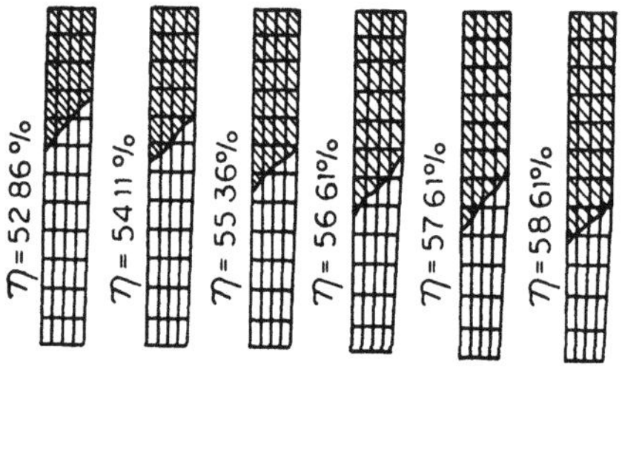

Figure 5

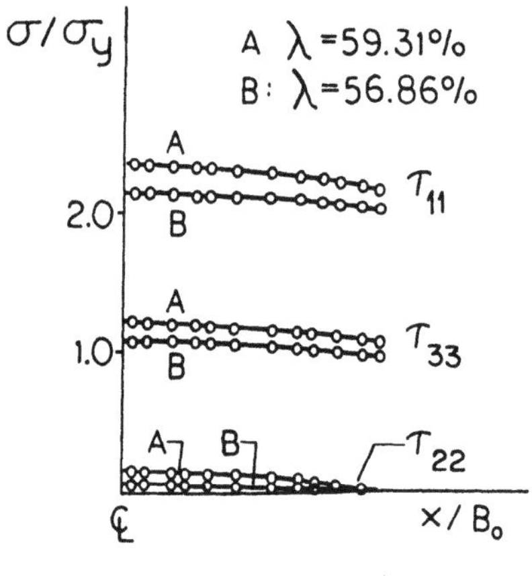

Figure 6

the problem of necking of an elastic-plastic bar, with initial imperfections, was also analysed by Osias [21], McMeeking and Rice [19], and Nemat-Nasser and Taya [21]. It is observed that the finite element meshes used in [19-21] are very much finer than the one used presently. Eventhough a precise mathematical statement as to this appears impossible, the above comparison of the mesh-sizes appears to indicate the relative advantages of the present complementary energy method.

Finite Deformation and Post-Buckling Analyses of a Thin Plate:

We consider large deformations (large rotations and large stretches) of a thin plate made of a semi-linear isotropic material (ie., a material exhibiting a linear relation between the stretch tensor $\underset{\sim}{h}$ and its conjugate stress-measure, the symmetrized Biot-Lure' or Jaumann tensor $\underset{\sim}{r}$). We invoke the well-known Kirchhoff-Love type plausible deformation hypotheses that a normal to the midplane of the initially flat undeformed plate remains normal to the deformed midplane and that there is no thickness stretch. In order to derive a consistent complementary energy principle for this constrained deformation problem, we start with the general (Hu-Washizu type) variational principle involving $\underset{\sim}{h}$, $\underset{\sim}{\alpha}$, $\underset{\sim}{u}$, and $\underset{\sim}{t}$ as variables.

By introducing the appropriate approximations to all these variables, we derive a general variational principle for the plate problem; from this we proceed to construct a consistent complementary energy principle for the plate problem. It is shown in [1] that a general functional, for the three-dimensional finite elasticity, whose stationary conditions lead to all the appropriate field equations, is given by:

$$\pi_G(\underline{u},\underset{\sim}{h},\underset{\sim}{\alpha},\underset{\sim}{t}) = \int_{V_o} \{W(\underset{\sim}{h}) - \rho_o\underline{g}\cdot\underline{u} + \underset{\sim}{t}^T : [(\underset{\sim}{I}+\underline{\nabla}\underline{u})^T - \underset{\sim}{\alpha}\cdot(\underset{\sim}{I}+\underset{\sim}{h})]\}dv - \int_{S_{uo}} t\cdot(\underline{u}-\bar{\underline{u}})ds - \int_{S_{\sigma o}} \bar{\underline{t}}\cdot\underline{u}ds \qquad (46)$$

where $W(\tilde{h})$ is the strain energy density per unit of initial volume V_o, as a function of pure stretch $\tilde{h}$; ρ_o is mass density/unit initial volume, and $\underline{g}$ are body forces/unit mass.

Let x_1, x_2 be cartesian coordinates in the mid plane, and x_3 normal to the midplane, of the plate. The position vector of an arbitrary material point in the undeformed plate is $\underline{x}=x_i\underline{e}_i$ (i=i,...3). Under the present deformation hypotheses, the position vector of the same particle after deformation is, $\underline{y} = (x_1+u_1^*)\underline{e}_1 + (x_2+u_2^*)\underline{e}_2 + u_3^*\underline{e}_3 + x_3\underline{N}$ where $\underline{N}$ is a unit normal to the deformed midplane. Further, the displacement u_i^* (i=1,2,3) are functions of x_1 and x_2 only. Thus the displacement of an arbitrary particle in the plate, is $\underline{y}-\underline{x} = u_1^*\underline{e}_1 + u_2^*\underline{e}_2 + u_3^*\underline{e}_3 + x_3(\underline{N}-\underline{e}_3)$. The base vector at an arbitrary point in the deformed plate are given by:

$$\partial\underline{y}/\partial x_\alpha = (\delta_{i\alpha} + u^*_{i,\alpha})\underline{e}_i + \underline{N},_\alpha x_3 \equiv G_\alpha \quad (\alpha=1,2;\ i=1,2,3) \qquad (47)$$

$$\partial\underline{y}/\partial x_3 = \underline{N}$$

The deformation gradient is

$$\tilde{F} = (\underline{\nabla}\underline{y})^T = \underline{G}_\alpha\underline{e}_\alpha + \underline{N}\underline{e}_3 \quad (\alpha=1,2) \qquad (48a)$$

Further, for the present kinematic hypotheses, we assume the stretch tensor to be:

$$\tilde{h} = h_{\alpha\beta}\underline{e}_\alpha\underline{e}_\beta;\ h_{\alpha\beta} = h_{\alpha\beta}(x_i) \quad [i=1,...3;\ \alpha,\ \beta=1,2] \qquad (48b)$$

From Eqs. (48a,b) it is seen:

$$\underline{N} = \tilde{F}\cdot\underline{e}_3 = [\tilde{\alpha}\cdot(\tilde{I}+\tilde{h})]\cdot\underline{e}_3 = \tilde{\alpha}\cdot\underline{e}_3 \qquad (49)$$

Thus the displacement vector can be written as:

$$\underline{u}(x_i) = u_i^*(x_\alpha)\underline{e}_i + (\tilde{\alpha}-\tilde{I})\cdot\underline{e}_3 x_3 \quad [i=1,2,3;\ \alpha=1,2] \qquad (50)$$

Further, we assume that $h_{\alpha\beta}(x_i)$ can be approximated as:

$$h_{\alpha\beta}(x_i) = h^*_{\alpha\beta}(x_\alpha) + x_3\chi_{\alpha\beta}(x_\delta) \quad [\alpha,\beta,\gamma=1,2] \qquad (51)$$

ie. $\tilde{h} = \tilde{h}^* + x_3\tilde{\chi}$

For the semilinear isotropic material we assume the constitutive law:

$$\tilde{r} = 2\mu\tilde{h} + \lambda(\tilde{h}:\tilde{I})\tilde{I} \qquad (52)$$

Since, for isotropy, $\underset{\sim}{h}$, $\underset{\sim}{g}$, and $\underset{\sim}{r}$ are coaxial, Eq. (3) becomes,

$$\underset{\sim}{r} = \underset{\sim}{t}.\underset{\sim}{\alpha} \tag{53}$$

The tensors $\underset{\sim}{t}$ and $\underset{\sim}{\alpha}$ are assumed to be:

$$\underset{\sim}{t} = t_{ij}\underline{e}_i\underline{e}_j;\ \underset{\sim}{\alpha}=\alpha_{ij}\underline{e}_i\underline{e}_j \quad [i,j=1,2,3;] \tag{54}$$

where, $t_{ij} = t_{ij}(x_1,x_2,x_3);\ \alpha_{ij} = \alpha_{ij}(x_1,x_2)$ (55)

Finally, the external forces distributed on the plate are assumed to be specified per unit area on the mid plane of the plate to be $g_i = g_i(x_\alpha)(i=1,..3,\ \alpha=1,2)$. When the assumptions in Eqs. (50-55) are substituted in Eq. (4-6), we find through atraight-forward algebra, that

$$\pi_G[u_i^*,\ h_{\alpha\beta}^*,\ \chi_{\alpha\beta},\ \alpha_{ij},\ T_{\alpha i},\ M_{\alpha i}]$$

$$= \int_{S_o} \{W^*(\underset{\sim}{h}^*,\underset{\sim}{\chi}) - \underline{g}.\underline{u}^* + \overset{\wedge}{\underset{\sim}{T}}{}^T:[\underline{e}_\alpha\underline{e}_\alpha + \underline{u}^*_{,\alpha}\underline{e}_\alpha$$

$$+ (\underset{\sim}{\alpha}.\underline{e}_3)\underline{e}_3 - \underset{\sim}{\alpha}.(\underset{\sim}{I}+\underset{\sim}{h})] + \overset{\wedge}{\underset{\sim}{M}}{}^T:[(\underset{\sim}{\alpha}.\underline{e}_3)_{,\alpha}\underline{e}_\alpha - \underset{\sim}{\alpha}.\underset{\sim}{\chi}]\}ds$$

$$- \int_{C_u} [\underline{T}.(\underline{u}^*-\bar{\underline{u}}^*) + \underline{M}.<(\underset{\sim}{\alpha}-\bar{\underset{\sim}{\alpha}}).\underline{e}_3>]dc - \int_{C_\sigma} [\bar{\underline{T}}.\underline{u}^*+\bar{\underline{M}}.(\underset{\sim}{\alpha}-\underset{\sim}{I}).\underline{e}_3]dc \tag{56}$$

where $\overset{\wedge}{\underset{\sim}{T}} = T_{\alpha i}\underline{e}_\alpha\underline{e}_i;\ \overset{\wedge}{\underset{\sim}{M}} = M_{\alpha i}\underline{e}_\alpha\underline{e}_i$ $(\alpha=1,2;\ i=1,...3)$, and,

$$T_{\alpha i} = \int_{x_3} t_{\alpha i}dx_3;\ M_{\alpha i} = \int_{x_3} t_{\alpha i}x_3dx_3$$

and
$$W^* = \int_{x_3} Wdx_3 \tag{57}$$

In Eq. (56), s_o is the area of the undeformed midplane, and C_u and C_σ are the displacement and traction prescribed boundaries of s. It is seen that only $t_{\alpha i}$ enter into the above energy expression due to the presently invoked deformation assumptions. The constitutive equations and LMB conditions obtainable from Eq. (57) are:

$$\partial W^*/\partial \underset{\sim}{h}^* = \overset{\wedge}{\underset{\sim}{T}}.\underset{\sim}{\alpha} \quad \underset{\sim}{R} \text{ and } \partial W^*/\partial\underset{\sim}{\chi} = \overset{\wedge}{\underset{\sim}{M}}.\underset{\sim}{\alpha} \equiv \underset{\sim}{N} \tag{58}$$

and
$$T_{\alpha i,\alpha} + g_i = 0 \quad (\alpha=1,2;\ i=1,...3) \tag{59}$$

When Eqs. (58, 59) and the appropriate traction boundary conditions on $T_{\alpha i}$ are satisfied a priori, one can eliminate from Eq. (56), (*i*) $\underset{\sim}{h}^*$ and χ through the usual contact transformations and by establishing a complementary energy density R^* such that $\partial R^*/\partial\underset{\sim}{R} = \underset{\sim}{h}^*$ and $\partial R^*/\partial\underset{\sim}{N} = \underset{\sim}{\chi}$, (*ii*) $\underline{u}^*$ through satisfying (53) a priori. When this is done, we obtain a complementary energy functional:

$$\pi_C(\underset{\sim}{\alpha}, \underset{\sim}{R}, \underset{\sim}{N}) = \int_s -R^*(\underset{\sim}{R},\underset{\sim}{N}) + \hat{\underset{\sim}{T}}^T:[\underline{e}_\alpha\underline{e}_\alpha + (\underset{\sim}{\alpha}\cdot\underline{e}_3)\underline{e}_3 - \underset{\sim}{\alpha}]$$

$$+ \underset{\sim}{M}^T:[(\underset{\sim}{\alpha}\cdot\underline{e}_3)_{,\alpha}\underline{e}_\alpha]]ds - \int_{C_u} (\underline{T}\cdot\bar{\underline{u}}^* + \underline{M}\cdot<(\underset{\sim}{\alpha}-\underset{\sim}{\alpha}^-)\cdot\underline{e}_3>)dc$$

$$- \int_{C_\sigma} \bar{\underline{M}}\cdot(\underset{\sim}{\alpha}-\underset{\sim}{I})\cdot\underline{e}_3 dc. \tag{60}$$

In Eq. (60), $\underset{\sim}{\alpha}$ is required to be orthogonal and further $\underset{\sim}{\alpha}$ is, as assumed in Eq. (55), a function only of x_1 and x_2. Also the variables $\underset{\sim}{R}$ and $\underset{\sim}{N}$ in Eq. (60) are assumed to be defined in terms of $\hat{\underset{\sim}{T}}$, $\hat{\underset{\sim}{M}}$, and $\underset{\sim}{\alpha}$ as in Eq. (58). To assume an orthogonal $\underset{\sim}{\alpha}(x_i,x_2)$, the concept of a finite rotation vector [23] is useful. Let ω be the finite angle of rotation around an arbitrarily oriented unit vector $\underline{e}$ in the midplane of the plate. The finite rotation vector is defined to be:

$$\underline{\Omega} = (\sin\omega)\underline{e} \tag{61}$$

The action of finite rotation $\underline{\Omega}$ on a vector $\underline{V}$ can be written [23] as the transformation of $\underline{V}$ to $\underline{V}^*$ as,

$$\underline{V}^* = \underline{V} + \underline{\Omega} \times \underline{V} + [\underline{\Omega}\times(\underline{\Omega}\times\underline{V}]/2\cos^2(\omega/2) \equiv \underset{\sim}{\alpha}\cdot\underline{V} \tag{62}$$

where $$\underset{\sim}{\alpha} = \underset{\sim}{I} + \underline{\Omega} \times \underset{\sim}{I} + [(\underline{\Omega}\times\underset{\sim}{I})\cdot(\underline{\Omega}\times\underset{\sim}{I})]/2\cos^2(\omega/2) \tag{63}$$

It can be shown that $\underset{\sim}{\alpha}$ of Eq. (63) is orthogonal, ie., $\underset{\sim}{\alpha}\cdot\underset{\sim}{\alpha}^T = \underset{\sim}{I}$.

The vector $\underline{e}$ in Eq. (61) can be written as:

$$\underline{e} = \underline{e}_1\cos\theta + \underline{e}_2 \sin\theta \tag{64}$$

Thus, the rotation tensor $\underset{\sim}{\alpha}$ of Eq. (63) is a function of two parameters: $\theta(x_1,x_2)$ and $\omega(x_1,x_2)$. The explicit expression for $\underset{\sim}{\alpha}$ can be shown to be:

$$\underset{\sim}{\alpha} = <1-(1-\cos\omega)\sin^2\theta>\underline{e}_1\underline{e}_1 + <(1-\cos\omega)\sin\theta\cos\theta>\underline{e}_1\underline{e}_2$$

$$+ (\sin\omega\sin\theta)\underline{e}_1\underline{e}_3 + (1-\cos\omega)\sin\theta\cos\theta\underline{e}_2\underline{e}_1$$

$$+ <1-(1-\cos\omega)\cos^2\theta>\underline{e}_2\underline{e}_2 - \sin\omega\cos\theta\underline{e}_2\underline{e}_3 - \sin\omega\sin\theta\underline{e}_3\underline{e}_1$$

$$+ \sin\omega\cos\theta\underline{e}_3\underline{e}_2 + \cos\omega\underline{e}_3\underline{e}_3.$$

In a Von-Karman type plate theory ω is assumed to be moderately large, such that $\cos\omega \approx 1-(\omega^2/2)$ and $\sin\omega = \omega$; while the angle θ can be assumed to be arbitrary.

Further details of the analysis of large rotations and stretches of thin plates using the complementary energy method sketched above, which are omitted here for space reasons, will be reported elsewhere. In the following we present some results using a hybrid-stress finite element method based on Eq. (60).

Post-Buckling of a Thin Plate:

The problem is that of a simply supported square plate $[-a/2 \leq x_1, x_2 \leq +a/2]$ that is subject to uniform uniaxial compression at edges $x_1 = \pm a/2$. Only a (2x2) mesh in a quarter of the plate was used. The pertinent geometrical and material-property data are indicated in Fig. 7.

In the lower half of Fig. 7, the maximum transverse displacement (W_{max}/t) is shown magnified for lower values of $(\bar{P}_x a^2/Et^2)$ where $\bar{P}_x$ is the axial compressive stress. In the present analysis a small uniform lateral load of $(Pa^2/Et^2) = 2x10^{-5}$ was applied along with the axial $\bar{P}_x$. The linear buckling load predicted by Levy [24] is $(\bar{P}_x a^2/Et^2) = 3.6$. The post-buckling displacement W_{max} as obtained by Levy [24] is also shown in both the parts of Fig. 7. It is seen that the present results agree excellently with those in [24] upto displacement levels of $(W_{max}/t) \simeq 3.5$.

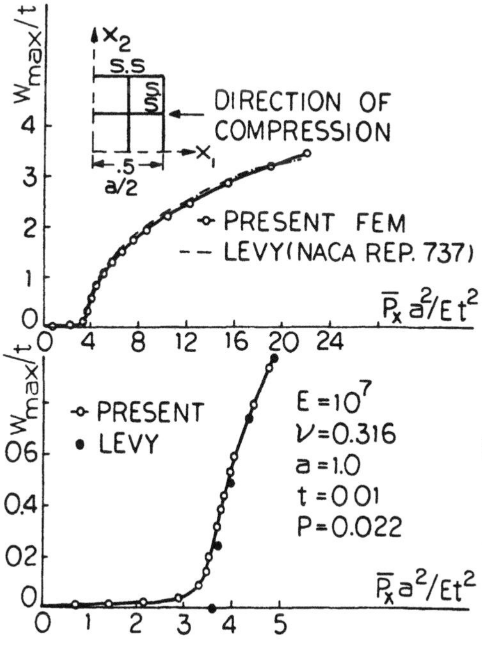

Figure 7

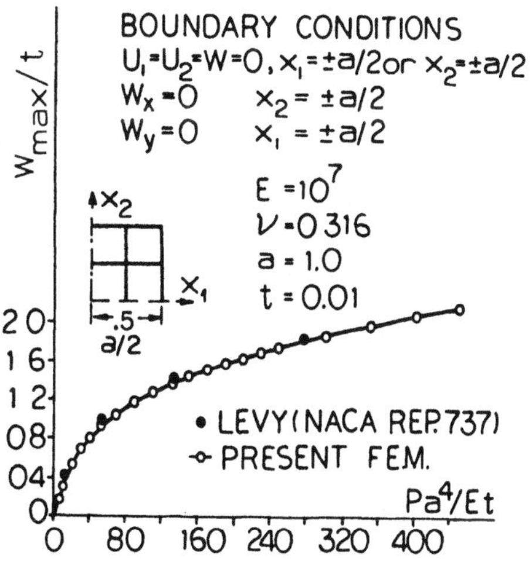

Figure 8

Large Deformation of a Transversley Loaded Plate:

The problem is that of a semilinear, isotropic thin elastic simply supported plate that is subject to uniform transverse pressure P. The pertinent geometric and material property data is shown in Fig. 8. The presently computed variation of the maximum transverse displacement W_{max} with P is seen to agree excellently with that of Levy [24].

The results presented in this paper, and further completed work to be reported elsewhere, appear encouraging to warrant us to pursue this line of thought in other areas of nonlinear continuum mechanics. Our current studies in this area will be reported on shortly.

Acknowledgements:

The research reported herein has been supported in parts by: NASA under grant NASA-NAG3-38; by ONR under contract number N00014-78-7636; and by AFOSR under contract number F49620-78-C-0085. The authors gratefully acknowledge these supports. The authors express their sincere thanks to Ms. Margarete Eiteman for her untiring efforts in the preparation of this typescript.

References:

[1] Atluri, S.N., and Murakawa, H., "On Hybrid Finite Element Models in Nonlinear Solid Mechanics", in Finite Elements in Nonlinear Mechanics, Vol 1 (Ed. P.G. Bergan, et al) Tapir Press, Norway, 1977, pp 3-41.

[2] Murakawa, H., "Incremental Hybrid Finite Element Methods for Finite Deformation Problems (With Special Emphasis on the Complementary Energy Principle) Ph.D. Thesis, Georgia Institute of Technology, Aug. 1978.

[3] Murakawa, H., and Atluri, S.N., "Finite Elasticity Solutions Using Hybrid Finite Elements Based on a Complementary Energy Principle" Journal of Applied Mechanics Trans. ASME, Vol 45, 1978, pp 539-548.

[4] Murakawa, H., and Atluri, S.N., "Finite Elasticity Solutions Using Hybrid Finite Elements Based on a Complementary Energy Principle - II. Incompressible Materials", Journal of Applied Mechanics, Vol 46, 1979, pp 71-78.

[5] Atluri, S.N., "On Rate Principles For Finite Strain Analysis of Elastic and Inelastic Nonlinear Solids", in Recent Research on Mechanical Behavior of Solids (Professor H. Miyamoto's 60th Anniversary Volume) Univ. of Tokyo Press, Tokyo, Japan 1979, pp 79-107.

[6] Atluri, S.N., "On Some New General and Complementary Energy Theorems for the Rate Problems of Finite Strain, Classical Elastoplasticity" Journal of Structural Mechanics, Vol. 8, No. 1, 1980, pp 61-92.

[7] Atluri, S.N., "Rate Complementary Energy Principles, Finite Strain Plasticity Problems, and Finite Elements" in Proc. IUTAM Symposium on Variational Methods Northwestern Univ., Sept. 1978 (S. Nemat-Nasser and K. Washizu, Eds.) Pergamon Press, 1980 (In Press).

[8] Murakawa, H., and Atluri, S.N., "Finite Element Solutions of Finite Strain Elastic-Plastic Problems, Based on a New Complementary Energy Rate Principle" in Advances in Computer Methods for Partial Differential Equations (R. Vishnevetsky and B. Stepleman, Eds.) IMAS, Rutgers Univ., 1979, pp 53-61.

[9] Atluri, S.N., "Finite-Strain Plasticity Computations Based on a New Complementary Rate Principle" Presented at Int. Conf. on Finite Element Methods, Chalmers Inst. of Tech. Goteborg, Sweden, Aug. 1979.

[10] Murakawa, H., Reed, K.W., Atluri, S.N., and Rubenstein, R., "Stability Analysis of Structures Via a New Complementary Energy Method" Proc. of Symp. on Computational Meth. in Nonlinear Structural and Solid Mechanics, (A.K. Noor, and H.C. McComb, Jr., Eds.) Washington, D.C., Oct. 6-8, 1980 (In Press).

[11] Hill, R., "Some Basic Principles in the Mechanics of Solids Without a Natural Time", J. Mech. & Phy. Solids, Vol 7, 1959, pp 209-225.

[12] Atluri, S.N., "On the Hybrid Stress Finite Element Model in Incremental Analysis of Large Deflection Problems", Int. Jnl. Solids and Structures, Vol 9, 1973, pp 1177-1191.

[13] Hill, R., "Eigenmodal Deformations in Elastic/Plastic Continua" Jnl. Mech. and Phys. Solids, Vol 15, 1967, pp 371-386.

[14] Perzyna, P., "Fundamental Problems in Visco-Plasticity" Jnl. Mech.and Phys. Solids, Vol 9, 1966, pp 243-377.

[15] Zienkiewicz, O.C., and Cormeau, I.C., "Viscoplasticity-Plasticity and Creep in Elastic Solids - A Unified Numerical Solution Approach" Int. J. Num. Methods in Engg., Vol 8, 1974, pp 821-845.

[16] Argyris, J.H., and Kleiber, M., "Incremental Formulation in Nonlinear Mechanica and Large Strain Elasto-Plasticity" Comp. Meth. in Appl. Mech. Engg., Vol 11, 1977, pp 215-247.

[17] Wang, H.C., "Finite Deformation Elastic-Viscoplastic Analysis By Finite Element Method" Ph.D. Thesis, Massachusetts Institute of Technology, Feb. 1980.

[18] Cowper, G.R., and Onat, E.T., "The Initiation of Necking and Buckling in Plane Plastic Flow", in Proc. 4th U.S. Congress of Applied Mech., ASME, N.Y., 1962, p 1023.

[19] McMeeking, R.M., and Rice, J.R., "Finite Element Formulations for Problems of Large Elastic-Plastic Deformations" Int. Jnl. Solids and Structures, Vol 11, 1975, pp 601-.

[20] Needleman, A., "A Numerical Study of Necking in Circular Cylindrical Bars", Jnl. Mech. Phys. Solids, Vol 20, 1972, p 111.

[21] Osias, J.R., "Finite Deformation of Elastic-Plastic Solids: The Example of Necking in Flat Tensile Bars", Ph.D. Thesis, Carnegie Mellon Univ. 1972.

[22] Nemat-Nasser, A., and Taya, M., "Model Studies of Ductile Fracture-Part II. Further Numerical Formulations", in Finite Elements in Nonlinear Mechanics, Vol 1 (P.G. Bergan et al Eds) Tapir, Norway, 1977, p 211.

[23] Lure', A.I., "Analytical Mechanics" (In Russian) Moscow, 1961.

[24] Levy, S., "Bending of Rectangular Plates with Large Deflections", NACA Tech. Note 846 and NACA Tech Report 737, 1942, p 139.

Formulation of Contact Problems by Assumed Stress Hybrid Elements

T. H. H. PIAN, K. KUBOMURA
Massachusetts Institute of Technology, Cambridge, USA

Summary

A modified complementary energy principle has been derived by introducing the continuity at contact surface as the condition of constraint and the reactions at contact surface as additional field variables. In the solution procedure an assumed contact surface is introduced and is divided into elements. An iterative procedure is used to obtain the correct location and extent of the contact surface. Example problems are limited only to elastic bodies under small deformations but are including extensive sliding between surfaces which may be either frictionless or frictional.

Introduction

The assumed stress hybrid method has been demonstrated to be very competitive to the conventional assumed displacement method in finite element analyses. One advantage of the approach is on its more accurate stress evaluation. Such advantage would be amplified for problems for which the solutions are dependent on accurate evaluation of stresses. The contact problems for which the behavior of frictional sliding depends on the contact pressure also belong to this category.

This paper presents an incremental finite element formulation of contact problems which may include extensive frictional sliding such that node-to-node contact is no longer maintained. The formulation is based on the introduction of contact reactions at the contact surface as independent variables in a Lagrange multiplier method.

INCREMENTAL FORMULATION

Variational principle for incremental finite element methods are described in Ref. 1. Modifications of variational principles for relaxing continuity conditions along interelement boundaries have been used for the derivation of many hybrid finite element models [2].

For the contact problem, consider two bodies A and B which are divided into finite elements V_n where n refers to the nth element. The boundary surface ∂V_n of a typical element may be composed of S_{σ_n}, S_{u_n}, S_n and S_{c_n}, which are respectively, the surface with prescribed tractions, the surface with prescribed displacements and the interelement surface and contact surface between A and B. The entire contact surface is discretized into contact element S_{c_m}. A modified variational principle which includes the conditions of continuity and equilibrium on the contact surface and for which the stresses satisfy the equilibrium equation is

$$\pi_{mc} = \sum_n \Big[\int_{V_n} \tfrac{1}{2} \Delta\underset{\sim}{\sigma}^T \underset{\sim}{S} \Delta\underset{\sim}{\sigma}\, dV - \int_{\partial V_n} (\underset{\sim}{T} + \Delta\underset{\sim}{T})^T \Delta\underset{\sim}{\tilde{u}}\, dS +$$

$$+ \int_{S_{\sigma_n}} (\underset{\sim}{\bar{T}} + \Delta\underset{\sim}{\bar{T}})^T \Delta\underset{\sim}{\tilde{u}}\, dS \Big]$$

$$- \sum_m \Big[\int_{S_{c_m}} \underset{\sim}{\lambda}^T (\Delta\underset{\sim}{\tilde{u}}^A - \Delta\underset{\sim}{\tilde{u}}^B)\, dS$$

$$+ \int_{S_{c_m}} \underset{\sim}{\lambda}^T \Big\{ (\underset{\sim}{\tilde{u}}^A + \underset{\sim}{x}^A) - (\underset{\sim}{\tilde{u}}^B + \underset{\sim}{x}^B) \Big\} dS \Big] = \text{stationary} \qquad (1)$$

in which

$\underset{\sim}{\sigma}$ = stresses

$\underset{\sim}{T}$ = element boundary tractions

$\tilde{\underset{\sim}{u}}$ = element boundary displacements

$\underset{\sim}{\lambda}$ = reactions at contact surface

$\underset{\sim}{x}$ = coordinate of material point prior to deformation

The superscripts A and B refer to bodies A and B respectively. With $\Delta \underset{\sim}{\sigma}$, $\Delta \tilde{\underset{\sim}{u}}$ and $\underset{\sim}{\lambda}$ as variables, and with $\Delta \underset{\sim}{\sigma}$ satisfying the equilibrium conditions, the Euler's equations of this variational principle provide, in addition to the compatibility conditions in V_n, the following equations along the element boundaries:

$$(\underset{\sim}{T} + \Delta \underset{\sim}{T})^a + (\underset{\sim}{T} + \Delta \underset{\sim}{T})^b = 0 \quad \text{on interelement boundary between } V_a \text{ and } V_b$$

$$\underset{\sim}{T} + \Delta \underset{\sim}{T} = \underset{\sim}{\bar{T}} + \Delta \underset{\sim}{\bar{T}} \quad \text{on } S_{\sigma_n}$$

$$\left.\begin{array}{l} (\underset{\sim}{T} + \Delta \underset{\sim}{T})^A + \underset{\sim}{\lambda} = 0 \\ (\underset{\sim}{T} + \Delta \underset{\sim}{T})^B - \underset{\sim}{\lambda} = 0 \\ \text{and} \quad \tilde{\underset{\sim}{u}}^A + \Delta \tilde{\underset{\sim}{u}}^A + \underset{\sim}{x}^A = \tilde{\underset{\sim}{u}}^B + \Delta \tilde{\underset{\sim}{u}}^B + \underset{\sim}{x}^B \end{array}\right\} \text{on } S_{c_m} \qquad (2)$$

These are, of course, the equilibrium conditions along the element boundaries and the condition of continuity along the contact surface. For a contact problem with frictional sliding, the following condition should also be satisfied:

$$|\lambda_s| \leq \mu |\lambda_n| \qquad (3)$$

where the reactions at the contact surface are resolved into the normal and tangential components, λ_n and λ_s, and μ is the coefficient of friction.

In the finite element implementation, the stress increments are expressed as

$$\Delta \underset{\sim}{\sigma} = \underset{\sim}{P} \Delta \underset{\sim}{\beta} \qquad (4)$$

from which

$$\Delta \underset{\sim}{T} = \underset{\sim}{R} \Delta \underset{\sim}{\beta} \qquad (5)$$

The boundary displacements are interpolated as

$$\tilde{\underset{\sim}{u}} = \underset{\sim}{L}\,\underset{\sim}{q} \tag{6}$$

and

$$\Delta\tilde{\underset{\sim}{u}} = \underset{\sim}{L}\,\Delta\underset{\sim}{q} \tag{7}$$

In the finite element implementation, the total contact reactions are actually resolved into their normal and tangential components and are interpolated in terms of their nodal values, $\underset{\sim}{t}$,

$$\underset{\sim}{\lambda} = \underset{\sim}{M}\,\underset{\sim}{t} \tag{8}$$

The displacements $\tilde{\underset{\sim}{u}}^A$ and $\tilde{\underset{\sim}{u}}^B$ over each contact element S_{c_m} are then interpolated in terms of the nearby nodal displacements $\underset{\sim}{q}^A$ and $\underset{\sim}{q}^B$ of the neighboring elements of the contacting bodies.

The functional π_{mc} is in the form of

$$\pi_{mc} = \sum_n \left[\frac{1}{2}\,\Delta\underset{\sim}{\beta}^T \underset{\sim}{H}_n\,\Delta\underset{\sim}{\beta} - \Delta\underset{\sim}{\beta}^T \underset{\sim}{G}_n\,\Delta\underset{\sim}{q} + \Delta\underset{\sim}{q}^T(\Delta\underset{\sim}{Q}_n + \underset{\sim}{R}_n)\right]$$

$$+ \sum_m \left[-(\Delta\underset{\sim}{q}^{AT}\,\underset{\sim}{J}_m^A - \Delta\underset{\sim}{q}^{BT}\,\underset{\sim}{J}_m^B)\,\underset{\sim}{t} - \underset{\sim}{t}^T\,\underset{\sim}{R}_c\right] \tag{9}$$

where

$$\underset{\sim}{H}_n = \int_{V_n} \underset{\sim}{P}^T \underset{\sim}{S}\,\underset{\sim}{P}\,dV \quad ; \quad \underset{\sim}{G}_n = \int_{\partial V_n} \underset{\sim}{R}^T\,\underset{\sim}{L}\,dS$$

$$\Delta\underset{\sim}{Q}_n = \int_{S_{\sigma_n}} \underset{\sim}{L}^T\,\Delta\bar{\underset{\sim}{T}}\,dS$$

$$\underset{\sim}{R}_n = -\int_{\partial V_n} \underset{\sim}{L}^T\,\underset{\sim}{T}\,dS + \int_{S_{\sigma_n}} \underset{\sim}{L}^T\,\bar{\underset{\sim}{T}}\,dS$$

$$\underset{\sim}{J}_m^A = \int_{S_{c_m}} \underset{\sim}{M}^T\,\underset{\sim}{L}^A\,dS \quad ; \quad \underset{\sim}{J}_m^B = \int_{S_{c_m}} \underset{\sim}{M}^T\,\underset{\sim}{L}^B\,dS$$

$$\underset{\sim}{R}_{c_m} = \int_{S_{c_m}} \underset{\sim}{M}^T[(\tilde{\underset{\sim}{u}}^A + \underset{\sim}{x}^A) - (\tilde{\underset{\sim}{u}}^B + \underset{\sim}{x}^B)]\,dS \tag{10}$$

In the above functional π_{mc} the incremental stress parameters $\Delta \underset{\sim}{\beta}$ are independent of those of other elements. Thus, by setting $\delta\, \pi_{mc} = 0$ one can obtain a set of equations which express $\Delta \underset{\sim}{\beta}$ in terms of $\Delta \underset{\sim}{q}$. Then, by eliminating $\Delta \underset{\sim}{\beta}$ and by summing up matrices over all elements one obtains

$$\pi_{mc} = -\frac{1}{2} \Delta \underset{\sim}{q}^T \underset{\sim}{K} \Delta \underset{\sim}{q} + \frac{1}{2} \Delta \underset{\sim}{q}^T (\Delta \underset{\sim}{Q} + \underset{\sim}{R})$$

$$- (\Delta \underset{\sim}{q}^{A^T} \underset{\sim}{J}^A - \Delta \underset{\sim}{q}^{B^T} \underset{\sim}{J}^B) \underset{\sim}{t} - \underset{\sim}{t}^T \underset{\sim}{R}_c \tag{11}$$

where

$$\underset{\sim}{K} = \sum_n \underset{\sim}{G}_n^T \underset{\sim}{H}_n^{-1} \underset{\sim}{G}_n$$

$$\Delta \underset{\sim}{Q} = \sum_n \Delta \underset{\sim}{Q}_n$$

$$\underset{\sim}{R} = \sum_n \underset{\sim}{R}_n \quad , \quad \underset{\sim}{R}_c = \sum_m \underset{\sim}{R}_{c_m}$$

$$\underset{\sim}{J}^A = \sum_m \underset{\sim}{J}_m^A \quad ; \quad \underset{\sim}{J}^B = \sum_m \underset{\sim}{J}_m^B \tag{12}$$

Setting $\delta\, \pi_{mc} = 0$ yields the matrix equation for the finite element solution

$$\begin{bmatrix} \underset{\sim}{K} & \begin{Bmatrix} \underset{\sim}{J}^A \\ -\underset{\sim}{J}^B \end{Bmatrix} \\ \lfloor \underset{\sim}{J}^{A^T} - \underset{\sim}{J}^{B^T} \rfloor & \underset{\sim}{0} \end{bmatrix} \begin{Bmatrix} \Delta \underset{\sim}{q} \\ \underset{\sim}{t} \end{Bmatrix} = \begin{Bmatrix} \Delta \underset{\sim}{Q} + \underset{\sim}{R} \\ - \underset{\sim}{R}_c \end{Bmatrix} \tag{13}$$

Iterative Procedure

Once the contact bodies are adequately constrained such that the inverse of the global stiffness matrix, $\underset{\sim}{K}^{-1}$ in Eq. (13) can be evaluated, it is used throughout the solution process. In locating the contact surface only $\underset{\sim}{J}^{A}$ and $\underset{\sim}{J}^{B}$ need to be recomputed in each iteration. Even in the case of material and/or geometrical nonlinearities, it is possible to use a modified Newton-Raphson method, hence to keep the global stiffness $\underset{\sim}{K}$ constant during the iterative process for each increment.

For a two-dimensional problem, before each iteration, the contact surface is a line fixed in the coordinate system, but not to the contacting bodies. Such a line is assumed known hence integrals over S_{cm} can be evaluated. The displacement increments are then solved using Eq. (13) and new range and location of the contact surface can be determined for further iteration. It has been found that, for best results, the length of the contact element should be the same as the length of the contacting elements.

The procedure for solving the contact problem is as follows: First an increment in the external load or prescribed displacement is applied. Second, a contact surface is assumed together with the points at which the nodes of the bodies are in contact. Also, the types of contact (sliding or nonsliding) at each contact point are assumed. For the initial calculation of the first loading increment the above choice is made simply by inspection, and for the first iteration after each new loading increment, the converged solution of the previous loading step is used. Third, at the i-th iteration of the N-th loading step, incremental displacements $\Delta \underset{\sim}{u}_i$ and contact reactions $\underset{\sim}{\lambda}_i$ are determined. Fourth, knowing the total displacement $\underset{\sim}{u}_{N-1}$ at the end of the previous loading step, the total displacements $\underset{\sim}{u}_{N-1}+\Delta \underset{\sim}{u}_i$ on the boundary and the contact reactions $\underset{\sim}{\lambda}_i$ are checked to determine if the conditions of contact are satisfied. If these conditions are not satisfied, the extent of the assumed contact surface is modified for further iteration.

To determine if the solution satisfies the conditions of contact, the following assurances are made:

(1) That nodes of either body, beyond the last contacting nodes from the previous iteration have not penetrated the other body.

(2) That tractions at the contacting nodes are compressive. These normal tractions can be calculated by three different methods; (a) from the stress coefficients, (b) from the equivalent nodal forces, and (c) from the contact reactions, $\underset{\sim}{\lambda}$. Here the last method was used.

(3) That the relationship between normal and tangential contact reactions,

$$|\lambda_s| \leq \mu \; |\lambda_n|$$

is satisfied.

Depending on which of the above checks, if any, is violated, one of the following procedures is employed to modify the assumed location of the contact surface.

(a) If (1) is violated, the contact surface may be extended to include the points at which penetration has occurred.

(b) If (2) is violated, the contact surface is reduced by excluding nodes at which the tractions are tensile.

(c) If (3) is violated, sliding is allowed to occur.

After the conditions of contact are satisfied, a test for convergence can be made by calculating the following quantity from a representing nodal displacement Δu,

$$R = \left| \frac{\Delta u_{i+1} - \Delta u_i}{\Delta u_i} \right|$$

If R is less than a prescribed quantity, say 0.01, the solution is considered as converged.

EXAMPLE SOLUTIONS

Example solutions have been carried out for problems of contact between a desk and a semi-infinite half-plane. The relevant dimensions and the finite element pattern are shown in Fig. 1a, with the area

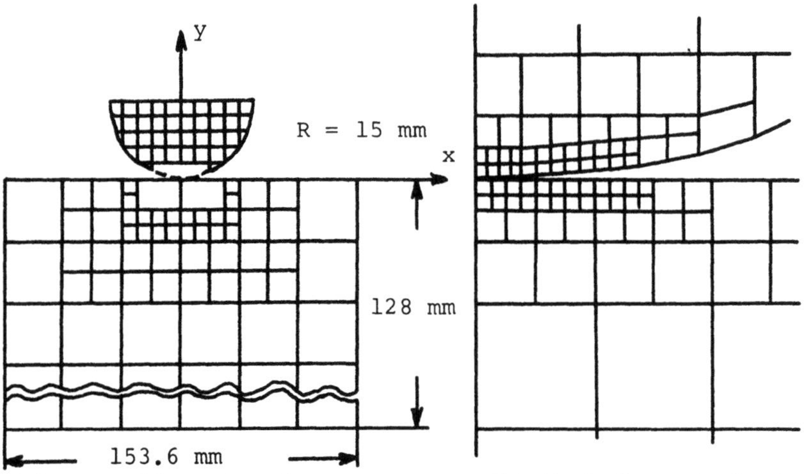

(a) Overall Mesh Pattern

(b) Mesh Near Contact Surface

Figure 1 Elastic Half-Disk and Half-Plane Contact ($E = 21{,}000$ Kg/mm^2, $\nu = 0.3$)

immediately surrounding the contact surface shown in great detail in Fig. 1b. The semi-infinite half-plane has been modeled by a finite one with overall dimensions much larger than those of the disk. The basic element used are four-node quadrilateral element derived by assuming seven β-parameters and linear displacement distribution along each edge. Five-node and six-node elements are also introduced in transition regions between coarser and finer meshes. Contact tractions along each contact element are approximated by linear interpolations.

Non-Sliding Contact

Problems are solved for the case with both applied loads and prescribed displacements at the top of the disk. In the problems, the ratio of Young's moduli are varied over a range from 1 to 10^{-4} and slightly different mesh patterns near the contact surface are used to accommodate for node-to-node and node-to-internode contacts. Loads or displacements are applied by three increments until the length of contact surface becomes about 2.4 mm. For each increment, the converged solutions are reached with three or four iterations. For these solutions, the best results for contact tractions are obtained when calculated from equivalent nodal forces and are compared excellently with the Hertz solution in all cases.

Frictionless Contact With Extensive Sliding

The half disk and the semi-infinite half-plane are also used to demonstrate the capability of this formulation to solve extensive sliding contact problems. Since the contact between the two bodies is frictionless, the solution is independent of the path, thus a Hertz solution is again available for comparison. Solutions are obtained for prescribed displacements at the top of the disk with both vertical and horizontal components. Stress distributions on the plane of contact, for two prescribed displacements, are plotted in Fig.2, where zero position represents the point of initial contact.

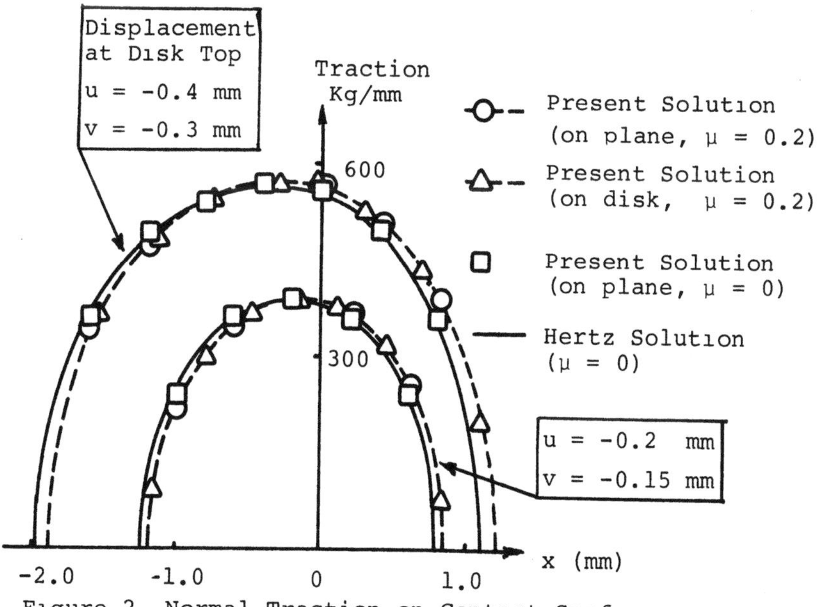

Figure 2 Normal Traction on Contact Surface

Four increments were used to obtain each solution. It is seen that the solution agrees almost exactly with that of the Hertz solution. It is noted that the center of symmetry of the stress distribution moves to the left as the half-disk slides in that direction.

Frictional Contact With Extensive Sliding

The same problem is again solved here with friction ($\mu = 0.2$) between the disk and the half-plane as an added consideration. The normal tractions at the contacting nodes between the disk and the half-plane for every other displacement increment are also shown in Fig. 2 and are compared with those of the frictionless case. Because of friction, it can be seen that the displacement of the contact surface is retarded. That the normal tractions of the plane and disk are equal in magnitude and opposite in sign is also evident in the figure. The contact surface of the two bodies and the locations of contacting nodes are shown in Fig. 3. It is seen that through averaging over the entire contact surface, the contact condition of no separation or penetration is satisfied.

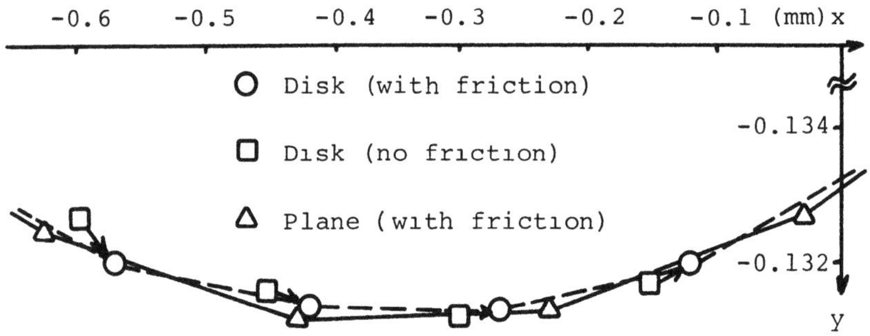

Figure 3 Locations of Contact Surface and Nodes of Contacting Elements (Frictional Sliding)

Conclusions

An incremental variational principle and a corresponding finite method have been formulated for contact problems based on the assumed stress hybrid model. The method is convenient for handling problems with node-to-internode contact, and has been verified by example solutions which involves only small deformation elasticity problems.

References

1. Pian, T.H.H., "Variational Principles for Incremental Finite Element Methods," J. Franklin Institute, Vol.302, 1976, pp 473-488.

2. Pian, T.H.H. and Tong, P., "Finite Element Methods in Continuum Mechanics" Advances in Applied Mechanics, Vol 12, Edited by C.S. Yih, Academic Press, 1972, pp. 1-58.

Part II:

Geometrically Nonlinear Problems – Rods and Shells

Strategies for Tracing the Nonlinear Response Near Limit Points

E. RAMM
Universitat Stuttgart, Germany

Abstract

For the prebuckling range an extensive literature of effective solution techniques exists for the numerical solution of structural problems but only a few algorithms have been proposed to trace nonlinear response from the pre-limit into the post-limit range. Among these are the simple method of suppressing equilibrium iterations, the introduction of artificial springs, the displacement control method and the "constant-arc-length method" of Riks/Wempner. It is the purpose of this paper to review these methods and to discuss the modifications to a program that are necessary for their implementation. Selected numerical examples show that a modified Riks/Wempner method can be especially recommended.

1. Introduction

Usually postcritical states are not tolerated in the design of a structure. However, the prediction of response in this range may still be of great value. A typical example is the imperfection sensitivity of certain structures which in general is directly related to the postcritical response. In particular this is true for structures exhibiting a decreasing post-limit characteristic. This may result in a dynamic snap-through or snap-back phenomenon depending on whether the load or the displacement controls the system. However, a static analysis traces the whole postcritical range allowing for a better judgement of the overall structural response.

It is well known that the usually applied Newton-Raphson iteration methods are not very efficient and often fail in the neighborhood of critical points. The stiffness matrix approaches singularity resulting in an increasing number of iterations and smaller and smaller load steps. Finally the solution diverges. In recent years several strategies have been proposed to overcome these problems and to trace the response beyond the critical point.

It is the purpose of this paper to describe some of the most commonly used techniques. These are the method of suppressing the equilibrium iterations in the neighborhood of the critical point, the method of artificial springs, the displacement control technique and the "constant - arc - length method" of Riks [1], [2] and Wempner [3]. In particular an attempt is made to show the correlation of the latter procedures. Special emphasis is given to some modifications of the Riks/Wempner method leading to an efficient iterative technique throughout the entire range of loading and not only near the critical point. Other methods for solving the same type of problem, e. g. the perturbation method or dynamic relaxation, are not studied.

The discussion refers to limit points only. Bifurcation problems may be included either by introducing a small perturbation in geometry or load (imperfect approach) or by superimposing on the displacement field of the critical load a part of the eigenmode (perfect approach). The procedures are described in conjunction with the Newton-Raphson method in its standard or modified versions. A combination with accelerated quasi Newton methods is possible. Proportional loading is assumed but few changes are necessary for non proportional loading.

2. Starting Point and Notation

The study is based on the incremental/iterative solution procedure in a nonlinear finite element analysis, i. e. the nonlinear problem is stepwise linearized and the linearization error is corrected by additional equilibrium iterations, see for instance [4]. A left superscript indicates the current configuration of the total displacements ${}^{m}\mathbf{u}$, the load vector ${}^{m}\mathbf{P}$, the internal forces ${}^{m}\mathbf{F}$ and the out-of-balance forces ${}^{m}\mathbf{R}$. For proportional loading the loads may be expressed by one load factor ${}^{m}\lambda$

$$ {}^{m}\mathbf{P} = {}^{m}\lambda \cdot \mathbf{P} \tag{1} $$

where $\mathbf{P}$ is a vector of reference loads. Within one increment from configuration m to m + 1, the positions i and j = i + 1, before and after an arbitrary iteration cycle, are distinguished (figure 1).

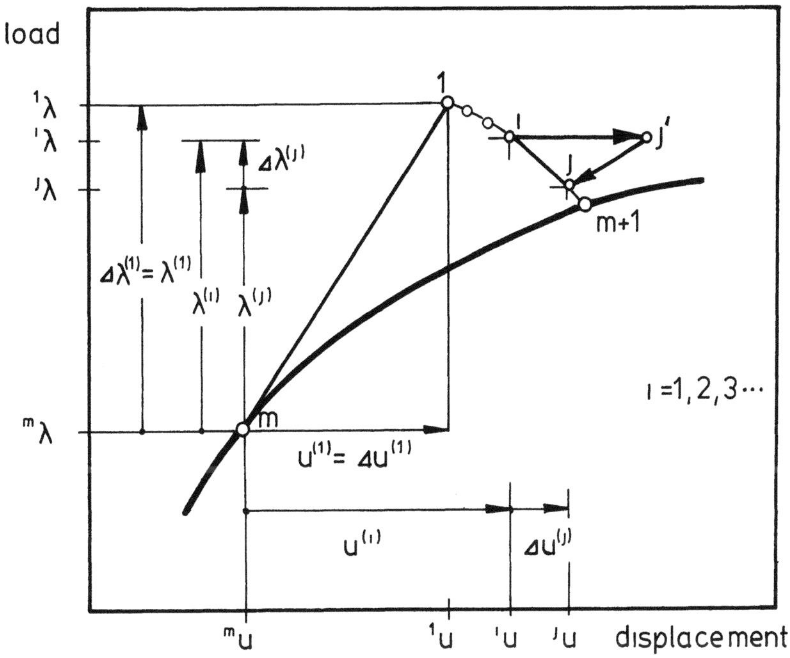

Figure 1: Notation

The total increments between positions m and i are denoted by $\mathbf{u}^{(i)}$, $\mathbf{P}^{(i)}$ and $\lambda^{(i)}$ whereas the changes in increments from i to j are denoted by $\Delta\mathbf{u}^{(j)}$, $\Delta\mathbf{P}^{(j)}$ and $\Delta\lambda^{(j)}$, respectively:

$$ {}^{j}\mathbf{P} = {}^{m}\mathbf{P} + \underbrace{\mathbf{P}^{(i)} + \Delta\mathbf{P}^{(j)}}_{\mathbf{P}^{(j)}} \quad \text{and} \quad {}^{j}\lambda = {}^{m}\lambda + \underbrace{\lambda^{(i)} + \Delta\lambda^{(j)}}_{\lambda^{(j)}} $$

$$ {}^{j}\mathbf{u} = {}^{m}\mathbf{u} + \underbrace{\mathbf{u}^{(i)} + \Delta\mathbf{u}^{(j)}}_{\mathbf{u}^{(j)}} \tag{2} $$

In view of the fact that iteration takes place in the displacement and load space the load level may change from one iterate to the other. In this case an intermediate position j' for the same load level $^{j'}\lambda = {}^{i}\lambda$ is introduced before the final state j is reached (figure 1).

Supposedly configuration i has already been determined and the incremental equilibrium equations may be expressed by the linearized stiffness expression.

$$^{i}\mathbf{K} \cdot \Delta\mathbf{u}^{(j)} = \Delta\mathbf{P}^{(j)} + {}^{i}\mathbf{P} - {}^{i}\mathbf{F} \tag{3 a}$$

If the out-of-balance forces $^{i}\mathbf{R} = {}^{i}\mathbf{P} - {}^{i}\mathbf{F}$ are inserted

$$^{i}\mathbf{K} \cdot \Delta\mathbf{u}^{(j)} = \Delta\lambda^{(j)} \cdot \mathbf{P} + {}^{i}\mathbf{R} \tag{3 b}$$

The tangent stiffness matrix $^{i}\mathbf{K}$ at position i may include all possible nonlinear effects. It may be kept unchanged through several iteration cycles following the modified Newton-Raphson technique. Eq. (3) is the basic relation used as the starting point for the different iterative techniques described below.

The static stability criterion indicates a limit or bifurcation point by

$$^{c}\mathbf{K} \cdot \Delta\mathbf{u}^{c} = \mathbf{0} \tag{4}$$

where $\Delta\mathbf{u}^{c}$ is the eigenmode of the critical point. The singularity is usually checked by the determinant

$$\det {}^{c}\mathbf{K} = 0 \tag{5}$$

The determinant can easily be calculated as the product of all diagonal terms in the triangularized matrix during Gaussian elimination. Note that a positive determinant is not a sufficient criterion for stable equilibrium. Rather, the signs of the diagonal terms should be monitored to detect negative eigenvalues. This is the point when the limit load is passed and unloading should start.

3. Description of Some Iterative Techniques

3.1 Suppressing Equilibrium Iterations

As mentioned the equilibrium iterations usually break down near the limit point even if the load increment is small. The simplest way of avoiding this difficulty is to suppress the iterations in the critical zone. This procedure is used with great success by Bergan [5] who introduced the "current stiffness parameter" to guide the algorithm (figure 2).

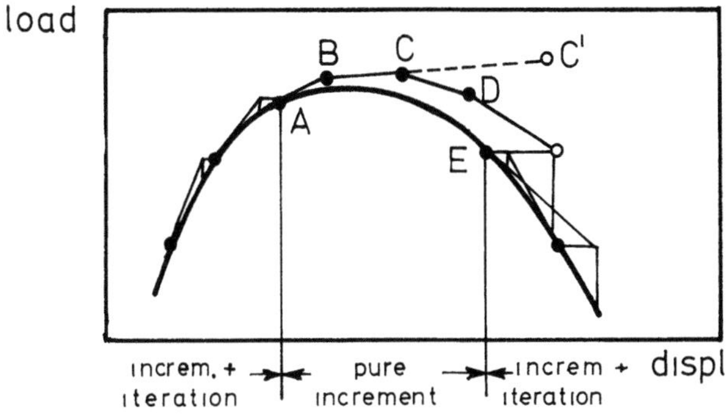

Figure 2: Suppressing iterations due to Bergan [5]

At a prescribed value of the stiffness parameter the iteration procedure is discontinued (point A). Then pure incrementation is used. If the Euclidean norm of the displacement increments exceeds a certain prescribed limit (point C') load and displacements are linearly scaled back (point C). Here negative diagonal elements may be detected in which case negative load increments are applied (point D). The iteration procedure is resumed when the stiffness parameter again reaches its prescribed value (point E). The limit point is located by a zero value of the stiffness parameter. The technique requires very small load increments to avoid drifting away from the equilibrium path.

3.2 Artificial - Spring - Method

This method was developed for frames by Wright and Gaylord [6] and has been applied to arch systems by Sharifi and Popov [7] and to shell structures by the author [8]. The technique is based on the observation that a snap-through problem may be transformed into one with a positive definite characteristic if linear artificial springs are added to the system (figure 3).

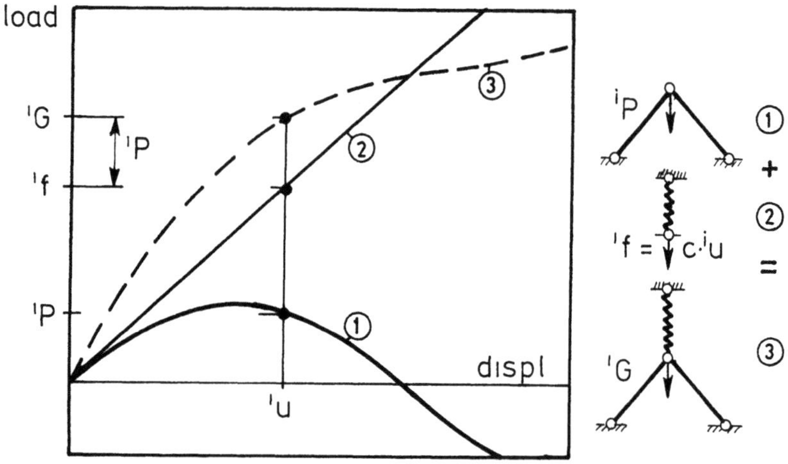

Figure 3: Artificial spring method

The method is described in detail in appendix I. It is an essential requirement that a separation of the real problem must be possible after the analysis of the stiffened system is obtained, i. e. for each stage only one load-reduction factor is defined. Furthermore the symmetry of the augmented stiffness matrix should be preserved. These requirements lead to springs at all loaded degrees of freedom, which are coupled, and depend on one single reference stiffness. This parameter has to be found by trial. The coupling of all artificial stiffnesses may destroy the banded nature of the stiffness matrix. In [8] the elements outside the band were omitted from the stiffness matrix but were retained on the right hand side to find the proper internal forces. Augmenting the spring stiffnesses on the band by a factor of three to five accelerates the convergence.

Because the "nonlinearity" of the system is diminished by the artificial springs the total number of iterations can nevertheless be reduced compared to the analysis without springs. Numerical experience shows that the method is successful only in real snap-through problems where the springs can keep the destabilizing structure alive. The method cannot be recommended for structures with local buckling or when a tendency to bifurcation is present.

3.3 Displacement - Control

The most often used method to avoid the singularity at the critical point is the interchange of dependent and independent variables. Here a single displacement component selected as a controlling parameter is prescribed and the corresponding load level is taken as unknown. The procedure was introduced first by Argyris [9] but in the meantime has been modified by several authors.

For simplicity let us assume that the stiffness expression, eq. (3), is reordered so that the prescribed component $\Delta u_2^{(j)} = \hat{u}_2$ is the last one in the displacement vector $\Delta \mathbf{u}^{(j)}$. Then equation (3) may be decomposed into two parts

$$ {}^{i}\begin{bmatrix} \mathbf{K}_{11} & \mathbf{K}_{12} \\ \mathbf{K}_{21} & \mathbf{K}_{22} \end{bmatrix} \begin{bmatrix} \Delta \mathbf{u}_1 \\ \Delta \mathbf{u}_2 \end{bmatrix}^{(j)} = \Delta\lambda^{(j)} \cdot \begin{bmatrix} \mathbf{P}_1 \\ \mathbf{P}_2 \end{bmatrix} + {}^{i}\begin{bmatrix} \mathbf{R}_1 \\ \mathbf{R}_2 \end{bmatrix} \qquad (6) $$

Interchanging the variables

$$ \begin{bmatrix} {}^{i}\mathbf{K}_{11} & -\mathbf{P}_1 \\ {}^{i}\mathbf{K}_{21} & -\mathbf{P}_2 \end{bmatrix} \begin{bmatrix} \Delta \mathbf{u}_1 \\ \Delta\lambda \end{bmatrix}^{(j)} = {}^{i}\begin{bmatrix} \mathbf{R}_1 \\ \mathbf{R}_2 \end{bmatrix} - {}^{i}\begin{bmatrix} \mathbf{K}_{12} \\ \mathbf{K}_{22} \end{bmatrix} \hat{u}_2 \qquad (7) $$

it is obvious that the loss of the symmetrical and banded structure of the stiffness matrix is a severe handicap. Later it was recognized that the solution of eq. (7) could be formed in two parts. The first line of eq. (7)

$$ {}^{i}\mathbf{K}_{11} \cdot \Delta \mathbf{u}_1^{(j)} = \Delta\lambda^{(j)} \cdot \mathbf{P}_1 + {}^{i}\mathbf{R}_1 - {}^{i}\mathbf{K}_{12} \cdot \hat{u}_2 \qquad (8) $$

is linear in the unknown increment at the load parameter $\Delta\lambda^{(j)}$.

Therefore its solution may be decomposed into (figure 4)

$$\Delta \mathbf{u}_1^{(j)} = \Delta \lambda^{(j)} \cdot \Delta \mathbf{u}_1^{(j)\,I} + \Delta \mathbf{u}_1^{(j)\,II} \tag{9}$$

corresponding to the two parts of the right hand side of eq. (8). That is, both solutions are obtained simultaneously using two different "load" vectors

$$'\mathbf{K}_{11}\ \Delta \mathbf{u}_1^{(j)I} = \mathbf{P}_1 \tag{10 a}$$

$$'\mathbf{K}_{11} \cdot \Delta \mathbf{u}_1^{(j)II} = {}'\mathbf{R}_1 - \underline{'\mathbf{K}_{12} \cdot \hat{\mathbf{u}}_2} \tag{10 b}$$

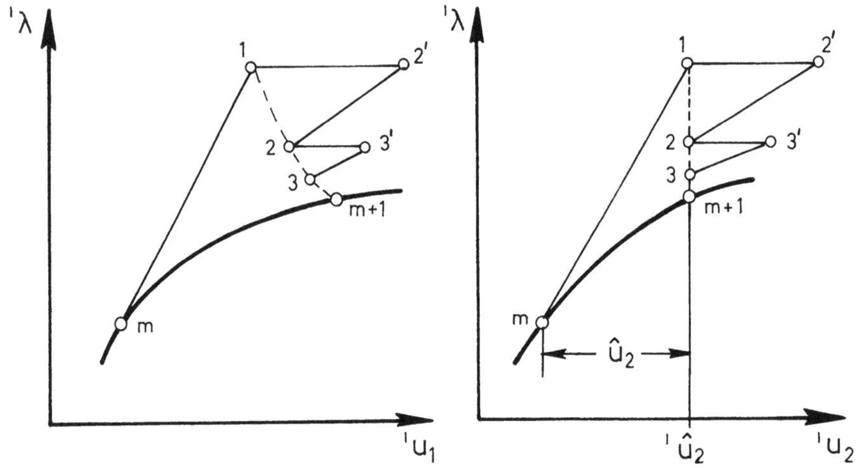

Figure 4: Displacement - Control Method

The displacement increment $\Delta \mathbf{u}_1^{(j)}$, eq. (9), is introduced into the second part of eq. (7). This allows the determination of the load parameter $\Delta\lambda^{(j)}$:

$$\Delta \lambda^{(j)} = \frac{-\,'\mathbf{R}_2 + {}'\mathbf{K}_{21}\ \Delta \mathbf{u}_1^{(j)II} + \underline{'\mathbf{K}_{22}\ \hat{\mathbf{u}}_2}}{\mathbf{P}_2 - {}'\mathbf{K}_{21} \cdot \Delta \mathbf{u}_1^{(j)I}} \tag{11}$$

Thus instead of solving an unsymmetrical equation the modified stiffness expression, eq. (8), is analysed for two right hand sides provided that ${}^1\mathbf{K}_{11}$ is not singular. Since the displacement $\hat{u}_2$ is held fixed during the iteration the underlined terms in equations (10 b) and (11) are omitted in all further iteration cycles.

This modified displacement control method was described first by Pian and Tong [10] without mentioning the out-of-balance terms. Zienkiewicz [11] refers to the standard programming technique and gives a physical interpretation of the two step method. Sabir and Lock [12] explicitly introduced the out-of-balance terms into the formulation. The method was also described in detail by Stricklin et al. [13]. A similar procedure has been applied by Nemat-Nasser and Shatoff [14] who used a direct substitution method instead of the Newton-Raphson technique.

A valuable simplification was utilized by Batoz and Dhatt [15]. Since the technique above described requires a modification of the stiffness matrix (${}^1\mathbf{K} \rightarrow {}^1\mathbf{K}_{11}$) the authors point out that it is not very likely to obtain exactly the singular point. Hence the original matrix ${}^1\mathbf{K}$ may still be used and equations (10) are replaced by

$$ {}^1\mathbf{K} \cdot \Delta\mathbf{u}^{(j)\mathrm{I}} = \mathbf{P} \tag{12 a} $$

$$ {}^1\mathbf{K} \cdot \Delta\mathbf{u}^{(j)\mathrm{II}} = {}^1\mathbf{R} \tag{12 b} $$

where the underlined term in eq. (10 b) is not required to be formed. Again both solutions are added:

$$ \Delta\mathbf{u}^{(j)} = \Delta\lambda^{(j)} \cdot \Delta\mathbf{u}^{(j)\mathrm{I}} + \Delta\mathbf{u}^{(j)\mathrm{II}} \tag{13 a} $$

The vector includes also the prescribed component

$$ \Delta u_2^{(j)} = \Delta\lambda^{(j)} \cdot \Delta u_2^{(j)\mathrm{I}} + \Delta u_2^{(j)\mathrm{II}} = \hat{u}_2 \tag{13 b} $$

This constraint equation used in the first iteration cycle ($m \rightarrow j = 1$)

allows the determination of the incremental load parameter

$$\Delta\lambda^{(1)} = \frac{\hat{u}_2 - \Delta u_2^{(1)\,II}}{\Delta u_2^{(1)\,I}} \tag{14}$$

Supposedly the structure is in an equilibrium state at the beginning of a step so the out-of-balance forces vanish and so does $\Delta u_2^{(j)\,II}$. Then $\Delta\lambda^{(1)}$ is simply a scaling factor providing the constraint $\Delta u_2^{(1)} = \hat{u}_2$. Batoz and Dhatt [15] even drop this first cycle. They update the displacement field only by its component $\Delta u_2^{(1)}$ and start to iterate.

For all further cycles $u_2^{(j)}$ does not change i. e. $\Delta u_2^{(j)}$ is zero and $\Delta\lambda^{(j)}$ is

$$\Delta\lambda^{(j)} = -\frac{\Delta u_2^{(j)\,II}}{\Delta u_2^{(j)\,I}} \qquad j = 2,3 \cdot \tag{15}$$

Applying the modified Newton-Raphson technique eq. (12 a) needs to be solved only when the stiffness matrix is updated. Then no additional computer time is required and the only additional vector stored is $\mathbf{\Delta u}^{(1)\,I}$. The iteration is continued until all other displacement components are adjusted and the new equilibrium position is found (fig. 4).

The displacement control method is usually used only in the neighborhood of the critical point although it may be applied throughout the entire load range. Obviously the method fails whenever the structure snaps back from one load level to a lower one (see example 5. 2). Some knowledge of the failure mode is required for a proper choice of the controlling displacement. It might even be necessary to change the prescribed parameter. Therefore an obvious modification is to relate the procedure to a measure including all displacements rather than to one single component. This is discussed in the next section.

3. 4 Modified Constant - Arc - Length - Method of Riks/Wempner

This iterative technique has been independently introduced by Riks [1], [2] and Wempner [3]. Both authors limit the load step $\Delta\lambda^{(1)}$

by the constraint equation

$$\Delta u^{(1)T} \Delta u^{(1)} + (\Delta\lambda^{(1)})^2 = ds^2 \tag{16}$$

That is, the generalized "arc length" of the tangent at m is fixed to a prescribed value ds. Then the iteration path follows a "plane" normal to the tangent (figure 5), so the scalar product of the tangent $\vec{t}^{(1)}$ and the vector $\Delta\vec{u}^{(j)}$ containing the unknown load and displacement increments must vanish:

$$\vec{t}^{(1)} \cdot \Delta\vec{u}^{(j)} = 0 \tag{17a}$$

or in matrix notation

$$\Delta u^{(1)T} \cdot \Delta u^{(j)} + \Delta\lambda^{(1)} \cdot \Delta\lambda^{(j)} = 0 \tag{17b}$$

$$j = 2,3$$

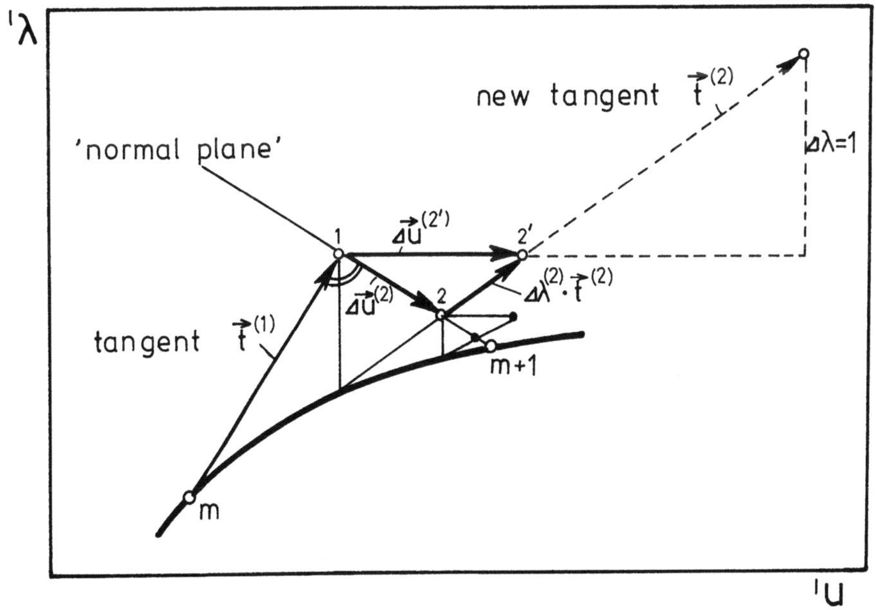

Figure 5: Constant - Arc - Length Method

The constraint equations originally were added to the incremental stiffness expression destroying symmetry and the banded structure of the matrix. It was realized by Wessels [16] based on geometrical considerations that these difficulties could be removed by a two step technique similar to that described in the previous section. It is this idea followed in this study. *)

Again the unknown vector $\Delta\vec{u}^{(j)}$ is formed in two parts

$$\Delta\vec{u}^{(j)} = \Delta\lambda^{(j)} \cdot \vec{t}^{(j)} + \Delta\vec{u}^{(j')} \tag{18 a}$$

or in matrix notation equivalent to eq. (13 a).

$$\Delta\mathbf{u}^{(j)} = \Delta\lambda^{(j)} \cdot \Delta\mathbf{u}^{(j)\mathrm{I}} + \Delta\mathbf{u}^{(j)\mathrm{II}} \tag{18 b}$$

Also here $\Delta\mathbf{u}^{(j)\mathrm{I}}$ and $\Delta\mathbf{u}^{(j)\mathrm{II}}$ are obtained by equations (12) using either the reference load vector $\mathbf{P}$ $(\Delta\lambda = 1)$ or the out-of-balance forces ${}^{1}\mathbf{R}$ as right hand sides. Then eq. (18) is inserted into the constraint eq. (17) and solved for the unknown load increment $\Delta\lambda^{(j)}$

$$\Delta\lambda^{(j)} = - \frac{\Delta\mathbf{u}^{(1)^{T}} \cdot \Delta\mathbf{u}^{(j)\mathrm{II}}}{\Delta\mathbf{u}^{(1)^{T}} \cdot \Delta\mathbf{u}^{(j)\mathrm{I}} + \Delta\lambda^{(1)}} \tag{19}$$

Geometrically this is the intersection j of the new tangent $\vec{t}^{(j)}$ with the "normal plane" (figure 5). Eq. (19) is equivalent to eq. (15) but contains the influence of all displacement components in an integral sense. The load increment $\Delta\lambda^{(1)}$ in the denominator, which obviously has another dimension, expresses the different scaling of the load axis with respect to the displacement space. It may be seen for the one degree-of-freedom system in figure 6 a that a low value $\Delta\lambda^{(1)}$ tends to a displacement control and a large value to a load control of the iteration. In many degree-of-freedom systems the value $\Delta\lambda^{(1)}$ in eq. (19) does not play an important role and may be suppressed.

---

*) During the preparation of this study the author became aware of the valuable paper by Crisfield [17] devoted to the same subject.

Again the modified Newton-Raphson technique simplifies the method because eq. (12 a) is solved only once at the beginning of the step and may even be replaced by the first solution $\Delta u^{(1)}$:

$$\frac{\Delta\lambda^{(j)}}{\Delta\lambda^{(1)}} = - \frac{\Delta u^{(1)^T} \cdot \Delta u^{(j)^{II}}}{\Delta u^{(1)T} \cdot \Delta u^{(1)} + (\Delta\lambda^{(1)})^2} \tag{20}$$

Instead of iterating in the "plane" normal to the tangent $\vec{t}^{(1)}$ it might be useful to define a "sphere" with a center at m and a radius ds [17] (see appendix II). Alternatively the "normal plane" may be updated in every iteration cycle (figure 6 b). That is, in eq. (19) $\Delta u^{(1)}$ is replaced by the total increment $u^{(1)}$. It was found that except for very large load steps the differences resulting from these formulations are minor.

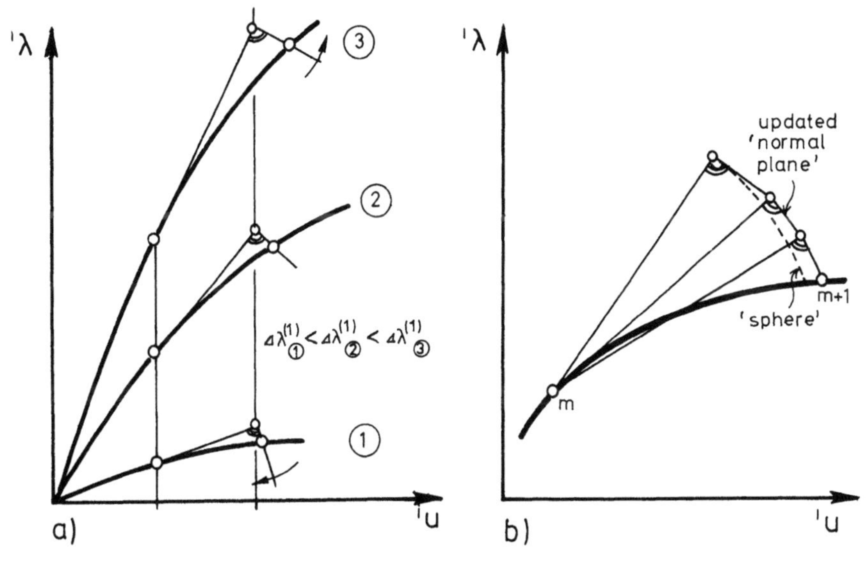

Figure 6: Modification of constant - arc - length method

Numerical experience has shown that this iterative technique is very efficient in the entire load range particularly when automatic load incrementation based on eq. (16) is used. The only additional storage required is the vector $\Delta u^{(1)}$. The extra computer time is negligible.

In addition to the "constant-arc-length" the step size may be scaled by relating the number of iterations, n_i, used in the previous step to a desired value, $\hat{n}_i$. It was found that a factor $\hat{n}_i/n_i$ results in oscillations in the number of iterations required from step to step so that $\sqrt{\hat{n}_i/n_i}$ is recommended. If material nonlinearities are involved smaller load steps should be defined to avoid drifting. Whenever a negative element in the triangularized matrix is encountered unloading is initiated. The convergence may be either monotonic or alternating and may in some cases be slow. Then relaxation factors may accelerate the iteration process. For instance, in the alternating case a cut-back of the next load change to 50 % resulted in a considerable improvement.

4. Summary of the Displacement Control and Modified Riks/Wempner Method

The algorithms for the displacement control method and the modified Riks/Wempner method differ only in the equation used for the evaluation of $\Delta\lambda^{(j)}$. The algorithm is summarized as follows:

1. Select a basic load increment as the reference load $\mathbf{P}$, thus defining the length ds in the first step (eq. 16).

2. In any step:
   a) Solve the equilibrium equations for $\mathbf{P}$ and linearly scale the load and displacements to produce the length ds. This determines $\Delta\lambda^{(1)}$, $\Delta\mathbf{u}^{(1)}$.
   b) Adjust the step size to the desired number of iterations $\hat{n}_i$, e. g. $\sqrt{\hat{n}_i/n_i}$.
   c) Check the triangularized matrix for unloading.

3. a)* Update the stiffness matrix ${}^{1}\mathbf{K}$
   b) and, simultaneously, determine the out-of-balance forces ${}^{1}\mathbf{R}$.
   c)* Solve for $\mathbf{P}$ to determine $\Delta\mathbf{u}^{(j)I}$.
   d) and, simultaneously, solve for the out-of-balance forces ${}^{1}\mathbf{R}$ to determine $\Delta\mathbf{u}^{(j)II}$.

Note: * indicates a step which is omitted in the modified Newton-Raphson procedure.

4. Use constraint eq. (15) or (19) to determine the load increment $\Delta\lambda^{(j)}$ and eq. (13 a) ≡ eq. (18 b) to determine displacement increments $\Delta u^{(j)}$. (If needed use acceleration factors.)

5. Update the load level and the displacement field.

6. Repeat steps 3 - 5 until the desired accuracy is achieved.

7. Reformulate the stiffness matrix and start a new step by returning to 2.

5. Numerical Examples

The examples have been analysed on CDC 6600/Cyber 174 computers using the nonlinear finite element code NISA [18]. The geometrical nonlinearity is based on the total Lagrangian formulation. For the arch example, an 8 node isoparametric plane stress element is used [4]. The plate and shell structures are idealized by degenerated isoparametric elements developed in [8], [19]. The modified Riks/Wempner method, in combination with the modified Newton-Raphson technique, has been applied exclusively. The ratio of the change of the incremental displacements to the total displacement increments, using Euclidean norms, is used for the convergence criterion.

5.1 Shallow Arch

The shallow circular arch under uniform pressure (figure 7) has already been analysed in [8] applying the artificial spring method (c_{11} = 28 lb/in), see also [7]. Ten 8 node isoparametric plane stress elements were used for one half of the arch. The analysis with a basic load of $\bar{p}$ = 0.3 and using the constant-arc-length constraint shows the typical step size reduction in the neighborhood of the limit point. Thirty steps with 1 to 2 iterations per step were needed. The analysis has been repeated for a basic load step of $\bar{p}$ = 1.0. The step size has been adjusted by the factor $\sqrt{\hat{n}_1/n_1}$ with a desired number of iterations $\hat{n}_1 = 5$. In addition, the load increment was reduced to 50 % whenever it alternated and the absolute value decreased. Now only 9 steps are sufficient. The number of iterations required are indicated in the figure. The diagram also shows the starting point in each step after the first Newton-Raphson iterate. Compared to the artificial spring technique considerable savings are achieved.

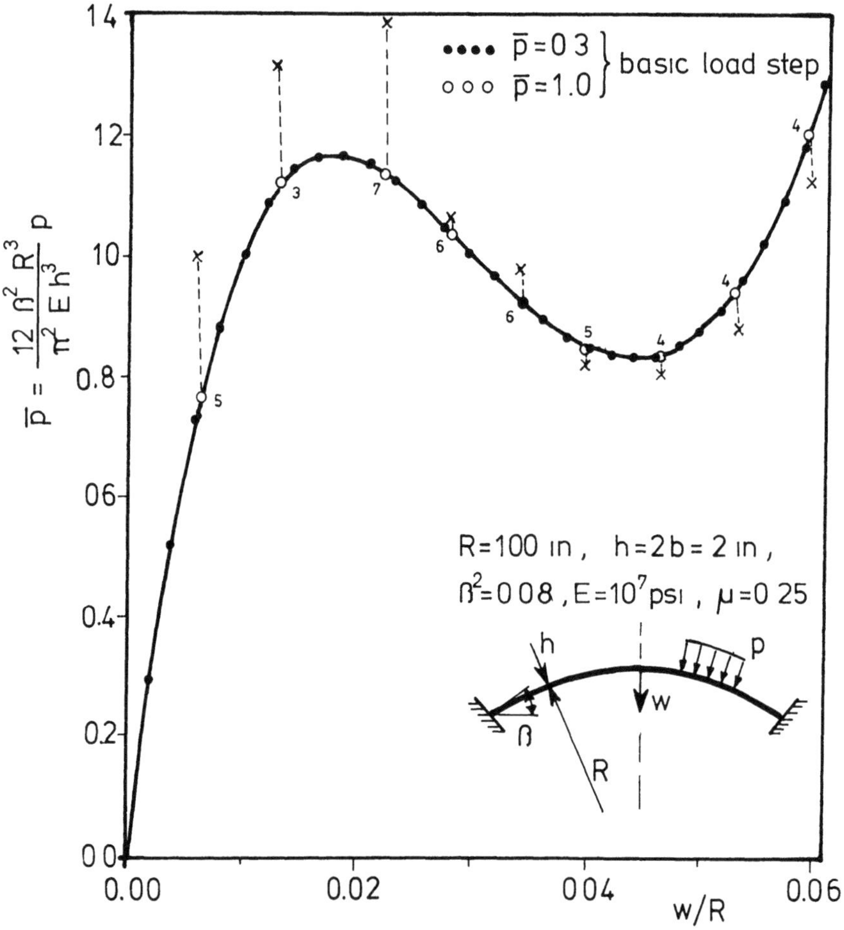

Figure 7: Shallow circular arch

5.2 Shallow Cylindrical Shell

The shallow cylindrical shell under one concentrated load (figure 8) is hinged at the longitudinal edges and free at the curved boundaries. The structure exhibits snap-through as well as snap-back phenomena with horizontal and vertical tangents. The shell has been analysed by

Sabir and Lock [20] who used a combination of the displacement and load control techniques. In the present study one quarter of the shell has been idealized by four 16 node bicubic degenerated shell elements. As the basic load step, P = 0.4 kN was chosen. Again the load steps were adjusted with $\sqrt{\hat{n}_1/n_1}$ and the acceleration scheme described for the arch was applied. The entire load deflection diagram is obtained in one solution with 15 steps and 3 to 9 iterations per step as indicated in the figure. If the acceleration technique was not used the number of iterations increased considerably especially at the minimum load.

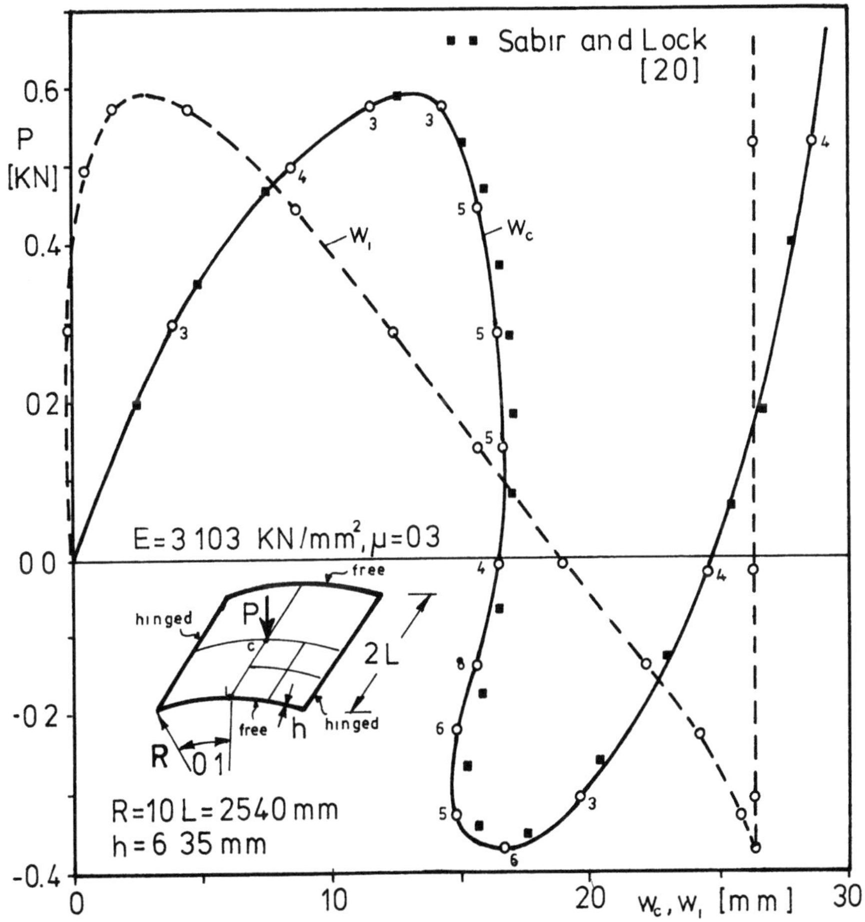

Figure 8: Shallow cylindrical shell

This part of the load-deflection curve is numerically difficult because of the abrupt changes of the response at, for instance, the point i at the free edge. The structure has also been analysed using 36 bilinear 4 node degenerated elements in combination with an uniform 1 x 1 reduced integration scheme. Approximately the same results have been obtained but at about 20 % of the CP-time.

5.3 Elastic - Plastic Buckling of a Plate

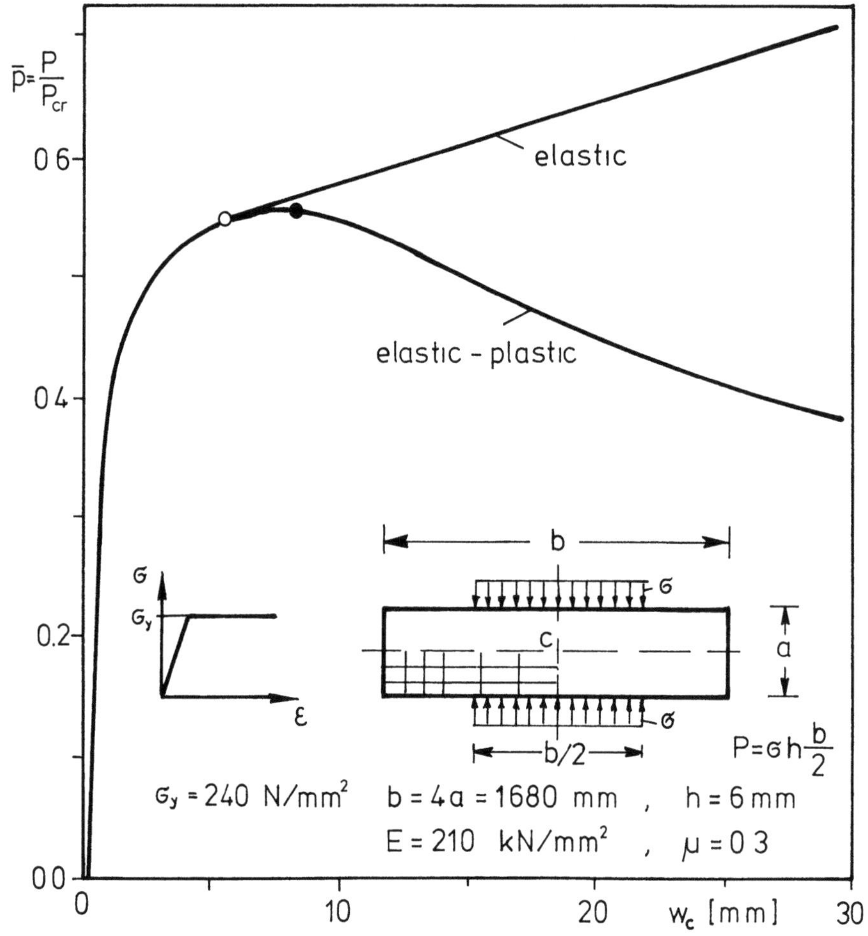

Figure 9: Buckling of a long plate

The simply supported plate shown in figure 9 has an aspect ratio of $\alpha = 1/4$ and is loaded only on its middle part. The plate has an initial geometrical imperfection, defined by a double sin-function, with a maximum amplitude of 0.294 mm. The yield limit σ_y of the elastic-ideally plastic steel is 240 N/mm². Eighteen bicubic degenerated elements unevenly spaced were used for one quarter of the plate. The thickness was divided into seven layers. The total load P is non-dimensionalized with the linear elastic buckling load P_{cr} of the plate with uniform load on the entire boundary:

$$P_{cr} = k \; b \; \frac{\pi^2 \cdot E \cdot h^3}{12\,(1-\mu^2)\,a^2} \; , \qquad k = (1 + \alpha^2)^2 \tag{21}$$

The basic load step chosen was $\bar{p}$ = 0.25. In figure 9 the normalized load is plotted versus the center lateral displacement. The plate fails under combined geometrical and material failure. The initial yield point at a deflection of about 6 mm is immediately followed by the limit point at about 8.3 mm. Thirty steps with 1 or 2 iterations per step were used. The elasto-plastic analysis was supplemented by a purely elastic solution also shown in the figure. Here the typical increasing postbuckling response of plates is recognized.

5.4 Cylindrical Shell under Wind Load

The buckling analysis of the closed cylindrical shell under wind load (figure 10) studied in [21] has been extended to the postbuckling range.

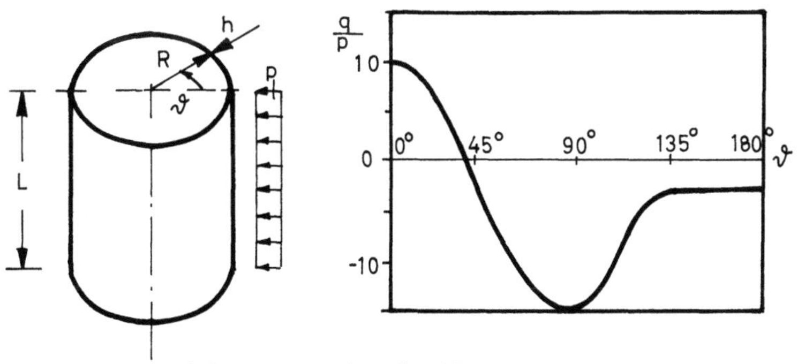

$R = L/2 = 220$ mm , $h = 0.105$ mm

$E = 6\;87\;10^4$ N/mm², $\mu = 0.3$

Figure 10. Geometry and load function of a cylindrical shell

The extremely thin structure with a radius to thickness ratio of over 2000 is simply supported at both ends. The variation of the wind load defined in figure 10 is taken as constant over the length of the cylinder. The maximal load p at the stagnation point is normalized to the linear buckling load of the shell under uniform pressure

$$p_{cl} = \frac{0.918 \; E\left(\frac{h}{R}\right)^2}{\frac{L}{R}\sqrt{\frac{R}{h}} - 0.657} \qquad \bar{p} = \frac{p}{p_{cl}} \tag{22}$$

One quarter of the shell is idealized by 2 x 18 bicubic 16 node elements. Two elements of unequal length are used in the axial direction, while the 18 elements in the circumferential direction are concentrated near the stagnation zone. The first load increment defined the basic step size as $\bar{p} = 0.25$. Both the perfect and an imperfect shell have been analysed. Figure 11 shows the displacement pattern of one quarter of

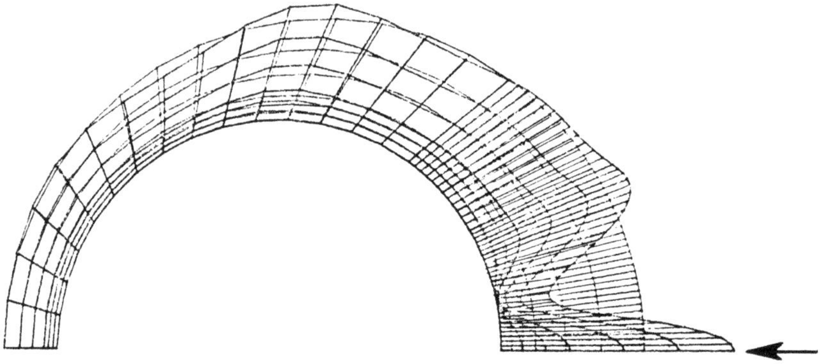

Figure 11: Displacement pattern

the shell near the limit point. A failure mode with one half a wave in the axial direction and a few buckling waves in the circumferential direction, located in the compression zone, is indicated. The post-buckling minimum of the load-deflection diagram (figure 12) is about 60 % of the limit point. The imperfection assumed for the second analysis corresponds to the failure mode of the perfect structure. The maximum imperfection amplitude is 2.5 times the wall thickness. The load deflection path (figure 12) indicates a reduction of the limit load to 68 % of that for the perfect shell. The postbuckling minima nearly coincide. It should be noted that the example is numerically very sensitive because of the extreme slenderness ratio and the local nature of the failure mechanism. In both cases over 60 steps were necessary.

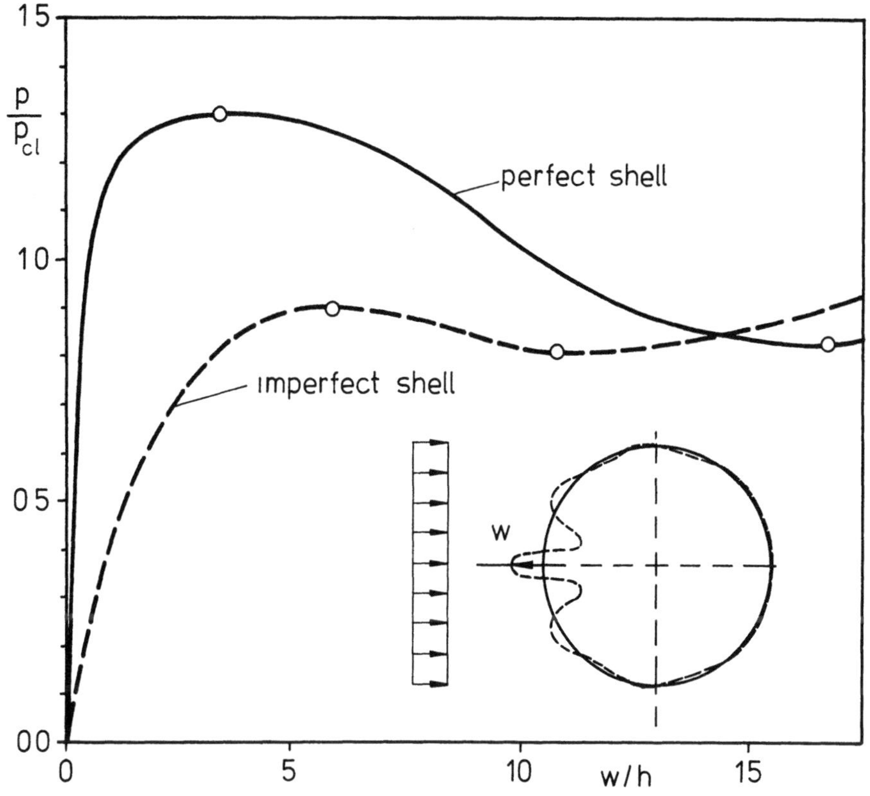

Figure 12: Load - deflection - diagram of a wind loaded shell

6. Conclusions

This study on iterative techniques for passing limit points allows the following conclusions:

* Suppression of equilibrium iterations near the limit point may be a useful procedure but requires very small load steps.

* The method of artificial springs is based on numerical experience and trial solutions. For local failure it may not be successful.

* The displacement control method requires a proper selection of the controlling parameter. It fails in snap-back situations.

* The constant - arc - length method of Riks /Wempner seems to be the most versatile technique, being advantageous in the entire load range.

  Due to modifications of the original method the constraint equation does not need to be solved simultaneously with the equilibrium equations.

  Automatic adjustment of the load step and acceleration schemes may further improve the performance. Only minor changes in coding are necessary. Applying the modified Newton-Raphson technique requires the storage of one additional vector. The extra computer time is negligible.

Acknowledgement

The author would like to thank Professor D. W. Murray, University of Edmonton, currently at the University of Stuttgart, for valuable discussions.

References

[1] Riks, E.: The Application of Newton's Method to the Problem of Elastic Stability. J. Appl. Mech. 39 (1972) 1060-1066.

[2] Riks, E.: An Incremental Approach to the Solution of Snapping and Buckling Problems. Int. J. Solids Struct. 15 (1979) 529-551.

[3] Wempner, G. A.: Discrete Approximations Related to Nonlinear Theories of Solids. Int. J. Solids Struct. 7 (1971) 1581-1599.

[4] Bathe, K.-J., Ramm, E., Wilson, E. L.: Finite Element Formulations for Large Deformation Dynamic Analysis. Int. J. Num. Meth. Engng. 9 (1975) 353-386.

[5] Bergan, P. G.: Solution Algorithms for Nonlinear Structural Problems. Int. Conf. on "Engng. Appl. of the F. E. Method", Høvik, Norway 1979, published by A. S. Computas.

[6] Wright, E. W., Gaylord, E. H.: Analysis of Unbraced Multi-Story Steel Rigid Frames. Proc. ASCE, J. Struct. Div. 94 (1968) 1143-1163.

[7] Sharifi, P., Popov, E. P.: Nonlinear Buckling Analysis of Sandwich Arches. Proc. ASCE, J. Engng. Div. 97 (1971) 1397-1412.

[8] Ramm, E.: Geometrisch nichtlineare Elastostatik und finite Elemente. Habilitationsschrift, Universitat Stuttgart, 1975.

[9] Argyris, J. H.: Continua and Discontinua. Proc. 1st Conf. "Matrix Meth. Struct. Mech.", Wright-Patterson A. F. B., Ohio 1965, 11-189.

[10] Pian, T. H. H., Tong, P.: Variational Formulation of Finite Displacement Analysis. IUTAM Symp. on "High Speed Computing of Elastic Structures", Liège 1970, 43-63.

[11] Zienkiewicz, O.C.: Incremental Displacement in Non-Linear Analysis. Int. J. Num. Meth. Engng. 3 (1971) 587-588.

[12] Lock, A. C., Sabir, A. B.: Algorithm for Large Deflection Geometrically Nonlinear Plane and Curved Structures. In "Mathematics of Finite Elements and Applications" (ed. J. R. Whiteman), Academic Press, N. Y. 1973, 483-494.

[13] Haisler, W., Stricklin, J., Key, J.: Displacement Incrementation in Nonlinear Structural Analysis by the Self-Correcting Methods. Int. J. Num. Meth. Engng. 11 (1977) 3-10.

[14] Nemat-Nasser, S., Shatoff, H. D.: Numerical Analysis of Pre- and Postcritical Response of Elastic Continua at Finite Strains. Comp. Struct. 3 (1973) 983-999.

[15] Batoz, J.-L., Dhatt, G.: Incremental Displacement Algorithms for Nonlinear Problems. Int. J. Num. Meth. Engng. 14 (1979) 1262-1267.

[16] Wessels, M.: Das statische und dynamische Durchschlagsproblem der imperfekten flachen Kugelschale bei elastischer rotationssymmetrischer Verformung. Dissertation, TU Hannover, 1977, Mitteil. Nr. 23 des Instituts für Statik.

[17] Crisfield, M. A.: A Fast Incremental/Iterative Solution Procedure that Handles "Snap-Through". Proc. Symp. on "Computational Methods in Nonlinear Structural and Solid Mech.", Washington, Oct. 1980.

[18] Brendel, B., Hafner, L., Ramm, E., Sattele, J. M.: Programmdokumentation - Programmsystem NISA. Bericht, Institut für Baustatik, Universitat Stuttgart, 1977.

[19] Ramm, E.: A Plate/Shell Element for Large Deflections and Rotations. Symp. "Formulations and Computational Algorithms in F. E. Analysis", Cambridge 1976, MIT Press 1977.

[20] Sabir, A. B., Lock, A. C.: The Application of Finite Elements to the Large Deflection Geometrically Nonlinear Behaviour of Cylindrical Shells. In "Variational Methods in Engng." (ed. C. A. Brebbia and H. Tottenham), Southampton, University Press (1972) 7/66 - 7/75.

[21] Brendel, B., Fischer, D., Ramm, E., Rammerstorfer, F.: Linear and Nonlinear Stability Analysis of Thin Cylindrical Shells under Windloads. To be published, J. Struct. Mech. 1981.

Appendix I: The Artificial Spring Method

According to figure 3 the vector of the total external loads ${}^{1}\mathbf{G}$ of the modified system is decomposed into the real load vector ${}^{1}\mathbf{P}$ and the part resisted by the springs ${}^{1}\mathbf{f}$.

$$ {}^{i}\mathbf{G} = {}^{i}\mathbf{P} + {}^{i}\mathbf{f} \qquad \text{(A 1)} $$

To retain the desired ratio of specified loads it is required that all components of the real load can be obtained by one common "load-reduction-factor" ${}^{1}\gamma$

$$ {}^{i}\mathbf{P} = {}^{i}\gamma \cdot {}^{i}\mathbf{G} \qquad \text{(A 2)} $$

That is, all components of configuration i have the same ratio

$$ \frac{{}^{i}f_k}{{}^{i}G_k} = 1 - {}^{i}\gamma \qquad i = 1,2,3 \quad n \qquad \text{(A 3)} $$

It follows that springs have to be attached to all loaded degrees-of-freedom and all spring stiffnesses are coupled. The spring stiffness matrix $\mathbf{c}$ is defined by

$$ {}^{i}\mathbf{f} = \mathbf{c} \cdot {}^{i}\mathbf{u} \qquad \text{(A 4)} $$

Energy principles require $\mathbf{c}$ to be a symmetrical matrix ($c_{k\ell} = c_{\ell k}$). Equation (A 3) allows the elements $c_{k\ell}$ of the matrix to be determined if one reference stiffness c_{11} is prescribed

$$ c_{kl} = \frac{{}^{i}G_k \cdot {}^{i}G_l}{({}^{i}G_1)^2} \, c_{11} \qquad \text{(A 5)} $$

or if the reference load vector $\mathbf{P}$ is introduced

$$ \mathbf{c} = \frac{c_{11}}{P_1^2} \cdot \mathbf{P} \cdot \mathbf{P}^T \qquad \text{(A 6)} $$

The iteration equation, eq. (3 a), is modified to

$$ ({}^{i}\mathbf{K} + \mathbf{c}) \cdot \Delta \mathbf{u}^{(j)} = \underbrace{{}^{j}\mathbf{P} + {}^{j}\mathbf{f}}_{{}^{j}\mathbf{G}} - {}^{i}\mathbf{F} - \mathbf{c}\,{}^{i}\mathbf{u} \qquad \text{(A 7)} $$

The right hand side expresses the out-of-balance forces. After iteration ($j \rightarrow m+1$) the real loads are determined by eq. (A 2):

$$ {}^{m+1}\mathbf{P} = {}^{m+1}\gamma \cdot {}^{m+1}\mathbf{G} \qquad \text{with} \quad {}^{m+1}\mathbf{G} = {}^{m+1}\lambda \cdot \mathbf{P} \tag{A 8} $$

The "load-reduction factor" is obtained by eq. (A 3):

$$ {}^{m+1}\gamma = 1 - \frac{c_{11}}{{}^{m+1}\lambda \cdot P_1^2} \; \mathbf{P}^T \cdot {}^{m+1}\mathbf{u} \tag{A 9} $$

It was found that an effective value of c_{11} is one which leads to $0 < {}^{m+1}\gamma < 0.6$ at the beginning of the analysis [7], [8].

Appendix II: Iteration on a "Sphere"

The "sphere" with the center at m and the radius ds of the initial tangent vector $\vec{t}^{(1)}$ (figure 13) is defined by

$$ \vec{r}^{(j)} \cdot \vec{r}^{(j)} - ds^2 = 0 \tag{A 10} $$

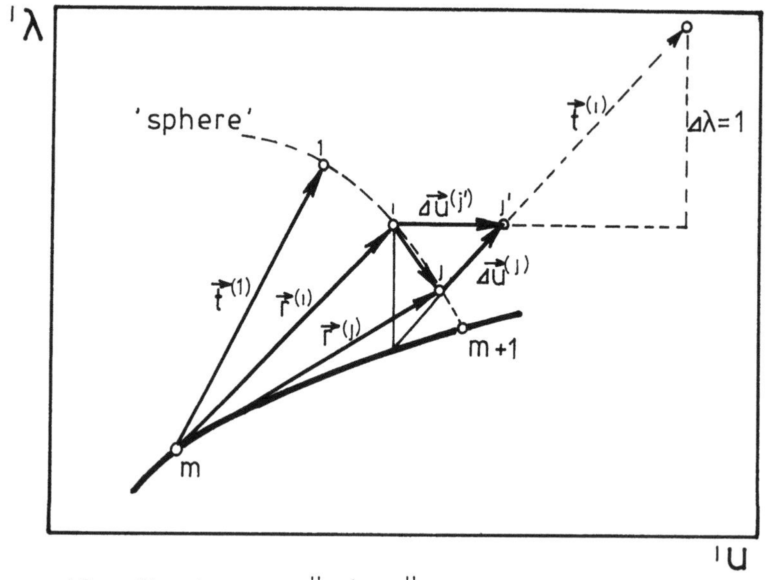

Figure 13: Iteration on a "sphere"

If the radius vector is replaced by

$$\vec{r}^{(j)} = \vec{r}^{(i)} + \vec{\Delta u}^{(j)} \tag{A 11}$$

and eq. (16) is taken into consideration eq. (10) results in

$$\vec{\Delta u}^{(j)} (\vec{\Delta u}^{(j)} + 2\vec{r}^{(i)}) = 0 \tag{A 12}$$

or in matrix notation

$$\mathbf{\Delta u}^{(j)T} (\mathbf{\Delta u}^{(j)} + 2 \cdot \mathbf{u}^{(i)}) + \Delta\lambda^{(j)} (\Delta\lambda^{(j)} + 2\,\lambda^{(i)}) = 0 \tag{A 13}$$

$\mathbf{\Delta u}^{(j)}$ is expressed by eq. (13 a). Then eq. (A 13) leads to a quadratic constraint equation for the load parameter $\Delta\lambda^{(j)}$ which is the equivalent to eq. (19)

$$a\,(\Delta\lambda^{(j)})^2 + 2\,b\,\Delta\lambda^{(j)} + c = 0 \tag{A 14}$$

with the coefficients

$$\begin{aligned} a &= 1 + (\mathbf{\Delta u}^{(j)II})^T \mathbf{\Delta u}^{(j)I} \\ b &= \lambda^{(i)} + (\mathbf{\Delta u}^{(j)I})^T (\mathbf{\Delta u}^{(j)II} + \mathbf{u}^{(i)}) \\ c &= (\mathbf{\Delta u}^{(j)II})^T (\mathbf{\Delta u}^{(j)II} + 2 \cdot \mathbf{u}^{(i)}) \end{aligned} \tag{A 15}$$

Parameter Sensitivity of Nonlinear Structures Concerning Stability Limit

F. G. RAMMERSTORFER, D. F FISCHER
Voest-Alpine AG, Linz, Austria

Summary

The stability of nonlinear structures is considered, especially having in view the parameter sensitivity of the stability limit load. It is shown that under special circumstances, for which criterions are derived, nonlinear structures behave with a sudden, i.e. non-continuous, change of the limit load if a certain system parameter is varied. Such phenomena belong to the field of the catastrophe theory which allows their classification and useful documentation.

1. Introduction

Considering the load displacement behaviour in combination with the variation of the determinant of the global tangent stiffness matrix and performing supplementary eigenvalue-analyses at several load steps, the stability behaviour of nonlinear structures can be determined. Some unexpected effects which a linear analysis would never reveal are found by those nonlinear investigations.

The paper shows how structures can undergo effects like a non-continuous parameter sensitivity. This means that an infinitesimal variation of a particular parameter of the physical system can cause a finite change of the stability limit. It is shown how those problems are embedded in the catastrophe theory.

As a particular example the described phenomena are discussed by the consideration of the nonlinear behaviour of an elastically supported initially curved beam which represents the behaviour of a wide-spanned, domed roof of a liquid storage tank.

2. Algorithms for nonlinear stability analysis

Assuming proportional conservative static loading the current load vector at the step n of the incremental analysis can be expressed by

$$ {}^{n}\underset{\sim}{R} = {}^{n}\lambda \, \underset{\sim}{R}_{ref} \tag{1} $$

with the reference load vector, $\underset{\sim}{R}_{ref}$, and the current load multiplier, ${}^{n}\lambda$. The linearized incremental equilibrium equation for the increment n → n+1 following the tangent stiffness concept (e.g. [1]) is

$$ {}^{n}\underset{\approx}{K} \, \Delta\underset{\sim}{u} = {}^{n+1}\lambda \, \underset{\sim}{R}_{ref} - {}^{n}\underset{\sim}{F}. \tag{2} $$

The current tangent stiffness matrix of the system, ${}^{n}\underset{\approx}{K}$, consists of

$$ {}^{n}\underset{\approx}{K} = \underset{\approx}{K}_{e} + {}^{n}\underset{\approx}{K}_{u} + {}^{n}\underset{\approx}{K}_{g} \tag{3} $$

with the constant linear elastic stiffness matrix, $\underset{\approx}{K}_e$, the initial displacement matrix, ${}^{n}\underset{\approx}{K}_u$, and the initial stress or geometric matrix ${}^{n}\underset{\approx}{K}_g$. ${}^{n}\underset{\sim}{F}$ denotes the vector of internal forces in configuration n. The solution of equ. (2) yields a first approximation of the incremental displacement vector $\Delta\underset{\sim}{u}$ which can be improved by equilibrium interation [2,3].

According to the static stability criterion, two adjacent equilibrium configurations exist at the critical load level ${}^{m^*}\lambda$. Hence,

$$ {}^{m^*}\underset{\approx}{K} \, \delta\underset{\sim}{u} = \underset{\sim}{0} \tag{4} $$

has a nontrivial solution at the stability limit. Equ. (4) leads to the condition

$$ \det {}^{m^*}\underset{\approx}{K} = \det \underset{\approx}{K}({}^{m^*}\lambda) = 0 \tag{5} $$

for the fundamental state m^* in which the equilibrium becomes instable. Therefore, the behaviour of the determinant of the load dependent tangent stiffness matrix, $\det \underset{\approx}{K}(\lambda)$, can be used as an indicator for the stability of the equilibrium. Denoting the critical load level by ${}^{m^*}\lambda = \lambda^*$, the following relations hold [4,5]:

a) $$ \lim_{\lambda \to \lambda^*} \det \underset{\approx}{K}(\lambda) = 0 \tag{6} $$

for buckling as well as for snap-through,

b) a sufficient but not necessary condition for snap-through is

$$ \lim_{\lambda \to \lambda^*} \frac{\partial}{\partial \lambda} \det \underset{\approx}{K}(\lambda) = -\infty, \tag{7} $$

c) buckling is indicated by a finite negative value of the left-hand-side of equ. (7).

All relations shown for det $\underset{\approx}{K}(\lambda)$ are also valid for the normalized determinant

$$\det_n \underset{\approx}{K}(\lambda) = \det \underset{\approx}{K}(\lambda) \;/\; \det \underset{\approx}{K}_e \tag{8}$$

with $\underset{\approx}{K}_e$ as explained in equ. (3). Due to practical reasons it is useful to consider the normalized rather than the original determinant value. Therefore, in the following part of this paper $\det_n \underset{\approx}{K}(\lambda)$ is used instead of det $\underset{\approx}{K}(\lambda)$.

As shown in detail in [4,6,7] the relations (3) and (4) can lead to an approximative (linearized) eigenvalue problem:

$$\left[\underset{\approx}{K}_e + {}^m\tilde{\eta}\,({}^m\underset{\approx}{K}_u + {}^m\underset{\approx}{K}_g)\right]\delta\underset{\sim}{u} = \underset{\sim}{0} \tag{9}$$

Solving the eigenvalue problem (9) at a certain load level ${}^m\lambda < \lambda^*$ renders an approximation of the critical load level:

$${}^m\tilde{\lambda}^* = {}^m\tilde{\eta}_j \; {}^m\lambda \;, \tag{10}$$

with ${}^m\tilde{\eta}_j$ the smallest eigenvalue, and $\delta\underset{\sim}{u}_j$ the corresponding eigenvector. In equ. (9) a linear relationship between the load dependent parts of the current stiffness matrix and the stiffness matrix at the stability limit is assumed. This approximative assumption becomes more and more accurate when the considered fundamental equilibrium state m respresented by ${}^m\lambda$, approaches to the stability limit configuration, represented by λ^*: In this case

$$\lim_{{}^m\lambda \to \lambda^*} {}^m\tilde{\eta}_j = {}^m\eta = 1 \tag{11}$$

with ${}^m\eta$ the exact "nonlinear" eigenvalue, for which

$$\lambda^* = {}^m\eta \; {}^m\lambda \tag{12}$$

is valid.

Equs. (9 - 12) describe the scheme of a supplementary eigenvalue analysis which renders - if it is applied at a sequence of incremental steps up to the limit load - a good insight into the nonlinear stability behaviour of the structure. Plotting the values of ${}^m\tilde{\lambda}^*$ (equ. (10)) versus an arbitrary component, ${}^m u^P$, of the displacement vector, ${}^m\underset{\sim}{u}$, yields a function $\tilde{\lambda}^*(u^P)$ which starts at $u^P = 0$ with the value λ^*_{lin}, the classical critical load level of the linear structure, and intersects the load-displacement path, $\lambda(u^P)$, at the stability limit point of the nonlinear structure (see fig. 2 of the following chapter).

The curve which is obtained by plotting ${}^m\tilde{\lambda}^*$ versus ${}^m\lambda$ intersects the line $f(\lambda) = \lambda$ at ${}^m\tilde{\lambda}^* = \lambda^*$. The behaviour of the $\tilde{\lambda}^*(\lambda)$ curve allows a distiction between buckling and snap-through, as shown in [6]:

$$\lim_{\lambda \to \lambda^*} \tilde{\lambda}^*(\lambda) = \lambda^* \tag{13}$$

holds in every case, but specially for snap-through

$$\lim_{\lambda \to \lambda^*} \frac{\partial}{\partial \lambda} \tilde{\lambda}^*(\lambda) = -\infty, \tag{14}$$

In the buckling case the left-hand side of equ. (14) converges to a finite value.

Further very useful algorithms for analysing the stability of nonlinear structures are to be found in Brendel's thesis [6]. Special attention will now be given to the "Current Stiffness Parameter", S_p, recently introduced by Bergan et al. [8 - 10] which represents the current stiffness of the structure as a relation between a load increment and the corresponding displacement increment. It indicates the snap-through limit load by

$$\lim_{\lambda \to \lambda^*_{SNAP}} S_p(\lambda) = 0 \tag{15}$$

and

$$\lim_{\lambda \to \lambda^*_{SNAP}} \frac{\partial}{\partial \lambda} S_p(\lambda) = -\infty \tag{16}$$

However $S_p(\lambda)$ cannot reveal the stability limit which is caused by bifurcation, i.e. buckling. Nevertheless the current stiffness parameter is in any case a very useful and easily calculated parameter which allows again in combination with another stability indicator, e.g. the determinant (equ. (6)), an investigation of the stability behaviour and the mechanism of stability loss as shown in [11].

3. A particular problem

Consider the circular arch, elastically restrained at both ends, in fig. 1. This simple structure represents a highly nonlinear system and has a very interesting stability behaviour as will be shown now.

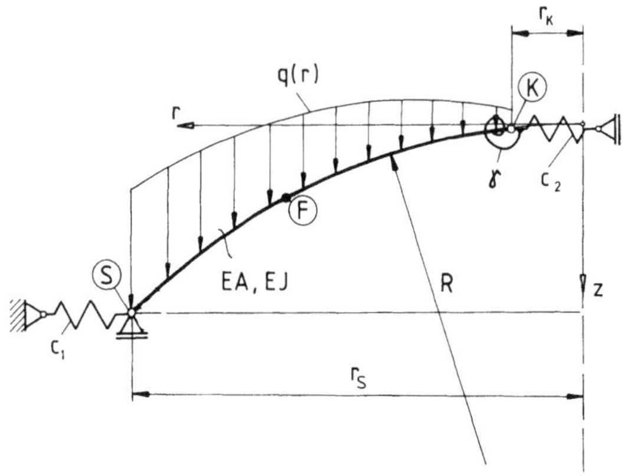

Fig. 1 The circular arch

The real background for the choice of this structure as an example is the consideration of the nonlinear behaviour of wide spanned roofs of very large storage tanks. The following data are used in the considered example: R = 105 m, r_k = 1.922 m, r_s = 34.6 m, EA = 1.741 x 10^9 N, EJ = = 4.765 x 10^7 Nm^2; the linear springs have the following stiffness coefficients c_{1ref} = 1.395 x 10^7 N/m, $c_2 \doteq 6.006 \times 10^8$ N/m, γ_{ref} = 2.708 x 10^7 Nm/rad. The stiffness coefficients c_1 and γ of the restraining springs are varied in order to calculate the sensitivity of the structure with regard to this variation:

$$c_1 = C\, c_{1ref}, \quad \gamma = \Gamma \gamma_{ref}, \tag{17}$$

C and Γ are used as restraint parameters.

The distributed load is location dependent:

$$q(r) = q_o \frac{r}{r_s}$$

with q_o = 5.04 x 10^4 N/m at load level λ = 1.

Only displacements in the y-z-plane are allowed (plane problem). The arch was modelled by 30 nonlinear beam elements. The elastic supports were realized with the aid of a combination of linear beam- and truss-elements. The calculations were performed using the NISA-program [12].

Fig. 2 shows the results of the nonlinear incremental-iterative analysis described above. The case with $C = 1$, $\Gamma = 1$ is considered.

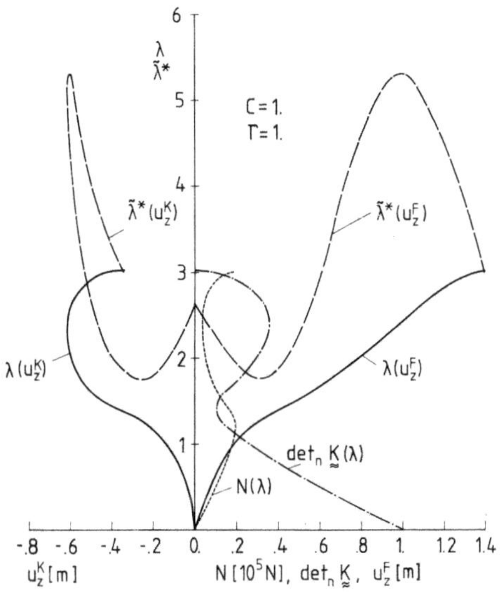

Fig. 2 The nonlinear behaviour of the arch ($C = 1$, $\Gamma = 1$)
—— load displacement path for nodes K and F
········ normal force $N(\lambda)$ in the cross section at F
—·— eigenvalue functions $\tilde{\lambda}^*(u)$
— — normalized determinant of the current tangent stiffness matrix $\det_n \underset{\approx}{K}(\lambda)$

A considerably nonlinear behaviour can be observed. Especially the motion of the support K which represents the motion of the central ring of the tank roof during load incrementation is remarkable. At the beginning of the load incrementation it moves upwards and after reaching a limit it moves downwards until the arch snaps through. This peculiar behaviour of the support motion is also reflected in the curious load dependence of the normal compression force $N(\lambda)$ which causes the alternating stiffness variation represented by the $\det_n \underset{\approx}{K}(\lambda)$ history. Considering the eigenvalue functions $\tilde{\lambda}^*(u)$ and the $\det_n \underset{\approx}{K}(\lambda)$ curve one gets the impression that the structure is almost at a stability limit at $\lambda \approx 1.4$ (the $\det_n \underset{\approx}{K}(\lambda)$ curve approaches the λ axis and the $\tilde{\lambda}^*$ curves tend to touch the $\lambda(u)$ paths) but a further load incrementation "stiffens" the structure until $\lambda \approx 2.5$. Continuation of the load incrementation leads finally to the snap-through at $\lambda \approx 3$: The $\det_n \underset{\approx}{K}$ curve reaches a limit with horizontal tangent. When the $\tilde{\lambda}^*(u)$ curves cross the load displacement paths, the determinant of the stiffness matrix behaves correspondingly to equs. (6) and (7). Special attention will now be given to the $\det_n \underset{\approx}{K}(\lambda)$ history. It is obvious that after a certain parameter variation the $\det_n \underset{\approx}{K}(\lambda)$ curve may cross the λ axis just before the "re-stiffening period". This would render a sudden change of the critical load level λ^*. In other words, an infinitesimal variation of this certain system parameter would cause a finite change of the stability limit load, i.e. a non-continuous parameter sensitivity concerning stability limit.

3.1 Parameter sensitivity

In order to find a sensitivity behaviour which was presumed in the previous chapter the spring stiffness c_1, expressed by C according to equ. (17) is varied. Some results are shown in fig. 3.

Fig. 3 shows the results of three typical cases (equ. (17)):

a) C = 1.0: Load incrementation leads to a stiffness-loss followed by a re-stiffening, a further stiffness-loss and finally the snap-through.

b) C = 0.5: It is remarkable that the determinant curve crosses the λ axis before re-stiffening takes place. Considering the $\lambda(u^K)$ behaviour one can observe that the support K snaps upwards quite contrary to the case C = 1 in which it snaps downwards. This value of C is already beyond the critical value of $C = C^*$ at which the limit load jumps (see also fig. 5).

c) C = 0.35: The system moves towards the snap-through in a direct manner just like the usual hinged-hinged arch.

The stability behaviour of these three typical cases can also be studied by consideration of fig. 4 which represents the $\tilde{\lambda}^*(\lambda)$ diagrams (see equs. (13) and (14)).

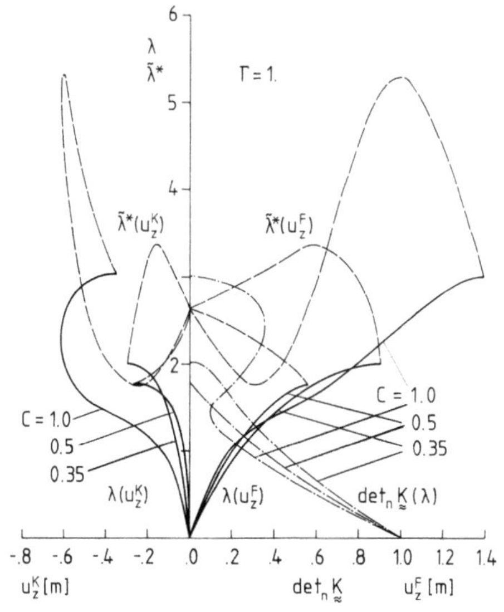

Fig. 3 Behaviour of the arch with different spring stiffness parameter C

——— load displacement paths
– – – eigenvalue functions
– — – determinant of the stiffness matrix

Corresponding to equ. (14) one can predict that in all the considered cases the stability loss is due to snap-through although the $\det_n \underset{\approx}{K}(\lambda)$ curve for the case C = 0.5 (see fig. 3) does not allow this prediction (refer to equ. (7)). This uncertainty in the use of the limit condition (7) (which is only a sufficient but not a necessary condition) is already pointed out in [6].

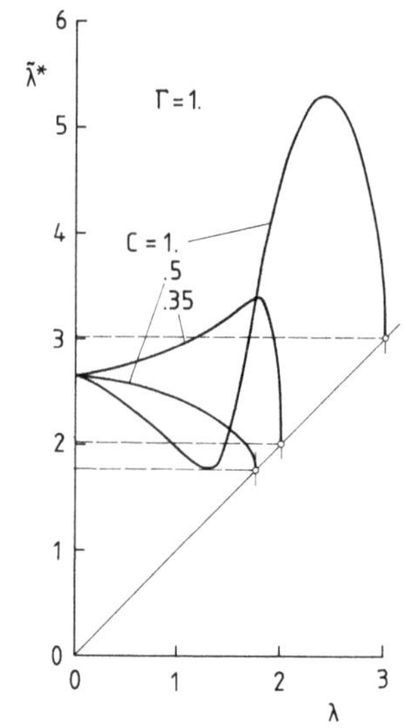

Fig. 4 Eigenvalue functions $\tilde{\lambda}^*(\lambda)$ for three different C parameter values

Further variation of the parameter C allows the presentation of the parameter sensitivity of the limit load, $\lambda^*(C)$, in fig. 5.

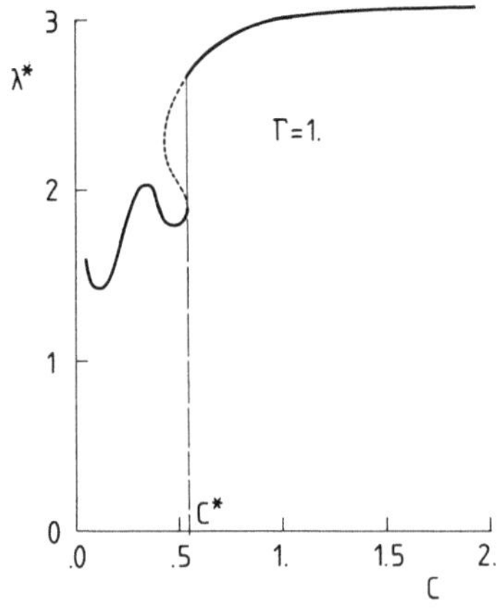

Fig. 5 Stability limit load as a function of the parameter C (dotted line is the expected physically not realizable part of the curve)

Fig. 5 shows clearly the considerable jump of $\lambda^*(C)$ at $C = C^* \approx 0.55$. This jump behaviour has of course serious consequences for the design of such structures and the criticism concerning their stability safety. Hence it is important to have a tool for classification of nonlinear structures with regard to the parameter sensitivity of the stability limit load. The catastrophe theory may be this tool as it is shown in the following chapter.

4. Embedding in the catastrophe theory

Systems like the considered one have catastrophic behaviour in a double sense: Firstly, the load displacement behaviour shows catastrophic states indicated by snap-through at $\lambda = \lambda^*$ and, secondly, the stability limit load itself has a catastrophic parameter sensitivity.

4.1 General remarks

The rather young mathematical theory of catastrophes has its origin in the endeavour to render mathematical models of natural processes having a certain jump behaviour, especially of such belonging to the field of biology [13]. In the meantime this theory has found useful application in a wide variety of other fields [14,15]. It is also used to classify phenomena concerning the stability of the equilibrium of mechanical structures [15 - 17].

4.2 A short insight into the catastrophe theory

Consider a mechanical system which has a parameter dependent behaviour. Points in the parameter domain at which the system changes its behaviour in a qualitative sense due to a variation of the parameters are called bifurcation points [17]. Assume for simplicity that the system has a real scalar potential function, $V(x,a,b)$, e.g. the complete potential energy of a loaded elastic structure with x the dependent variable, describing the state of the system, and a and b the independent system parameters (control parameters). The surface in the a, b, x domain which is defined by

$$\frac{\partial V}{\partial x} = 0 \qquad (18)$$

represents the behaviour surface, e.g. the equilibrium states of the elastic structure. The parameter domain is in this case the plane a - b and is named the control surface. The system is structurally stable if a small variation of the system parameters does not change the qualitative character of the potential function, otherwise the system is structurally unstable [17]. Applying the more general classification theorem of R. Thom [13] to this special case the following statement holds: The behaviour surface is - provided the system is structurally stable - a smooth surface, whose projection into the parameter domain (control surface) has only fold curves and cusps, see fig. 6. The fold curves are the set of the bifurcation points (a,b)* in the control surface for which the condition

$$\left.\frac{\partial^2 V}{\partial x^2}\right|_{(a,b)^*} = 0 \tag{19}$$

is met, e.g. snap-through conditions for the elastic structure.

In the considered special case the potential, V(a,b,x), can be described in the vicinity of the cusp qualitatively by one of the following polynoms [13,17]:

$$V = \frac{1}{3}\, x^3 + ax \tag{20}$$

or

$$V = \frac{1}{4}\, x^4 + \frac{1}{2}\, ax^2 + bx \tag{21}$$

In case of equ. (21) the behaviour surface is locally qualitatively defined by equ. (18) which renders now:

$$\frac{\partial V}{\partial x} = x^3 + ax + b = 0 \tag{22}$$

Application of equ. (19) and elimination of x yields under consideration of equ. (22) the bifurcation set in the control surface:

$$4a^3 + 27b^2 = 0, \tag{23}$$

which represents parabolas forming a cusp.

Catastrophes defined by a potential corresponding to equ. (21) belongs to the class of cusp catastrophes.

A simple example taken from [17], the v. Mises truss, may demonstrate the statements given above. Fig. 6 shows the behaviour surface of this structure, provided the Euler-buckling of the single truss element does not appear before the snap-through takes place, otherwise the system would be structurally unstable.

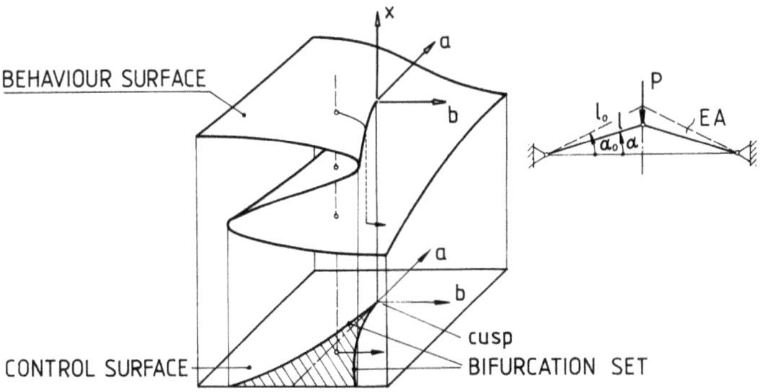

Fig. 6 The behaviour surface and the bifurcation set of the v. Mises truss (taken from [17] with permission)

For this example the potential is given by [17]:

$$V = \frac{1}{4}\alpha^4 - \frac{\alpha_o^2}{2}\alpha^2 + \frac{P}{EA}\alpha + (\frac{\alpha_o^4}{4} - \frac{P}{EA}\alpha_o) \tag{24}$$

In order to correspond with equs. (20 - 23) the following quantities must be introduced:

$$x = \alpha$$
$$a = -\alpha_o^2 \tag{25}$$
$$b = P/EA$$

Thus, the catastrophic behaviour of the v. Mises truss is described by the equs. (22) and (23) concerning equilibrium states and stability limits, respectively. The shaded area in fig. 6 is limited by the fold curves; it has the following significance: If a parameter value crosses one of the fold curves coming from the shaded area, then a continuous parameter variation renders a non-continuous system behaviour, i.e. snap-through.

4.3 Application of the catastrophe theory to the problem under consideration

Let us now come back to the problem described in chapter 3.

Choosing the load multiplier, λ, and the restraining parameter, C, as independent control parameters and the vertical displacement, u_z^K, of the support K as dependent behaviour variable, the results of the above described analysis (see chapter 3) renders the behaviour surface as shown qualitatively in fig. 7.

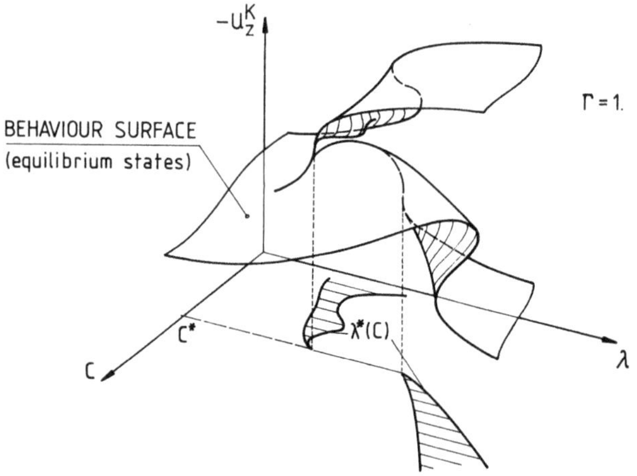

Fig. 7 The behaviour surface of the arch with regard to the equilibrium states

The potential V describing this behaviour is, naturally, the total potential energy. Considering the behaviour surface in fig. 7 one observes that it is cut into two different surfaces at a line along $C = C^*$. Hence, Thom's theorem concerning the existence of a smooth behaviour surface (see chapter 4.2) is only valid for two separated parameter sub-domains which are connected along the line $C = C^*$. Thus, the structure has a structural instability at $C = C^*$. This fact corresponds to the observation that for $C < C^*$ the support K snaps through upwards and for $C > C^*$ downwards.

It should be mentioned that the algorithm used allows the calculation of the system behaviour only for load incrementation up to the limit load. Therefore the continuation of the behaviour surface for $\lambda > \lambda^*(C)$, i.e. the post-buckling behaviour, has only qualitatively predicted character.

Concerning the fold curve $\lambda^*(C)$ in the control surface which is relevant for the stability loss due to increased load, one should notice the correspondence with fig. 5.

The described behaviour $u_z^K(\lambda, C)$ is the first catastrophic aspect, i.e. the snap-through of the structure. The second one is the non-continuous variation of the stability limit load for some parameter variation. Fig. 8 shows schematically the dependence of the critical load multiplier, $\lambda = \lambda^*$, on the system parameters (control parameters) C and Γ. It will be shown by the relevant potential function $\tilde{V}(C, \Gamma, \lambda)$ that we again are dealing with a cusp catastrophe.

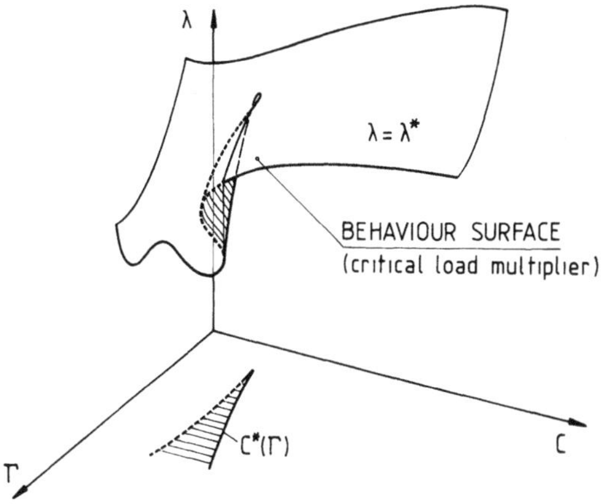

Fig. 8 The catastrophic behaviour of the critical load level $\lambda^*(C, \Gamma)$

In fig. 8 the behaviour surface is shown for $\Gamma \leq 1$. The cross section through the behaviour surface $\lambda^*(C, \Gamma)$ at $\Gamma = 1$ corresponds with the $\lambda^*(C)$ curve in fig. 5. Furthermore, some points on the behaviour surface for $C = 1.0$, i.e. $\lambda^*(\Gamma, C=1.)$ are calculated. From these informations the behaviour surface was drawn qualitatively. There is one important fact to be noted: Although the behaviour surface is a smooth surface, the folded part has only mathematical but no physical meaning.

Always the lowest critical value λ^* is the stability limit. Hence, the relevant part of the behaviour surface is again slotted and only one branch of the bifurcation set in the control surface, namely the $C^*(\Gamma)$ curve, contains the physically relevant jump conditions.

What potential function $\tilde{V}(\lambda, C, \Gamma)$ belongs to this catastrophic behaviour? - V has to satisfy:

$$\left.\frac{\partial \tilde{V}}{\partial \lambda}\right|_{\lambda=\lambda^*} = 0 \quad , \tag{26}$$

$$\left.\frac{\partial^2 V}{\partial \lambda^2}\right|_{\lambda=\lambda^*,\ C=C^*(\Gamma)} = 0. \tag{27}$$

A potential function which fulfills these requirements is

$$\tilde{V} = - \int_0^{\lambda} \det_n \underset{\approx}{K} (\bar{\lambda})\, d\bar{\lambda} + D, \tag{28}$$

with $\det_n K(\bar{\lambda})$ denoting the normalized determinant of the tangent stiffness matrix at the load level $\bar{\lambda}$. D is an arbitrary real constant.

The potential function (28) renders with equ. (26) the necessary condition (5) for the stability limit. Hence, $\tilde{V}$ in equ. (28) yields the behaviour surface $\lambda = \lambda^*$(fig. 8). The condition for the jump of λ^*, equ. (27), leads in combination with equ. (28) to that situation in which the $\det_n \underset{\approx}{K}(\lambda)$ curve touches the λ axis without crossing it. This is exactly the case if C approaches C^* (see figs. 9 - 11).

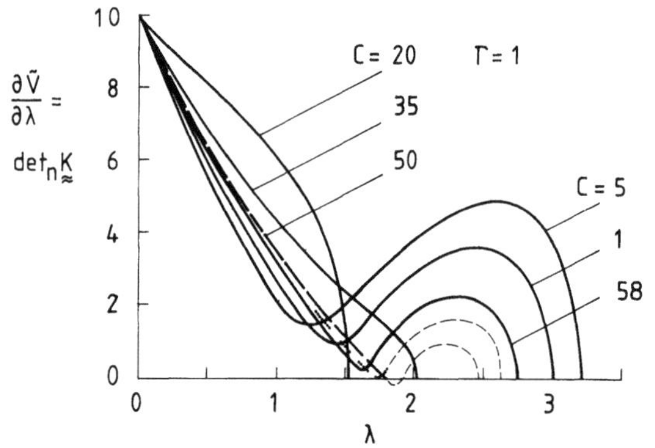

Fig. 9 The normalized determinant of the tangent stiffness matrix as a function of the load level (thin dotted lines are expected curves)

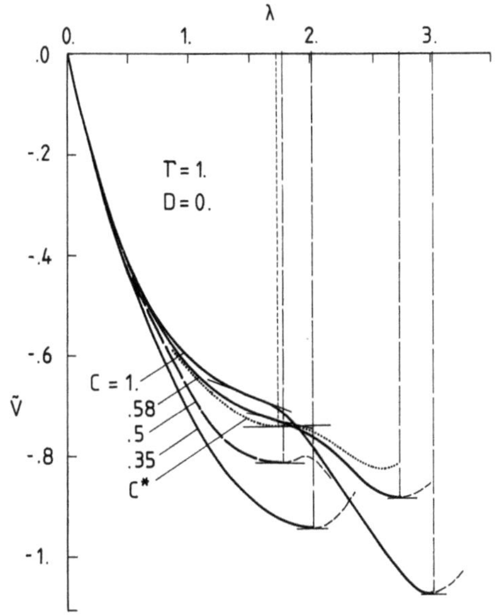

Fig. 10 The potential function Ṽ (thin dotted lines are expected curves)

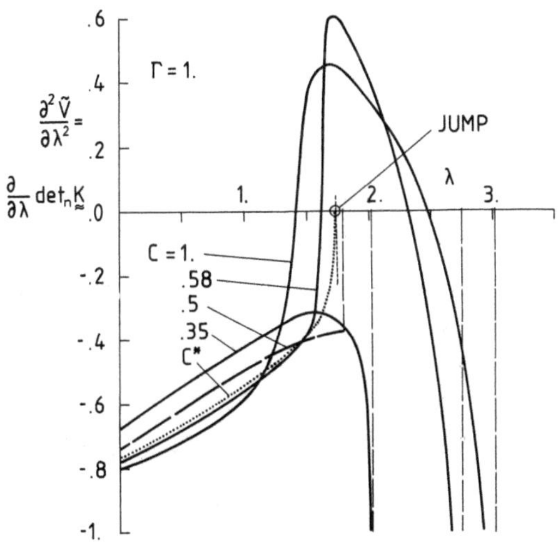

Fig. 11 Second derivative of the potential function

Fig. 10 shows that the potential $\tilde{V}$ behaves like a polynominal with order four which is in correspondence with equ. (21). Referring to Thom's theorem one can conclude that the assumption of a cusp catastrophe as shown schematically in fig. 8 is justified. Furthermore fig. 10 shows clearly for decreasing C that for values $C > C^*$ indeed a point of inflection in the $\tilde{V}(\lambda)$ curve exists for a λ value smaller than λ^* but not till $C \leq C^*$ a second minimum of $\tilde{V}$ appears prior the original one. The appearance of a second minimum of $\tilde{V}$ renders the jump of $\lambda^*(C)$.

In order to generalize the results and considerations described above following theorems are stated:

Theorem 1: If a nonlinear structure which tends to a stability loss, i.e.

$$\exists \lambda^*: \lim_{\lambda \to \lambda^*} \det_n \underset{\approx}{K}(\lambda) = 0, \tag{29}$$

has any one system parameter, Q, (control parameter) for which is valid that

$$\exists \tilde{Q} \in \mathcal{S}: \begin{cases} \exists \tilde{\lambda} \leq \lambda^*_{\tilde{Q}}: \lim_{\lambda \to \tilde{\lambda}} \left[\dfrac{\partial}{\partial \lambda} \det_n \underset{\approx}{K}(\lambda) \right]_{Q=\tilde{Q}} = 0 & (30) \\ \text{with} \\ \left[\dfrac{\partial^2}{\partial \lambda^2} \det_n \underset{\approx}{K}(\lambda) \right]_{\lambda=\tilde{\lambda}} > 0 & (31) \end{cases}$$

with $\mathcal{S}$ denoting the feasible region of Q, then it is possible that the stability limit has a non-continuous parameter sensitivity with regard to a variation of the parameter Q. For explanation, see the curve for C = 5.0 in fig. 9.

Theorem 2: If the assumptions of theorem 1 are met and $\tilde{\lambda}$ fulfills the condition

$$\lim_{\lambda \to \tilde{\lambda}} \left[\det_n \underset{\approx}{K}(\lambda) \right]_{Q=\tilde{Q}} = 0, \tag{32}$$

then the structure has a non-continuous parameter sensitivity at

$$Q^* = \tilde{Q}, \tag{33}$$

i.e. a sudden jump of the critical load level $\lambda^*(Q)$.

The proof of these theorems is obvious by considering equs. (6), (8) and (26 - 28). Equ. (31) means that only minima at load levels smaller or equal to the stability limit load have to be considered because of the fact that usually a practical physical structure behaves in the unloaded state, i.e. $\lambda = 0$, with

$$\det_n \underset{\approx}{K}(\lambda=0) > 0 \tag{34}$$

and maxima of $\det_n \underset{\sim}{K}(\lambda)$ at λ values smaller than λ^* may only point to a possible jump of the stability limit load if a minimum exists prior the maximum (see fig. 9).

5. Conclusions

Using theorems 1 and 2 a numerical procedure - which calculates in each incremental step the behaviour of the determinant of the current tangent stiffness matrix and its first and second derivatives with respect to λ - allows a classification of the considered structure concerning its parameter sensitivity with regard to the stability limit. In a future paper the use of theorem 2 in combination with a steepest descent method will be presented which renders a procedure which can be used to calculate the bifurcation set in the control surface (see e.g. fig. 8). This procedure will help to find out critical combinations of system parameters at which a sudden jump of the stability limit load takes place. The knowledge of such a jump is of great importance for the design of supporting structures especially in the field of lightweigt construction.

6. References

1 Bathe, K.-J.; Ramm, E; Wilson, E.L: Finite Element Formulations for Large Deformation Analysis. Int.J.Numer.Meth.Engng. 9 (1975), 353 - 386.

2 Bathe, K.-J.; Bolourchi, S.; Ramaswamy, S.; Snyder, M.D.: Some Computational Capabilities for Nonlinear Finite Element Analysis. Nucl. Eng.Design 46 (1978), 429 - 455.

3 Matthies, H.; Strang, G.: The Solution of Nonlinear Finite Element Equations. Int.J.Numer.Meth.Engng. 14 (1979), 1613 - 1626.

4 Ramm, E.: Geometrisch nichtlineare Elastostatik und finite Elemente. Habilitationsschrift. Univ. Stuttgart. 1975.

5 Gallagher, R.H.; Mau, S.: A Method of Limit Point Calculation in Finite Element Structural Analysis, NASA CR-2115. 1972.

6 Brendel, B.: Geometrisch nichtlineare Elastostabilität. Bericht Nr. 79-1, Inst. f. Baustatik, Univ. Stuttgart. 1979.

7 Brendel, B.; Ramm, E.; Fischer, D.F.; Rammerstorfer, F.G.: Linear and Nonlinear Stability Analysis of Thin Cylindrical Shells Under Wind Loads. To be published in: J.Struct.Mech. 9 (1981).

8 Bergan, P.G.; Horrigmoe, G.; Kråkeland, B.; Søreide, T.H.: Solution Techniques for Non-Linear Finite Element Problems. Int.J.Numer. Meth.Engng. 12 (1978), 1677 - 1696.

9 Bergan, P.G.: Solution Algorithms for Nonlinear Structural Problems. Proc.Int.Conf.Engng.Appl. FEM, Høvik, 1979.

10 Bergan, P.G.; Holand, I.; Søreide, T.H.: Use of the Current Stiffness Parameter in Solution of Nonlinear Problems. Energy Methods in Finite Element Analysis. R. Glowinski, E.Y. Rodin, O.C. Zienkiewicz (editors). John Wiley & Sons, Chichester - New York - Brisbane - Toronto: 1979

11 Rammerstorfer, F.G. ; Fischer D.F.: Nonlinear Elastic-Plastic Stability and Contact Problems Concerning the Straightening of a Wave-Like Deformed Strip Plate. To appear in Proc.1st Int.Conf.Num. Meth. for Non-Linear Problems. Swansea: 1980.

12 Brendel, B.; Häfner, L.; Ramm, E.; Sättele, J.M.: Programmdokumentation - Programmsystem NISA. Univ. Stuttgart, Inst. f. Baustatik. 1977.

13 Thom, R.: Topological Models in Biology. Topology 8, (1969),313-335.

14 Thom, R.: Structural Stability and Morphogenesis. New York: Benjamin 1975.

15 Zeeman, E.C.: Catastrophe Theory. Scientific American, April 1976, 65 - 83.

16 Troger, H.: Zur Einteilung von Sprungeffekten in mechanischen Systemen. ZAMM 54 (1974), T177 - T179.

17 Troger, H.: Ein Beitrag zum Durchschlagen einfacher Strukturen. Acta Mechanica 23 (1975), 179 - 191.

Behaviour of Finite Element Solutions Near a Bifurcation Point

F. BREZZI
Università di Pavia, Italy

Summary

We analyze the behaviour of finite dimensional approximation of Galerkin type in a neighborhood of a simple critical point. Error bounds of optimal type are derived. Some computational aspects are also treated.

1. Introduction

The aim of this paper is to present some recent results in the theory of approximation of nonlinear problems, with a particular accent on the behaviour of the approximate solutions in a neighborhood of singular points, such as normal limit points and bifurcation points. Although some computational aspects will be briefly sketched, the main interest will be focused on the problem of error bounds. The well known definitions of the singularities will be given first, for the sake of simplicity, in finite dimension. In the sequel we shall briefly present the continuous problem and the abstract hypotheses on the finite dimensional approximation. Then the main results on the error estimate will be stated. An example of application to the model problem of the von Kármán plate equations will also be given

2. Regular and singular points in finite dimension

Let $\underline{f}(\underline{x},\lambda)=(f_1(x_1,\ldots,x_n,\lambda),\ldots,f_n(x_1,\ldots,x_n,\lambda))$ be a C^r-mapping ($r\geqslant 3$) from $R^{n+1}=R^n \times R$ into R^n and consider the problem

$$\begin{cases} \text{find pairs } (\underline{x},\lambda) \text{ such that} \\ \qquad \underline{f}(\underline{x},\lambda)=0. \end{cases} \tag{2.1}$$

Let $(\underline{x}^0,\lambda^0)$ be a solution of (2.1), that is

$$\underline{f}(\underline{x}^0,\lambda^0)=0. \tag{2.2}$$

The total and partial Fréchet derivatives of $\underline{f}$ at the point $(\underline{x}^0,\lambda^0)$ are defined as follows (M_{ms} being the space of $m\times s$ matrices and using the convention $\lambda=x_{n+1}$).

$$Df^0 \in M_{n,n+1} \quad ; \quad (D\underline{f}^0)_{ij} = \left(\frac{\partial f_i}{\partial x_j}\right)_{\underline{x}=\underline{x}^0,\lambda=\lambda^0} \quad \begin{cases} i=1,\dots,n \\ j=1,\dots,n+1 \end{cases} \tag{2.3}$$

$$D_x\underline{f}^0 \in M_{n,n} \quad ; \quad (D_x\underline{f}^0)_{ij} = (D\underline{f}^0)_{ij} \quad i,j=1,\dots,n \tag{2.4}$$

$$D_\lambda\underline{f}^0 \in M_{n,1} \quad ; \quad (D_\lambda\underline{f}^0)_{i} = (D\underline{f}^0)_{i,n+1} \quad i=1,\dots,n. \tag{2.5}$$

Remark - In the general case of a mapping F from a Banach space E_1 into another Banach space E_2 the Fréchet derivative of F at a point $e^0\in E_1$ is defined as the unique linear mapping DF^0 from E_1 into E_2 that verifies

$$||F(e^0+e)-F(e^0)-DF^0.e||_{E_2} \leqslant o(||e||_{E_1}) \quad e\in E_1 \tag{2.6}$$

It is easy to check that (2.3)-(2.5) satisfy the general definition (2.6). In its turn the map $x^0\to DF^0$ (from E_1 into $\mathcal{L}(E_1,E_2)$) can be differentiated (if F is smooth enough); this leads to higher order Fréchet derivatives. In our finite dimensional example we would have, for instance, $D_{xx}\underline{f}^0 : R^n\times R^n\to R^n$ given by

$$D_{xx}\underline{f}^0: \underline{y}\to(\underline{z}\to \sum_{j,r=1}^{n} \frac{\partial^2 f_i^0}{\partial x_j \partial x_r} z_r\, y_j) \quad \underline{y},\underline{z}\in R^n \tag{2.7}$$

or equivalently

$$D_{xx}\underline{f}^0: (\underline{y},\underline{z}) \to \sum \frac{\partial^2 f_i^0}{\partial x_j \partial x_r} z_r\, y_j \qquad \underline{y},\underline{z}\in R^n \tag{2.8}$$

Definition - A point $(\underline{x}^0,\lambda^0)$ solution of (2.1) is called a regular point if the Jacobian matrix $D_x\underline{f}^0$ is nonsingular.

The well known implicit function theorem ensures that if

$(\underline{x}^0,\lambda^0)$ is a regular point then there exists in a neighborhood $X\times\Lambda$ of $(\underline{x}^0,\lambda^0)$ a unique branch $\lambda\to\underline{x}(\lambda)$ such that:

$$f(\underline{x}(\lambda),\lambda) \equiv 0 \quad \forall\lambda\in\Lambda. \tag{2.9}$$

Definition - A point $(\underline{x}^0,\lambda^0)$ solution of (2.1) is called a simple critical point of $\underline{f}$ if zero is a simple eigenvalue of $D_{\underline{x}}\underline{f}^0$.

Assume now that $(\underline{x}^0,\lambda^0)$ is a simple critical point of $\underline{f}$, and let $(D_{\underline{x}}\underline{f}^0)^*$ be the transposed matrix of $D_{\underline{x}}\underline{f}^0$. It is well known that there exist a unique (up to a sign) pair of vectors $\underline{\phi}^0,\underline{\phi}^*$ such that

$$D_{\underline{x}}\underline{f}^0\cdot\underline{\phi}^0 = 0 \qquad ||\underline{\phi}^0|| = 1 \tag{2.10}$$

$$(D_{\underline{x}}\underline{f}^0)^*\underline{\phi}^* = 0 \qquad (\underline{\phi}^0,\underline{\phi}^*)=1. \tag{2.11}$$

Among the simple critical points there are points for which the criticality is, in some sense, more apparent than real. These are points which belong to some branch of solutions that cannot be parametrized with respect to λ.

Definition - A simple critical point is called a normal limit point if $D_\lambda\underline{f}^0$ is not in the range of $D_{\underline{x}}\underline{f}^0$, or, equivalently, if

$$(D_\lambda\underline{f}^0,\phi_0^*)\neq 0. \tag{2.12}$$

In order to simplify the notations, we shall assume from now on that

$$\underline{\phi}^0 \equiv \underline{\phi}^* = (0,0,\ldots,0,1) \tag{2.13}$$

so that the last row and column of $D_{\underline{x}}\underline{f}^0$ are identically zero. This will not result in a loss of generality; in fact, since zero is a simple eigenvalue, we can always make a change of coordinates (like in the Jordan normal form) such that (2.13) is satisfied. It is also easy to see that in such case the submatrix

$$S_{ij} = (D_{\underline{x}}\underline{f}^0)_{ij} \qquad i,j=1,\ldots,n-1 \tag{2.14}$$

has to be nonsingular.

If $(\underline{x}^0,\lambda^0)$ is a normal limit point condition (2.12) implies $(D_\lambda \underline{f}^0)_n \neq 0$; interchanging x_n and $\lambda = x_{n+1}$, we are now back to the case of regular points.

Remark - In other words we could say that a normal limit point is a point $(\underline{x}^0,\lambda^0)$ such that the unique branch of solutions through it has a tangent that is "orthogonal" to the λ axis. If in addition one has

$$(D_{xx}\underline{f}^0(\underline{\phi}^0,\underline{\phi}^0),\underline{\phi}^*)>0 \quad (\text{resp} < 0)$$

then $(\underline{x}^0,\lambda^0)$ is called a nondegenerated turning point; it is easy to see that in such cases the branch through $(\underline{x}^0,\lambda^0)$ verifies, in a neighborhood, $\lambda \leq \lambda_0$ ($\lambda \geq \lambda_0$ respectively).

Let us deal finally with simple bifurcation points. Assume that $(\underline{x}^0,\lambda^0)$ is a simple critical point and that (2.12) is not satisfied. Writing

$$\underline{x}=\underline{x}'+\alpha\underline{\phi}^0 \qquad \begin{cases} x' = (x_1,\ldots,x_{n-1},0) \\ \alpha = x_n \end{cases} \tag{2.16}$$

equation (2.1) can be splitted as

$$\begin{cases} f_i(\underline{x}'+\alpha\underline{\phi}^0,\lambda)=0 \qquad i=1,\ldots,n-1, \\ f_n(\underline{x}'+\alpha\underline{\phi}^0,\lambda)=0 \end{cases} \tag{2.17}$$

Since the matrix S_{ij} defined in (2.14) is nonsingular, the implicit function theorem can be applied to deduce, from the first n-1 equation of (2.17), the existence of a unique mapping $\underline{x}'=\underline{x}'(\alpha,\lambda)$, defined in a neighborhood of (x_n^0,λ^0) such that

$$f_i(\underline{x}'(\alpha,\lambda)+\alpha\underline{\phi}^0,\lambda)\equiv 0 \qquad i=1,\ldots,n-1 \tag{2.18}$$

Hence (2.17) is reduced to the last scalar equation

$$f_n(\underline{x}'(\alpha,\lambda)+\alpha\underline{\phi}^0,\lambda)=0 \tag{2.19}$$

in the unknowns α,λ. We set for simplicity

$$g(\alpha,\lambda)=f_n(\underline{x}'(\alpha,\lambda)+\alpha\underline{\phi}^0,\lambda). \tag{2.20}$$

It is an easy matter to check that

$$g = \frac{\partial g}{\partial \alpha} = \frac{\partial g}{\partial \lambda} = 0 \text{ at } \alpha = x_n^0, \lambda = \lambda^0 \tag{2.21}$$

and therefore for the study of the solutions of (2.19) we have to look to the second derivatives. It is well known for instance that if

$$\det \begin{pmatrix} \frac{\partial^2 g}{\partial \alpha^2} & \frac{\partial^2 g}{\partial \alpha \partial \lambda} \\ \frac{\partial^2 g}{\partial \lambda \partial \alpha} & \frac{\partial^2 g}{\partial \lambda^2} \end{pmatrix} < 0 \quad \text{at } (x_n^0, \lambda^0), \tag{2.22}$$

the solutions of (2.19) lie on two C^{r-2} branches crossing transversally at (x_n^0, λ^0). If on the opposite the determinant in (2.22) is positive, then (x_n^0, λ^0) is an isolated solution of (2.19). The case det=0 corresponds to a higher degeneracy and will not be studied here.

Definition - Let $(\underline{x}^0, \lambda^0)$ be a simple critical point of $\underline{f}$ such that

$$(D_\lambda \underline{f}^0, \underline{\phi}^*) = 0.$$

We set

$$g(\alpha, \lambda) = f_n(\underline{x}'(\alpha, \lambda) + \alpha \underline{\phi}^0, \lambda)$$

where $\underline{x}'(\alpha, \lambda)$ is the implicit function defined in (2.18). If the determinant of the Hessian matrix of g at (x_n^0, λ^0) is negative then $(\underline{x}^0, \lambda^0)$ is called a simple bifurcation point.

At a simple bifurcation point the directions tangent to the branches can be computed as follows: let $\underline{\sigma}^0$ be the unique solution of

$$\begin{cases} D_{\underline{x}} \underline{f}^0 \underline{\sigma}^0 = -D_\lambda \underline{f}^0 \\ (\underline{\sigma}^0, \underline{\phi}^*) = 0. \end{cases} \tag{2.23}$$

If $(\dot{\underline{x}}, \dot{\lambda})$ $R^n \times R$ is a tangent to a branch through $(\underline{x}^0, \lambda^0)$, we must have

$$\|\dot{\underline{x}}\|^2 + \dot{\lambda}^2 = 1, \tag{2.24}$$

$$D_{\underline{x}} \underline{f}^0 \dot{x} + D_\lambda \underline{f}^0 \dot{\lambda} = 0. \tag{2.25}$$

Hence $\dot{\underline{x}}$ must be of the form

$$\dot{\underline{x}} = \xi_1 \underline{\phi}^0 + \xi_2 \underline{\sigma}^0 \tag{2.26}$$

and we have only to compute ξ_1 and ξ_2. An easy computation shows that

$$(D_{xx}\underline{f}^0(\dot{\underline{x}},\dot{\underline{x}})+2\dot{\lambda}D_{x\lambda}\underline{f}^0\dot{\underline{x}}+\dot{\lambda}^2D_{\lambda\lambda}\underline{f}^0,\underline{\phi}^*)=0 . \qquad (2.27)$$

Substituting (2.26) into (2.27) and using (2.24) and (2.25) we may easily compute ξ_1 and ξ_2 and hence $\dot{\underline{x}}$ and $\dot{\lambda}$. For more details see [7].

3. The infinite dimensional case

From now on we shall restrict ourselves to nonlinear problems of the following form:

$$F(u,\lambda)=0 \qquad F:V\times R\to V \qquad (3.1)$$

where F can be written as

$$F(u,\lambda)=u+TG(u,\lambda); \qquad (3.2)$$

we make the following assumptions:

V is a Banach space, (3.3)

G(u,λ) is an r-times continuously differentiable mapping from V×R into another Banach space W(r≥4), (3.4)

T is a linear compact operator from W into V. (3.5)

Example - Let us consider the classical von Kármán nonlinear plate equations:

$$\begin{cases}\Delta\Delta\phi = -\frac{1}{2}[w,w] & \text{in } \Omega,\\ \Delta\Delta w = [\phi+\lambda\bar{\phi},w] & \text{in } \Omega,\end{cases} \qquad (3.6)$$

$$w = \frac{\partial w}{\partial n} = \phi = \frac{\partial\phi}{\partial n} = 0 \quad \text{on } \partial\Omega . \qquad (3.7)$$

with w=transversal displacement, ϕ=stress function, $[u,v]=u_{xx}v_{yy}+u_{yy}v_{xx}-2u_{xy}v_{xy}$ and $\bar{\phi}$=given function. Setting

$$V=H_0^2(\Omega)\times H_0^2(\Omega) \qquad W=L^1(\Omega)\times L^1(\Omega) \qquad (3.8)$$

$$G:((w,\phi),\lambda)\to(\tfrac{1}{2}[w,w],-[\phi+\lambda\bar{\phi},w]) \qquad (3.9)$$

$$T : (p_1,p_2) \to (\psi_1,\psi_2) \text{ solution of} \qquad (3.10)$$

$$\Delta\Delta\psi_i = p_i \ , \ \psi_i \ H_0^2(\Omega) \quad i=1,2$$

and applying T to both sides of (3.6) we are in the framework (3.2)-(3.5).

Let now go back to (3.2) and let (u_0,λ_0) be a solution of (3.1). We denote by $D_uF^0, D_\lambda F^0, D_{uu}F^0, D_{u\lambda}F^0, D_{\lambda\lambda}F^0$ the first and second Fréchet derivatives of F at the point (u_0,λ_0). We recall the following definitions, which are the natural extension of the finite dimensional case.

<u>Definition</u> - The point (u_0,λ_0) is called a regular point if D_uF^0 is an isomorphism from V onto itself.

<u>Definition</u> - The point (u_0,λ_0) is called a simple critical point if zero is an eigenvalue of D_uF^0 with algebraic multiplicity 1.

In analogy with the finite dimensional case at a simple critical point the two eigenvectors ϕ^0 and ϕ^* will be uniquely determined by the equations

$$\begin{cases} D_uF^0\phi^0 = 0 & \|\phi^0\|_V = 1 \\ (D_uF^0)^*\phi^* = 0 & (\phi^0,\phi^*)=1 \end{cases} \qquad (3.11)$$

<u>Definition</u> - The point (u_0,λ_0) is a normal limit point if it is a simple critical point and

$$(D_\lambda F^0,\phi^*) \neq 0 \ . \qquad (3.12)$$

If in addition one has

$$(D_{uu}F^0(\phi^0,\phi^0),\phi^*) \neq 0 \qquad (3.13)$$

then (u_0,λ_0) is a nondegenerate turning point.

<u>Definition</u> - The point (u_0,λ_0) is a simple bifurcation point if it is a simple critical point where

$$(D_\lambda F^0,\phi^*)=0 \qquad (3.14)$$

and

$$A_0C_0-B_0^2 < 0 \qquad (3.15)$$

with

$$\begin{cases} A_0=(D_{uu}F^0(\phi^0,\phi^0),\phi^*) \\ B_0=(D_{u\lambda}F^0\phi^0+D_{uu}F^0(\phi^0,\sigma^0),\phi^*) \\ C_0=(D_{\lambda\lambda}F^0+2D_{u\lambda}F^0\sigma^0+D_{uu}F^0(\sigma^0,\sigma^0),\phi^*) \end{cases} \tag{3.16}$$

and σ^0 is the unique solution of :

$$\begin{cases} D_uF^0\sigma^0=-D_\lambda F^0 \\ (\sigma^0,\phi^*)=0 \ . \end{cases} \tag{3.17}$$

4. Finite dimensional approximation

Assume again that hypotheses (3.2)-(3.5) of the previous section are satisfied. For the sake of simplicity we shall deal only with approximations "of Galerkin type", although the theory in [2],[3],[4] is much more general. We assume then that V is a Hilbert space and that the operator T is associated with a bilinear continuous V-elliptic form $a(u,v)$ in such a way that for any f in V' we have

$$a(Tf,v)=(f,v) \quad \forall v\in V \qquad (W\subseteq V'). \tag{4.1}$$

We also assume that we are given a sequence $\{V_h\}_h$ of finite dimensional subspaces of V and we define the operators $T_h:V'\to V_h$ by

$$a(T_hf,v_h)=(f,v_h) \quad \forall v_h\in V_h. \tag{4.2}$$

We set the approximate problem:

$$F_h(u_h,\lambda)\equiv u_h+T_hG(u_h,\lambda)=0. \tag{4.3}$$

Example - In the case of the von Kármán equations for a clamped nonlinear plate we may set

$$a((w,\phi),(\tilde{w},\tilde{\phi}))=\int_\Omega(\Delta w\Delta\tilde{w}+\Delta\phi\Delta\tilde{\phi})\,d\Omega. \tag{4.4}$$

As a finite dimensional subspace V_h of $V=H_0^2(\Omega)\times H_0^2(\Omega)$ we could take, for the simplest case of a polygonal domain Ω, pairs of some usual C^1-conforming finite elements, like the composite Hsieh-Clough Tocher triangular element (12 d.o.f.) or the complete P_5 element (21 d.o.f.) or the reduced Bell element

(18 d.o.f.) or any other that we may like. For a nonpolygonal Ω isoparametric elements should be considered. We notice that the operator T_h defined in (4.2) associates to a given pair (p_1,p_2) the approximate finite element solution of the linear problem (3.10) so that the formulation (4.3) is the one that is used in practice. We also point out that $(T-T_h)(p_1,p_2)$ measures the error in the finite element approximation of the linear problem (3.10).

We define an "optimal order of convergence" by assuming that there exist a subspace $\mathcal{V} \subseteq V$ (of smooth functions) a positive constant c and an integer k>0 such that

$$\|(T-T_h)p\|_V \leqslant ch^k \|Tp\|_{\mathcal{V}} \qquad p\in V' \tag{4.5}$$

whenever the right hand side of (4.5) is finite.

<u>Example</u> - In the case of the von Kármán equations we have k=2 for the HCT 12 d.o.f. element, k=4 for the complete P_5 and k=3 for the Bell element (cfr. e.g. [5]). In any case $\mathcal{V}=H^{k+2}(\Omega)\times H^{k+2}(\Omega)$.

It is well known that if (3.1) has a regular solution (u^0,λ^0) there exists a neighborhood U of u^0 such that for h small enough (4.3) has a unique solution (u_h^0,λ^0) with $u_h^0\in U$; moreover

$$\|u^0-u_h^0\|_V \leqslant O(h^k)$$

provided that $u^0\in\mathcal{V}$. For a similar result in a more general context see [2].

Let us now state some results on the behaviour of the solutions of (4.3) in a neighborhood of a simple critical point.

<u>Theorem 1</u> - Let (u^0,λ^0) be a normal limit point of (3.1). There exists a neighborhood $U\times\Lambda$ of (u^0,λ^0) such that for h small enough both (3.1) and (4.3) have a unique branch of solutions in $U\times\Lambda$. Moreover these branches can be parametrized in such a way that

$$\|u(\alpha)-u_h(\alpha)\|_V + |\lambda(\alpha)-\lambda_h(\alpha)| \leqslant ch^k \|u(\alpha)\|_{\mathcal{V}} \tag{4.7}$$

for $\alpha_1\leqslant\alpha\leqslant\alpha_2$ with c,α_1,α_2 independent of h.

Theorem 2 - If (u^0,λ^0) is a nondegenerate turning point of (3.1) there exists a neighborhood $U\times\Lambda$ of (u^0,λ^0) such that far h small enough (4.3) has a unique nondegenerate turning point (u_h^0,λ_h^0) in $U\times\Lambda$ and we have

$$h^k \|u^0-u_h^0\|_V + |\lambda^0-\lambda_h^0| \leq ch^{2k} \tag{4.8}$$

with c independent of h, provided that $u_0, T^*\phi^*$ and the derivative of u in the direction of ϕ^0 at (u^0,λ^0) (tangent to the branch) belong to $\mathcal{V}$.

We point out that we have a double order of convergence for the critical values λ^0 and λ_h^0. The proof of theorem 1 and 2 can be found, in a more general framework, in [3]. Previous results in this direction where obtained for instance in [8],[6].

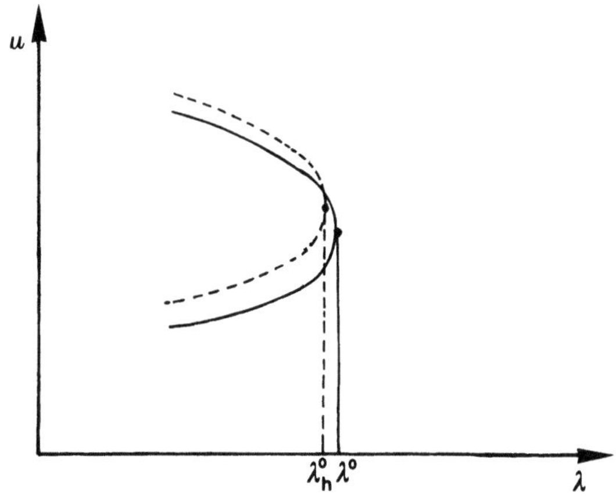

The difficulties connected with the actual computations of the branch of approximated solutions near a normal limit point can be overcome with various techniques. See for instance [10],[12] and the references therein contained.

Theorem 3 - If (u^0,λ^0) is a simple bifurcation point then there exists a neighborhood $U\times\Lambda$ of (u^0,λ^0) such that for h small enough the set S_h of the solutions of (4.3) in $U\times\Lambda$ is composed of two C^{r-2} branches which may not intersect (and, in general, do not). In any case, denoting by S the set of solutions of (3.1) in $U\times\Lambda$ we have

$$\mathrm{dist}(S,S_h)^{(1)} \leq O(h^k) \tag{4.9}$$

provided that $u^0, T^*\phi^*$ and the two derivatives of u in the direction of the branches at (u^0,λ^0) belong to $\mathcal{V}$.

For the proof and for more general results we refer to [4]. Theorem 3 states that the bifurcation pattern is not (in general) stable with respect to the finite element discretization. Exceptional cases are for instance the bifurcation from the trivial branch and the symmetry-breaking bifurcation. See [4] for more details. Previous results in this direction can be found for instance in [9], [6].

From the computational point of view two classes of problems arise when we follow a branch: one is due to the fact that, if we proceed too fast along the branch, we may jump over the bifurcation point without noticing it, and keep following an unstable branch. This usually results in a change in the sign of $\det(D_uF_h)$; if the determinant is computed at each step we are able to go back with a smaller size step and try to locate the bifurcation point. This is easy but not cheap in general. Another problem arises when we come too close to the bifurcation point: then D_uF_h becomes (numerically) singular and the "direction of the branch" does not exist anymore; we say that we are in a "black box". In order to go out of the black box in the direction of the branches we may use the following trick: assume that we are at a point $(\tilde{u}_h,\tilde{\lambda})$ in the balck box and let ν_h be the eigenvalue of $D_u\tilde{F}_h$ which is smallest in modulus. Let $\tilde{\phi}$ and $\tilde{\phi}^*$ be the corresponding eigenvectors:

$$D_u\tilde{F}_h\tilde{\phi} = \nu_h\tilde{\phi} \qquad ||\phi||=1, \tag{4.10}$$

$$(D_u\tilde{F}_h)^*\tilde{\phi}^*=\bar{\nu}_h\phi^* \quad (\tilde{\phi},\tilde{\phi}^*)=1, \tag{4.11}$$

We consider the following problem:

---

(1) - The distance dist(A,B) of two closed sets is defined for instance as

$$\mathrm{dist}(A,B)=\max\ \{\sup_{x\in A}\inf_{y\in B}\ ||x-y||\ ,\ \sup_{y\in B}\inf_{x\in A}\ ||x-y||\}.$$

$$\Psi_h(u_h,\lambda) = 0 \tag{4.12}$$

where

$$\Psi_h(u_h,\lambda)=F_h(u_h,\lambda)-F_h(\tilde{u}_h,\tilde{\lambda})-\nu_h(u_h-\tilde{u}_h)$$
$$-(\lambda-\tilde{\lambda})(D_\lambda\tilde{F}_h,\tilde{\phi}^*)\tilde{\phi}. \tag{4.13}$$

Clearly Ψ_h has a simple bifurcation point at $(\tilde{u}_h,\tilde{\lambda})$ (artificial bifurcation). The tangent directions to the branches of solutions of (4.12) through $(\tilde{u}_h,\tilde{\lambda})$ can be computed with the procedure (2.24)-(2.27). If $(\tilde{u}_h,\tilde{\lambda})$ is close enough to the simple bifurcation point (u^0,λ^0) the previous procedure works and the tangent directions to the artificial bifurcation are good approximations to the tangent directions at (u^0,λ^0). In particular if $\|(\tilde{u}_h,\tilde{\lambda})-(u^0,\lambda^0)\| \leq 0(h^k)$ then it is easy to prove (using the techniques of [4]) that the tangent directions are computed with an error of $0(h^k)$.
For other considerations on the computational aspects see for instance the references of [10], [12].
For some extensions to the case of Hopf bifurcation or multiple bifurcations see [1] and [11] respectively.

References

[1] Bernardi C. (to appear)

[2] Brezzi F. - Rappaz J. - Raviart P.A. - "Finite dimensional approximation of nonlinear problems. Part I: Branches of nonsingular solutions" - Rapport interne n. 52 (1979) Centre de Mathématiques Appliquées, Ecole Polytechnique, Palaiseau, (submitted to Num. Math.)

[3] ——— " ——— Part II: Limit points." Rapport interne n. 64 (1980) Centre de Mathématiques Appliquées, Ecole Polytechnique, Palaiseau, (submitted to Num. Math.)

[4] ——— " ——— Part III: Simple bifurcation points". Rapport interne n. 67 (1980) Centre de Mathématiques Appliquées, Ecole Polytechnique, Palaiseau (submitted to Num. Math.)

[5] Ciarlet P.G. "The finite element method for elliptic problems" North Holland (Amsterdam) 1978.

[6] Fujii H. - Yamaguti M. "Structure of singularities and its numerical realization in nonlinear elasticity" Research Report KSU/ICS 78-06. Kyoto Sangyo University (to appear in J. Math. Kyoto Univ.).

[7] Keller H.B. "Numerical solutions of bifurcation and nonlinear eigenvalue problems" Applications of Bifurcation Theory (P.H. Rabinowitz ed.) Academic Press (New York) 1977

[8] Kikuchi F. "Finite element approximations to bifurcation problems of turning point type" Theoretical and Applied Mechanics 27 (1979), 99-114.

[9] Kikuchi F. "An iterative finite element scheme for bifurcation analysis of semilinear elliptic equations" Report Inst. Space Aero. Sc. n. 542 (1976), Tokyo University.

[10] Mittelmann H.D. - Weber H. "Numerical methods for bifurcation problems - A survey and classification" Report n. 45 (1980) Universitat Dortmund, Lehrstuhl Mathematik III.

[11] Rappaz J. - Raugel G. "Finite-dimensional approximation of bifurcation problems at a double eigenvalue" (to appear).

[12] Reinhart L. "Sur la résolution numérique del problèmes aux limites nonlinéaires par des méthodes de continuation" Thése de 3 ème cycle - Université P. et M. Curie 1980.

Some Results in the Analysis of Thin Shell Structures

K.-J. BATHE, L.W. HO
Massachusetts Institute of Technology, Cambridge, USA

Abstract

This paper represents a progress report of some of the research that we are conducting in the finite element analysis of thin shell structures. We consider our isoparametric displacement-rotation thin shell element and our discrete-Kirchhoff-theory (DKT) plate/shell element, which we are continuously refining for accurate and effective geometric and materially nonlinear analysis. In the paper we briefly discuss the locking phenomenon of the isoparametric element, the use of this element as a transition element between shell surfaces and in shell-solid transitions, and we give some results using the DKT element.

1. INTRODUCTION

Much research has been conducted during the last two decades in the development of thin shell finite elements. However, despite the large amount of research effort there do not exist as yet what may be considered to be cost-effective, reliable and general thin shell analysis capabilities. We believe that an effective thin shell element should satisfy the following criteria:

1) The element should yield accurate solutions when modeling any shell geometry and under all boundary and loading conditions. In particular, the element should not contain any spurious zero energy modes, so that reliable results can always be expected. The theory of the element formulation must be well-understood and should not contain any "numerical fudge factors." These considerations are most important and many elements that have been published do not satisfy this criterion. Such element developments can represent

interesting research but should not be used in actual engineering analyses, because the generated analysis results cannot be interpreted with confidence.

2) We should be able to use the element in the modeling of general shell structures with beam stiffeners, cut-outs, intersections, and so on.

3) The element should be cost-effective in linear as well as in nonlinear, static and dynamic analysis. In nonlinear analysis, the element should be applicable to large displacement, large rotation, and materially nonlinear conditions.

Various approaches have been advocated for the development of thin shell elements [1-2]; however, based on the above objectives we have concentrated our research efforts onto two procedures:

a) The use of displacement/rotation isoparametric elements that can be employed with a variable number of nodes, but are usually used with 9 or 16 nodes.

b) The use of a simple triangular element which is obtained by superimposing the bending and membrane behaviors.

We have reported our formulations and first experiences with these elements in previous publications [3-5]. Our objective in this contribution is to present additional results that we have obtained recently in our continued research on the element formulations and implementations. The aim in our research is to improve and refine the performance of the elements to the maximum extent possible but always subject to the constraints summarized in 1) to 3) above.

2. SOME RECENT RESULTS WITH OUR ISOPARAMETRIC DISPLACEMENT-ROTATION ELEMENT

The use of an isoparametric displacement-rotation shell element was earlier proposed by Ahmad et al. [6] for linear analysis and Ramm [7] and Kråkeland [8] for nonlinear analysis. We have used the same concepts but refined the formulation of the element in some details to enhance its practical usage. The element formulation is very general because no specific shell theory is used; instead, only the two following basic assumptions are employed on the shell behavior:

(i) Particles originally on a straight line in the direction of the "normal vector" to the midsurface of the shell element remain on a straight line during the deformation of the shell.

(ii) The stress in the direction of the "normal vector" to the shell midsurface is zero.

The "normal vector" at any point of the midsurface of the element is used in the description of the shell geometry and displacements and is defined by the direction cosines of the "normal vectors" at the nodal points. These direction cosines are interpolated over the midsurface of the shell elements. As previously discussed [3], the element formulation reduces in the analysis of plates to the finite element discretization of the Mindlin plate theory. The formulation includes shear deformations which, however, are assumed to be constant over the plate thickness.

We have implemented the element with a variable number of nodes, but in practical analysis the element is usually most effective using the 9-node (parabolic) or 16-node (cubic) Lagrangian interpolations, as discussed further below. The element can also be employed as a transition element between solid and shell element idealizations or different shell surfaces. Figures 1 and 2 illustrate some of the features of the element.

The major advantage of the element is its general applicability, but it is also recognized that the use of the higher-order elements can be expensive. For this reason

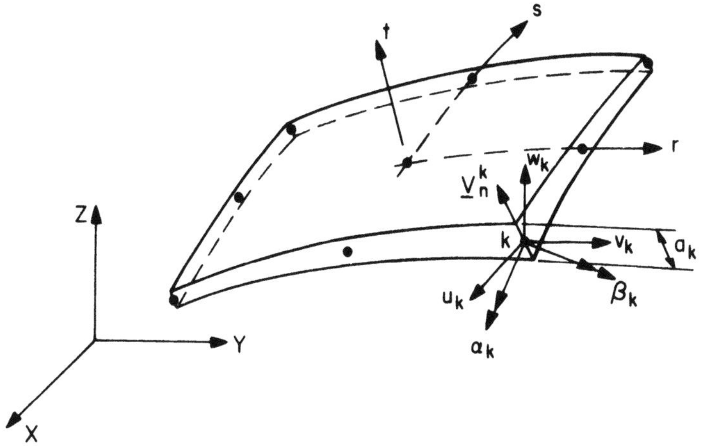

Figure 1 Nine-Node Isoparametric Shell Element

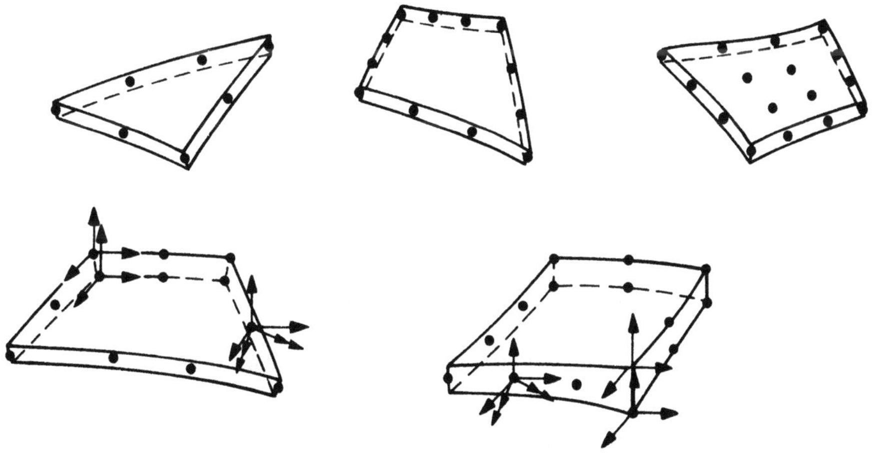

Figure 2 Shell Elements with Variable Number of Nodes and Transition Elements

much attention has been given to the use of the low-order elements. However, if the stiffness matrices of these elements are fully integrated, the elements display a "locking phenomenon" and therefore some reduced or selective integration techniques have been proposed [1,2,9,10]. These techniques, which can also be related to mixed formulations [11, 12, 5], relieve the "locking behavior" of the elements but because of the various difficulties encountered have not resulted as yet into a shell element that satisfies our three criteria of an effective practical shell solution capability. Indeed, based on the experience available so far it appears that for the development of low-order shell elements the approach used in the discrete Kirchhoff theory is more effective [5].

In the following sections we consider two important ingredients of the isoparametric shell element: the convergence of the higher-order elements (when fully integrated) to a finite element discretization of the Kirchhoff plate theory and the use of the transition element.

Although the results of a number of investigations have been published in which the locking phenomenon in conjunction with reduced integration procedures was considered, there is still a need for further insight into the problem of whether a mesh will lock or not. The objective in our work was to address this question first for the simpler case when using fully integrated elements. We are currently extending the results obtained to elements that are not fully integrated.

2.1 On the Use of the Higher-Order Elements

As shown earlier the low-order elements and the higher-order serendipity elements can greatly overestimate the stiffness of a thin plate when the element stiffness matrices are fully integrated [9, 13]. This phenomenon referred to as "element locking" can be explained by considering the variational indicator of the Mindlin plate theory,

$$\Pi = \frac{h^3}{2}\left[\underbrace{\int_A \underline{\kappa}^T \underline{C}_b \underline{\kappa}\, dA}_{\text{TERM 1}} + \alpha \underbrace{\int_A \underline{\gamma}^T \underline{C}_s \underline{\gamma}\, dA}_{\text{TERM 2}}\right] \quad (1)$$

$$-(\text{potential of external loads})$$

where

$$\underline{\kappa} = \begin{bmatrix} \frac{\partial \beta_x}{\partial x} \\ -\frac{\partial \beta_y}{\partial y} \\ \frac{\partial \beta_x}{\partial y} - \frac{\partial \beta_y}{\partial x} \end{bmatrix} ; \quad \underline{\gamma} = \begin{bmatrix} \frac{\partial w}{\partial y} - \beta_y \\ \frac{\partial w}{\partial x} + \beta_x \end{bmatrix} \quad (2)$$

$$\underline{C}_b = \frac{E}{12(1-\nu^2)}\begin{bmatrix} 1 & \nu & 0 \\ \nu & 1 & 0 \\ 0 & 0 & \frac{1-\nu}{2} \end{bmatrix} ; \quad \underline{C}_s = \frac{Ek}{2(1+\nu)}\begin{bmatrix} 1 & 0 \\ 0 & 1 \end{bmatrix} \quad (3)$$

and

w = transverse displacement of plate

β_y, β_x = section rotations about x and y-axes, respectively

h = thickness of plate (assumed constant)

k = shear correction factor

E, ν = Young's modulus and Poisson ratio

$\alpha = \left(\frac{L}{h}\right)^2$; L = characteristic length

The expression labelled "TERM 1" corresponds to the bending strain energy and the expression labelled "TERM 2" corresponds to the shear strain energy. We notice that for thin plates the constant α is large and can be regarded as a penalty parameter, which enforces that the shear strains are small and converge to zero as $\frac{h}{L}$ decreases. Hence, a Mindlin plate theory analytical solution converges to the Kirchhoff plate theory solution when the thickness of the plate becomes very small.

Considering now a finite element discretization of the variational indicator, we can directly conclude that for the finite element scheme to be applicable to the analysis of thin plates, we must have that, with the interpolations used on w, β_x and β_y, the two expressions for the shear strains

$$\gamma_{zy} = \frac{\partial w}{\partial y} - \beta_y \tag{4}$$

$$\gamma_{zx} = \frac{\partial w}{\partial x} + \beta_x \tag{5}$$

admit very small values. Also, since we want to use full numerical integration--to avoid the problem of generating spurious zero energy modes in the element formulation--we must have that the shear strain expressions in Eqs. (4) and (5) can be very small throughout the element. Hence, our criterion on whether an element or a patch of elements will lock is simply that the finite element interpolations with the available degrees of freedom must enable the conditions

$$\gamma_{zy} = 0 \tag{6}$$

$$\gamma_{zx} = 0 \tag{7}$$

throughout the element or patch of elements. However, this means that a simple condition for a mesh of rectangular plate elements not to lock is:

$$2q - k > np \tag{8}$$

where

q = total number of nodal points,

k = number of constrained degrees-of-freedom, maximum corresponding to rotations about x and y-axes,

n = number of finite elements,

p = number of nodes (basis functions) per element.

The relation in Eq. (8) is derived by considering Eqs. (6) and (7) when using the finite element displacement and rotation interpolations. The maximum number of individual polynomial terms that can occur in the shear expressions is equal to np. If the relation in Eq. (8) is satisfied, the coefficients of these terms can all be individually equal to zero, which represents the worst case. Hence, the relation in Eq. (8) is a sufficient but not necessary condition for the element mesh not to lock. Figures 3 and 4 illustrate the application of the condition in Eq. (8) for various analysis cases and give computed results using ADINA [14] (in each case one quarter of the plate was modeled).

The relation in Eq. (8) is a sufficient but not necessary condition because the np equations derived as described above may be linearly dependent. In such case the patch of elements may not lock although Eq. (8) is violated. Figure 5 summarizes some results obtained using the parabolic Lagrangian element, for which the condition in Eq. (8) predicted locking but the element meshes proved adequate in the finite element solutions using ADINA. Hence, in summary, the relation in Eq. (8) is a conservative rule for telling whether a mesh locks.

We should emphasize that the simple rule in Eq. (8) is strictly only applicable to rectangular (undistorted) plate elements. The development of a similar criterion for general plate and shell elements appears to be much more difficult, since the element distortions (measured on the natural element coordinates) enter into the calculation of the shear strains.

2.2 On the Use of the Transition Element

The transition element can be employed to model shell-solid transitions and the intersections of different shell

ELEMENT TYPE	NO OF D.O F (2q - k)	NO OF POLYNOMIAL TERMS (np)	ADINA RESULTS
	5	8	LOCKS
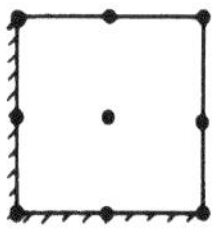	7	9	LOCKS
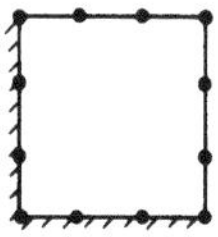	9	12	LOCKS
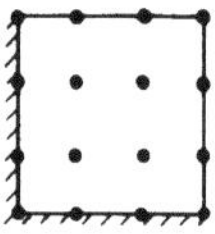	17	16	DOES NOT LOCK

(a) Analysis of a Clamped Square Plate Modeled with a Single Shell Element

ELEMENT MESH CONSIDERED	CASE	NO OF DOF (2q - k)	NO OF POLYNOMIAL TERMS (np)	ADINA RESULTS
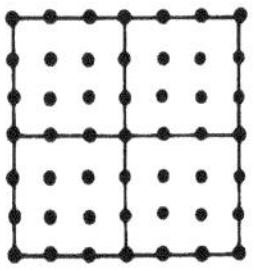	SIMPLY SUPPORTED	78	64	DOES NOT LOCK
	CLAMPED	71	64	DOES NOT LOCK
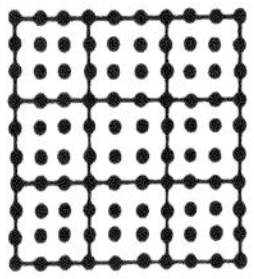	SIMPLY SUPPORTED	171	144	DOES NOT LOCK
	CLAMPED	161	144	DOES NOT LOCK

(b) Analysis of a Square Plate Modeled with Cubic Elements

Figure 3 Application of the Locking Criterion

ELEMENT MESH CONSIDERED	NO OF D O F (2q-k)	NO OF POLYNOMIAL TERMS (n p)	ADINA RESULTS
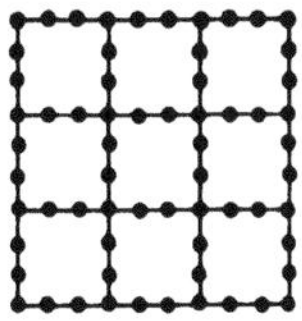	89	108	LOCKS
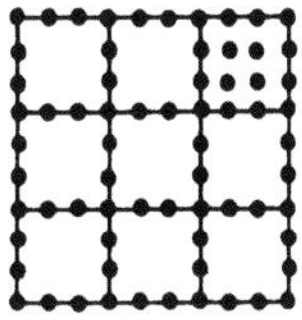	97	112	LOCKS (12-NODE ELEMENTS ARE LOCKING)
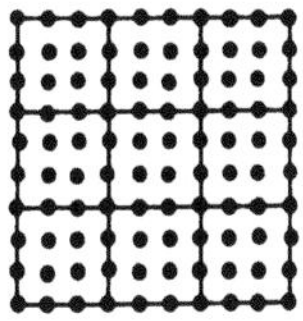	161	144	DOES NOT LOCK

Figure 4 Application of the Locking Criterion to the Analysis of a Clamped Square Plate Modeled with Different Elements

ELEMENT MESH CONSIDERED	CASE	NO OF D O F (2q-k)	NO OF POLYNOMIAL TERMS (np)	ADINA RESULTS
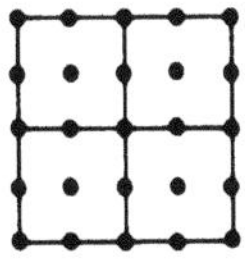	SIMPLY SUPPORTED	36	36	DOES NOT LOCK
	CLAMPED	31	36	DOES NOT LOCK
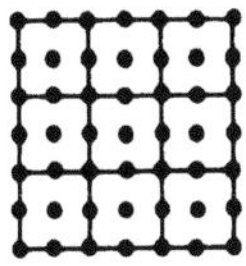	SIMPLY SUPPORTED	78	81	DOES NOT LOCK
	CLAMPED	71	81	DOES NOT LOCK

Figure 5 Application of the Locking Criterion to the Analysis of a Square Plate Modeled with Parabolic Elements

surfaces[2]. The use of the element is relatively easy because no special constraint equations need be written at the transition regions and yet full compatibility between the elements is preserved. However, an important question must be whether the element predicts displacements and stresses accurately enough in the transition regions.

Figure 6 shows a simple folded roof structure that we analyzed using the transition element. Figure 7 lists the various finite element idealizations used. Since no analytical or accurate numerical solution of the stresses in the transition region could be located we used the idealization in Fig. 7(a) to obtain an accurate prediction of the stress components. The model in Fig. 7(b) corresponds to the usual procedure of modeling shell intersections, whereas the model in Fig. 7(c) may be attractive because of ease of program input of only one normal. The definition of the model in Fig. 7(d) is equally effective, but we need to question the accuracy of the results of the stress predictions.

Figures 8 and 9 give the stresses predicted in the analyses. It is interesting to note that the results with an average normal at the shell intersection are not far from the results using the usual modeling procedure for shell intersections (the model in Fig. 7(b)). The results obtained with the transition element are different near the transition region, but they show the trend of the 2-D fine mesh results. Also, they indicate that the transition element should be used to model only the actual transition region.

Figure 10 shows the model used in the next analysis of the folded roof, which was established based on the above considerations. Figures 11 and 12 give the results obtained in the analysis of this model. It is seen that the predicted stresses are significantly more accurate than those calculated with the usual shell idealization (in Fig. 7(b)). We should note that the number of degrees-of-freedom in the analysis was about $\frac{1}{5}$ th of those used in the two-dimensional model. Hence, although more detailed studies of the

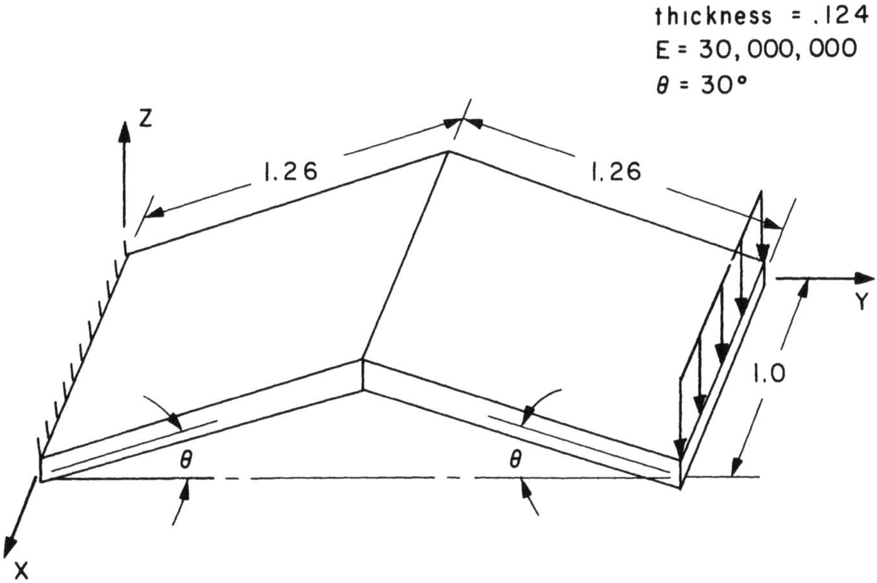

Figure 6 Folded Cantilever Plate Subjected to a Line Load at Its Tip

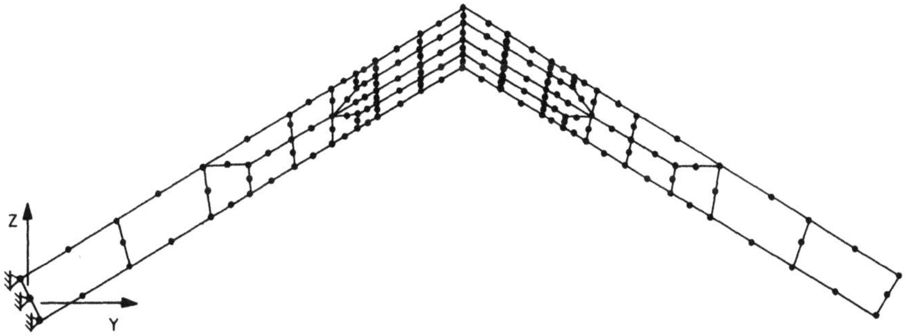

(a) Two-Dimensional Finite Element Mesh

Figure 7 Models Used for the Analysis of the Folded Cantilever Plate

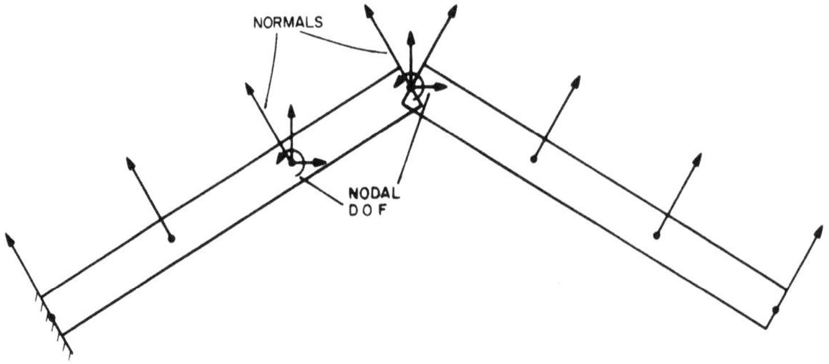

(b) Shell Elements Using Constraint Equations

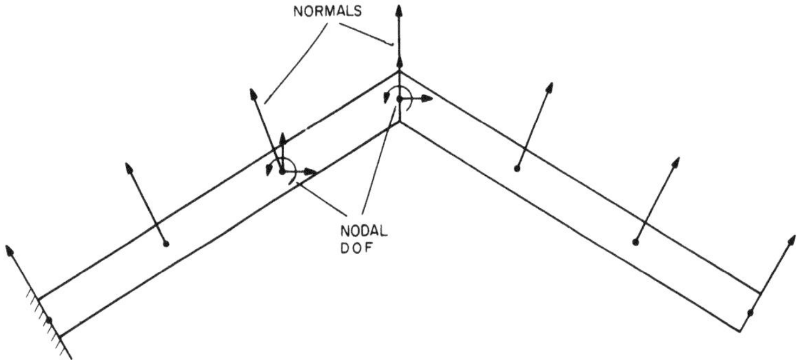

(c) Shell Elements Using the Average Normal Technique

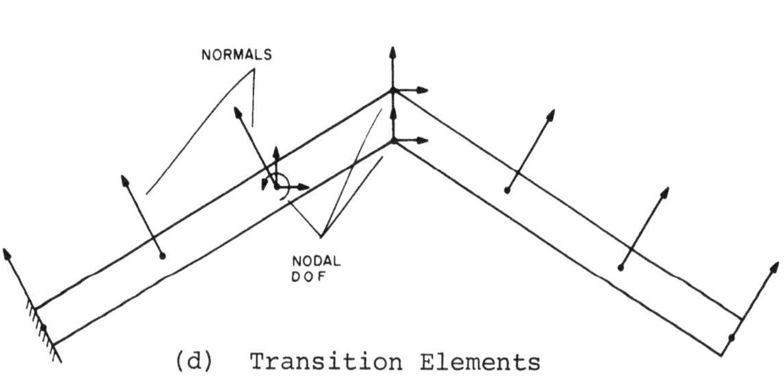

(d) Transition Elements

Figure 7 (Continued)

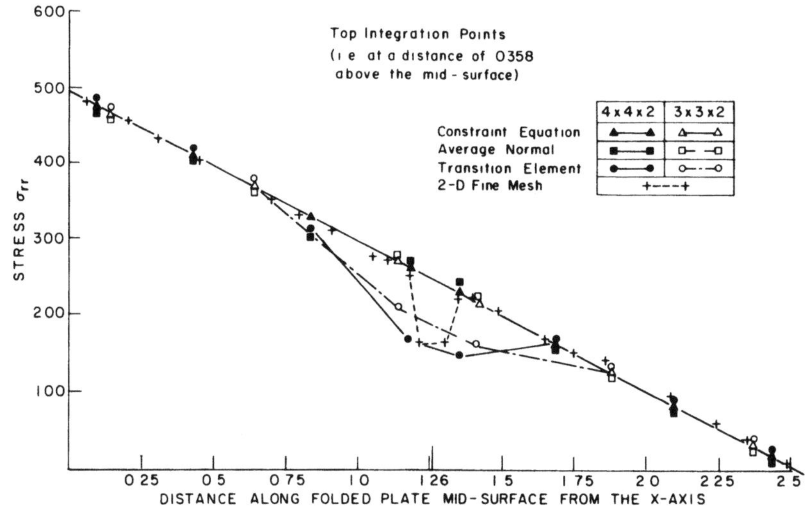

Figure 8 Predicted Stresses with Folded Plate Models of Fig. 7 (Gauss Integration Used)

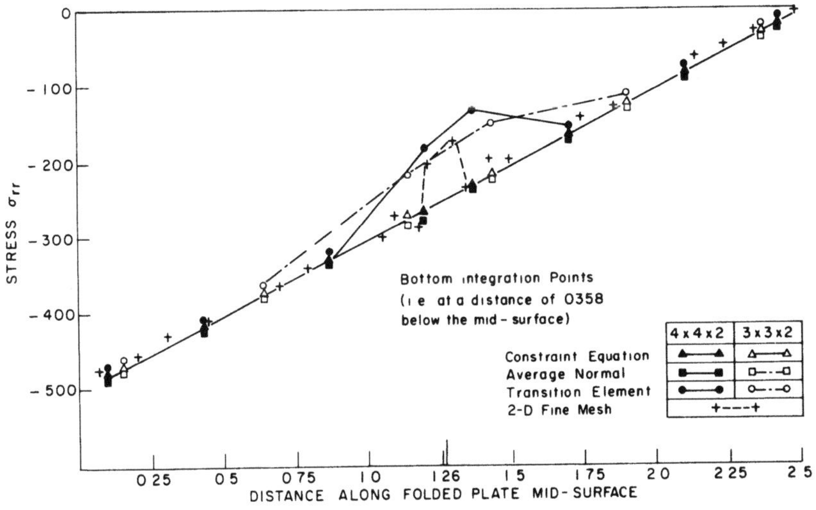

Figure 9 Predicted Stresses with Folded Plate Models of Fig. 7 (Gauss Integration Used)

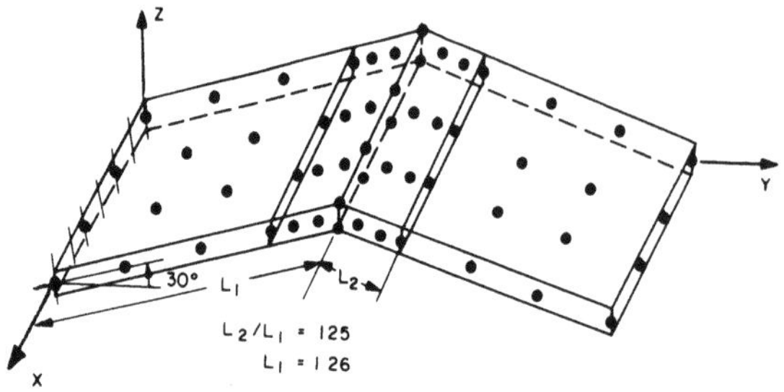

Figure 10 A Refined Model of the Folded Plate with Transition and Shell Elements

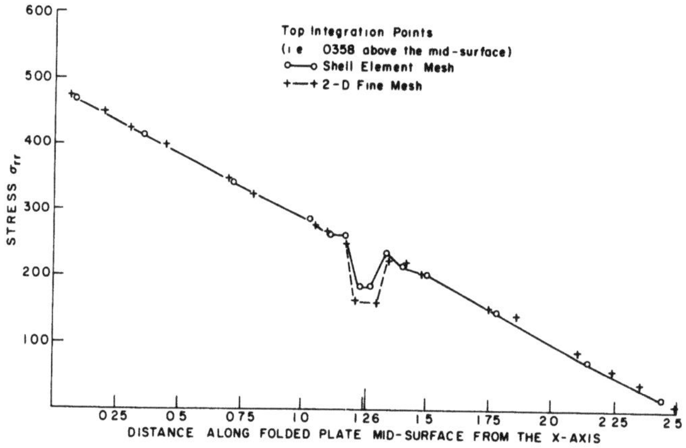

Figure 11 Predicted Stresses with Folded Plate Models of Figs. 7(a) and 10

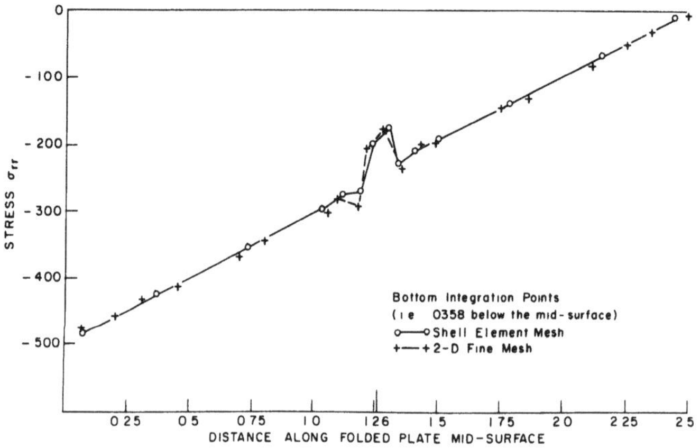

Figure 12 Predicted Stresses with Folded Plate Models of Figs. 7(a) and 10

transition element are still necessary, this example analysis shows the potential in the practical use of the element.

3. SOME RECENT RESULTS WITH OUR DISCRETE-KIRCHHOFF-THEORY TRIANGULAR ELEMENT

The triangular element developed has three corner nodes and six degrees-of-freedom per node as shown in Fig. 13. The element is flat and therefore a shell surface is modeled with the element as an assemblage of flat facets in the way some of the earliest shell analyses were carried out [1,2,15]. The primary objective in the development of the element was to have a very simple and cost-effective element of good accuracy as an alternative to the higher-order isoparametric elements. The three-node DKT shell element stiffness matrix is constructed as follows [16]:

1) The bending behavior is described by a discrete Kirchhoff formulation. This description has been found to yield the most effective three-node plate bending element of a large number of elements considered in a survey study [4, 5]. The element has high accuracy, has no spurious zero energy modes, is computationally very inexpensive and the programming is very simple (less than 100 Fortran statements). The element is very likely the most effective simple plate bending element currently available.

2) The membrane behavior is currently described by the simple constant strain triangle. However, it would be desirable to increase the order of accuracy of the membrane description.

3) The above stiffnesses are combined in a local coordinate system and the normal rotational stiffness is arbitrarily set to $(\frac{1}{10000})$ x (smallest bending stiffness) in order to obtain 6 stiffness degrees-of-freedom per node.

The element can be used very effectively in geometric nonlinear analysis because it remains flat during the response history. Also, for elastic-plastic analysis it is effective

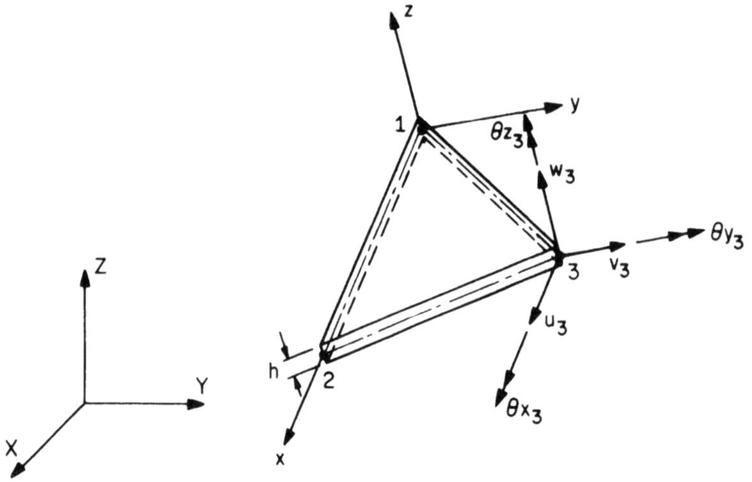

(a) Configuration and Degrees-of-Freedom

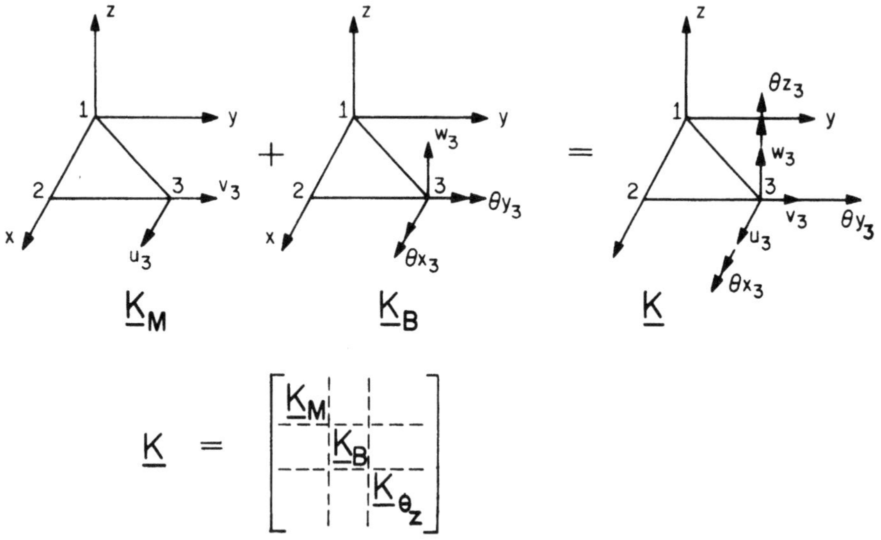

(b) Element Stiffness Matrix Components

Figure 13 The DKT Three-Node Shell Element

to use in the material description stress resultants, which, as in elastic analysis, circumvents the numerical integration through the element thickness.

To indicate the effectiveness of the element we are presenting in the following the results obtained in the analyses of three problems.

3.1 Analysis of a Square Plate

A square plate of side lengths 2a with clamped and simply-supported edges was analyzed [5]. Because of the symmetry conditions only one quarter of the plate was considered. Figure 14 gives the finite element meshes used in the analyses for $N = 2, 4$ and 8. Figure 15 summarizes the results obtained for the centre deflection of the plate. We may note that, considering the computational effort in the solution, these results should be compared with 2x2, 4x4 and 8x8 mesh idealizations using four-node square elements or 1x1, 2x2 and 4x4 meshes with nine-node square elements. Such evaluation and a comparison with other 3-node elements show that the 3-node DKT element is indeed very effective.

3.2 Analysis of a Pinched Cylindrical Shell

The structure analyzed and a typical finite element idealization used are shown in Fig. 16. In the figure the 10x10 mesh used is shown, but the analysis was also carried out with 4x4, 6x6, 8x8 and 16x16 mesh topologies. Figures 17 to 19 give calculated displacement and stress resultant distributions along the lines DC, BC and AD of the shell, respectively. It is seen that the finite element predictions converge rapidly to the analytical solution as a reasonable number of shell elements is employed in the structural idealization. Table 1 summarizes the solution times used in the analysis of the shell structure.

3.3 Large Displacement Analysis of a Simply - Supported Plate

The plate was subjected to a uniform pressure loading q. Figure 20 shows the finite element idealization used for the plate and

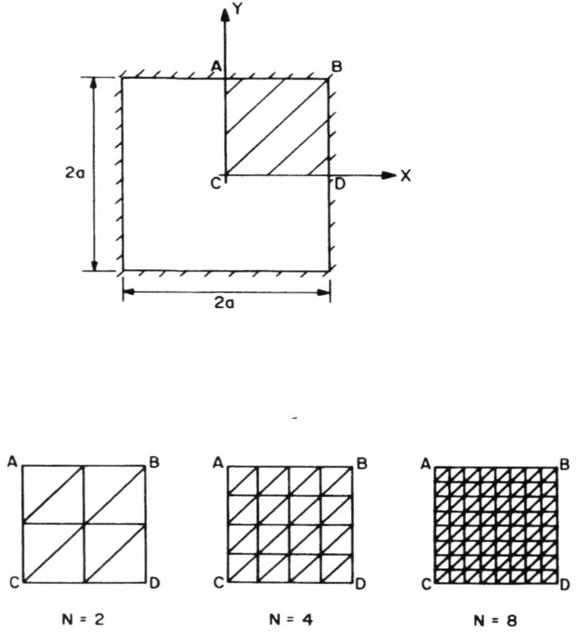

Figure 14 Element Meshes for the Analysis of a Square Plate Using DKT Elements

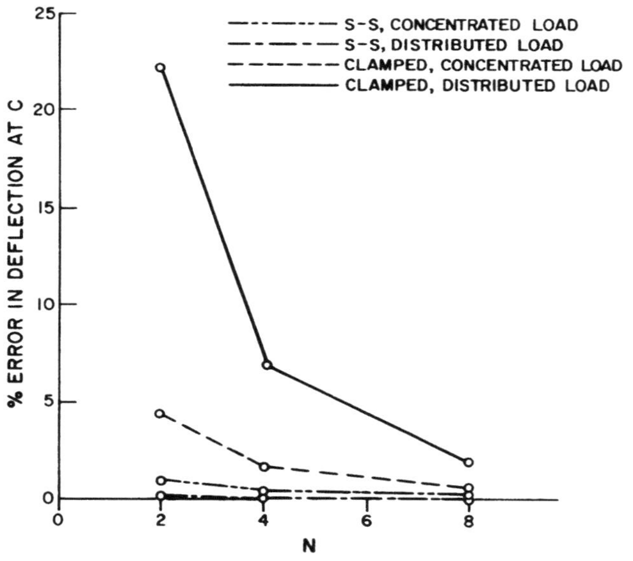

Figure 15 Percentage Error in the Predicted Centre Deflection of the Square Plate Using Element Meshes of Fig. 14

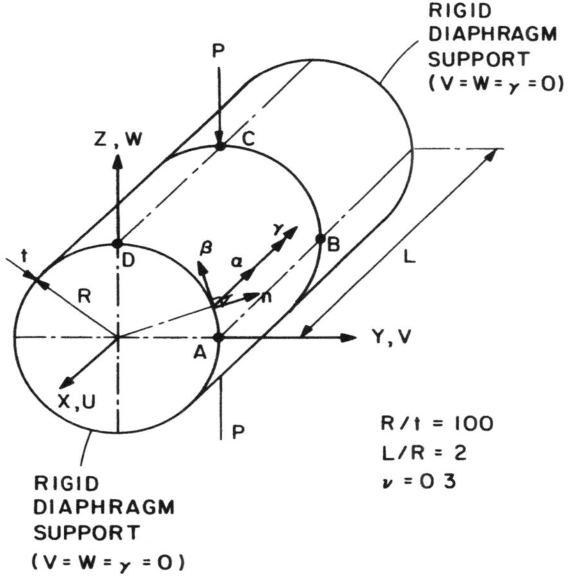

(a) Structure Considered

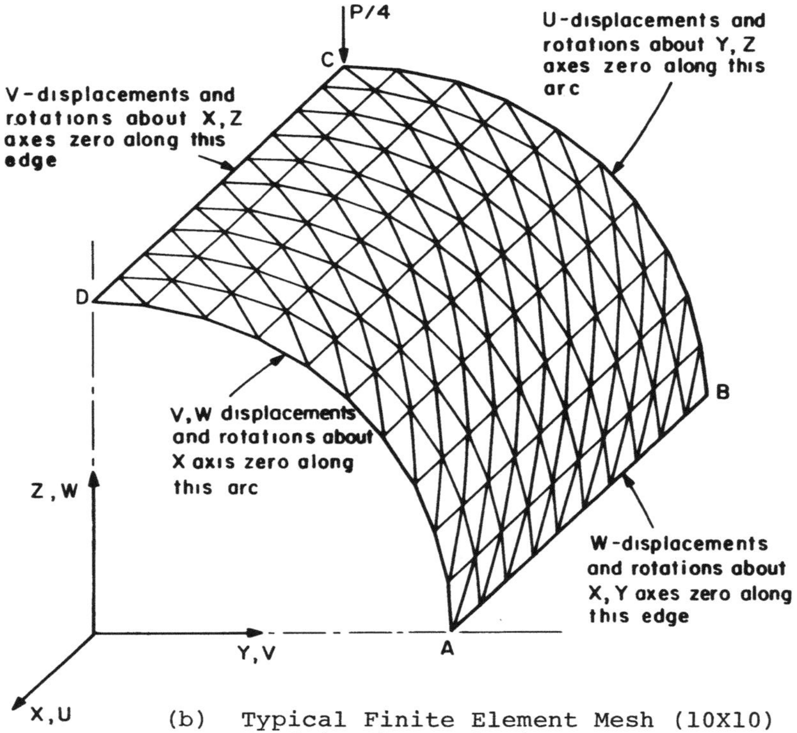

(b) Typical Finite Element Mesh (10X10) Used in the Analysis

Figure 16 Analysis of a Pinched Cylindrical Shell Structure Using DKT Elements

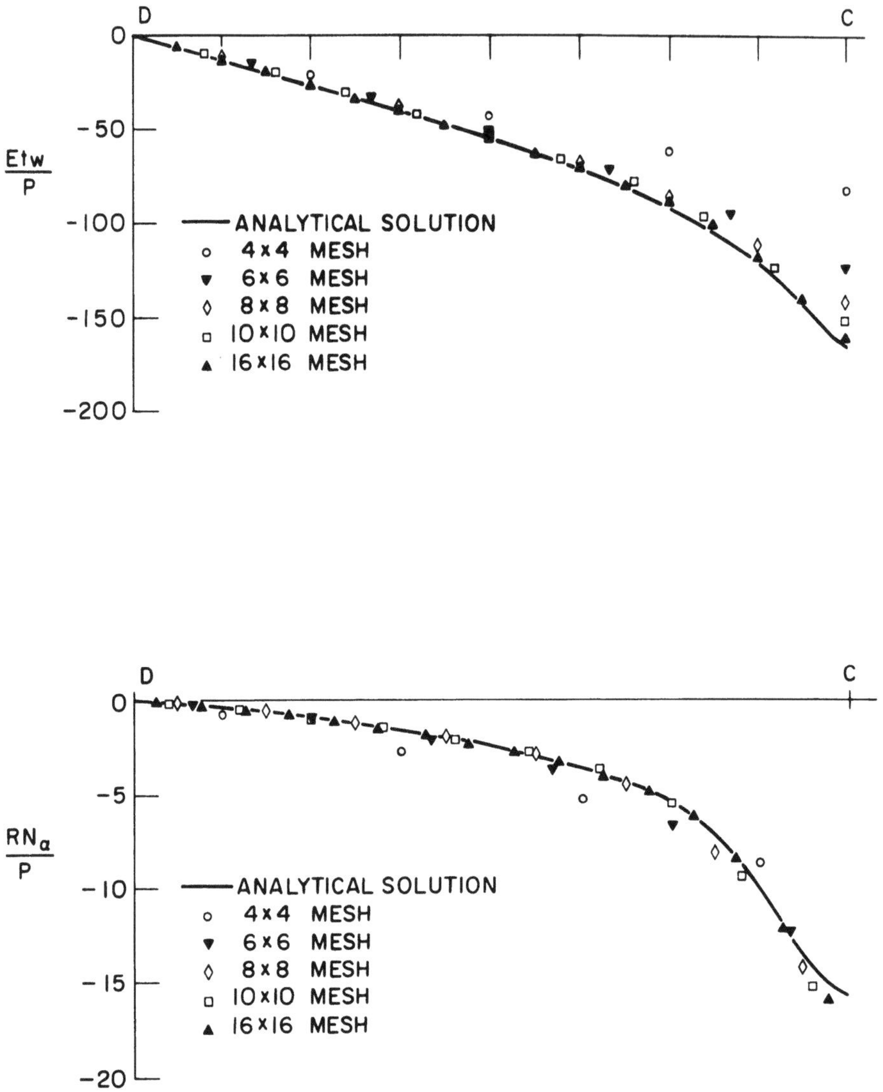

Figure 17 Predicted Displacement and Stress Distributions along DC of Shell in Fig. 16

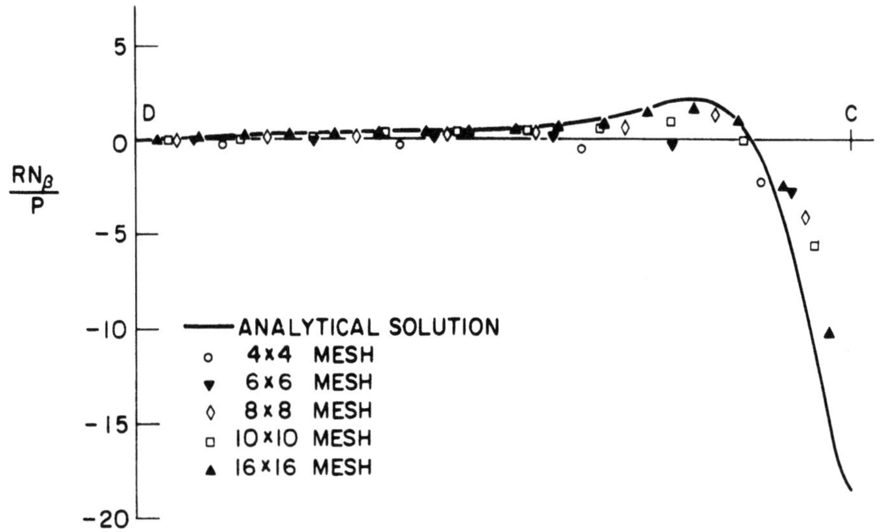

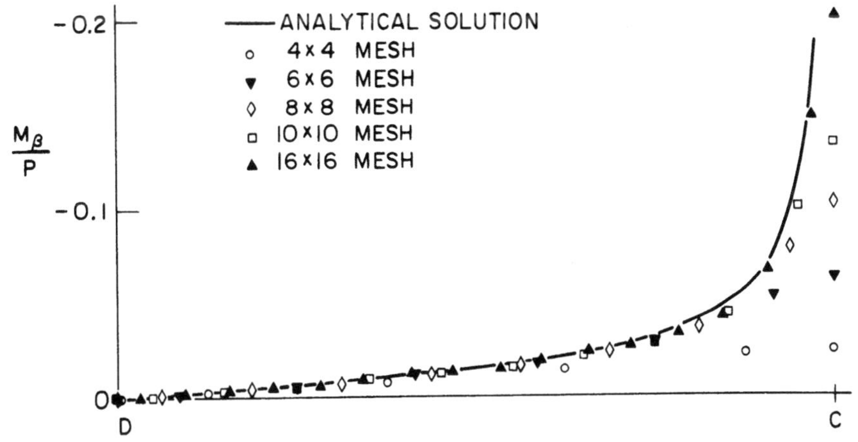

Figure 17 (Continued)

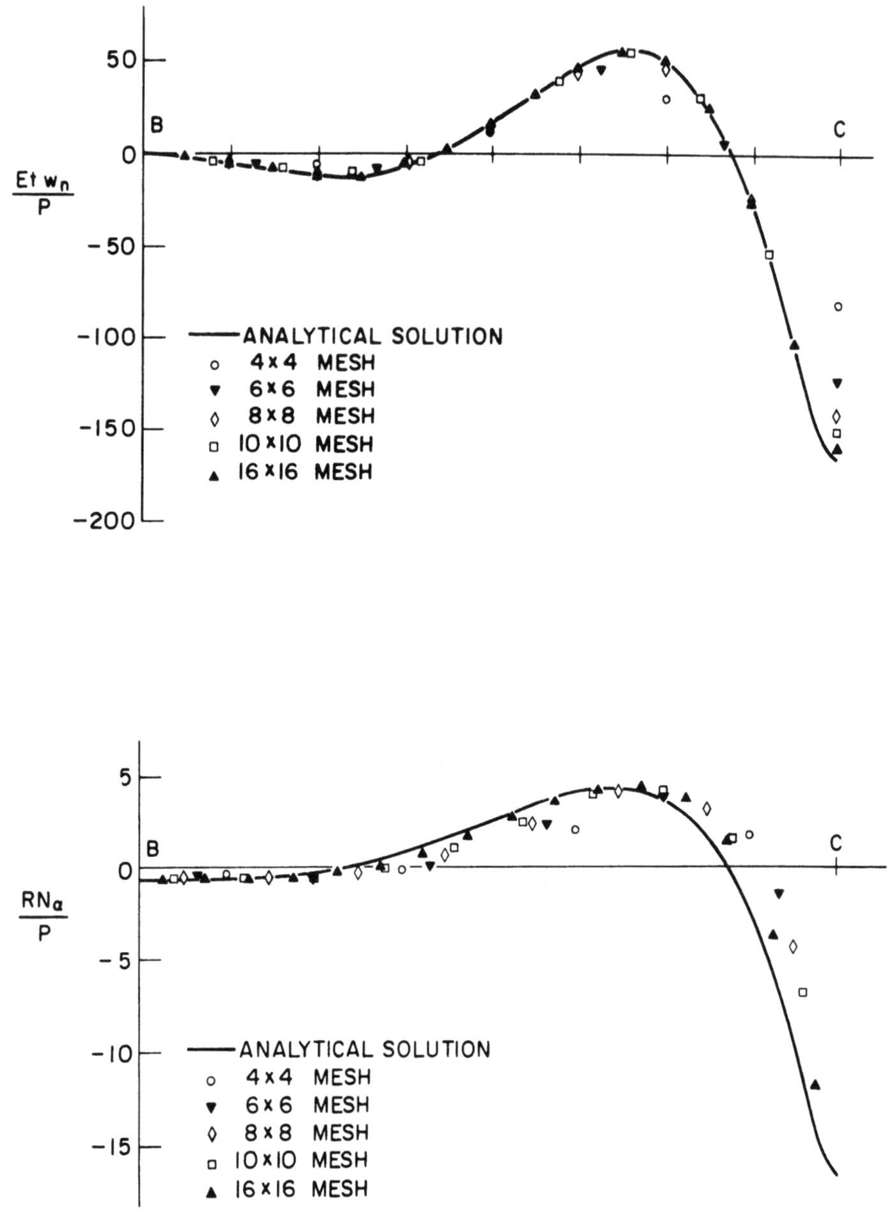

Figure 18 Predicted Displacement and Stress Distributions along BC of Shell in Fig. 16

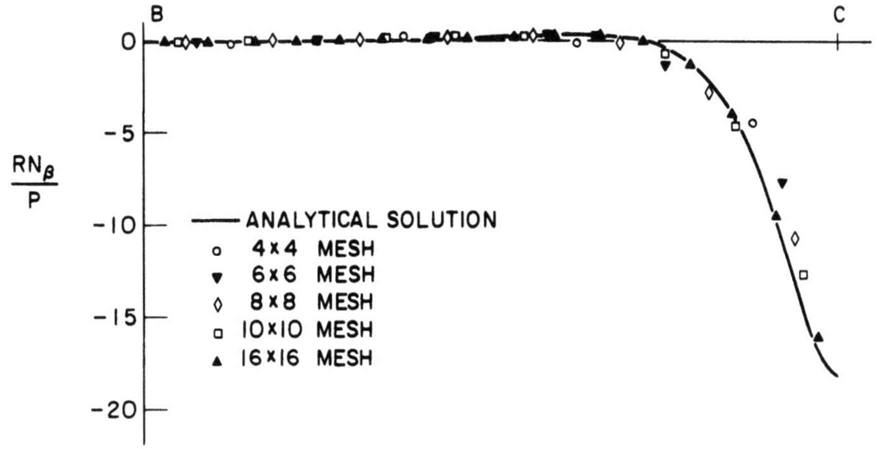

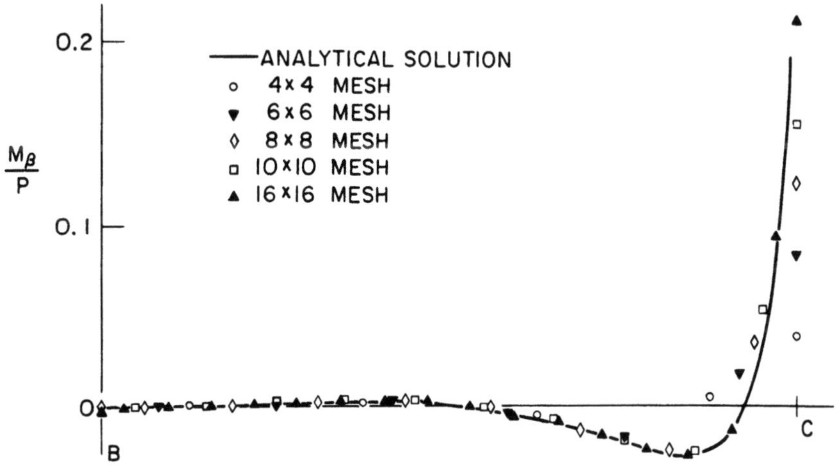

Figure 18 (Continued)

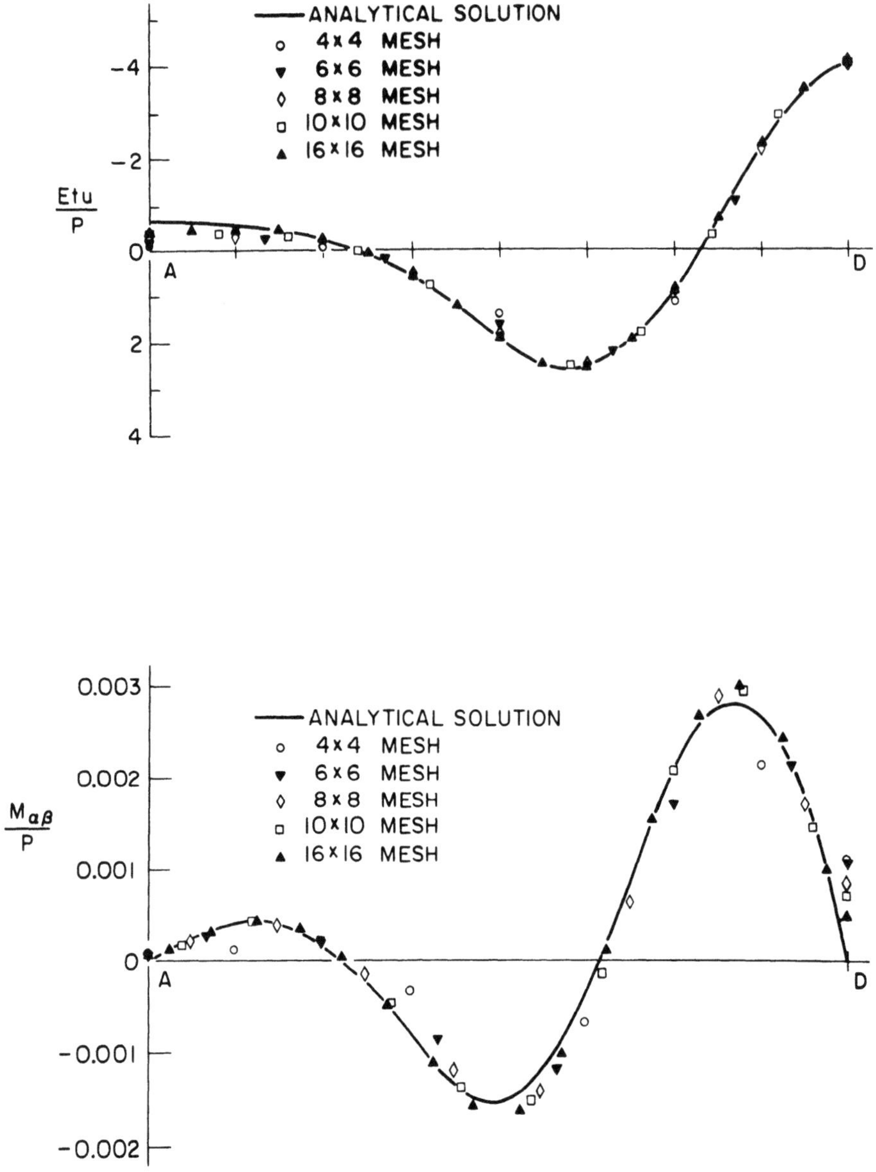

Figure 19 Predicted Displacement and Stress Distributions along AD of Shell in Fig. 16

the response predicted. In a first analysis, the edges were constrained not to move in the plane of the plate, and in a second analysis, the u and v edge displacements were left free. The predicted responses in these analyses are compared in Fig. 20 with other solutions reported earlier.

4. CONCLUDING REMARKS

In this paper we have reported some recent results of our research in finite element thin shell analysis capabilities. Our present analysis capabilities can be employed in linear analysis, in large displacement, large rotation and materially nonlinear analysis, and for static and dynamic solutions. A few demonstrative analysis results have been presented in the paper, and the reader may refer to ref. [3-5, 14, 16] for additional analysis results.

We are conducting our research to obtain increasingly more cost-effective and general solution procedures but we noted that primary emphasis is directed towards the reliability of the analysis methods. This is so, because we are convinced that an ultimate extensive usage of nonlinear finite element analysis capabilities will depend primarily on the reliability of the solution procedures available and much research work should be concentrated in this area.

Acknowledgment

We are grateful to the ADINA users group for supporting financially our research and development efforts, and we thank C. Keilers and U. Tsach for having carried out some of the analyses with ADINA.

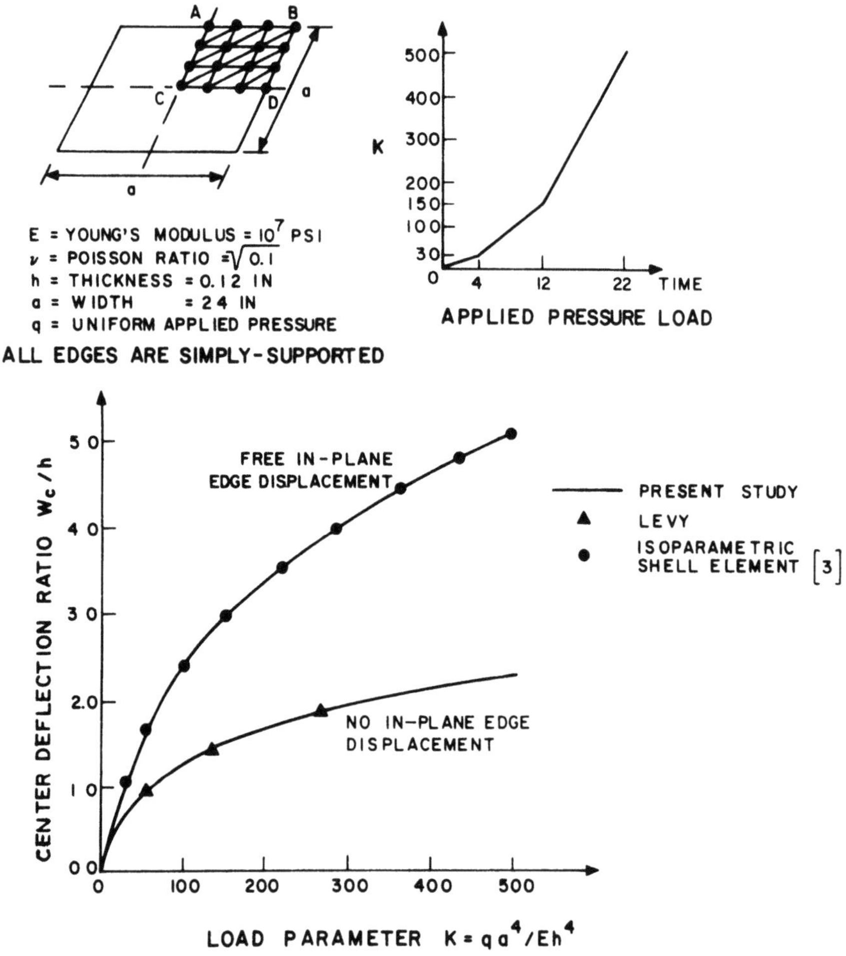

Figure 20 Large Displacement Analysis of a Simply-Supported Square Plate Subjected to Pressure Loading Using DKT Elements

References

[1] Zienkiewicz, O.C., The Finite Element Method, McGraw-Hill, 1977.

[2] Bathe, K.J., Finite Element Procedures in Engineering Analysis, Prenctice-Hall, Inc. in press.

[3] Bathe, K.J., and Bolourchi, S., "A Geometric and Material Nonlinear Plate and Shell Element," J. Computers and Structures, Vol. 11, pp. 23-48.

[4] Batoz, J.L., Bathe, K.J., and Ho, L.W., "A Search for the Optimum Three-node Triangular Plate Bending Element," Acoustics and Vibration Lab., Report 82448-8, Dept. of Mech. Eng., M.I.T., Dec. 1978, 1980.

[5] Batoz, J.L., Bathe, K.J., and Ho, L.W., "A Study of Three-Node Triangular Plate Bending Elements," Int. J. Num. Meth. in Eng., in press.

[6] Ahmad, S., Irons, B.M., and Zienkiewicz, O.C., "Analysis of Thick and Thin Shell Structures by Curved Elements," Int. J. Num. Meth. in Eng., Vol. 2, pp. 419-451, 1970.

[7] Ramm, E., "A Plate/Shell Element for Large Deflections and Rotations," in Formulations and Computational Algorithms in Finite Element Analysis, Bathe, K.J., Oden, J.T., and Wunderlich, W., (eds.), M.I.T. Press, 1977.

[8] Kråkeland, B., "Large Displacement Analysis of Shells Considering Elastic-Plastic and Elasto-Viscoplastic Materials," Report No. 776, The Norwegian Institute of Technology, The University of Trondheim, Norway, Dec. 1977.

[9] Hughes, T.J.R., Taylor, R.L., and Kanoknukulchai, W., "A Simple and Efficient Finite Element for Plate Bending," Int. J. Num. Meth. in Eng., Vol. 11, 1977, pp. 1529-1543.

[10] Zienkiewicz, O.C., Bauer, J., Morgan, K., and Onate, E., "A Simple and Efficient Element for Axisymmetric Shells," Int. J. Num. Meth. in Eng., Vol. 11, 1977, pp. 1545-1558.

[11] Lee, S.W., and Pian, T.H.H., "Improvement of Plate and Shell Finite Elements by Mixed Formulations," AIAA J., Vol. 16, Jan. 1978, pp. 29-34.

[12] Malkus, D.S., and Hughes, T.J.R.,"Mixed Finite Element Methods--Reduced and Selective Integration Techniques: A Unification of Concepts," Computer Methods in Appl. Mech. and Eng., Vol. 15, 1978, pp. 63-81.

[13] Pugh, E.D., Hinton, E., and Zienkiewicz, O.C., "A Study of Quadrilateral Plate Bending Elements with Reduced Integration," Int. J. Num. Meth. in Eng., Vol. 12, 1978, pp. 1059-1078.

[14] Bathe, K.J., "ADINA - A Finite Element Program for Automatic Dynamic Incremental Nonlinear Analysis," Report AVL 82448-1, Mech. Eng. Dept., M.I.T., Sept. 1975 (rev. Dec. 1978).

[15] Clough, R.W., and Wilson, E.L., "Dynamic Finite Element Analysis of Arbitrary Thin Shells," J. Computers and Structures, Vol. 1, pp. 33-56, 1971.

[16] Bathe, K.J., and Ho, L.W. "A Simple and Effective Element for Analysis of General Shell Structures," paper to be presented at the ADINA Conference, June 1981, M.I.T.

Table 1 Total Solution Times in Analysis of Pinched Cylindrical Shell (on a CDC Cyber 175)

Grid	Solution time (sec.)
4x4	1.00
6x6	1.86
8x8	3.19
10x10	5.73
16x16	21.94

Nonlinear Dynamic Finite Element Analysis of Shells

T. J. R. HUGHES*, W. K. LIU**, I. LEVIT**
California Institute of Technology, Pasadena, USA

Summary

Dynamical aspects of a general nonlinear finite element shell analysis procedure are described. The work extends previous endeavors of the authors on quasistatic plate and shell analysis. Several sample problems of a two- and three-dimensional nature are presented which demonstrate the applicability of the methodology.

1. Introduction

In this paper we are primarily concerned with the nonlinear dynamic finite element analysis of shells. The work described herein is a sequel to a general nonlinear static formulation presented in [20] and [21] and referred to in this paper as "Part I" and "Part II", respectively. Due to the extensiveness of Parts I and II, and in an effort to keep the present work as brief as possible, the details of the static aspects of the theory will not be repeated here. However, to enable the reader to better appreciate the "big picture" we will briefly mention the salient features of Parts I and II.

Part I concerns itself with three-dimensional and fundamental aspects of nonlinear finite element shell analysis. Large strain and finite rotation effects are accounted for. In addition, a fairly general class of nonlinear finite-deformation constitutive equations may be accommodated. The elements are generalizations of reduced/selective integration Mindlin plate elements which have been described in previous works of the senior author and colleagues [15-17,

* Associate Professor of Structural Mechanics

** Graduate Research Assistant

,24]. The extension of the selective integration procedure to the nonlinear case is novel, the basic idea having been described earlier for a related situation [12,13]. This in particular facilitates the development of a "heterosis-type" nonlinear shell element which exhibits high accuracy without ostensible defects. It has been noted in the past that the reduced/selective integration process is equivalent to a mixed formulation [29] and thereby has some implementational advantages by virtue of the elimination ab initio of the auxiliary field. However, some new thoughts on the matter have recently appeared [30] and advantages of directly implementing the mixed formulation are argued. This and other aspects of shell element behavior will no doubt remain active areas of research for the coming years. Another noteworthy aspect of Part I is the development of a constitutive algorithm which is "objective" for large rotation increments (i.e., appropriately treats finite rotation effects) and maintains the zero normal-stress condition in a rotating stress coordinate system. Lastly, the treatment of finite-rotational nodal degrees-of-freedom precludes the appearance of zero-energy in-plane rotational modes and is felt to be especially simple and clean.

In Part II, two-dimensional shell situations are dealt with. The formulation applies to the practically important cases of rings, tubes, beams, frames, and shells of revolution. The point of view adopted is to deduce the two-dimensional special cases from the three-dimensional formulation of Part I. This results in a conceptual economy and permits much of the three-dimensional software to be used for the two-dimensional cases, particularly constitutive routines. Thus the implementational phase is considerably simplified, however, some computational overhead results due to the degree of generality attempted in one program package. Details are presented of how to modify kinematic quantities so that the particular constraints of the various two-dimensional cases are manifested (e.g., in- and out-of-plane zero normal stress conditions). The elements employed are the uniform reduced integration Lagrange elements, which have very nice properties for this class of problems. A number of numerical problems are presented in Parts I and II and the interested reader is urged to consult these references for further information.

In this work we begin where Parts I and II leave off. In Section 2 we describe the construction of element mass matrices. (The construction of all other element arrays is described in Parts I and II.) Particular attention is paid to the development of lumped mass matrices which yield large critical time steps for explicit transient analysis. The basic idea employed is that originally presented by Key and Beisinger [25]. In Section 3 we present aspects of the transient algorithm. For the most part the presentation is very similar to our other recent works on this topic [14,18, 19,22,23] and is only briefly sketched here. However, we do point out the specific modifications which are necessary for a successful transient shell formulation, namely, those affecting rotation-like nodal degrees-of-freedom. In Section 4, several numerical examples are presented which demonstrate the capabilities. In addition to some dynamic elastic-plastic calculations, we present some static elastic buckling calculations which are of some interest.

2. Element Mass Matrices

Consistent mass

The element consistent mass matrix is defined by

$$\underset{\sim}{m} = [\underset{\sim}{m}_{ab}]$$

$$\underset{\sim}{m}_{ab} = \underset{\sim}{S}_a^T \int_{\square} \int_{-1}^{+1} \underset{\sim}{N}_a^T \underset{\sim}{N}_b \rho_o j_o \, d\zeta \, d\square \, \underset{\sim}{S}_b \tag{2.1}$$

where $1 \leq a, b \leq n_{en}$, the number of element nodes; $\underset{\sim}{S}_a$ is a transformation matrix which maps rotational degrees-of-freedom into so-called "fiber" directions; ρ_o is the mass density in the initial configuration; j_o is the Jacobian determinant for the initial configuration; and

$$\int_{\square} \cdots \, d\square = \begin{cases} \displaystyle\int_{-1}^{+1}\int_{-1}^{+1} \cdots \, d\xi \, d\eta & \text{(three dimensions)} \\ \displaystyle\int_{-1}^{+1} \cdots \, d\eta & \text{(two dimensions)} \end{cases} \tag{2.2}$$

$$\underset{\sim}{N}_a = \begin{cases} \begin{bmatrix} N_a & 0 & 0 & N_a z_a & 0 & 0 \\ 0 & N_a & 0 & 0 & N_a z_a & 0 \\ 0 & 0 & N_a & 0 & 0 & N_a z_a \end{bmatrix} & \text{(three dimensions)} \\ \begin{bmatrix} N_a & 0 & N_a z_a & 0 \\ 0 & N_a & 0 & N_a z_a \end{bmatrix} & \text{(two dimensions)} \end{cases} \tag{2.3}$$

in which N_a is the lamina shape function; and $z_a = z_a(\zeta)$ is a thickness function (see Part I for further details).

<u>Lumped mass</u>

Element lumped mass matrices may be computed by performing the following steps:

1. Calculate

$$\bar{J}_o = \int_{-1}^{+1} J_o \, d\zeta \tag{2.4}$$

$$m_a^{rot} = \int_{\square} N_a^2 \, \rho_o \bar{J}_o \, d\square \tag{2.5}$$

(ρ_o is assumed to be independent of ζ).

2. For elements other than heterosis:

$$m_a^{disp} = m_a^{rot} \, , \qquad a = 1,2,\ldots,n_{en} \tag{2.6}$$

For heterosis.

$$m_a^{disp} = \int_{\square} (N_a + N_a^{ser}(0,0)N_9)^2 \, \rho_o \bar{J}_o d\square \, , \qquad a = 1,2,\ldots,8 \tag{2.7}$$

$$m_9^{disp} = 0 \tag{2.8}$$

(N_a^{ser} is the 8-node serendipity shape function, see e.g., [34].)

3. $$V = \int_{\square} \bar{j}_o \, d\square \quad , \quad M = \int_{\square} \rho_o \bar{j}_o \, d\square \tag{2.9}$$

4. Normalization

$$\tilde{M}^{rot} = \sum_{a=1}^{n_{en}} m_a^{rot} \tag{2.10}$$

$$\tilde{M}^{disp} = \sum_{a=1}^{n_{en}} m_a^{disp} \tag{2.11}$$

$$m_a^{rot} \leftarrow (M/\tilde{M}^{rot}) \, m_a^{rot} \tag{2.12}$$

$$m_a^{disp} \leftarrow (M/\tilde{M}^{disp}) \, m_a^{disp} \tag{2.13}$$

5. Adjustment to rotational inertia:

$$\langle z_a \rangle = (z_a^+ + z_a^-) / 2 \tag{2.14}$$

$$[z_a] = z_a^+ - z_a^- \tag{2.15}$$

$$\alpha_a = \int_{-1}^{+1} z_a^2 \, d\zeta / 2 = \langle z_a \rangle^2 + \frac{1}{12} [z_a]^2 \tag{2.16}$$

$$h = \sum_{a=1}^{n_{en}} [z_a] / n_{en} \quad , \quad A = V/h \tag{2.17}$$

$$\alpha_a = \max\{\alpha_a, A/8\} \tag{2.18}$$

$$m_a^{rot} = m_a^{rot} \alpha_a \qquad \text{(no sum)} \tag{2.19}$$

Remarks

a) In the axisymmetric case, j_o needs to be replaced throughout by $r j_o$, where r is the radial coordinate.

b) The rationale behind the lumped mass matrices has been discussed more fully in Hughes et al. [17]. The original idea is due to Key and Beisinger [25]. Although ad hoc in nature, this technique can dramatically increase the allowable time step in explicit transient calculations without adverse effect on accuracy (see e.g., [11]).

3. Transient Algorithm

The transient algorithm is basically the implicit-explicit predictor/multi-corrector scheme which has been described previously (Hughes et al. [12,14,18,19,22,23]). However, a slight reorganization is required to appropriately treat rotational degrees-of-freedom. Let i be the iteration counter; n be the time step number; Δt be the time step; $\underset{\sim}{d}_n, \underset{\sim}{v}_n$, and $\underset{\sim}{a}_n$ be the (generalized) displacement, velocity, and acceleration vectors at step n, respectively; and define the Newmark "predictors" by

$$\tilde{\underset{\sim}{d}}_{n+1} = \underset{\sim}{d}_n + \Delta t \underset{\sim}{v}_n + \frac{1}{2}(1-2\beta)\,\Delta t^2\,\underset{\sim}{a}_n \tag{3.1}$$

$$\tilde{\underset{\sim}{v}}_{n+1} = \underset{\sim}{v}_n + (1-\gamma)\,\Delta t\,\underset{\sim}{a}_n \tag{3.2}$$

where β and γ are the Newmark parameters. The global effective mass matrix is defined by

$$\underset{\sim}{M}^* = \begin{cases} \underset{\sim}{M} + \gamma \Delta t\,\underset{\sim}{C} + \beta \Delta t^2 \underset{\sim}{K} & \text{implicit} \\ \underset{\sim}{M} & \text{explicit} \end{cases} \tag{3.3}$$

where $\underset{\sim}{C}$ and $\underset{\sim}{K}$ are the global tangent damping and stiffness matrices, respectively. Various "mesh partitions" may also be employed (see Hughes et al. [14,22] for further details). In each time step the following calculations need be performed:

1. Set $i = 0$.
2. Predictor phase:

$$\underset{\sim}{d}_{n+1}^{(o)} = \tilde{\underset{\sim}{d}}_{n+1} \tag{3.4}$$

$$\underset{\sim}{v}_{n+1}^{(o)} = \tilde{\underset{\sim}{v}}_{n+1} \tag{3.5}$$

$$\underset{\sim}{a}_{n+1}^{(o)} = \underset{\sim}{0} \tag{3.6}$$

3. Equation solution:

$$\underset{\sim}{M}^{*}\Delta\underset{\sim}{a} = \underset{\sim}{R}^{(i)} \qquad \text{(residual or out-of-balance force)} \tag{3.7}$$

4. In geometrically linear analysis go to step 5. In geometrically nonlinear analysis go to step 6.

5. Corrector phase for geometrically linear analysis:

$$\underset{\sim}{a}_{n+1}^{(i+1)} = \underset{\sim}{a}_{n+1}^{(i)} + \Delta\underset{\sim}{a} \tag{3.8}$$

$$\underset{\sim}{v}_{n+1}^{(i+1)} = \tilde{\underset{\sim}{v}}_{n+1} + \gamma\Delta t\, \underset{\sim}{a}_{n+1}^{(i+1)} \tag{3.9}$$

$$\underset{\sim}{d}_{n+1}^{(i+1)} = \tilde{\underset{\sim}{d}}_{n+1} + \beta\Delta t^{2}\, \underset{\sim}{a}_{n+1}^{(i+1)} \tag{3.10}$$

If further iterations are required, increment i and go to step 3. Otherwise, increment n and go to step 1.

6. Corrector phase for geometrically nonlinear analysis: Employ (3.8) and (3.10) to obtain $\underset{\sim}{d}_{n+1}^{(i+1)}$. The rotational components of $\underset{\sim}{d}_{n+1}^{(i+1)}$ are, in this case, viewed as tentative, since they will not generally maintain the fiber inextensibility condition. A radial projection procedure is used to appropriately redefine these components (see Part I, eqs. 2.21-2.23 for details). The corrected $\underset{\sim}{d}_{n+1}^{(i+1)}$ is then used to calculate $\underset{\sim}{a}_{n+1}^{(i+1)}$ via (3.10). With this in hand, (3.9) may be employed to define $\underset{\sim}{v}_{n+1}^{(i+1)}$. At this point, if further iterations are required, increment i and go to step 3. Otherwise, increment n and go to step 1.

Remark

For static analysis purposes, the algorithm may be reduced to a standard Newton-Raphson iterative procedure by setting $\gamma = 0$, $\beta = \Delta t^{2}$, $\tilde{\underset{\sim}{d}}_{n+1} = \underset{\sim}{d}_{n}$, and ignoring all equations in which the velocity appears, i.e., (3.2), (3.5), and (3.9).

4. Numerical Examples

The calculations described herein were performed on the California Institute of Technology IBM 3032 computer in double precision (64 bits/floating point word). In three-dimensional situations 4-node quadrilaterals were employed and in two-dimensional cases 2-node elements were used. These elements are described in detail in Parts I and II, respectively. In cases in which the material was completely elastic, we used 2-point Gaussian quadrature through the thickness ("fiber integration"). This has always been sufficient in our experience. However, in elastic-plastic situations we found it necessary to use 4-point Gaussian quadrature through the thickness. In the calculations employing 4-node quadrilaterals, we used both selective (S1) and uniform (U1) reduced Gaussian rules for the lamina integration (see Part I for further details). For the two-node element we used only the 1-point Gaussian rule in the lamina direction (U1). The plasticity theory employed was that due to Krieg and Key [27] which allows any arbitrary linear combination of isotropic and kinematic hardening. A plane stress radial return implementation was employed [26]. In all cases the corotational hypothesis [1-3,5,7-9,33] was adopted in the context of the constitutive algorithm presented in Part I. Newmark parameters were taken to be $\beta = 1/4$ and $\gamma = 1/2$ throughout.

4.1 Transient elastic-plastic response of a simply-supported plate

This problem was solved previously by Liu and Lin [28] who used small deflection theory and ignored membrane effects. In our calculations we assumed full geometric as well as material nonlinearity. One would anticipate some stiffening in this case due to the tensile stresses developed as the plate deforms. Since the center displacement is of the order of half the plate thickness, one would expect small, but not insignificant differences. As may be seen from the results of Figure 1, the expected behavior is produced by our calculations which tend to be slightly stiffer than Liu and Lin's with respect to peak response.

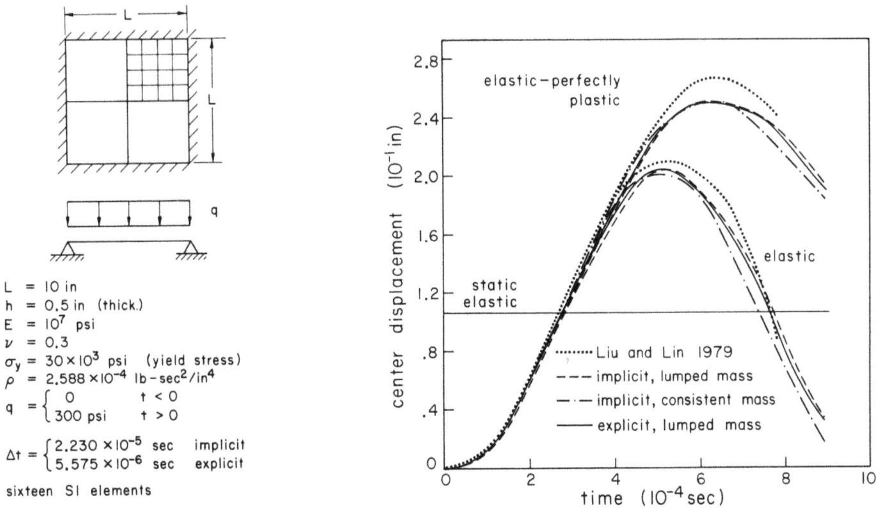

Fig 1 Transient response of a simply-supported plate

This is more pronounced in the elastic perfectly plastic case than in the purely elastic case as may be seen. This is also reasonable, since larger membrane effects would tend to develop due to the smaller bending resistance in plasticity.

Comparisons between lumped and consistent mass matrices, and implicit and explicit algorithms tend to be quite good. Despite the fact that the explicit time step was taken to be one-fourth the implicit, an economy of approximately a factor of seven was noted for the explicit calculations in the elastic-plastic case. In the purely elastic case this was reduced to a factor of four, favoring the explicit technique.

4.2 <u>Impulsively loaded elastic-plastic strip</u>

Experimental results for this problem were given by Balmer and Witmer [6], and results of a finite element calculation were presented by Belytschko and Marchertos [10] who used an elastic perfectly plastic model with artificial viscosity. In our calculations we experimented with different types of hardening, but did not include any viscous effects.

Another difference between the Belytschko-Marchertos calculation and ours was transverse shear effects which were not included in theirs, but were included in both the kinematics and constitution of ours. A comparison of results is presented in Figure 2. The displacement is quite large, being

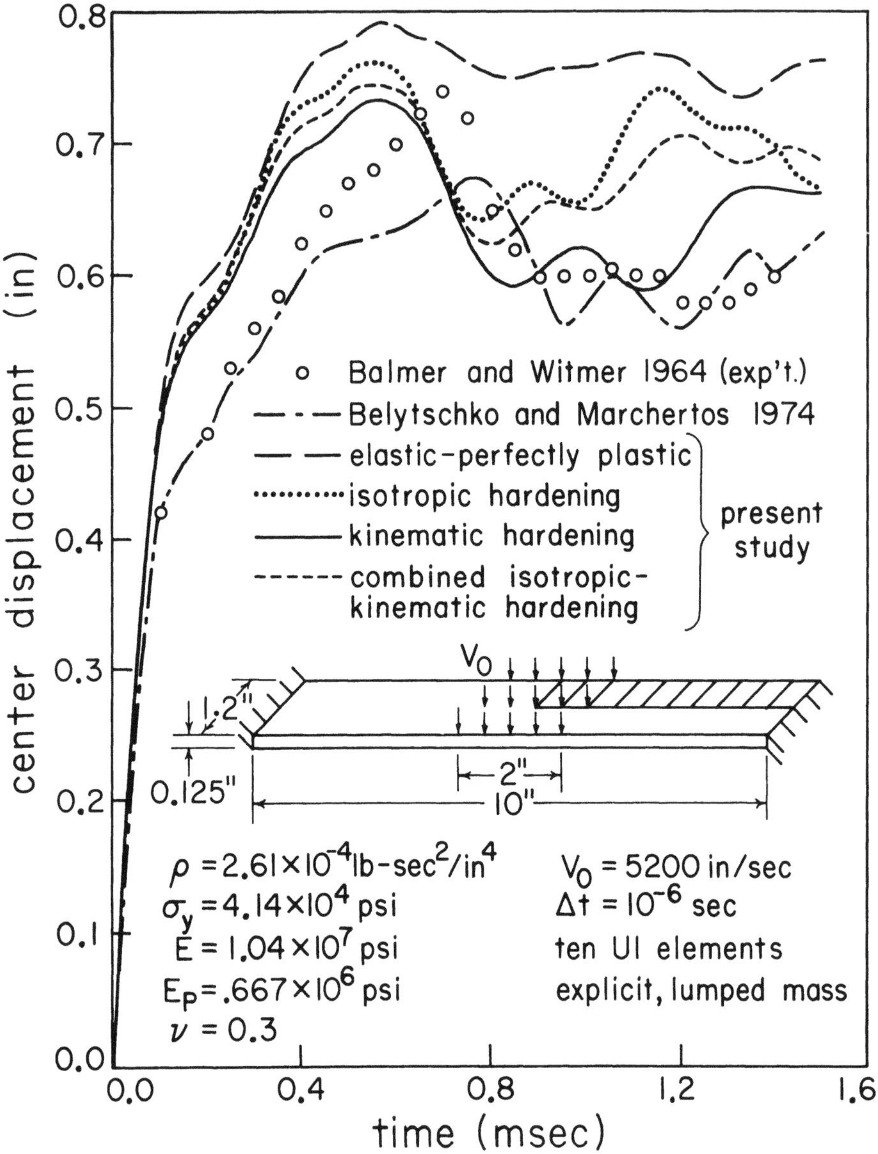

Fig. 2. Impulsively loaded elastic-plastic strip.

approximately six times the strip thickness. Our peak responses tend to be in good agreement with the experimental value, although they tend to occur at somewhat earlier times. The closest agreement we are able to get is attained with purely kinematic hardening, as may be seen. Considering the ambiguities in material modeling, we feel the correlation is quite reasonable.

4.3 Static axisymmetric buckling of elastic cylindrical shells

Three types of boundary conditions were employed in this study:

a) bottom fixed--top free;

b) bottom fixed--top fixed with respect to horizontal displacement and rotation; and

c) bottom simply-supported--top fixed with respect to horizontal displacement.

Load versus displacement data and deformed profiles are shown in Figure 3 up to the buckling loads at which time the calculations were terminated. Comparison is made with analytical results based upon eigenvalue analysis (see e.g., [32]).

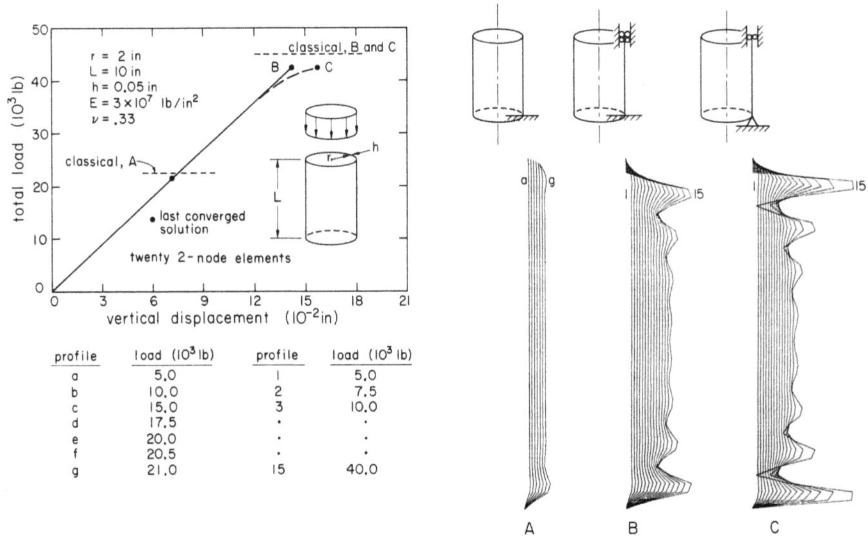

profile	load (10^3 lb)	profile	load (10^3 lb)
a	5.0	1	5.0
b	10.0	2	7.5
c	15.0	3	10.0
d	17.5	·	·
e	20.0	·	·
f	20.5	·	·
g	21.0	15	40.0

Fig 3 Static axisymmetric buckling of elastic cylindrical shells

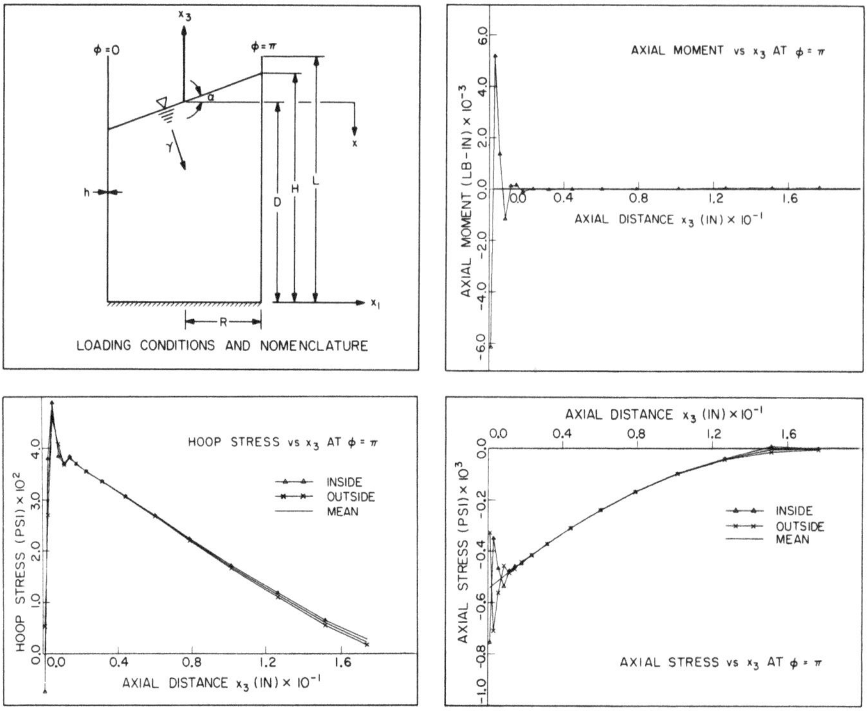

Fig 4. Tilted cylindrical shell subjected to internal hydrostatic loading. Stress results in the linear regime.

4.4 Cylindrical shell containing a fluid

An analysis was performed of a tilted cylindrical shell subjected to internal hydrostatic loading. The problem statement is shown in Figure 4 along with stress results prior to any significant nonlinear phenomena. Data employed in the analysis are given as follows:

$E = 7.35 \times 10^5$ lb/in^2 (Young's modulus)

$\nu = 0.3$ (Poisson's ratio)

$D = 16$ in , $L = 20$ in

$R = 4$ in , $h = 0.01$ in

$\alpha = 30°$, $\gamma = 0.0361$ lb/in^3 (specific gravity)

The effective specific gravity, γ_{eff}, was defined to be

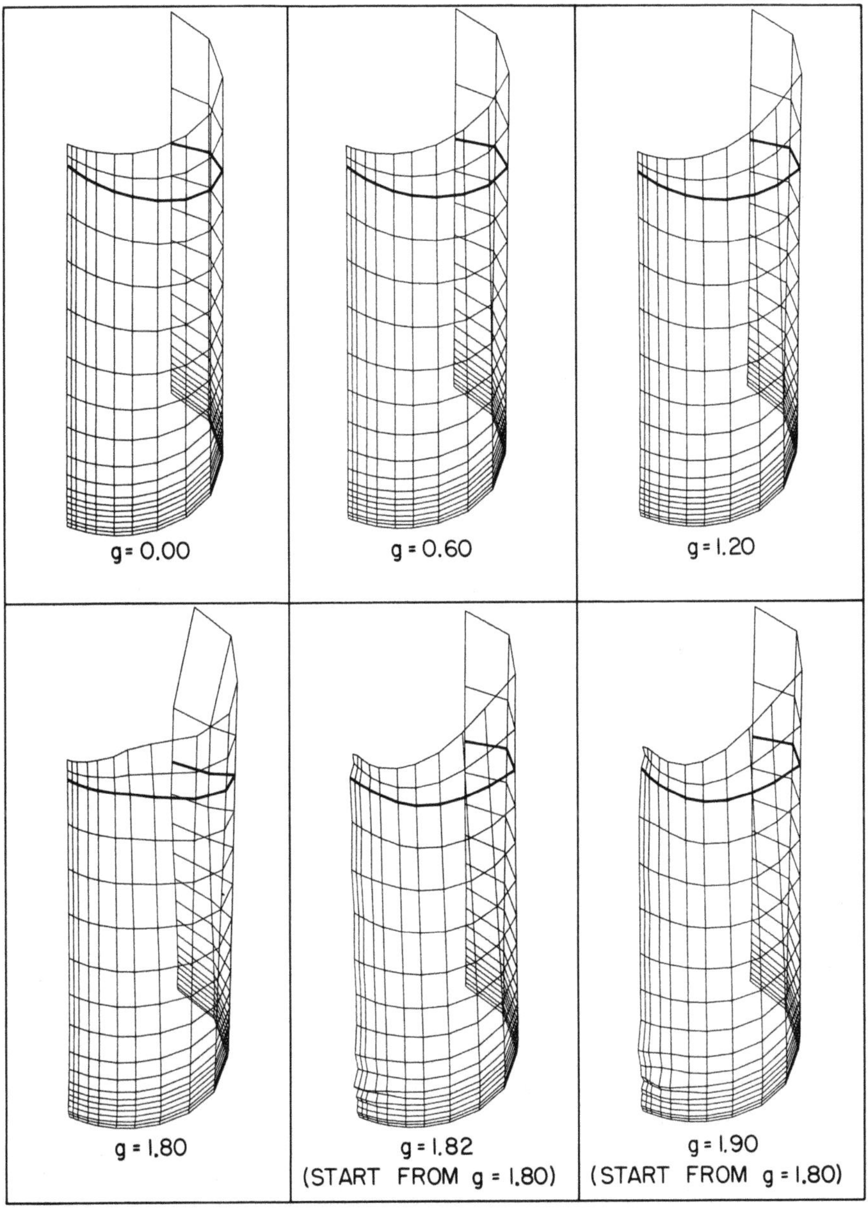

Fig 5. Deformed configurations for the tilted cylindrical shell.

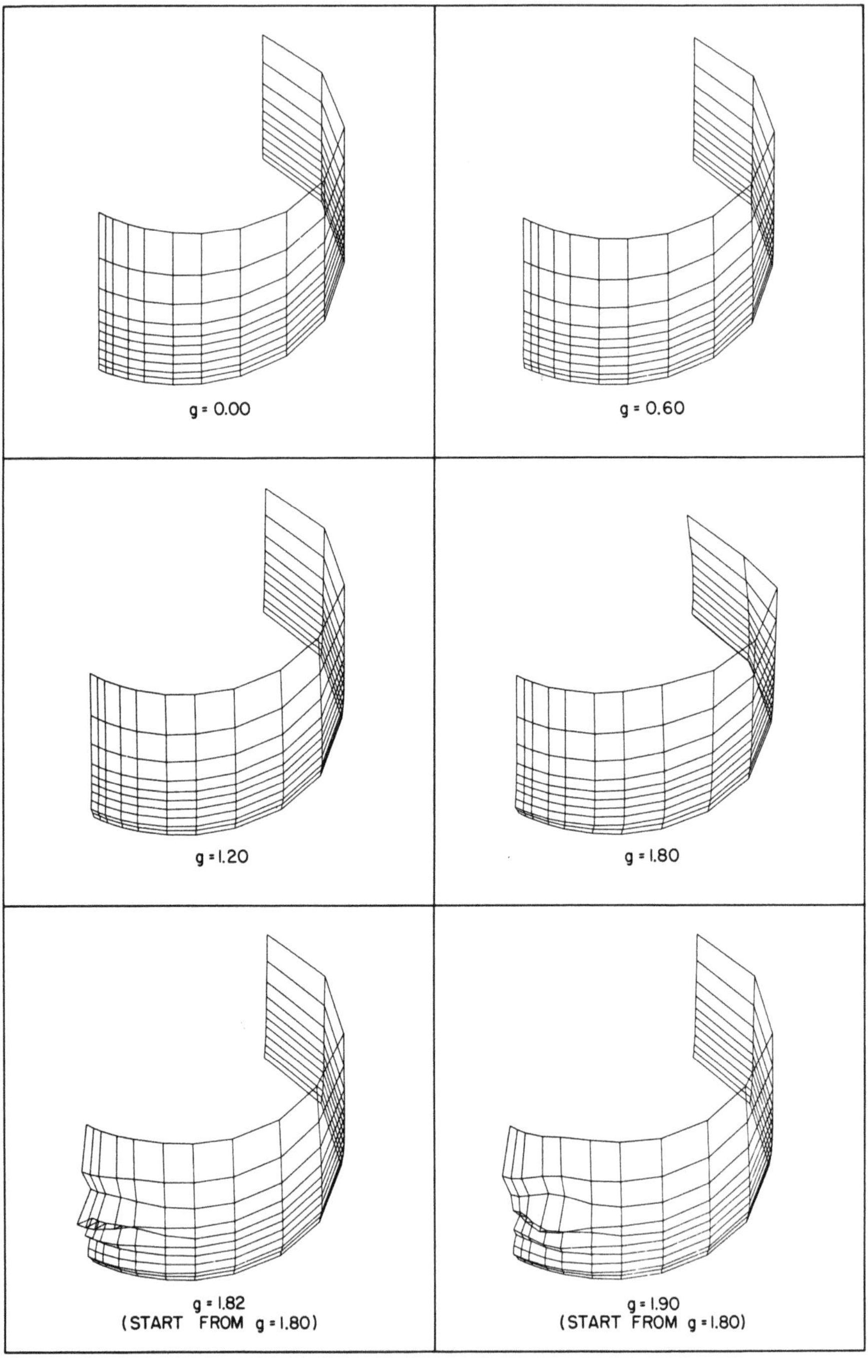

Fig. 6. Detail of lower regions for the tilted cylindrical shell.

$\gamma_{eff} = \gamma g$. The value of g used in obtaining the stresses shown in Figure 4 was 1.8. Comparison with a membrane theory solution [31] confirmed the correctness of the numerical results outside of the boundary layer at the fixed end. Buckling calculations were performed by incrementing g. Results are shown in Figures 5 and 6. As may be seen, the initial bifurcation occurred in attempting to increment g beyond 1.8. This corresponds to a maximum axial membrane stress of 547.53 psi at $x = L$, $\phi = \pi$, which is approximately half the classical value. Although this result was not anticipated, it bears a great deal of similarity to results obtained by Argyris and Dunne [4] in their penetrating study of a compressed cylindrical panel. They too found an initial bifurcation of approximately half classical. When they pursued the analysis into the post-buckled regime, a drop in load of only 3% was noted at which time the load was able to be increased again to a value in the vicinity of classical. At this point a second, and more significant, bifurcation ensued, and this was followed by several more bifurcations. An important point in considering the results of Argyris and Dunne was that a preliminary eigenvalue analysis, employing the same mesh used in the finite deformation case, yielded a buckling load approximating the classical value on the high side. This lends credence to the analysis, but a full understanding of the initial bifurcation in this case, and in ours also, does not yet seem to exist. The results shown were obtained using S1 elements. Similar results were also obtained with heterosis elements (see Part I).

5. Conclusions

In this paper we have presented dynamical aspects of a general nonlinear finite element shell capability. This work is a sequel to statical aspects which have been reported upon in [20,21]. In particular, we have described the construction of mass matrices and presented details of the transient algorithms employed. Several sample problems of a dynamic and static nature have been included, which demonstrate the applicability of the methodology.

Acknowledgments

I would like to thank the following individuals and organizations for their interest and support of the research presented herein J. Crawford, Civil Engineering Laboratory, Port Hueneme, California, The Electric Power Research Institute, Palo Alto, California; and the National Science Foundation.

References

1 Argyris, J.H.: Recent advances in matrix methods of structural analysis. Progress in Aeronautical Sciences 4. Oxford· Pergamon Press, 1964.

2 Argyris, J.H.; Continua and discontinua. Proceedings of the Conference on Matrix Methods in Structural Mechanics. Ohio Wright-Patterson Air Force Base, Oct. 1965.

3 Argyris, J.H., Balmer, H., St. Doltsinis, J., Dunne, P.C., Haase, M.; Kleiber, M.; Malejannakis, G.A.; Mlejnek, H.-P., Müller, M., Scharpf, D.W., Finite element method--the natural approach. Computer Methods in Applied Mechanics and Engineering 17/18 (1979) 1-106.

4 Argyris, J.H.; Dunne, P.C.. Post-buckling, finite element analysis of circular cylinders under end load. Report 224, Institut für Statik und Dynamik der Luft und Raumfahrtkonstruktionen, University of Stuttgart, Germany, 1977.

5 Argyris, J.H., Dunne, P.C., Malejannakis, G.A., Scharpf, D.W.: On large displacement—small strain analysis of structures with rotational degrees of freedom. Computer Methods in Applied Mechanics and Engineering 14 (1978) 401-451, 15 (1978) 99-135.

6 Balmer, H.A., Witmer, E.A.: Theoretical-experimental correlation of large dynamic and permanent deformation of impulsively loaded simple structures. Report FDL-TDR-64-108, Air Force Flight Dynamics Laboratory, Wright-Patterson Air Force Base, Ohio, July 1964.

7 Belytschko, T., Glaum, L.: Applications of higher-order corotational formulations for nonlinear finite element analysis. Computers and Structures 10 (1979) 175-182.

8 Belytschko, T.; Hsieh, B.J., Nonlinear transient finite element analysis with convected coordinates. International Journal for Numerical Methods in Engineering 7 (1973) 255-271.

9 Belytschko, T., Schwer, L., Klein, M.J.: Large displacement transient analysis of space frames. International Journal for Numerical Methods in Engineering 11 (1977) 64-84.

10 Belytschko, T., Marchertas, A.H.: Nonlinear finite element method for plates and its application to dynamic response of reactor fuel subassemblies. Journal of Pressure Vessel Technology (1974) 251-257.

11 Hinton, E., Pica, A.: Efficient transient dynamic plate bending analysis with Mindlin elements. Earthquake Engineering and Structural Dynamics, in press.

12 Hughes, T.J.R.: Recent developments in computer methods for structural analysis. Nuclear Engineering and Design, in press.

13 Hughes, T.J.R.. Generalization of selective integration procedures to anisotropic and nonlinear media. International Journal for Numerical Methods in Engineering, in press.

14 Hughes, T.J.R. Implicit-explicit techniques for symmetric and nonsymmetric systems. Proceedings of the Conference on Nonlinear Problems in Mechanics, Swansea, U.K., Sept. 1980.

15 Hughes, T.J.R., Cohen, M.. The "heterosis" finite element for plate bending. Computers and Structures 9 (1978) 445-450.

16 Hughes, T.J.R., Cohen, M.: The "heterosis" family of plate finite elements. Proceedings of the ASCE Electronic Computations Conference, St. Louis, Missouri, Aug. 1979.

17 Hughes, T.J.R., Cohen, M. Haroun, M.: Reduced and selective integration techniques in the finite element analysis of plates. Nuclear Engineering and Design 46 (1978) 203-222.

18 Hughes, T.J.R., Liu, W.K.: Implicit-explicit finite elements in transient analysis. stability theory. Journal of Applied Mechanics 45 (1978) 371-374.

19 Hughes, T.J.R., Liu, W.K.: Implicit-explicit finite elements in transient analysis; implementation and numerical examples. Journal of Applied Mechanics 45 (1978) 375-378.

20 Hughes, T.J.R., Liu, W.K.: Nonlinear finite element analysis of shells: Part I: Three-dimensional shells. Computer Methods in Applied Mechanics and Engineering, in press.

21 Hughes, T.J.R.; Liu, W.K.; Nonlinear finite element analysis of shells; Part II; Two-dimensional shells. Computer Methods in Applied Mechanics and Engineering, in review.

22 Hughes, T.J.R.; Pister, K.S., Taylor, R.L.: Implicit-explicit finite elements in nonlinear transient analysis. Computer Methods in Applied Mechanics and Engineering 17/18 (1979) 159-182.

23 Hughes, T.J.R.; Stephenson, R.A.; Convergence of implicit-explicit algorithms in nonlinear transient analysis. International Journal of Engineering Science, in press.

24 Hughes, T.J.R.; Taylor, R.L.; Kanoknukulchai, W.: A simple and efficient element for plate bending. International Journal for Numerical Methods in Engineering 11 (1977) 1529-1543.

25 Key, S.W.; Beisinger, Z.E.; The transient dynamic analysis of thin shells in the finite element method. Proceedings of the 3rd Conference on Matrix Methods in Structural Mechanics, Wright-Patterson Air Force Base, Ohio, 1971.

26 Krieg, R.D. Private communication, 1979.

27 Krieg, R.D.; Key, S.W.: Implementation of a time independent plasticity theory into structural computer programs, pp. 125-137 in Constitutive Equations in Viscoplasticity: Computation and Engineering Aspects (eds. Stricklin, J.A., Saczalaski, K.J.) AMD-Vol. 20. New York: ASME, 1976.

28 Liu, S.C.; Lin, T.H.: Elastic-plastic dynamic analysis of structures using known elastic solutions. Earthquake Engineering and Structural Dynamics 7 (1979) 147-159.

29 Malkus, D.S.; Hughes, T.J.R.: Mixed finite element methods--reduced and selective integration techniques; a unification of concepts. Computer Methods in Applied Mechanics and Engineering 15 (1978) 63-81.

30 Noor, A.K.; Peters, J.M.; Mixed models and reduced/selective integration displacement models for nonlinear analysis of curved beams. International Journal for Numerical Methods in Engineering, in press.

31 Shih, C.; Babcock, Jr., C.D.: Private communication, 1979.

32 Timoshenko, S.P.; Gere, J.M.: Theory of Elastic Stability. New York: McGraw-Hill, 1961.

33 Wempner, G.: Finite elements, finite rotations and small strains of flexible shells. International Journal of Solids and Structures 5 (1969) 117-153.

34 Zienkiewicz, O.C.: The Finite Element Method. London: McGraw-Hill, 1977.

Elastic Collapse Analysis of Shells Via Global-Local Approach

A.K. NOOR, J.M. PETERS
George Washington University Center at
NASA Langley Research Center, Hampton,USA

Summary

A two-stage global-local approach is presented for predicting the collapse behavior of shells. The first stage is that of spatial discretization wherein the shell is discretized by using finite elements (or finite differences) which cover the entire region of the shell. In the second stage the vector of unknown nodal parameters is expressed as a linear combination of small number of global functions (or basis vectors). A Rayleigh-Ritz (or Bubnov-Galerkin) technique is then used to approximate the nonlinear equations of the discretized shell by a reduced system of nonlinear algebraic equations. For the case of loading applied by means of axial end shortening, a scalar function is introduced which measures the degree of nonlinearity of the structure. Also, a quantitative measure for the error of the reduced system of equations is proposed. The effectiveness of the proposed technique for predicting the collapse behavior of shells is demonstrated by means of a numerical example of the elastic collapse of a cylindrical shell with a rectangular cutout.

Introduction

In spite of the significant advances made in numerical discretization procedures for nonlinear problems, the collapse analysis of complex shell structures, having thousands of degrees of freedom, is still not economically feasible on present-day computers. Hence, increasing interest has been shown in the application of a two-stage, global-local approach to these problems (see, for example, Refs. 1 to 7).

The first stage in this hybrid approach is that of *spatial discretization* wherein the shell is discretized by using finite elements (or finite differences) with local support for the approximation functions of the individual elements. In the second stage a *significant reduction in the number of degrees of freedom* is achieved by expressing the vector of unknown nodal parameters as a linear combination of a small number of global functions (or basis vectors). A Rayleigh-Ritz technique is then used to approximate the finite element equations of the discretized shell by a reduced system of nonlinear algebraic equations.

As would be expected, the effectiveness of the approach depends, to a great extent, on the proper choice of the global functions (or reduced basis vectors). Various choices for global functions were proposed in the literature. These include linear bifurcation buckling modes (Refs. 3 and 4); linear solution and corrections to it (Ref. 5); and nonlinear solution and its various order path derivatives (Refs. 6 to 8). Numerical examples presented in Refs. 6 and 7 demonstrated that the latter choice of a nonlinear solution, and its various order path derivatives (which are commonly used in the static perturbation technique) as global functions results in an effective global-local approach. However, in all the cited references, the structures were subjected to externally applied loads. The case of loading applied by means of axial end shortening, as would occur in a laboratory compression test was not considered.

The present study focuses on the application of the global-local approach to the elastic collapse analysis of shells subjected to prescribed edge displacements (or end shortening). Specifically, the objectives of this paper are: a) to outline the modifications in the global-local approach of Refs. 6 and 7 which are required in the case of prescribed (nonzero) edge displacements, b) to demonstrate the effectiveness of the proposed technique by means of a numerical example, and c) to give some insight into why and when the reduced basis technique works.

The analytical formulation is based on a form of the geometrically nonlinear shallow shell theory with the effects of transverse shear deformation included. A displacement formulation is used with the fundamental unknowns consisting of the three displacement components and the two rotations at each point of the middle surface of the shell. The numerical solutions were obtained by using 16-node quadrilateral elements with bicubic Lagrangian interpolation functions for each of the displacement and rotation components. The computational algorithm presented herein is a modification and an extension of that presented in Refs. 6 and 7.

Mathematical Formulation

Stage I - Spatial Discretization:

A total Lagrangian formulation is used and the shell is discretized by using a displacement finite element model. The governing finite element equations, for the case of a loading applied by means of axial end shortening, can be cast in the following compact form:

$$[K]\{X\} + \{G(X)\} - \{Q\} = 0 \tag{1}$$

where [K] is the nxn linear global stiffness matrix; n is the total number of displacement degrees of freedom; {X} is the vector of nodal displacements; {G(X)} is the vector of nonlinear terms which is cubic in {X} (see Ref. 7); and {Q} is the vector of constraint and applied forces.

It is convenient to partition the vector of nodal displacements as follows:

$$\{X\} = \begin{Bmatrix} X_f \\ X_c \end{Bmatrix} \tag{2}$$

where subscripts f and c refer to free and constrained (prescribed nonzero) displacements, respectively. The constrained zero displacements and their associated equations are not included in Eqs. 1. For simplicity, the constrained nonzero displacements are assumed to be proportional to a single parameter λ, i.e.

$$\{X_c\} = \lambda\{Z\} \tag{3}$$

As the edge displacements are incremented, the value of the parameter λ changes, but the components {Z} remain constant.

Equations 1 can be conveniently partitioned into two sets of matrix equations as follows:

$$\begin{bmatrix} K_{ff} & K_{fc} \\ K_{cf} & K_{cc} \end{bmatrix} \begin{Bmatrix} X_f \\ X_c \end{Bmatrix} + \begin{Bmatrix} G_1(X_f, X_c) \\ G_2(X_f, X_c) \end{Bmatrix} - \begin{Bmatrix} Q_f \\ Q_c \end{Bmatrix} = 0 \qquad (4)$$

In the absence of externally applied loading (case of prescribed displacements only), $\{Q_f\}=0$ and $\{Q_c\}$ equals the vector of constraint forces associated with the prescribed nonzero displacements $\{X_c\}$. The first set of Eqs. 4 can be used to determine $\{X_f\}$ and the second set is then used to evaluate the constraint forces $\{Q_c\}$.

Stage II - Reduction of the Number of Degrees of Freedom:
In the second stage of the proposed technique, the vector of nodal displacements $\{X\}$ of the discretized structure is approximated over a range of values of λ, by a linear combination of $\{X\}$ corresponding to a particular value of λ and a number of its path derivatives (derivatives of $\{X\}$ with respect to the path parameter λ) at the same value of λ. This can be expressed as follows:

$$\begin{Bmatrix} X_f \\ X_c \end{Bmatrix} = [\Gamma]\{\psi\} \qquad (5)$$

where

$$[\Gamma]_{n,r} = \left[\begin{Bmatrix} 0 \\ Z \end{Bmatrix} \begin{Bmatrix} X_f \\ 0 \end{Bmatrix} \frac{\partial}{\partial\lambda} \begin{Bmatrix} X_f \\ 0 \end{Bmatrix} \cdots \frac{\partial^{r-2}}{\partial\lambda^{r-2}} \begin{Bmatrix} X_f \\ 0 \end{Bmatrix} \right] \qquad (6)$$

and $\{\psi\}$ is a vector of reduced unknowns (with r components only) which are functions of the path parameter λ. Note that r is considerably smaller than the total number of degrees of freedom n. The columns of the matrix $[\Gamma]$ are the global approximation functions.

The Rayleigh-Ritz technique is now applied to replace the n equations obtained in the first stage, Eqs. 4, by the following reduced system of r nonlinear equations:

$$[\tilde{K}]\{\psi\} + \{\tilde{G}(\psi)\} - \{\tilde{Q}\} = 0 \qquad (7)$$

subject to the condition $\psi_1 = \lambda$.

The coefficients of the reduced equations, Eqs. 7, are given by:

$$[\tilde{K}] = [\Gamma]^T[K][\Gamma] \tag{8}$$

$$\{\tilde{G}(\psi)\} = [\Gamma]^T\{G(\psi)\} \tag{9}$$

and

$$\{\tilde{Q}\} = [\Gamma]^T \begin{Bmatrix} 0 \\ Q_c \end{Bmatrix} \tag{10}$$

where the superscript T denotes transposition and $\{G(\psi)\}$ is obtained from $\{G(X)\}$ by replacing $\{X\}$ by its expression in terms of $\{\psi\}$, Eqs. 5. Note the $\{\tilde{Q}\}$ has only one nonzero component, namely, $\tilde{Q}_1$.

For a given value of λ, the last r-1 equations of Eqs. 7 can be solved for $\psi_2, \psi_3, \ldots \psi_r$. The first equation is then used to evaluate $\tilde{Q}_1$.

The global functions in Eqs. 6 are obtained by successive differentiation of the finite element equations of the discretized shell, Eqs. 4, and solving the resulting system of linear algebraic equations. Also, as shown in Refs. 6 and 7, the evaluation of all global approximation functions requires only one matrix factorization.

The criterion for selecting the number of basis vectors proposed in Ref. 6 was adopted in the present study. This criterion is based on monitoring the condition number of the Gram matrix of the global functions, and terminating the generation of these vectors when the condition number exceeds a prescribed value. Also, upper and lower limits for the number of global functions were chosen to be seven and two, respectively.

The range of λ for which Eqs. 7 provides an acceptable approximation for the original discrete system, Eqs. 4, depends on the degree of nonlinearity of the problem. This range can be identified by monitoring the accuracy of the solutions of the reduced system, Eqs. 7. When the error exceeds a prescribed tolerance, the vector $\{X\}$ generated by the reduced system of equations is used as a predictor and the Newton-Raphson iterative technique is used in conjunction with the original equations, Eqs. 4, to obtain a corrected (improved) solution. Then a new (updated) set of global functions is generated.

Computational Procedure

The computational procedure outlined in Ref. 7 for the efficient evaluation of the global functions, generation of the reduced system of equations, automatic selection of the displacement increments, sensing and controlling the errors of the reduced equations is adopted in the present study. However, the scalar S (current stiffness parameter) used to characterize the nonlinear response (see Ref. 9) and the error norm used in Ref. 7 are not applicable to the case of prescribed edge displacements, and therefore, are modified as described subsequently.

Current Stiffness Parameter:

In the case of loading applied by means of prescribed edge displacement, the following scalar S is introduced for characterizing the nonlinear response:

$$S_{(i)} = \frac{1}{\left\{\frac{\partial Q_c}{\partial \lambda}\right\}^T_{(i)} \{Z\}} \Big/ S_o \quad \text{for the full system} \tag{11}$$

$$= \frac{1}{\frac{\partial \tilde{Q}_1}{\partial \lambda}} \Big/ S_o \quad \text{for the reduced system} \tag{12}$$

where

$$\left\{\frac{\partial Q_c}{\partial \lambda}\right\} = \left[[K_{cf}] + \left[\frac{\partial G_{2_i}}{\partial X_{f_j}}\right]\right]\left\{\frac{\partial X_f}{\partial \lambda}\right\} + \left[[K_{cc}] + \left[\frac{\partial G_{2_i}}{\partial X_{c_j}}\right]\right]\{Z\} \tag{13}$$

$$\frac{\partial \tilde{Q}_1}{\partial \lambda} = \sum_{j=1}^{r} \left[\tilde{K}_{1j} + \frac{\partial \tilde{G}_1}{\partial \psi_j}\right] \frac{\partial \psi_j}{\partial \lambda} \tag{14}$$

and

$$S_o = \frac{1}{\left\{\frac{\partial Q_c}{\partial \lambda}\right\}_o^T \{Z\}} \tag{15}$$

The subscript (i) in Eqs. 11 and 12 refers to point i of the solution path and the subscript o refers to the point at $\lambda=0$.

The current stiffness parameter defined in Eqs. 11 is similar to that introduced in Ref. 9 for the case of externally applied loads. It has the major advantage of being easily computed from the reduced system of equations. The selection of displacement increments and the frequency of error sensing are then related to changes in this parameter (see Ref. 6).

The parameter $S_{(i)}$ provides a global measure for the stiffness of the structure at point i of the solution path. It has an initial value of 1.0, increases when the structure stiffens and decreases when the structure softens. For stable equilibrium paths, S is positive, for unstable paths, S is negative, and at limit or collapse points, S is zero.

Error Norm:
To check the accuracy of the solution obtained by the reduced system of equations, Eqs. 7, the following error norm is used:

$$e = \sqrt{\{R_f\}^T\{R_f\}/\{Q_c\}^T\{Q_c\}} \tag{16}$$

where

$$\{R_f\} = [K_{ff}]\{X_f\} + [K_{fc}]\{X_c\} + \{G_1(X_f,X_c)\} \tag{17}$$

If the error norm e is less than a prescribed tolerance, the solution is continued; otherwise a corrected (or improved) estimate of $\{X_f\}$ is obtained using Newton-Raphson technique in conjunction with Eqs. 4. Then a new set of global functions is generated using the technique outlined in Ref. 7.

Numerical Studies

To test and evaluate the effectiveness of the proposed global-local approach, a number of nonlinear shell problems were solved by this approach. Comparisons were made with solutions based on the full system of equations of the finite element model. Typical results are presented herein for the elastic collapse analysis of the cylindrical shell with rectangular cutout shown in Fig. 1. The problem is one of three problems used in Ref. 10 to assess the capability of various programs to analyze shell structures. The load is applied to the cylinder by means of a uniform axial end shortening which is increased incrementally until the cylinder collapses.

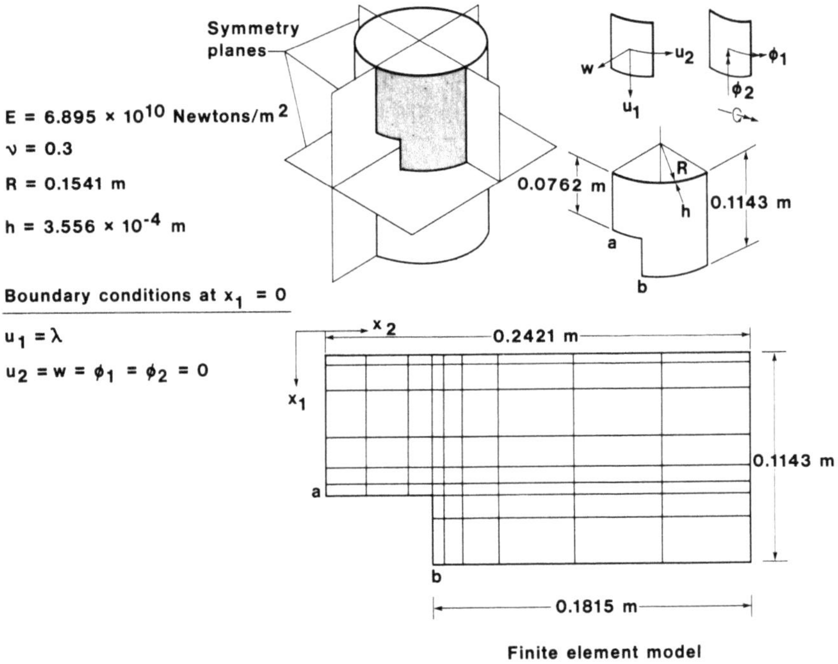

Figure 1.-Cylindrical shell with cutout and finite element model used in the present study.

Due to symmetry, only one octant of the shell was modeled using the grid of shear-flexible elements shown in Fig. 1. Bicubic Lagrangian interpolation functions were used to approximate each of the displacement and rotation components (a total of 2996 nonzero displacement degrees of freedom). Finite difference solutions to this problem using the STAGS

(STructural Analysis of General Shells) computer code are presented in Refs. 10 and 11. Finite element solutions for the same problem are presented in Ref. 12

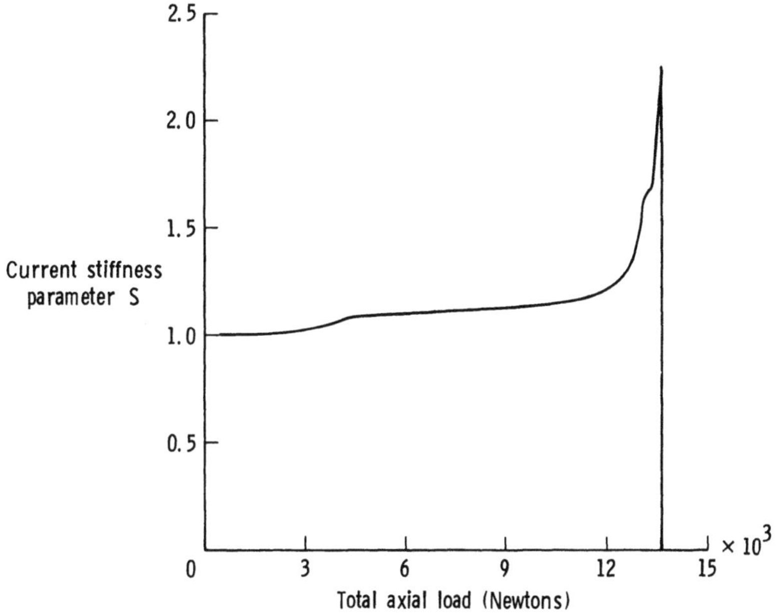

Figure 2.-Variation of current stiffness parameter with loading for the cylindrical shell with cutout shown in Fig. 1.

Figure 2 shows the variation of the current stiffness parameter S with loading. Figure 3 gives an indication of the accuracy of the normal displacement w, at the centers of two sides of the rectangular cutout, obtained by the global-local approach. Figure 4 gives an indication of the accuracy of the total strain energy and Figure 5 shows the error norms of the global-local approach with seven basis vectors at various load levels. Figure 6 shows contour plots of the normal displacements w at four different load levels, each normalized by dividing by w_{max} for that load level.

In the present study the path parameter λ was chosen to be equal to the axial edge displacements, and therefore, the components of the vector $\{Z\}$ are equal to unity. The global functions were first computed for the unloaded shell ($\lambda=0$, $\{X_f\}=0$). An error tolerance $e \leq 0.05$ was prescribed. The seven global functions were used to advance the solution to

$\lambda=3.048 \times 10^{-5}$ m. at which value the error norm was checked and was found to exceed the prescribed tolerance (see Fig. 5). New (updated) set of seven global functions were generated and used to advance the solution until $\lambda=9.398 \times 10^{-5}$ m. The predicted collapse load of the cylinder was 13.656×10^3 Newtons corresponding to $\lambda=9.70 \times 10^{-5}$ m. The collapse loads predicted in Refs. 11 and 12 are 10.008×10^3 and 12.792×10^3 Newtons, respectively.

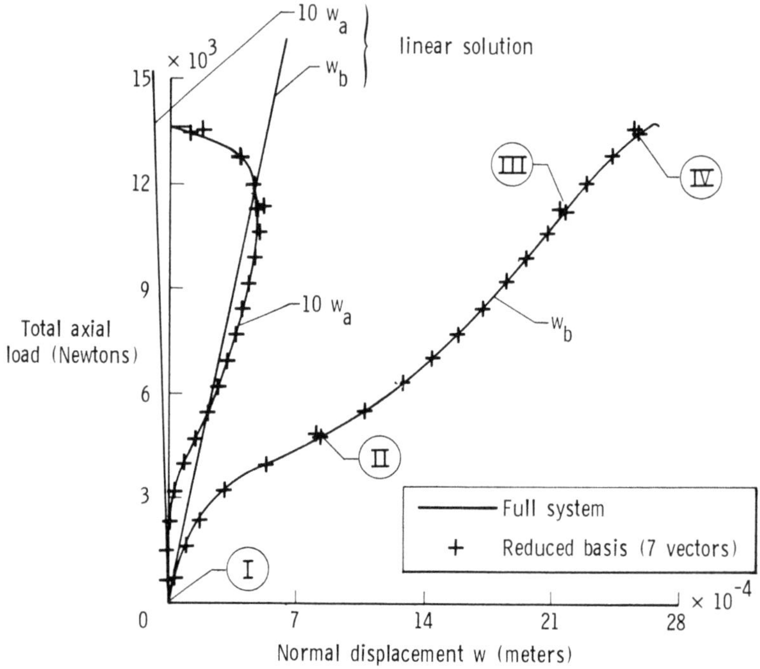

Figure 3.-Accuracy of normal displacements obtained by reduced basis technique at various load levels. Cylindrical shell with cutout shown in Fig. 1. Roman numerals indicate points of generating basis vectors.

The high accuracy of the normal displacements and strain energies obtained by the reduced system of equations is demonstrated in Figs. 3 and 4. At $\lambda=9.398 \times 10^{-5}$ m. the errors in the maximum normal displacement w_b and the total strain energy U obtained by using seven basis vectors were 0.53% and 0.023%, respectively.

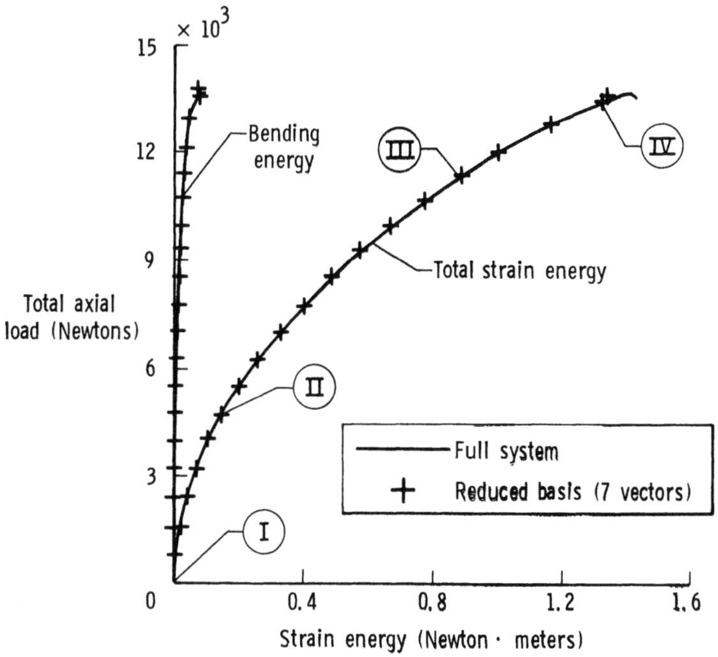

Figure 4.-Accuracy of strain energies obtained by reduced basis technique at various load levels. Cylindrical shell with cutout shown in Fig. 1. Roman numerals indicate points of generating basis vectors.

Higher accuracy of the reduced solutions can be achieved by backtracking the equilibrium path every time a new (updated) set of global functions is generated. This amounts to effectively reducing the error norm well below the prescribed tolerance. When this technique was used in the present problem, the maximum value of the error norm using seven functions reduced to less than 0.003 (see Fig. 5). The computational expense involved in the backtracking process was insignificant.

In summary, the generation of the whole solution path up to collapse of the cylinder involved a) generation of an initial set of global functions at $\lambda=0$ and b) updating the global functions three times.

The use of the global-local approach in this problem resulted in reducing the number of degrees of freedom by a factor of over 400 (from

2996 degrees of freedom for the original finite element model to seven degrees of freedom for the reduced system).

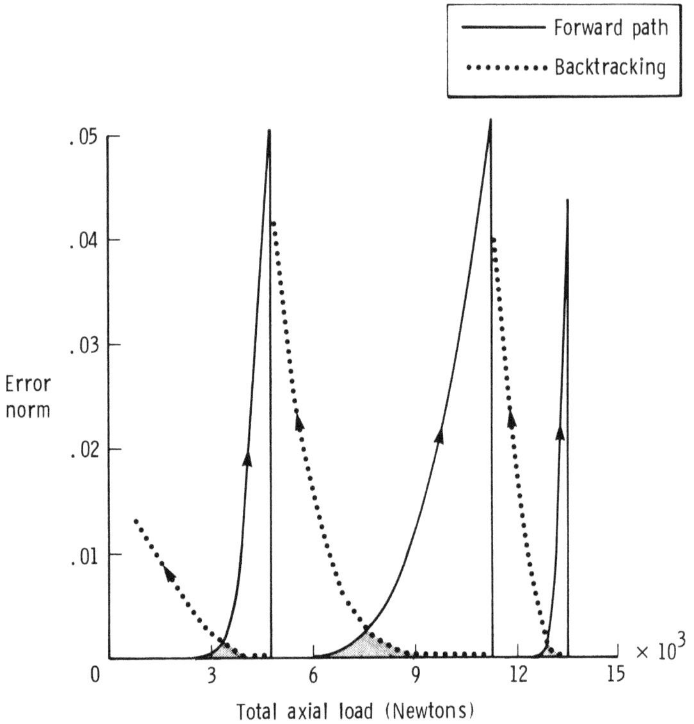

Figure 5.-Error norms of reduced basis technique (using 7 vectors) at various load levels. Cylindrical shell with cutout shown in Fig. 1.

<u>Comments on the Potential of the Proposed Approach</u>

The following comments concerning why and when the global-local approach works seem to be in order:

1. The success of the proposed global-local approach can be mainly attributed to the separation of spatial distribution of the response quantities at any load level from the variation of these quantities with loading. The form of the spatial distribution of the fundamental unknowns is given by the discrete finite element model in the first stage. The use of the path derivatives as global functions in the second stage permits the known information about the nonlinear response in the neighborhood of a point to be brought into the analysis.

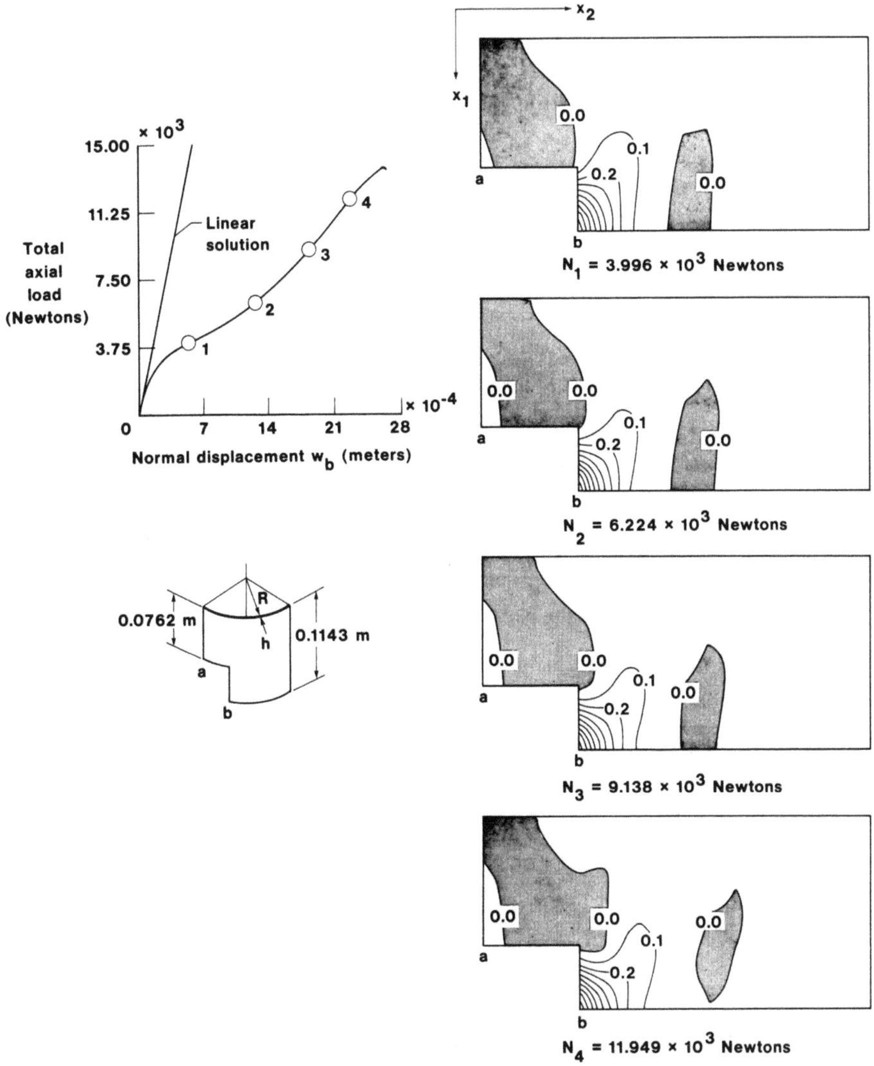

Figure 6.-Contour plots for the normal displacements w at various load levels (each normalized by dividing by w_{max} for that load level). Cylindrical shell with cutout (see Fig. 1).

In this manner substantially fewer degrees of freedom will be required to achieve a desired overall accuracy in comparison with that based on the finite element technique.

2. Numerical experiments have shown that the spatial distribution of stresses and displacements varies slowly

with the loading. This can be seen from the normalized contour plots of Fig. 6, wherein large changes in loading are associated with small changes in the contour plots. On the other hand, large changes in the contour plots require updating the global functions.

3. If the proposed global-local approach is contrasted with the static perturbation technique the following can be noted. In both methods the vector of unknown nodal parameters is approximated by a linear combination of a small number of path derivatives, Eqs. 5 and 6. However, the coefficients of the linear combination $\{\psi\}$ in the static perturbation technique are fixed and are equal to: $\lambda+\Delta\lambda$, 1, $\Delta\lambda$, $\frac{(\Delta\lambda)^2}{2!}$, ... $\frac{(\Delta\lambda)^{r-2}}{(r-2)!}$. By contrast, the coefficients $\{\psi\}$ in the proposed global-local approach are left as free parameters, and are determined by applying the Rayleigh-Ritz technique. Numerical experiments indicate that the use of free parameters leads to accurate solutions not only within the radius of convergence of the Taylor series but also well beyond it.

4. If both external and internal forces are conservative, the efficiency of the global-local approach can be increased by increasing the error tolerance and maintaining the accuracy of the reduced system of equations through backtracking the solution path every time a new (updated) set of basis vectors is generated.

5. In cases where rapid changes in the spatial distribution occur in a small region of the shell, the efficiency of the global-local approach can be enhanced by using this approach in conjunction with a substructuring technique (or a partitioning scheme). The global functions in the region of rapid change are updated more frequently than the global functions in other regions.

Concluding Remarks

A two-stage global-local approach is used to predict the collapse behavior of shells. The computational algorithm presented previously by the authors is modified to handle the case of loading applied by means of axial end shortening, as would occur in a laboratory compression test. A scalar function is introduced for measuring the degree of nonlinearity of the structure. Also, a quantitative measure for the error of the reduced system of equations is proposed.

The potential of the proposed technique for predicting the collapse behavior of shells is demonstrated by means of a numerical example of elastic collapse of an axially compressed cylindrical shell with a cutout, wherein reduction in the number of degrees of freedom by a factor of over 400 was achieved. The example also provides an insight into why and when the global-local approach works.

References

1 Besseling, J. F.: Nonlinear analysis of structures by the finite element method as a supplement to a linear analysis. Computer Methods in Applied Mechanics and Engineering 3 (1976) 173-194.

2 Besseling, J. F.: Post-buckling and nonlinear analysis by the finite element method as a supplement to a linear analysis. ZAMM 55 (1975) T3-16.

3 Nagy, D. A.: Modal representation of geometrically nonlinear behavior by the finite element method. Computers and Structures 10 (1979) 683-688.

4 Nagy, D. A.; König, M.: Geometrically nonlinear finite element behavior using buckling mode superposition. Computer Methods in Applied Mechanics and Engineering 19 (1979) 447-484.

5 Almroth, B. O.; Stern, P.; Brogan, F. A.: Automatic choice of global shape functions in structural analysis. AIAA Journal 16 (1978) 525-528.

6 Noor, A. K.; Peters, J. M.: Reduced basis technique for nonlinear analysis of structures. AIAA Journal 18 (1980) 455-462.

7 Noor, A. K.; Andersen, C. M.; Peters, J. M.: Global-local approach for nonlinear shell analysis. Proc. Seventh ASCE Conference on Electronic Computation, St. Louis, Mo., Aug. 6-8 (1979) 634-657.

8 Noor, A. K.; Peters, J. M.: Nonlinear analysis via global-local mixed finite element approach. To appear in the International Journal for Numerical Methods in Engineering.

9 Bergan, P. G.; Horrigmoe, G.; Kråkeland, B.; Søreide, T. H.: Solution techniques for nonlinear finite element problems. International Journal for Numerical Methods in Engineering 12 (1978) 1677-1696.

10 Hartung, R. F.; Ball, R. E.: A comparison of several computer solutions to three structural shell analysis problems. AFFDL-TR-73-15 (1973).

11 Almroth, B. O.; Holmes, A. M. C.: Buckling of shells with cutouts, experiments and analysis. International Journal of Solids and Structures 8 (1972) 1057-1071.

12 Argyris, J. H.; Hilpert, O.; Hindenlang, U.; Malejannakis, G. A.; Schelkle, E.: Flächentragwerke im Konstruktiven Ingenieurbau. ISD-Bericht Nr. 263, Universität Stuttgart (1979).

Large Spatial Deformations of Rods Using Generalized Variational Principles

W. WUNDERLICH, H. OBRECHT
Ruhr-Universität Bochum, Germany

Summary

General equations and corresponding incremental formulations for the analysis of rods undergoing large spatial deformations are presented. The derivation is based on a generalized three-dimensional variational principle with the increments of the Lagrangian stresses, the deformed position vector and the rotation as independent variables. "Engineering" strains and conjugate "Jaumann"-stresses are related by a semilinear material law. The principle allows assumptions for the stresses and for the kinematical variables simultaneously. This is used in the reduction to the one-dimensional description for general rods. Warping effects and arbitrary cross sections are considered in the fully spatial and geometrically nonlinear description. The resulting generalized principle in terms of seven stress resultants and conjugate displacements and rotations is given in its general and incremental forms. It could be used for a mixed or hybrid finite element approach. In the paper "exact" stiffness matrices are obtained via the integration of the local equations of the principle in the form of a system of first-order ordinary differential equations. Numerical results of sample problems are given.

1. Introduction

To evaluate the actual load carrying capacity of slender rods and of frame structures, nonlinear calculations which account for large deflections and moderate or large rotations are often necessary. This is particularly true if such structures may fail by buckling into complex modes (e.g. by interactive sidesway buckling) or if, as a result of initial imperfections the structure, or invidual members, show a pronounced nonlinear behavior. Then a relevant decrease in overall stiffness may occur before a "critical" load level has been reached. In this respect the effect of coupling between bending and torsion as well as the influence of arbitrary cross sections (thin walled, open, and closed) have to be taken into account to realistically model the actual behavior. Warping constraints and the cross-sectional shape sometimes play an important role. It is obvious that the proper formulation of this complex problem and its solution requires special attention.

The finite element approach to this subject has been treated in a number of papers (see e.g. the lists of references in [1,2,3]). It is mostly based on the incremental form of the principle of virtual work and a displacement formulation. Curved beams are sometimes approximated by straight elements, and the nonlinear coupling of bending and torsion as well as the influence of arbitrary cross-sections is rarely mentioned. In the approach of [2,4] to the treatment of large deflections and moderate rotations of spatial rods a generalized variational principle with Lagrangian stresses, displacements and rotations was used as the starting point, and the effect of warping was included. The derivation in this paper is similar in spirit, but embodies several important generalizations. It employs a more consistent formulation, and is also applicable to large rotation problems. The use of a generalized variational principle, with rotations as unknowns, as the basis for the reduction of the three-dimensional equations to the one-dimensional theory for rods facilitates the formulation significantly. Moreover, since in slender structures - like rods, plates, and shells - moderate and large rotations play an important role, it appears quite natural to include them in the description explicitly, rather than implicitly through the displacement gradient. This is accomplished here through the use of the polar decomposition theorem and a generalized variational principle of Hellinger-Reissner type which, for the continuum, was first given by Fraeijs de Veubeke [5] and has been recast in incremental form by Atluri and Murakawa [6]. In this paper the principle is modified further. The three-dimensional principle, written in terms of so-called "Jaumann" stresses and conjugate "engineering" strains, is reduced to the corresponding principle for rods. It is expressed in terms of components of the position vector of the deformed configuration, the angles of rotation about principal axes, and the corresponding (Lagrangian) components of the "Jaumann"-stress-resultants as fundamental unknowns. The general principle is given as well as its total (T.L.) and updated Lagrangian (U.L.) forms. The independent variables chosen seem to be the most general and consistent ones, and are also most appropriate for a formulation which applies to arbitrarily loaded general rods.

Due to its generalized character, the principle allows assumptions on the distribution of stresses over the cross-sections and, simultaneously, kinematic assumptions. In the latter the deformed position of points outside the axis are related to the corresponding position and rotation of the reference line in such a way that arbitrary cross-sections can be taken into account. For the warping effect a kinematic unknown and the corresponding bimoment are additional parameters in the formulation. A somewhat lengthy, but straightforward, procedure of reduction to the one-dimensional beam problem finally leads to a generalized variational principle with seven kinematic variables and the same number of stress resultants as unknowns. This principle could be taken as the basis of a usual mixed-type finite element discretization. For one-dimensional problems, however, a proven algorithm which is based on the integration of the corresponding Euler-equations, written in the form of a system of first-order ordinary differential equations [11] , is more advantageous, because it results in highly accurate element stiffness matrices. In this manner, the rapidly decaying functions which account for warping effects do not need special treatment. The initial displacement and initial stress matrices of the nonlinear problem are treated in the same way as the linear parts of the problem. Numerical calculations of sample problems have been performed to demonstrate the effect of combined bending and torsion

2. Basic general relationships (continuum)

2.1 Reference configurations

The configurations of a body undergoing deformations may be referred to the undeformed initial state with coordinates X^A and base vectors $\underline{G}_A$, to a current, deformed state with coordinates x^i and base vectors $\underline{g}_i$, or to a global cartesian coordinate system with unit vectors $\underline{e}_A$ (see Fig. 1). For the description of curved members it is useful to use convected, curvilinear coordinates, such that in all configurations a material particle is identified by the same set of values θ^1, θ^2, θ^3. The notation employed here follows that of [7,8].

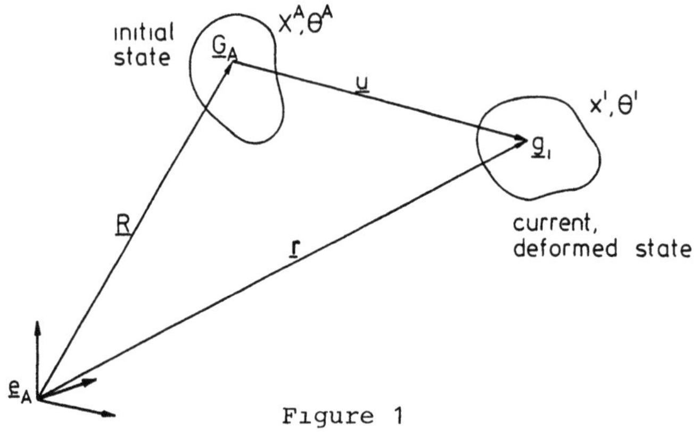

Figure 1

Denote the position vector of a particle and its neighborhood in the undeformed body by $\underline{R} = X^A \underline{e}_A$, and $d\underline{R} = dX^A \underline{e}_A$ while its location and neighborhood in the current, deformed state may be given by $\underline{r} = x^i \underline{e}_i$ and $d\underline{r} = dx^i \underline{e}_i$. The respective base vectors and metrics are $\underline{G}_A$, $\underline{g}_i$ and $G_{AB} = \underline{G}_A\,\underline{G}_B$, $g_{ij} = \underline{g}_i\,\underline{g}_j$. The use of capital and lower case indices is an aid which helps to differentiate whether a tensor is associated with initial or deformed base vectors, or both, in which case it is a two point tensor. The orthogonal triads of the principal strain directions are denoted by $\underline{N}_\alpha = N^M_\alpha \underline{G}_M$ in the initial and by $\underline{n}_\alpha = n^i_\alpha \underline{g}_i$ in the deformed state, respectively.

The deformation of the body causes the material lines of principal strain to stretch and to rotate. The amount and orientation of the rigid rotation is given by the orthonormal two-point rotation tensor $\underline{\alpha}$ whose components satisfy

$$\underline{\alpha}\,\underline{\alpha}^T = \underline{I} \quad , \quad \alpha^{j}_{\cdot M}\,\alpha_i^{\cdot M} = \alpha_N^{\cdot j}\,\alpha^N_{\cdot i} = \delta^j_i \quad . \tag{1}$$

Note that for vanishing rotation, $\underline{\alpha}$ becomes unity. The initial and current principal strain triads are then related by

$n^k_\alpha = \alpha^k_{.M} \, N^M_\alpha$ and the following identities hold

$$\alpha^k_{\cdot M} = n^k_\alpha \, N^\alpha_M \, , \qquad N^\alpha_K \, N^K_\beta = \delta^\alpha_\beta \, , \qquad N^\alpha_K \, N^M_\alpha = \delta^M_K \tag{2}$$

A one point rotation tensor with components $\alpha^K_{.M}$ may be defined by introducing the shifter $g^K_i = \underline{G}^K \, \underline{g}_i$ to give $\alpha^K_{\cdot M} = (g^K_i \, n^i_\alpha) N^\alpha_M$.

2.2 Deformation quantities

As a result of the deformation the neighborhood of a particle is transformed according to

$$dx^i = F^i_{.A} \, dX^A \, , \tag{3}$$

where $F^i_{.A}$ denotes the components of the deformation gradient $\underline{F}$. It is a two point tensor and may be decomposed as follows

$$\underline{F} = \underline{\alpha} \, \underline{U} = \underline{\alpha}(\underline{I} + \underline{h}) \quad , \qquad F^i_{\cdot A} = \alpha^i_{\cdot C}(\delta^C_A + h^C_A) \quad . \tag{4}$$

In (4) $\underline{\alpha}$ is the rotation tensor defined above and $\underline{U} = \underline{I} + \underline{h}$ is the symmetric right stretch tensor. In the terminology introduced in [5] $\underline{h}$ is called the engineering strain tensor. As above, a related one-point tensor may be defined as

$$F^K_{\cdot A} = g^K_i \, \alpha^i_{\cdot C}(\delta^C_A + h^C_A) = \alpha^K_{\cdot C}(\delta^C_A + h^C_A) \quad . \tag{4a}$$

The base vectors of the deformed and undeformed configurations are related in a similar fashion

$$\underline{g}^i = F^i_{\cdot A} \, \underline{G}^A = \alpha^i_{\cdot C}(\delta^C_A + h^C_A)\underline{G}^A = \alpha^i_{\cdot C} \, \hat{\underline{G}}^C \quad . \tag{5}$$

This shows that the rotation tensor can also be written as a shifter $\underline{g}^i \, \hat{\underline{G}}_A = \alpha^i_{\cdot A}$ and thus describes the rigid rotation of the deformed base $\underline{g}_i$ to the stretched base $\hat{\underline{G}}_A$. Recall, that in terms of the usual displacement gradient the base vectors $\underline{g}_i$ are given by

$$d\underline{r} = \underline{g}_i \, d\Theta^i = (\delta^C_A + u^C|_A)\underline{G}_C \, \delta\Theta^A \quad , \tag{6}$$

where convected coordinates have been used and $(\quad)|_A$ denotes covariant differentation with respect to the undeformed metric.

2.3 Incremental relationships

To describe the change of quantities in going from a current state to a closely neighboring state, the notation given below will be used (see[7] and Figure 2).

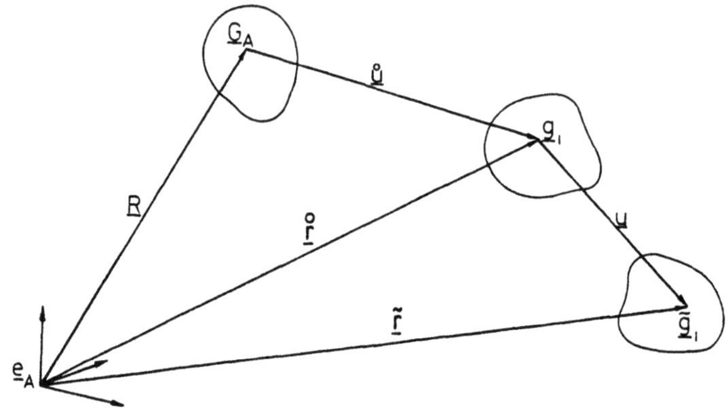

Figure 2

All quantities which characterize the current state are presumed known and will be denoted by the supercript (^{o}), all quantities of the neighboring state by (~), whereas incremental quantities will not be labeled (e.g. $\overset{o}{\underline{u}}$ in Figure 2 denotes the current, $\underline{u}$ the incremental displacement vector). The pertinent relationships between the principal strain triad $\tilde{\underline{n}}_\alpha$, the rotation tensor $\tilde{\underline{\alpha}}$ the position vector $\tilde{\underline{r}}$, and the deformation gradient $\underline{\nabla\tilde{r}}$ of the adjacent state, and the quantities characterizing the current state as well as their increments are summarized below in their T.L. and U.L. form:

T.L.: U.L.:

$$\left.\begin{array}{ll}
\tilde{\underline{n}}_\alpha = (\overset{o}{\underline{\alpha}} + \underline{\alpha})\underline{N}_\alpha & \tilde{\underline{n}}_\alpha = (\underline{I} + \underline{\alpha})\overset{o}{\underline{n}}_\alpha \\
\tilde{\underline{r}} = \underline{R} + \overset{o}{\underline{u}} + \underline{u} & \tilde{\underline{r}} = \overset{o}{\underline{r}} + \underline{u} \\
d\tilde{\underline{r}} = (\underline{G}_A + \overset{o}{\underline{u}}_{,A} + \underline{u}_{,A})d\Theta^A & d\tilde{\underline{r}} = (\underline{g}_i + \underline{u}_{,i})d\Theta^i \\
\underline{\nabla\tilde{r}} = (\overset{o}{\underline{\alpha}} + \underline{\alpha})\cdot(\underline{I} + \overset{o}{\underline{h}} + \underline{h}) & \underline{\nabla\tilde{r}} = (\underline{I} + \underline{\alpha})\cdot(\underline{I} + \underline{h}) \\
\underline{\nabla}(\tilde{\underline{r}} - \overset{o}{\underline{r}}) = \overset{o}{\underline{\alpha}}\cdot\underline{h} + \underline{\alpha}\overset{o}{\underline{h}} + \underline{\alpha}(\underline{I} + \underline{h}) & \nabla(\tilde{\underline{r}} - \overset{o}{\underline{r}}) = \underline{h} + \underline{\alpha}(\underline{I} + \underline{h}) \\
y^B|_A = \overset{o}{\alpha}{}^B_{\cdot C}\, h^C_A + \alpha^B_{\cdot C}\, \overset{o}{h}{}^C_A + \alpha^B_{\cdot A} & y^j|_i = h^j_i + \alpha^j_{\cdot i} \quad .
\end{array}\right\} \quad (7)$$

The components of the tensor $\overset{\circ}{\underline{\alpha}}$ which describes the general rotation of an orthogonal triad, may be expressed in terms of trigonometric functions of suitably defined rotation angles. Euler angles may be used, as was done in [1] but a formulation in terms of angles about the axes of a reference triad are more appropriate for the description of beam deformations. The full expression of $\overset{\circ}{\underline{\alpha}}$ is given in the Appendix (equ. A1) While $\overset{\circ}{\underline{\alpha}}$ satisfies the ortho-normality relation (1), the increments $\underline{\alpha}$ do not. Rather, from

$$\underline{\tilde{\alpha}}^T \, \underline{\tilde{\alpha}} = (\overset{\circ}{\underline{\alpha}} + \underline{\alpha})^T \; (\overset{\circ}{\underline{\alpha}} + \underline{\alpha}) = \underline{I}$$

it follows that for small increments $\underline{\alpha}$ a skew symmetric rotation increment tensor $\underline{\omega}$ may be defined as follows:

$$\underline{\omega} = \underline{\alpha}^T \, \overset{\circ}{\underline{\alpha}} = -\overset{\circ}{\underline{\alpha}}{}^T \, \underline{\alpha} = -\underline{\omega}^T \quad . \tag{8}$$

For the case of a rotation through an increment φ about one axis only, these relations take the special form

$$\underline{\tilde{\alpha}} = \overset{\circ}{\underline{\alpha}} + \underline{\alpha} = \begin{bmatrix} \cos(\overset{\circ}{\varphi} + \varphi) & \sin(\overset{\circ}{\varphi} + \varphi) \\ -\sin(\overset{\circ}{\varphi} + \varphi) & \cos(\overset{\circ}{\varphi} + \varphi) \end{bmatrix} = \begin{bmatrix} 1 & \varphi \\ -\varphi & 1 \end{bmatrix} \begin{bmatrix} \cos\overset{\circ}{\varphi} & \sin\overset{\circ}{\varphi} \\ -\sin\overset{\circ}{\varphi} & \cos\overset{\circ}{\varphi} \end{bmatrix} ,$$

or
$$\underline{\tilde{\alpha}} = \overset{\circ}{\underline{\alpha}} + \underline{\alpha} = (\underline{\omega} + \underline{I})^T \, \overset{\circ}{\underline{\alpha}} \tag{9}$$

The corresponding general expressions for $\underline{\alpha}$ and $\underline{\omega}$ in terms of $\overset{\circ}{\underline{\varphi}}$ and increments $\underline{\varphi}$ are also given in the Appendix (equ. A2a). Since $\underline{\omega}$ is skew symmetric, it may be expressed in terms of a rotation increment vector with components ω^M, such that

$$\omega_{AB} = \varepsilon_{MBA} \, \omega^M \quad , \qquad \omega_M = -\frac{1}{2} \, \varepsilon_{MKL} \, \omega^{KL} \quad ,$$

where ε_{MKL} denotes the permutation tensor. In turn, the rotation rate of a corotating coordinate system may be expressed in terms of the increments φ_1 about the axes of the reference triad by:

$$\underline{\underline{\omega}} = \begin{bmatrix} 0 & -\omega^3 & \omega^2 \\ \omega^3 & 0 & -\omega^1 \\ -\omega^2 & \omega^1 & 0 \end{bmatrix} ; \qquad \underline{\omega} = \begin{bmatrix} \omega^1 \\ \omega^2 \\ \omega^3 \end{bmatrix} = \overset{\circ}{\underline{b}} \begin{bmatrix} \varphi_1 \\ \varphi_2 \\ \varphi_3 \end{bmatrix} = \overset{\circ}{\underline{b}}\, \underline{\varphi} \quad . \tag{10}$$

Note that $\overset{\circ}{\underline{b}}$ becomes unity in the reference state. The matrix $\overset{\circ}{\underline{b}}$ is given in the Appendix (equ. A3). A summary of the above relationships follows:

T.L.:

$$\omega_{AB} = \overset{\circ}{\alpha}{}^{K}_{\cdot A}\, \alpha_{KB} = \varepsilon_{MBA}\, \omega^M$$

$$\alpha_{KM} = \varepsilon_{LMA}\, \alpha_K^{\cdot A}\, \omega^L$$

U.L.:

$$\omega_{ij} = \alpha_{ij} = -\alpha_{ji}$$

$$\alpha_{ij} = \varepsilon_{mji}\, \varphi^m$$

$$\omega^i = \varphi^i \quad .$$

2.4 Strains and stresses

The Lagrangian strain tensor $\overset{\circ}{\varepsilon}_{AB}$ of the current configuration is defined by

$$ds^2 - dS^2 = 2\overset{\circ}{\varepsilon}_{AB}\, dX^A\, dX^B \quad ,$$

where ds and dS are the deformed and undeformed lengths of a material line element. It can be expressed in terms of the deformation gradient $F^{i}_{\cdot A}$ or the displacement gradient $u_{A\,|\,B}$ by

$$2\overset{\circ}{\varepsilon}_{AB} = (\overset{\circ}{F}{}^{i}_{\cdot A}\, \overset{\circ}{F}{}^{\cdot C}_{i} - \delta^C_A)\, G_{CB} = \overset{\circ}{u}_{A|B} + \overset{\circ}{u}_{B|A} + \overset{\circ}{u}{}^{C}|_A\, \overset{\circ}{u}_{C|B} \quad . \tag{11}$$

Substitution of the polar decomposition relationship (4) into the first expression of (11) and use of the orthogonality property (1) of $\overset{\circ}{\alpha}{}^{i}_{\cdot M}$ leads to the known expression of $\overset{\circ}{\varepsilon}_{AB}$ in terms of the "engineering strain" tensor $\overset{\circ}{h}_{AB}$

$$\overset{\circ}{\varepsilon}_{AB} = \overset{\circ}{h}_{AB} + \frac{1}{2}\, \overset{\circ}{h}{}^{C}_{A}\, \overset{\circ}{h}_{CB} \quad . \tag{12}$$

Alternatively $\overset{\circ}{h}_{AB}$ may be expressed in terms of the displacement gradient and $\overset{\circ}{\alpha}{}^{\cdot A}_{C}$:

$$\overset{\circ}{h}{}^{A}_{B} = \overset{\circ}{\alpha}{}^{\cdot A}_{C}\, \overset{\circ}{y}{}^{C}|_B - \delta^A_B = \overset{\circ}{\alpha}{}^{\cdot A}_{C} (\delta^C_B + \overset{\circ}{u}{}^{C}|_B) - \delta^A_B \quad . \tag{12a}$$

The related incremental quantities are given below in their T.L. and U.L. forms. Note that the underlined terms are usually omitted in a linearized incremental formulation. They may, however, be of significance in numerical procedures designed to satisfy the full set of nonlinear equations iteratively.

T.L.: U.L.:

$$\varepsilon_{AB} = (\delta_A^C + \overset{o}{h}{}_A^C + \underline{\tfrac{1}{2}\, h_A^C})\, h_{CB} \qquad \varepsilon_{ik} = h_{ik} + \underline{\tfrac{1}{2}\, h_i^j\, h_{jk}}$$

$$h_B^A = \alpha_C^{\cdot A}(\delta_B^C + \overset{o}{u}{}^C|_B + \underline{u^C|_B}) + \overset{o}{\alpha}{}_C^{\cdot A}\, u^C|_B \qquad h_j^i = \alpha_k^{\cdot i}(\delta_j^k + \underline{u^k|_j}) + u^i|_j \; .$$

As appropriate stress measures the so-called "Jaumann" stresses r^{AB} [5,8] are introduced. They are related to the "Lagrangian" stresses t^{AB}, to the 1. Piola-Kirchhoff stresses π^{AK}, and to the 2. Piola-Kirchhoff stresses s^{AB} by

$$\overset{o}{r}{}^{AB} = \overset{o}{t}{}^{AC}\, \overset{o}{\alpha}{}_C^{\cdot B} = \overset{o}{\pi}{}^{Ak}\, \overset{o}{\alpha}{}_K^{\cdot B} = \overset{o}{s}{}^{AC}(\delta_C^B + \overset{o}{h}{}_C^B) \quad . \tag{13}$$

The symmetric part of $\overset{o}{r}{}^{AB}$:

$$\overset{o}{\bar{r}}{}^{AB} = \frac{1}{2}(\overset{o}{r}{}^{AB} + \overset{o}{r}{}^{BA}) = \frac{1}{2}(\overset{o}{t}{}^{AC}\, \overset{o}{\alpha}{}_C^{\cdot B} + \overset{o}{t}{}^{BC}\, \overset{o}{\alpha}{}_C^{\cdot A}) \tag{14}$$

has been termed "engineering stress". Only this part contributes to the work expressions and the complementary energy density of the functionals which are used for the derivation. It arises as the conjugate quantity of the "engineering strain" tensor $\overset{o}{h}{}_{AB}$ [5].

From (1) it follows that the first equation in (13) may be inverted uniquely to give

$$\overset{o}{t}{}^{AB} = \overset{o}{r}{}^{AC}\, \overset{o}{\alpha}{}_{\cdot C}^{B} \quad . \tag{15}$$

The respective incremental expressions are given below where, for completeness, the relationship between r^{AB} and s^{AB} has been included.

T.L.: U.L.:

$$r^{AB} = \overset{o}{t}{}^{AC}\, \alpha_C^{\cdot B} + t^{AC}\, \overset{o}{\alpha}{}_C^{\cdot B} \qquad r^{ik} = t^{ik} + \overset{o}{\tau}{}_r^{i}\, \varepsilon^{mkr}\, \varphi_m$$

$$t^{AB} = \overset{o}{r}{}^{AC}\, \alpha_{\cdot C}^{B} + r^{AC}\, \overset{o}{\alpha}{}_{\cdot C}^{B} \qquad t^{ik} = r^{ik} - \overset{o}{\tau}{}_r^{i}\, \varepsilon^{mkr}\, \varphi_m$$

$$r^{AB} = s^{AB} - s^{AC}\, \overset{o}{h}{}_C^B + \overset{o}{s}{}^{AC}\, h_C^B \qquad r^{ik} = s^{ik} + \overset{o}{\tau}{}^{ir}\, h_r^k \quad .$$

Here $\overset{\circ}{\tau}^{ir}$, $\overset{\circ}{\tau}{}^{i}_{r}$ denote components of Cauchy stress.

2.5 Constitutive relations

In nonlinear problems the choice of constitutive relations is of special significance. Generally speaking, such relations are only meaningful if they relate objective measures of stress and their conjugate strain measures. The following relations, usually called hyper- and hypo-elastic, are appropriate if strains remain small:

$$\overset{\circ}{\varepsilon}{}^{B}_{A} = \frac{1+\nu}{E}\{\overset{\circ}{s}{}^{B}_{A} - \frac{\nu}{1+\nu}\overset{\circ}{s}{}^{K}_{K}\,\delta^{B}_{A}\} \tag{16}$$

$$\varepsilon^{B}_{A} = \frac{1+\nu}{E}\{s^{B}_{A} - \frac{\nu}{1+\nu}s^{K}_{K}\,\delta^{B}_{A}\} \quad . \tag{17}$$

A linear relation between engineering stresses and strains is often called "semilinear":

$$\overset{\circ}{h}{}^{B}_{A} = \frac{1+\nu}{E}\{\overset{\circ}{\bar{r}}{}^{B}_{A} - \frac{\nu}{1+\nu}\overset{\circ}{\bar{r}}{}^{K}_{K}\,\delta^{B}_{A}\} \quad . \tag{18}$$

Since the engineering measures $\overset{\circ}{\bar{r}}{}^{AB}$ and $\overset{\circ}{h}_{AB}$ are objective a Legendre transformation

$$Q^*(\overset{\circ}{\underline{r}}) = \overset{\circ}{\bar{r}}{}^{A}_{B}\,\overset{\circ}{h}{}^{B}_{A} - Q(\overset{\circ}{\underline{h}}) \tag{19}$$

exists [5], where $Q(\overset{\circ}{\underline{h}})$ and $Q^*(\overset{\circ}{\underline{r}})$ are potential functions corresponding to the strain energy and complementary energy densities, respectively. For isotropic materials it may be shown that

$$Q^*(\overset{\circ}{\underline{\bar{r}}}) = Q^*(\overset{\circ}{\underline{r}}) .$$

The corresponding incremental relationship employed in this paper relates the increments in an analogous manner:

$$h^{B}_{A} = \frac{1+\nu}{E}\{\bar{r}^{B}_{A} - \frac{\nu}{1+\nu}\bar{r}^{K}_{K}\,\delta^{B}_{A}\} \quad . \tag{20}$$

This strain-stress-relation implies that a transformation analogous to (19) exists, and that the incremental complementary energy density has the simple form

$$W^*(\underline{\bar{r}}) = \frac{1+\nu}{E}\{\bar{r}^{A}_{B}\,\bar{r}^{B}_{A} - \frac{\nu}{1+\nu}\bar{r}^{K}_{K}\,\bar{r}^{L}_{L}\} = W^*(\underline{r}) \quad . \tag{21}$$

It is noted that the use of the simple (nonobjective) increments $\bar{\underline{r}}$ in (20,21) is limited to problems with small strains $\underline{h}$. This is assumed in the present formulation for rods. For large strains appropriate corotational increments of stress should be used in the respective constitutive laws.

Recall that the corresponding Legendre transformation relating the increments of the 2nd Piola Kirchhoff stress and Green's strain tensor reads [6] :

$$W^*(\underline{s},\underline{u}) = s^A_B\,\varepsilon^B_A + 2\overset{o}{s}{}^A_B\,\eta^B_A - W(\underline{\varepsilon}) \quad , \tag{22}$$

where $\eta_{AB} = u^C{}_{|A}\, u_{C|B}$ and $W(\underline{\varepsilon})$ is again the incremental strain energy density. Substitution of (17) gives

$$W^*(\underline{s},\underline{u}) = \overset{o}{s}{}^A_B\,\eta^B_A + \frac{1+\nu}{E}\{s^A_B\,s^B_A - \frac{\nu}{1+\nu}\,s^K_K\,s^L_L\} \tag{23}$$

It is seen that in this case W* may not be written in terms of s^{AB} alone. Similar problems arise in the attempt to formulate W and W* in terms of Lagrangian stresses and displacement gradients. Although t^{AB} and $u_{B|C}$ are not objective measures of stress and strain, the sum of W and W^* can simply be written as

$$W + W^* = t^A_{\cdot B}\,u^B|_A = s^A_B\,\varepsilon^B_A + 2s^A_B\,\eta^B_A \quad . \tag{24}$$

2.6 Generalized variational principle

A variational principle of Hellinger-Reissner type using the rotation-tensor of the polar decomposition explicitly has been derived by Fraeijs de Veubeke [5] . In [8] Wempner used it and a related principle in the context of buckling problems. An incremental version was presented by Atluri and Murakawa on several occasions, e.g. in [6] . The generalized principle used in this paper as the basis for the reduction to the beam problem employs the stresses $\overset{o}{\underline{r}}$ and the conjugate strains $\overset{o}{\underline{h}}$ as unknowns, whereby the latter is expressed in terms of the rotation and the deformation gradient (see 12a):

$$\delta J_{CB} = \delta\{\int_{V_o} [-Q^*(\overset{o}{\underline{r}}) + \overset{o}{r}{}^A_B \overset{o}{h}{}^B_A - \rho_o \overset{o}{f}{}^A \overset{o}{u}_A]\, dV - \int_{S_p} \overset{o}{\bar{t}}{}^A \overset{o}{u}_A\, dS - \int_{S_u} \overset{o}{t}{}^A (\overset{o}{u}_A - \overset{o}{\bar{u}}_A)\, dS\}$$

$$= \delta\{\int_{V_o} [-Q^*(\overset{o}{\underline{r}}) + \overset{o}{r}{}^{AB} \overset{o}{\alpha}{}^C_{\cdot B} \overset{o}{y}_{C|A} - \overset{o}{r}{}^A_A - \rho_o \overset{o}{f}{}^A \overset{o}{u}_A]\, dV + \text{boundary terms}\} = 0 \quad . \tag{25}$$

After the definition of stress-resultants and reduction, the final principle for rods will be written in terms of the deformation gradient, the rotations, and nominal resultants, which are components of the cross-sectional forces and moments in the direction of the reference configuration. They are analogous to t^{AB} in three-dimensional problems.

The incremental versions of (25) follow after substitution of $\underline{\alpha}$ by $\underline{\omega}$ using (8). For the "total Lagrangian" form one obtains

T.L.:

$$\begin{aligned}\delta J_{CB} = \delta\{\int_{V_o} [&-W^*(\underline{r}) + \overset{o}{\alpha}{}^L_{\cdot B}\, r^{AB}\, y_{L|A} + \varepsilon_{MBL} \overset{o}{r}{}^{AB} \overset{o}{\alpha}{}^{\cdot L}_K\, y^K|_A\, \omega^M \\ &+ \varepsilon_{MBL} \overset{o}{\alpha}{}^{\cdot L}_K \overset{o}{y}{}^K|_A\, r^{AB}\, \omega^M - \frac{1}{2} \overset{o}{y}{}^K|_A \overset{o}{r}{}^A_B \overset{o}{\alpha}{}^{\cdot B}_K\, \omega_M\, \omega^M \\ &+ \frac{1}{2} \overset{o}{y}{}^K|_A \overset{o}{\alpha}{}^{\cdot L}_K \overset{o}{r}{}^{AB}\, \omega_L\, \omega_B - \rho_o f^A y_A]\, dV \\ &- \int_{S_p} \bar{t}^A y_A\, dS - \int_{S_u} t^A (y_A - \bar{y}_A)\, dS\} = 0 \quad .\end{aligned} \tag{26}$$

The "updated Lagrangian" form reads

U.L.:

$$\begin{aligned}\delta J_{CB} = \delta\{\int_V [&-W^*(\underline{r}) + r^{ij}\, y_{j|i} + \varepsilon_{lkj} \overset{o}{r}{}^{ik}\, y^j|_i\, \varphi^l \\ &- \frac{1}{2} (\overset{o}{r}{}^i_i\, \varphi^j \varphi_j - \overset{o}{r}{}^{il}\, \varphi_i \varphi_l) - \rho f^j y_j]\, dV \\ &- \int_{S_p} \bar{t}^j y_j\, dS - \int_{S_u} t^j (y_j - \bar{y}_j)\, dS\} = 0 \quad ,\end{aligned} \tag{27a}$$

which can alternatively be expressed in terms of t^{ij} instead of r^{ij} to give

U.L.:

$$\delta J_{CB} = \delta\{\int_V [-W^*(\underline{r}) + t^{ij}\, y_j|_i - \varepsilon_{kij}\, t^{ij}\, \varphi^k + \tfrac{1}{2}(\overset{\circ}{\tau}{}^k_k\, \varphi^r \varphi_r - \overset{\circ}{\tau}{}^{ks}\, \varphi_k \varphi_s) - \rho f^j\, y_j]\, dV - \int_{S_p} \bar{t}^j\, y_j\, dS - \int_{S_u} t^j (y_j - \bar{y}_j)\, dS\} = 0 \tag{27b}$$

The Euler equations resulting from (27a,b) are the force equilibrium (28a or 29a) and momentum balance (28b or 29b) equations, the polar decomposition relation (28c or 29c) as well as all boundary conditions:

U.L.:

$$r^{ij}|_i + \varepsilon^{mkj}\, \overset{\circ}{r}{}^i_k|_i\, \varphi_m + \varepsilon^{mkj}\, \overset{\circ}{r}{}^i_k\, \varphi_m|_i + \rho f^j = 0 \tag{28a}$$

$$r^{ij} + \overset{\circ}{\tau}{}^{kj}\, h^i_k \quad : \quad \text{symmetric} \tag{28b}$$

$$y_j|_i = h_{ji} + \varepsilon_{mij}\, \varphi^m \quad , \tag{28c}$$

or

$$t^{ij}|_i - \rho f^j = 0 \tag{29a}$$

$$t^{ij} + \overset{\circ}{\tau}{}^{jk}\, h^i_k + \overset{\circ}{\tau}{}^{ik}\, \alpha^j_{\cdot k} \quad : \quad \text{symmetric} \tag{29b}$$

$$y_j|_i = h_{ji} + \alpha_{ji} \quad . \tag{29c}$$

The indices CB of the functional should not be confused with the capital tensorial indices used above. They refer to a classification of generalized variational principles introduced by the first author (see e.g. [7]). Some differences in comparison with the versions given by Murakawa and Atluri [6] should be noted. Firstly, the deformation gradient, rather than the displacement gradient appears. Although this difference is minor in incremental problems it seems more consistent to employ the deformation gradient since it appears in the polar decomposition theorem. Secondly, the increments of the rotation angles have been introduced explicitly which helps to further clarify the role played by the rotation of principal axes in large displacement problems. With fifteen independent variables the proposed forms also contain fewer unknowns. Moreover, this choice of variables is particularly well suited to the formulation of beam theories, since for beams the orientation of principal axes is known and, with appropriate definitions of stress resultants, the rotation angles about these axes arise as natural unknowns.

3. Generalized principle for slender rods

3.1 General remarks

In the deformation of slender structures, such as rods, plates, and shells, large displacements and moderate to large rotations may occur. It is thus quite natural to include rotational degrees of freedom in the description explicitly, rather than implicitly through the displacement gradient. In the context of shell theory this was recognized by Wempner [10] who derived appropriate kinematic relations using the polar decomposition theorem . Here the theorem is also used for the description of the kinematical behavior of a rod. But unlike in [10] it is inserted into the variational principle (25) from which, together with additional assumptions on stresses, a consistent and general principle for rods undergoing large displacements and rotations is derived. Its U.L. incremental form gives local equations which directly reduce to those of well known small displacement theories.

3.2 Kinematic assumptions

In the initial, undeformed configuration the position vector $\underline{R}$ of an arbitrary point in a plane cross section which is perpendicular to a suitably chosen reference axis may be expressed by

$$\underline{R} = \overset{c}{\underline{R}} + \eta \underline{A}_2 + \zeta \underline{A}_3 \quad , \qquad (30)$$

where ζ and η are the distances from the axis, $\overset{C}{\underline{R}}$ is the position vector of the point on the axis, and $\underline{A}_k$ are the unit vectors of the orthogonal reference triad at this point (see Fig. 3). Vector $\underline{A}_1$ is tangent to the reference axis. The base vectors $\underline{G}_K$ are then given in terms of $\underline{A}_k$ by

$$\underline{G}_1 = (\delta_1^K + \Gamma_{11}^K)\underline{A}_K \ ; \quad \underline{G}_2 = \underline{A}_2 \ ; \quad \underline{G}_3 = \underline{A}_3 \ , \tag{31}$$

where the Christoffel symbol Γ_{1l}^K accounts for the axial curvature and twist of an initially curved rod.

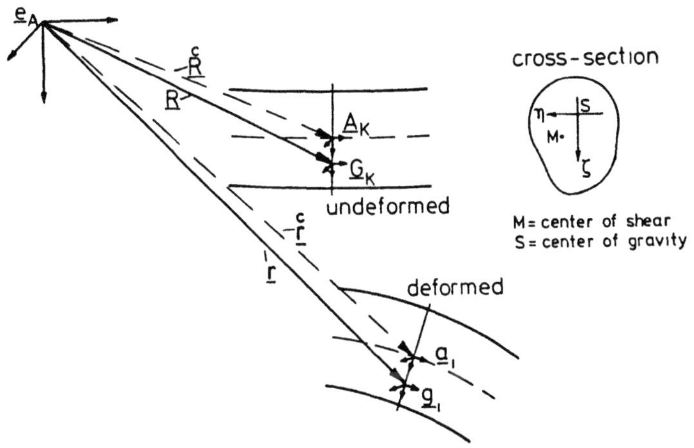

Figure 3

The number of the stiffness parameters of the cross section can be minimized by referring the torsional quantities to the center of shear, with coordinates $\overset{M}{\zeta}$, $\overset{M}{\eta}$, the bending quantities to the center of gravity, and by orienting $\underline{A}_2$ and $\underline{A}_3$ such that they point in the direction of the principal axes of inertia. This choice is most advantageous if complex, nonsymmetric cross-sections are considered.

After deformation the position vector, the vector of the origin of the center triad and the base vectors will be denoted ba $\underline{r}$,

$\overset{C}{\underline{r}}$, $\underline{a}_i$ and $\underline{g}_i$ respectively. From (5) it follows that in general

$$\underline{a}_i = \alpha_i^{\cdot C}(\delta_C^B + h_C^B)\,\underline{A}_B = \alpha_i^{\cdot C}\,\hat{\underline{A}}_C \quad , \tag{32}$$

where $\hat{\underline{A}}_C$ denotes the stretched reference triad. Employing now the usual assumption that in bending normals to the axis remain normal and unstretched we obtain

$$\hat{\underline{A}}_1 = (1+h_1^1)\underline{A}_1, \quad \hat{\underline{A}}_2 = \underline{A}_2 \;, \quad \hat{\underline{A}}_3 = \underline{A}_3 \quad , \tag{33}$$

from which it follows that during deformation points normal to the axis are only rotated but not stretched. In accordance with (30) one obtains

$$\underline{r} = \overset{C}{\underline{r}} + \eta\underline{a}_2 + \zeta\underline{a}_3 - \omega(\eta,\zeta)\,\Psi\,\underline{a}_1 \;, \tag{34a}$$

$$\underline{r} = [\overset{C}{y}{}^{B} + \eta\alpha_{\cdot 2}^{B} + \zeta\alpha_{\cdot 3}^{B} - \omega\Psi\alpha_{\cdot 1}^{B}]\,\underline{A}_B = y^B\,\underline{A}_B \quad . \tag{34b}$$

The last term in (34a) accounts for warping of the cross section due to torsion and $\omega(\eta,\zeta)$ depends on the cross-sectional shape. Thus, the deformation of any point perpendicular to the reference axis is completely described by that of the respective point on the axis. Moreover, as a consequence of (33) and (32) the vectors a_i are seen to coincide with the triad of principal strains and the angles $\underline{\omega}$ and $\underline{\varphi}$ appearing in (26,27) may be identified with the increments of the rotation angles about the local coordinate vectors $\underline{a}_i$.

In the U.L. formulation expression (34b) reduces to the linear relationship

$$\begin{bmatrix} y_1 \\ y_2 \\ y_3 \end{bmatrix} = \begin{bmatrix} \overset{c}{y}_1 \\ \overset{c}{y}_2 \\ \overset{c}{y}_3 \end{bmatrix} + \begin{bmatrix} 0 & \zeta & -\eta \\ -(\zeta-\overset{M}{\zeta}) & 0 & 0 \\ \eta-\overset{M}{\eta} & 0 & 0 \end{bmatrix} \begin{bmatrix} \overset{c}{\varphi}_1 \\ \overset{c}{\varphi}_2 \\ \overset{c}{\varphi}_3 \end{bmatrix} - \begin{bmatrix} \omega\Psi \\ 0 \\ 0 \end{bmatrix} . \tag{35}$$

It is noted that in (35) χ, $\overset{c}{\chi}$, and $\overset{c}{\varphi}$ denote incremental quantities, and that the same relationship is applicable in the T.L. description if rotations remain moderate. Furthermore, (35) differs slightly from (34) in that (35) provides for the fact that the centers of shear and of gravity may not coincide. The appropriate modifications of (34) have, however, been made in all general expressions which follow.

3.3 Stress assumptions and stress resultants

From the assumptions implicit in (33) and from the inverses of the constitutive relationship (18) it is seen that the stress state is essentially uniaxial, and that the components of the Jaumann stress corresponding to the vanishing strain components must also vanish

$$\overset{\circ}{r}{}^{22} = \overset{\circ}{r}{}^{33} = \overset{\circ}{r}{}^{23} = \overset{\circ}{r}{}^{32} = 0 \quad . \tag{36}$$

The remaining components $\overset{\circ}{r}{}^{1K}$ are the normal and shear stresses acting in the directions of the vectors $\underline{a}_k$ of the displaced and rotated cross section (see Fig. 3).

The reduction of the volume integrals in (25) to one-dimensional integrals along the rod axis requires assumptions on the distribution of the components $\overset{\circ}{r}{}^{1K}$ across the cross-section. Consistent with the assumptions of beam theory the distribution of the axial stress $\overset{\circ}{r}{}^{11}$ is taken to consist of a constant term, two linear terms in the coordinates η and ζ along $\underline{a}_2$ and $\underline{a}_3$, and a bilinear term resulting from the "unit warping function" $\omega(\eta,\zeta)$:

$$\overset{\circ}{r}{}^{11} = \frac{\overset{\circ}{N}}{A} - \frac{\overset{\circ}{M}_3}{F_{\eta\eta}}\,\eta + \frac{\overset{\circ}{M}_2}{F_{\zeta\zeta}}\,\zeta - \frac{\overset{\circ}{M}_\omega}{F_{\omega\omega}}\,\omega(\eta,\zeta) \quad . \tag{37}$$

The distribution of the shear stresses is taken as

$$\overset{\circ}{r}{}^{12} = \frac{\overset{\circ}{Q}_2}{A} - \left[(\zeta-\overset{M}{\zeta}) + \omega,_{\eta}\right]\frac{\overset{\circ}{M}_{1p}}{J_T} - \overset{\circ}{r}{}^{11}\,(\zeta-\overset{M}{\zeta})\,\overset{\circ}{\Psi} \quad , \tag{38a}$$

$$\overset{\circ}{r}{}^{13} = \frac{\overset{\circ}{Q}_3}{A} - \left[-(\eta-\overset{M}{\eta}) + \omega,_{\zeta}\right]\frac{\overset{\circ}{M}_{1p}}{J_T} + \overset{\circ}{r}{}^{11}\,(\eta-\overset{M}{\eta})\,\overset{\circ}{\Psi} \quad . \tag{38b}$$

In (37) A is the area of the cross-section, $F_{\eta\eta}$ and $F_{\zeta\zeta}$ are the principal moments of inertia. $F_{\omega\omega}$ is the principal warping stiffness, and J_T is the torsional stiffness. It may be noted that the representation (37) can also be deduced by substituting the deformation gradient from (34b) into (12a) and then using the inverse of the constitutive law (18).

Relations (38a,b) differ from those usually assumed in beam theory. The third terms are due to warping of the cross section. They account for the fact that the actual normal and shear stresses acting on a warped cross-section are no longer parallel to the unit vectors $\underline{a}_i$. This difference is most pronounced so far as shear stresses are concerned. The effect on the axial stresses can be shown to be small.

The quantities $\overset{o}{N}$, $\overset{o}{M}_2$, $\overset{o}{M}_3$, $\overset{o}{M}_\omega$ and $\overset{o}{Q}_2$, $\overset{o}{Q}_3$, $\overset{o}{M}_{1p}$ in (37,38) are identical with the usual stress resultants. This may be seen from the variation of the surface integral of (26). Noting that the initial normal vector $\underline{N}$ is identical with $\underline{A}_1$ (Fig. 3) one obtains

$$\int_A \overset{o}{t}{}^B \, \delta y_B \, dA = \int_A \overset{o}{t}{}^{AB} N_A \, \delta y_B \, dA = \overset{o}{\alpha}{}^B_{\cdot K} \int_A \overset{o}{r}{}^{1K} \, \delta y_B \, dA \quad . \tag{39}$$

By substituting (34) or (37) into (39), carrying out the integrations over the thickness, and comparing coefficients one finds

$$\int_A \begin{bmatrix} \overset{o}{r}{}^{11} \\ \overset{o}{r}{}^{12} \\ \overset{o}{r}{}^{13} \end{bmatrix} dA = \begin{bmatrix} \overset{o}{N} \\ \overset{o}{Q}_2 \\ \overset{o}{Q}_3 \end{bmatrix} = \underline{\overset{o}{Q}} \quad , \tag{40a}$$

$$\int_A \begin{bmatrix} 0 & -(\zeta-\overset{M}{\zeta}) & (\eta-\overset{M}{\eta}) \\ \zeta & 0 & 0 \\ -\eta & 0 & 0 \end{bmatrix} \begin{bmatrix} \overset{o}{r}{}^{11} \\ \overset{o}{r}{}^{12} \\ \overset{o}{r}{}^{13} \end{bmatrix} dA = \begin{bmatrix} \overset{o}{M}_1 \\ \overset{o}{M}_2 \\ \overset{o}{M}_3 \end{bmatrix} = \underline{M} \quad , \tag{40b}$$

$$\int_A \omega \, \overset{o}{r}{}^{11} \, dA = \overset{o}{M}_\omega \quad . \tag{40c}$$

$\overset{o}{N}$ is thus the resultant axial force, $\overset{o}{M}_2$ and $\overset{o}{M}_3$ are the bending moments about $\underline{a}_3$ and $\underline{a}_2$, and $\overset{o}{M}_\omega$ is the "warping bi-moment". It is noted that the torsional moment $\overset{o}{M}_1$ in (40b) differs from $\overset{o}{M}_{1p}$ in (38a,b) by terms which are due to warping. Similar terms influencing $\overset{o}{Q}_2$, $\overset{o}{Q}_3$ and $\overset{o}{M}_2$, $\overset{o}{M}_3$ have been omitted in relations (40a,b) but are taken into account in the development of the next chapter. In matrix notation (39) then reads

$$\int_A \overset{o}{t}{}^B \, \delta y_B \, dA = \overset{o}{\underline{\alpha}} \cdot \overset{o}{\underline{Q}} \, \delta \overset{c}{\underline{y}} + \overset{o}{\underline{M}} \, \delta \overset{c}{\underline{\omega}} + \overset{o}{M}_\omega \, \delta \Psi \quad . \tag{41}$$

It is also seen that $\overset{o}{\underline{y}}{}^c$, $\overset{o}{\underline{\omega}}{}^c$, and $\overset{o}{\psi}$ are the kinematic variables which are conjugate to $\overset{o}{\underline{\alpha}}\,\overset{o}{\underline{Q}}$, $\overset{o}{\underline{M}}$ and $\overset{o}{M}_\omega$. By further substituting $\delta \overset{c}{\underline{\omega}} = \overset{o}{\underline{b}} \cdot \delta \overset{c}{\underline{\varphi}}$ from (10), where $\overset{c}{\underline{\varphi}}$ are the rotation angle increments about the initial reference triad, one may rewrite (41) as follows:

$$\begin{aligned} \int_A \overset{o}{t}{}^B \, \delta y_B \, dA &= \overset{o}{\underline{\alpha}} \, \overset{o}{\underline{Q}} \, \delta \overset{c}{\underline{y}} + \overset{o}{\underline{M}} \, \overset{o}{\underline{b}} \, \delta \overset{c}{\underline{\varphi}} + \overset{o}{M}_\omega \, \delta \Psi \\ &= \overset{o}{\underline{V}} \, \delta \overset{c}{\underline{y}} + \overset{o}{\underline{B}} \, \delta \overset{c}{\underline{\varphi}} + \overset{o}{M}_\omega \, \delta \Psi \quad , \end{aligned} \tag{42}$$

where $\quad \overset{o}{\underline{V}} = \overset{o}{\underline{\alpha}} \, \overset{o}{\underline{Q}} \quad , \qquad \overset{o}{\underline{B}} = \overset{o}{\underline{b}}{}^T \, \overset{o}{\underline{M}} \quad . \qquad (42a)$

Their conjugate kinematic variables are $\overset{oc}{\underline{y}}$ and $\overset{oc}{\underline{\varphi}}$. Equation (42) may thus be interpreted as the virtual work performed by $\overset{o}{\underline{V}}$, $\overset{o}{\underline{B}}$ and $\overset{o}{M}_\omega$ in the initial configuration. Figuratively speaking, $\overset{o}{\underline{V}}$ consists of the sum of the components of $\overset{o}{\underline{Q}}$ pointing in the direction of the initial reference system. The simple form of (42) implies that the general field equations take on a particularly simple form if expressed in terms of these conjugate quantities.

3.4 Generalized variational principle for beams and corresponding local equations

Introducing the assumptions made in the previous chapters into (25) results, after lenghty but straightforward operations, in a generalized variational principle for beams in terms of the deformed position vector of the beam axis, the rotation angles and the warping parameter ψ as well as the stress resultants given above.

One obtains from (25):

$$\delta J_{CB} = \delta\{\int_x [-Q^*(\overset{\circ}{\underline{Q}},\overset{\circ}{\underline{M}},\overset{\circ}{M}_\omega) + V^A \overset{c}{y}_A|_1 - Q_1 + B^A \overset{c}{\varphi}_A|_1$$
$$-M_1 \Psi + M_4 \Psi,_1]dx + \text{boundary terms}\} = 0, \quad (43)$$

where

$$Q^*(\overset{\circ}{\underline{Q}},\overset{\circ}{\underline{M}},\overset{\circ}{M}_\omega) = \frac{1}{2}\{\frac{\overset{\circ}{N}^2}{EA} + \frac{\overset{\circ}{M}_2^2}{EF_{\zeta\zeta}} + \frac{\overset{\circ}{M}_3^2}{EF_{\eta\eta}} + \frac{\overset{\circ}{M}_\omega^2}{EF_{\omega\omega}} + \frac{\overset{\circ}{Q}_2^2}{GA} + \frac{\overset{\circ}{Q}_3^2}{GA}\} \quad (43a)$$

which may be expressed in terms of the transformed resultants $\overset{\circ}{\underline{V}}$ and $\overset{\circ}{\underline{B}}$ by using (42a). Similar expressions for the T.L. and U.L. formulation may be derived from (26,27) or directly from (43). This completes the reduction to the one-dimensional problem.

The incremental versions associated with (43) are represented schematically below. Writing them in this form most clearly displays their common structure.

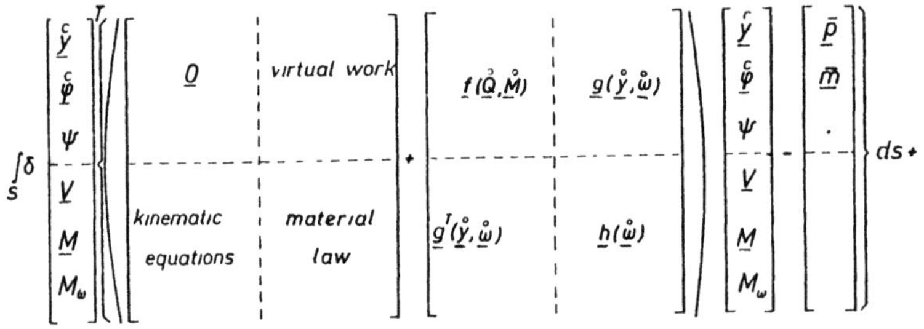

$$\int_S \delta \begin{bmatrix} \overset{c}{\underline{y}} \\ \overset{c}{\varphi} \\ \psi \\ \hline \underline{V} \\ \underline{M} \\ M_\omega \end{bmatrix}^T \left\{ \begin{bmatrix} \underline{0} & | & \text{virtual work} \\ & | & \\ \hline \text{kinematic} & | & \text{material} \\ \text{equations} & | & \text{law} \end{bmatrix} + \begin{bmatrix} & | & \\ f(\overset{\circ}{\underline{Q}},\overset{\circ}{\underline{M}}) & | & g(\overset{\circ}{\underline{y}},\overset{\circ}{\omega}) \\ \hline g^T(\overset{\circ}{\underline{y}},\overset{\circ}{\omega}) & | & h(\overset{\circ}{\omega}) \\ & | & \end{bmatrix} \right\} \begin{bmatrix} \overset{c}{\underline{y}} \\ \overset{c}{\varphi} \\ \psi \\ \hline \underline{V} \\ \underline{M} \\ M_\omega \end{bmatrix} - \begin{bmatrix} \overline{\underline{p}} \\ \overline{m} \\ \cdot \\ \hline \\ \\ \end{bmatrix} ds + \quad (44)$$

or $\quad \int_S \{\delta\underline{z}^T (\underline{D}_L + \underline{D}_G) \underline{z} - \delta\underline{z}^T \overline{\underline{p}}\} ds + \ldots \quad (44a)$

In (44) the components of $\underline{z}$ are the incremental kinematical quantities and stress resultants, and $\overline{\underline{p}}$ denotes the vector of external forces and moments. Matrix $\underline{D}_L$ contains constant material and geometric parameters as well as differential operators which operate on the components of $\underline{z}$. All terms containing quantities of the current state are grouped together in $\underline{D}_G$.

It is noted that the components of $\underline{z}$ representing stress resultants are the increments of the quantities $\underline{V}$ and $\underline{M}$ rather than $\underline{B}$. With this choice of incremental variables the field equations are particularly amenable to the numerical solution scheme outlined in the next chapter. For the case of the T.L. formulation and an initially straight beam, expression (44) is given in detail in the Appendix (A4, A4a). In the respective expressions for curved beams additional terms accounting for the initial curvatures and twist appear [4] .

The Euler equations associated with (44) are obtained by integration by parts along the axis. They are again expressed in terms of $\underline{z}$ as defined above, which leads to a system of first order ordinary differential equations which is again given schematically below:

$$\frac{d}{dx}\begin{bmatrix} \overset{c}{\underline{y}} \\ \overset{c}{\underline{\varphi}} \\ \psi \\ \hdashline \underline{V} \\ \underline{M} \\ M_\omega \end{bmatrix} = \left(\left[\begin{array}{c:c} \begin{array}{c}\textit{kinematical}\\ \textit{relations}\end{array} & \begin{array}{c}\textit{material}\\ \textit{law}\end{array} \\ \hdashline \multicolumn{2}{c}{\begin{array}{c}\textit{force - }\\ \textit{moment-}\end{array}\ \textit{equilibrium}} \end{array} \right] + \left[\begin{array}{c:c} -\underline{g}^T(\mathring{\underline{y}},\mathring{\underline{\omega}}) & -\underline{h}(\mathring{\underline{\omega}}) \\ \hdashline \underline{f}(\mathring{\underline{Q}},\mathring{\underline{M}}) & \underline{g}(\mathring{\underline{y}},\mathring{\underline{\omega}}) \end{array} \right] \right) \begin{bmatrix} \overset{c}{\underline{y}} \\ \overset{c}{\underline{\varphi}} \\ \psi \\ \hdashline \underline{V} \\ \underline{M} \\ M_\omega \end{bmatrix} - \begin{bmatrix} \bar{\underline{p}} \\ \bar{\underline{m}} \\ \\ \\ \cdot \\ \\ \end{bmatrix} \tag{45}$$

or

$$\underline{z}' = (\underline{A}_L + \underline{A}_G) \, \underline{z} - \bar{\underline{p}} \ , \tag{45a}$$

As before all nonlinear terms are written separately in $\underline{A}_G$. Due to the particular choice of independent variables the matrices $\underline{f}$, $\underline{g}$ and $\underline{h}$ are the same as those appearing in $\underline{D}_G$ of (44).

Finally it is noted that the corresponding expressions in the U.L. description follow from (44) and (45) if kinematic variables of the current state vanish, and the covariant derivatives are taken with respect to the updated, deformed geometry. Frequently this is not done in practice. Instead, the deformed configuration is often approximated by straight elements. In the full expression (45) which is given in the Appendix(A5) these influences are incorporated via the terms k_2, k_3 and h which represent the initial (current) values of the curvatures and the twist of the beam axis [4] .

4. Numerical solution technique

The variational principle (44) and the corresponding local equations (45) can now be used as the basis of the discretization process, either in the T.L. or in the U.L. description.

One way is to approximate the unknowns by interpolation functions, resulting in a finite element approach of mixed type. Another possibility is to start from the local equations (45). In this approach the system of first-order ordinary differential equations (45) is integrated numerically over each element, and the resulting integral matrix is transformed to a stiffness matrix. It can then be treated in the same way as in the usual displacement finite element approach. This method is particularly well suited to the treatment of the one-dimensional rod problem and has been shown to give very accurate results. It has the additional advantage that a certain relative error-bound can be given a-priori for the element matrix. It also appears to better account for the rapidly decaying warping functions, and their sometimes significant effect on the load-carrying behavior of slender beams, than stiffness matrices developed from a usual displacement formulation. For further details of the approach, which has been used successfully by the first author and his coworkers, see [2, 4, 11] . The examples given below have been solved in this way.

5. Numerical examples

The examples were mainly chosen to demonstrate the interaction of bending and torsion during large spatial deformations. It is assumed that the material is sufficiently elastic so that material nonlinearities need not be considered. The calculations are based on the U.L. expression of (45), i.e. (A5).

5.1 Cantilever under end moment

In this standard example an initially straight cantilever is deformed into a full circle by an increasing bending moment at the free end. The numerical results given in Figure 4 show good agreement with the analytical solution, and demonstrate the usefulness of the above approach.

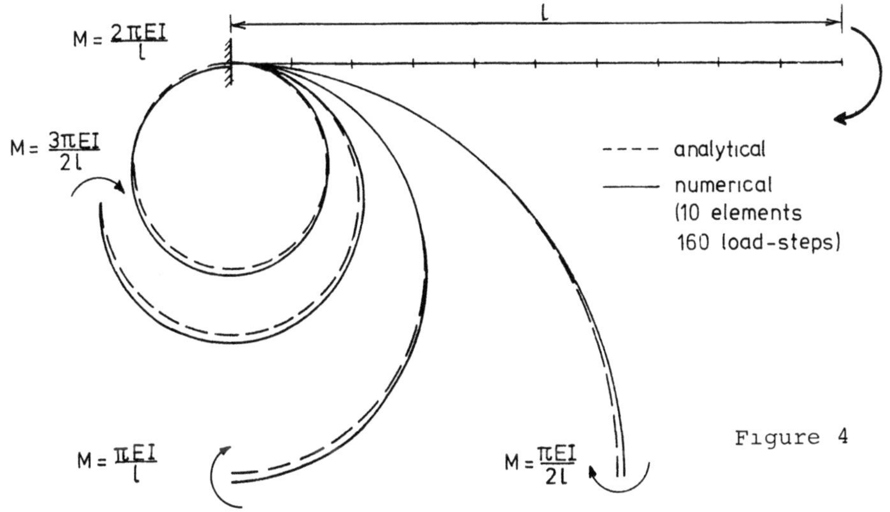

Figure 4

5.2 Two span beam under bending and torsion

In this example a two-span beam loaded by a single load in midspan, a constant distributed load as well as a distributed torsional moment is analyzed. It was chosen to demonstrate that warping constraints over the central support may be important, and that near the load causing sidesway buckling the torsional rotations are no longer small. The configuration is shown in Figure 5. The cross sectional parameters were taken as follows: $\overset{M}{\zeta} = -0.088\,m$, $EF = 1.008 \cdot 10^6 kN$, $EF_{\eta\eta} = 1575\ kN\,m^2$, $EF_{\zeta\zeta} = 16902.9\ kN\,m^2$, $EF_{\omega\omega} = 15.057\ kN\,m^4$, $GJ_T = 9.882\ kN\,m^2$. The boundary conditions are obvious from the figure and all loads were increased proportionally, i.e. $\bar{\underline{p}} = \lambda \cdot \overset{o}{\underline{p}}$, where the values of $\overset{o}{\underline{p}}$ were taken as: $\overset{o}{p}_{\zeta 1} = 5kN/m$, $\overset{o}{p}_{\zeta 2} = 1kN/m$, $\overset{o}{m}_1 = 0.2\ kNm/m$ $\bar{p}_\zeta = 10\ kN$.

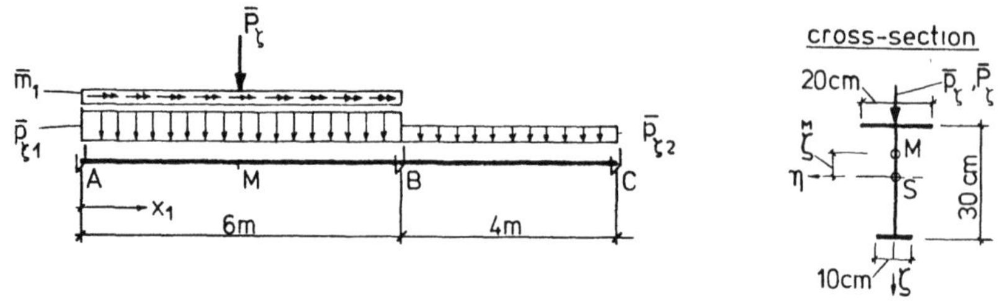

Figure 5

The numerical results are given in Figs. 5a-d. In Fig. 5a the load parameter λ is plotted against the vertical and lateral deflections at the midpoint M. It shows a clear nonlinear behavior as the critical load is approached. The dashed lines in Fig. 5a and all subsequent figures indicate the results obtained from a version of the above equations which is restricted to moderate rotations. It is seen that the discrepancies are small at low load levels but that they may become considerable as higher loads are approached. It is interesting to note that, depending on the loading range, these simplified equations may over- as well as underestimate displacements and moments.

Figure 5b shows that since failure is due to torsion the rotation angle φ_1 may become quite large at higher load levels. Finally, the torsional and bending moments M_1, M_2, M_3 as well as M_ω at points A, B, and M are plotted in Figs. 5c, d as functions of λ. From Fig. 5c it is seen that the bi-moment M_ω at the center support is of the same order of magnitude as the bending moment, and that at M the bending moment M_2 about the η axis even decreases after a certain load level has been reached, while the vertical deflection continues to increase rapidly.

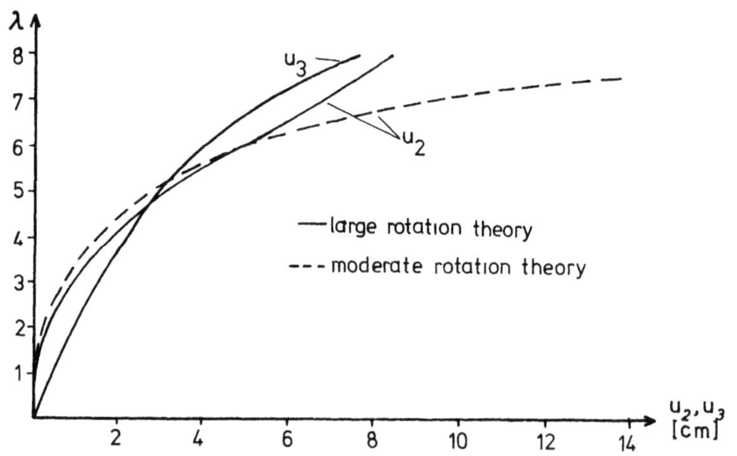

Figure 5a: Displacements u_2, u_3 at M

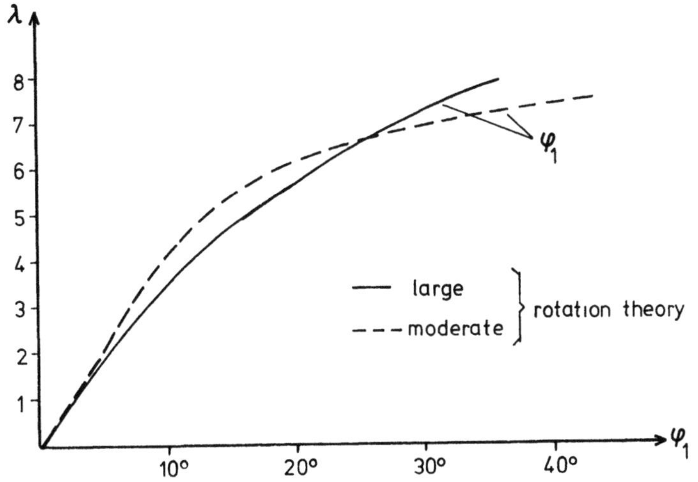

Figure 5b : Torsion angle φ at M

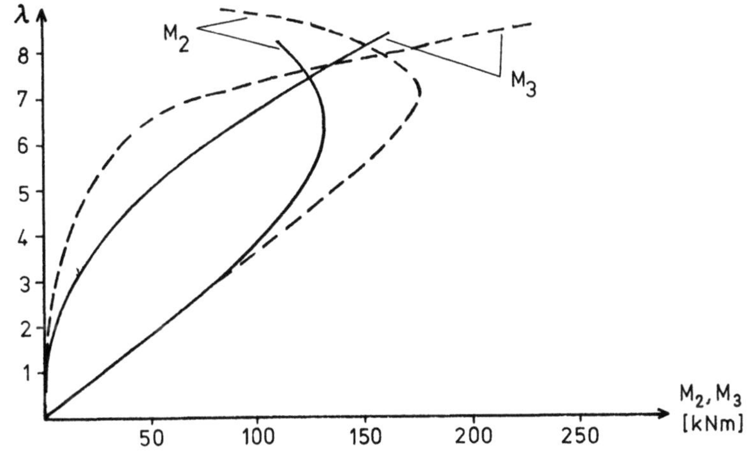

Figure 5c: Bending Moments M_2, M_3 at M

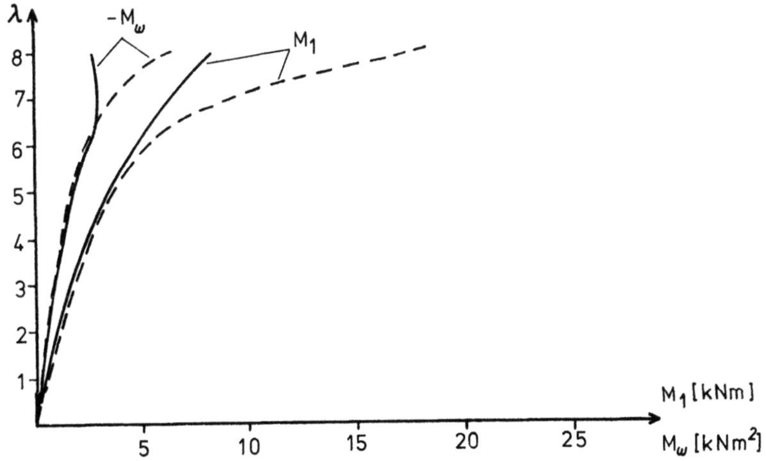

Figure 5d:Torsional moment M_1 and warping bi-moment M_ω at B

5.3 Cantilever under bending and torsion

The cantilever shown in Figure 6 has the same cross section as the beam in the previous example. It was again loaded proportionally by a vertical and a lateral end load as well as an end moment with relative magnitudes $\overset{\circ}{P}_\zeta = 2.5$ kN, $\overset{\circ}{P}_\zeta = 0.8$ kN, $\overset{\circ}{M}_1 = 0.2$ kN m.

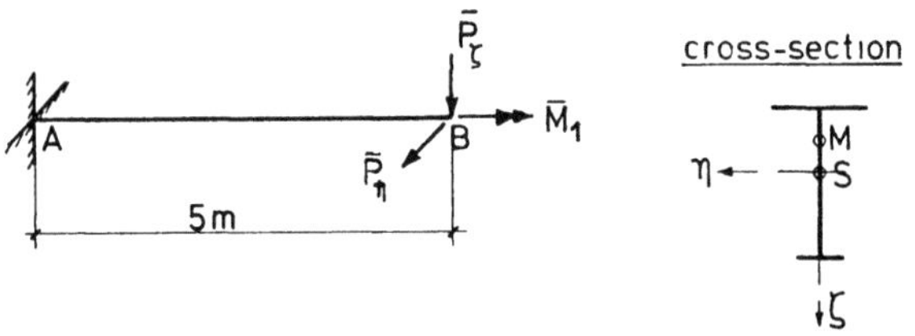

Figure 6

In Figs. 6 a-d the same quantities as before are plotted versus the load parameter λ. It is seen that as the failure load is approached, both vertical and lateral displacements increase very rapidly. The torsion angle however, increases much faster and becomes quite large at loads of less than 50 % of the failure load. Fig. 6d shows that at the clamped end the warping constraint is so severe that its influence may not be neglected.

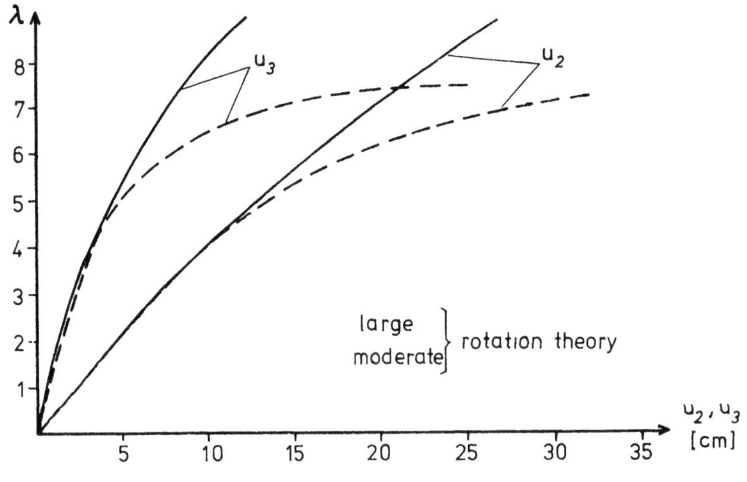

Figure 6a: Displacements u_2, u_3 at B

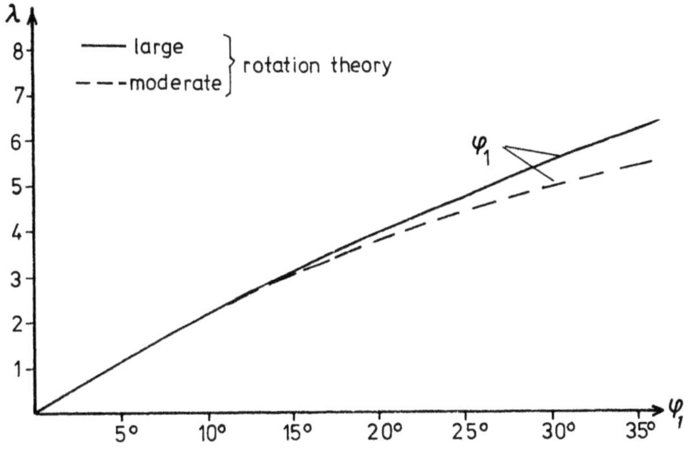

Figure 6b: Torsion angle φ_1 at B

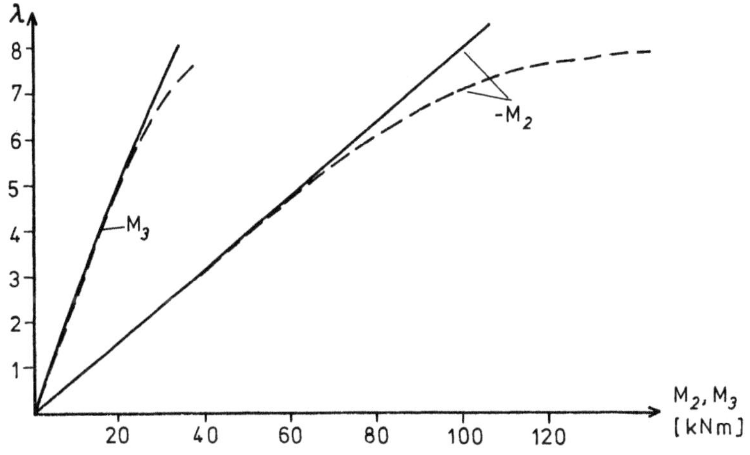

Figure 6c: Bending moments M_2, M_3 at A

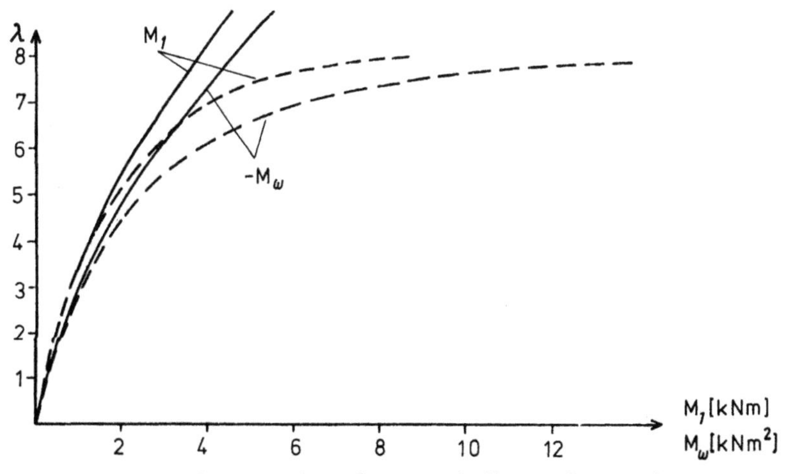

Figure 6d: Torsional moment M_1 and warping bi-moment M_ω at A

6. References

1 Bathe, K.J. and Bolourchi, S: Large displacement analysis of three-dimensional beam structures. Int. J. Num. Meth. Eng. 14 (1979) 961-986.

2 Wunderlich, W. and Beverungen, G.: Geometrically nonlinear theory and analysis of curved rods (in German). Bauingenieur 52 (1977) 225-237.

3 Argyris, J.H. et al.: On large displacement-small strain analysis of structures with rotational degrees of freedom. Comp. Meth.Appl. Mech. Eng. 14 (1978) 401-451, 15 (1978) 99-135.

4 Beverungen, G.: Geometrically nonlinear stress- and stability analysis of spatially curved rods (in German). Mitteilung Nr. 76-13 (1978),Techn.-wiss. Mitteilungen, Inst. f. Konstruktiven Ingenieurbau, Ruhr-Universität Bochum.

5 Fraeijs de Veubeke, B.: A new variational principle for finite elastic displacements. Int. J. Eng. Sci. 10 (1972) 745-763.

6 Murakawa, H. and Atluri, S.N.: Finite elasticity solutions using hybrid finite elements based on a complementary energy principle. J. Appl. Mech. 45 (1978) 539-547.

7 Wunderlich, W.: Incremental formulation for geometrically nonlinear problems. In "Formulations and Computational Algorithms in Finite Element Analysis". K.J. Bathe, J.T. Oden, W. Wunderlich, Eds. MIT Press (1977) 193-240.

8 Malvern, L.E.: Introduction to the mechanics of a continuous medium. Prentice-Hall, Englewood Cliffs (1969).

9 Wempner, G.: Complementary theorems of solid mechanics. In "Variational Methods in the Mechanics of Solids". S. Nemat-Nasser, Ed. Pergamon Press (1980).

10 Wempner, G.: Finite elements, finite rotations and small strains of flexible shells. Int. J. Solids Structures 5 (1969) 117-153.

11 Wunderlich, W.: Calculation of transfer matrices, applied to the bending theory of shells of revolution. In "The Use of Electronic Digital Computers in Structural Engineering". Proc. Int. Symp. Newcastle upon Tyne (1966).

Acknowledgement - The support of the "Deutsche Forschungsgemeinschaft" and the valuable help of Dr.-Ing. H.G.Liekweg and Dipl.-Ing. V. Schrödter in carrying out the numerical calculations is greatfully acknowledged.

Appendix

$$\overset{o}{\underline{\alpha}} = \begin{bmatrix} C_2C_3 & S_1S_2C_3-C_1S_3 & S_1S_3+C_1S_2C_3 \\ C_2S_3 & C_1C_3+S_1S_2S_3 & C_1S_2S_3-S_1C_3 \\ -S_2 & S_1C_2 & C_1C_2 \end{bmatrix} \quad \text{(A1)}$$

where $C_i = \cos \overset{o}{\varphi}_i \quad ; \quad S_i = \sin \overset{o}{\varphi}_i \quad (i = 1,2,3)$

$$\underline{\alpha} = \begin{bmatrix} \begin{matrix}-S_2C_3\varphi_2 \\ -C_2S_3\varphi_3\end{matrix} & \begin{matrix}(C_1S_2C_3+S_1S_3)\varphi_1 \\ +S_1C_2C_3\ \varphi_2 \\ -(C_1C_3+S_1S_2S_3)\varphi_3\end{matrix} & \begin{matrix}(C_1S_3-S_1S_2C_3)\varphi_1 \\ +C_1C_2C_3\varphi_2 \\ +(S_1C_3-S_2S_3C_1)\varphi_3\end{matrix} \\ \begin{matrix}-S_2S_3\varphi_2 \\ +C_2C_3\varphi_3\end{matrix} & \begin{matrix}(C_1S_2S_3-S_1C_3)\varphi_1 \\ +S_1S_3C_2\varphi_2 \\ +(S_1S_2C_3-S_3C_1)\varphi_3\end{matrix} & \begin{matrix}(-C_1C_3-S_1S_2S_3)\varphi_1 \\ +S_3C_1C_2\varphi_2 \\ +(S_1S_3+S_2C_1C_3)\varphi_3\end{matrix} \\ -C_2\varphi_2 & C_1C_2\varphi_1-S_1S_2\varphi_2 & -S_1C_2\varphi_1-S_2C_1\varphi_2 \end{bmatrix} \quad \text{(A2a)}$$

$$\underline{\omega} = \begin{bmatrix} 0 & S_1\varphi_2-C_1C_2\varphi_3 & C_1\varphi_2+S_1C_2\varphi_3 \\ -S_1\varphi_2+C_1C_2\varphi_3 & 0 & -\varphi_1+S_2\varphi_3 \\ -C_1\varphi_2-S_1C_2\varphi_3 & \varphi_1-S_2\varphi_3 & 0 \end{bmatrix} \quad \text{(A2b)}$$

$$\overset{o}{\underline{b}} = \begin{bmatrix} 1 & 0 & -S_2 \\ 0 & C_1 & S_1C_2 \\ 0 & -S_1 & C_1C_2 \end{bmatrix} \quad \text{(A3)}$$

214

$$
\left\{ \overset{c}{y}_1 \;\middle|\; \overset{c}{y}_2 \;\middle|\; \overset{c}{y}_3 \;\middle|\; \overset{c}{\omega}_1 \;\middle|\; \overset{c}{\omega}_2 \;\middle|\; \overset{c}{\omega}_3 \;\middle|\; \psi \;\middle|\; V_1 \;\middle|\; V_2 \;\middle|\; V_3 \;\middle|\; M_1 \;\middle|\; M_2 \;\middle|\; M_3 \;\middle|\; M_\omega \right\}
$$

$$
\delta \underline{z}^T \cdot
\left[\begin{array}{ccccccc|ccccccc}
 & & & & & & & {}_{,}\partial & & & & & & \\
 & & & & & & & & {}_{,}\partial & & & & & \\
 & & & & & & & & & {}_{,}\partial & & & & \\
 & & & \underline{L} & & & & & & & {}_{,}\partial & \mathring{\varkappa}_3 & -\mathring{\varkappa}_2 & \\
 & & & & & & & \frac{\mathring{\alpha}_{11}\mathring{Q}_3}{EA}+\mathring{\alpha}_{13} & \frac{\mathring{\alpha}_{21}\mathring{Q}_3}{EA}+\alpha_{23} & \frac{\mathring{\alpha}_{31}\mathring{Q}_3}{EA}+\mathring{\alpha}_{33} & -\mathring{\varkappa}_3 & {}_{,}\partial & \mathring{\psi} & \\
 & & & & & & & -\frac{\mathring{\alpha}_{11}\mathring{Q}_2}{EA}+\mathring{\alpha}_{12} & -\frac{\mathring{\alpha}_{21}\mathring{Q}_2}{EA}-\mathring{\alpha}_{22} & -\frac{\mathring{\alpha}_{31}\mathring{Q}_2}{EA}-\mathring{\alpha}_{32} & \mathring{\varkappa}_2 & -\mathring{\psi} & {}_{,}\partial & \\
 & & & & & & & & & & -1 & & & {}_{,}\partial \\
\hline
\partial_1 & & & & \frac{\mathring{\alpha}_{11}\mathring{Q}_3}{EA}+\mathring{\alpha}_{13} & -\frac{\mathring{\alpha}_{11}\mathring{Q}_2}{EA}-\mathring{\alpha}_{12} & & -\frac{\mathring{\alpha}_{11}^2}{EA} & -\frac{\mathring{\alpha}_{11}\mathring{\alpha}_{21}}{EA} & -\frac{\mathring{\alpha}_{11}\mathring{\alpha}_{31}}{EA} & & & & \\
 & \partial_1 & & & \frac{\mathring{\alpha}_{21}\mathring{Q}_3}{EA}+\alpha_{23} & -\frac{\mathring{\alpha}_{21}\mathring{Q}_2}{EA}-\mathring{\alpha}_{22} & & -\frac{\mathring{\alpha}_{11}\mathring{\alpha}_{21}}{EA} & -\frac{\mathring{\alpha}_{21}^2}{EA} & -\frac{\mathring{\alpha}_{21}\mathring{\alpha}_{31}}{EA} & & & & \\
 & & \partial_1 & & \frac{\mathring{\alpha}_{31}\mathring{Q}_3}{EA}+\mathring{\alpha}_{33} & -\frac{\mathring{\alpha}_{31}\mathring{Q}_2}{EA}-\mathring{\alpha}_{32} & & -\frac{\mathring{\alpha}_{11}\mathring{\alpha}_{31}}{EA} & -\frac{\mathring{\alpha}_{21}\mathring{\alpha}_{31}}{EA} & -\frac{\mathring{\alpha}_{31}^2}{EA} & & & & \\
 & & & \partial_1 & -\mathring{\varkappa}_3 & \mathring{\varkappa}_2 & -1 & & & & & & & \\
 & & & \mathring{\varkappa}_3 & \partial_1 & -\mathring{\psi} & & & & & & -\frac{1}{EF_{\zeta\zeta}} & & \\
 & & & -\mathring{\varkappa}_2 & \mathring{\psi} & \partial_1 & & & & & & & -\frac{1}{EF_{\eta\eta}} & \\
 & & & & & & \partial_1 & & & & & & & -\frac{1}{EF_{\omega\omega}}
\end{array}\right]
\qquad (A4)
$$

$$
\begin{array}{c}
\{\ \overset{c}{y}_1 \ \ \overset{c}{y}_2 \ \ \overset{c}{y}_3 \ \ \overset{c}{\omega}_1 \ \ \overset{c}{\omega}_2 \ \ \overset{c}{\omega}_3 \ \ \psi\ \}
\end{array}
$$

$$
\underline{L} = \left[\begin{array}{c|c|c|c|c|c|c}
 & & & & & & \\
\hline
 & & & & & & \\
\hline
 & & & & & & \\
\hline
 & & & \mathring{M}_1 \mathring{\psi} & \begin{array}{c} \mathring{M}_1 \mathring{\varkappa}_2 + \frac{1}{2} \mathring{M}_2 \mathring{\psi} \\ - \frac{1}{2} \mathring{M}_\omega \mathring{\psi} \mathring{\varkappa}_3 - \mathring{Q}_2 \end{array} & \begin{array}{c} \mathring{M}_1 \mathring{\varkappa}_3 + \frac{1}{2} \mathring{M}_3 \mathring{\psi} \\ + \frac{1}{2} \mathring{M}_\omega \mathring{\psi} \mathring{\varkappa}_2 - \mathring{Q}_3 \end{array} & -\overset{M}{\eta} \mathring{Q}_2 + \overset{M}{\zeta} \mathring{Q}_3 \\
\hline
 & & & \begin{array}{c} \mathring{M}_1 \mathring{\varkappa}_2 + \frac{1}{2} \mathring{M}_2 \mathring{\psi} \\ - \frac{1}{2} \mathring{M}_\omega \mathring{\psi} \mathring{\varkappa}_3 - \mathring{Q}_2 \end{array} & \begin{array}{c} -\mathring{M}_1 \mathring{\psi} + \mathring{M}_2 \mathring{\varkappa}_2 + \mathring{M}_3 \mathring{\varkappa}_3 \\ + \mathring{M}_\omega \mathring{\varkappa}_\omega - \mathring{Q}_3^2/EA + \mathring{Q}_1 \end{array} & & \mathring{M}_3 + \mathring{M}_\omega \mathring{\varkappa}_2 \\
\hline
 & & & \begin{array}{c} \mathring{M}_1 \mathring{\varkappa}_3 + \frac{1}{2} \mathring{M}_3 \mathring{\psi} \\ + \frac{1}{2} \mathring{M}_\omega \mathring{\psi} \mathring{\varkappa}_2 - \mathring{Q}_3 \end{array} & & \begin{array}{c} -\mathring{M}_1 \mathring{\psi} + \mathring{M}_2 \mathring{\varkappa}_2 + \mathring{M}_3 \mathring{\varkappa}_3 \\ + \mathring{M}_\omega \mathring{\varkappa}_\omega - \mathring{Q}_2^2/EA + \mathring{Q}_1 \end{array} & -\mathring{M}_2 + \mathring{M}_\omega \mathring{\varkappa}_3 \\
\hline
 & & & -\overset{M}{\eta} \mathring{Q}_2 + \overset{M}{\zeta} \mathring{Q}_3 & \mathring{M}_3 + \mathring{M}_\omega \mathring{\varkappa}_2 & -\mathring{M}_2 + \mathring{M}_\omega \mathring{\varkappa}_3 & GJ_T + i_M^2 \mathring{Q}_1
\end{array}\right] \qquad (A4a)
$$

$$\{\ \overset{c}{y}_1 \ \ \overset{c}{y}_2 \ \ \overset{c}{y}_3 \ \ \overset{c}{\varphi}_1 \ \ \overset{c}{\varphi}_2 \ \ \overset{c}{\varphi}_3 \ \ \psi \ \ V_1 \ \ V_2 \ \ V_3 \ \ M_1 \ \ M_2 \ \ M_3 \ \ M_\omega\ \}$$

$$\frac{d}{dx}\begin{bmatrix} \overset{c}{y}_1 \\ \overset{c}{y}_2 \\ \overset{c}{y}_3 \\ \overset{c}{\varphi}_1 \\ \overset{c}{\varphi}_2 \\ \overset{c}{\varphi}_3 \\ \psi \\ \hline V_1 \\ V_2 \\ V_3 \\ M_1 \\ M_2 \\ M_3 \\ M_\omega \end{bmatrix} = \left[\begin{array}{ccccccc|ccccccc}
 & k_3 & -k_2 & & -\frac{\mathring{Q}_3}{EA} & \frac{\mathring{Q}_2}{EA} & & \frac{1}{EA} & & & & & & \\
-k_3 & & h & & & 1 & & & & & & & & \\
k_2 & -h & & & -1 & & & & & & & & & \\
 & & & & +k_3 & -k_2 & 1 & & & & & & & \\
 & & & -k_3 & & h & & & & & & \frac{1}{EF_{\zeta\zeta}} & & \\
 & & & k_2 & h & & & & & & & & \frac{1}{EF_{\eta\eta}} & \\
 & & & & & & & & & & & & & \frac{1}{EF_{\omega\omega}} \\ \hline
 & & & & & & & & k_3 & -k_2 & & & & \\
 & & & & & & & -k_3 & & h & & & & \\
 & & & & & & & k_2 & -h & & & & & \\
 & & & & -\mathring{Q}_2 & -\mathring{Q}_3 & -\overset{M}{\eta}\mathring{Q}_2 + \overset{M}{\zeta}\mathring{Q}_3 & & & & & k_3 & -k_2 & \\
 & & & -\mathring{Q}_2 & \mathring{Q}_1 & & \mathring{M}_3 & \frac{\mathring{Q}_3}{EA} & & 1 & -k_3 & & -h & \\
 & & & -\mathring{Q}_3 & & \mathring{Q}_1 & -\mathring{M}_2 & -\frac{\mathring{Q}_2}{EA} & -1 & & k_2 & -h & & \\
 & & & -\overset{M}{\eta}\mathring{Q}_2 + \overset{M}{\zeta}\mathring{Q}_3 & \mathring{M}_3 & -\mathring{M}_2 & GJ_T + i_M^2\mathring{Q}_1 & & & & -1 & & &
\end{array}\right] - \begin{bmatrix} \\ \\ \\ \\ \\ \\ \\ \hline \bar{p}_1 \\ \bar{p}_2 \\ \bar{p}_3 \\ \bar{m}_1 \\ \bar{m}_2 \\ \bar{m}_3 \\ \bar{m}_\omega \end{bmatrix} \qquad (A5)$$

Large Deflection Finite Element Analysis of Pre- and Postcritical Response of Thin Elastic Frames

D. KARAMANLIDIS, A. HONECKER, K. KNOTHE
Technische Universitat Berlin, Germany

Abstract

A finite element variational approach and a computational algorithm are presented for predicting the nonlinear static, pre- and postcritical response of elastic framed structures. An updated Lagrangian formulation is used and the structure is discretized by using curved, mixed-hybrid beam elements. Throughout the development correction terms are maintained in the incremental energy functional to reduce the drifting of the approximate solution from the true solution. These correction terms correspond to checks on the stress resultants equilibrium and compatibility in the reference state. To tracing the unstable postcritical load deflection path in snap-through and bifurcation problems a generalized incrementation procedure has been developed and implemented. The high accuracy and effectiveness of the proposed approach is demonstrated by means of well-selected numerical examples.

1. Introduction

Large deflection nonlinear analysis has become in the last decade the focus of intensive efforts. Considerable progress has been made in the development of versatile and powerful finite element discretization methods as well as of improved numerical methods and programming techniques for nonlinear analysis of structures. Since the first applications of computers to nonlinear structural analysis a number of nonlinear beam elements have been presented. The large number of publications on this topic is due to the fact that various kinematic formulations as well as different energy functionals can be employed.

In the development of a geometrically nonlinear finite element procedure an updated Lagrangian or a total Lagrangian formulation can be employed. Although both formulations lead to identical results (as pointed out by Dupuis et al [1] and Bathe and Bolourchi [2]) their computational efficiency is different. In the case of geometrically nonlinear framework analysis the updated Lagrangian appears to be more effective than the total Lagrangian formulation.

The most popular finite element procedure for geometrically nonlinear analysis is still the so-called assumed displacement method based either on the principle of minimum potential work or the principle of virtual work (see, for example, Refs. [1 - 5]). Pirotin [6] was propably the first investigator to study the nonlinear response of a sinusoidal arch utilizing curved beam elements. In that work three different models have been implemented: an assumed displacement model, a mixed Reissner model and what he called a modified stress hybrid model. The idea underlying the latter model has been introduced by Pian's pioneering work [7]. Pirotin's arch problem was being analysed later by Atluri [8] and Boland [9]. Geometrically nonlinear analyses of plane frames using mixed variational procedures have been carried out also by Bäcklund [10] and more recently by Noor et al [11], Wunderlich and Beverungen [12] and Karamanlidis et al [13].

The objective of this paper is to present a consistent large rotation updated Lagrangian formulation for two-dimensional curved beams. The finite element procedure is based on an incremental hybrid mixed variational principle Π_{mR} containing both displacements and stress resultants as independent variables. The beam elements are assumed to be curved and the conventional beam displacement functions are employed as shape functions formulated in local, convective co-ordinates. The deformed element geometry has been for convenience interpolated in the same fashion as the beam lateral deflection. Finally, stress resultants have been approximated by trial functions which a priori satisfy the incremental homogeneous part of the equilibrium equations for the curved beam. The element has been implemented for use in elastic and elastic-plastic analysis in the pre- and postbuckling range (see, Refs. [14 - 16]). In the paper some demonstrative sample solutions are presented.

2. A geometrically nonlinear theory for plane curved beams

The equations of the two-dimensional, arbitrarily curved plane beam element are formulated based on the general continuum mechanics equations, using an incremental updated Lagrangian approach. These equations have been outlined by various authors previously (see, for example, Refs. [2,4,9]).

Fig. 1 shows a single curved beam element in three different states, in the initial configuration $\Omega^{(o)}$ in the reference configuration $\Omega^{(N)}$ and in the incrementally adjacent current configuration $\Omega^{(N+1)}$.

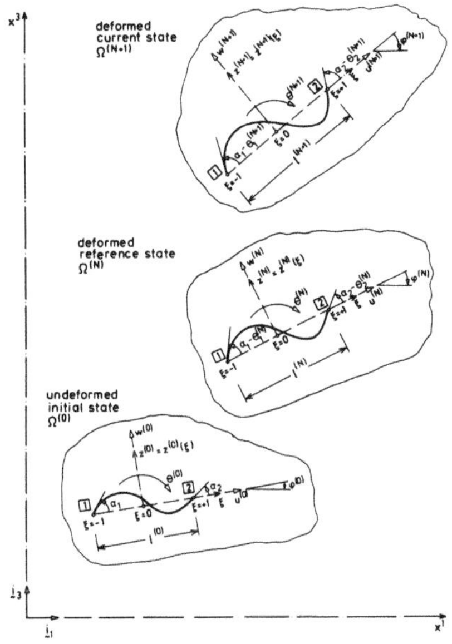

Figure 1. Local, convective co-ordinate-systems

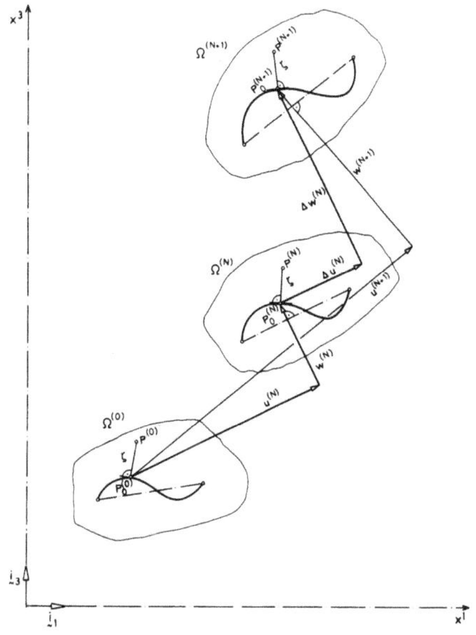

Figure 2. Description of motion by the updated Lagrangian formulation

The updated Lagrangian formulation of the governing equations of a deformable body requires the use of Euler stresses ${}^{(E)}\sigma_{ij}^{(N)}$ as initial stresses, and Truesdell stresses $\Delta^{*}\sigma_{ij}^{(N)}$ as incremental stresses. Consistent to these are the Almansi strains $\alpha_{ij}^{(N)}$ as initial strains and updated Green strains $\Delta^{*}\alpha_{ij}^{(N)}$ as incremental strains. In the case of a two-dimensional curved beam, these tensor quantities are reduced to the scalar state variables ${}^{(E)}\sigma^{(N)}$, $\Delta^{*}\sigma^{(N)}$, $\alpha^{(N)}$ and $\Delta^{*}\alpha^{(N)}$.

Following Washizu [17] the updated Green strain increment $\Delta^* \alpha^{(N)}$ can be for convenience decomposed into a linear and into a quadratic part. The linear part as well as the Almansi initial strain $\alpha^{(N)}$ can be further decomposed in their extensional and curvature parts:

$$\Delta^*\alpha^{(N)} = \Delta^* e^{(N)} + \zeta\Delta^*\kappa^{(N)} + \frac{1}{2}(\Delta w,_x^{(N)})^2 \qquad (1a),$$

$$\alpha^{(N)} = e^{(N)} + \zeta\kappa^{(N)} \qquad (1b).$$

where (see Fig. 2)

$$\Delta^* e^{(N)} = \Delta u,_x^{(N)} + z,_x^{(N)} \Delta w,_x^{(N)} \qquad (2a),$$

$$e^{(N)} = u,_x^{(N)} + z,_x^{(N)} \; w,_x^{(N)} - \frac{1}{2}(w,_x^{(N)})^2 \qquad (2b),$$

$$\Delta^*\kappa^{(N)} = - \Delta w,_{xx}^{(N)} \qquad (2c),$$

$$\kappa^{(N)} = - w,_{xx}^{(N)} \qquad (2d).$$

To obtain these strain displacement relations the assumptions of Marguerre's shallow shell theory have been adopted (see, for example, Refs. [9,13,14]).

Integration of the normal stress ${}^{(E)}\sigma^{(N)}$ resp. $\Delta^* \sigma^{(N)}$ over the depth h of the beam leads to the stress resultants (see Fig.3):

$${}^{(E)}N^{(N)} + \Delta^* N^{(N)} = \int_{-h/2}^{h/2} \left[{}^{(E)}\sigma^{(N)} + \Delta^*\sigma^{(N)}\right] d\zeta \qquad (3a),$$

$${}^{(E)}M^{(N)} + \Delta^* M^{(N)} = \int_{-h/2}^{h/2} \left[{}^{(E)}\sigma^{(N)} + \Delta^*\sigma^{(N)}\right] \zeta d\zeta \qquad (3b).$$

In this paper stress resultants and beam strain quantities are assumed to satisfy the linear constitutive law:

$$\left\{\begin{bmatrix} {}^{(E)}N^{(N)} + \Delta^* N^{(N)} \\ {}^{(E)}M^{(N)} + \Delta^* M^{(N)} \end{bmatrix}\right\} = \begin{bmatrix} EA & o \\ o & EI \end{bmatrix} \left\{\begin{bmatrix} e^{(N)} + \Delta^* e^{(N)} \\ \kappa^{(N)} + \Delta^*\kappa^{(N)} \end{bmatrix}\right\} \qquad (4),$$

where E A and E I are the extensional and flexural beam stiffness, respectively.

Finally, the following stress resultants equilibrium equations can be formulated

$$^{(E)}N^{(N)}_{,x} + \Delta^{*}N^{(N)}_{,x} + \bar{p}^{(N)}_{1} + \Delta\bar{p}^{(N)}_{1} = 0 \qquad (5a),$$

$$^{(E)}S^{(N)}_{,x} + \Delta^{*}S^{(N)}_{,x} + \bar{p}^{(N)}_{3} + \Delta\bar{p}^{(N)}_{3} = 0 \qquad (5b),$$

$$^{(E)}M^{(N)}_{,x} + \Delta^{*}M^{(N)}_{,x} + {}^{(E)}N^{(N)}\Delta w^{(N)}_{,x} + \left[{}^{(E)}N^{(N)}+\Delta^{*}N^{(N)}\right]z^{(N)}_{,x} = {}^{(E)}S^{(N)}+\Delta^{*}S^{(N)} \qquad (5c),$$

where ${}^{(E)}S^{(N)}+\Delta^{*}S^{(N)}$ denotes the shear force and $\bar{p}_1{}^{(N)}+\Delta\bar{p}_1{}^{(N)}$ and $\bar{p}_3{}^{(N)}+\Delta\bar{p}_3{}^{(N)}$ are the external distributed loads (see Fig.4).

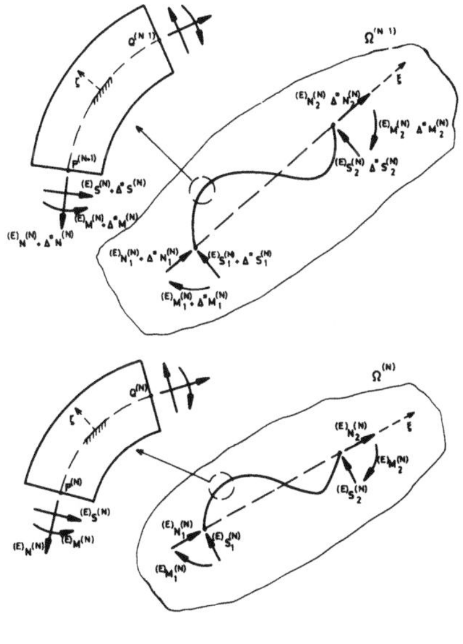

Figure 3. Beam stress resultants within the updated Lagrangian formulation

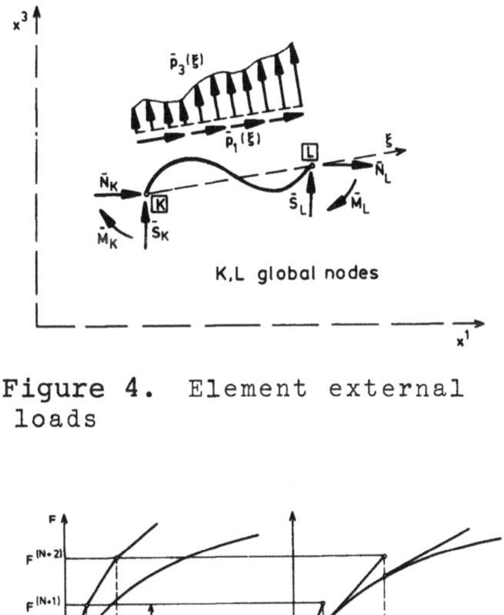

Figure 4. Element external loads

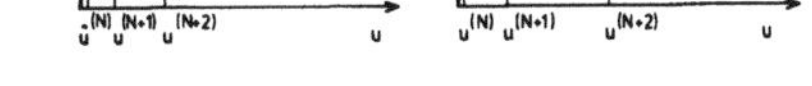

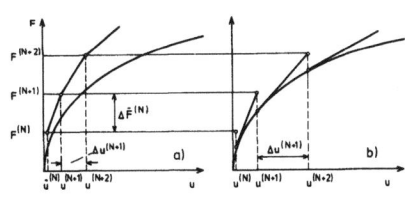

Figure 5. Incremental solution schemes, a) Euler method
b) self-correcting method

3. Finite element variational approach

The finite element procedure utilized in this paper is based on a generalized incremental variational principle in which the displacements Δu and Δw and the stress resultants $\Delta^{*}N$ and $\Delta^{*}M$ are the independent variables⁺. Assuming the two-dimensional frame to be discretized into (n) individual elements the generalized energy functional Π_{mR} is defined as follows [14]:

$$
\begin{aligned}
\Pi_{mR} \equiv \sum_{n} \Big\{ \int_{l(N)} \Big[-\frac{(\Delta^{*}N)^{2}}{2EA} - \frac{(\Delta^{*}M)^{2}}{2EI} + {}^{(E)}N\,\tfrac{1}{2}(\Delta w,_{x})^{2} \Big]\,dx \\
+ \Big[\Delta^{*}N\,\Delta u + \Delta^{*}S\,\Delta w + \Delta^{*}M\,\Delta\Theta \Big]_{o}^{l(N)} \Big\} \\
\underbrace{\begin{aligned}
- \sum_{n} \int_{l(N)} (\Delta\bar{p}_{1}\Delta u + \Delta\bar{p}_{3}\Delta w)\,dx - \sum_{k} (\Delta\bar{N}\Delta u + \Delta\bar{S}\Delta w + \Delta\bar{M}\Delta\Theta) \\
+ \sum_{n} \int_{l(N)} \Big[{}^{(E)}N(\Delta u,_{x} + z,_{x}\Delta w,_{x}) - {}^{(E)}M\Delta w,_{xx} - \bar{p}_{1}\Delta u - \bar{p}_{3}\Delta w \Big]\,dx \\
- \sum_{k} (\bar{N}\Delta u + \bar{S}\Delta w + \bar{M}\Delta\Theta)
\end{aligned}}_{①} \\
\underbrace{\begin{aligned}
- \sum_{n} \int_{l(N)} \Big\{ \Delta^{*}N \Big[\frac{{}^{(E)}N}{EA} - u,_{x} - z,_{x}\,w,_{x} + \tfrac{1}{2}(w,_{x})^{2} \Big] \\
+ \Delta^{*}M \Big(\frac{{}^{(E)}M}{EI} + w,_{xx} \Big) \Big\}\,dx
\end{aligned}}_{②} = \text{stationary} \qquad (6)
\end{aligned}
$$

Eqs. (6) is subject to the following constraint conditions

$$\Delta^{*}N,_{x} = 0 \qquad \text{in } l^{(N)} \qquad (7a),$$

$$\Delta^{*}M,_{xx} + (\Delta^{*}N\,z,_{x}),_{x} = 0 \qquad \text{in } l^{(N)} \qquad (7b),$$

which represent the linear incremental homogeneous parts of the stress resultants equilibrium equations (5). On the other hand the kinematic boundary conditions and the interelement displacement compatibility must be satisfied exatly.

⁺ Since no confusion is possible the state label 'N' will be dropped in the following.

The functional Π_{mR} , eqs. (6), contains two types of correction terms, an equilibrium correction term (1) and a compatibility correction term (2) . These correction terms are due to the fact that equilibrium equations for the initial stresses as well as strain displacement relations for the initial strains are nonlinear equations which cannot be fulfiled completely by a sequence of linearized incremental equations. During the incremental procedure presented here the correction terms in the load step (N) are taken into account in the following step (N+1). According to the nomenclature introduced by Pian [18] and Stricklin et al [19] the geometrically nonlinear procedure presented in this paper can be thereafter identified as a self-correcting incremental procedure (see Fig. 5).

4. Interpolation functions

It is convenient to introduce matrix notation in this chapter to replace the scalar notation above. Also, the independent variables of the functional Π_{mR} shall be interpolated in terms of trial functions appropriately chosen to satisfy the aforementionend constraint conditions of the functional.

The displacement interpolation functions are constructed assuming cubic variations for Δw and linear variations for Δu

$$\Delta \underset{\sim}{u} = \underset{\sim}{N} \, \Delta \hat{\underset{\sim}{u}}_n \tag{8}$$

where $\Delta \hat{\underset{\sim}{u}}_n$ is the element nodal displacement vector. Integration of the ordinary differential equations (7) with respect to the local, convective co-ordinate x leads to the stress resultants interpolation functions

$$\Delta \underset{\sim}{\sigma} = \begin{bmatrix} \Delta^* N \\ \Delta^* M \end{bmatrix} = \begin{bmatrix} 1 & 0 & 0 \\ -z & 1 & x \end{bmatrix} \begin{bmatrix} \beta_1 \\ \beta_2 \\ \beta_3 \end{bmatrix} = \underset{\sim}{P} \, \underset{\sim}{\beta}_n \tag{9}$$

Introducing eqs. (8) and (9) in eqs. (6) yields

$$\Pi_{mR} \equiv \sum_n \{ -\frac{1}{2} \beta_n^T H_n \beta_n + \frac{1}{2} \Delta\hat{u}_n^T K_n^\sigma \Delta\hat{u}_n$$

$$- \Delta\hat{u}_n^T \Delta\bar{q}_n + \beta_n^T G_n \Delta\hat{u}_n + \Delta\hat{u}_n^T g_n - \beta_n^T c_n \}$$

$$- \sum_k \Delta u_k^T \Delta\bar{T}_k \qquad = \text{stationary} \qquad (10)$$

where H is the flexibility matrix
K^σ is the initial stress stiffness matrix
$\Delta\bar{q}$ is the incremental element lumped load vector
g is the equilibrium imbalance term and
c is the compatibility mismatch term.

$\Delta u^T_k = \{\Delta u_k, \Delta w_k, \Delta\Theta_k\}$ and $\Delta\bar{T}^T_k = \{\Delta\bar{N}_k, \Delta\bar{S}_k, \Delta\bar{M}_k\}$ denote the displacement and load vector of the global nodal point k, respectively.

Taking the variation of Π_{mR}, eqs. (10), with respect to the independent coefficients β_n yields a solution for the β_n in terms of the nodal displacement vectors $\Delta\hat{u}_n$:

$$\beta_n = H_n^{-1} G_n \Delta\hat{u}_n - H_n^{-1} c_n \qquad (12).$$

Placing this into Π_{mR} results

$$\Pi_{mR} \equiv \sum_n \{\frac{1}{2}\Delta\hat{u}_n^T(G_n^T H_n^{-1} G_n + K_n^\sigma)\Delta\hat{u}_n - \Delta\hat{u}_n^T(\Delta\bar{q}_n$$

$$+ G_n^T H_n^{-1} c_n - g_n)\} - \sum_k \Delta u_k^T \Delta\bar{T}_k$$

$$= \text{stationary} \qquad (13).$$

Finally, taking the variation of Π_{mR} with respect to the global nodal displacement vector $\Delta\hat{u}$ yields the matrix equation

$$K \Delta\hat{u} = \Delta\bar{q} - k$$

where

K is the assembled (global) tangent stiffness matrix
$\Delta\bar{q}$ is the global load vector and
k is the consistently generated equivalent correction load vector.

5. Algorithms for tracing the complete deformation path

Eqs. (13) is usually applied in an inversed form such that the incremental displacement vector $\Delta\hat{\underset{\sim}{u}}$ is obtained by premultiplying the right hand side vector $\Delta\bar{\underset{\sim}{q}} - \underset{\sim}{k}$ by the inverse of the matrix $\underset{\sim}{K}$. However, difficulty arises if the load displacement relation presents a snap-through or bifurcation behaviour. In these cases, in the neighborhood of the critical points the matrix $\underset{\sim}{K}$ becomes singular, so that its inversion is impossible.

In the past, several schemes have been suggested for tracing the postcritical deformation path of instability problems. The most popular of these procedures are:

- displacement incrementation (Argyris, Pian/Tong, et al),
- ficticious springs concept (Whitman/Gaylord, Sharifi),
- " arc length" incrementation (Riks, Wempner, Wessels), and
- current stiffness parameter concept (Bergan).

The common feature of the first three methods is to introduce modifications or constraints that render the matrix $\underset{\sim}{K}$ positive definite in the postbuckling range, while the procedure by Bergan simply "jumps" over the instability points.

In solution of practical problems it is highly desirable that an algorithm tracing the complete deformation path fulfils the following appropriate requirements:

- The method should not fail in the case of load displacement curves having vertical tangents.
- The method must be capable of detecting that a bifurcation point has been reached in order to trace the secondary equilibrium path.
- Symmetry and sparse population (bandwidthness) of the coefficient matrix $\underset{\sim}{K}$ must be preserved.
- In order to keep the truncation error of the incremental procedure nearly the same for each loading step, the load steps should be varied automatically according to the local curvature of the load displacement curve.

In a recent work [20] by the junior author (A.H.) a new incrementation scheme for tracing the complete deformation path has been developed and implemented. The numerical results for several test examples lead to the conclusion, that the new incrementation scheme, to be outlined and discussed in the following, satisfies nearly all of the requirements above.

The basic idea behind this new step-by-step algorithm in to impose, in each load step, external work increments $\Delta S^{(N)}$ (see Fig. 6)

$$\Delta S^{(N)} = \Delta \hat{\underset{\sim}{u}}^{(N)^T} \Delta \bar{\underset{\sim}{q}}^{(N)} \tag{16}$$

rather than load or displacement increments. For proportional loading an arbitrary load increment may be expressed by scaling an appropriately choised reference load $\bar{\underset{\sim}{q}}_{ref}$ (e.g. the estimated buckling load):

$$\Delta \bar{\underset{\sim}{q}}^{(N)} = \Delta\lambda^{(N)} \bar{\underset{\sim}{q}}_{ref} \tag{17}$$

where λ denotes the load parameter. It is convenient to scale the incremental displacement vector $\Delta\hat{\underset{\sim}{u}}^{(N)}$ in the same manner as the incremental load vector, i.e.:

$$\Delta \hat{\underset{\sim}{u}}^{(N)} = \Delta\lambda^{(N)} \hat{\underset{\sim}{u}}^{(N)}_{ref} \tag{18}$$

where the reference displacement vector $\hat{\underset{\sim}{u}}^{(N)}_{ref}$ is defined by (see Fig. 6)

$$\underset{\sim}{K}^{(N)} \hat{\underset{\sim}{u}}^{(N)}_{ref} = \bar{\underset{\sim}{q}}_{ref} \tag{19}$$

Placing eqs. (17) and (18) into eqs. (16) yields:

$$\Delta S^{(N)} = (\Delta\lambda^{(N)})^2 \bar{\underset{\sim}{q}}^T_{ref} \hat{\underset{\sim}{u}}^{(N)}_{ref} \tag{20}$$

so that

$$\Delta\lambda^{(N)} = \pm \sqrt{\frac{\Delta S^{(N)}}{\bar{\underset{\sim}{q}}^T_{ref} \hat{\underset{\sim}{u}}^{(N)}_{ref}}} \tag{21}$$

In our numerical studies, carried out by the new incrementation algorithm the following choices have been made:

$$\Delta S^{(N)} = \rho \left| \bar{\underset{\sim}{q}}^T_{ref} \hat{\underset{\sim}{u}}^{(1)}_{ref} \right| \tag{22}$$

$$\rho = 0.025 \div 0.100 \ \% \tag{23}$$

$$\mathrm{sign}\{\Delta\lambda^{(N)}\} = \mathrm{sign}\{S^{(N)}\} \cdot \mathrm{sign}\{S^{(N)} \cdot \Delta\lambda^{(N-1)}\} \qquad (24).$$

In the neighborhood of instability points, application of eqs. (21) is still possible, producing very small load increments and in some cases local drifting from the true load deflection path. On the other hand, it is seen from eqs. (21) that the algorithm fails in regions of the load deflection path where the denominator $\bar{q}^T_{ref}\ \hat{u}^{(N)}_{ref}$ becomes close to zero. For problems where the loading consists of a single concentrated load only, this situation arises if the load deflection curve has a nearly vertical tangent. To overcome this difficulty, in the numerical studies to be presented in the next chapter, it has been found to be convenient to replace in these regions (A → B and C → D, Fig. 6) eqs. (21) by

$$\Delta\lambda^{(N)} = \mathrm{sign}\{\Delta\lambda^{(N-1)}\} \cdot \Delta\lambda^{(1)} \qquad (25).$$

The applicability of the proposed incrementation algorithm will be demonstrated in the next chapter.

Finally, it is worthnoting that Bergan's current stiffness parameter concept appears to fail not only in the case of the load deflection curve having a vertical tangent, but also for problems where the current stiffness parameter S_p is nearly constant (e.g. Euler buckling problems).
The reasons for these shortcommings are:

- In the case of a vertical tangent the current stiffness parameter S_p becomes infinite (see, for example, eqs. (6) in Ref. [24]).
- If S_p is nearly constant ($\Delta S_p^{(N)}$ close to zero) eqs. (10) in [24] becomes meaningless.

Indeed, inspection of the available literature on the current stiffness parameter concept [21-26, 28] reveal that nonlinear structural problems having one of the properties above (i.e. vertical tangent or S_p = constant) have never been considered.

6. Numerical test studies

In order to test and evaluate the beam element resp. the step-by-step algorithm proposed, a large number of arch and frame problems were solved. In this paper results of some selected problems are presented and discussed. These problems are: 1) Symmetrical buckling of Williams'toggle frame, 2) Shallow sinusoidal arch under gravitiy loading, 3) Bifurcation of a right-angle frame, and 4) Nonlinear response of a nonshallow clamped-hinged circular arch.

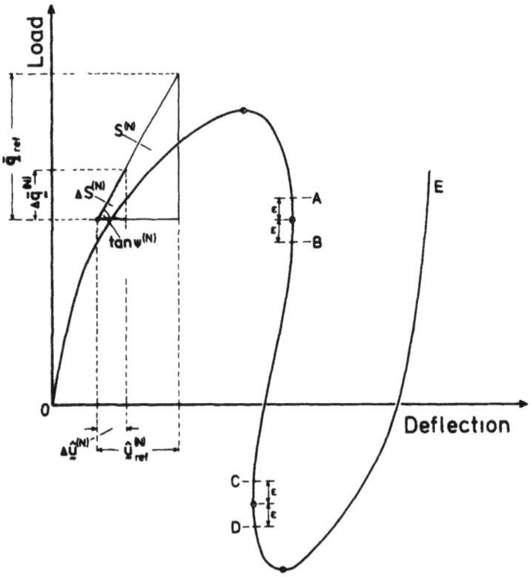

Figure 6. Idealized load deflection curve with horizontal and vertical tangents

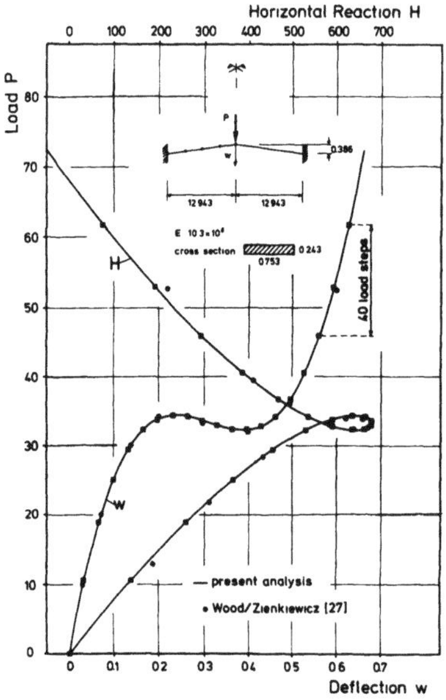

Figure 7. Load deflection curves of Williams'toggle frame

6.1 Williams'toggle frame

This problem has been treated by Frey [4] and Wood and Zienkiewicz [27] previously. In both references the assumed displacement finite element approach together with a modified Newton/Raphson method have been used. The finite element results using the mixed-hybrid beam element and the self-correcting step-by-step procedure, are shown in Fig. 7, where very good agreement with the numerical results of Ref. [27] is obtained. The variation of Bergan's current stiffness parameter S_p as function of load level is shown in Fig. 8.

6.2 Shallow sinusoidal arch under gravity loading

The shallow sinusoidal arch shown in Fig. 9 has been frequently used to demonstrate the applicability of Bergan's current stiffness parameter concept (see Refs. [21-24]). As it is seen from Figs. 7, 8, 9, 10 and 11 the nonlinear response of this structure is very similar to that exhibited by William's toggle. The analysis presented here has been carried out using the new incrementation algorithm. Fig. 9 indicates very good agreement of the obtained solution with the numerical results by Bergan et al [22]. The variations of the axial thrust N and the current stiffness parameter S_p as functions of load level are shown in Figs. 10 and 11, respectively.

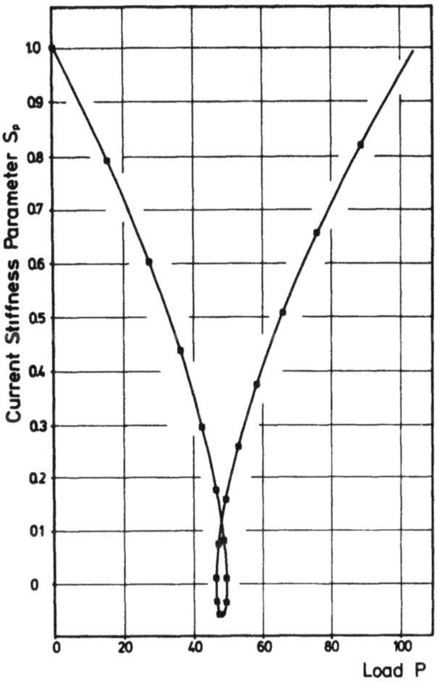

Figure 8. Variation of S_p for Williams' toggle frame

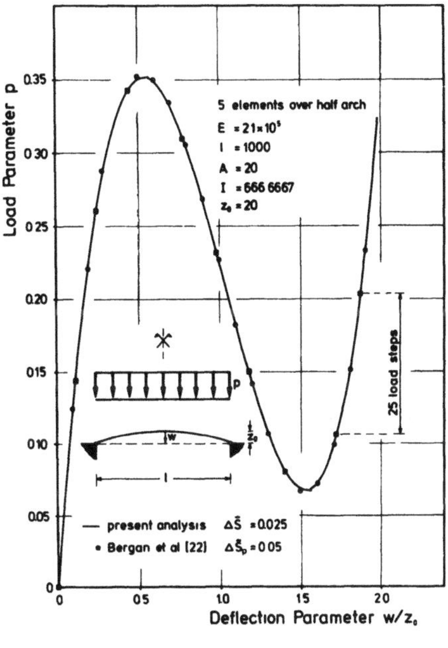

Figure 9. Load deflection curve of shallow arch

6.3 Right-angle frame

The right-angle frame under point load shown in Fig. 12 has become nearly a classical case for demonstrating experimentally or analytically structural imperfection sensitivity. The problem has been analysed previously by Argyris and Dunne

[5] and by Frey [4] using the finite element displacement method together with a modified Newton/Raphson iterative procedure.

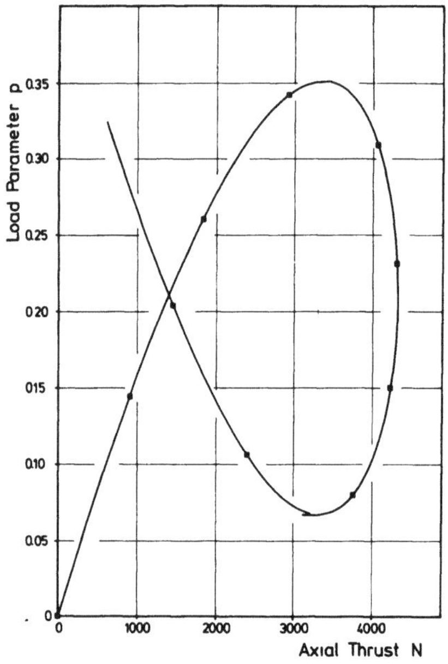

Figure 10. Variation of axial thrust N for a shallow arch

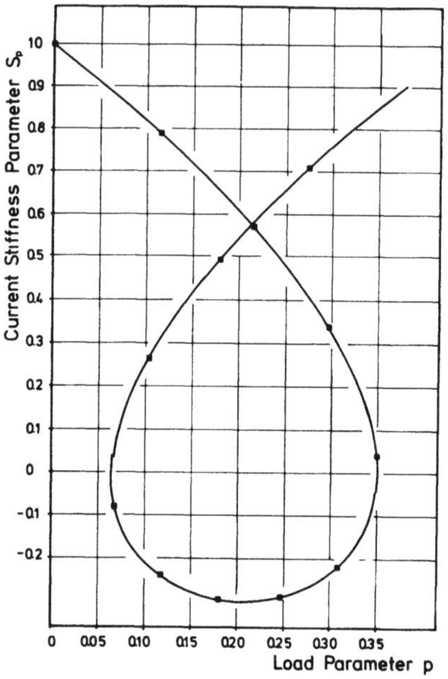

Figure 11. Sinusoidal shallow arch: Variation of S_p

The finite element results of the present analysis compared with the reference solutions [5] and [4] (see Figs. 12 and 13) clearly indicate the following:

- The proposed nonlinear mixed-hybrid beam element produces accurate results even if the deflections arising in the problem under consideration are very large.

- Using the pure incremental solution algorithm large deviations of the approximate solution from the true solution can arise, even if the loading steps used have been kept very small.

- The step-by-step algorithm proposed in this paper is capable to trace the complete nonlinear load deflection path of this complex structure. It should be remarked in this connection that according the variation of S_p as function of load level (not shown in this paper), it seems to be not possible to analyse this problem completely using Bergan's current stiffness parameter concept.

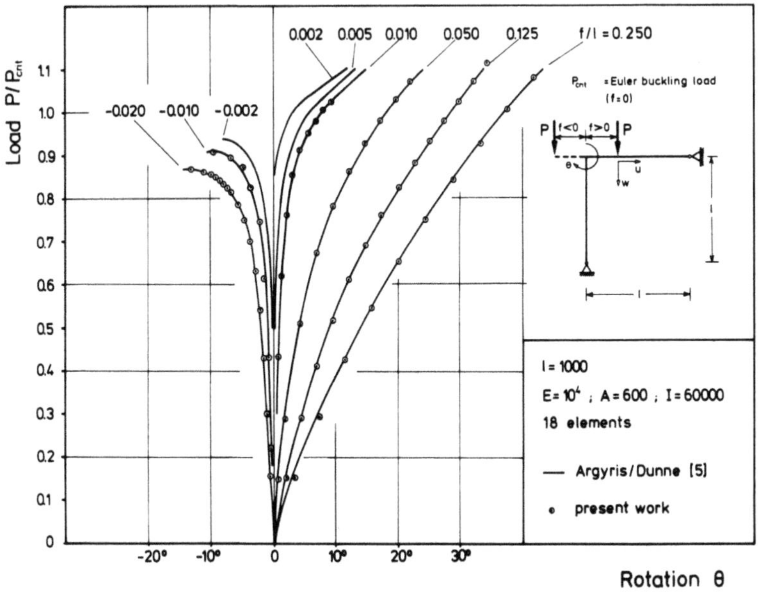

Figure 12. Right-angle frame: geometry and imperfection sensitivity

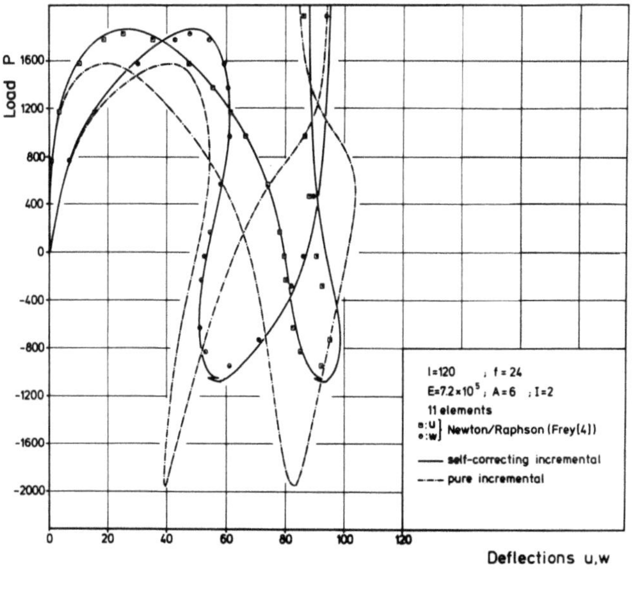

Figure 13. Complete load deflection curves of right-angle frame

6.4 Nonshallow clamped-hinged circular arch

Because of its extreme complexity, the complete nonlinear pre- and postcritical response of the clamped-hinged circular arch shown in Fig. 14 has not been studied previously. In this paper the example was selected to test the usefulness of the proposed new algorithm to trace a very complex load deflexion path. An illustration of the nonlinear behaviour of this arch is provided in Fig. 14 which shows a plot of vertical central deflection versus applied load. It is seen that the clamped-hinged arch with single load at apex displays a bifurcation as well as a snap-through singularity. In the example under consideration it was possible to trace the two braches of the

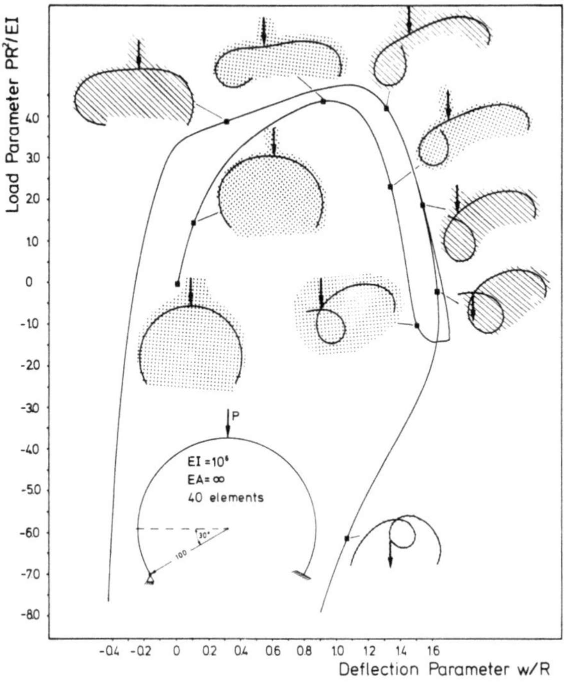

Figure 14. Clamped-hinged circular arch: load deflection curves and deformed configurations

load deflection curve by off-loading after passing the bifurcation point. It should be noted that as no reference solution exists, accuracy of the numerical results presented here can be indicated by realising that for all load steps the bending moment at left end as well as the norm of the correction vector $\underset{\sim}{k}$ are very small (ca. 10^{-4}:1) in comparison with the stress resultants. Regardless of that, there is no doubt that further studies using an iterative scheme are necessary. Only then, one has the possibility to make the numerical drifting error of the approximate solution from the true solution as small as possible.

7. Conclusions

In this paper it has been demonstrated that

- the mixed-hybrid beam element provides an adequate tool for geometrically nonlinear analyses even if displacements as well as rotations are very large,
- the proposed self-correcting, step-by-step procedure is capable to trace automatically the complete pre- and post-critical equilibrium path even in very complex nonlinear situations. Nevertheless, further studies are needed to improve the capability of the scheme to trace the passage from the primary to the secondary path in general cases of bifurcation instability.

References

1 Dupuis, G.A.; Hibitt, H.D.; McNamara, S.F.; Marcal, P.V.: Nonlinear material and geometric behavior of shell structures. Comp. & Struct. 1 (1971) 223-239.

2 Bathe, K.-J.; Bolourchi, S.: Large displacement analysis of three -dimensional beam structures. Int. J.Num. Meth. Engng. 14 (1979) 961-986.

3 Ramm, E.: Geometrisch nichtlineare Elastostatik und finite Elemente. Habilitationsschrift, Universität Stuttgart, 1976.

4 Frey, F.: L'analyse statique non lineaire des structures par la methode des elements finis et son application a la construction metallique. Ph. D. Thesis, Universite de Liege, 1978.

5 Argyris, J.H.; Dunne, P.C.: On the application of the natural mode technique to small strain large displacement problems. Proc. World Congress on Finite Element Methods in Structural Mechanics, Bournemouth/England, 1975.

6 Pirotin, S.D.: Incremental large deflection analyses of elastic structures. Ph. D. Thesis, M.I.T., 1971.

7 Pian, T.H.H.: Derivation of element stiffness matrices by assumed stress distributions, AIAA J.2 (1964)1333-1336.

8 Atluri, S.: On the hybrid stress finite element model for incremental analysis of large deflection problems. Int. J. Solids Struct. 9 (1973) 1177-1191.

9 Boland, P.L.: Large deflection analysis of thin elastic structures by the assumed stress hybrid finite element method. Ph. D. Thesis, M.I.T., 1975.

10 Bäcklund, J.: Finite element analysis of nonlinear structures. Ph. D. Thesis, Göteborg, 1973.

11 Noor, A.K.; Greene, W.H.; Hartley, S.J.: Nonlinear finite element analysis of curved beams. Comp. Meth.Appl.Mech. Engng. 12 (1977) 289-307.

12 Wunderlich, W.; Beverungen, G.: Geometrisch nichtlineare Theorie eben gekrümmter Stäbe. Bauingenieur 52 (1977) 225-237.

13 Karamanlidis, D.; Tsuzuki, O.; Knothe, K.: A study of the geometrically nonlinear behaviour of plane curved beam structures using a mixed hybrid finite element procedure. ILR Mitt. 54, Berlin, 1978.

14 Karamanlidis, D.: Finite Elementmodelle zur numerischen Berechnung des geometrisch nichtlinearen Verhaltens ebener Rahmentragwerke im unter- und überkritischen Bereich. Fortschr. - Ber. VDI - Z 1/63, Düsseldorf, 1980.

15 Mast, St.; Karamanlidis, D.: Elastoplastische Berechnungen von ebenen Rahmentragwerken nach einem gemischt-hybriden Finite-Element-Verfahren. DER STAHLBAU (in press).

16 Karamanlidis, D.; Knothe, K.: Geometrisch nichtlineare Berechnung von ebenen Stabwerken auf der Grundlage eines gemischt-hybriden Finite-Element-Verfahrens. To appear in: Ingenieur Archiv.

17 Washizu, K.: Variational methods in elasticity and plasticity, 2nd Edition. Pergamon Press, 1975.

18 Pian, T.H.H.: Variational principles for incremental finite element methods. J. Franklin Inst. 302 (1976) 473-488.

19 Stricklin, J.A.; Haisler, W.E.; von Riesemann, W.A.; Evaluation of solution procedures for material and/or geometric nonlinear structural analysis. AIAA J.11 (1973) 292-299.

20 Honecker, A.: Entwicklung, Implementierung und Austestung von Lösungsalgorithmen für Gleichungssysteme mit singulär werdender Funktionalmatrix. Diploma Thesis, Berlin, 1980.

21 Bergan, P.G.; Horrigmoe, G.; Krakeland, B.; Söreide, T.: Solution techniques for nonlinear finite element problems. Int. J. num. Meth. Engng. 12 (1978) 1677-1696.

22 Bergan, P.G.; Holand, I.; Söreide, T.H.: Use of the current stiffness parameter in solution of nonlinear problems. Energy Methods in Finite Element Analysis,edited by R. Glowinski, E.Y. Rodin and O.C. Zienkiewicz. John Wiley & Sons Ltd., 265 - 282,1979.

23 Bergan, P.G.: Solution algorithm s for nonlinear structural problems. Proc. Int. Conf. Engineering Application of the finite Element Method, 13.1 - 13.38, Hövik/Norway, 1979.

24 Bergan, P.G.; Soreide, T.H.: Solution of Large displacement and instability problems using the current stiffness parameter. Proc. Int. Conf. Finite Elements in Nonlinear Solid and Structural Mechanics, Geilo/Norway, 1977.

25 Remseth, S.N.; Holthe, K.; Bergan, P.G.; Holand, S.: Tube buckling analysis by the finite element method. Ibidem.

26 Krakeland, B.: Nonlinear analysis of shells using degenerate isoparametric element. Ibidem.

27 Wood, R.D.: Zienkiewicz, O.C.: Geometrically nonlinear finite element analysis of beams, frames, arches and axissymmetric shells. Comp. & Struct. 7 (1977) 725-735.

28 Noor, A.K.; Peters, J.M.: Reduced basis technique for nonlinear analysis of structures. AIAA J. 18 (1980) 455 - 462.

Geometrically Correct Formulations for Curved Finite Bar Elements Under Large Deformations

K. BRINK*, W. B. KRÄTZIG**
Ruhr-Universität Bochum, Germany

* Dr.-Ing., Hochtief AG Essen, formerly Research Group Leader, Ruhr-University Bochum, Germany

** Dr.-Ing., Professor of Structural Engineering, Ruhr-University Bochum, Germany

Summary

Based on an energy-consistent, geometrically nonlinear theory of rods several curved finite bar elements are derived. Large deformations and large rotations are considered in a kinematically rigorous sense. The solution-procedure - a modified Newton-Raphson-method within the modular software-system FEMAS - demonstrates the excellent convergence of the derived elements compared with standard examples of the literature.

1. Introduction

The state of the art of computer-hardware and computational techniques enables the analysis of arbitrarily complicated structural responses. The finite element method as the most frequently applied approximative solution technique is often combined with further (sometimes very tough) mechanical or mathematical approximations. In this contribution the authors intend to demonstrate the advantage and simplicity of the use of approved mechanical modells and geometrically complete formulations.

For the subsequent treatment we use the following assumptions within the framework of the classical theory of curved rods [6, 16]:

a) The thickness h of the rod is small compared with its length.

b) Stresses and strains normal to the axis are neglected, thus the cross-section preserves its shape during deformation.

c) Any straight normal to the undeformed axis will remain straight and retain its rectangularity during deformation.

d) Homogeneous, isotropic and hyper-elastic material will be assumed.

e) Only plane deformations are considered in this paper.

f) Self-equilibrating stresses over the cross-section and pre-imperfections of the axis of the structure will not be considered.

2. Basic nonlinear equations

Within these assumptions we now derive an energy-consistent theory of plane curved rods subjected to arbitrary large displacements and rotations. Due to Fig. 1 any bar element will be described by the system of convected co-ordinates z^1 in the axial and z^3 in the normal direction. The undeformed configuration $\mathring{P}$, $\mathring{P}^*$ of the structure, fixed by the spatial co-ordinate system x^1, x^3, shall be used as the initial state of reference for the future Total-Lagrange-Formulation. Introducing for z^1 the natural curve length $\mathring{s}$ of the undeformed state of reference we achieve a formulation, which contains only physical tensor components: $/\mathring{a}_1/ = /\mathring{a}_3/ = 1$.

In order to relate the base vectors of the deformed state to those of the undeformed initial state of the structure we start with the position vectors of any point $P(P^*)$ lying on (outside) the deformed axis (see Fig. 1):

$$\underset{\sim}{r} = \underset{\sim}{\mathring{r}} + \underset{\sim}{u} \; , \quad \underset{\sim}{r}^* = \underset{\sim}{\mathring{r}}^* + \underset{\sim}{u}^* \; . \tag{1}$$

Considering

$$\underset{\sim}{u}^* = \underset{\sim}{u} + z\underset{\sim}{w} , \quad z = z^3 \tag{2}$$

we find for the components of the displacement vectors

$$\underset{\sim}{u} = \overset{\circ}{u}{}^i \, \overset{\circ}{\underset{\sim}{a}}_i = \overset{\circ}{u}{}^1 \, \overset{\circ}{\underset{\sim}{a}}_1 + \overset{\circ}{u}{}^3 \, \overset{\circ}{\underset{\sim}{a}}_3 ,$$

$$\underset{\sim}{u}^* = (\overset{\circ}{u}{}^1 + z\overset{\circ}{w}{}^1) \, \overset{\circ}{\underset{\sim}{a}}_1 + (\overset{\circ}{u}{}^3 + z\overset{\circ}{w}{}^3) \, \overset{\circ}{\underset{\sim}{a}}_3 , \tag{3}$$

where the superscript $\circ$ marks the decomposition with respect to the initial state. With the help of

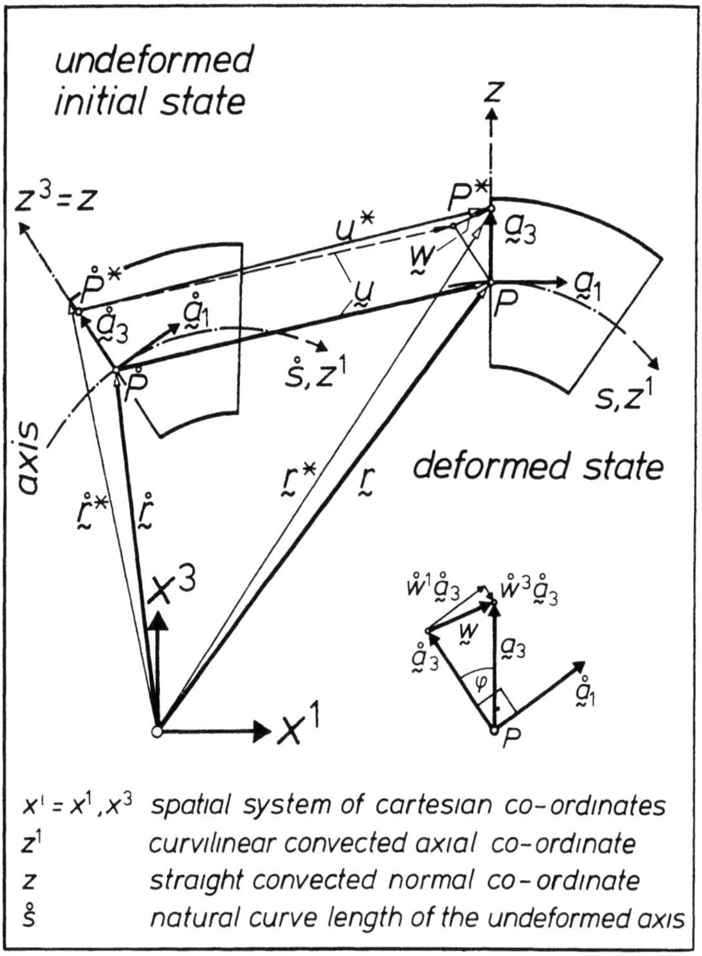

Fig. 1: Undeformed and deformed bar element

$$\frac{\partial \overset{\circ}{\underset{\sim}{r}}}{\partial z^1} = \frac{\partial \overset{\circ}{\underset{\sim}{r}}}{\partial \overset{\circ}{s}} = \overset{\circ}{\underset{\sim}{r}}{}_{,1} = \overset{\circ}{\underset{\sim}{a}}{}_1 \; , \quad \underset{\sim}{r}{}_{,1} = \underset{\sim}{a}{}_1 \tag{4}$$

we further find from (1) the relation

$$\underset{\sim}{a}{}_1 = \overset{\circ}{\underset{\sim}{a}}{}_1 + \underset{\sim}{u}{}_{,1} \; , \tag{5}$$

which can be transcribed by virtue of Frenet's equations

$$\begin{aligned} \overset{\circ}{\underset{\sim}{a}}{}_{1,1} &= \overset{\circ}{b}_{11} \overset{\circ}{\underset{\sim}{a}}{}_3 = \overset{\circ}{\varkappa} \overset{\circ}{\underset{\sim}{a}}{}_3 \; , \\ \overset{\circ}{\underset{\sim}{a}}{}_{3,1} &= -\overset{\circ}{b}{}^1_1 \overset{\circ}{\underset{\sim}{a}}{}_1 = -\overset{\circ}{\varkappa} \overset{\circ}{\underset{\sim}{a}}{}_1 \end{aligned} \tag{6}$$

into the transformation for the base vector $\underset{\sim}{a}{}_1$:

$$\begin{aligned} \underset{\sim}{a}{}_1 &= (1 + \overset{\circ}{u}{}^1{}_{,1} - \overset{\circ}{u}{}^3 \overset{\circ}{b}{}^1_1) \overset{\circ}{\underset{\sim}{a}}{}_1 + (\overset{\circ}{u}{}^3{}_{,1} + \overset{\circ}{u}{}^1 \overset{\circ}{b}_{11}) \overset{\circ}{\underset{\sim}{a}}{}_3 \\ &= \overset{\circ}{a}{}^1_1 \overset{\circ}{\underset{\sim}{a}}{}_1 + \overset{\circ}{a}{}^3_1 \overset{\circ}{\underset{\sim}{a}}{}_3 \; . \end{aligned} \tag{7}$$

This expression holds for arbitrary large displacements. A similar treatment for the normal base vector a_3 leads to

$$\begin{aligned} \underset{\sim}{a}{}_3 &= - \frac{\overset{\circ}{u}{}^3{}_{,1} + \overset{\circ}{u}{}^1 \overset{\circ}{b}_{11}}{\sqrt{a_{11}}} \overset{\circ}{\underset{\sim}{a}}{}_1 + \frac{1 + \overset{\circ}{u}{}^1{}_{,1} - \overset{\circ}{u}{}^3 \overset{\circ}{b}{}^1_1}{\sqrt{a_{11}}} \overset{\circ}{\underset{\sim}{a}}{}_3 \\ &= - \frac{\overset{\circ}{a}{}^3_1}{\sqrt{a_{11}}} \overset{\circ}{\underset{\sim}{a}}{}_1 + \frac{\overset{\circ}{a}{}^1_1}{\sqrt{a_{11}}} \overset{\circ}{\underset{\sim}{a}}{}_3 \end{aligned} \tag{8}$$

using the abbreviation:

$$a_{11} = (\overset{\circ}{a}{}^1_1)^2 + (\overset{\circ}{a}{}^3_1)^2 \; . \tag{9}$$

In order to evaluate the vector components of w for arbitrary large rotations we decompose (see Fig. 1):

$$w = \overset{\circ}{w}{}^1 \overset{\circ}{a}_1 + \overset{\circ}{w}{}^3 \overset{\circ}{a}_3 = \sin \varphi \, \overset{\circ}{a}_1 + (\cos \varphi - 1) \, \overset{\circ}{a}_3 . \tag{10}$$

Using the orthogonality condition

$$\underset{\sim}{a}{}_1 \cdot \underset{\sim}{a}{}_3 = (\overset{\circ}{\underset{\sim}{a}}{}_1 + \underset{\sim}{u}{}_{,1}) \cdot (\overset{\circ}{\underset{\sim}{a}}{}_3 + \underset{\sim}{w}) = \overset{\circ}{\underset{\sim}{a}}{}_1 \cdot \overset{\circ}{\underset{\sim}{a}}{}_3 = 0 \tag{11}$$

we find after some additional transformations for the exact angle of rotation φ

$$\varphi = \arctan\left[-\frac{\overset{\circ}{u}{}^3{}_{,1} + \overset{\circ}{u}{}^1\, \overset{\circ}{b}_{11}}{1 + \overset{\circ}{u}{}^1{}_{,1} - \overset{\circ}{u}{}^3\, \overset{\circ}{b}{}^1_1}\right] = \arctan\left[-\frac{\overset{\circ}{a}{}^3_1}{\overset{\circ}{a}{}^1_1}\right],$$

$$\varphi_{,1} = \frac{1}{a_{11}}\left[-(\overset{\circ}{a}{}^3_1)_{,1}\, \overset{\circ}{a}{}^1_1 + \overset{\circ}{a}{}^3_1\, (\overset{\circ}{a}{}^1_1)_{,1}\right]. \tag{12}$$

The adequate relations for small rotations (index L: linearized values) take the form ($\overset{\circ}{w}{}^3 \approx 0$):

$$\varphi^L = -\overset{\circ}{a}{}^3_1 , \quad \varphi^L{}_{,1} = -(\overset{\circ}{a}{}^3_1)_{,1} . \tag{13}$$

Green's well-known three-dimensional strain tensor

$$\varepsilon_{ij} = \frac{1}{2}\left(\underset{\sim}{r}^*{}_{,i} \cdot \underset{\sim}{r}^*{}_{,j} - \underset{\sim}{\overset{\circ}{r}}{}^*{}_{,i} \cdot \underset{\sim}{\overset{\circ}{r}}{}^*{}_{,j}\right), \quad i,j = 1,2,3 \tag{14}$$

furnishes with the assumptions made only one (axial) component ε_{11}, which can be broken down into a constant part α_{11} and another one, varying linearly over the thickness h:

$$\varepsilon_{11} = \alpha_{11} + z\beta_{11} . \tag{15}$$

By virtue of the previously derived formulae and under the abbreviations $\overset{\circ}{u}{}^1 = \overset{\circ}{u}$, $\overset{\circ}{u}{}^3 = \overset{\circ}{w}$ we find for the complete expressions of the 1. and 2. strain tensor:

$$\begin{aligned}\alpha_{11} = {} & \underline{\overset{\circ}{u}_{,1} - \overset{\circ}{w}\,\overset{\circ}{\varkappa}} + \frac{1}{2}(\overset{\circ}{u}_{,1})^2 + \frac{1}{2}(\overset{\circ}{w}_{,1})^2 + \frac{1}{2}(\overset{\circ}{u}\,\overset{\circ}{\varkappa})^2 \\ & + \frac{1}{2}(\overset{\circ}{w}\,\overset{\circ}{\varkappa})^2 - \overset{\circ}{u}_{,1}\,\overset{\circ}{w}\,\overset{\circ}{\varkappa} + \overset{\circ}{u}\,\overset{\circ}{w}_{,1}\,\overset{\circ}{\varkappa} ,\end{aligned}$$

$$\begin{aligned}\beta_{11} = {} & \underline{\overset{\circ}{\varkappa} + \cos\varphi\,(-\overset{\circ}{\varkappa} + \varphi_{,1} - \overset{\circ}{u}_{,1}\,\overset{\circ}{\varkappa} + \overset{\circ}{w}\,\overset{\circ}{\varkappa}{}^2} + \varphi_{,1}\,\overset{\circ}{u}_{,1} \\ & - \varphi_{,1}\,\overset{\circ}{w}\,\overset{\circ}{\varkappa}) + \sin\varphi\,(\overset{\circ}{w}_{,1}\,\overset{\circ}{\varkappa} + \overset{\circ}{u}\,\overset{\circ}{\varkappa}{}^2 - \varphi_{,1}\,\overset{\circ}{w}_{,1} \\ & - \varphi_{,1}\,\overset{\circ}{u}\,\overset{\circ}{\varkappa}).\end{aligned} \tag{16}$$

The approximation for small rotations changes the 2. strain tensor into

$$\beta_{11}^{SR} = \underline{\varphi,_1 - \mathring{u},_1 \mathring{\varkappa} + \mathring{w} \mathring{\varkappa}^2} + \varphi,_1 \mathring{u},_1 - \varphi,_1 \mathring{w} \mathring{\varkappa}$$

$$+ \varphi\mathring{w},_1 \mathring{\varkappa} + \varphi\mathring{u} \mathring{\varkappa}^2, \tag{17}$$

while the assumption of infinitesimal small deformations leads to the most far-reaching simplification (underlined terms).

Defining the hyper-elastic potential energy function

$$U = \int_{\mathring{s}} \left(\frac{EF}{2} \alpha_{11} \alpha_{11} + \frac{EI}{2} \beta_{11} \beta_{11}\right) d\mathring{s}, \tag{18}$$

the global rate of energy equation, in which inertia terms are neglected, can be broken down as [6, 9]:

$$\frac{D}{Dt} \int_{\mathring{s}} \left(\frac{EF}{2} \alpha_{11} \alpha_{11} + \frac{EI}{2} \beta_{11} \beta_{11}\right) d\mathring{s}$$

$$= \frac{D}{Dt} \int_{\mathring{s}} (\underset{\sim}{f} \cdot \underset{\sim}{u} + \underset{\sim}{t} \cdot \underset{\sim}{u}) \, d\mathring{s} \; . \tag{19}$$

Carrying out the time derivatives in (19) and combining body forces $\underset{\sim}{f}$ with surface forces $\underset{\sim}{t}$ into a general load vector $\underset{\sim}{p}$ this statement transforms into

$$\int_{\mathring{s}} (EF\, \alpha_{11} \dot{\alpha}_{11} + EI\, \beta_{11} \dot{\beta}_{11}) \, d\mathring{s} = \int_{\mathring{s}} \underset{\sim}{p} \cdot \dot{\underset{\sim}{u}} \, d\mathring{s} \; , \tag{20}$$

where dotted quantities represent material time derivatives or increments.

With this functional we conclude our summary of non-linear equations as a basis of the intended finite element approximation. For more complete informations the presentation [3] may be recommended.

3. Incremental nonlinear relations

These just derived nonlinear equations can be used only in iterative solution techniques. To include the ability of incremental procedures we now define in Fig. 2 a neighbouring position $\bar{P}$, $\bar{P}^*$ of our previously considered deformed state P, P^* by adding an infinitesimal small state of deformation. This neighbouring state shall be reached from P, P^* by a linear step of computation. Under this assumption the incremental angle of rotation $\bar{\varphi}$, superposed upon φ, has to be considered sufficiently small, e.g.: $\sin\bar{\varphi} \approx \bar{\varphi}$, $\cos\bar{\varphi} \approx 1$.

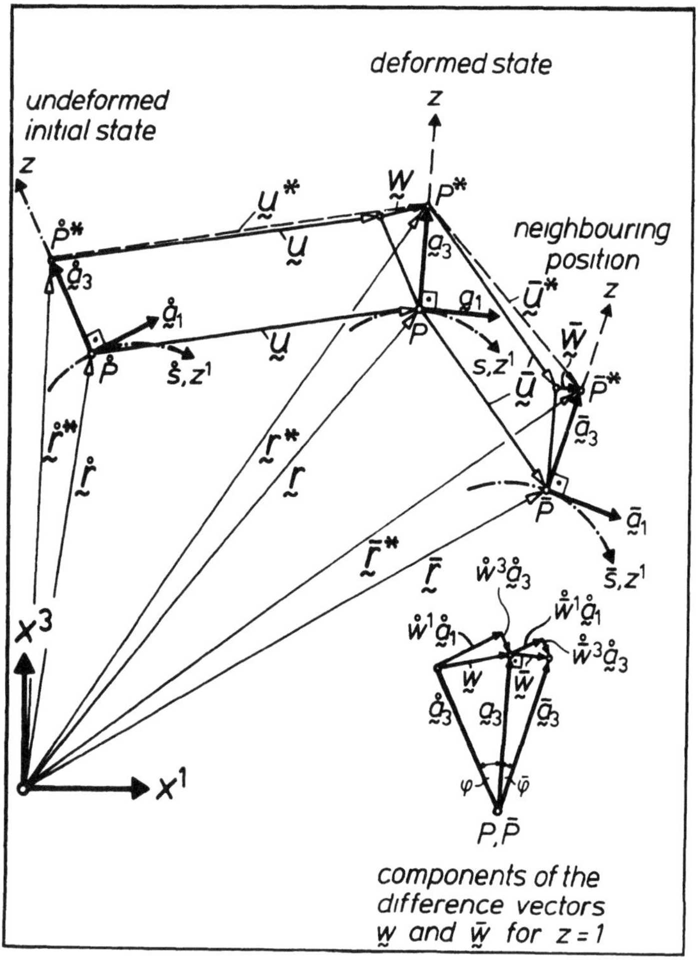

Fig. 2: Initial state, deformed state and neighbouring position

With this linearization we find for the incremental difference vector $\bar{\underset{\sim}{w}}$ (Fig. 2), decomposed again with respect to the initial state:

$$\bar{\underset{\sim}{w}} = \bar{\varphi} \cos \varphi \; \overset{\circ}{\underset{\sim}{a}}_1 + \bar{\varphi} \sin \varphi \; \overset{\circ}{\underset{\sim}{a}}_3 \; . \tag{21}$$

From the orthogonality condition, applied to the neighbouring position

$$\bar{\underset{\sim}{a}}_1 \cdot \bar{\underset{\sim}{a}}_3 = (\overset{\circ}{\underset{\sim}{a}}_1 + \underset{\sim}{u},_1 + \bar{\underset{\sim}{u}},_1) \cdot (\overset{\circ}{\underset{\sim}{a}}_3 + \underset{\sim}{w} + \bar{\underset{\sim}{w}}) = 0 \; , \tag{22}$$

a series of transformations similar to those following (11) for the incremental rotation leads to:

$$\bar{\varphi} = \frac{1}{a_{11}} \left[- (\overset{\circ}{\bar{w}},_1 + \overset{\circ}{\bar{u}} \overset{\circ}{\varkappa}) \overset{\circ}{a}{}^1_1 + (\overset{\circ}{\bar{u}},_1 - \overset{\circ}{\bar{w}} \overset{\circ}{\varkappa}) \overset{\circ}{a}{}^3_1 \right],$$

$$\begin{aligned} \bar{\varphi},_1 = \frac{1}{a_{11}} \Big[& - (\overset{\circ}{\bar{w}},_1 + \overset{\circ}{\bar{u}} \overset{\circ}{\varkappa}),_1 \overset{\circ}{a}{}^1_1 - (\overset{\circ}{\bar{w}},_1 + \overset{\circ}{\bar{u}} \overset{\circ}{\varkappa}) \overset{\circ}{a}{}^1_1,_1 \\ & + (\overset{\circ}{\bar{u}},_1 - \overset{\circ}{\bar{w}} \overset{\circ}{\varkappa}),_1 \overset{\circ}{a}{}^3_1 + (\overset{\circ}{\bar{u}},_1 - \overset{\circ}{\bar{w}} \overset{\circ}{\varkappa}) \overset{\circ}{a}{}^3_1,_1 \Big] \\ & - \frac{a_{11},_1}{(a_{11})^2} \left[- (\overset{\circ}{\bar{w}},_1 + \overset{\circ}{\bar{u}} \overset{\circ}{\varkappa}) \overset{\circ}{a}{}^1_1 + (\overset{\circ}{\bar{u}},_1 - \overset{\circ}{\bar{w}} \overset{\circ}{\varkappa}) \overset{\circ}{a}{}^3_1 \right] \end{aligned} \tag{23}$$

Again we have introduced the abbreviations $\overset{\circ}{\bar{u}}{}^1 = \overset{\circ}{\bar{u}}$, $\overset{\circ}{\bar{u}}{}^3 = \overset{\circ}{\bar{w}}$. The incremental displacement vectors $\bar{\underset{\sim}{u}}$, $\bar{\underset{\sim}{w}}$ are defined by the difference of the position vectors $\bar{\underset{\sim}{r}}$, $\bar{\underset{\sim}{r}}^*$ and (1):

$$\begin{aligned} \underset{\sim}{r} &= \overset{\circ}{\underset{\sim}{r}} + \overset{\circ}{\underset{\sim}{u}} + \bar{\underset{\sim}{u}} \; , \\ \underset{\sim}{r}^* &= \overset{\circ}{\underset{\sim}{r}}{}^* + \underset{\sim}{u}^* + \bar{\underset{\sim}{u}}^* = \overset{\circ}{\underset{\sim}{r}}{}^* + \underset{\sim}{u} + \bar{\underset{\sim}{u}} + z\,(\underset{\sim}{w} + \bar{\underset{\sim}{w}}) \; . \end{aligned} \tag{24}$$

Continuing with the incremental strain tensor (14)

$$\bar{\varepsilon}_{11} = \frac{1}{2} (\bar{\underset{\sim}{r}}{}^*,_1 \cdot \bar{\underset{\sim}{r}}{}^*,_1 - \underset{\sim}{r}^*,_1 \cdot \underset{\sim}{r}^*,_1) \tag{25}$$

and the corresponding strain energy increment

$$\dot{\bar{U}} = \int_{\overset{\circ}{s}} \Big[EF\,(\alpha_{11} + \overset{+}{\alpha}_{11} + \overset{++}{\alpha}_{11})\,(\dot{\overset{+}{\alpha}}_{11} + \overset{++}{\alpha}_{11}) + EI\,(\beta_{11} + \overset{+}{\beta}_{11} + \overset{++}{\beta}_{1})\,(\overset{+}{\beta}_{11} + \dot{\overset{++}{\beta}}_{11}) \Big]\, d\overset{\circ}{s}\,, \tag{26}$$

in which all products up to quadratic nonlinear terms are considered consistently:

$$\dot{\bar{U}} = \int_{\overset{\circ}{s}} \Big[EF\,(\alpha_{11}\,\overset{+}{\alpha}_{11} + \alpha_{11}\,\overset{++}{\alpha}_{11} + \overset{+}{\alpha}_{11}\,\overset{+}{\alpha}_{11}) + EI\,(\beta_{11}\,\overset{+}{\beta}_{11} + \beta_{11}\,\overset{++}{\beta}_{11} + \overset{+}{\beta}_{11}\,\overset{+}{\beta}_{11}) \Big]\, d\overset{\circ}{s}\,, \tag{27}$$

we can find after considerable evaluations all basic equations of the neighbouring position, related to and decomposed with respect to the undeformed initial state, in Table 1. This consistently linearized set of equations forms the fundament of the future finite translation.

The global rate of energy equation in Table 1 contains on the left hand side - in the terms $\overset{+}{\alpha}_{11}\,\dot{\overset{+}{\alpha}}_{11}$, $\overset{+}{\beta}_{11}\,\dot{\overset{+}{\beta}}_{11}$ - the contributions of the incremental and original displacements, in the terms $\alpha_{11}\,\dot{\overset{++}{\alpha}}_{11}$, $\beta_{11}\,\dot{\overset{++}{\beta}}_{11}$ the contributions of the initial strains and the quadratic displacement-increments. The right hand side contains the energy rates of the load-increments and - emphazised by underlining - the rate of energy equation of the original deformed state. In any state of equilibrium these underlined terms have to vanish; the remaining statement would correspond with a purely incremental solution technique. During a certain solution process, if equilibrium has been reached not yet, they can be interpreted as unbalanced external of internal forces. This leads to the advantage of a very efficient coupling of incremental and iterative solution techniques.

Incremental global rate of energy equation (homogeneous hyper-elastic material, conservative loads)

$$\int_{\mathring{s}} [EF(\alpha_{11}\overset{++}{\alpha}_{11} + \overset{+}{\alpha}_{11}\overset{\dot{+}}{\alpha}_{11}) + EI(\beta_{11}\overset{++}{\beta}_{11} + \overset{+}{\beta}_{11}\overset{\dot{+}}{\beta}_{11})]\, d\mathring{s} =$$

$$\int_{\mathring{s}} [(\underline{\mathring{p}^i} + \overset{\dot{+}}{p}{}^i)\,\dot{\mathring{u}}_i + (\underline{m} + \overset{+}{m})\,\dot{\bar{\varphi}} \underbrace{- EF\alpha_{11}\overset{\dot{+}}{\alpha}_{11} - EI\beta_{11}\overset{\dot{+}}{\beta}_{11}}_{(i = 1\ 3,\ \mathring{u}_1 = \mathring{u},\ \mathring{u}_3 = \mathring{w})}]\, d\mathring{s}$$

Kinematic equations for incremental strain variables

for large rotations

$$\overset{+}{\alpha}_{11} = \mathring{\bar{u}}\,\mathring{\kappa}\,\mathring{a}_1^3 + \mathring{\bar{u}}_{,1}\,\mathring{a}_1^1 - \mathring{\bar{w}}\,\mathring{\kappa}\,\mathring{a}_1^1 + \mathring{\bar{w}}_{,1}\,\mathring{a}_1^3$$

$$\overset{++}{\alpha}_{11} = \frac{1}{2}(\mathring{\bar{u}}\,\mathring{\bar{u}}\,\mathring{\kappa}^2 + \mathring{\bar{u}}_{,1}\mathring{\bar{u}}_{,1} + \mathring{\bar{w}}\,\mathring{\bar{w}}\,\mathring{\kappa}^2 + \mathring{\bar{w}}_{,1}\mathring{\bar{w}}_{,1}) + \mathring{\bar{u}}\,\mathring{\bar{w}}_{,1}\,\mathring{\kappa} - \mathring{\bar{u}}_{,1}\,\mathring{\bar{w}}\,\mathring{\kappa}$$

$$\begin{aligned}\overset{+}{\beta}_{11} = {} & \mathring{\bar{u}}\,\mathring{\kappa}\sin\varphi(\mathring{\kappa} - \varphi_{,1}) - \mathring{\bar{u}}_{,1}\cos\varphi(\mathring{\kappa} - \varphi_{,1}) + \mathring{\bar{w}}\,\mathring{\kappa}\cos\varphi(\mathring{\kappa} - \varphi_{,1}) \\ & + \mathring{\bar{w}}_{,1}\sin\varphi(\mathring{\kappa} - \varphi_{,1}) + \bar{\varphi}(\mathring{\kappa} - \varphi_{,1})(\cos\varphi\,\mathring{a}_1^3 + \sin\varphi\,\mathring{a}_1^1) \\ & + \bar{\varphi}_{,1}(\cos\varphi\,\mathring{a}_1^1 - \sin\varphi\,\mathring{a}_1^3)\end{aligned}$$

$$\begin{aligned}\overset{++}{\beta}_{11} = {} & \mathring{\bar{u}}\,\bar{\varphi}\,\mathring{\kappa}\cos\varphi(\mathring{\kappa} - \varphi_{,1}) - \mathring{\bar{u}}\,\bar{\varphi}_{,1}\,\mathring{\kappa}\sin\varphi + \mathring{\bar{u}}_{,1}\,\bar{\varphi}\sin\varphi(\mathring{\kappa} - \varphi_{,1}) \\ & + \mathring{\bar{u}}_{,1}\,\bar{\varphi}_{,1}\cos\varphi - \mathring{\bar{w}}\,\bar{\varphi}\,\mathring{\kappa}\sin\varphi(\mathring{\kappa} - \varphi_{,1}) - \mathring{\bar{w}}\,\bar{\varphi}_{,1}\,\mathring{\kappa}\cos\varphi \\ & + \mathring{\bar{w}}\,\bar{\varphi}\cos\varphi(\mathring{\kappa} - \varphi_{,1}) - \mathring{\bar{w}}_{,1}\,\bar{\varphi}_{,1}\sin\varphi\end{aligned}$$

for small rotations

$$\begin{aligned}\overset{+}{\beta}_{11} = {} & \mathring{\bar{u}}\,\varphi\,\mathring{\kappa}^2 - \mathring{\bar{u}}_{,1}(\mathring{\kappa} - \varphi_{,1}) + \mathring{\bar{w}}\,\mathring{\kappa}(\mathring{\kappa} - \varphi_{,1}) + \mathring{\bar{w}}_{,1}\,\varphi\,\mathring{\kappa} \\ & + \bar{\varphi}\,\mathring{\kappa}\,\mathring{a}_1^3 + \bar{\varphi}_{,1}\,\mathring{a}_1^1\end{aligned}$$

$$\overset{++}{\beta}_{11} = \mathring{\bar{u}}\,\bar{\varphi}\,\mathring{\kappa}^2 + \mathring{\bar{u}}_{,1}\,\bar{\varphi}_{,1} - \mathring{\bar{w}}\,\bar{\varphi}_{,1}\,\mathring{\kappa} + \mathring{\bar{w}}_{,1}\,\bar{\varphi}\,\mathring{\kappa}$$

Coefficients of transformation

$$\mathring{a}_1^1 = 1 + \mathring{u}_{,1} - \mathring{w}\,\mathring{\kappa}$$

$$\mathring{a}_1^3 = \mathring{w}_{,1} + \mathring{u}\,\mathring{\kappa}$$

Table 1: Incremental basic equations of the neighbouring position, related to the initial state

4. Matrix formulation of basic equations

For the numerical treatment we select as primary kinematic variable the vector $\overset{\circ}{\underline{u}}$ of the complete incremental displacements, given in the first line of Table 2. This vector contains the displacements $\overset{\circ}{\bar{u}}$, $\overset{\circ}{\bar{w}}$ as independent quantities, the rotation $\overset{\circ}{\bar{\varphi}}$ and all first derivatives as dependent ones. Additionally Table 2 shows the linear parts of the incremental strain-displacement relations; the non-linear parts may be found in Table 3. In both cases small and arbitrary large rotations are distinguished. For details of the evaluation procedure of these matrices and in connection with their numerical computation see [3]. The specific discretizing prozess of the incremental matrix equations is similar to that one of any linear theory and may therefore be passed over here.

5. Circular bar elements

In order to check the derived incremental theory of curved rods, arbitrary arches are - to a large extend - the most suitable simple structures. Their ability to bifurcation as well as to snaps-through combines two completely different non-linear response phenomena. The encouraging results of our start, a 2-node circular bar element with a total of 6 DOF, finally led to the systematic development of several element families [8], which will be reviewed briefly.

In the case of (straight) beam elements the interpolation functions for the axial and normal displacements in general can be chosen easily such that rigid body modes are described properly and will cause no strains. For curved bar elements, e.g. circular bar elements, both requirements are difficult to fullfill. Three different ways are known to solve this problem:

a) A set of n functions, which describe the rigid body modes correctly, is superposed upon the m original in-

Complete incremental displacement variables

$$\underline{\mathring{u}} = \left\{ \mathring{u} \;\middle|\; \mathring{u}_{,1} \;\middle|\; \mathring{w} \;\middle|\; \mathring{w}_{,\cdot} \;\middle|\; \bar{\varphi} \;\middle|\; \bar{\varphi}_{,1} \right\}$$

Matrix of incremental strains

$$\underline{\bar{\varepsilon}} = \begin{bmatrix} \bar{\alpha}_{11} \\ \bar{\beta}_{11} \end{bmatrix} = \underline{\overset{+}{\varepsilon}} + \underline{\overset{++}{\varepsilon}} = \begin{bmatrix} \overset{+}{\alpha}_{11} \\ \overset{+}{\beta}_{11} \end{bmatrix} + \begin{bmatrix} \overset{++}{\alpha}_{11} \\ \overset{++}{\beta}_{11} \end{bmatrix}$$

Linearized incremental strain-displacement relation.

$$\underline{\overset{+}{\varepsilon}} = (\underline{\mathring{D}} + \underline{\mathring{D}}_v) \, \underline{\mathring{u}}$$

Operator matrices.

$$\underline{\mathring{D}} = \left[\begin{array}{c|c|c|c|c|c} 0 & 1 & -\mathring{\varkappa} & 0 & 0 & 0 \\ \hline 0 & -\mathring{\varkappa} & \mathring{\varkappa}^2 & 0 & 0 & 1 \end{array}\right]$$

for large rotations

$$\underline{\mathring{D}}_v = \left[\begin{array}{c|c|c|c|c|c} \mathring{\varkappa}\,\mathring{a}_1^3 & \mathring{a}_1^1 - 1 & \mathring{\varkappa}(1-\mathring{a}_1^1) & \mathring{a}_1^3 & 0 & 0 \\ \hline \mathring{\varkappa}(\mathring{\varkappa}-\varphi_{,1}) & \mathring{\varkappa}-(\mathring{\varkappa}-\varphi_{,1}) & \mathring{\varkappa}(\mathring{\varkappa}-\varphi_{,1}) & (\mathring{\varkappa}-\varphi_{,1}) & (\mathring{\varkappa}-\varphi_{,1}) & \mathring{a}_1^1\cos\varphi - 1 \\ \sin\varphi & \cos\varphi & \cos\varphi - \mathring{\varkappa}^2 & \sin\varphi & (\mathring{a}_1^3\cos\varphi + \mathring{a}_1^1\sin\varphi) & -\mathring{a}_1^3\sin\varphi \end{array}\right]$$

for small rotations

$$\underline{\mathring{D}}_v = \left[\begin{array}{c|c|c|c|c|c} \mathring{\varkappa}\,\mathring{a}_1^3 & \mathring{a}_1^1 - 1 & \mathring{\varkappa}(1-\mathring{a}_1^1) & \mathring{a}_1^3 & 0 & 0 \\ \hline \varphi\,\mathring{\varkappa}^2 & \varphi_{,1} & -\varphi_{,\cdot}\,\mathring{\varkappa} & \varphi\,\mathring{\varkappa} & \mathring{\varkappa}\,\mathring{a}_1^3 & \mathring{a}_1^1 - 1 \end{array}\right]$$

Table 2: Linear incremental strain-displacement relations, related to the initial state

Non-linear incremental strain-displacement relation

$$\overset{++}{\underline{\varepsilon}} = \begin{bmatrix} \mathring{\underline{D}}_s & \mathring{\underline{u}} \end{bmatrix}^T \begin{bmatrix} \mathring{\underline{D}}_s & \mathring{\underline{u}} \end{bmatrix}, \quad \dot{\overset{++}{\underline{\varepsilon}}} = 2 \begin{bmatrix} \mathring{\underline{D}}_s & \dot{\mathring{\underline{u}}} \end{bmatrix}^T \begin{bmatrix} \mathring{\underline{D}}_s & \mathring{\underline{u}} \end{bmatrix}$$

Non-linear operator matrix.

for large rotations

$$\mathring{\underline{D}}_s = \begin{bmatrix} \frac{1}{2}\mathring{\varkappa} & 0 & 0 & \frac{1}{2} & \frac{1}{2}(\mathring{\varkappa}-\varphi_{,1})\cos\varphi & -\frac{1}{2}\sin\varphi \\ 0 & \frac{1}{2} & -\frac{1}{2}\mathring{\varkappa} & 0 & \frac{1}{2}(\mathring{\varkappa}-\varphi_{,1})\sin\varphi & \frac{1}{2}\cos\varphi \\ 0 & \frac{1}{2} & -\frac{1}{2}\mathring{\varkappa} & 0 & \frac{1}{2}(\mathring{\varkappa}-\varphi_{,1})\sin\varphi & \frac{1}{2}\cos\varphi \\ \frac{1}{2}\mathring{\varkappa} & 0 & 0 & \frac{1}{2} & \frac{1}{2}(\mathring{\varkappa}-\varphi_{,1})\cos\varphi & -\frac{1}{2}\sin\varphi \end{bmatrix}$$

for small rotations

$$\mathring{\underline{D}}_s = \begin{bmatrix} \frac{1}{2}\mathring{\varkappa} & 0 & 0 & \frac{1}{2} & \frac{1}{2}\mathring{\varkappa} & 0 \\ 0 & \frac{1}{2} & -\frac{1}{2}\mathring{\varkappa} & 0 & 0 & \frac{1}{2} \\ 0 & \frac{1}{2} & -\frac{1}{2}\mathring{\varkappa} & 0 & 0 & \frac{1}{2} \\ \frac{1}{2}\mathring{\varkappa} & 0 & 0 & \frac{1}{2} & \frac{1}{2}\mathring{\varkappa} & 0 \end{bmatrix}$$

Matrix product $\mathring{\underline{D}}_s^T \; \mathring{\underline{D}}_s$

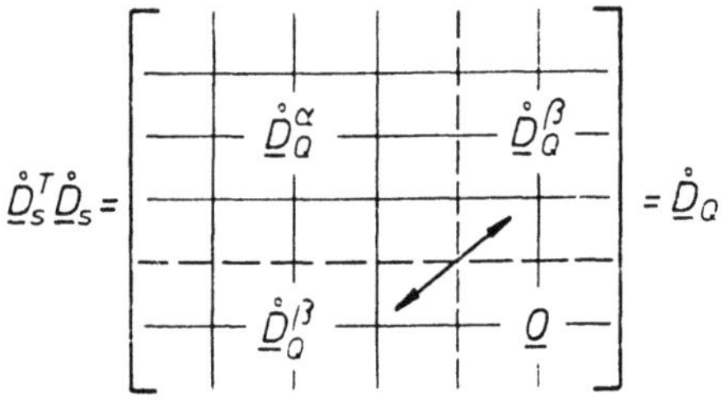

$\mathring{\underline{D}}_Q^{\alpha}$. *Matrix for initial stresses due to* α_{11}

$\mathring{\underline{D}}_Q^{\beta}$. *Matrix for initial stresses due to* β_{11}

$\underline{0}$ *neglected sub-matrix (elements depend on* z^2*)*

Table 3: Non-linear incremental strain-displacement relations, related to the initial state

terpolation functions. The (m+n) free parameters then are determined such that boundary as well as intermediate conditions are fullfilled and rigid body motions do not lead to strains.

b) The requirement of strainless rigid body motions transfers the (linearized) kinematic equations for $\overset{+}{\alpha}_{11}$, $\overset{+}{\beta}_{11}$ in Table 1 into a homogeneous differential equation of order 2. The solutions describe rigid body modes; they have to be completed by additional interpolation functions, which satisfy the inhomogeneous kinematic relations.

c) The third way is the separation of the displacement field into rigid body modes and strain-producing displacements [14].

Finally, for the interpolation functions we may select arbitrary polynomials or Hermite-polynominals; the latter safe a time-consuming matrix inversion.

The interested reader may find 13 different circular bar elements in [8], further more in [3]. In the present paper we sketch the evaluation of one of the best elements containing 5 nodes and 12 DOFs (see Fig. 3). The requirement of vanishing strains transforms the (linearized) kinematic equations (16)

$$\alpha_{11} = 0 = \overset{\circ}{u}_{,1} - \overset{\circ}{w}\,\overset{\circ}{\varkappa}\,,$$

$$\beta_{11} = 0 = \varphi_{,1} - \overset{\circ}{u}_{,1}\,\overset{\circ}{\varkappa} + \overset{\circ}{w}\,\overset{\circ}{\varkappa}{}^2 \qquad (28)$$

under consideration of (23)

$$\varphi_{,1} = -\overset{\circ}{w}_{,1} - \overset{\circ}{u}\,\overset{\circ}{\varkappa} \qquad (29)$$

and for constant (circular) curvature $\overset{\circ}{\varkappa} = -1/R$ into the harmonic differential equation:

$$\overset{\circ}{w}_{,11} + \frac{1}{R^2}\,\overset{\circ}{w} = 0\ . \qquad (30)$$

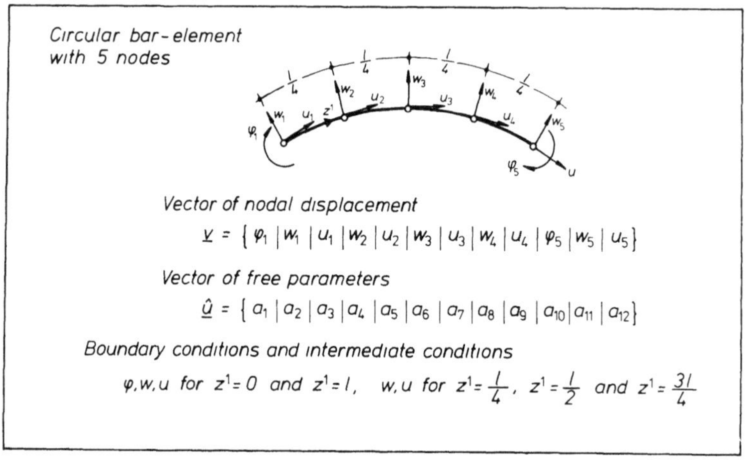

Fig. 3: Circular bar element with 2 external and 3 internal nodes

As a first solution of (30)

$$\overset{\circ}{w}^s(z^1) = a_1 \sin\frac{z^1}{R} + a_2 \cos\frac{z^1}{R},$$
$$\overset{\circ}{u}^s(z^1) = a_1 \cos\frac{z^1}{R} - a_2 \sin\frac{z^1}{R} + a_3 R,$$
$$\varphi^s(z^1) = a_3 \qquad (31)$$

we confirm the rigid body modes found by [13]. A slightly extended solution of (30) is formed by:

$$\overset{\circ}{w}^s(z^1) = a_1 \sin\frac{z^1}{R} + a_2 \cos\frac{z^1}{R} - a_3 R \sin\frac{z^1}{R},$$
$$\overset{\circ}{u}^s(z^1) = a_1 \cos\frac{z^1}{R} - a_2 \sin\frac{z^1}{R} + a_3 R\left(1 - \cos\frac{z^1}{R}\right),$$
$$\varphi^s(z^1) = a_3 . \qquad (32)$$

Furthermore, as interpolation functions for the strain variables α_{11}, β_{11} we select polynomials of order 3 and 4 with a total of 9 free parameters. Assuming trial solutions for $\overset{\circ}{u}$, $\overset{\circ}{w}$ their undetermined constants can be evaluated by comparison from the corresponding relations of Table 1. After suitable combinations of all parameters we end up with the interpolation functions of Table 4. Based on the vector of the complete incremental displacement variables in Table 2, the edge and intermediate conditions of Fig. 3 we conclude this brief information with the basic matrices $\underline{\emptyset}$, $\hat{\underline{\emptyset}}$ in Table 5. For further details concerning the derived elements see [3, 8].

Interpolation functions describing rigid body modes

$$u^s(z^1) = a_1 \cos\frac{z^1}{R} - a_2 \sin\frac{z^1}{R} + a_3 R\left(1 - \cos\frac{z^1}{R}\right)$$

$$w^s(z^1) = a_1 \sin\frac{z^1}{R} + a_2 \cos\frac{z^1}{R} - a_3 R \sin\frac{z^1}{R}$$

$$\varphi^s(z^1) = a_3$$

Interpolation functions describing strains

$$u^\varepsilon(z^1) = a_8 \frac{z^1}{R} + a_9 \frac{(z^1)^2}{2R^2} + a_{10} \frac{(z^1)^3}{3R^3} + a_{11} \frac{(z^1)^4}{4R^4} + a_{12} \frac{(z^1)^5}{5R^5}$$

$$w^\varepsilon(z^1) = a_4 + a_5 \frac{z^1}{R} + a_6 \frac{(z^1)^2}{R^2} + a_7 \frac{(z^1)^3}{R^3} - a_{12} \frac{(z^1)^4}{R^4}$$

$$\varphi^\varepsilon(z^1) = -a_5 \frac{1}{R} - a_6 \frac{2z^1}{R^2} - a_7 \frac{3(z^1)^2}{R^3} + a_8 \frac{z^1}{R^2} + a_9 \frac{(z^1)^2}{2R^3} + a_{10} \frac{(z^1)^3}{3R^4} + a_{11} \frac{(z^1)^4}{4R^5} + a_{12}\left(\frac{4(z^1)^3}{R^4} + \frac{(z^1)^5}{5R^6}\right)$$

Corresponding strains for the 5 node circular bar-element

$$\overset{*}{\alpha}_{11} = (a_4 + a_8)\frac{1}{R} + (a_5 + a_9)\frac{z^1}{R^2} + (a_6 + a_{10})\frac{(z^1)^2}{R^3} + (a_7 + a_{11})\frac{(z^1)^3}{R^4}$$

$$\overset{*}{\beta}_{11} = (a_4 - 2a_6 + 2a_8)\frac{1}{R^2} + (a_5 - 6a_7 + 2a_9)\frac{z^1}{R^3} + (a_6 + 2a_{10} + 12a_{12})\frac{(z^1)^2}{R^4} + (a_7 + 2a_{11})\frac{(z^1)^3}{R^5} + a_{12}\frac{(z^1)^4}{R^6}$$

Combined interbolation functions for element displacements

$$u(z^1) = u^s(z^1) + u^\varepsilon(z^1)$$

$$w(z^1) = w^s(z^1) + u^\varepsilon(z^1)$$

$$\varphi(z^1) = \varphi^s(z^1) + \varphi^\varepsilon(z^1)$$

Table 4: Interpolation functions for circular bar element

$\underline{u} = \underline{\Phi}\,\underline{\hat{u}}$:

$\underline{\Phi} =$

$\cos\frac{z'}{R}$	$\sin\frac{z'}{R}$	$R(1-\cos\frac{z'}{R})$					$\frac{z}{R}$	$\frac{(z')^2}{2R^2}$	$\frac{(z')^3}{3R^3}$	$\frac{(z')^4}{4R^4}$	$\frac{(z')^5}{5R^5}$
$-\frac{1}{R}\sin\frac{z'}{R}$	$-\frac{1}{R}\cos\frac{z'}{R}$	$\sin\frac{z'}{R}$					$\frac{1}{R}$	$\frac{z'}{R^2}$	$\frac{(z')^2}{R^3}$	$\frac{(z')^3}{R^4}$	$\frac{(z')^4}{R^5}$
$\sin\frac{z'}{R}$	$\cos\frac{z'}{R}$	$-R\sin\frac{z'}{R}$	1	$\frac{z'}{R}$	$\frac{(z')^2}{R^2}$	$\frac{(z')^3}{R^3}$					$-\frac{(z')^4}{R^4}$
$\frac{1}{R}\cos\frac{z'}{R}$	$-\frac{1}{R}\sin\frac{z'}{R}$	$-\cos\frac{z'}{R}$		$\frac{1}{R}$	$\frac{2z'}{R^2}$	$\frac{3(z')^2}{R^3}$					$-\frac{4(z')^3}{R^4}$
		1	.	$-\frac{1}{R}$	$-\frac{2z'}{R^2}$	$-\frac{3(z')^2}{R^3}$	$\frac{z'}{R^2}$	$\frac{(z')^2}{2R^3}$	$\frac{(z')^3}{3R^4}$	$\frac{(z')^4}{4R^5}$	$\frac{4(z')^3}{R^4}+\frac{(z')^5}{5R^5}$
					$-\frac{2}{R^2}$	$-\frac{6z'}{R^3}$	$\frac{1}{R^2}$	$\frac{z'}{R^3}$	$\frac{(z')^2}{R^4}$	$\frac{(z')^3}{R^5}$	$\frac{12(z')^2}{R^4}+\frac{(z')^4}{R^6}$

$\underline{v} = \hat{\underline{\Phi}}\,\underline{\hat{u}}$:

$\hat{\underline{\Phi}} =$

		1		$-\frac{1}{R}$							
	1		1								
1											
$\sin\frac{s}{4R}$	$\cos\frac{s}{4R}$	$-R\sin\frac{s}{4R}$	1	$\frac{s}{4R}$	$\frac{s^2}{16R^2}$	$\frac{s^3}{64R^2}$					$-\frac{s^4}{256R^4}$
$\cos\frac{s}{4R}$	$-\sin\frac{s}{4R}$	$R(1-\cos\frac{s}{4R})$					$\frac{s}{4R}$	$\frac{s^2}{32R^2}$	$\frac{s^3}{192R^3}$	$\frac{s^4}{1024R^4}$	$\frac{s^5}{5120R^5}$
$\sin\frac{s}{2R}$	$\cos\frac{s}{2R}$	$-R\sin\frac{s}{2R}$	1	$\frac{s}{2R}$	$\frac{s^2}{4R^2}$	$\frac{s^3}{8R^3}$					$-\frac{s^4}{16R^4}$
$\cos\frac{s}{2R}$	$-\sin\frac{s}{2R}$	$R(1-\cos\frac{s}{2R})$					$\frac{s}{2R}$	$\frac{s^2}{8R^2}$	$\frac{s^3}{24R^3}$	$\frac{s^4}{64R^4}$	$\frac{s^5}{160R^5}$
$\sin\frac{3s}{4R}$	$\cos\frac{3s}{4R}$	$-R\sin\frac{3s}{4R}$	1	$\frac{3s}{4R}$	$\frac{9s^2}{16R^2}$	$\frac{27s^3}{64R^3}$					$-\frac{81s^4}{256R^4}$
$\cos\frac{3s}{4R}$	$-\sin\frac{3s}{4R}$	$R(1-\cos\frac{3s}{4R})$					$\frac{3s}{4R}$	$\frac{9s^2}{32R^2}$	$\frac{27s^3}{192R^3}$	$\frac{81s^4}{1024R^4}$	$\frac{243s^5}{5120R^5}$
		1		$-\frac{1}{R}$	$-\frac{2s}{R}$	$-\frac{3s^2}{R^3}$	$\frac{s}{R^2}$	$\frac{s^2}{2R^3}$	$\frac{s^3}{3R^4}$	$\frac{s^4}{4R^5}$	$\frac{4s^3}{R^4}+\frac{s^5}{5R^6}$
$\sin\frac{s}{R}$	$\cos\frac{s}{R}$	$-R\sin\frac{s}{R}$	1	$\frac{s}{R}$	$\frac{s^2}{R^2}$	$\frac{s^3}{R^3}$					$-\frac{s^4}{R^4}$
$\cos\frac{s}{R}$	$-\sin\frac{s}{P}$	$R(1-\cos\frac{s}{R})$					$\frac{s}{R}$	$\frac{s^2}{2R^2}$	$\frac{s^3}{3R^3}$	$\frac{s^4}{4R^4}$	$\frac{s^5}{5R^5}$

Table 5: Additional matrices for circular bar element

6. Software implementation and examples

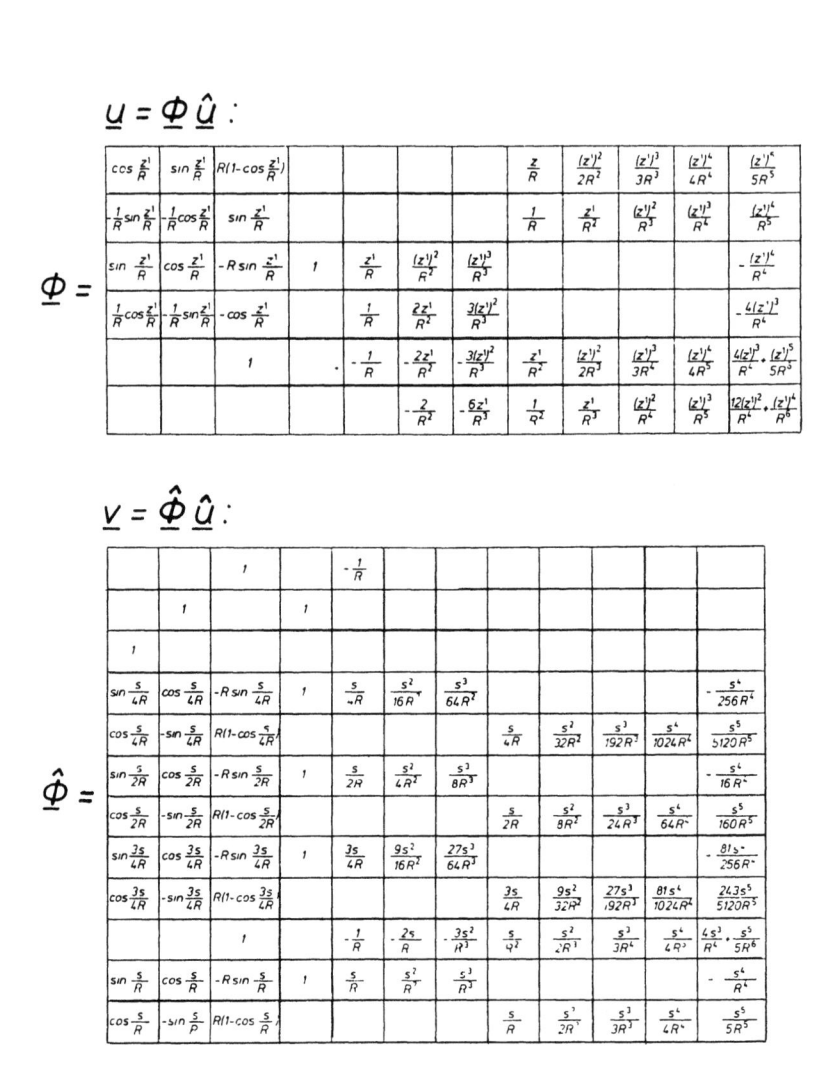

The previously described basic incremental relations have been implemented into the FEMAS-system [2] together with several circular bar elements forming a combined incremental-iterative solution procedure. FEMAS, an in-core modular software system, has been designed for the handling of banded, symmetric matrices; it has been proved to be a highly efficient tool for the present purpose.

It should be emphasized, that the internal control of any future load-step in FEMAS has been carried out automatically - for every state of equilibrium - by the computed increment of the mechanical work. We could show, that this physical quantity permits a far better estimate of the future stiffness of the structure than a certain displacement component: the computed work-increment controls the magnitude of the next load-increment. By this option different load-increments easily and automatically can be chosen as the state of deformation demands.

Finally, two often cited examples of the literature will be recomputed and their results compared. The first one, given in Fig. 4, shows a very shallow circular arch, clamped at both ends and loaded by a point-load P in the apex. The structure deformes continuously into a symmetric, stable post-buckling configuration. During this path very small rotations appear, thus small and large rotation theories lead to identical results. Our computation has been carried out with only 2 bar elements for one half of the structure. Comparable results were obtained with 4 3-dimensional elements [10], 12 8-node plane stress elements [1, 11]and 4 to 32 beam elements [4] as shown in Fig. 4. Introducing a small imperfection pattern of 0.05 in of amplitude also the experimental result of [5] could be verified.

The second example - a steep, pin-jointed circular arch bearing again a point-load P in the apex - shows an unstable symmetric bifurcation: The stable post-buckling configuration appears as the first unsymmetric buckling mode, reached at a considerable lower limit load. This structure (Fig. 5) has been investigated analytically in [7] and numerically by [12, 15], both using 20 plane stress elements. In order to force the iteration into the unsymmetric

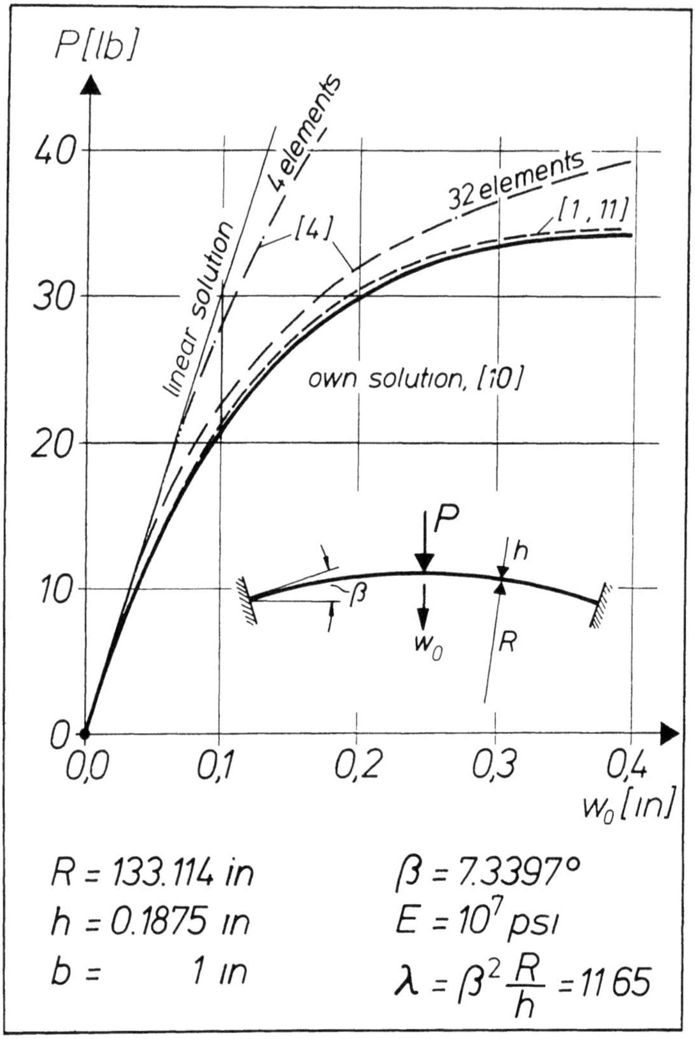

Fig. 4: Shallow clamped arch

buckling path, a horizontal perturbation load P/1000 has been applied successfully.

Using the same remedy in our computation the analytical solution could be verified with 6 bar elements and 60 load-increments. As Fig. 5 demonstrates for this highly nonlinear response any small rotation theory, leading to an error-margin of about 40%, is completely insufficient. These and many further results prove the applicability of the

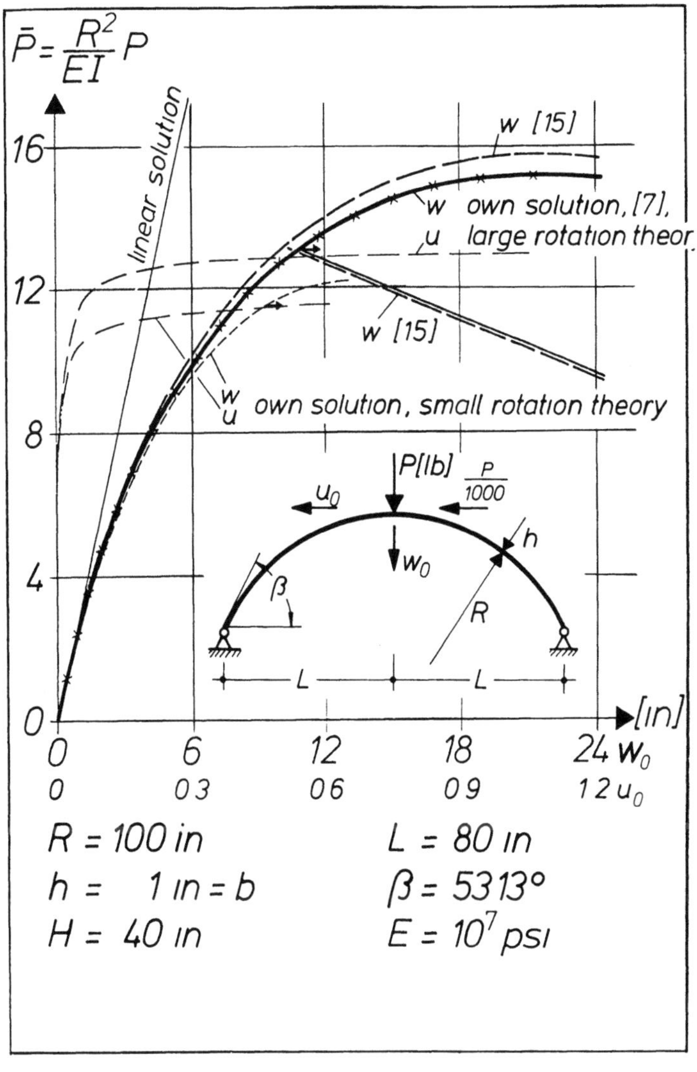

Fig. 5: Steep pin-jointed arch

derived theoretical concept and its software implementation.

Acknowledgement: Conceptual ideas and substantial support in the FEMAS implementation of our colleagues Beate Lechtleitner and Hermann Beem are gratefully acknowledged.

7. References

[1] Bathe, K.J. / Ozdemır, H. / Wilson, E.L.: Static and Dynamic Geometric and Material Nonlinear Analysis. UC-SESM Report Nr. 74-4, UC-Berkeley 1974.
[2] Beem, H. / Brink, K. / Krätzig, W.B.: FEMAS - Benutzeranweisung und Programmdokumentation (in print).
[3] Brink, K.: Theorie und Berechnung in einer Ebene gekrümmter Stäbe unter großen Verformungen. Institut für Konstr. Ingenieurbau, Ruhr-University Bochum, Techn. Rep. No. 79-1.
[4] Dupuis, G.A. et al.: Nonlinear Material and Geometrıc Behavior of Shell Structures. Comp. a. Struct. 1 (1971) 223-239.
[5] Gjelsvik, A. / Bodner, S.R.: The Energy Crıterion and Snap Bucklıng of Arches. Journ. Eng. Mech. Div. (1962) 87-134.
[6] Green, A.E. / Naghdi, P.M.: Non-isothermal Theory of Rods, Plates and Shells. Int. J. Sol. Struct. 6 (1970) 209-244.
[7] Huddleston, J.V.: Finıte Deflection and Snap-Through of High Circular Arches. Journ. Appl. Mech. (1968) 763-769.
[8] Krätzıg, W.B. et al.: Elementbibliothek, Lehrstuhl III des Instituts für Konstruktıven Ingenıeurbau, Ruhr-Universıtät Bochum.
[9] Krätzig, W.B.: Eınführung ın die Thermodynamik der Deformationen I. Seminarbericht Nr. 73-1, Lehrstuhl für Baumechanık, T.U. Hannover, 95-115.
[10] Mallet, R.H. / Berke, L.: Automated Method for the Large Deflection and Instability Analysıs of 3-Dımensional Truss and Frame Assemblıes. AFFDL-TR-66-102 (1966).
[11] Papenhausen, V.: Eıne energiegerechte, inkrementelle Formulierung der geometrischen nıchtlinearen Theorie elastischer Kontinua. Inst. f. Konstr. Ingenıeurbau, Ruhr-University Bochum, Techn. Rep. No. 75-13.
[12] Ramm, E.: Geometrısch nıchtlineare Elastostatık und finite Elemente. Habilitatıonsschrift Stuttgart 1976.
[13] Sabir, A.B. / Lock, A.C.: Large Deflection, Geometrically Nonlinear Finıte Element Analysis of Cırcular Arches. Int. J. Mech. Sci., 15 (1973), 43-47.
[14] Schrader, K.-H.: Genauıngkeitsprobleme beı Ansätzen mit Starrkörperverschıebungen. Finite Elemente ın der Baupraxis, Verlag W. Ernst & Sohn, Berlin 1978, 231-245.
[15] Sharifi, P. / Popov, E.P.: Nonlinear Bucklıng Analysıs of Sandwich Arches. Journ. Eng. Mech. Div. (1971) 1397-1412.
[16] Wunderlich, W. / Beverungen, G.: Geometrısch nıchtlıneare Theorie und Berechnung eben gekrümmter Stäbe. Bauingenieur 52 (1977) 225-237.

Part III:

Physically Nonlinear Problems

Some Nonlinear Problems of Soil Statics and Dynamics

O. C. ZIENKIEWICZ
University College of Swansea, U K.

Summary

This brief presentation discusses the appropriate formulation for dynamic, quasi-static and static problems in soil mechanics. Some constitutive relations of plasticity are presented and numerical computation illustrated by an example of an earth dam subject to earthquake forces.

Introduction

Soil, and indeed rock and concrete are porous materials with the voids generally filled with water which exerts a pressure on the solid particles and migrates through the material. Any numerical computation must therefore allow for the essentially two phase behaviour - and on occasion when the saturation is not complete for a possible three phase situation.

In this paper we shall confine our attention to the purely two phase situation and discuss the formulation necessary to obtain the static or dynamic response of 'soil structures' up to the point of collapse. Practical problems to which the numerical solution must be applicable involve such typical situations as the settlement of buildings, incremental collapse of offshore platform structures due to the action of sea waves or the possible earthquake collapse of earth dams or nuclear power station foundations. For many problems in such categories quite large displacements and local strains can be tolerated so appropriate consideration has to be given to both material and geometric nonlinearity.

The effective stress concept

If we define the total, Cauchy, stress in terms of all the forces acting on an isolated portion of the solid-fluid ensable as σ_{ij} , then the "effective" stress is defined as

$$\sigma'_{ij} = \sigma_{ij} + \delta_{ij}\, p \tag{1}$$

where the symbol p stands for the fluid pressure in the pores of the solid skeleton (Fig. 1).

In above the usual convention of positive tensile stresses is maintained although in soil practice it is often useful to reverse this.

The idea of using the separation given in (1) stems from Terzhagi [1] who observed that uniform pressure changes alone could not deform the solid material appreciably or lead to its failure. The only noticeable effect of such pressure changes can be to cause a volumetric strain change associated with the compression of solid grains - this we can write as

$$d\varepsilon_{ii}^{p} = -dp/K_S \tag{2}$$

where K_S is the average bulk modulus of the solid phase.

The volumetric strain ε_{ij}^{p} is usually neglected in analysis of soils but may be of importance in rock or concrete mechanics where the total strains of the solid skeleton are small. We shall thus retain it in subsequent derivations.

With the very limited effect of uniform pressure changes it is quite clear that the major deformations are associated with the effective stress σ'. Whatever the specific form of the constitutive law we can write the incremental (or rate) relationship as

$$d\sigma_{ij}' = D_{ijkl}\ (\ d\varepsilon_{kl} - d\varepsilon_{kl}^{\circ} - d\varepsilon_{kl}^{p}/3) + \sigma_{ip}' d\Omega_{pj} + \sigma_{jp}' d\Omega_{pi} \tag{3}$$

where the first term corresponds to the constitutive law associated with the Jaumann, corrotational, stress increment and the second term accounts for effective stress changes due to 'rigid body' rotation of an element of volume. In above $d\varepsilon_{ij}^{\circ}$ stands for any 'initial' or creep strain increments and

$$d\varepsilon_{ij}^{p} = \frac{1}{3}\, d\varepsilon_{ii}^{p} \tag{4}$$

The strain and 'spin' increments, $d\varepsilon_{ij}$ and $d\Omega_{ij}$, are defined in the usual manner as

$$d\varepsilon_{ij} = \frac{1}{2}\ (du_{i,j} + du_{j,i}\ ) \tag{5}$$

and

$$d\Omega_{ij} = \frac{1}{2}\ (du_{i,j} - du_{j,i}\ ) \tag{6}$$

in which du_i are the displacement increments.

At this stage we should note that quite generally the tangent modulus D_{ijkl} will depend on the current effective stress levels $\underset{\sim}{\sigma}'$, accumulated plastic strains or other history or damage parameters and in the case of anisotropic materials on the orientation of the material elements. In the last case it is necessary in computation to keep account of this orientation by a suitable integration of elementary rotations $d\underset{\sim}{\Omega}$.

Equilibrium relations

Following Biot we shall use the vector w_i to denote the mean displacement of the fluid relative to the solid skeleton. This is defined by taking the total quantity of fluid and dividing by the total cross section. Thus the average displacement in pores is w_i/n where n is the porosity.

With this we can write the total equilibrium of the solid-fluid volume as

$$\sigma_{ij,j} + \rho g_i = \rho \ddot{u}_i + \rho_f \ddot{w}_i \tag{7}$$

with g_i being body force acceleration, ρ_f fluid density and ρ density of the total 'mixture'.

To above equation the fluid equilibrium conditions have to be added. Here the essential effect is that of the viscous forces exerted by the solid boundaries on the flowing fluid. These can be defined in terms of a permeability tensor, k_{ij} and we can write

$$-p_{,i} + \rho_f g_i = \rho_f \ddot{u}_i + \frac{\rho_f}{n} \ddot{w}_i + k_{ij}^{-1} \dot{w}_i \tag{8}$$

where k_{ij}^{-1} are the 'resistivity' coefficients which are the inverse of the permeability matrix.

Mass balance

The final equation coupling the behaviour of two phase materials is that of mass balance. Clearly during straining of the solid matrix the volume of pores changes and the resulting change has to be compensated by the outflow and compressibility of the fluid.

$$\dot{\varepsilon}_{ii} + \dot{w}_{i,i} = -\dot{p}\, n/K_f - \dot{\varepsilon}_{ii}^{p}(1-n) \tag{9}$$

In the above equation we have neglected a very small term arising due to the change of pore shapes when the material is subject to pure effective stress - assuming thus that during such action the solid grains are essentially incompressible. The inclusion of this term in a similar framework of analysis is possible as shown by Zienkiewicz et al [4] but represents a negligible effect.

Alternative derivations of relationships governing the two phase behaviour are possible using a mixture theory. Some examples of such formulations are available in literature [5] - generally for linear constitutive relations - but except for ultra-high frequency phenomena in practice these reduce to the above format.

An approximation

Before attempting a finite element solution of the equations 1-8 it is convenient but not essential to introduce an approximation. In this we take the effect of the additional fluid acceleration $\ddot{w}_{ii}$ as negligible. Such an approximation allows us to use u_i and p as primary variables rather than u_i and w_i thus reducing in general the number of unknowns. With that approximation it is easy to eliminate w from (8) and (9) giving now in their place the well known equation of flow in porous media

$$\dot{\varepsilon}_{ii} - (k_{ij}\, p_{,i} - k_{ij}\, \rho_f\, g_i + k_{ij} \rho_f \ddot{u}_i)_{,i} = -\dot{p}\, n/K_f - \dot{\varepsilon}^{p}_{ii}(1-n) \qquad (10)$$

The above equation together with (1) through (7) in which effects of $\ddot{w}_i$ have been omitted governs the solution for all static and dynamic problems of soil/rock or concrete behaviour.

Though the above formulation disregards one of the dynamic effects it is applicable for most engineering dynamics situations associated with soil behaviour even if permeability values are relatively high and the frequencies are of the order of those occurring in earthquake analysis.

In Fig. 2 we show results of a study on a linear problem of a soil layer in which the zone of analysis in which the approximation is not valid is indicated [6] in terms of two non dimensional parameters.

The same figure indicates zones in which the effects of permeability are of no importance i.e. the zone of undrained behaviour; and a zone in which dynamic effects can be totally disregarded i.e. the zone of consolidation behaviour.

Large displacements and deformation

The basic equations presume that the difference between the natural coordinates of a material point i.e. $x_i = X_i + u_i$ and those spatial coordinates X_i is such that

$$\frac{\partial \underset{\sim}{x}}{\partial \underset{\sim}{X}} = \underset{\sim}{I}$$

If displacements are large this assumption is still valid providing an updating of the spatial coordinates is carried out in reasonably small increments of displacement u_i . Computation procedures which we shall now describe can follow such a pattern and the changes of domain are taken care of automatically.

Solution of dynamic and quasi-static problems

As in purely static situations the coupling of the two phases disappears and the fluid flow and solid equations can be solved separately there is little to be said about such solutions for which any of the 'standard' finite element procedures available can be used. Interest therefore focusses on dynamic problems of the kind involved in earthquake analysis of foundations or in a special case of problems in which dynamic effects are negligible and the transient nature of the phenomena is due entirely to the seepage of the fluid. It is on such classes of problems that the methodology expounded here is focussed.

Proceeding in a standard manner of finite element discretisation for displacements u_i and pressures p we can write (now using a vectorial notation)

$$\underset{\sim}{u} = \underset{\sim}{N}\bar{\underset{\sim}{u}} \qquad \underset{\sim}{p} = \tilde{\underset{\sim}{N}}\bar{\underset{\sim}{p}} \tag{11}$$

where $\underset{\sim}{N}$ and $\tilde{\underset{\sim}{N}}$ stand for appropriate shape functions and $\bar{\underset{\sim}{u}}$ and $\bar{\underset{\sim}{p}}$ for nodal quantities. In above

$$\underset{\sim}{u}^T = (u, v, w)^T$$

where u, v, w are the Cartesian displacement components. Eq.7 can now be written using standard Galerkin methods and substituting (1) as

$$\int_\Omega \underset{\sim}{B}^T \underset{\sim}{\sigma}' d\Omega - \underset{\sim}{Q}\bar{\underset{\sim}{p}} + \underset{\sim}{M}\ddot{\bar{\underset{\sim}{u}}} + \underset{\sim}{f} = 0 \tag{12}$$

where $M = \int_\Omega \underset{\sim}{N}^T \rho \underset{\sim}{N} d\Omega$

is the usual mass matrix, $Q = \int_\Omega \underset{\sim}{B}^T \underset{\sim}{m} \tilde{\underset{\sim}{N}} \, d\Omega$

and $\underset{\sim}{f}$ the force vector including body forces $\rho\underset{\sim}{g}$ and boundary tractions [7]. In above $\underset{\sim}{B}$ is the standard small strain matrix arising from (5).

Similarly (10) is discretised to give

$$\underset{\sim}{H}\bar{\underset{\sim}{p}} + \underset{\sim}{S}\dot{\bar{\underset{\sim}{p}}} + \underset{\sim}{Q}^T\dot{\bar{\underset{\sim}{u}}} - \hat{\underset{\sim}{M}}\ddot{\bar{\underset{\sim}{u}}} = \underset{\sim}{q} \tag{13}$$

where

$$\underset{\sim}{H} = \int_\Omega (\underset{\sim}{\nabla}\tilde{\underset{\sim}{N}})^T \underset{\sim}{k} \, \underset{\sim}{\nabla}\tilde{\underset{\sim}{N}} \, d\Omega$$

$$\underset{\sim}{S} = \int_\Omega \tilde{\underset{\sim}{N}}^T (n/K_f + (1-n)/K_s) \tilde{\underset{\sim}{N}} \, d\Omega$$

$$\hat{\underset{\sim}{M}} = \int_\Omega \tilde{\underset{\sim}{N}}^T \underset{\sim}{\nabla}^T \underset{\sim}{k} \rho_f \underset{\sim}{N} d\Omega$$

In all of the above equations Ω represents the <u>current</u> domain and if displacements are large an updating of coordinates is necessary at each step of the computation.

Equations (12) and (13) are supplemented by the constitutive relationship (3) and an appropriate definition of the spin vector of (6). In the previous expressions the "standard" strain matrices were used. Now we have various alternatives and the writer prefers to retain the notation consistent with writing the stresses in a vector form. We thus define following Nayak (in 3 dimensions)

$$d\underset{\sim}{\Omega} = \begin{bmatrix} 0 & 0 & 0 & 0 & -\omega_y & +\omega_z \\ 0 & 0 & 0 & +\omega_x & 0 & -\omega_z \\ 0 & 0 & 0 & -\omega_x & +\omega_y & 0 \\ 0 & -\frac{1}{2}\omega_x & +\frac{1}{2}\omega_x & 0 & -\frac{1}{2}\omega_z & +\frac{1}{2}\omega_y \\ +\frac{1}{2}\omega_y & 0 & -\frac{1}{2}\omega_y & +\frac{1}{2}\omega_z & 0 & -\frac{1}{2}\omega_x \\ -\frac{1}{2}\omega_z & +\frac{1}{2}\omega_z & 0 & -\frac{1}{2}\omega_y & +\frac{1}{2}\omega_x & 0 \end{bmatrix} \qquad (14)$$

where the angular velocities are

$$\omega_x = \frac{\partial(dv)}{\partial z} - \frac{\partial(dw)}{\partial y}$$

$$\omega_y = \frac{\partial(dw)}{\partial x} - \frac{\partial(du)}{\partial z}$$

$$\omega_z = \frac{\partial(du)}{\partial y} - \frac{\partial(dv)}{\partial z} \qquad (15)$$

reducing to

$$d\Omega = \begin{bmatrix} 0 & 0 & 1 \\ 0 & 0 & 1 \\ -\frac{1}{2} & \frac{1}{2} & 0 \end{bmatrix} \omega_z \qquad (16)$$

in 2 dimensional plane problems. Now the stress increment in vectorial form is written as

$$d\underset{\sim}{\sigma}' = \underset{\sim}{D}(\underset{\sim}{B}\, d\underset{\sim}{u} - d\underset{\sim}{\varepsilon}^{\circ} - \underset{\sim}{m}\, d\underset{\sim}{\varepsilon}_v^p / 3) + d\underset{\sim}{\Omega}\, \underset{\sim}{\sigma}' \qquad (17)$$

with

$$d\varepsilon_v^p = -dp/K_s \quad ; \quad \underset{\sim}{m}^T = (1,1,1,0,0,0)$$

The above equations with a specified constitutive relation complete the discretised solution which has to be solved now in the time domain.

Time stepping algorithms

The nonlinear system of ordinary differential equations (12) to (17) can be solved by a variety of algorithms. However in all it is convenient to use some form of staggering or partitioning to avoid very large equation systems. In such staggered solutions one set of equations is stepped forward at a time using extrapolated values of the second variable. Thus for instance we can apply a central difference algorithm to the dynamic equation (12). Now we have for the n-th step

$$\left(\int_{\Omega} \underset{\sim}{B}^T \underset{\sim}{\sigma}' d\Omega\right)_n - \underset{\sim}{Q}\underset{\sim}{\bar{p}}_n + \underset{\sim}{M}(\underset{\sim}{\bar{u}}_{n+1} - 2\underset{\sim}{\bar{u}}_n + \underset{\sim}{\bar{u}}_{n-1})/\Delta t^2 + \underset{\sim}{f}_n \qquad (18)$$

from which $\underset{\sim}{\bar{u}}_{n+1}$ can be found explicity if a lumped matrix form is used and $\underset{\sim}{\sigma}'_n$, $\underset{\sim}{\bar{p}}_n$, $\underset{\sim}{\bar{u}}_n$, $\underset{\sim}{\bar{u}}_{n-1}$ etc. are known.

The stress increment can be evaluated using eqs. 16 and 17. Thus first we compute $\Delta\underset{\sim}{\Omega}$ using a difference form and then find

$$\Delta\underset{\sim}{\sigma}' = \underset{\sim}{D}_n\,(\underset{\sim}{B}_n \Delta \underset{\sim}{u} - \Delta\underset{\sim}{\varepsilon}^{\circ} - \underset{\sim}{m}\,\Delta\varepsilon_v^p/3) + \Delta\underset{\sim}{\Omega}\,\underset{\sim}{\sigma}' \qquad (19)$$

(Note that in above calculations we have used a simple Euler rule of forward integration although with little additional effort average values of $\underset{\sim}{D}_{n+1/2}$, $\underset{\sim}{B}_{n+1/2}$ etc. could have been inserted.

Once $\underset{\sim}{\bar{u}}_{n+1}$ is known the pore pressure changes can be found using an independent algorithm on the seepage flow equation inserting appropriate approximations for $\dot{\underset{\sim}{\bar{u}}}$ and $\ddot{\underset{\sim}{\bar{u}}}$.

$$\dot{\underset{\sim}{\bar{u}}}_{n+1} = (\underset{\sim}{\bar{u}}_{n+1} - \underset{\sim}{\bar{u}}_n)/\Delta t$$

$$\ddot{\underset{\sim}{\bar{u}}}_{n+1} = (\underset{\sim}{\bar{u}}_{n+1} - 2\underset{\sim}{\bar{u}}_n + \underset{\sim}{\bar{u}}_{n-1})/\Delta t^2 \qquad (20)$$

We choose here to use an unconditional stable algorithm $1/2 < \theta \leqslant 1$ (generally with the upper value)

$$\underset{\sim}{H}_n((1-\theta)\underset{\sim}{\bar{p}}_n + \theta\underset{\sim}{\bar{p}}_{n+1}) + \underset{\sim}{S}_n(\underset{\sim}{\bar{p}}_{n+1} - \underset{\sim}{\bar{p}}_n)/\Delta t + \underset{\sim}{Q}^T\dot{\underset{\sim}{\bar{u}}}_{n+1} - \hat{\underset{\sim}{M}}\,\ddot{\underset{\sim}{\bar{u}}}_{n+1} \quad \underset{\sim}{q} = 0$$

From this the new set $\underset{\sim}{\bar{p}}_{n+1}$ is easily obtained and computation can proceed further.

Stability of such staggered algorithms has recently been investigated thoroughly by Park-Fellipa.

In the present case we note that

(a) as the permeability increases i.e. $\underset{\sim}{k} \to \infty$ the equation (13) becomes approximate

$$\underset{\sim}{H}\underset{\sim}{\bar{p}} - \hat{\underset{\sim}{M}}\dot{\underset{\sim}{\bar{u}}} + \underset{\sim}{q} = 0$$

and if further the frequencies involved are such that $\hat{\underset{\sim}{M}}\ddot{\underset{\sim}{\bar{u}}} = 0$ then uncoupling of the two systems occurs. In such a case the stability of computation of the explicit scheme depends on the wave speed in the desired medium

and

(b) as the permeability decreases i.e. $\underline{k} \to 0$ then the seepage equation reduces to

$$\underline{S}\dot{\underline{p}} + \underline{Q}^T\dot{\underline{u}} = 0$$

Now we can evaluate $\bar{\underline{p}}$ as

$$\bar{\underline{p}} = -\underline{S}^{-1}\underline{Q}^T\bar{\underline{u}}$$

and eliminating this from eq. 12 results in a stiffness term

$$\underline{Q}\,\underline{S}^{-1}\underline{Q}^T\bar{\underline{u}}$$

from which we conclude that stability is governed by undrained conditions.

As generally the ratio of 'drained' to 'undrained' wave speeds is of the order of 1/10, the advantages of using an undrained assumption disappear.

In practical computation we generally find that for conditions not corresponding to the above limiting cases an intermediate assumption for critical Δt needs to be used. A study is now being completed on the critical stability of the combined algorithm.

The explicit-implicit, staggered scheme just outlined is advantageous and practical for relatively short time span phenomena such as explosions or earthquakes. For longer term computation such as those involving consolidation or wave loads on offshore structures, implicit-implicit schemes must be used. We shall not discuss the detail of such schemes except to mention that now, if large displacement form of equations is to be used, the computation will involve the calculation of the derivatives

$$\frac{\partial}{\partial \bar{\underline{u}}}\left(\int_\Omega \underline{B}^T \underline{\sigma}'\, d\Omega\right)$$

These yield tangent stiffness matrices due to geometric (initial stress) effects and details of such calculations can be found elsewhere.

It should be noted that here generally one determination of such stiffness matrices will only be necessary if quasi-Newton (B.F.G.S.) methods are used.

Constitutive law

The brief scope of this paper permits only a very brief discussion of the constitutive relations applicable to soils under dynamic or quasi-static conditions.

The basic models used generally fall into one or another of the two general classes:

(1) elasto-plastic or elasto-viscoplastic

(2) endochronic.

For static or quasi-static monotonic loading problems it is generally conceded that models of the first category are more efficient requiring the least number of parameters for their description. Here the best models are those of the "critical state" family developed by Roscoe in which a hardening process occurs with increasing plastic compaction. However, quite adequate models are those with a non associative ideal elasto plastic behaviour. Here a yield surface corresponding with the Mohr-Coulomb failure is usually used.

A survey of some models of this type has recently been presented by the author and their performance compared on some typical problems.

The need to examine the behaviour under load reversals and to incorporate rate effects has caused much effort to be devoted to the subject recently. In a conference devoted to the subject of behaviour of soils under transient and cyclic load [12] various approaches have been analysed.

It appears that once again elasto-plastic models alone or in combination with a "deterioration parameter" of an endochronic type are most advantageous. In the first the concepts of anisotropic kinematic hardening of the type first introduced by Mroz [13] combined with plastic modulus interpolation suggested by Popov and Defalias [14] are effective [15]. Here the critical state ellipse is used as a "bounding" surface with an interior ellipse of smaller size representing a kinematic yield surface.

With such models the phenomenon of "densification" which occurs in cyclically loaded soils and which is responsible for pore pressure rises and occasionally liquefaction of the soil can be modelled. Fig. 3 shows such a model in the stress invariant space and the resulting volumetric strain increase.

In the second approach an established elasto-plastic, isotropic model is used and the densification is added as an initial strain $\underset{\sim}{\varepsilon}^{o}$ dependent on the 'endochronic' deterioration parameter

$$d\xi = \sqrt{de_{ij} \quad de_{ij}}$$

where e_{ij} is the deviatoric strain.

Such a model was used with success by the author in various practical computations. Fig. 4 shows for instance the computation carried out by the scheme described to simulate the partial collapse of San Fernando dam [16][17].

Other uses of the deterioration parameter have been reported and some very successful models are now being produced, though much work in this area is yet necessary.

Concluding remarks

The field of problems described in this paper is one in which both the formulation as well as physical modelling present some of the greatest challenges. It is hoped that this interim statement of the state of the art will stimulate further contributions which will increase the efficiency and width of application of the now existing methodology.

Acknowledgements

The author is indebted to research co-workers and students in the various phases of this work. Dr. D. Naylor, Dr. E. Hinton, Dr. G. Pande, Dr. C. T. Chang, Mr. K. H. Leung and Mr. V. Norris have made particularly important contributions. Thanks are also due to the Science Research Council (U.K.) for supporting some of the members of the team.

References

1. Terzhagi, K.; Erdboumechanik, Franz Deuticke, Vienna, 1925.

2. Biot, M. A.; Theory of three-dimensional consolidation, Journal of Applied Physics, Vol. 12, pp. 155-164, 1941.

3. Biot, M. A.; Mechanics of deformation and acoustic propagation in porous media, J. Appl. Physics, Vol. 33, pp. 1483-98, 1960.

4. Zienkiewicz, O. C., Humpheson, C. and Lewis, R. W.; A unified approach to soil mechanics problems, Chapter 4 of Finite Elements in Geomechanics, pp. 151-178, ed. G. Gudehus, J. Wiley, 1977.

5. Bowen, R. M.; "Theory of mixtures", Continuum Physics, Vol. III, editor Eringen; Academic Press, New York, 1976.

6. Zienkiewicz, O. C. and Bettess, P.; Soil and other saturated porous material under transient, dynamic condition, general formulation and validity of various simplified assumptions. Soils under Cyclic and Transient Loading, editors O. C. Zienkiewicz and G. N. Pande, John Wiley (to be published).

7. Zienkiewicz, O. C.; The Finite Element Method, third edition, McGraw-Hill, 1977.

9. Park, K. C.; Partitioned transient analysis procedures for coupled field problems - stability analysis - to be published.

10. Park, K. C. and Felippa, C. A. Partitioned transient analysis procedures for coupled field problems - accuracy analysis - to be published.

11. Zienkiewicz, O. C.; Constitutive laws and numerical analysis for soil foundation under static, transient or cyclic loading. Proceedings of Second International Conference on Behaviour of Offshore Structures (BOSS), London 1979.

12. Pande, G. N. and Zienkiewicz, O. C. Proceedings Conference on Soils under Cyclic and Transient loading, Swansea 1980. Balkema Press 1980. (See also special volume to be published by J. Wiley & Sons, 1981).

13. Mroz, Z. On the description of anisotropic work hardening, J. Mech. Phys. Solids, 15, pp. 163-75, 1967.

14. Dafalias, Y. F. and Popov, E. P. Plastic internal variables formalism of cyclic plasticity. J. Appl. Mech. 98 (4), 645-650, 1976.

15a Mroz, Z., Norris, V. A. and Zienkiewicz, O. C.; An anisotropic hardening model for soils and its application to cyclic loading. Int. J. Num. and Analytical Meth. Geomech. 1978, 2, 203-221.

15b Mroz, Z., Norris, V. A. and Zienkiewicz, O. C.; Application of an anisotropic hardening model in the analysis of elastic-plastic deformation of soils. Geotechnique, 1979, 29, 1-34.

16. Seed, H. B.; Consideration in the earthquake-resistant design of earth and rockfill dams. 19th Rankine Lecture of the British Geotechnical Society, Geotechnique, 1979, 29, 215-263.

17. Zienkiewicz, O. C., Leung, K. H., Hinton, E. and Chang, C. T.; Earth dam analysis for earthquakes. Proceedings Conference on 'Design of Dams to resist Earthquakes', Inst. Civ. Eng. London, October 1-2, 1980.

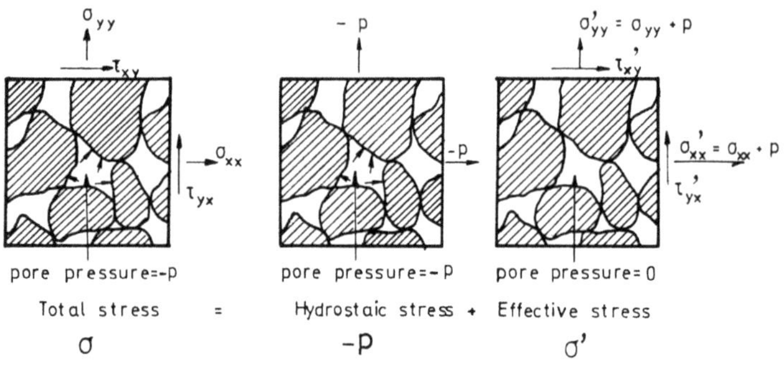

Figure 1 Total and effective stress in a porous material

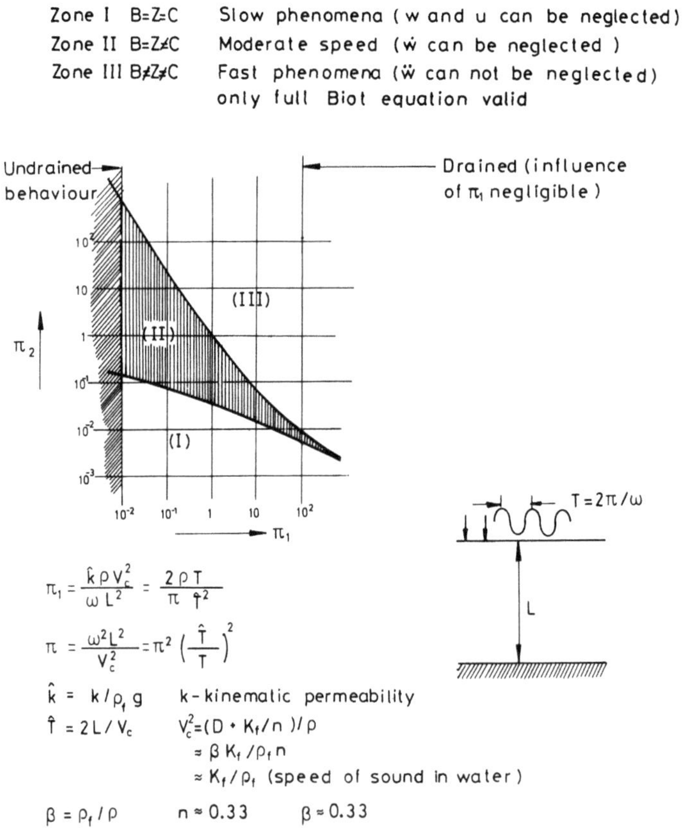

Figure 2 A linear elastic semi-infinite space subject to a periodic surface loading -

Zones of applicability of various assumptions

B: Biot's equation, Z: Zienkiewicz's approximation,

C: Conventional (Terzhagi) equation

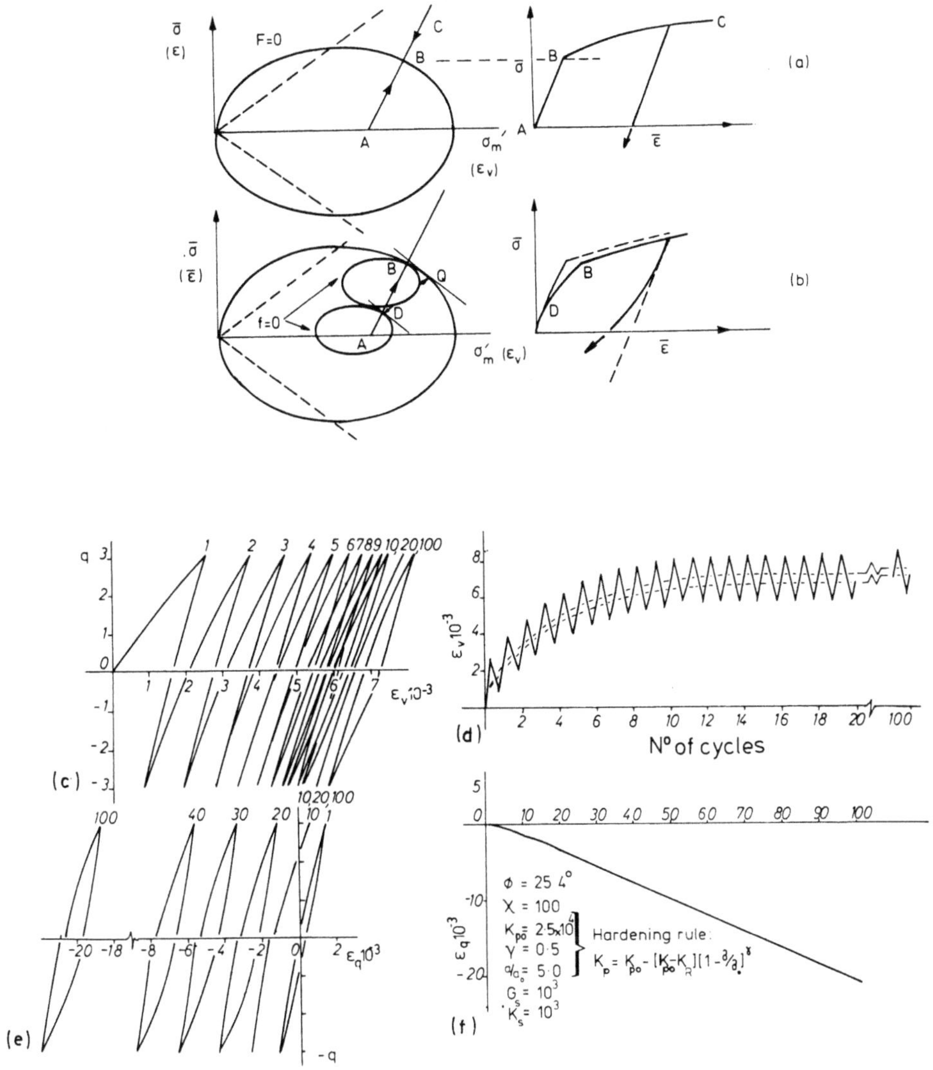

Figure 3 Critical state plasticity
a Single surface b two surface

Cyclic loading under drained triaxial conditions
(c) shear stress vs. volumetric strain
(d) volumetric strains vs. number of cycles
(e) shear stress vs. shear strain
(f) shear strain (at end of cycle) vs. number of cycles

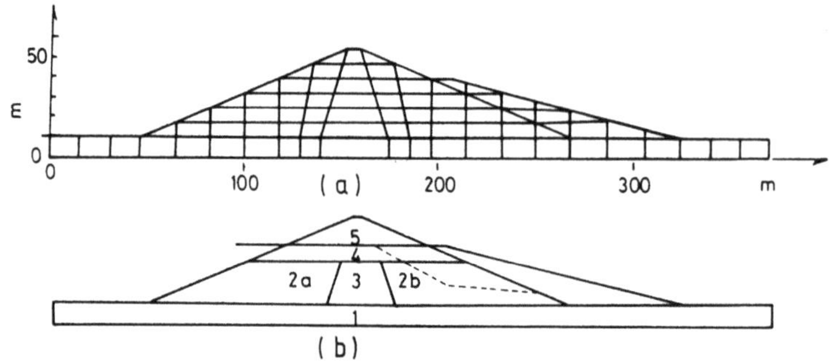

	Material zone	Elastic modulus E MN/M²	ν	Φ' Degree	C' KN/M²	Unit weight T/M³	$\hat{p}=\frac{A}{B}(1+Be^{\gamma\theta})$ A	B	Y
1	Alluvium	200	0.4	38	10	2.09			
2a	Hydraulic fill - sand	90	0.41	37	10	2.02	2914	1904	4.6
2b	" " "	110	0.41	37	10	2.02-sat 1.71-dry	2914	1904	4.6
3	Clay core	90	0.41	37	10	2.02			
4	Ground shale-hydraulic fill	90	0.41	37	10	2.02sat 1.71dry			
5	Rolled fill	60	0.3	25	126	2.0			

(a)

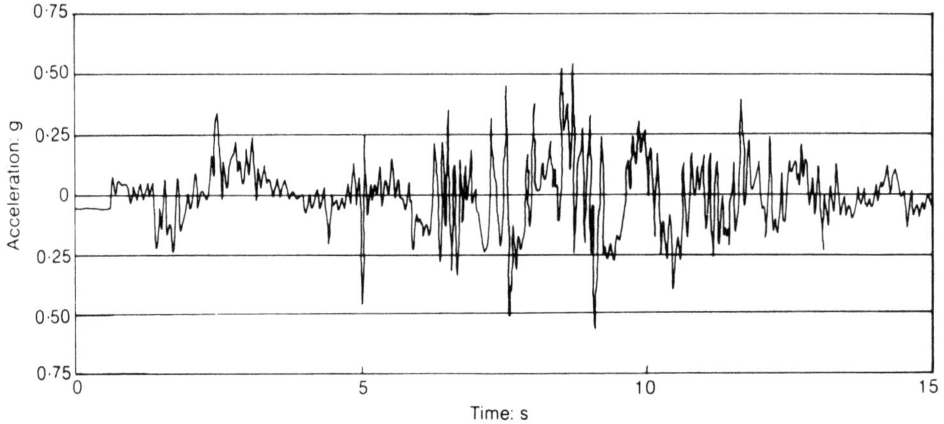

(b)

Figure 4 The Lower San Fernando Dam analysis

(a) Data assumed for the analysis and the mesh of quadratic finite element

(b) Base motion input

(c) Deformed mesh plot (displacement x 2.0)

(d) Contours of excess pore water pressure build up

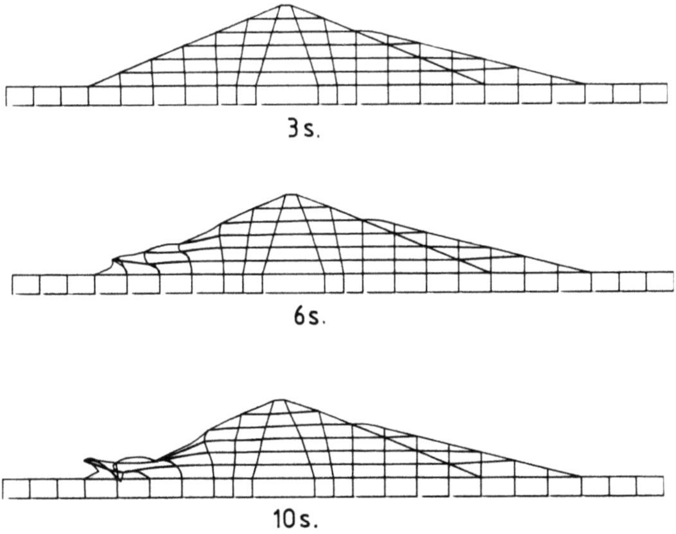

(c)

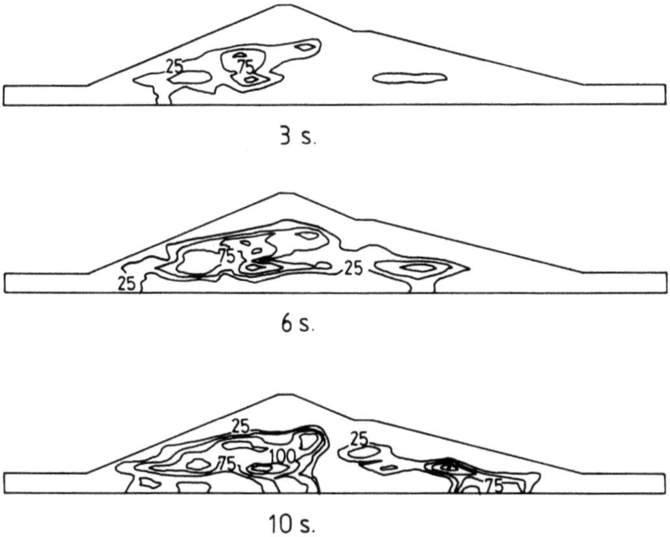

Contour unit = kN/m^2

Contour spacing= 25 kN/m^2

(d)

Numerical Methods in Elasto-Plasticity - A Comparative Study

A. SAMUELSSON, M. FRÖIER
Chalmers University of Technology, Göteborg, Sweden

Summary

Problems involving nonviscid, quasistatic, small strain elastoplasticity are discussed in different aspects: the role of an objective rate of stress, the choice of variational formulation, the choice of time step procedure.

Introduction

The development of the mathematical theory of nonviscid, quasistatic plasticity started with Tresca, Lévy and St. Venant in the middle of the 19th century. In the first part of our century v Mises, Prandtl and Reuss established a theory for small strain plasticity for an isotropic material with isotropic hardening. Their work is still the basis of the mathematical theory of plasticity. After the second world war the theory was generalized and extended in many directions by among others Hill, Prager, Drucker, and Koiter.

Lately, the theory was redefined in the framework of functional analysis, especially the theory of variational inequalities by among others Duvaut and Lions [1], and Johnson [2]. It was, within this theory, possible to prove the existence of a solution to the cases with elastic - perfectly plastic and elastic - linear hardening properties.

Before the development of the finite element method very few practical solutions were obtained. In late 60ies the first finite element applications to twodimensional boundary value problems governed by Prandtl-Reuss equations appeared in literature. An important contribution here was due to Yamada, Yoshimura, and Sakurai [3] who deduced an explicit tangent stiffness matrix according to the Prandtl-Reuss theory. The solution was obtained in a step by step method such that in each step a linear problem with an elastoplastic stiffness matrix was solved. It was found that due to the step approximation the yield condition became violated. The method was then improved by projection to the yield surface and iteration. Important contributions from this period were due to Marcal et al [4] and Zienkiewicz et al [5].

Following the development of the application of functional analysis to the Prandtl-Reuss problem equilibrium and mixed finite element procedures suggested by the analysis were proposed by Johnson [6] and others. A mixed method has been applied to problems with plane stress by the present authors [7], [8]. Linear programming techniques used for limit load determinations have by Maier and others been shown to be useful also for the solution of elastoplastic boundary value problems [9].

The Prandtl-Reuss theory has been extended to problems with finite strain by Hill [10], Lee [11], Hutchinson [12], and others. In this theory stress and rate of stress have to be defined in a strict and useful way. When specializing this general theory to small strain elastoplasticity the result is found to differ from Prandtl-Reuss theory.

In this paper some questions concerning small strain plasticity for isotropic material and monotonous loading will be discussed.

Questions

1. Is it of practical importance to use objective rate of stress also in small strain plasticity?

2. Which are the essential differences between the conventional finite element method with an elastoplastic stiffness matrix and the method based on a variational inequality?

3. The transient problem is solved in the two methods with a forward respectively a backward Euler method, so the order of error is $O(\Delta t)$. Small time steps are certainly needed. Would it be worth trying a higher order method?

<u>The tangent modulus and tangent matrix in small strain plasticity</u>

Let an infinitesimal unit element be loaded in principal direction 1 with a force P so that the nominal stress vector is $\{\sigma_0, 0, 0\}$, $\sigma_0 = P$. After deformation the side lengths of the elements are $1+e_1$, $1+e_2$, $1+e_3$. Then the Cauchy stress vector is $\{\sigma_1, 0, 0\} = \{\sigma_0/(1+e_2)(1+e_3), 0, 0\}$, or with use of volume constancy, $\{\sigma_0(1+e_1), 0, 0\}$. Derivation with respect to time of $\sigma_0 = \sigma_1/(1+e_1)$ gives

$$\frac{d\sigma_0}{dt} = \frac{1}{1+e_1}\frac{d\sigma_1}{dt} - \frac{\sigma_1}{(1+e_1)^2}\frac{de_1}{dt} \tag{1}$$

For small strain, $e_1 \ll 1$, then

$$\frac{\dot{\sigma}_1}{\dot{e}_1} = \frac{\dot{\sigma}_0}{\dot{e}_1} + \sigma_1 \tag{2}$$

The nominal and real tangent modulus thus differ by the amount of σ_1. For low or disappearing tangent modulus the effect of σ_1 is significant.

Hutchinson [12] gives the following elastoplastic constitutive relation in stiffness form for the large strain case

$$\overset{*}{\tau}{}^{ij} = \overset{*}{S}{}^{ijkl}\dot{\eta}_{kl} \tag{3}$$

with

$$\overset{*}{S}{}^{ijkl} = 2G\{\tfrac{1}{2}(g^{ik}g^{jl} + g^{il}g^{jk}) + \tag{4}$$

$$+ [\nu/(1-2\nu)]g^{ij}g^{kl} - (\alpha/g)s^{ij}s^{kl}\}$$

where $\overset{*}{\tau}{}^{ij}$ is the Jaumann rate of a contravariant component of the Kirchhoff stress, $\dot{\eta}_{kl}$ is the rate of a covariant component of the Lagrangian strain, $\underset{\sim}{g}$ is the metric tensor in the deformed body and

$$s^{ij} = \tau^{ij} - g^{ij}g_{kl}\tau^{kl}/3 \tag{5}$$

Further

$$\alpha = 1 \text{ if } s^{kl}\dot{\eta}_{kl} \geq 0 \text{ and} \tag{6}$$

$$J_2 = g_{ik}g_{jl}s^{ij}s^{kl}/2 = (J_2)_{max}$$

$$\alpha = 0 \text{ else,}$$

and g is a generalized hardening modulus

$$1/g = (E/E_t - 1)/[E/E_t - (1-2\nu)/3]2J_2 \tag{7}$$

with E_t and J_2 taken from a uniaxial case.
The Jaumann rate of the Kirchhoff stress $\underset{\sim}{\tau}$ is related to the convected rate by

$$\overset{*}{\tau}{}^{ij} = \dot{\tau}^{ij} + \frac{1}{2}(g^{ik}\tau^{jl} + g^{jk}\tau^{il} + g^{il}\tau^{jk} + \\ + g^{jl}\tau^{ik})\dot{\eta}_{kl} \tag{8}$$

For small strain $g^{ik} \sim \delta^{ik}$, $\eta_{kl} \sim \varepsilon_{kl}$, $\tau^{ik} \sim \sigma_0^{ik}$, the nominal stress, here denoted by σ^{ik}. Combination of (3) and (8) and restriction to small strain and plane stress gives the following tangent elastoplastic relation

$$\begin{bmatrix} \dot{\sigma}_x \\ \dot{\sigma}_y \\ \dot{\tau}_{xy} \end{bmatrix} = (2G/g) \begin{bmatrix} A_{11} & A_{12} & A_{13} \\ & A_{22} & A_{23} \\ \text{SYM} & & A_{33} \end{bmatrix} \begin{bmatrix} \dot{\varepsilon}_x \\ \dot{\varepsilon}_y \\ \dot{\gamma}_{xy} \end{bmatrix} \tag{9}$$

where G is the shear modulus and

$$A_{11} = A - s_x^2 - (B-s_x s_z)^2/(A-s_z^2) - \sigma_x g/G \tag{10}$$

$$A_{22} = A - s_y^2 - (B-s_y s_z)^2/(A-s_z^2) - \sigma_y g/G$$

$$A_{12} = B - s_x s_y - (B-s_x s_z)(B-s_y s_z)/(A-s_z^2)$$

$$A_{13} = (B-s_x s_z) s_z \tau_{xy}/(A-s_z^2) - s_x \tau_{xy} - \tau_{xy} g/2G$$

$$A_{23} = (B-s_y s_z) s_z \tau_{xy}/(A-s_z^2) - s_y \tau_{xy} - \tau_{xy} g/2G$$

$$A_{33} = g/2 - \tau_{xy}^2 - s_z^2 \tau_{xy}^2/(A-s_z^2) - (\sigma_x+\sigma_y) g/4G$$

with

$$A = g(1-\nu)/(1-2\nu), \quad B = g\nu/(1-2\nu)$$

The numerical experiment reported below indicates that the effect of using the more correct stress rate definition is significant also for small strain if the tangent modulus also is small. There seems to be no practical reason not to include the extra terms in the tangent matrix.

Constitutive equations in two formulations

In the Prandtl-Reuss theory with isotropic linear hardening the admissible set is a convex

$$P = \{(\sigma_{ij}, \varepsilon^p) : F(\sigma_{ij}, \varepsilon^p) \leq 0\} \tag{11}$$

where

$$F(\sigma_{ij}, \varepsilon^p) = \sqrt{3/2}\,|s_{ij}| - (\sigma_y + H\varepsilon^p) \tag{12}$$

and $\dot{\varepsilon}^p = \sqrt{2/3}\,|\dot{\varepsilon}^p_{ij}|$ is proportional to the Euclidean length of the 9-dimensional plastic strain rate vector.
It is assumed that during plastic flow $\dot{F} = 0$, that is

$$\frac{\partial F}{\partial \sigma_{ij}} \dot{\sigma}_{ij} - H\dot{\varepsilon}^p = 0 \tag{13}$$

and that the rate of plastic strain takes place in the direction normal to F in the point σ_{ij}:

$$\dot{\sigma}_{ij} = A^{-1}_{ijkl}(\dot{\varepsilon}_{kl} - \dot{\varepsilon}^p \frac{\partial F}{\partial \sigma_{kl}}) \tag{14}$$

The rate $\dot{\varepsilon}^p$ of effective plastic strain can be eliminated between (13) and (14) giving an elastoplastic relation valid for $E_t >> \sigma$

$$\dot{\sigma}_{ij} = S_{ijkl}\dot{\varepsilon}_{kl} \tag{15}$$

In the alternative formulation the constitutive equation is given in two equations,

$$\dot{\varepsilon}_{ij} = A_{ijkl}\dot{\sigma}_{kl} + \varepsilon^p_{ij} \tag{16}$$

$$-\dot{\varepsilon}^p_{ij}(\tau_{ij}-\sigma_{ij}) + H\dot{\varepsilon}^p(\eta-\varepsilon^p) \geq 0, \quad \forall(\tau_{ij},\eta) \in P \tag{17}$$

The latter equation expresses for $F(\sigma_{ij}, \varepsilon^p) = 0$ that the vector $\{\dot{\varepsilon}^p_{ij}, -H\dot{\varepsilon}^p\}$ is a normal vector to F at $(\sigma_{ij}, \varepsilon^p)$ so

$$\begin{bmatrix} \dot{\varepsilon}^p_{ij} \\ -H\dot{\varepsilon}^p \end{bmatrix} = \lambda \begin{bmatrix} \partial F/\partial \sigma_{ij} \\ \partial F/\partial \varepsilon^p \end{bmatrix} \tag{18}$$

Now $|\partial F/\partial \sigma_{ij}| = \sqrt{3/2}$ and $\partial F/\partial \varepsilon^p = -H$ so

$$\begin{bmatrix} |\dot{\varepsilon}^p_{ij}| \\ -H\dot{\varepsilon}^p \end{bmatrix} = \begin{bmatrix} \sqrt{3/2}\ \dot{\varepsilon}^p \\ -H\dot{\varepsilon}^p \end{bmatrix} = \lambda \begin{bmatrix} \sqrt{3/2} \\ -H \end{bmatrix} \tag{19}$$

It follows that $\lambda = \dot{\varepsilon}^p$ so (14) is deduced from (16), (17).

Variational equations in two formulations

Let a regular region $\Omega \subset R^3$ be loaded at point x_i and time t by volume load U_i. At the part $\partial\Omega_T$ of the boundary $\partial\Omega$ the region is loaded by $\bar{T}_i$ and at the remaining part $\partial\Omega_u$ of the boundary the displacement u_i is equal to zero. An initial-boundary value problem is then defined as:

Find the displacements $u_i(t)$ and stresses $\sigma_{ij}(t)$ for a given volume load $U_i(t)$ and surface load $\bar{T}_i(t)$ satisfying

$$\left.\begin{array}{lll} \text{(a)} & -\sigma_{ji,j} = U_i & \\ \text{(b)} & \varepsilon_{ij} = \frac{1}{2}(u_{i,j}+u_{j,i}) & \\ \text{(c)} & \dot{\varepsilon}_{ij} = A_{ijkl}\dot{\sigma}_{kl} + \dot{\varepsilon}^p_{ij} & \\ \text{(d)} & -\dot{\varepsilon}^p_{ij}(\tau_{ij}-\sigma_{ij}) + H\dot{\varepsilon}^p(\eta-\varepsilon^p) \geq 0, \ \forall(\tau_{ij},\eta) \in P & \\ \text{(e)} & u_i(0) = \sigma_{ij}(0) = 0 & \end{array}\right\} \text{on } \Omega \qquad (20)$$

$$\text{(f)} \quad u_i(t) = 0 \qquad \text{on } \Omega_u$$

$$\text{(g)} \quad T_i = n_j\sigma_{ji} = \bar{T}_i(t) \quad \text{on } \Omega_T$$

The constitutive relations (c), (d) may be substituted by the stiffness relation (15)

$$\text{(c')(d')} \quad \dot{\sigma}_{ij} = S_{ijkl}\dot{\varepsilon}_{kl} \qquad (20)$$

In both procedures (20a) is replaced by a virtual work equation

$$\int_\Omega \varepsilon_{ij}(u_i)\sigma_{ij}dV = \int_\Omega u_iU_idV + \int_{\partial\Omega_T} u_i\bar{T}_idS, \ \forall u_i \in V \qquad (21)$$

where V is a space of functions that satisfy (20f) and are regular enough. A regular finite element mesh is then chosen and a subspace $V_h \subset V$ of finite element functions is defined. Let a general such function be

$$u = \phi^T\tilde{u} \qquad (22)$$

where ϕ is a rectangular matrix of basis functions and $\tilde{u}$ is a column matrix of node values. With the functions in ϕ as u_i in (21) the following system of equations is obtained

$$\int_\Omega (\tilde{\nabla}\phi^T)^T\sigma dV = \int_\Omega \phi U dV + \int_{\partial\Omega_T} \phi\bar{T}dS \qquad (23)$$

where σ, U, and $\bar{T}$ are matrices with components σ_{ij}, U_i and $\bar{T}_i$ and $\tilde{\nabla}$ is the linear strain operator.

In the <u>first procedure</u> (23) is time differentiated and combined with (20c',d') giving

$$\{\int_{\Omega}(\tilde{\nabla}\phi^T)^T S(\sigma,H)\nabla\phi^T dV\}\dot{\tilde{u}} = \int_{\Omega}\phi\dot{U}dV + \int_{\partial\Omega_T}\phi\dot{\bar{T}}dS \quad (24)$$

This transient system is to be integrated under constraint of initial conditions, variational equilibrium (23) and yield condition (11).

In the <u>second procedure</u> (20c) is scalarly multiplied by $(\tau_{ij} - \sigma_{ij})$, integrated over the domain and combined with (20d) giving a variational inequality

$$\int_{\Omega}A_{ijkl}\dot{\sigma}_{kl}(\tau_{ij}-\sigma_{ij})dV + H\int_{\Omega}\dot{\varepsilon}^p(\eta-\varepsilon^p)dV - \quad (25)$$

$$- \int_{\Omega}\dot{\varepsilon}_{ij}(\tau_{ij}-\sigma_{ij})dV \geq 0, \quad \forall(\tau_{ij},\eta) \in P$$

Let in a finite element mesh the stress components and the hardening functions be discretized to values at the Gauss points. Collect all these values in column matrices $\tilde{\sigma}$ and $\tilde{\varepsilon}^p$. The variational inequality (25) then becomes a set of uncoupled equations

$$(\tilde{\tau}-\tilde{\sigma})^T A\dot{\tilde{\sigma}} + H(\tilde{\eta}-\tilde{\varepsilon}^p)^T\dot{\tilde{\varepsilon}}^p - (\tilde{\tau}-\tilde{\sigma})^T\tilde{\nabla}\phi^T\dot{\tilde{u}} \geq 0, \quad (26)$$

$$\forall(\tilde{\tau},\tilde{\eta}) \in P$$

Together with (23) a transient system of equations is defined which should be integrated under constraint of initial conditions.

We observe that the equilibrium equation is in rate form in the first formulation and in total form in the second and that the hardening function ε^p is eliminated in the first formulation by use of the consistency condition (13) but is kept as unknown in the second.

Time discretization in two formulations

In the first formulation, (24), it is natural to use a simple forward Euler procedure. Starting at a known state at time t, $\{\tilde{u}_t, \tilde{\sigma}_t\}$, set

$$\tilde{u}_{t+\Delta t} = \tilde{u}_t + \Delta t \cdot \dot{\tilde{u}}_t \qquad (27)$$

where Δt is a short time-step. In (24) both the coefficient matrix and the right hand side are supposed to be known so $\dot{\tilde{u}}_t$ can be solved. Then

$$\dot{\tilde{\sigma}}_t = S(\sigma_t, H_t)\tilde{\nabla}\phi^T\dot{\tilde{u}}_t \qquad (28)$$

and

$$\tilde{\sigma}_{t+\Delta t} = \tilde{\sigma}_t + \Delta t\dot{\tilde{\sigma}}_t \qquad (29)$$

For a given space discretization this procedure tends to the correct solution for $\Delta t \to 0$. For a finite Δt the consistency condition (13) becomes violated so stress points may fall outside P. The values of (27) and (29) therefore need to be adjusted. Points outside P are projected onto $F = 0$ in an approximate way. Since with the projected stresses the equilibrium equation is not satisfied the residual and its effect upon $\dot{\tilde{u}}_t$ and $\dot{\tilde{\sigma}}_t$ have to be calculated.

With the <u>second formulation</u>, (23), (26), a backward Euler method may be used. With

$$\dot{\tilde{\sigma}}_t = (\tilde{\sigma}_t - \tilde{\sigma}_{t-\Delta t})/\Delta t, \quad \dot{\tilde{\varepsilon}}^p_t = (\tilde{\varepsilon}^p_t - \tilde{\varepsilon}^p_{t-\Delta t})/\Delta t \qquad (30)$$

at a time-step Δt, (26) becomes

$$(\tilde{\tau}-\tilde{\sigma}_t)^T A(\tilde{\sigma}_t - \tilde{\sigma}_{t-\Delta t}) + H(\tilde{\eta}-\tilde{\varepsilon}^p_t)(\tilde{\varepsilon}^p_t - \tilde{\varepsilon}^p_{t-\Delta t}) - \qquad (31)$$

$$- \Delta t(\tilde{\tau}-\tilde{\sigma}_t)^T\tilde{\nabla}\phi^T\dot{\tilde{u}}_t \geq 0, \quad \forall(\tilde{\tau},\tilde{\eta}) \in P$$

The solution to (31) is the minimum of a functional, namely

$$L(\tilde{\tau},\tilde{\eta}) = (\tilde{\sigma}^f-\tilde{\tau})^T A(\tilde{\sigma}^f-\tilde{\tau}) + H(\tilde{\varepsilon}^p_{t-\Delta t}-\tilde{\eta})^2, \qquad (32)$$

$$\forall(\tilde{\tau},\tilde{\eta}) \in P$$

where

$$\tilde{\sigma}^f = \tilde{\sigma}_{t-\Delta t} + \Delta t A^{-1}(\tilde{\nabla}\phi^T)\dot{\tilde{u}}_t \qquad (33)$$

is a fictitious stress obtained by adding to $\tilde{\sigma}_{t-\Delta t}$ an elastic stress increment.

For plane stress the variables of (32) are τ_{11}, τ_{22}, τ_{12}, η. In order to satisfy also the yielding condition this is augmented according to Lagrange's method with a multiplier λ. The following non-linear system of equations is obtained

$$2\tau_{11} - 2\nu\tau_{22} + 2(\nu\sigma^f_{22}-\sigma^f_{11}) + \lambda(2\tau_{11}-\tau_{22}) = 0 \qquad (34)$$

$$2\tau_{22} - 2\nu\tau_{11} + 2(\nu\sigma^f_{11}-\sigma^f_{22}) + \lambda(2\tau_{22}-\tau_{11}) = 0$$

$$4(1+\nu)\tau_{12} - 4(1+\nu)\sigma^f_{12} + 6\lambda\tau_{12} = 0$$

$$-2H(\varepsilon^p-\eta)E - \lambda\cdot 2H(\sigma_Y+H\eta) = 0$$

$$\tau^2_{11} - \tau_{11}\tau_{22} + \tau^2_{22} + 3\tau^2_{12} - (\sigma_Y+H\eta)^2 = 0$$

This system is easily solved by Newton-Raphson's method for known σ^f_{11}, σ^f_{22}, σ^f_{12}. These are obtained from (33) where, however, $\dot{\tilde{u}}_t$ is unknown. The problem has in each time-step to be solved by iteration between the equilibrium equation and (34). Stresses obtained from (34) give a residual R_j when put into (23). This corresponds to a change $\Delta t(\dot{u}_j-\dot{u}_{j-1})$ in displacement which can be obtained approximately from an elastic calculation

$$[\int_\Omega(\tilde{\nabla}\phi^T)^T A^{-1}\tilde{\nabla}\phi^T dV]\Delta t(\dot{\tilde{u}}_j-\dot{\tilde{u}}_{j-1}) = \rho R_j \qquad (35)$$

where ρ is a constant that should be chosen in a way that speeds up the calculation.

Small time steps Δt are needed in order to describe the transient problem in a sufficiently accurate way, particularly in the highly non-linear region near the limit load.

Small time steps are also needed in order to get satisfactorily good convergence in the iterative procedures. It should be observed that higher order time step procedures than the simple Euler method leads to complicated solution agorithms.

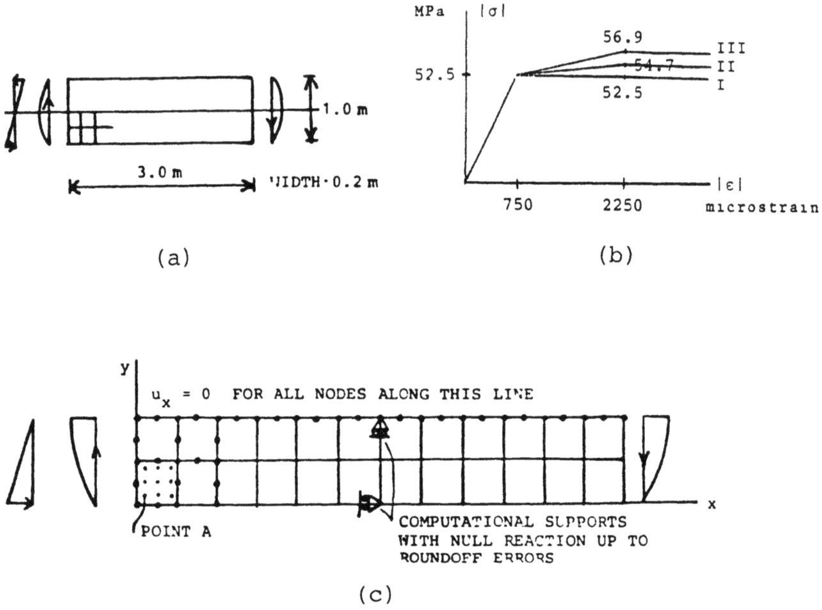

Fig 1. a) Selfequilibrating load on cantilever
b) Assumed uniaxial stress-strain relations
c) Finite element model

Numerical experiment

A cantilever beam with self-equilibrating load, Fig. 1a, made of an isotropic material with a uniaxial stress-strain relation according to Fig. 1b, is analyzed according to the conventional method with an elastoplastic stiffness matrix on an IBM-computer with double precision, [13]. The results obtained with an objective stress is compared with the results obtained with a nominal stress. Due to the antisymmetric conditions only half of the structure is studied. The elements are eight-node isoparametric with 3 x 3 Gausspoints according to Fig. 1c.

The strain ε_x in the indicated Gauss-point A is calculated with the two stress definitions, the three stress-strain relations I, II, III, according to Fig. 1b, with uniform load-steps $\Delta P/P_E = 0.025$ after the elastic limit load P_e is obtained, Table 1.

P/P_E		1 025	1 050	1 075	1 100	1.125	1 150	1 175	1 200	1.225	1 250	1.275
I	OBJECTIVE STRESS	784	882	982	1152	1803	4335					
	NOMINAL STRESS	784	882	982	1152	1803	4328					
II	OBJECTIVE STRESS	782	874	966	1071	1352	2019	2830	4473	12803		
	NOMINAL STRESS	782	874	966	1071	1352	2019	2814	4376	11742		
III	OBJECTIVE STRESS	781	866	952	1038	1249	1583	2056	2549	3434	5220	11846
	NOMINAL STRESS	781	866	952	1038	1249	1583	2056	2540	3403	5045	10687

Table 1. Strain $-\varepsilon_x$ in point A in microstrain for three stress-strain relations and two different theories, $\Delta P/P_E =$ $= 0.025$

In Table 2 results from calculations with three different but uniform load-steps, 0.100, 0.050, and 0.025, are displayed.

The solutions were in each load case obtained by solving the equilibrium equation and satisfaction of the yield condition iteratively. The procedure was stopped when no monotonous convergence was obtained.

	P/P_E	1.025	1.050	1.075	1.100	1.125	1.150	1.175	1.200	1.225	1.250
$\frac{\Delta P}{P_E}$	0.1000				1029				3865		
	0.0500		857		1055		2006		3823		14133
	0.0250	782	874	966	1071	1352	2019	2814	4376	11742	
	0.0125	791	883	975	1080	1362	2019	3001	4947	13321	

Table 2. Strain $-\varepsilon_x$ in point A in microstrain for three different load increments. Nominal stress is used.

In Table 3 and Table 4 results from calculations with the two methods described above are compared. The calculations are made for the problem in Fig 1 with the strain-stress relation II, Fig 1b, but changed to a bilinear form by extension of the second line of the graph. The obtained values of the effective strain ε^p in point A are displayed in Table 3. As shown in this Table the convergence of the time-step procedure is fast for this bilinear case. In Table 4 the development of the plastic zones can be followed. The labels of the integration points are given in Fig 2.

As can be seen from Tables 3 and 4 the numerical values of the strain in point A as well as the development of the plastic zones differ considerably between the methods after 10% increase of the elastic limit load. The different time step procedures in the methods do not seem to effect this conclusion.

	P/P_E	1 025	1 050	1 075	1 100	1 125	1 150	1 175	1 200	1 225	1 250	1 275	1 300
First method	0 0500		105		302		1240		2893		4780		7308
	0 0250	32	121	213	317	595	1253	2035	2908	3942	5000	6132	7320
	0 0125	40	130	221	325	604	1253	2038	2910	3946	5004	6135	7317
Second method	0 0500		127		295		787		1702		2729		3811
	0 0250	45	127	210	295	49[illegible]	798	1254	1732	2208	2733	3286	3819
	0 0125	45	127	210	295	493	799	1264	1732	2206	2735	3287	3817

Table 3. Effective strain ε^p in point A in microstrain for three load-steps and two methods

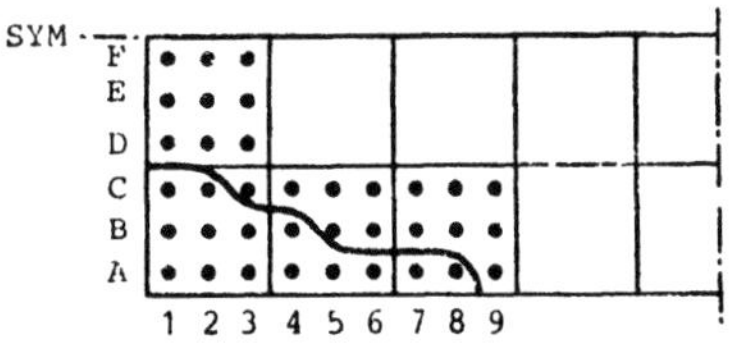

Fig 2. Order of yielding in integration points

P/P_E	1 025	1 050	1 075	1 100	1 125	1 150	1 175	1 200	1 225	1 250	1 275	1 300
First method	1A 2A		3A	1B	4A	2B 5A		1C	6A	7A	2C 4B 8A	3B
Second method	1A 2A		3A		1B 4A	2B 5A			6A 7A	1C	4B 8A	3B

Table 4. Order of yielding in integration points for the two methods

Conclusions

It is important to use the more correct stress-strain definition obtained with a Jaumann rate of stress for low tangent modulus. It can easily be implemented in a finite element program.

In the conventional method for elastoplastic numerical analysis the constitutive relation is given in form of an equation between rates of stress and strain. The yield condition is utilized and the rate of the hardening parameter ε^p is eliminated at the deduction of this equation by use of the consistency condition.

In the method with use of a variational inequality the constitutive equation is satisfied by a restricted minimization of a functional with ε^p kept as a variable.

Our numerical experiments show that large differences in the development of the plastic zones can be obtained in finite element calculations with the two methods.

It does not seem to be practical to use a higher order difference method than the simple Euler method. This means that small time steps are needed particularly near the limit load. The number of iterations to obtain equilibrium in each time step will in general be low when short time steps are used.

References

1 Duvaut, G.; Lions, S.L.: Les inéquations en Mécanique et en Physique, Dunod, Paris, 1973

2 Johnson, C.: Existence theorems for plasticity problems, J. de Math. Pures et Appl., Vol. 55, 1976, pp. 431-444

3 Yamada, Y.; Yoshimura, N.; Sakurai, T.: Plastic stress-strain matrix and its application for the solution of elastic-plastic problems by the finite element method, Int. J. Mech. Sci., Vol. 10, 1968, pp. 343-354

4 Marcal, P.V.; King, L.P.: Elasto-plastic analysis of two-dimensional stress systems by the finite element method, Int. J. Mech. Sci., Vol. 9, 1967, pp. 143-155

5 Zienkiewicz, O.C.; Valliappan, S.; King, I.P.: Elasto-plastic solution of engineering problems; 'inital stress', finite element approach, Int. J. num. Meth. Engng, Vol. 1, 1969, pp. 75-100

6 Johnsson, C.: A mixed finite element method for plasticity with hardening, SIAM J. Numer. Anal., Vol. 14, 1977, pp. 575-584

7 Fröier, M.; Samuelsson, A.: Variational inequalities in plasticity, recent developments, Finite Elements in Non-linear Mechanics, Tapir, Trondheim, 1978, pp. 63-85

8 Samuelsson, A; Fröier, M.: Finite elements in plasticity - A variational inequality approach. The Matematics of Finite Elements and Applications III, Academic Press, London, 1979, pp. 105-115

9 Maier, G.: A minimum principle for incremental elastoplasticity with non-associated flow-laws, J. Mech. Phys. Solids, Vol. 18, 1970, pp 319-330

10 Hill, R.: On the classical constitutive relations for elastic-plastic solids, Recent Progress in Applied Mechanics, the Folke Odqvist Volume, Almqvist & Wiksell, Stockholm, 1967, pp. 241-249

11 Lee, E.H.: Elastic-plastic deformation at finite strains, J. Appl. Mechs, Vol. 36, 1969, pp. 1-6

12 Hutchinson, J.W.: Finite strain analysis of elastic-plastic solids and structures, Numerical Solution of Nonlinear Structural Problems, ASME, AMD, Vol. 6, 1973, pp. 17-29

13 Bäcklund, J.; Wennerström, H.; Axelsson, K.: PIFEM, Computer program for elastic-plastic structures (in Swedish), Skrift LiTH-IKP-S-067, Dept. of Mech. Eng., University of Technology, Linköping, Sweden, 1976

Finite Element Elastoplastic and Limit Analysis: Some Consistency Criteria and Their Implications

L. CORRADI, G. MAIER
Politecnico di Milano, Italy

Summary

The formulation of element elastic-plastic laws is considered in the framework of the displacement approach. The finite element model is regarded as an actually discrete system, composed of a finite number of parts whose individual behaviour is described in terms of generalized stresses and strains. A general procedure for formulating the element laws is proposed, and it is shown that some commonly used formulations of finite element elastic-plastic analysis can be recovered on the basis of particular assumptions. It is pointed out that these formulations may violate some "consistency" requirements, which are discussed in the paper; these violations may explain some of the inaccuracies experienced in computations. A fairly general method for restoring consistency is proposed. The implications of some of the approximations involved are also discussed, with particular reference to the limit analysis problem. Simple examples illustrate the effects of lack of consistency.

Introduction

When material non-linearities began to be considered in finite element analyses, rather straightforward adjustments and generalizations of the well established elastic methods were first adopted. Because of the clear impossibility of enforcing the non-linear material law throughout, this law was imposed only at a selected number of points within each element. The "Gauss integration points" used to compute numerically the element matrices proved to be convenient choice |1| . However, inaccurate solutions were occasionally experienced. In order to avoid such occurrences, more or less empirical rules, based on computational experience, were proposed in the literature (see, e.g. |2|).

In the author's opinion, the implications of this approach were not completely explored so far. Recent papers, e.g. |3| , have re-considered the properties of finite element models within the framework of the algebra of finite dimensional spaces. In this way is was possible to assess precisely the number and kind of independent strain and stress distributions to assume in association with a given displacement model. The approach gave evidence to certain "unconsistencies" which may arise with traditional procedures, e.g. when an attempt is made to enforce the material law at an excessive number of points |4|. Actually, these concepts are not completely new; within the framework of the displacement method they can be regarded as a more precise and rigorous formulation of the ideas on which the so-called "natural" approach is based |5|; as such, they were known since long and exploited in order to achieve operative and computational advantages in elastic analyses |6|.

When this point of view is adopted, the finite element model is regarded as an aggregate of constituents each of which has an individual behavior described in terms of generalized (element) variables. Thus, the non-linear material law has to be replaced by an "element" law concerning generalized quantities associated with the assumed displacement field. However, difficulties may arise in the explicit derivation of the non-linear constitutive relations for high order elements. With reference to piecewiselinear elastic-plastic laws |7| a general procedure was proposed in |8|, and generalized in |9,10| to regular yield functions. In particular, in the latter paper |10| it was shown that the values assumed by strains and stresses at Gauss points within the element can be regarded as element generalized variables, provided that certain "consistency requirements" be complied with. In some instances, consistency requires the use of "reduced" integration procedures which, in fact, had turned out to be computationally advantageos in some cases; however, the use of reduced integration imposes some restrictions on the finite element models which can be adopted in plastic analysis.

This paper provides a review and a generalization of the aforementioned results and proposes a fairly general method for formulating "consistent" elastic-plastic laws for finite elements. For the sake of brevity, only perfectly plastic material behavior is considered. Reference is made to the displacement approach alone. The approximations implied by the procedure proposed are discussed, with particular reference to the limit analysis (collapse load) problem. It is shown, in particular, how an upper bound to the collapse load is not always obtained, in spite of the use of compatible models.

Elastic-plastic element laws

In the displacement formulation of finite element analysis, the displacement field within each element is expressed as a function of the nodal values by means of suitably defined interpolation functions. Then, the strain-displacement relations and the principle of virtual work provide the following equilibrium and compatibility equations

$$\underline{\varepsilon}_e(\underline{x}) = \underline{B}(\underline{x})\ \underline{u} \qquad ; \qquad \underline{f} = \int_V \underline{B}^t(\underline{x})\ \underline{\sigma}_e(\underline{x})\ dV \tag{1a,b}$$

where: $\underline{\varepsilon}_e$ and $\underline{\sigma}_e$ are S-vectors of the strain and stress distributions, respectively, within the element; $\underline{u}$ and $\underline{f}$ are I-vectors of element nodal displacements and forces, respectively; V is the element volume. Subscript e denotes quantities variable over the element and will later permit to drop the specification $(\underline{x})$ in order to simplify the symbology. Eqs.(1) must be supplemented by the material constitutive law. For brevity, only elastic-perfectly plastic behaviors will be considered here and the material properties will be assumed as constant within each element. The constitutive laws first imply that

$$\underline{\sigma}_e(\underline{x}) = \underline{D}_e\ \underline{e}_e(\underline{x}) \tag{2}$$

$$\underline{\varepsilon}_e(\underline{x}) = \underline{e}_e(\underline{x}) + \underline{p}_e(\underline{x}) \tag{3}$$

where: $\underline{D}_e$ is the elastic matrix; $\underline{e}_e$ and $\underline{p}_e$ denote the elastic and plastic portion, respectively, of the total strain. The plasticity condition (here assumed to be "regular", i.e. to define a smooth yield locus) and the associated incremental flow rule can be written as follows |7|:

$$\phi_e(\underline{\sigma}_e) \leqq 0 \qquad ; \qquad \dot{\underline{p}}_e = \{\partial\phi_e/\partial\underline{\sigma}_e\}\ \dot{\lambda}_e \tag{4a,b}$$

$$\dot{\lambda}_e \geqq 0 \qquad ; \qquad \phi_e\ \dot{\lambda}_e = 0 \qquad ; \qquad \dot{\phi}_e\ \dot{\lambda}_e = 0 \tag{4c-e}$$

Since for high order (nonconstant strain) elements it is difficult to enforce the constitutive relations everywhere, they are usually imposed only at some selected points within the element.

An alternative approach to formulation of numerical (finite element) elastoplastic analysis is based on the definition of element generalized strains $\underline{\varepsilon}$ and stresses $\underline{\sigma}$, in such a way that the "virtual work equivalence" requirement is fulfilled |11|:

$$\underline{\sigma}^t\ \underline{\varepsilon} = \int_V \underline{\sigma}_e^t(\underline{x})\ \underline{\varepsilon}_e(\underline{x})\ dV \tag{5}$$

In this case, the role of constitutive laws is played by "element" laws, which can be written as follows

$$\underline{\sigma} = \underline{D}\ \underline{e} \tag{6}$$

$$\underline{\varepsilon} = \underline{e} + \underline{p} \tag{7}$$

$$\underline{\phi}(\underline{\sigma}) \leqq \underline{0} \quad ; \quad \dot{\underline{p}} = [\partial\underline{\phi}/\partial\underline{\sigma}]\ \dot{\underline{\lambda}} \tag{8a,b}$$

$$\dot{\underline{\lambda}} \geqq \underline{0} \quad ; \quad \underline{\phi}^t\ \dot{\underline{\lambda}} = 0 \quad ; \quad \dot{\underline{\phi}}^t\ \dot{\underline{\lambda}} = 0 \tag{8c-e}$$

In the above relations, symbols have the same meaning as in eqs.(2-4), except that they refer now to generalized (element) variables.
In order to give explicit expression to the quantities involved in eqs.(6-8), a relation between generalized strains and stresses and the corresponding field values is to be established. This relation can be written in the general form:

$$\underline{\varepsilon}_e(\underline{x}) = \underline{b}(\underline{x})\ \underline{\varepsilon} \quad ; \quad \underline{\sigma} = \int_V \underline{b}^t(\underline{x})\ \underline{\sigma}_e(\underline{x})\ dV \tag{9a,b}$$

Equations similar to (9a), with the same interpolation function matrix $\underline{b}(\underline{x})$, hold for elastic and plastic strain fields. Eq.(9b) is a consequence of eqs.(9a) and (5). Eqs.(9) can be inverted, to give

$$\underline{\sigma}_e(\underline{x}) = \underline{s}(\underline{x})\ \underline{\sigma} \quad ; \quad \underline{\varepsilon} = \int_V \underline{s}^t(\underline{x})\ \underline{\varepsilon}_e(\underline{x})\ dV \tag{10a,b}$$

with

$$\underline{s}(\underline{x}) = \underline{b}(\underline{x})\ \underline{W}^{-1} \quad ; \quad \underline{W} = \int_V \underline{b}^t\ \underline{b}\ dV \tag{11a,b}$$

An explicit expression for the element elastic matrix $\underline{D}$ is readily obtained from eqs.(9) and (2) and reads:

$$\underline{D} = \int_V \underline{b}^t\ \underline{D}_e\ \underline{b}\ dV \tag{12}$$

In order to formulate the plastic portion of the element law, let a row matrix $\underline{\Lambda}(\underline{x})$ of interpolation functions be defined such that

$$\dot{\lambda}_e(\underline{x}) = \underline{\Lambda}(\underline{x})\ \dot{\underline{\lambda}} \quad ; \quad \underline{\phi} = \int_V \underline{\Lambda}^t(\underline{x})\ \phi_e(\underline{x})\ dV \tag{13a,b}$$

Then, by applying to eq.(13b) the chain rule of differentiation and making use of eq.(10a), one obtains

$$[\partial \underline{\phi}/\partial \underline{\sigma}] = \int_V \underline{s}^t \{\partial \phi_e/\partial \underline{\sigma}_e\} \underline{\Lambda} \, dV \tag{14}$$

Thus, as soon as matrix $\underline{\Lambda}(\underline{x})$ is chosen, all the element matrices appearing in eqs.(8) can be defined. Different element laws are obtained with different $\underline{\Lambda}(\underline{x})$, each involving different approximations |9|. The implications of some assumptions will be discussed subsequently. However, it is worth noting here that some formulations earlier proposed for the element plastic laws can be regarded as particular cases of the present approach; e.g. the "yield criterion of the mean", successfully used for limit analysis and shakedown problems |12|, is recovered from the preceeding equations when matrix $\underline{\Lambda}(\underline{x})$ reduces to the unit scalar.
The element properties must next be expressed in terms of nodal variables, as required for the assemblage. To this purpose, the equilibrium and compatibility equations for the element have to be used. They read:

$$\underline{\varepsilon} = \underline{C}\,\underline{u} \quad ; \quad \underline{f} = \underline{C}^t \underline{\sigma} \tag{15}$$

where $\underline{C}$ is a constant matrix. Eqs.(6),(7) and (15) permit to express the element elastic properties in the form:

$$\underline{f} = \underline{k}_{uu}\,\underline{u} - \underline{k}_{up}\,\underline{p} \tag{16}$$

having set:

$$\underline{k}_{uu} = \underline{C}^t \underline{D}\,\underline{C} \quad ; \quad \underline{k}_{up} = \underline{C}^t \underline{D} \tag{17a,b}$$

These matrices represent the elastic stiffness matrix and the initial strain matrix, respectively, consistent with the present approach.

The "natural" formulation of the displacement method

Vector $\underline{\varepsilon}$ of generalized strain parameters cannot be chosen independently of the assumed displacement model. As formally shown in |4|, in a finite element whose displacement field is governed by I nodal displacement components $\underline{u}$, there are precisely D = I - R independent strain states, R being the number of the element rigid modes. A most spontaneous choice is to assume these independent straining modes as generalized strains for the element.
As first pointed out by Argyris |5|, vector $\underline{u}$ can always be expressed as a linear combination of the R-vector $\underline{\rho}$ of element rigid body motions

and the D-vector $\underline{\delta}$ of the "natural" or straining modes, through a non-singular transformation

$$\underline{u} = [\underline{A}_\rho \ \underline{A}_\delta] \begin{Bmatrix} \underline{\rho} \\ \underline{\delta} \end{Bmatrix} \quad ; \quad \begin{Bmatrix} \underline{\rho} \\ \underline{\delta} \end{Bmatrix} = \begin{bmatrix} \underline{G}_\rho \\ \underline{G}_\delta \end{bmatrix} \underline{u} \tag{18a,b}$$

Let $\underline{r}$ and $\underline{d}$ denote the force quantities corresponding to $\underline{\rho}$ and $\underline{\delta}$. The expressions dual to eqs.(18) read

$$\underline{f} = [\underline{G}_\rho^t \ \underline{G}_\delta^t] \begin{Bmatrix} \underline{r} \\ \underline{d} \end{Bmatrix} \quad ; \quad \begin{Bmatrix} \underline{r} \\ \underline{d} \end{Bmatrix} = \begin{bmatrix} \underline{A}_\rho^t \\ \underline{A}_\delta^t \end{bmatrix} \underline{f} \tag{19a,b}$$

If vectors $\underline{\delta}$ and $\underline{d}$ are chosen as strain and stress parameters, respectively, it follows from eqs.(15) and (18)(19) that $\underline{C} = \underline{G}_\delta$. Since $\underline{G}_\delta$ is a matrix of full row rank, it follows from eq.(15b) that $\underline{d} \neq \underline{0}$ implies $\underline{f} \neq \underline{0}$, i.e. the element generalized stress vector $\underline{d}$ does not contain self-equilibrating (or "redundant") stress states. Moreover, as by the very definition of rigid body motion it must be $\underline{B}(\underline{x}) \ \underline{A}_\rho = \underline{0}$, from eqs. (1a),(18a) and (9a) one obtains

$$\underline{b}(\underline{x}) = \underline{B}(\underline{x}) \ \underline{A}_\delta \tag{20a}$$

A comparison of eq.(1a) to eqs.(9a)(15a) shows that the following equality holds

$$\underline{B}(\underline{x}) = \underline{b}(\underline{x}) \ \underline{G}_\delta \tag{20b}$$

Eqs.(20) provide the relations between matrices $\underline{B}(\underline{x})$ and $\underline{b}(\underline{x})$, which must be fulfilled when $\underline{\varepsilon} = \underline{\delta}$ if eqs.(1a) and (9a)(15a) have to represent the same strain field within the element.

However, the formulation of the element elastic-plastic behaviour requires also some connections between matrices $\underline{b}(\underline{x})$ and $\underline{\Lambda}(\underline{x})$, since they both contribute to the representation of the strain field within the element. In fact, on the basis of the relations of the preceeding section, one can write both the following expressions for the plastic strain rate field over the element

$$\dot{\underline{p}}_e(\underline{x}) = \underline{b}(\underline{x}) \ \dot{\underline{p}} = \underline{b}(\underline{x}) \ [\partial\underline{\phi}/\partial\underline{\sigma}] \ \dot{\underline{\lambda}} \tag{21a}$$

$$\dot{\underline{p}}_e(\underline{x}) = \{\partial\phi_e/\partial\underline{\sigma}_e\} \ \dot{\lambda}_e(\underline{x}) = \{\partial\phi_e/\partial\underline{\sigma}_e\} \ \underline{\Lambda}(\underline{x}) \ \dot{\underline{\lambda}} \tag{21b}$$

Clearly, it would be desirable to choose $\underline{\Lambda}(\underline{x})$ so that the alternative distributions of $\dot{\underline{p}}_e(\underline{x})$ provided by eqs.(20) are as close as possible to each other. Unfortunately, the rather formal definition of the natural modes makes it difficult to discuss the implications of different choices of $\underline{\Lambda}(\underline{x})$ when $\underline{\varepsilon} = \underline{\delta}$ is assumed. In order to overcome this difficulty, alternative definitions of the generalized variables will be proposed.

Alternative definitions and consistency requirements

Let vectors $\dot{\underline{\varepsilon}}$ and $\dot{\underline{\lambda}}$ collect the corresponding field values at J suitably chosen "strain points" within the element, i.e.

$$\dot{\underline{\varepsilon}} = \{\ldots \dot{\underline{\varepsilon}}_e^t(\underline{x}_j) \ \ldots\}^t \ ; \ \dot{\underline{\lambda}} = \{\ldots \dot{\lambda}_e(\underline{x}_j) \ \ldots\}^t \tag{22a,b}$$

where $\underline{x}_j$ $(j = 1,\ldots,J)$ are the strain point coordinates. Then, the following interpolation matrices can be assumed

$$\underline{b}(\underline{x}) = [\ldots \underline{I}_S \, b_j(\underline{x}) \ \ldots] \quad ; \quad \underline{\Lambda}(\underline{x}) = [\ldots \, b_j(\underline{x}) \ldots] \tag{23a,b}$$

$$b_j(\underline{x}_i) = \delta_{ji} \qquad , \ i,j = 1,\ldots,J \tag{23c}$$

Here, $\underline{I}_S$ denotes the S×S identity matrix and δ_{ji} the Kronecker symbol. With the above assumptions, the dimension of vectors $\underline{\varepsilon}$, $\underline{\sigma}$ is SJ and that of $\dot{\underline{\lambda}}$, $\underline{\phi}$, $\dot{\underline{\phi}}$ is J.

A complete discussion of consistency between matrices $\underline{b}(\underline{x})$ and $\underline{\Lambda}(\underline{x})$, as defined by eqs.(23), cannot be presented in this paper because of space limitations. However, a certain degree of consistency can be reasonably expected, since both $\dot{\underline{p}}_e(x)$ and $\dot{\lambda}_e(\underline{x})$ are governed by their values at the same points. As shown in | 9 | , complete consistency in the sense that both eqs.(20) lead to the same distribution of $\dot{\underline{p}}_e(\underline{x})$, can be achieved only if ϕ_e is a linear function of stresses (or a piecewise linear one, an assumption which could be easily be included in the present formulation by generalizing it to singular yield loci). Some approximations are implied in other cases; nevertheless, the above assumptions lead to element laws which represent the most natural extension to high order elements of the formulation first proposed by Hodge | 13| .

Some restrictions on the number and location of strain points must be complied with in order to ensure that the strain field defined as function of $\underline{\varepsilon}$ through eq.(22a) is consistent with the one expressed by eq.(1a) in terms of $\underline{u}$. To this purpose, note first that from eqs.(1a), (15a) and (22a) one obtains

$$\underline{C} = [\ldots \underline{B}^t(\underline{x}_J) \ldots]^t \tag{23}$$

On the other hand, if eqs.(1a) and (9a)(15a) have to provide the same strain field, the following equality must hold

$$\underline{B}(\underline{x}) = \underline{b}(\underline{x}) \, \underline{C} \tag{24}$$

Since $\underline{C} \, \underline{A}_\rho = \underline{0}$, from eqs.(15) and (18)(19) it follows that:

$$\underline{\varepsilon} = \underline{M} \, \underline{\delta} \quad ; \quad \underline{d} = \underline{M}^t \, \underline{\sigma} \tag{25a,b}$$

where

$$\underline{M} = \underline{C} \, \underline{A}_\delta \tag{26}$$

is a SJ×D constant matrix. When SJ = D and $\underline{M}$ is non-singular, there is a one-to-one correspondance between $\underline{\varepsilon}$ and $\underline{\delta}$. Then, eq.(22a) can be considered as a particular choice of natural modes and a valid definition of generalized strains. It can be easily shown that in this case eqs.(20) are replaced by eq.(24) and by the following relation

$$\underline{b}(\underline{x}) = \underline{B}(\underline{x}) \, \underline{A}_\delta \, \underline{M}^{-1} \tag{27}$$

For linear elements, S = 1 and it is always possible to select J = D strain points where generalized strains are defined. However, in plane or space elements it seldom happens that D is a multiple of S . Then two alternatives are possible.

(a) SJ > D (rank of $\underline{M}$ = D). In this situation, eq.(25a) imposes SJ-D constraints among the components of $\underline{\varepsilon}$. Hence, these components cannot be conceived as generalized strains for the element. Moreover, eq.(25b) shows that a "fictitious redundancy" (inconsistency of the first type) is generated, in the sense that a (D-SJ)-dimensional subspace of the stress parameters $\underline{\sigma}$ corresponds to $\underline{d} = \underline{0}$ (and, hence, to $\underline{f} = \underline{0}$).

(b) SJ< D (rank of $\underline{M}$ = SJ). In this case, if the element strain state has to be governed by vector $\underline{\varepsilon}$, D-SJ fictitious modes of rigid body motion (in the sense of deformations without strain energy changes) must be attributed to the element, in addition to the R true rigid modes. The fictitious, additional rigid modes are represented by these vectors $\underline{\delta}$ for which $\underline{\varepsilon} = \underline{0}$ is obtained from eq.(25a) and can be generated by replacing in eqs.(1) the original expression of matrix $\underline{B}(\underline{x})$ with the one obtained in terms of $\underline{b}(\underline{x})$ from eq.(24). In this way, a consistent element is arrived at; however, its elastic stiffness matrix $\underline{k}_{uu}$, eq.(17a), has only rank SJ , instead of D . If the additional rigid modes are not introduced, two different element behaviors are obtained from eqs.(1) and (9), the latter being "stiffer" than the former. This second type of inconsistency will be called "artificial stiffening".

The preceeding discussion shows that, at least in principle, consistent element formulation can be achieved starting from the assumptions (22) (23) , provided $SJ \leq D$. Note that, while the components of vectors $\underline{\varepsilon}$ and $\underline{\lambda}$ coincide with the relevant field values at the strain points, the same is not true, in general, for the components of $\underline{\sigma}$ and $\underline{\phi}$. However, this coincidence can be enforced by selecting as strain point locations $\underline{x}_j$ the coordinates of the points that would be used to evaluate numerically by Gaussian quadrature an integral over the element volume. This result can be readily arrived at if eqs.(9b) and (13b) are numerically integrated, account taken of eqs.(23).

The above special choice of $\underline{x}_j$ turns out to be particularly convenient. In fact, the element constitutive law, eqs.(6-8), can be shown to reduce, in this case, to the material law, eqs.(2-1), enforced at the strain (Gauss) points |10|. Thus, the element properties defined by eqs.(13b) and (14) need not to be evaluated, with a substantial saving in computational effort. It can also be shown that the element stiffness matrix $\underline{k}_{uu}$ coincides with the customary expression, provided a numerical integration based on the J Gauss (strain) points is used to evaluate it |10|. When $SJ < D$, a "reduced integration" is performed, which automatically generates the required additional rigid modes in the element behaviour.

Thus, the familiar procedure for elastic-plastic finite element analysis which enforces the material law at the Gauss integration points |1|, can be regarded as a special case of the present approach, based on the assumptions (22) (23) and on a particular choice of the strain point coordinates.

The consistency requirements previously discussed might explain some aspects of the numerical experience described in the literature. For instance, the inaccurate solutions sometimes obtained when using an excessive number of Gauss points (such that $SJ > D$) could be justified by the fictitions redundancy implicitly introduced in the element behaviour. Also the advantages provided by the use of reduced integration |2| can be understood as due to the fact that consistency is restored by making the rank of matrix $\underline{k}_{uu}$ equal to the number of implicitly assumed generalized strain parameters. Practical limitations to the applicability of this procedure arise from the fact that D is seldom exactly a multiple of S and, hence, reduced integration is usually required in order to achieve consistency. Reduced integration, even if always possible in principle, must be used with some care; in fact the reduced rank of matrix $\underline{k}_{uu}$ may cause singularity of the assembled stiffness matrix |2| or decrease the possibility of stress redistribution in the plastic range due to the diminished degree of redundancy of the assembled system |10|.

On the other hand, inconsistency of the first type cannot be avoided with assumptions (22)(23) when SJ > D; in fact, no matter how refined is the numerical integration, the rank of the resulting stiffness matrix cannot exceed D. In this case, however, it is possible to avoid fictitious redundancy by assuming $\underline{\varepsilon} = \underline{\delta}$ and $\dot{\underline{\lambda}}$ according to eqs.(22b), (23b,c). If $\underline{x}_j$ are Gauss point coordinates, vector $\underline{\phi}$ will still contain as components the yield function values at these points; however, now these values are functions of the element generalized stresses $\underline{\sigma} = \underline{d}$; eq.(10a) permits to write

$$\phi_e(\underline{x}_j) = \phi_e(\underline{\sigma}_e(\underline{x}_j)) \quad , \quad \underline{\sigma}_e(\underline{x}_j) = \underline{s}(\underline{x}_j)\ \underline{\sigma} \tag{28a,b}$$

where matrix $\underline{s}(\underline{x})$ is evaluated, through eqs.(11), on the basis of the expression (20a) of $\underline{b}(\underline{x})$.

In this case, the elastic element matrices in eqs.(17) are the "exact" ones. However, matrix $[\partial\underline{\phi}/\partial\underline{\sigma}]$, eq.(14), must now be computed on the basis of the current values of the material state variables. Therefore, whereas in principle the plasticity condition can be enforced to a high level of accuracy by selecting for J a large number, the computational effort involved in evaluating the element properties will increase appreciably with increasing J.

The present procedure cannot ensure consistency when SJ < D. In this case, the plastic strain rate field would be actually governed by a subset of the set of available parameters and the element will "artificially stiffen" as plastic flow occurs.

Assemblage relations

The equations governing the behaviour of the entire system, composed of N finite elements, are derived below. Subscript n will denote variables pertaining to element n (n = 1,...,N); e.g. vectors $\underline{u}_n$ and $\underline{f}_n$ will collect the nodal displacement and force components of element n. Let $\underline{U}$ and $\underline{F}$ indicate the vectors of the nodal displacement and force components, respectively, for the assembled system. For each element, the following customary connectivity relations can be written

$$\underline{u}_n = \underline{L}_n \underline{U} \quad ; \quad \underline{F} = \sum_{n=1}^{N} \underline{L}_n^t \underline{f}_n \tag{29a,b}$$

If $\underline{u}_n$ and $\underline{U}$ are defined in the same reference frame, $\underline{L}_n$ is a Boolean matrix, containing at most a single unit entry in each row. The compatibility and equilibrium equations for the discrete system, relating the nodal variables of the discretized structure to the element generalized strains and stresses, can be obtained from eqs.(29) and (15). They read

$$\underline{\varepsilon}_n = \underline{C}_n \underline{L}_n \underline{U} \quad , \quad n = 1,\dots,N \tag{30a}$$

$$\underline{F} = \sum_{n=1}^{N} \underline{L}_n^t \underline{C}_n^t \underline{\sigma}_n \tag{30b}$$

Eqs.(30) and the element law, eqs.(6-8), written for each element n, provide a complete formulation of the incremental elastic-plastic problem. It is worth mentioning that, as shown in |9|, the present formulation fulfills all the requirements for the validity of some extremum properties which, in turn, permit the use of numerical methods based on mathematical programming (specifically quadratic programming) for the solution of the analysis problem |7,14|. This aspect will not be discussed here; attention will be focused on the limit analysis problem alone.

Limit analysis

Let the finite element model in point be subjected to a given distribution of nodal loads $\underline{F}_o$. For perfectly plastic systems, the "collapse multiplier" s is defined as the maximum load multiplier ψ such that loads $\psi\underline{F}_o$ can be carried by the structure. The static theorem of limit analysis |11| states that s is the maximum value of ψ for which in all elements generalized stresses $\underline{\sigma}_n$ can be found which are in equilibrium with loads $\psi \underline{F}_o$ and do not violate the plasticity condition:

$$s = \max \psi \ , \quad \text{subject to :} \quad \psi \underline{F}_o = \sum_{n=1}^{N} \underline{L}_n^t \, \underline{C}_n^t \, \underline{\sigma}_n \quad ;$$

$$\underline{\phi}_n(\underline{\sigma}_n) \leq \underline{0} \quad \forall \, n \tag{31}$$

The kinematic theorem of limit analysis can be stated as follows: s is the minimum of the ("kinematically admissible") load factors such that the rate of external work equals the internal dissipation rate for some compatible field of velocities and strain rates. The explicit formulation of limit analysis by the kinematic theorem as a minimization problem is generally more complicate than that by the static theorem, unless piecewiselinear yield conditions are assumed |7,14|.

Eq.(31) provides the collapse load for the discrete system. It is of interest to compare the value of s obtained in this way to the collapse load of the original continuous structure which was modelled by finite elements. Since the present approach concerns displacement models, in which compatibility is ensured throughout, s would be expected to represent an upper bound to the actual collapse load. This condition does not necessarily hold when reduced integration techniques are used to impose consistency, since in this case rigorous interelement compatibility may not be ensured. However, even when compatibility is strictly imposed, the upper bound nature of s cannot be "a-priori" guaranteed, because of the approximations introduced when writing the plastic portion of the element constitutive law.

In fact, except for very simple (constant strain) elements, eqs.(8) enforce the material law only in an "average" sense. In order to clarify this point, consider the expression (23b) of $\underline{\Lambda}(\underline{x})$. The functions $b_j(\underline{x})$ are usually negative over part of the element volume; thus, eqs.(8a) and (8c) do not prevent possible local positive values of $\phi_e(\underline{x})$ or negative values of $\dot{\lambda}_e(\underline{x})$. Moreover, eqs.(4d,e) are imposed only globally by their element counterparts, eqs.(8d,e), since from eqs.(13) it merely follows that

$$\underline{\phi}^t \, \dot{\underline{\lambda}} = \int_V \phi_e(\underline{x}) \, \dot{\lambda}_e(\underline{x}) \, dV \tag{32}$$

In general, when $\underline{x}_j$ represent Gauss point coordinates, upper bounds on the collapse load of the continuous structure are obtained, even if s is less than the exact value of the kinematically admissible load multiplier associated with the assumed displacement model. This remark explains why accurate estimates of the collapse load are often obtained even by means of comparatively "stiff" finite element models.

Examples

Some very simple examples are considered below in order to illustrate some of the concepts discussed in what preceeds. Consider first a cantilever beam of uniform cross section (with flexural rigidity EI and limit moment M_p). The beam is acted upon by a concentrated force F at the free end. The beam is modelled with a single finite element, with a cubic displacement field; thus, the strain (curvature) distribution in the beam is linear.

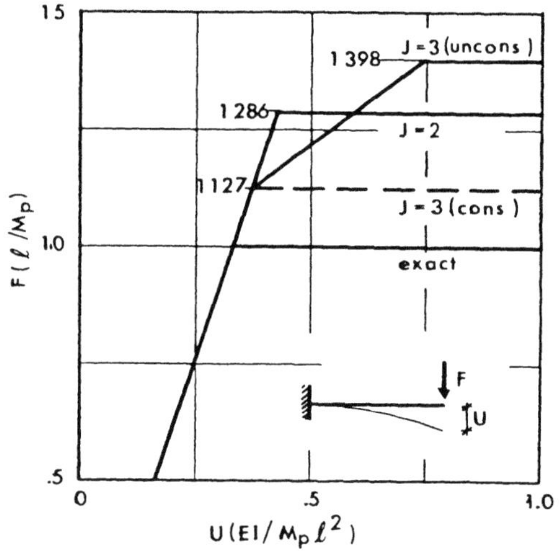

Fig.1. Force-displacement diagram for a statically determinate, perfectly plastic cantilever beam and different element laws.

Since the sectional behaviour was assumed as perfectly plastic, the "exact" load-displacement relationship results in a linear plot up to the value $F_L = M_p/\ell$; at this point, the load cannot further increase and collapse occurs. This relation is depicted in fig.1, together with the diagrams obtained with different element laws.
Since I = 4 and R = 2, a consistent element is obtained with the use of J = 2 strain points. They were located at the Gauss point coordinates and matrices $\underline{b}(\underline{x})$ and $\underline{\Lambda}(\underline{x})$ were defined according to eqs.(22)(23). The plot indicates the same behaviour as the exact one, even if the collapse load is overestimated by about 29%.

When J = 3 is adopted in connection with eqs.(22)(23), the effects of the inconsistency of the first type become apparent. For F = 1.127 (M_p/ℓ) the first strain (Gauss) point yields; however, the fictitious redundancy generated in the model prevents the beam from collapsing. Collapse occurs for F = 1.398 (M_p/ℓ), when the secon strain point also yields (fig.1). This result is obtained with traditional procedures, by enforcing the "material" constitutive law at three Gauss points within the element.

For J = 3, consistency can be restored either by assuming a fourth order displacement distribution through the addition of a displacement type nodal variable (I = 5), or by using the natural modes as generalized strains for the element, while retaining the expression (23b,c) for $\underline{\Lambda}(\underline{x})$ and expressing the element plasticity condition through eqs.(28). The two procedures are fully equivalent and the dashed line in fig.1 shows that the fictitious redundancy has thus disappeared.

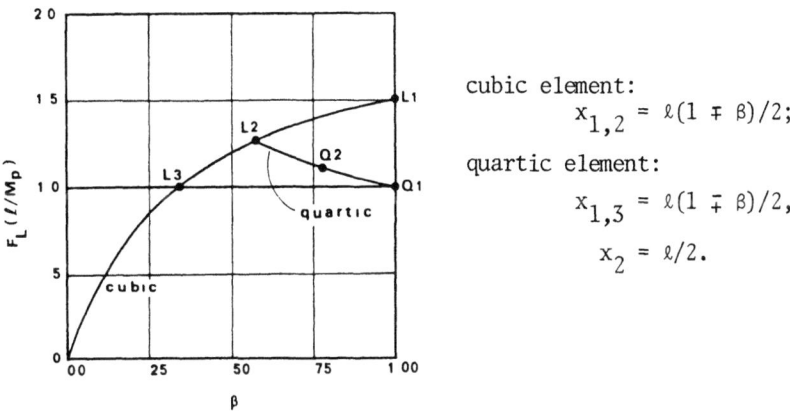

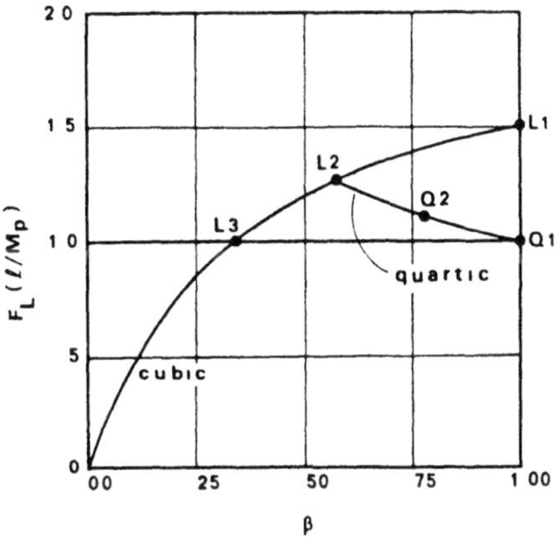

cubic element:

$$x_{1,2} = \ell(1 \mp \beta)/2;$$

quartic element:

$$x_{1,3} = \ell(1 \mp \beta)/2,$$

$$x_2 = \ell/2.$$

Fig.2. Collapse load for the cantilever beam of fig.1, as a function of the strain point locations.

Since Gauss point coordinates were chosen as strain point locations, an upper bound on the actual collapse load was obtained. However, it is not always so with either choices of strain point coordinates, in spite of full compatibility. This circumstance is illustrated in fig.2 (from ref. |8|), where the collapse loads obtained for consistent

cubic (I = 4, J = 2) and quartic (I = 5, J = 3) elements is depicted. In the latter case, unsound constitutive laws are obtained for $\beta < 1/\sqrt{3}$, since negative dissipation becomes possible. It appears that with cubic elements, the finite element collapse load is below the exact value of $F_L = M_p/\ell$ when $\beta < 1/3$. Points L2 and Q2 indicate the values obtained when strain points and Gauss points coincide.
For linear elements, there are no difficulties in determining, on the basis of the displacement model, the number of Gauss points where the constitutive law is to be enforced without violating consistency. This is not always possible with more complicated elements and the procedure based on eqs.(28) should be used to avoid inconsistencies. As an example, consider an axisymmetric, tronco-conical thin shell element (fig.3) |15|. By neglecting transverse shear deformations, the strain distribution within the element is represented by the 4-component vector

$$\underline{\varepsilon}_e(r) = \{\varepsilon_s \quad \varepsilon_\Phi \quad \chi_s \quad \chi_\Phi\}^t \tag{32}$$

where ε_s, ε_Φ denote, respectively, the meridian and circumferential membrane strains and χ_s, χ_Φ the corresponding curvatures; r is the radial distance from the shell axis.
The simplest displacement model able to ensure interelement compatibility consists of a linear expression for $u_e(r)$ and a cubic one for $w_e(r)$. Thus, the displacement field depends on I = 6 parameters, which can be identified with the nodal components indicated in fig.3. Since the only rigid mode for the element is a vertical translation, we have D = 5. A possible expression for matrix $\underline{b}(r)$, in terms of natural displacements, is (θ denoting the angle between generatrix and shell axis):

$$\underline{b}(r) = \begin{bmatrix} \dfrac{2\cos\theta\,\sin\theta}{r} & 1 & \mathrm{ctg}\theta & r\,\dfrac{\mathrm{ctg}\theta}{\sin\theta} & r^2\,\dfrac{\mathrm{cgt}\theta}{\sin^2\theta} \\ 0 & 1 & 0 & 0 & 0 \\ 0 & 0 & -\dfrac{\sin\theta}{r} & -2 & -3\,\dfrac{r}{\sin\theta} \\ 0 & 0 & 0 & -2 & -6\,\dfrac{r}{\sin\theta} \end{bmatrix} \tag{34}$$

From eqs.(34) and (11), the expression of $\underline{s}(r)$ can be readily derived.

Two strain points can be assumed in the element and matrix $\underline{\Lambda}(r)$ can be defined accordingly. If the constitutive law were directly imposed at the strain points, a fictitious redundancy would show up, since SJ = 8 is greater than D. However, the inconsistency can be removed through the use of eqs.(28); if strain points are located at Gauss points, the element yield function will result in the following two-component vector

$$\begin{aligned} \phi_1 &= \phi_e(r_1) = \phi_e(\underline{s}(r_1)\ \underline{\sigma}) \\ \phi_2 &= \phi_e(r_2) = \phi_e(\underline{s}(r_2)\ \underline{\sigma}) \end{aligned} \tag{35}$$

where $\underline{\sigma}$ is the 5-vector of generalized stresses associated to the assumed natural modes.

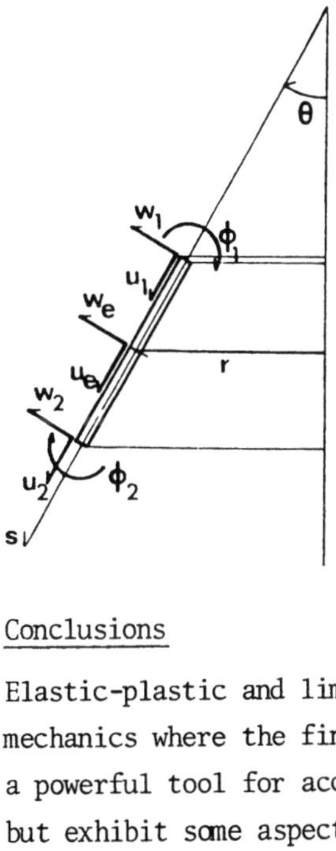

Fig.3. Tronco-conical thin shell element.

Conclusions

Elastic-plastic and limit analysis are areas of non-linear structural mechanics where the finite element method turns out to be, once again, a powerful tool for accurate numerical solutions in complex situations, but exhibit some aspects and requirements quite distinct and partly novel with respect to elastic analysis. Some of these peculiarities, considered in the paper, basically concern non trivial, sometimes overlooked "consistency" conditions, to be fulfilled in modelling the element fields, or, equivalently, in formulating element plastic laws in (element) generalized variables; it has been shown that these should be taken into account in order to avoid sistematic and possibly significant source of inaccuracy.

Aknowledgment

The financial support of C.N.R. (National (Italian) Research Council) is gratefully aknowledged.

References

1 Nayak, G.C.; Zienkiewicz, O.C.: Elastic-plastic stress analysis. A Generalization for various constitutive relations including strain softening. Int. J. Num. Meth. Engrg., 5, (1972), 113-135.

2 Zienkiewicz, O.C.: The finite element method. McGraw-Hill, London, 1977.

3 Besseling, J.F.: The force method and its application in plasticity problems. Computers and Structures, 8, (1978) 323-330.

4 Besseling, J.F.: Finite element method. 53-78, in: Trends in Solid Mechanics, Besseling and van der Heijden edts., Sijthoff & Noordhoff, 1979.

5 Argyris, J.H.: Continua and discontinua. 11-189, in: Proc. 1st Conf. on Matrix Methods in Structural Mechanics, Dayton, Ohio, AFFDL TR. 66.80, 1966.

6 Scharpf, D.W.: A new method of stress calculation in the matrix displacement analysis. Computers and Structures, 8, (1978) 465-477.

7 Maier, G.: A matrix structural theory of piecewiselinear plasticity with interacting yield planes. Meccanica, 5 (1970) 54-66.

8 Corradi, L.: On compatible finite element models for elastic plastic analysis. Meccanica, 13 (1978) 133-150.

9 Corradi, L.: On some "consistent" finite element approximations in non-linear structural analysis. Proc. 5th AIMETA Conference, Palermo, Italy, Oct. 1980.

10 Corradi, L.; Gioda, G.: On the finite element modelling of elastic-plastic behaviour with reference to geotechnical problems. 1st. Int. Conf. on Numerical Methods for Non-Linear Problems, Swansea, U.K. Sept.1980.

11 Martin, J.: Plasticity. MIT Press, 1975.

12 Dang Hung, N.; König, J.A.: A finite element formulation of the shakedown problem using a yield criterion of the mean. Comp. Meth. Appl. Mech. Engrg., 8 (1976) 179-192.

13 Hodge, P.G. Jr.: A Consistent finite element model for the two-dimensional continuum. Ingenieur Archiv, 39 (1970) 375-383.

14 Cohn, M.Z.; Maier, G. Eds.: Engineering Plasticity by Mathematical Programming. Pergamon Press, New York, 1979.

15 Dang Hung, N.; Trapletti, M.; Ransart, D.: Bornes quasi inférieures et bornes supérieures de la pression de ruine de coques de révolution par la méthode des éléments finis et par la programmation nonlinéare Int. J. Nonlinear Mech., 13 (1979) 79-102.

A Finite Element Method for Problems in Perfect Plasticity Using Discontinuous Trial Functions

C. JOHNSON, R. SCOTT
Chalmers University of Technology, Göteborg, Sweden

Introduction. It is known (see e.g. [3],[5]) that the displacements in an elastic perfectly-plastic body may be discontinuous. Conventional finite element methods (displacement methods) for plasticity problems are based on using continuous trial functions and are thus not particularly well adapted to the nature of the true solution. In this note we propose a finite element method of displacement type for problems in perfect plasticity where we use a finite element space V_h of piecewise polynomial functions with no requirement on inter-element continuity. In order to be able to approximate a discontinuous solution u of a plasticity problem accurately with functions in V_h, the finite element mesh will have to fit the discontinuities of u. Thus, since the location of these discontinuities is in general not known in advance, one would like to use some kind of adaptive technique where according to the results of computations the finite element mesh is succesively modified. In this note we do not consider this more general problem but concentrate on analyzing the proposed method in the case of a given mesh.

An outline of the note is as follows: In Section 2 we formulate the problem, in Section 3 we introduce the finite element method and prove a convergence result, and finally in Section 3 we give the results of some numerical experiments in one space dimension.

1. The plasticity problem

For simplicity we shall consider a stationary (static) problem corresponding to the so-called Henky's law. The quasi-static rate problem corresponding to Prandtl-Reuss' flow rule does not present any new serious difficulties and leads to a problem of Henky type at each incremental step.

Let Ω be a bounded region in $\mathbb{R}^3$ occupied by a body $\mathcal{E}$ made up of an elasto-plastic material. The *displacements* of $\mathcal{E}$ is given by $u = (u_i)$, $i = 1,2,3$, where u_i is the displacement in coordinate direction x_i. The *strains* corresponding to u is given by the *strain tensor* $\varepsilon(u) = \{\varepsilon_{ij}(u)\}$, $i,j = 1,2,3$, where

$$\varepsilon_{ij}(u) = \tfrac{1}{2}\{\frac{\partial u_i}{\partial x_j} + \frac{\partial u_j}{\partial x_i}\}.$$

Further, let the *stresses* in $\mathcal{E}$ be given by the *symmetric stress tensor* $\sigma = \{\sigma_{ij}\}$, $\sigma_{ij} = \sigma_{ji}$, $i,j = 1,2,3$. Let the elastic properties of the material, which we assume to be independent of x for simplicity, be given by the symmetric positive definite transformation $A: \mathbb{R}^9 \to \mathbb{R}^9$ so that

$$\varepsilon = A\sigma \qquad \text{(Hooke's law)},$$

is the *elastic strain* corresponding to the stress σ. The set of *plastically admissible stresses* is given by the closed convex set $B \subset \mathbb{R}^9_s$ with $0 \in$ interior of B, where $\mathbb{R}^9_s$ is the set of symmetric 3×3 tensors $\tau = \{\tau_{ij}\}$. For instance we may take $B = \{\tau \in \mathbb{R}^9_s: |\tau| \leq 1\}$, where $|\cdot|$ is the Euclidean norm in $\mathbb{R}^9$, i.e.,

$$|\tau| = (\sum_{i,j=1}^{3} \tau_{ij}^2)^{\frac{1}{2}}.$$

For simplicity, let us assume that B is bounded. Together,

A and B define the properties of the elasto-plastic material. For simplicity, we assume below that A = Identity.

We can now formulate (in a non-precise way) the plasticity problem as follows: Find σ and u such that

$$\begin{cases} \sigma = \Pi\varepsilon(u) & \text{in } \Omega, \quad (1.1a) \\ -\mathrm{div}\sigma = f & \text{in } \Omega, \quad (1.1b) \\ u = 0 & \text{on } \Gamma, \quad (1.1c) \end{cases}$$

where $\Pi:\mathbb{R}^9_s \to B$ is the projection onto B with respect to the usual scalar product in $\mathbb{R}^9$, $f = (f_1,f_2,f_3)$ is a given volume load and $\mathrm{div}\,\sigma = ((\mathrm{div}\,\sigma)_i)$ with

$$(\mathrm{div}\,\sigma)_i = \frac{\partial\sigma_{ij}}{\partial x_j}.$$

Here and below we use the following summation convention: repeated indices indicate summation from 1 to 3 so that in particular

$$\frac{\partial\sigma_{ij}}{\partial x_j} = \sum_{j=1}^{3} \frac{\partial\sigma_{ij}}{\partial x_j}.$$

For simplicity, we assume that the boundary conditions are given by (1.1c). Note that (1.1a) can also be written

$$\varepsilon(u) = (\varepsilon(u) - \Pi\varepsilon(u)) + \sigma \equiv \varepsilon^p + \varepsilon^e,$$

which corresponds to a splitting of the total deformation ε into an elastic part ε^p and plastic part ε^e.

In order to give a precise formulation of (1.1) let us introduce some function spaces. Let

$$H = \{\tau = (\tau_{ij}) : \tau_{ij} = \tau_{ji} \in L_2(\Omega)\},$$

with scalar product and norm given by

$$(\sigma,\tau) = \int_\Omega \sigma_{ij}\tau_{ij}dx, \qquad \|\tau\| = (\sigma,\sigma)^{\frac{1}{2}},$$

$$P = \{\tau \in H : \tau(x) \in B \quad \text{a.e. } x \in \Omega\},$$

$V = [H_0^1(\Omega)]^3$ and finally $W = [L_2(\Omega)]^3$ with the usual scalar product $(\cdot,\cdot)$ and norm $\|\cdot\|$.

The stress σ can, according to the Principle of the complementary energy, be characterized as the unique solution of the problem

$$\sup_{\substack{\tau\in P\\ -\operatorname{div}\tau=f}} -\tfrac{1}{2}\|\tau\|^2, \tag{1.2}$$

where $f \in W$. Recalling Green's formula:

$$\int_\Omega \tau_{ij}\varepsilon_{ij}(v)dx = \int_\Gamma v_i\tau_{ij}n_j ds - \int_\Omega v_i\tau_{ij,j}dx ,$$

where $n = (n_j)$ denotes the outward unit normal to Γ , we have if $v \in V$ and $\tau \in H$ with $\operatorname{div}\tau \in W$,

$$(\tau,\varepsilon(v)) = -(\operatorname{div}\tau , v).$$

Thus

$$\inf_{v\in V} [(\tau,\varepsilon(v))-(f,v)] = \begin{cases} 0 & \text{if } -\operatorname{div}\tau = f, \\ -\infty & \text{otherwise,} \end{cases}$$

and hence (1.2) can be written

$$\sup_{\tau\in P} [\inf_{v\in V} L(\tau,v)] , \tag{1.3}$$

where

$$L(\tau,v) = -\tfrac{1}{2}\|\tau\|^2 + (\tau,\varepsilon(v))-(f,v).$$

The corresponding dual problem obtained by interchanging the order of Inf and Sup reads

$$\inf_{v\in V} [\sup_{\tau\in P} L(\tau,v)]. \tag{1.4}$$

Here it is easy to evaluate the Sup: We have

$$\begin{aligned}\sup_{\tau\in P} L(\tau,v) &= \sup_{\tau\in P}[-\tfrac{1}{2}\|\tau-\varepsilon(v)\|^2 + \tfrac{1}{2}\|\varepsilon(v)\|^2 - (f,v)]\\ &= -\tfrac{1}{2}\|\varepsilon(v)-\Pi\varepsilon(v)\|^2 + \tfrac{1}{2}\|\varepsilon(v)\|^2 - (f,v) \qquad (1.5)\\ &= (\varepsilon(v),\Pi\varepsilon(v))-\tfrac{1}{2}\|\Pi\varepsilon(v)\|^2 - (f,v) \equiv F(v).\end{aligned}$$

Thus the dual problem can be formulated

$$\operatorname*{Inf}_{v \in V} F(v), \tag{1.6}$$

with F defined in (1.5). In the case $B = \{\tau \in \mathbb{R}^9_s : |\tau| \leq 1\}$ one has

$$F(v) = \int_\Omega g(|\varepsilon(v)|)dx - (f,v),$$

where $g: R \to \mathbb{R}$ is given by

$$g(r) = \begin{cases} \frac{1}{2}r^2 & \text{if } |r| \leq 1, \\ r-\frac{1}{2} & \text{if } |r| > 1. \end{cases}$$

The functional $F: V \to \mathbb{R}$ is not coercive on V and thus the problem (1.6) does not have a solution in V in general. In order to find a solution we have to extend the space V and the natural space is then the space of "functions of bounded deformation" $\overset{\circ}{BD}(\Omega)$ defined as follows ([1] , [5]):

$$\overset{\circ}{BD}(\Omega) = \{v \in [L_1(\Omega)]^3 : \varepsilon(v) \in M(Q)_s\},$$

where Q is a cube containing $\bar{\Omega}$, v has been extended by zero outside $\bar{\Omega}$, $M(Q)$ is the set of bounded measures on Q and $M(Q)_s$ is the set of symmetric tensors with components in $M(Q)$. The norm in $\overset{\circ}{BD}(\Omega)$ is given by

$$||v||_{BD} = ||\varepsilon(v)||_{M(Q)_s} = \sum_{i,j=1}^{3} ||\varepsilon_{ij}(v)||_{M(Q)} .$$

Furthermore, one can prove (cf. [3], [4], [5]) that for $1 \leq p \leq 3/2$,

$$||v||_{L_p(\Omega)} \leq c||\varepsilon(v)||_{M(Q)_s},$$

and that ([5]) $\overset{\circ}{BD}(\Omega)$ is compactly imbedded in $[L_p(\Omega)]^3$ if $1 \leq p < 3/2$.

One can prove that under a safe load hypothesis, to be stated precisely below, the functional F is coercive on $\overset{\circ}{BD}(\Omega)$ and since $\overset{\circ}{BD}(\Omega)$ is the dual of a Banach space (cf.[1]) the problem

$$\operatorname*{Inf}_{v\in \overset{\circ}{BD}(\Omega)} F(v), \qquad (1.7)$$

admits a solution $u \in \overset{\circ}{BD}(\Omega)$. Here the domain of F is extended to $\overset{\circ}{BD}(\Omega)$ by defining for $v \in \overset{\circ}{BD}(\Omega)$

$$F(v) = \operatorname*{Sup}_{\tau\in P\cap C(Q)_s} [\,-\tfrac{1}{2}\|\tau\|^2 + \langle\tau,\varepsilon(v)\rangle - (f,v)\,], \qquad (1.8)$$

where $C(\Omega)_s$ denotes the set of symmetric tensors with components in $C(Q)$ = {continuous functions on Q} and $\langle\cdot,\cdot\rangle$ denotes the duality between $C(Q)_s$ and $M(Q)_s$ extending the scalar product $(\tau,\varepsilon(v))$.

The safe load hypothesis reads as follows:

$$\begin{cases} \exists\ \chi\in P \text{ and } \delta > 0 \text{ such that } -\operatorname{div}\chi = f \text{ in } \Omega \text{ and} \\ \operatorname{dist}(\chi(x),\partial B) \geq \delta,\ x \in \Omega\ , & (1.9a) \\ \chi \in C(\bar{\Omega})_s, & (1.9b) \end{cases}$$

where ∂B denotes the boundary of B in $\mathbb{R}^9_s$.

The finite element scheme to be considered below may be thought of as a discrete analogue of (1.7) obtained by replacing the space $\overset{\circ}{BD}(\Omega)$ by a finite-dimensional space consisting of piecewise polynomial discontinuous functions. In the analysis of this scheme below we shall use the following alternative formulation of the plasticity problem in both stresses and displacements where the somewhat non-standard space $\overset{\circ}{BD}(\Omega)$ is not utilized. Define

$$\tilde{V} = [L_{3/2}(\Omega)]^3,$$

$$\tilde{P} = \{\tau \in P: \operatorname{div}\tau \in [L_3(\Omega)]^3\},$$

$$\tilde{L}(\tau,v) = -\tfrac{1}{2}\|\tau\|^2 - (\operatorname{div}\tau\ ,v) - (f,v),$$

and assume that (1.9a) holds and $f \in [L_3(\Omega)]^3$. Then, one can prove (see [3], [5]) that the functional $\tilde{L}:\tilde{P}\times\tilde{V}\to\mathbb{R}$ admits a saddlepoint $(\sigma,u)\in\tilde{P}\times\tilde{V}$, i.e. $\exists(\sigma,u)\in\tilde{P}\times\tilde{V}$ such that

$$\tilde{L}(\tau,u) \leq \tilde{L}(\sigma,u) \leq \tilde{L}(\sigma,v) \qquad \forall(\tau,v) \in \tilde{P} \times \tilde{V}, \tag{1.10}$$

or, equivalently,

$$(\sigma,\tau-\sigma)+(u,\operatorname{div}\tau-\operatorname{div}\sigma) \geq 0 \qquad \forall\tau \in \tilde{P}, \tag{1.11a}$$

$$-(\operatorname{div}\sigma,v) = (f,v) \qquad \forall v \in \tilde{V}. \tag{1.11b}$$

Moreover, $u \in \overset{\circ}{BD}(\Omega)$ and we note that σ is uniquely determined. From (1.10) it follows that $\sigma \in \tilde{P}$ and $u \in \tilde{V}$ are solutions of the following analogues of (1.3) and (1.4):

$$\operatorname*{Sup}_{\tau\in\tilde{P}} [\operatorname*{Inf}_{v\in\tilde{V}} \tilde{L}(\tau,v)], \tag{1.12}$$

$$\operatorname*{Inf}_{v\in\tilde{V}} [\operatorname*{Sup}_{\tau\in\tilde{P}} \tilde{L}(\tau,v)]. \tag{1.13}$$

To see the connection with our original formulation (1.1), we note that if $u \in V$ then by Green's formula, (1.11a) can be written

$$(\sigma-\varepsilon(u),\tau-\sigma) \geq 0 \qquad \forall\tau \in P,$$

i.e.,

$$\sigma = \Pi\varepsilon(u).$$

Thus, if $u \in V$ (1.11a) is equivalent to (1.1a). However, in general $u \notin V$ and then it is not immediately clear how to interpret (1.1a) (or 1.11a)).

2. The finite element method

Let us for simplicity consider a problem in two space dimensions. We use the same notation as in the previous section with the obvious modifications (the only notable difference is that the spaces $L_{3/2}(\Omega)$ and $L_3(\Omega)$ in the definition of $\tilde{P}$ and $\tilde{V}$ may be replaced by $L_2(\Omega)$ in the two-dimensional case). Thus, Ω is a bounded region in the plane which we assume to be polygonal to simplify the presentation. Let now $\mathcal{C}_h = \{K\}$ be a triangulation of Ω and introduce

$$W_h = \{v \in [L_2(\Omega)]^2 : v|_K \in \mathbb{P}_k(K),\ K \in \mathcal{C}_h\},$$

where $\mathbb{P}_k(K)$ is the set of polynomials of degree at most k defined on K. Thus, W_k consists of piecewise polynomial functions with no continuity requirement on interelement boundaries. We consider a family of triangulations $\{\mathcal{C}_h\}$ indexed by the positive parameter h such that $\bigcup_h W_h$ is strongly dense in V. Clearly, $W_h \subset \overset{\circ}{BD}(\Omega)$ and we now pose the following discrete analogue of (1.7):

$$\operatorname*{Inf}_{v \in W_h} F(v). \tag{2.1}$$

Let us evaluate F(v) for $v \in W_h$ according to (1.8). We have for τ smooth, $v \in W_h$ using Green's formula

$$\langle \tau, \varepsilon(v) \rangle = -\int_\Omega v \operatorname{div} \tau \, dx$$

$$= \sum_{K \in \mathcal{C}_h} (\tau, \varepsilon(v))_K - \sum_S \int_S [v_i] \tau_{ij} n_j ds \ ,$$

where we sum over all sides S of the triangulation $\mathcal{C}_h$, $[v_i]$ is the jump of v_i across S and

$$(\tau, \varepsilon(v))_K = \int_K \tau_{ij} \varepsilon_{ij}(v) dx.$$

Thus, by varying τ in the interior of each element K and in a thin strip around each side S, it follows that for $v \in W_h$

$$F(v) = \sum_K ((\varepsilon(v), \Pi\varepsilon(v))_K - \tfrac{1}{2}\|\Pi\varepsilon(v)\|_K^2) + \sum_S j_S(v) - (f,v) \tag{2.2}$$

where

$$j_S(v) = \int_S \sup_{\tau \in B} [v_i] \tau_{ij} n_j ds \ .$$

The functional $F:W_h \to \mathbb{R}$ is clearly continuous and by the safe load hypothesis (1.9) it follows replacing f by $-\mathrm{div}\,\chi$ and using Green's formula that

$$F(v) = \sum_K ((\varepsilon(v),\Pi\varepsilon(v))_K - (\varepsilon(v),\chi)_K - \tfrac{1}{2}\|\Pi\varepsilon(v)\|_K^2)$$

$$+ \sum_S \int_S (\sup_{\tau\in B}[v_i]\tau_{ij}n_j - [v_i]\chi_{ij}n_j)ds.$$

Using the fact that $\mathrm{dist}(\chi(x),\partial B) \geq \delta$ it is then easy to see that

$$F(v) \geq \frac{C}{\delta}\|v\|_{BD},$$

for $\|v\|_{BD}$ sufficiently large and thus $F(v) \to \infty$ as $\|v\|_{BD} \to \infty$. This proves that (2.1) admits a solution $u_h \in W_h$. The functional $F:W_h \to \mathbb{R}$ is convex but not strictly convex and thus u_h may not be uniquely determined.

Let us notice that in the case $B = \{\tau \in \mathbb{R}^4_s : |\tau| \leq 1\}$ we have

$$F(v) = \sum_K \int_K g(|\varepsilon(v)|)dx + \sum_S |[v]| - (f,v).$$

2.2 Convergence

To prove convergence of the method (2.2) it is convenient to use the following characterization of the solution $u_h \in W_h$. Let us introduce the Lagrangian $\mathcal{L}:\mathcal{P}_h \times W_h \to \mathbb{R}$ defined by

$$\mathcal{L}(\tau,v) = -\tfrac{1}{2}\|\bar{\tau}\|^2 + [\tau,\varepsilon(v)]_h - (f,v),$$

where

$$[\tau,\varepsilon(v)]_h = \sum_K (\bar{\tau},\varepsilon(v))_K - \sum_S \int_S [v_i]\,\tilde{\tau}_{ij}n_j ds,$$

$$\mathcal{P}_h = \{\tau = (\bar{\tau},\tilde{\tau}) : \bar{\tau} \in P, \tilde{\tau} \in \prod_S L_2(S)_s$$
with $\tilde{\tau}(x) \in B$ a.e. on S, $\forall S\}$.

Thus, a $\tau \in \mathcal{P}_h$ has an "interior component" $\bar{\tau}$ and a "boundary component" $\tilde{\tau}$ and these components are independent. Consider now the problems

$$\operatorname*{Inf}_{v\in W_h} [\operatorname*{Sup}_{\tau\in\mathcal{P}_h} \mathcal{L}(\tau,v)], \tag{2.3}$$

$$\operatorname*{Sup}_{\tau\in\mathcal{P}_h} [\operatorname*{Inf}_{v\in W_h} \mathcal{L}(\tau,v)]. \tag{2.4}$$

The problem (2.3) is clearly the same as our original discrete problem (2.1) and thus has a solution $u_h \in W_h$. Since B is bounded in $\mathbb{R}^4_s$, $\mathcal{P}_h$ is bounded and thus the problem (2.4) has a solution $\sigma_h \in \mathcal{P}_h$ (in fact we may consider $\mathcal{P}_h$ to be a finite dimensional since W_h is finite-dimensional). Thus (cf. [2]) $\mathcal{L}$: $\mathcal{P}_h \times W_h \to \mathbb{R}$ has a saddle point $(\sigma_h,u_h) \in \mathcal{P}_h \times W_h$ and the extremality relations can be written:

$$(\bar{\sigma}_h,\bar{\tau} - \bar{\sigma}_h) - [\tau - \sigma_h,\varepsilon(u_h)]_h \geq 0 \qquad \forall\,\tau \in \mathcal{P}_h, \tag{2.5a}$$

$$[\sigma_h,\varepsilon(v)]_h = (f,v) \qquad \forall\, v \in W_h. \tag{2.5b}$$

We shall prove the following result:

Theorem 1. If the safe load hypothesis (1.9) is satisfied, then for a subsequence $\{h\}$ tending to zero

$\bar{\sigma}_h \to \sigma$ weakly in H,

$u_h \to u$ weak* in $\overset{\circ}{BD}(\Omega)$,

$u_h \to u$ in $[L_p(\Omega)]^2$ for $1 \leq p < 2$,

where $(\sigma,u) \in \tilde{P} \times \overset{\circ}{BD}(\Omega)$ satisfies (1.11).

Proof. Since $-\operatorname{div} \chi = f$ where χ is given by (1.9) we have by Green's formula

$$[\chi,\varepsilon(v)]_h = (f,v) \qquad \forall\, v \in W_h, \tag{2.6}$$

where we consider χ to be an element of $\mathcal{P}_h$ in the natural way i.e. by letting

$$\bar{\chi} = \chi|_{\underset{K}{\cup} \text{int } K}, \quad \text{int } K = \text{interior of } K.$$

$$\tilde{\chi} = \chi|_{\underset{S}{\cup} S}.$$

Now, taking $\tau = \chi$ in (2.5a) we obtain

$$0 \leq (\bar{\sigma}_h, \bar{\chi} - \bar{\sigma}_h) - [\chi - \sigma_h, \varepsilon(u_h)]_h = (\bar{\sigma}_h, \bar{\chi} - \bar{\sigma}_h), \tag{2.7}$$

so that

$$||\bar{\sigma}_h|| \leq C, \tag{2.8}$$

with C independent of h. Next taking $\tau = \chi + \delta\tau_1$, where $|\tau_1| \leq 1$, in (2.5a) we get using (2.5b) and (2.6)

$$[\delta\tau_1, \varepsilon(u_h)]_h \leq (\bar{\sigma}_h, \bar{\chi} + \sigma\bar{\tau}_1 - \bar{\sigma}_h) - [\chi - \sigma_h, \varepsilon(u_h)]_h$$

$$= (\bar{\sigma}_h, \bar{\chi} + \delta\bar{\tau}_1 - \bar{\sigma}_h) \leq C,$$

which shows that

$$||u_h||_{BD(\Omega)} \leq C. \tag{2.9}$$

By (2.8) it follows that $\exists\, \sigma \in P$ such that for a subsequence $\{h\}$ tending to zero

$$\bar{\sigma}_h \to \sigma \qquad \text{weakly in } H.$$

Further, by (2.9) it follows, using the fact that $\overset{o}{BD}(\Omega)$ is compactly contained in $[L_p(\Omega)]^2$ for $1 \leq p < 2$ (see [5]), that $\exists u \in \overset{o}{BD}(\Omega)$ such that

$$u_h \to u \qquad \text{weak* in } \overset{o}{BD}(\Omega),$$

$$u_h \to u \qquad \text{in } [L_p(\Omega)]^2.$$

It remains to pass to the limit in (2.5). Given $w \in V$ there exists a sequence $w_h \in W_h \cap V$ such that $w_h \to w$ strongly in V. If $w_h \in W_h \cap V$ then

$$[\sigma_h, \varepsilon(w_h)]_h = (\bar{\sigma}_h, \varepsilon(w_h)),$$

and thus

$$[\sigma_h, \varepsilon(w_h)]_h \to (\sigma, \varepsilon(w)).$$

This proves that

$$(\sigma, \varepsilon(w)) = (f,w) \qquad \forall w \in V,$$

i.e.

$$-(\operatorname{div}\sigma, w) = (f,w) \qquad \forall w \in \tilde{V}.$$

Finally, by choosing τ in (2.5a) smooth we find since then

$$[\tau, \varepsilon(u_h)]_h = -(\operatorname{div}\tau, u_h),$$

that

$$0 \le \underline{\lim}\{-\tfrac{1}{2}\|\bar{\sigma}_h\|^2 + (\bar{\sigma}_h, \tau) + (\operatorname{div}\tau, u_h) + (f, u_h)\}$$

$$\le -\tfrac{1}{2}\|\sigma\|^2 + (\sigma,\tau) - (\operatorname{div}\tau, u) + (f,u)$$

$$= (\sigma, \tau-\sigma) - (u, \operatorname{div}\tau - \operatorname{div}\sigma).$$

By density we then obtain (1.11a). Thus $(\sigma,u) \in \tilde{\mathbb{P}} \times \overset{\circ}{BD}(\Omega)$ satisfies (1.11) and the proof is complete. ■

3. A numerical experiment

Let us consider the following analogue of (1.7) in the case of one space dimension:

$$\inf_{v \in \overset{\circ}{BD}(I)} F(v), \tag{3.1}$$

where $I = (0,1)$ and

$$F(v) = \int_I g(|v'|)dx - \int_I fv\,dx, \qquad v \in H_0^1(I),$$

with g as in section 2, corresponding to taking $B = \{\tau \in \mathbb{R}: |\tau| \le 1\}$. Alternatively this problem can be formulated as follows (c.f. (1.11)): Find $(\sigma,u) \in \tilde{\mathbb{P}} \times \tilde{\mathbb{V}}$ with $\tilde{\mathbb{P}} = \{\tau \in L_2(I): \tau' \in L_2(I),\ |\tau(x)| \le 1 \text{ a.e. } x \in \Omega\}$ and

$\tilde{V} = L_2(I)$ such that

$$\begin{cases} (\sigma,\tau-\sigma) + (u,\tau' - \sigma') \geq 0 & \forall\tau \in \tilde{P}, \quad (3.2a) \\ (\sigma',v) = (f,v) & \forall v \in \tilde{V}. \quad (3.2b) \end{cases}$$

Here $\tau' = \frac{d\tau}{dx}$ and $(\cdot,\cdot)$ denotes the scalar product in $L_2(I)$. Let us choose the load f to be

$$f(x) = \lambda\bar{f}(x), \qquad x \in I,$$

where

$$\bar{f}(x) = \begin{cases} x & \text{for } 0 \leq x \leq \frac{1}{2}, \\ x-1 & \text{for } \frac{1}{2} \leq x \leq 1. \end{cases}$$

It is easy to check that the (unique) solution of (3.2) is as follows for $0 \leq x \leq \frac{1}{2}$ ($\sigma(x)$ is symmetric and $u(x)$ antisymmetric around $x = \frac{1}{2}$):

$$\begin{cases} u(x) = \frac{\lambda}{6}(x^3 - \frac{x}{4}) \\ \sigma(x) = \frac{\lambda}{2}(x^2 - \frac{1}{12}) \end{cases} \quad \text{for } 0 \leq \lambda \leq 12,$$

$$\begin{cases} u(x) = \frac{\lambda x^3}{6} + (1 - \frac{\lambda}{8})x, \\ \sigma(x) = \frac{\lambda x^2}{2} + 1 - \frac{\lambda}{8}, \end{cases} \quad \text{for } 12 \leq \lambda \leq 16,$$

and that no solution exists for $\lambda > 16$. We note that the solution is elastic (i.e. $|\sigma(x)| < 1$ for $x \in I$) for $0 \leq \lambda < 12$ and elasto-plastic for $12 \leq \lambda \leq 16$ with the plastic region concentrated at the point $x = \frac{1}{2}$ ($\sigma(\frac{1}{2}) = 1$ while $|\sigma(x)| < 1$ for $x \neq \frac{1}{2}$). Further, we note that the displacement u has a jump at $x = \frac{1}{2}$ of magnitude $\frac{\lambda}{12} - 1$ for $12 < \lambda \leq 16$ (see Fig. 1).

We shall now compare the conventional finite element method for (3.1) using continuous trial functions with the finite element method proposed above using discontinuous trial functions. Let

$$0 = x_0 < x_1 < \dots < x_n = 1, \quad I_j = (x_{j-1}, x_j),$$

be a partition of I, introduce the spaces

$$V_h = \{v: v \text{ is continuous on } I \text{ and } v|_{I_j} \text{ linear}, \; j = 1, \dots n\},$$

$$W_h = \{v: v|_{I_j} \text{ linear}, \; j = 1, \dots n\},$$

and consider the following discrete analogues of (3.1):

$$\operatorname*{Inf}_{v \in V_h} F(v), \tag{3.3}$$

$$\operatorname*{Inf}_{v \in W_h} F(v), \tag{3.4}$$

Here

$$F(v) = \sum_{j=1}^{n} \int_{I_j} g(|v'|)dx + \sum_{j=0}^{n} |[v(x_j)]| - (f,v)$$

for $v \in W_h$, where $[v(x_j)]$ denotes the jump of v at node x_j. To solve (3.3) numerically we used Newton's method on the non-linear equation $F'(u_h) = 0$ characterizing the exact solution u_h. In order to apply the same method for (3.4) the non-differentiable functional F was first regularized by replacing the term $|[v(x_j)]|$ by

$$\varepsilon g\left(\frac{|[v(x_j)]|}{\varepsilon}\right),$$

with ε small. The problem (3.4) was then solved approximately by applying Newton's method to the regularized problem. In the computations we used $\varepsilon = 10^{-7}$ and the iterations were terminated when the relative error between two consecutive iterations was less than 10^{-7}. The number of iterations required was then in the range 5 - 10.

We considered some different partitions. By $\delta(N)$ we denote a uniform partition with N intervals. Then if N is odd the jump discontinuity in the displacement at $x = \frac{1}{2}$ will fall in the middle of an interval in $\delta(N)$. Given a partition $\delta(N)$ with N odd we introduce four partitions $\delta_k = \delta_k(N)$, $k = 0,\dots,3$ obtained by adding to $\delta(N)$ the mesh point

$$y_k = \tfrac{1}{2} + k\,\frac{1}{12N}\,, \qquad k = 0,\dots,3.$$

Thus, δ_0 has a mesh point at the jump in u and δ_k, $k = 1,\dots,3$ are modifications of δ_0 obtained by shifting the mesh point at the jump slightly (c.f. Fig. 2).

The error between the exact solution (u,σ) given above and the solution (u_h,σ_h) of the discrete problems (3.3) and (3.4) (here $\sigma_h = u'_h$ in each subinterval) is as follows for $\lambda = 15$:

	$\|u-u_h\|_{L_2(I)}$				$\|\sigma-\sigma_h\|_{L_2(I)}$			
	cont. (V_h)		discont. (W_h)		cont. (V_h)		discont. (W_h)	
$\delta(9)$	0.64	-1	0.64	-1	0.23	0	0.23	0
δ_0	0.64	-1	0.46	-2	0.23	0	0.12	0
δ_1	0.58	-1	0.31	-1	0.22	0	0.13	0
δ_2	0.55	-1	0.48	-1	0.21	0	0.18	0
δ_3	0.55	-1	0.55	-1	0.21	0	0.21	0
$\delta(19)$	0.34	-1	0.34	-1	0.12	0	0.12	0
δ_0	0.34	-1	0.11	-2	0.12	0	0.-2	-1
δ_1	0.32	-1	0.18	-1	0.11	0	0.70	-1
δ_2	0.30	-1	0.28	-1	0.10	0	0.90	-1
δ_3	0.30	-1	0.29	-1	0.10	0	0.10	0
$\delta(39)$	0.19	-1	0.19	-1	0.57	-1	0.57	-1
δ_0	0.19	-1	0.28	-3	0.57	-1	0.31	-1
δ_1	0.18	-1	0.12	-1	0.53	-1	0.35	-1
δ_2	0.17	-1	0.18	-1	0.50	-1	0.45	-1
δ_3	0.17	-1	0.17	-1	0.50	-1	0.50	-1

We note that in the case of discontinuous trial functions we have for the partition δ_0 with a meshpoint exactly at the jump of u at x = ½ an $\mathcal{O}(h^2)$ convergence in displacements and $\mathcal{O}(h)$ in stresses. We also note that the reduction in the error in displacements is considerable when adding the mesh point at x = ½ (that is when passing from $\delta(N)$ to $\delta_0(N)$) while the reduction in the error in stresses is less pronounced. However the error in displacements is very sensitive to the location of the mesh point close to the jump in u as is clear from partitions δ_0 to δ_3.

Let us also give the jump in u_h at the meshpoint closest to x = ½ for the partitions δ_0 to δ_3. The jumps in u_h at other meshpoints are very small. The jump in the exact solution u at x = ½ is 0.25.

	N = 9				N = 19				N = 39			
	δ_0	δ_1	δ_2	δ_3	δ_0	δ_1	δ_2	δ_3	δ_0	δ_1	δ_2	δ_3
jump in u_h	0.25	0.18	0.11	.0	0.25	0.22	0.18	0	0.25	0.23	0.21	.0

Thus, we see that the discrete solution u_h will have a jump at the meshpoint closest to ½ if this meshpoint is sufficiently close to ½.

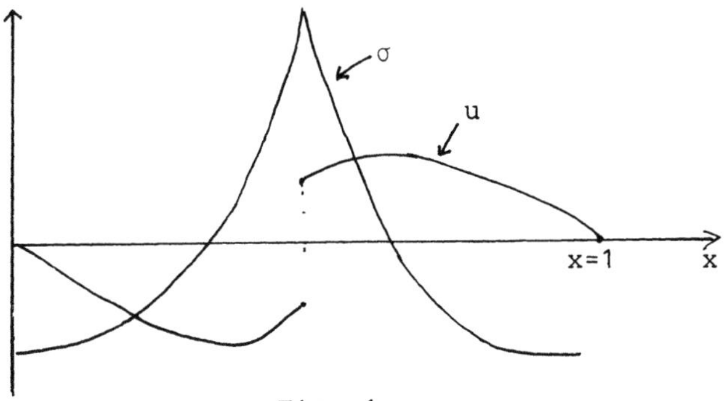

Fig. 1

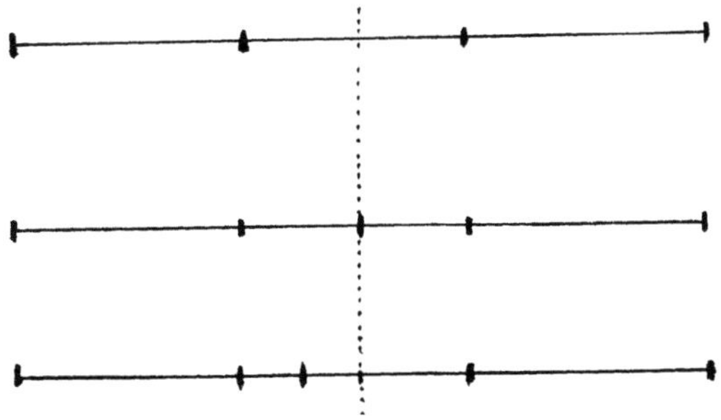

Fig. 2

Remark. The limit load problem (cf. [1]) can be formulated as follows:

$$\inf_{\substack{v \in \overset{\circ}{BD}(\Omega) \\ (f,v)=1}} G(v), \tag{3.5}$$

where

$$G(v) = \sup_{\tau \in P \cap C(Q)_s} \langle \tau, \varepsilon(v) \rangle.$$

If B is the unit ball and $v \in V$ we have

$$G(v) = \int_\Omega |\varepsilon(v)| dx.$$

A natural approach for the numerical solution of (3.5) is as follows

$$\inf_{\substack{v \in W_h \\ (f,v)=1}} G(v), \tag{3.6}$$

with W_h consisting of piecewise polynomial functions of degree k (k = 0,1,...). If B is the unit ball we have if $v \in W_h$

$$G(v) = \sum_K \int_K |\varepsilon(v)|dx + \sum_S \int_S |[v]|ds.$$

For k = 0 (which is probably the most interesting case) this method corresponds to the yield line method for limit load problems, a classical method in engineering computations (note that in the elastic-plastic case considered above we must have $k \geq 1$ to achieve convergence). ∎

References

1 E. Christiansen, H. Matthies and G. Strang: The saddle point of a differential problem in "Energy methods in finite element analysis" ed. Rodin, Zienkiewicz and Glowinski, Wiley (1979).

2 I. Ekeland and R. Temam: Convex Analysis and Variational Problems, North Holland, Amsterdam (1976).

3 C. Johnson: Existence theorems for plasticity problems, J. Math. pures et appl. 55 (1976), 431-444.

4 G. Strang and R. Temam: Functions of Bounded Deformation, Université de Paris-Sud (1979).

5 P. Suquet: Existence et regularité de solutions des equations de la plasticité parfaite, Thèse 3me cycle, Université Paris VI (1978).

Finite Element Analysis for Combined Material and Geometric Nonlinearities

M. A. CRISFIELD
Transport and Road Research Laboratory, Crowthorne, Berkshire, U. K

Introduction

For a number of years, the author has used the finite element method to investigate the collapse strength of thin plated steel structures [1-3]. The work has been directed primarily towards steel bridges which are usually fabricated from engineering steel for which the stress-strain curve exhibits a significant plateau. The collapse behaviour usually involves an interaction between material and geometric non-linearities and is influenced by initial geometric imperfections and residual welding stresses. The present communication describes a number of numerical techniques that the author has developed in order to analyse such structures. The topics covered include approximate yield criteria, accelerated iterative methods and incremental solutions using a 'length constraint'.

Yield criteria involving stress resultants

When analysing steel plates and shells, the most accurate way of treating plasticity is to apply von Mises yield criterion in conjunction with a numerical integration through the thickness [4-6]. Up to eleven layers of integration stations have been used [4]. Unfortunately such an approach can lead to large requirements for computer storage and time. Since only modest computer facilities were available, the author adopted an alternative approach and applied an approximate yield criterion that is a direct function of the six stress resultants [1]. The savings in computer storage are obvious but there was also found to be an appreciable saving in computer time [1]. This followed partly from the reduced number of 'disc accesses' but also from the reduced number of iterations required to achieve equilibrium. The latter reduction was caused by the smaller number of 'stresses' moving on the yield surface during the iterations.

Using the von Mises yield criterion and assuming that the equivalent stress is at yield throughout the depth of the plate or shell, Ilyushin [7] derived a yield surface involving the six stress resultants $M_x, M_y, M_{xy}, N_x, N_y$ and N_{xy}. To this end, he defined the following non-dimensional quadratic stress intensities.

$$\left.\begin{aligned} Q_t &= \frac{1}{N_o^2}(N_x^2 + N_y^2 - N_x N_y + 3 N_{xy}^2) = \frac{\overline{N}}{N_o^2} \\ Q_m &= \frac{1}{M_o^2}(M_x^2 + M_y^2 - M_x M_y + 3 M_{xy}^2) = \frac{\overline{M}}{M_o^2} \\ Q_{tm} &= \frac{1}{N_o M_o}(M_x N_x + M_y N_y - \tfrac{1}{2} M_x N_y \\ &\quad - \tfrac{1}{2} M_y N_x + 3 M_{xy} N_{xy}) = \frac{\overline{MN}}{M_o N_o} \end{aligned}\right\} \qquad \ldots (1)$$

where N_o is the uniaxial yield force $\sigma_o t$ and M_o is the uniaxial yield moment $\sigma_o t^2/4$. The stress intensities were related by means of complex parametric relationships of the form:

$$Q_t = f_1(\phi,\mu), \quad Q_m = f_2(\phi,\mu), \quad Q_{tm} = f_3(\phi,\mu) \qquad \ldots (2)$$

where the two parameters ϕ and μ are

$$\phi = \frac{e_{i2}}{e_{i1}}, \quad \mu = \frac{e_{i0}}{e_{i1}} \qquad \ldots (3)$$

and e_{i1} is the equivalent strain on the top surface, e_{i2} the equivalent strain on the bottom surface and e_{io} the minimum value of e_i. Unfortunately, equations (2) are not in a form that are suitable for computation. Consequently a number of attempts have been made to approximate equations (2) by eliminating the parameter ϕ and μ. Ilyushin proposed the approximation:

$$F_1 = Q_t + Q_m + \frac{1}{\sqrt{3}} |Q_{tm}| = 1 \qquad \dots (4)$$

Robinson [8] has shown that this is the best of the various approximations that are linear in Q_t, Q_m, Q_{tm} space. The author [1] invoked normality and treated the yield function in a flow sense to derive elasto-plastic modular matrices relating the increments of the stress resultants to the increments of the total 'strain'. The matrices were incorporated in an incremental/iterative finite element formulation and applied to a number of problems involving the collapse analysis of imperfect steel plates. Reasonable agreement was obtained with both alternative solutions using a layered von Mises approach and with experimental results [1].

For problems involving significant elements of compressive loading, the computed collapse loads tended to be too large. This follows from the basic assumption that the equivalent stress is at yield throughout the full section. Consequently, no allowance is made for the softening induced by 'fibre plasticity' before the attainment of 'full section yield'. For bending dominant situations, this 'full section yield' can strictly only occur as the equivalent plastic curvature, χ_{ps}, tends to infinity. Consequently, an approximate method was proposed [9] to allow for this 'fibre yield' by involving the equivalent plastic curvature in the yield function. To this end, the M_o term in equation (4) was replaced by $\alpha(\chi_{ps})M_o$ where α was chosen so that $\alpha(\chi_{ps})M_o$ follows the uniaxial moment plastic curvature relationship with increasing equivalent plastic curvature. i.e. equation (4) is replaced by

$$F_2 = Q_t + \frac{Q_m}{\alpha^2} + \frac{1}{\sqrt{3}\alpha}|Q_{tm}| = 1 \qquad \ldots (5)$$

The following expression for α was proposed in Ref. [2].

$$\alpha = 1.0 - 0.4\ \mathrm{EXP}(-\ 2.6\sqrt{\bar{\chi}_{ps}}) \qquad \ldots (6)$$

where $\bar{\chi}_{ps}$ is the non-dimensional equivalent plastic curvature obtained by summing the incremental equivalent plastic curvatures

$$\Delta\bar{\chi}^2_{ps} = \left(\frac{Et}{3\sigma_o}\right)^2 \Delta\chi^2_{ps}$$

$$= \left(\frac{Et}{3\sigma_o}\right)^2 \cdot \frac{4}{3}\left(\Delta\chi^2_{px} + \Delta\chi^2_{py} + \Delta\chi_{px}\Delta\chi_{py} + \frac{\Delta\chi^2_{pxy}}{4}\right) \ldots (7)$$

E is Young's modulus and $\Delta_{px}, \Delta\chi_{py}$ and $\Delta\chi_{pxy}$ are the incremental plastic curvatures.
As $\chi_{ps} \to \infty$, $\alpha \to 1$ and F_2 (equation (5)) coincides with the original Ilysushin approximation of equation (4). When $\chi_{ps} = 0$, $\alpha = 0.6$ and equation (5) gives

$$\left(F_2\right)\chi_{ps=0} = Q_t + \frac{9}{4}Q_m + \frac{\sqrt{3}}{2}|Q_{tm}| = 1 \qquad \ldots (8)$$

which is a reasonable approximation to the true expression for first fibre yield:

$$F_3 = Q_t + 9/4\ Q_m + 3|Q_{tm}| = 1 \qquad \ldots (9)$$

For imperfect plates subject to uniaxial compression, application of this modified yield criterion gave computed collapse loads that were about 5% lower than those obtained using the original Ilyushia yield function. The maximum loads were very similar to those obtained using a layered von Mises approach. However, following collapse, the computed load/shortening curves dropped too sharply. In an attempt to improve this situation, the author replaced Ilyushin's approximate yield surface (eqn. (4)) by a more accurate approximation due to Ivanov [10]. Ivanov's yield surface is given by

$$F_4 = Q_t + \tfrac{1}{2}Q_m + \sqrt{\tfrac{1}{4}Q_m^2 + Q_{tm}^2}$$

$$- \tfrac{1}{4}\,\frac{(Q_t\,Q_m - Q_{tm}^2)}{(Q_t + 0.48\,Q_m)} = 1 \qquad \ldots (10)$$

The equivalent plastic curvature may be introduced into the yield function [11] using a similar technique to that described for Ilyushin's yield criterion. The method has been used to study the collapse behaviour of simply supported imperfect steel plates subject to uniaxial compression. The results are compared in Fig. 1 with alternative solutions obtained [6] using a five layered von Mises approach. The agreement is very satisfactory.

Accelerated iterative solution procedures

In his early work, the author used the standard solution procedure combining incremental loading with modified Newton-Raphson (mNR) iterations. Recently, the mNR method has been replaced by accelerated iterative techniques [13-14] that are related to quasi-Newton methods [15,16]. Two such procedures are related to the BFGS up-date [16,17] and will be discussed here.

Using the BFGS method, the displacements $\underset{\sim}{p}$ are up-dated according to

$$\underset{\sim}{p}_{i+1} = \underset{\sim}{p}_i + \eta_i\,\underset{\sim}{\delta}_i \qquad \ldots (11)$$

where $$\underset{\sim}{\delta}_i = -\underset{\sim}{K}_i^{-1}\,\underset{\sim}{g}_i \qquad \ldots (12)$$

$\underset{\sim}{g}_i$ is the current gradient (of the total potential energy) or out-of-balance force vector and $\underset{\sim}{K}_i^{-1}$ is up-dated from

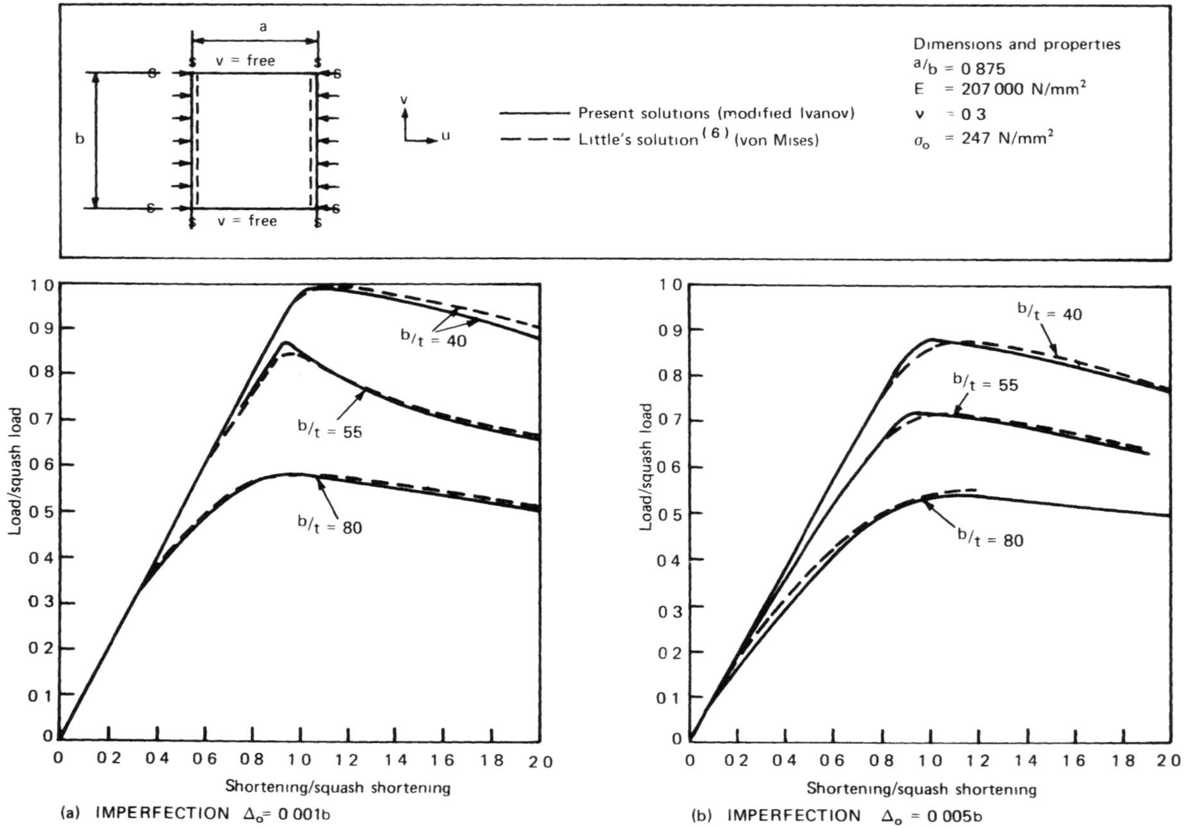

Fig 1 NON–DIMENSIONAL LOAD SHORTENING RELATIONSHIPS FOR UNIAXIALLY COMPRESSED SIMPLY SUPPORTED PLATES

$\underset{\sim}{K}_{i-1}^{-1}$ according to :

$$\underset{\sim}{K}_i^{-1} = \underset{\sim}{K}_{i-1}^{-1} - \frac{\underset{\sim}{\delta}_{i-1}\,\underset{\sim}{\gamma}_i^T\,\underset{\sim}{K}_{i-1}^{-1}}{\underset{\sim}{\delta}_{i-1}^T\,\underset{\sim}{\gamma}_i} - \frac{\underset{\sim}{K}_{i-1}\,\underset{\sim}{\gamma}_i\,\underset{\sim}{\delta}_{i-1}^T}{\underset{\sim}{\delta}_{i-1}^T\,\underset{\sim}{\gamma}_i}$$

$$+\left(1 + \frac{\underset{\sim}{\gamma}_i^T\,\underset{\sim}{K}_{i-1}^{-1}\,\underset{\sim}{\gamma}_i}{\eta_{i-1}\,\underset{\sim}{\delta}_{i-1}^T\,\underset{\sim}{\gamma}_i}\right)\frac{\eta_i\,\underset{\sim}{\delta}_{i-1}\,\underset{\sim}{\delta}_{i-1}^T}{\underset{\sim}{\delta}_{i-1}^T\,\underset{\sim}{\gamma}_i} \quad .. (13)$$

where $\underset{\sim}{\gamma}_i = \underset{\sim}{g}_i - \underset{\sim}{g}_{i-1}$ ... (14)

η_i is a 'step length parameter' which may be found using a line search [15] but has usually been taken as unity in the author's work. If

$$\overset{*}{\underset{\sim}{\delta}}_i = - \underset{\sim}{K}_{i-1}^{-1}\,\underset{\sim}{g}_i \quad ... (15)$$

is the standard mNR iteration neglecting any up-dating of $\underset{\sim}{K}$, substitution of equation (13) into equation (12) gives

$$\underset{\sim}{\delta}_i = h_i\overset{*}{\underset{\sim}{\delta}}_i + f_i\eta_{i-1}\underset{\sim}{\delta}_{i-1} + (1 - h_i)\overset{*}{\underset{\sim}{\delta}}_{i-1} \quad ... (16)$$

where $\overset{*}{\underset{\sim}{\delta}}_{i-1}$ is given by

$$\overset{*}{\underset{\sim}{\delta}}_{i-1} = - \underset{\sim}{K}_{i-1}^{-1}\,\underset{\sim}{g}_{i-1} = - \underset{\sim}{K}_a^{-1}\,\underset{\sim}{g}_{i-1} \quad ... (17)$$

and $\underset{\sim}{K}_{i-1}$ becomes a fixed matrix $\underset{\sim}{K}_a$. The scalars h_i and f_i are given by

$$h_i = - \frac{a_i}{b_i}$$
$$f_i = - \frac{d_i}{b_i} + (h_i - 1)\,(1 - \frac{t_i}{b_i}) \quad ... (18)$$

a_i, b_i, d_i and t_i are scalars given by the inner products

$$\left.\begin{aligned} a_i &= \eta_{i-1}\underset{\sim}{\delta}_{i-1}^T\underset{\sim}{g}_{i-1} \\ b_i &= \eta_{i-1}\underset{\sim}{\delta}_{i-1}^T\underset{\sim}{\gamma}_i \\ d_i &= (\underset{\sim}{\overset{*}{\delta}}_i - \underset{\sim}{\overset{*}{\delta}}_{i-1})^T\underset{\sim}{g}_i \\ t_i &= (\underset{\sim}{\overset{*}{\delta}}_i - \underset{\sim}{\overset{*}{\delta}}_{i-1})^T\gamma_i \end{aligned}\right\} \qquad \ldots (19)$$

Equation (16) defines a memoryless 'one-step' version of a vectorised BFGS method given by Mathies and Strang [17]. The procedure is also a scaled version of a new conjugate gradient method due to Shanno [18].
By making the approximation

$$\eta_{i-1}\underset{\sim}{\delta}_{i-1} \triangleq - \underset{\sim}{K}_{i-1}^{-1}\, \underset{\sim}{g}_{i-1} \qquad \ldots (20)$$

substitution of equation (13) into equation (12) gives

$$\underset{\sim}{\delta}_i = h_i\underset{\sim}{\overset{*}{\delta}}_i + e_i\eta_{i-1}\underset{\sim}{\delta}_{i-1} \qquad \ldots (21)$$

where h_i has already been defined (equation (18)) and

$$e_i = h_i(1 - \frac{c_i}{b_i}) - 1 \qquad \ldots (22)$$

and $$c_i = \underset{\sim}{\overset{*}{\delta}}_i^T\underset{\sim}{\gamma}_i \qquad \ldots (23)$$

Equation (21), which is Crisfield's faster mNR iteration [12], requires the storage of one less vector than does equation (16).

The two methods (equations (16) and (21)) have been applied to a number of elastic and elasto-plastic large-deflection problems [13]. Both techniques have been found to require substantially less iterations than the standard mNR method. The extra computation per iteration is almost negligable, the only penalty being the extra storage. In general, equation (16) has been found to be marginally more efficient than equation (21) but this slight advantage is compensated by the reduced storage required by the latter equation. The two accelerated methods and the standard mNR procedure have each been used to analyse a clamped cylindrical shell subject to uniform pressure loading. A five-by-five mesh was used to idealise a quarter of the shell and line searches were ommitted (i.e. η_i, equation (11) was set to unity). The number of iterations required for the accelerated method of equation (21) is compared with the number of iterations required by the standard mNR technique in Fig. 2 (fixed load increments). Clearly the accelerated procedure has led to a very significant improvement in the convergence characteristics. The performance of the accelerated method of equation (16) was found to be very similar [13] to that of equation (21).

<u>Incremental procedures using 'length control'</u>

Generally in structural analysis, the iterative solution procedures are not used to find a single equilibrium position for a given fixed load. Instead, a complete equilibrium path is often traced. Clearly advantage should be taken from the adjacent equilibrium positions. This can be achieved by automatically adjusting the sizes of the load increments to account for the degree of non-linearity [3,19]. Alternatively, the load level can be continuously adjusted as the iterations proceed. Such approaches have been proposed by Bergan [19] and Riks [20]. The author has had considerable success in adopting a modified version [14] of Riks's procedure [20]. The method is not only faster than the standard fixed load mNR procedure but also allows limit points to be passed [14,20]. The technique can be used in conjunction with the standard mNR method or with the accelerated techniques that have previously been discussed [13,14]. The former application will be briefly described.

$\underset{\sim}{q}$ will be taken as the total fixed applied load vector while the scalar λ will represent the 'load level'. The

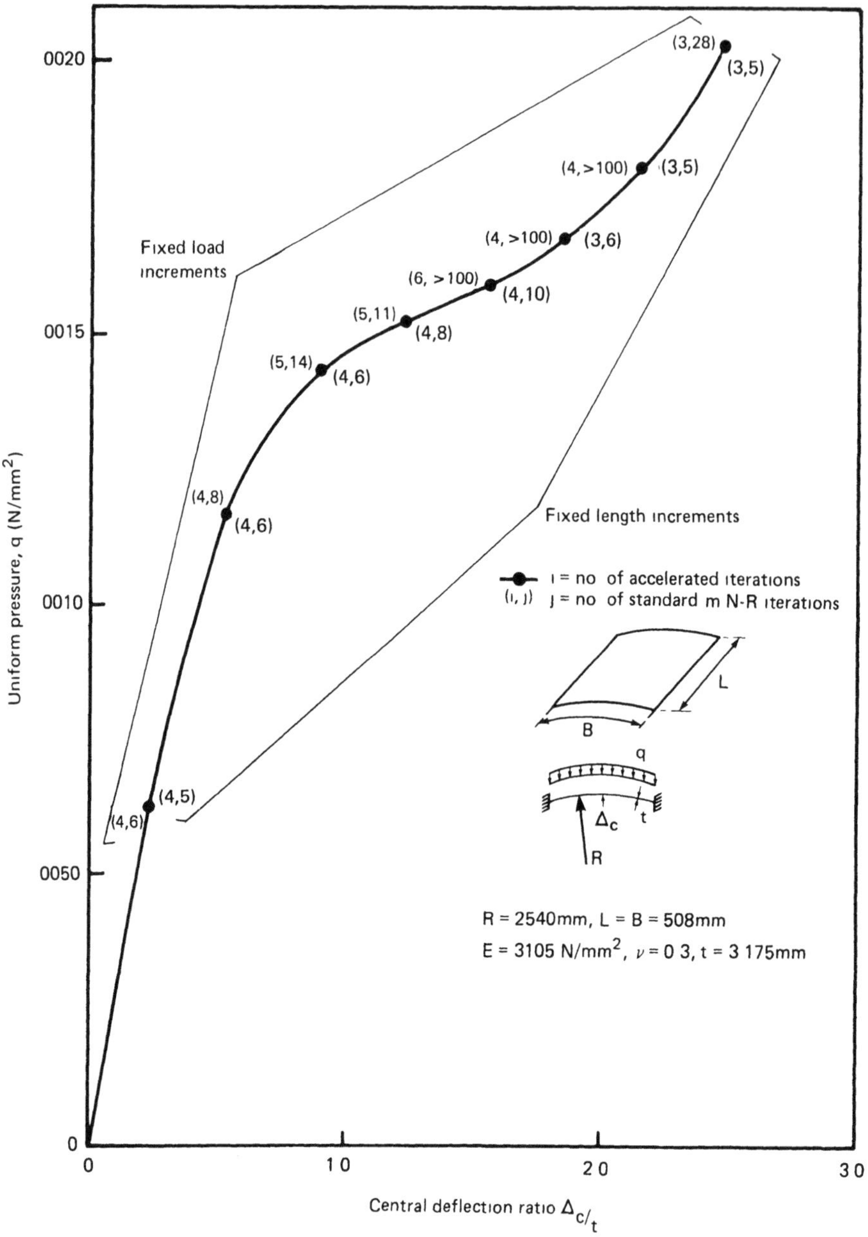

Fig. 2 FULLY CLAMPED CYLINDRICAL SHELL WITH UNIFORM PRESSURE LOADING

standard equilibrium equations can then be expressed as

$$\underset{\sim}{g}_{int}(\underset{\sim}{p},\lambda) - \lambda \, q = 0 \qquad \ldots (24)$$

If λ is now considered to be an additional variable, a further governing equation is required. Riks provided this equation by means of the incremental length constraint

$$\Delta\underset{\sim}{p}^T \, \Delta\underset{\sim}{p} + \Delta\lambda^2 \underset{\sim}{q}^T \, \underset{\sim}{q} = \Delta\ell^2 \qquad \ldots (25)$$

where $\Delta\ell$ is a given fixed length in N+1 dimensional space. (N is the number of displacement variables). The author has found it advantageous to replace equation (25) by the simpler expression

$$\Delta\underset{\sim}{p}^T \, \Delta\underset{\sim}{p} = \Delta\ell^2 \qquad \ldots (26)$$

Riks directly added the constraint of equation (25) to the N equilibrium equations of equation (24). He then solved the complete system using the Newton-Raphson procedure. Unfortunately this destroyed both the symmetry and the banded nature of the original equilibrium equations. An alternative approach is possible and follows from a different presentation of the equilibrium equations. i.e.

$$\underset{\sim}{g}(\lambda + \delta\lambda) = \underset{\sim}{g}(\lambda) - \delta\lambda \, \underset{\sim}{q} = 0 \qquad \ldots (27)$$

Following from equation (27), the iterative change given by the mNR method (for the unknown load level $\lambda_i + \delta\lambda_i$ where i is the iteration number) is

$$\underset{\sim}{\delta}_i = -\underset{\sim}{K}^{-1} \, \underset{\sim}{g}(\lambda_i + \delta\lambda_i) = \underset{\sim}{\delta}_i(\lambda_i) + \delta\lambda_i \, \underset{\sim}{\delta}_T \qquad \ldots (28)$$

where $\underset{\sim}{\delta}_i(\lambda_i) = -\underset{\sim}{K}^{-1} \, \underset{\sim}{g}_i(\lambda_i)$ ... (29)

is the standard (fixed load level) mNR change and δ_T is the fixed (and stored) tangential displacement

$$\underset{\sim}{\delta}_T = \underset{\sim}{K}^{-1} \underset{\sim}{q} \qquad \ldots (30)$$

$\delta\lambda_i$ is as yet unknown. However,

$$\Delta \underset{\sim}{p}_{i+1} = \Delta \underset{\sim}{p}_i + \underset{\sim}{\delta}_i \qquad \ldots (31)$$

and substitution from equation (28) into equations (31) and (26) (for $\Delta \underset{\sim}{p}_{i+1}$) leads to the following simple scalar quadratic equation for $\delta\lambda_i$.

$$a_1 \, \delta\lambda_i^{\,2} + a_2 \, \delta\lambda_i + a_3 = 0 \qquad \ldots (32)$$

$$a_1 = \underset{\sim}{\delta}_T^{\,T} \, \underset{\sim}{\delta}_T$$

$$a_2 = 2 \, (\Delta \underset{\sim}{p}_i + \underset{\sim}{\delta}_i(\lambda_i))^T \, \underset{\sim}{\delta}_T \qquad \ldots (33)$$

$$a_3 = (\Delta \underset{\sim}{p}_i + \underset{\sim}{\delta}_i(\lambda_i))^T \, (\Delta \underset{\sim}{p}_i + \underset{\sim}{\delta}_i(\lambda_i)) - \Delta \ell^2$$

The appropriate root is chosen with a view to maintaining a positive 'angle' between the incremental displacement vector at the present and past iteration. Details are given in [14] which also describes the application of the length constraint to the accelerated iterative methods.

As an example, the clamped shell of Fig. 2 has been re-analysed using 'fixed length' rather than 'fixed load' increments. The lengths were chosen to coincide with those obtained from the previous 'fixed load increment' analysis. The results are shown below the curve in Fig. 2. Clearly the imposition of the 'length constraint' to the mNR method has dramatically improved its convergence characteristics. Further improvements are gained (Fig. 2) by adding the acceleration (of equation (21)). However, the acceleration is relatively less beneficial once the length constraint has been added. The addition of the length constraint only marginally increases the computer time required for each iteration.

Further work is required before the most efficient solution strategy can be defined. However, the simplicity and effectiveness of the standard mNR method, when combined with the 'length constraint', make this technique a strong contender.

Acknowledgements

The work described in this Paper forms part of the programme of the Transport and Road Research Laboratory and the Paper is published by permission of the Director.

References

1 Crisfield, M.A.: Large deflection elasto-plastic buckling analysis of plates using finite elements. TRRL Report LR 598, Crowthorne, England, (1973).

2 Crisfield, M.A. and Puthli, R.S.: Approximations in the non-linear analysis of thin plated structures. Finite elements in non-linear mechanics, Vol. 1, Tapir Press, Trondheim, (1978) 373-392.

3 Crisfield, M.A.: The automatic nonlinear analysis of stiffened plates and shallow shells using finite elements. Proc. Instn. Civ. Engrs., Part 2, Paper 8335, to be published (Sept. 1980).

4 Marcal, P.V.: Large deflection analysis of elastic-plastic shells of revolution. AIAA Journal 8 (1970) 1627-1634.

5 Backlund, J. and Wennestrom, H.: Finite element analysis of elasto-plastic shells. Int. J. Num. Meth. in Engng. 8 (1974) 415-425.

6 Little, G.H.: Rapid analysis of plate collapse by live energy minimisation. Int. J. Mech, Sci. 19 (1977) 725-744.

7 Ilyushin, A.: Plasticité. Editions Eyrolles, Paris, (1965).

8 Robinson, M.A.: A comparison of yield surfaces for thin shells. Int. J. Mech. Sci. 13 (1971) 345-354.

9 Crisfield, M.A.: On an approximate yield criterion for thin shells. TRRL Report LR 658, Crowthorne, England (1974).

10 Ivanov, E.V.: Inzhenernyi Zhurnal Mekhanika Tverdogo Tela. 6 (1967) 74-75.

11 Crisfield, M.A.: Ivanov's yield criterion for thin plates and shells using finite elements. TRRL Report LR 919, Crowthorne, England (1979).

12 Crisfield, M.A.: A faster modified Newton-Raphson iteration. Comp. Meth. in Appl. Mech. and Engng. 20 (1979) 267-278.

13 Crisfield, M.A.: Incremental/iterative solution procedures for nonlinear structural analysis. Int. Conf. Num. Meth. for Nonlinear Problems, Swansea (Sept. 1980).

14 Crisfield, M.A.: A fast incremental/iterative solution procedure that handles 'snap through'. Symp. on Comp. Meth. in Nonlinear Struct. and Solid Mech., Washington, (Oct.1980).

15 Dennis, J. Jnr. and More, J.: Quasi-Newton methods, motivation and theory. SIAM Review 19 (1977) 46-84.

16 Broyden, C.G.: The convergence of a double-rank minimisation 2: The new algorithm. J. Inst. Math. Appl. 6 (1970) 222-231.

17 Matthies, H. and Strang, G.: The solution of nonlinear finite element equations. Int. J. for Num. Meth. in Engng. 14 (1979) 1613-1626

18 Shanno, D.F.: Conjugate gradient methods with inexact searches. Math. of Operations Res. 3 (1978) 244-256.

19 Bergan, P.G.: Solution algorithms for nonlinear structural problems. Engineering Appls. of the Finite Element Method, Computas, Hovik, Norway, (1979) 13.1-13.39.

20 Riks, E.: An incremental approach to the solution of snapping and buckling problems. Int. J. Solids & Structs. 15 (1979) 529-551

Incremental Elastic-Plastic Deformations of Stiffened Plates in Compression and Bending

J PAULUN, E. STEIN*
Technische Universitat Hannover, Germany

Summary

The deformation process of stiffened plates (e.g. box girders for bridges) with increasing static compression loads until critical states is investigated in the frame of geometrical nonlinear theory. Instead of geometrical imperfections a constant normal load is applied on the plate. The elastic-plastic material is described according to v. Mises with an isotropic hardening. Partly plastifying over the thickness of the plate is considered in the FE-description by using elements with the full plate thickness. So, neither layered elements nor smeared elements for plate and stiffeners are used. Two methods for the numerical treatment of incremental elastic-plastic compression and bending are developed and compared, the first following a proposal of Pflüger and the second improving the first by an equilibrium iteration in order to reduce the errors with respect to the Prandtl-Reuss equations. Up to now a Tresca-material was mostly used for plate bending. Concering the FEM, rectangular elements (12-parametric for bending and 8-parametric for plane stress) are used for the plate and beam elements for the stiffeners. Calculations with various stiffness parameters give insight into the load carrying behaviour near critical deformation states.

Introduction

It is wellknown from tests that so-called orthotropic plates have large load-capacities in the region of geometrically and physically nonlinear deformations. The postcritical states and especially the state of ultimate load should be known in order to estimate a construction in the frame of safety-analysis. Those constructions are very complex because of the stiffeners which are necessary for stabiliza-

---

* In honour of Prof. Dr.-Ing. Th. Lehmann on his 6oth birthday on August 1o, 198o

tion, and the statical systems change during the deformation process.

Because of the nonlinearities tests need to be large scale, and they are very expensive. Therefore numerical methods - especially FE-methods - are necessary for the engineering treatment. The discretized system must be investigated by an incremental and iterative technique with special consideration of critical, postcritical and ultimate loadings. The first problem arises from the discretization itself. Due to elastic-plastic deformation different directions of main stress axis are present in the plate-layers for combined membrame and bending forces. Therefore layered plate elements would be recommendable but too expensive for practical calculations. On the other hand fully integrated elements for the plate including the stiffeners are only possible for dense patterns of stiffeners. So, integrated plate elements over the full thickness of the plate and seperate beam elements for the stiffeners seem to be recommendable. Those integrated plate elements for small elastic-plastic strains but moderately large rotations are developed in this paper. The second problem concerns the incremental calculation of critical and postcritical states because of complicated interactions of local buckling and plastifying. So, local unloading is possible for increasing total loading.

2. Geometrically nonlinear treatment of deformation states

The geometrically nonlinear description of thinwalled plates and shells allows in most cases - even for elasto-plastic deformations - the assumption of small strains combined with large rotations and displacements. For the elastic part a linear relation of the spatial strain rate tensor $\underset{\sim}{D}$ and the Jaumann stress flux $\overset{\circ}{\underset{\sim}{T}}$ is oftenly chosen, see Lehmann [1]. For isotropic material it has the form

$$\overset{\circ}{\underset{\sim}{T}} = 2G\left[\underset{\sim}{D} + \frac{\nu}{1-2\nu}\,\mathrm{tr}(\underset{\sim}{D})\,\underset{\sim}{1}\right] \qquad (2.1)$$

(G: shear modulus, ν: Poisson's ratio).
For small strains it becomes a linear relation between the Green strain tensor $\underset{\sim}{E}$ and 2. Piola-Kirchhoff stress tensor $\underset{\sim}{S}$

$$\dot{\underset{\sim}{S}} = 2G\left[\dot{\underset{\sim}{E}} + \frac{\nu}{1-2\nu}\,\mathrm{tr}(\dot{\underset{\sim}{E}})\underset{\sim}{1}\right] . \qquad (2.2)$$

Correspondingly a flow rule for large plastic deformations with plastic incompressibility (compare Paulun [8])

$$\dot{\underset{\sim}{E}}_p = \dot{\lambda}\left[\underset{\sim}{F}_p^T\underset{\sim}{F}_p\underset{\sim}{S}\;\underset{\sim}{F}_p^T\underset{\sim}{F}_p - \frac{1}{3}\,\mathrm{tr}(\underset{\sim}{F}_p\underset{\sim}{S}\;\underset{\sim}{F}_p^T)\underset{\sim}{F}_p^T\underset{\sim}{F}_p\right] \qquad (2.3)$$

simplifies for small strains into

$$\dot{\underset{\sim}{E}}_p = \dot{\lambda}\left[\underset{\sim}{S} - \frac{1}{3}\,\mathrm{tr}(\underset{\sim}{S})\underset{\sim}{1}\right] . \qquad (2.4)$$

For proofing the above statement the material deformation gradient

$$\underset{\sim}{F} = \frac{d\underset{\sim}{x}}{d\underset{\sim}{X}} := \frac{\partial x_i}{\partial X_k}\,\underset{\sim}{N}_i \otimes \underset{\sim}{N}_k$$

(with: $\underset{\sim}{N}_i$ fixed orthonormal base vectors in space, X_k Lagrangian and x_i Eulerian coordinates) is split by polar decomposition

$$\underset{\sim}{F} = \underset{\sim}{R}\;\underset{\sim}{U} \qquad (2.5)$$

into the rotation tensor $\underset{\sim}{R}$ and the stretch tensor $\underset{\sim}{U}$. For small strains it holds

$$\underset{\sim}{U} = \underset{\sim}{1} + \varepsilon\underset{\sim}{A} , \qquad (2.6)$$

with $\|\underset{\sim}{U}\| \approx \|\underset{\sim}{1}\| \approx \|\underset{\sim}{A}\|$ and $\varepsilon \to 0$. Then, because of the transformation

$$\underset{\sim}{F} = \underset{\sim}{R} = (\underset{\sim}{R}^T)^{-1} = (\underset{\sim}{F}^T)^{-1} \; ; \; \det \underset{\sim}{F} = 1 \qquad (2.7)$$

the objective stress flux $\overset{\Delta}{\underset{\sim}{T}}$ of Truesdell [2]

$$\overset{\Delta}{\underset{\sim}{T}} := (\det \underset{\sim}{F})^{-1} \underset{\sim}{F}\, \dot{\underset{\sim}{S}}\, \underset{\sim}{F}^T \tag{2.8}$$

has the relation with the Jaumann stress flux $\overset{\circ}{\underset{\sim}{T}}$

$$\overset{\Delta}{\underset{\sim}{T}} = \overset{\circ}{\underset{\sim}{T}} - \underset{\sim}{T}\,\underset{\sim}{D} - \underset{\sim}{D}\,\underset{\sim}{T} + \mathrm{tr}(\underset{\sim}{D})\underset{\sim}{T} \, . \tag{2.9}$$

Introducing the material law (2.1) into (2.9) it follows

$$\overset{\Delta}{\underset{\sim}{T}} = (G\underset{\sim}{1} - \underset{\sim}{T})\underset{\sim}{D} + \underset{\sim}{D}(G\underset{\sim}{1} - \underset{\sim}{T}) + \left[\frac{2G\nu}{1-2\nu}\underset{\sim}{1} + \underset{\sim}{T}\right]\mathrm{tr}(\underset{\sim}{D}) \, . \tag{2.1o}$$

Due to the above assumptions, $\underset{\sim}{T}$ is of higher order small compared with $G\underset{\sim}{1}$ so that (2.1o) simplifies to

$$\overset{\Delta}{\underset{\sim}{T}} = 2G\left[\underset{\sim}{D} + \frac{\nu}{1-2\nu}\,\mathrm{tr}(\underset{\sim}{D})\,\underset{\sim}{1}\right] = \overset{\circ}{\underset{\sim}{T}}. \tag{2.11}$$

On the other hand from (2.7) and (2.8) follows

$$\overset{\Delta}{\underset{\sim}{T}} = \underset{\sim}{R}\,\dot{\underset{\sim}{S}}\,\underset{\sim}{R}^T \, , \tag{2.12}$$

and from the known relation

$$\dot{\underset{\sim}{E}} = \underset{\sim}{F}^T \underset{\sim}{D}\,\underset{\sim}{F} \tag{2.13}$$

we finally get

$$\underset{\sim}{D} = \underset{\sim}{R}\,\dot{\underset{\sim}{E}}\,\underset{\sim}{R}^T. \tag{2.14}$$

Introducing (2.12) and (2.14) into the material equation (2.11) the prediction (2.2) comes out immediately. Correspondingly, the assumption of small plastic strains $\underset{\sim}{F}_p^T\underset{\sim}{F}_p = \underset{\sim}{R}_p^T\underset{\sim}{R}_p = \underset{\sim}{1}$ is used in (2.3), and leads to the material equation (2.4).

The tensors $\underset{\sim}{E}$ and $\underset{\sim}{S}$ allow a Lagrangian representation of large deformations. A linearization of Green strain tensor $\underset{\sim}{E}$ is only possible incrementally as described below. Otherwise large rotations can not be considered. For bending of a plane plate in the initial configuration Kirchhoff hypothesis is used. Then the kinematic variables are reduced to the displacements and the rotations of the middle surface of the plate. The numerical treatment of plates in compression and bending uses the FE-method. The kinematic variables of the discretization process are the nodal displacements, row matrix $\underline{V}$. The corresponding load vector resulting from the given load is $\underline{\bar{R}}$.

3. Incremental representation for discretization methods

The total load, given by the column matrix $\underline{\bar{R}}$, is applied with increments $\Delta\underline{\bar{R}}^N$ (N = 1, 2, ..., n). A total strong nonlinearity is mostly weak in the increments, so that the deformation process only needs few iterations for a single load step. Furthermore load history of an elastic-plastic material can be better described. The material tensor is changed after each load step and improved during the iteration process. For the calculation of the incremental displacement vector $\Delta\underline{V}$ from N to N + 1 we postulate that the whole configuration is known after step N. The principle of virtual work for calculating the incremental displacement vector $\Delta\underline{V}$ can be written in the form

$$\int_{V_o} \mathrm{tr}\left[\delta\Delta\underset{\sim}{E}(\underset{\sim}{S}^N + \Delta\underset{\sim}{S})\right] dV - \delta\Delta\underline{V}^T(\underline{\bar{R}}^N + \Delta\underline{\bar{R}}) = 0 \tag{3.1}$$

This"total Lagrangian representation (T.L)" is wellknown from Bathe, Ramm, Wilson [3] and was used also by Bathe, Borlourchi [4] , Washizu [5] , Atluri [6] , Larsen, Popov [7] a.o. For shorter representation matrix notation is used for the virtual work of the given load in (3.1). With

$$\begin{aligned} \Delta\underset{\sim}{E} &= \mathrm{sym}\,\{\Delta\underset{\sim}{F}^T\underset{\sim}{F}^N + \tfrac{1}{2}\,\Delta\underset{\sim}{F}^T\Delta\underset{\sim}{F}\} \\ \delta\Delta\underset{\sim}{E} &= \mathrm{sym}\,\{\delta\Delta\underset{\sim}{F}^T\underset{\sim}{F}^N\} + \mathrm{sym}\,\{\delta\Delta\underset{\sim}{F}^T\Delta\underset{\sim}{F}\} \end{aligned} \tag{3.2}$$

we get from (3.1) by using the symmetry of $\underset{\sim}{S}^N$ and $\Delta\underset{\sim}{S}$

$$\int_{V_o} \mathrm{tr}(\delta\Delta\underset{\sim}{E}\ \Delta\underset{\sim}{S})\,dV + \int_{V_o} \mathrm{tr}(\delta\Delta\underset{\sim}{F}^T\Delta\underset{\sim}{F}\ \underset{\sim}{S}^N)\,dV$$

$$= \left[\delta\Delta\underline{V}^T\underline{\bar{R}}^N - \int_{V_o} \mathrm{tr}(\delta\Delta\underset{\sim}{F}^T\underset{\sim}{F}^N\underset{\sim}{S}^N)\,dV\right] + \delta\Delta\underline{V}^T\Delta\underline{\bar{R}}\ . \tag{3.3}$$

The two terms in brackets on the righthand side of (3.3) vanish as follows. We replace in the principle of virtual work for state N, namely

$$\int_{V_o} \mathrm{tr}(\delta\underset{\sim}{E}^N\underset{\sim}{S}^N)\,dV - \delta\underline{V}^{N^T}\underline{\bar{R}}^N = 0 \tag{3.4}$$

with $\delta\underset{\sim}{E}^N = \mathrm{sym}\ \{\delta\underset{\sim}{F}^{N^T}\underset{\sim}{F}^N\}$, the variation of the displacements by the variation of the increments, $\delta\underline{V}^N \rightarrow \delta\Delta\underline{V}$, and correspondingly $\delta\underset{\sim}{E}^N \rightarrow \mathrm{sym}\ \{\delta\Delta\underset{\sim}{F}^T\underset{\sim}{F}^N\}$. From (3.4) follows the simplification in (3.3) as was shown by Klee, Paulun [9] .

The collection of (2.2) and (2.4) leads to the Prandtl-Reuß material law which is given here in incremental form

$$\Delta\underset{\sim}{S} = 2G\Delta\underset{\sim}{E} + \frac{2G\nu}{1-2\nu}\ \mathrm{tr}(\Delta\underset{\sim}{E})\ \underset{\sim}{1} + \frac{1}{\alpha}\ \mathrm{tr}(\underset{\sim}{S}^D\Delta\underset{\sim}{E})\underset{\sim}{S}^D \tag{3.5}$$

with the hardening parameter $\alpha = \frac{\zeta k^2}{G^2} + \mathrm{tr}(\underset{\sim}{S}^D\underset{\sim}{S}^D)$, the plastic tangent modulus ζ and the stress deviator $\underset{\sim}{S}^D := \underset{\sim}{S} - \frac{1}{3}\ \mathrm{tr}(\underset{\sim}{S})\underset{\sim}{1}$. This constitutive equation is a linear dependence between the components of $\Delta\underset{\sim}{E}$ and $\Delta\underset{\sim}{S}$ if the quantities ζ, k and $\underset{\sim}{S}$ can be kept constant by suitable average values. For relatively small load steps we can use the initial values ζ^N, k^N, $\underset{\sim}{S}^N$. In the case of pure elastic deformations one gets $\frac{1}{\alpha} \rightarrow 0$. Introducing the linear incremental material law (3.5) into the principle (3.3) and discretizing the system by finite displacement elements we get a cubic system of algebraic equations for the displacement increments $\Delta\underline{V}$ ($\delta\Delta\underset{\sim}{E}$ contains $\Delta\underline{V}$ linearly, and $\Delta\underset{\sim}{S}$ contains $\Delta\underline{V}$ quadraticly). A linearization of the strain increments (3.2) in the form

$$\overset{L}{\Delta\underset{\sim}{E}} = \mathrm{sym}\ \{\Delta\underset{\sim}{F}^T\ \underset{\sim}{F}^N\}\ ;\ \ \delta\overset{L}{\Delta\underset{\sim}{E}} = \mathrm{sym}\ \{\delta\Delta\underset{\sim}{F}^T\ \underset{\sim}{F}^N\} \tag{3.6}$$

leads to discrepances for large strains which can be corrected by iteration of the equilibrium conditions with a modified Newton Raphson process

$$\int_{V_o} tr(\delta\Delta \underset{\sim}{F}^T \underset{\sim}{F}^N \Delta\tilde{\underset{\sim}{S}})dV + \int_{V_o} tr(\delta\Delta \underset{\sim}{F}^T \Delta\tilde{\underset{\sim}{F}} \underset{\sim}{S}^N)dV$$
$$= \int_{V_o} \Delta\underline{V}^T(\underline{\bar{R}}^N + \Delta\underline{\bar{R}}) - \int_{V_o} tr(\delta\Delta\underset{\sim}{F}^T \tilde{\underset{\sim}{F}}^{N+1}\tilde{\underset{\sim}{S}}^{N+1})dV. \qquad (3.7)$$

Diverging from (3.3) we have to replace $\underset{\sim}{F}^N$ and $\underset{\sim}{S}^N$ by the unknown tensors F^{N+1} and $\underset{\sim}{S}^{N+1}$ on the righthand side. One can chose $\underset{\sim}{F}^N$, $\underset{\sim}{S}^N$ as initial values and gets iteratively improved tensors $\tilde{\underset{\sim}{F}}^{N+1}$, $\tilde{\underset{\sim}{S}}^{N+1}$ by the correction terms $\Delta\tilde{\underset{\sim}{F}}$ and $\Delta\tilde{\underset{\sim}{S}}$. At the end of the iteration process the righthand side of (3.7) vanishes. Then principle of virtual work - represented by the lefthand side of (3.7) - describes the equilibrium conditions of state N+1 exactly. This is corresponding with the vanishing bracket terms for state N in (3.3).

Within a theory of small strains - as presumed in this paper - the correction values due to equilibrium iteration for linearized strain-increments are small of higher order. Then (3.3) can be reduced to

$$\int_{V_o} tr(\delta\Delta\overset{L}{\underset{\sim}{E}} \overset{L}{\Delta\underset{\sim}{S}})dV + \int_{V_o} tr(\delta\Delta\underset{\sim}{F}^T\Delta\underset{\sim}{F}\,\underset{\sim}{S}^N)dV = \delta\Delta\underline{V}^T\Delta\underline{\bar{R}}. \qquad (3.8)$$

For purely elastic deformations also the second term in (3.8) vanishes. In the case of elasto-plastic behaviour this second term in (3.8) can only be neglected if the achieved yield stress Y is small compared with the plastic tangent modulus ζ

$$\int_{V_o} tr(\delta\Delta\underset{\sim}{F}^T\Delta\underset{\sim}{F}\,\underset{\sim}{S}^N)dV \to 0 \quad \text{for} \quad Y << \zeta. \qquad (3.9)$$

For the following we restrict the formulation to plane plates in the undeformed state. After N load increments the resulting rotations of the middle surface are presumed to be large. The FE-discretization is supposed to be adapted in such a way that the rotation tensor $\underset{\sim}{R}^N$ corresponding to (2.5) and (2.7) is approximately constant within each element. Then we can introduce local orthonormal base vectors

$$\underset{\sim}{n}_i = \underset{\sim}{R}^N \underset{\sim}{N}_i \tag{3.10}$$

which nearly coincide with the base vectors of the convected material coordinates X_i. We define $\underset{\sim}{N}_3$ as the unit normal vector of the undeformed middle surface and correspondingly $\underset{\sim}{n}_3$ of the deformed middle surface at state N, see fig. 1.

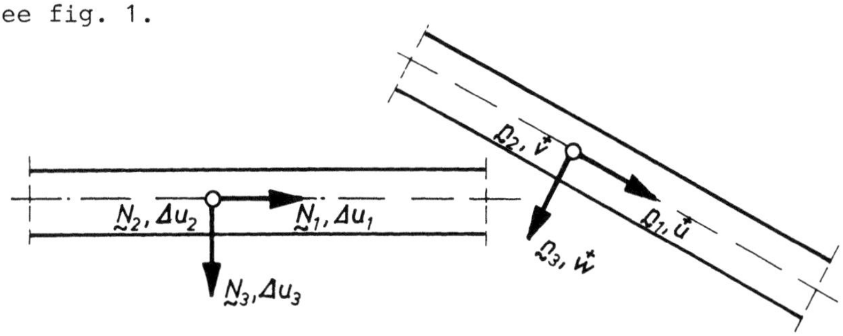

fig. 1. Global and local base vectors and displacements for a load increment from state N to state N+1

Then, the displacement increments

$$\Delta\underset{\sim}{u} = \underset{\sim}{x}^{N+1} - \underset{\sim}{x}^N \tag{3.11}$$

can be represented in directions of the local base vectors. In the first step, $\Delta\underset{\sim}{u}$ is decomposed into the undeformed directions

$$\overset{+}{\underset{\sim}{u}} = \overset{+}{u}\underset{\sim}{N}_1 + \overset{+}{v}\underset{\sim}{N}_2 + \overset{+}{w}\underset{\sim}{N}_3 = \underset{\sim}{R}^{N^T} \Delta\underset{\sim}{u}. \tag{3.12}$$

For the local displacement gradient $\overset{+}{\underset{\sim}{H}} = \frac{d\overset{+}{\underset{\sim}{u}}}{d\underset{\sim}{X}} := \frac{\partial \overset{+}{u}_i}{\partial X_j} \underset{\sim}{N}_i \otimes \underset{\sim}{N}_j$ it follows

$$\Delta\underset{\sim}{F} := \underset{\sim}{F}^{N+1} - \underset{\sim}{F}^N = \underset{\sim}{R}^N \overset{+}{\underset{\sim}{H}} , \tag{3.13}$$

and with the assumption of a linear material behaviour in the increments $\Delta \underset{\sim}{S} = \mathbb{L}[\Delta \underset{\sim}{E}]$, principle of virtual work takes the form

$$\sum_i \left[\int_{V_o^i} \mathrm{tr}(\delta \overset{+}{\underset{\sim}{H}}{}^T \mathbb{L}[\overset{+}{\underset{\sim}{H}}])\,dV + \int_{V_o^i} \mathrm{tr}(\delta \overset{+}{\underset{\sim}{H}}{}^T \overset{+}{\underset{\sim}{H}}\, \underset{\sim}{S}^N)\,dV \right] = \delta \Delta \underline{V}^T \Delta \underline{\overline{R}} \quad (3.14)$$

with summation over elements i.

Using Kirchhoff plate theory for combined bending and plane stress we get from the first term in (3.14) the wellknown matrix representation of the geometrically linear problem, namely the expression $\delta \overset{+}{\underline{v}}{}^{i\,T} \overset{+}{\underline{k}}{}^i \overset{+}{\underline{v}}{}^i$ for element i as will be explained in chapter 5. The row matrix $\overset{+}{\underline{v}}{}^i$ contains the local nodal displacements and rotations. The transformation into global coordinates

$$\overset{+}{\underline{v}}{}^i = \underline{T}^i \Delta \underline{v}^i \quad (3.15)$$

leads to the element stiffness matrix $\underline{k}^i$ in global coordinates

$$\underline{k}^i = \underline{T}^{i\,T}\, \overset{+}{\underline{k}}{}^i\, \underline{T}^i .$$

The calculation of transformation matrix $\underline{T}$ needs special considerations. The nodal displacements $\overset{+}{u}_k$ are transformed according to (3.12) by using

$$\underset{\sim}{R}^N = \underset{\sim}{n}_k \otimes \underset{\sim}{N}_k = R^N_{jk}\, \underset{\sim}{N}_j \otimes \underset{\sim}{N}_k \text{ with } R^N_{jk} = \underset{\sim}{N}_j \cdot \underset{\sim}{n}_k .$$

One gets

$$\Delta u_j = R^N_{jk}\, \overset{+}{u}_k \text{ or } \overset{+}{u}_k = R^N_{kj}\, \Delta u_j . \quad (3.16)$$

The local incremental derivatives in the nodes are assumed to be small so that we approximately get, see fig. 2,

$$\frac{\partial u_3}{\partial X_\alpha} = \tan \overset{+}{\phi}_{3\alpha} \doteq \cos(\tfrac{\pi}{2} - \overset{+}{\phi}_{3\alpha}) = \underset{\sim}{n}_3 \cdot \overset{+}{\underset{\sim}{n}}_\alpha \,(\alpha = 1,2). \quad (3.17)$$

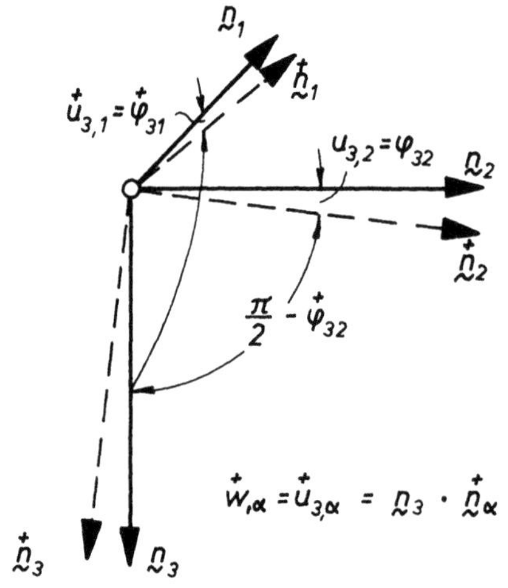

fig. 2. Local base vectors in states N and N+1

For the local rotational tensor $\overset{+}{\underset{\sim}{R}}$ it yields

$$\underset{\sim}{1} + \overset{+}{\underset{\sim}{H}} = \begin{bmatrix} 1 & \overset{+}{\phi} & -\overset{+}{w}_{,1} \\ -\overset{+}{\phi} & 1 & -\overset{+}{w}_{,2} \\ \overset{+}{w}_{,1} & \overset{+}{w}_{,2} & 1 \end{bmatrix} \underset{\sim}{N}_i \otimes \underset{\sim}{N}_j \approx \overset{+}{\underset{\sim}{R}} = (\underset{\sim}{n}_i \cdot \overset{+}{\underset{\sim}{n}}_j)\underset{\sim}{N}_i \otimes \underset{\sim}{N}_j$$

with the average inplane rotation $\overset{+}{\phi} = \frac{1}{2}(\overset{+}{u}_{1,2} - \overset{+}{u}_{2,1}) \approx 0$.
For the increments with respect to global directions we get

$$\underset{\sim}{R}^{N+1} - \underset{\sim}{R}^{N} = \begin{bmatrix} \cdot & \Delta a_1 & \cdot \\ \cdot & \cdot & \Delta a_2 \\ \Delta a_3 & \cdot & \cdot \end{bmatrix} \underset{\sim}{N}_i \otimes \underset{\sim}{N}_j$$

with $\Delta a_1 = \underset{\sim}{N}_1 \cdot (\overset{+}{\underset{\sim}{n}}_2 - \underset{\sim}{n}_2)$, $\Delta a_2 = \underset{\sim}{N}_2 \cdot (\overset{+}{\underset{\sim}{n}}_3 - \underset{\sim}{n}_3)$ and $\Delta a_3 = \underset{\sim}{N}_3 \cdot (\overset{+}{\underset{\sim}{n}}_1 - \underset{\sim}{n}_1)$. These rotational increments are calculated with the transformation

$$\underset{\sim}{R}^{N+1} - \underset{\sim}{R}^{N} = \underset{\sim}{R}^{N}\,\overset{+}{\underset{\sim}{H}}. \tag{3.18}$$

The components are

$$\begin{aligned}
\Delta a_1 &= R^N_{11}\stackrel{+}{\phi} + R_{13}\stackrel{+}{w}_{,2} \\
\Delta a_2 &= -R^N_{21}\stackrel{+}{w}_{,1} - R^N_{22}\stackrel{+}{w}_{,2} \qquad (3.19)\\
\Delta a_3 &= -R^N_{32}\stackrel{+}{\phi} + R^N_{33}\stackrel{+}{w}_{,1} \;.
\end{aligned}$$

So, the transformation matrix $\underline{T}^i$ for each element i can be computed. The initial values for state N+1 are obtained using the nodal displacements $\underset{\sim}{u}^N + \Delta\underset{\sim}{u}$ for the calculation of $\underset{\sim}{R}^{N+1}$, not (3.18).

The solution of systems of linear algebraic equations for the increments of displacements and rotations needs special techniques for postcritical states. Riks [1o, 11] has given a modification of Newton-Raphson iteration using a normal to the tangent in each load-displacement diagram. So the equation-matrix remains regular beyond the snapping points. Wessels [12] has developed a similar method with symmetric matrices, and he has obtained good results for postcritical buckling of spherical shells.

4. The elasto-plastic deformation of a plate in bending

A closed solution of elasto-plastic bending of plates for ductile metals (J_2-material with isotropic hardening) is only possible for special cases. This results from the fact that the yield condition etc are formulated in stresses and strains whereas in the plate theory we need integrated quantities like bending moments and curvatures. Former investigations are restricted either to the simpler problem of brittle metal materials (Mohr, Coulomb, Tresca) or to approximated yield conditions for the bending moments (Bäcklund [13], Eggers, Kröplin [14] e.a.). In the first of the following calculation methods for ductile materials we use a proposal of Pflüger [15, 16] where a "smearing" over the thickness of the plate is also used. In the second method given here, the first method is improved in a sense that the Prandtl-Reuß material law is fulfilled by a post-iteration of equilibrium conditions.

4.1 The assumption of Pflüger is that the plastic strain-rates are proportional to the total deviatoric strain-rates

$$\dot{\tilde{E}}^{pl} = \eta \dot{\tilde{E}}^{D} \tag{4.1}$$

with

$$\dot{\tilde{E}}^{D} := \dot{\tilde{E}} - \frac{1}{3}\,\mathrm{tr}(\dot{\tilde{E}})\,\tilde{1}.$$

The Prandtl-Reuß material equation is fulfilled by Pflüger's assumption if the stress deviator is proportonial to its own rate, $\dot{\tilde{S}}^D = \kappa \tilde{S}^D$. For the example of a rectangular plate, see fig. 3, the directions of main stresses and the stress rates coincide in regions which are responsible for the buckling behaviour.

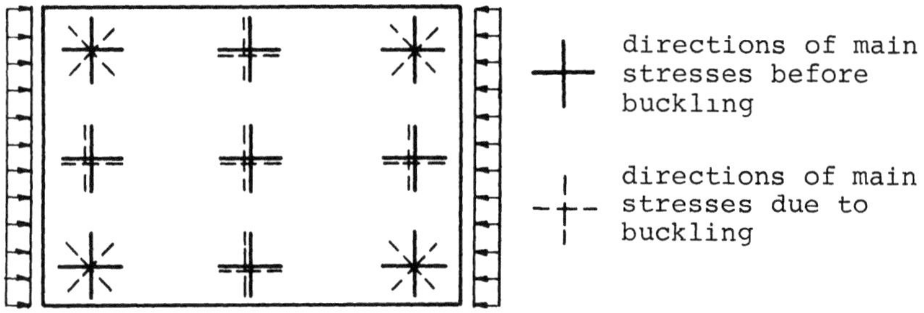

fig. 3. Buckling of a simply supported rectangular plate

The proportional factor η in (4.1) can be derived from the relations

$$\eta = \eta_o \Phi \ , \tag{4.2}$$

$$\Phi = \frac{\mathrm{tr}(\tilde{S}^D \dot{\tilde{S}}^D)}{\left[\mathrm{tr}(\tilde{S}^D \tilde{S}^D)\ \mathrm{tr}(\dot{\tilde{S}}^D \dot{\tilde{S}}^D)\right]^{1/2}} = \cos\alpha \ , \tag{4.3}$$

$$\eta_o = \frac{3(E - T)}{3E - (1-2\nu)T} \ . \tag{4.4}$$

Here E is Young's modulus, T the tangent modulus in the uniaxial tension test, and α the angle between the change of stress and the outer normal at the yield surface, see fig. 4.

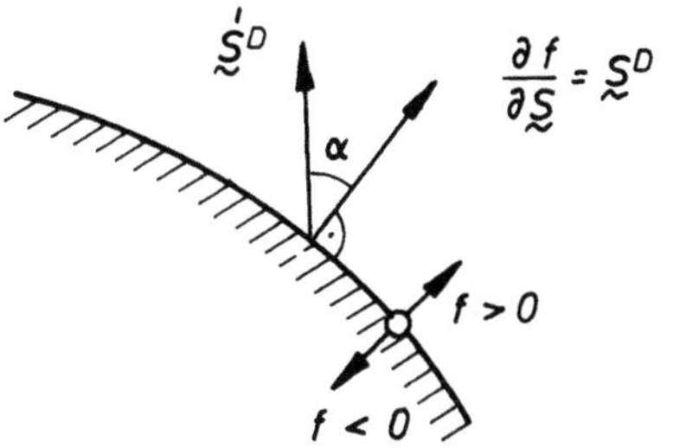

yield surface f = 0

fig. 4. Yield surface, outer normal and angle α

For a neutral change of stresses we get $\alpha = \frac{\pi}{2}$, and therefore $\dot{\tilde{E}}^{pl} = \tilde{0}$. The relations (4.2 - 4.4) lead after some transformations to modified elasticity constants, compare Pflüger [15] ,

$$\overset{*}{E} = \frac{3(1-\eta)}{2(1+\nu) + (1-\eta)(1-2\nu)} E , \qquad (4.5)$$

$$\overset{*}{\nu} = \frac{1+\nu - (1-\eta)(1-2\nu)}{2(1+\nu) + (1-\eta)(1-2\nu)}$$

These constants change for each load increment so that the plastifying plate is calculated as an elastic system in each step from N to N+1. An evaluation of (4.4) is given in fig.5. One can see the weak change of η_o with respect to ν. For $\nu = 0,5$, (4.4) is reduced to $\eta_o = 1 - \frac{T}{E}$. In case of an affine change of stress ($\Phi = 1$) we finally get

$$\eta = \eta_o = 1 - \frac{T}{E}. \qquad (4.7)$$

For the case $\Phi = 1$ we also refer to a representation of Lehmann [17] .

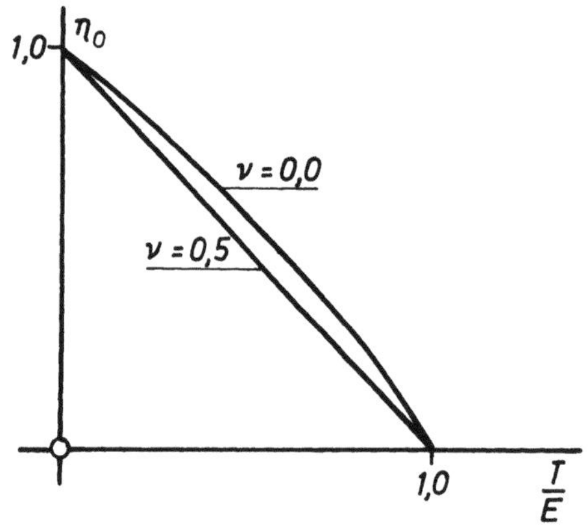

fig. 5. η_o in dependence of $\frac{T}{E}$ and ν

The values $\overset{*}{E}$ and $\overset{*}{\nu}$ are variable over the thickness and the coordinates of the middle surface as well.

Using FE-techniques a Gaussian-point-integration is recommendable. The average values for a Gaussian-point can be calculated by the help of n layer-points, see fig. 6. From the values $\overset{*}{E}$ and $\overset{*}{\nu}$ for each layer-point one can calculate a weighted average. The weighting coefficients can be obtained from a beam model.

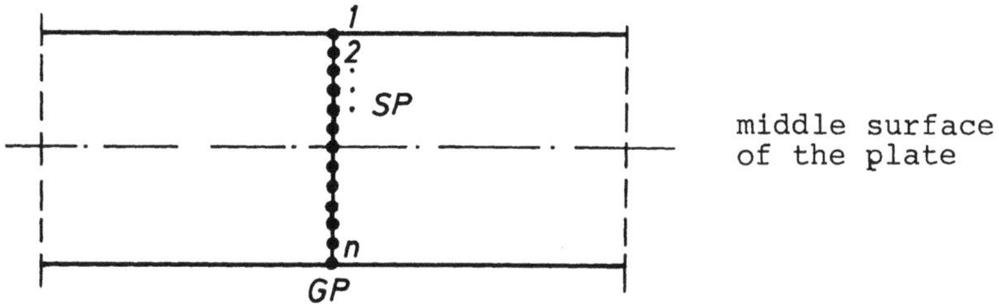

fig. 6. n layer-points, LP's, at the normal of the middle surface for one Gaussian-point GP

4.2 The second method is based on Prandtl-Reuß's constitutive equation which is fulfilled iteratively using the first method. For pure bending this procedure was described by Paulun [18] within a geometrically linear theory. We want the fictitious load increment $\Delta\underline{\overset{*}{R}}$ which effects the same displacement increments $\Delta\underline{V}$ for the "elastic" plate with $\overset{*}{E}$ and $\overset{*}{\nu}$ as the real load increment $\Delta\underline{\bar{R}}$ acting on the elasto-plastic plate. $\Delta\underline{\overset{*}{R}}$ is calculated with the following iteration scheme (j: iteration index):

a) $\Delta\underline{\overset{*}{R}}_j$ from h), initial values $\Delta\underline{\bar{R}}$,
b) $\Delta\underline{V}_j$ with $\Delta\underline{\overset{*}{R}}_j$ using the first method according to 4.1.
c) strain- and curvature increments of the middle surface, using $\Delta\underline{V}_j$ and initial values depending from the interpolation functions of the FE-model ,
d) strain increments in the layer-points from displacement increments of the middle surface,
e) stress increments from strain increments and initial values due to Prandtl-Reuß equation,
f) bending moment- and normal force increments by numerical integration of stress increments over the layer-points,
g) load increment $\Delta\underline{\bar{R}}_j$ by Gaussian integration of the increments f), transformation and assembling of the element terms,
h) correction of fictitious load increments

$$\Delta\underline{\overset{*}{R}}_{j+1} = \Delta\underline{\overset{*}{R}}_j - (\Delta\underline{\bar{R}} - \Delta\underline{\bar{R}}_j),$$

and back to a).

The iteration is finished if $\|\Delta\underline{\bar{R}} - \Delta\underline{\bar{R}}_j\| < \varepsilon$ for given $\varepsilon<<1$ is fulfilled. After the iteration we calculate the final increments of comparative plastic strains in each layer and add them to the previous values.

5. Comprehensive representation of the incremental- iterative process

We use matrix notation for the example of a 12-parameter rectangular plane bending element and a 8-parameter rectangular plane stress element. The incremental nodal displacements and rotations are denoted in fig. 7. We consider the "elastic" plate with $\overset{*}{E}$ and $\overset{*}{\nu}$ and use the work principle in the form (3.14). Matrix notation of the lefthandside yields

$$\delta \Delta \underline{V}^T \underline{K}\ \Delta \underline{V} = \sum_i \delta \overset{+}{\underline{v}}{}^{i\,T} \overset{+}{\underline{k}}{}^{i}\ \overset{+}{\underline{v}}{}^{i} \tag{5.1}$$

with the global stiffness matrix $\underline{K}$.

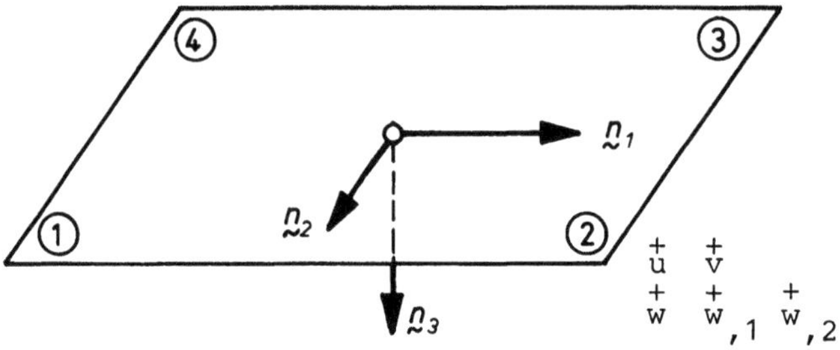

fig. 7. Rectangular element with 24 parameters (nodal displacements and rotations)

The element oriented inner virtual works $\delta \overset{+}{\underline{v}}{}^{i\,T} \overset{+}{\underline{k}}{}^{i}\ \overset{+}{\underline{v}}{}^{i}$ can be developed very similar to the geometrically linear elastic case, see Stein, Wunderlich [19], e.g. Putting all kinematic variables together in vector $\overset{+}{\underline{h}}$ we get the integral of the inner virtual work for an element i $\int \delta \overset{+}{\underline{h}}{}^{T} \overset{*}{\underline{E}}\ \overset{+}{\underline{h}}\ dA$, with $\overset{+}{\underline{h}} = \{\overset{+}{u}_{,1}\ \overset{+}{v}_{,2}\ \overset{+}{u}_{,2}\ \overset{+}{v}_{,1}\ w_{,11}\ w_{,22}\ \overset{+}{w}_{,12}\ \overset{+}{w}_{,21}\}$,

$$\overset{*}{\underline{E}} = \begin{bmatrix} t\overset{*}{\underline{C}}_1 & \underline{O} \\ \underline{O} & \frac{t^3}{12}\overset{*}{\underline{C}}_2 \end{bmatrix}, \quad \overset{*}{\underline{C}}_1 = \overset{*}{\underline{C}}_2 = \overset{*}{\underline{C}}_P + \overset{*}{\underline{C}}_G \tag{5.2}$$

and the plate thickness t.

$$\overset{*}{\underline{C}}_P = \frac{\overset{*}{E}}{1-\overset{*}{\nu}} \begin{bmatrix} 1 & \overset{*}{\nu} & 0 & 0 \\ \overset{*}{\nu} & 1 & 0 & 0 \\ 0 & 0 & \frac{1-\overset{*}{\nu}}{2} & 0 \\ 0 & 0 & 0 & \frac{1-\overset{*}{\nu}}{2} \end{bmatrix}; \quad \overset{*}{\underline{C}}_G = \begin{bmatrix} \sigma^N_{11} & 0 & \sigma^N_{12} & 0 \\ 0 & \sigma^N_{11} & 0 & \sigma^N_{21} \\ \sigma^N_{21} & 0 & \sigma^N_{22} & 0 \\ 0 & \sigma^N_{12} & 0 & \sigma^N_{22} \end{bmatrix}$$

influence of physical nonlinearity — influence of geometrical nonlinearity

From the decoupling of bending and plane stress follows that $\underline{\overset{*}{C}}$ does not change along the plate thickness. The fictitious "elastic" plate only yields the initial displacements for the iteration so that no mistake comes out from the above separation.

The discretization $\underline{\overset{+}{u}} = \underline{\Omega}(X_1,X_2)\underline{\overset{+}{v}}$ with the unit displacement states $\underline{\Omega}$ and 20 nodal displacements in $\underline{\overset{+}{v}} = \{\underline{\overset{+}{v}}_1\ \underline{\overset{+}{v}}_2\ \underline{\overset{+}{v}}_3\ \underline{\overset{+}{v}}_4\}$; $\underline{\overset{+}{v}}_k = \{\overset{+}{u}_k\ \overset{+}{v}_k\ \overset{+}{w}_k\ \overset{+}{w}_{,1}\ \overset{+}{w}_{,2}\}$ leads to the relation

$$\underline{\overset{+}{h}}^i = \underline{H}^i\underline{\overset{+}{v}}^i \tag{5.3}$$

with submatrices for bending and plane stress in $\underline{H}^i$ according to [19]. The transformation into global coordinates, namely

$$\underline{\overset{+}{v}}^i = \underline{T}^i\Delta\underline{v} \tag{5.4}$$

follows from (3.16) and (3.19). For a node k the transformation matrix has the form, see (3.16) and (3.19)

$$\begin{bmatrix} \overset{+}{u} \\ \overset{+}{v} \\ \overset{+}{w} \\ \overset{+}{w}_{,1} \\ \overset{+}{w}_{,2} \end{bmatrix} = \left[\begin{array}{ccc|ccc} R^N_{11} & R^N_{21} & R^N_{31} & & & \\ R^N_{12} & R^N_{22} & R^N_{32} & & \underline{O} & \\ R^N_{13} & R^N_{23} & R^N_{33} & & & \\ \hline & & & \dfrac{R^N_{22}\,R^N_{32}}{N} & \dfrac{R^N_{13}\,R^N_{32}}{N} & \dfrac{R^N_{11}\,R^N_{22}}{N} \\ & \underline{O} & & \dfrac{-R^N_{21}\,R^N_{32}}{N} & \dfrac{-R^N_{11}\,R^N_{33}}{N} & \dfrac{-R^N_{11}\,R^N_{21}}{N} \end{array}\right] \begin{bmatrix} \Delta u_1 \\ \Delta u_2 \\ \Delta u_3 \\ \Delta a_1 \\ \Delta a_2 \\ \Delta a_3 \end{bmatrix} \tag{5.5}$$

with $N = R^N_{11}\,R^N_{22}\,R^N_{33} - R^N_{13}\,R^N_{32}\,R^N_{21}$.

N can vanish in special cases of large rotations. Then, instead of Δa_i, other relevant components of the incremental rotation tensor (3.18) with the corresponding transformation matrix are chosen.

Additional nodal variables, especially higher kinematic quantities (like $\overset{+}{w}_{,12}$ for the 16-parameter bending element) need further consideration for transformation and assembling. One can average or introduce torsional springs, e.g.

Considering plastic material behaviour according to chapter 4.1, we have to store the stresses $\underline{\sigma}^N = \{\sigma_{11}^N \ \sigma_{22}^N \ \sigma_{12}^N\}$ of each layer of the plate, see fig. 6., furthermore the comparative plastic strain ε_v^{pl} and the actual yield stress Y of state N. With the tangent modulus $T = T(\varepsilon_v^{pl})$ out of the characteristic material line we get $\overset{*}{E}$ with (4.5) and $\overset{*}{\nu}$ with (4.6) for each layer point. For one Gaussian point with n layer points we can calculate the average values

$$(E,\nu) = \sum_{k=1}^{n} (\kappa_k E_k \ , \ \kappa_k \nu_k) \tag{5.6}$$

with the weighting factors κ_k $(k = 1,\dots,n)$. The weighting factors are determined seperately for plane stress and bending and lead to the matrices $\overset{*}{\underline{C}}_1$ and $\overset{*}{\underline{C}}_2$, see (5.2), for each Gaussian point (GP).

After calculating the element stiffness matrices, assembling and solution of the system of algebraic equations we have the incremental nodal displacements and rotations, and finally with (3.15), (5.2), (5.3) the local incremental displacement gradient, the changes of curvatures and the membrane forces and bending moments for each Gaussian point.

With the strain increments for a layer-point k

$$\Delta\underline{\varepsilon}_k = \begin{bmatrix} \Delta\varepsilon_{11} \\ \Delta\varepsilon_{22} \\ 2\Delta\varepsilon_{12} \end{bmatrix}_k = \begin{bmatrix} 1 & 0 & 0 & 0 & X_{3k} & 0 & 0 & 0 \\ 0 & 1 & 0 & 0 & 0 & X_{3k} & 0 & 0 \\ 0 & 0 & 1 & 1 & 0 & 0 & X_{3k} & X_{3k} \end{bmatrix} \overset{+}{\underline{h}}_{GP} \tag{5.7}$$

we get the stress increments using Prandtl-Reuß equation

$$\Delta\underline{\sigma}_k = \left[\underline{E} - \frac{\underline{E}\ \underline{s}\ \underline{s}^T\underline{E}}{\frac{4\zeta Y^2}{9} - \underline{s}^T\underline{E}\ \underline{s}}\right]_k \Delta\underline{\varepsilon}_k \tag{5.8}$$

with the elasticity matrix $\underline{E}$, the plastic tangent modulus $\zeta = \frac{ET}{E-T}$ and the row matrix $\underline{s}$ of the deviatoric stresses

$$\underline{s} = \underline{A}\ \underline{\sigma}\ ;\ \underline{A} = \frac{1}{3}\begin{bmatrix} 2 & -1 & 0 \\ -1 & 2 & 0 \\ 0 & 0 & 6 \end{bmatrix}. \tag{5.9}$$

The plastic strain increment $\Delta\varepsilon^{pl} = \{\Delta\varepsilon^{pl}_{11}\ \Delta\varepsilon^{pl}_{22}\ \Delta\varepsilon^{pl}_{12}\}$ is calculated as

$$\Delta\underline{\varepsilon}^{pl} = \frac{9}{4\zeta Y^2}\ \underline{s}\ \underline{s}^T\Delta\underline{\sigma} \tag{5.1o}$$

and yields the increment of comparative plastic strain

$$\Delta\varepsilon^{pl}_v = \left[\frac{4}{3}(\Delta\varepsilon^{pl^2}_{11} + \Delta\varepsilon^{pl^2}_{22} + \Delta\varepsilon^{pl}_{11}\Delta\varepsilon^{pl}_{22} + \Delta\varepsilon^{pl^2}_{12})\right]^{1/2}. \tag{5.11}$$

With these datas the determination of initial values for the next load increment is possible.

Following the more accurate method according to chapter 4.2, we have to compute the membrane and bending forces $\overset{+}{\underline{n}} = \int_{(t)} \Delta\underline{\sigma}dz$ and $\overset{+}{\underline{m}} = \int_{(t)} \Delta\underline{\sigma}zdz$ by numerical integration over the plate thickness. Furthermore fictitious nodal forces $\overset{+}{\underline{r}}{}^i$ of element i (acting according to the nodal displacements $\overset{+}{\underline{v}}{}^i$) are determined with principle of virtual work as follows

$$\delta\overset{+}{\underline{v}}{}^{i^T}\overset{+}{\underline{r}}{}^i = \delta\overset{+}{\underline{v}}{}^{i^T}\int_{A^i}\underline{H}^{i^T}\begin{bmatrix}\overset{+}{\underline{n}}\\ \overset{+}{\underline{m}}\end{bmatrix}dA \tag{5.12}$$

with a Gaussian integration on the righthand side. After transformation into global directions

$$\Delta\underline{r}^i = \underline{T}^{i^T}\overset{+}{\underline{r}}{}^i \tag{5.13}$$

and assembling the element terms in each node we get the global load increment $\Delta\bar{R}_J$ corresponding to step g) in the iteration procedure, chapter 4.2. During the iteration one can simultaneously improve the material matrix in (5.8), using adequate average values for $\underline{s}$, ζ, Y from the last iteration loop.

6. Examples

In the first example, fig. 8, membrane stresses and horizontal displacements of the middle surface of a simply supported square plate under uniform load are calculated. 144 square elements in a quarter of the plate were applied. These elements have 4 nodes with 5 DOF (degrees of freedom) in each node. The compression stresses in a ringlike outer region effect the supporting of the inner tensioned plate in states of large deflections. Combined with this behaviour we see horizontal displacements of the support lines which is known as the pillow effect.

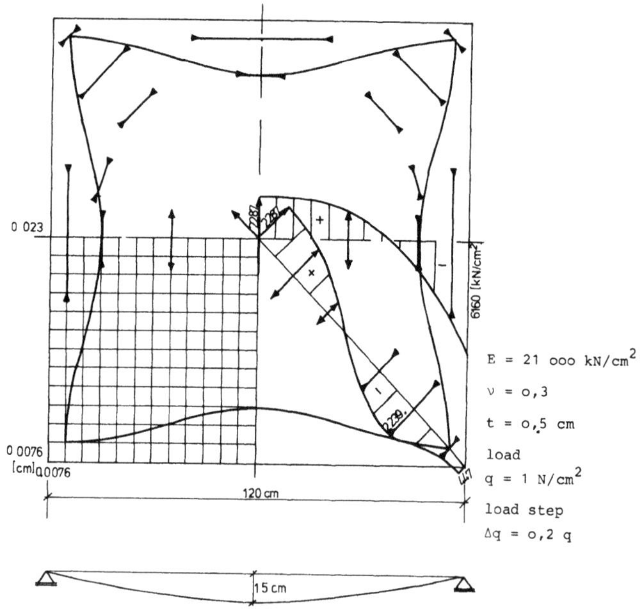

fig. 8: simply supported square plate under uniform load

In the second example, fig. 9, a simply supported rectangular plate under in-plane load and dead weight is considered. 1o8 square elements (6 x 18) as used in the first example were applied. The deflections of point A and B of the obtained buckling shape, fig. 1o, are compared with results by Girkmann [21] , see fig. 11.

$E = 21000\ kN/cm^2$	$a = 300$ cm
$\nu = 0.3$	$b = 100$ cm
$I_{St} = 19.0\ cm^4$ girder	$t = 1.4$ cm
$p = \sigma_o/25000$ bending load	
$\sigma_o t = 7.0 - 20.3$ kN/cm in-plane load	

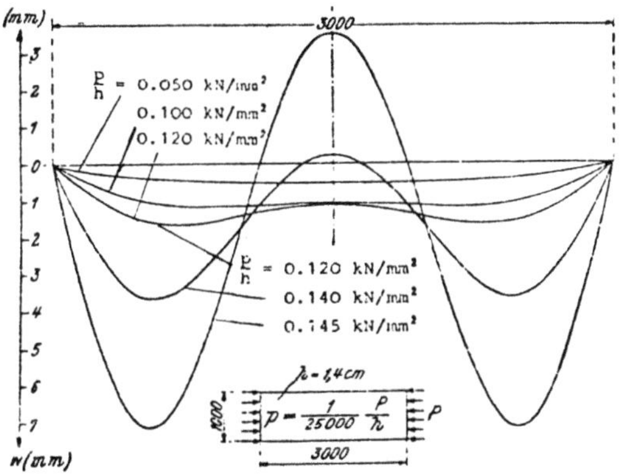

fig. 9: Example and results by Girkmann for the unstiffened plate

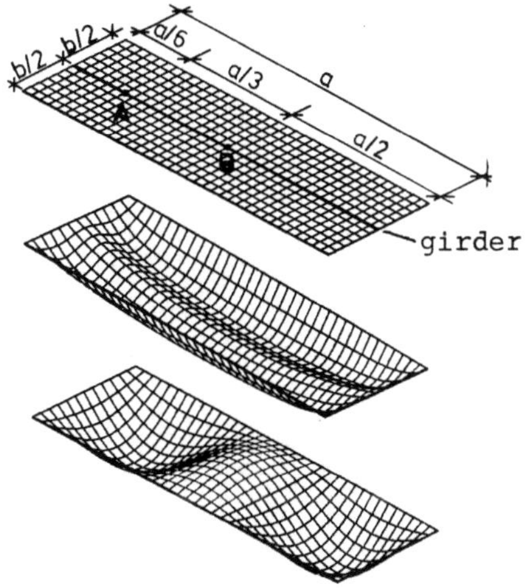

fig. 1o: Plots of the undeformed the deformed stiffened and the buckled unstiffened plate

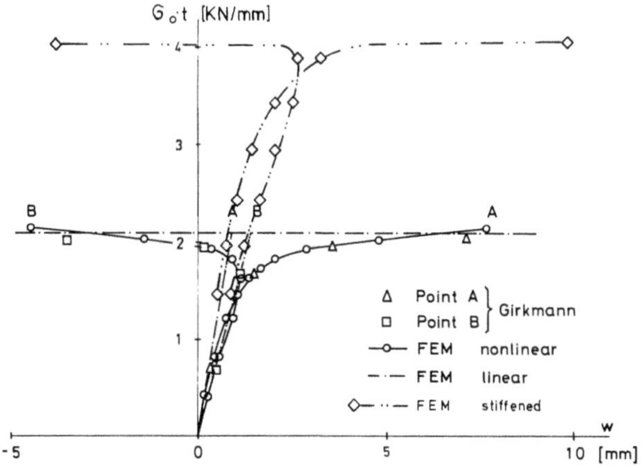

fig. 11: Deflection of point A and B (fig. 1o) under given loads $\sigma_o t$

The improved plastic calculation concept is applied in the third example, fig. 12. Triangular 21-parameter elements with 6 DOF in the corners and 4 layer-points over half of the thickness were applied. The increasing of yielding domains over the plate area and thickness is calculated due to uniform load normal to the plate applying 8 load steps. The lower part of fig. 12 shows the increasing of yielding zones over the thickness (in the upper half) for a diagonal section. One can recognize the wellknown fact that plastifying begins in the corners of the plate and from the 4th load step on also in the middle. The results correspond to those given by Laudiero, Tralli [2o].

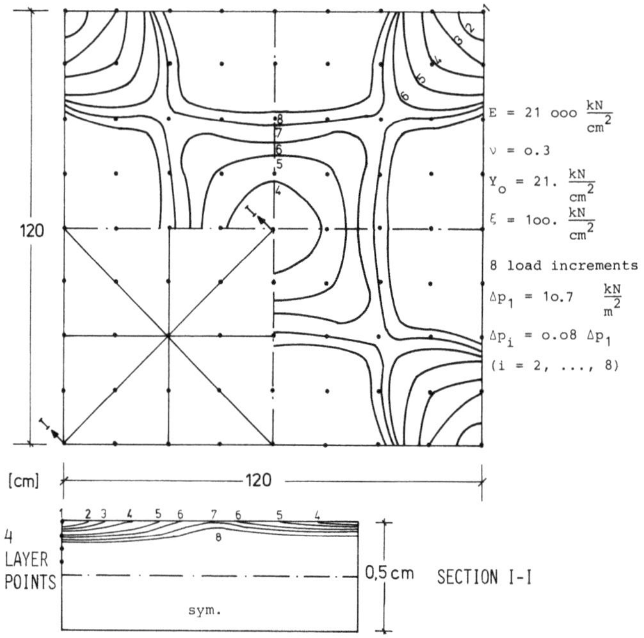

fig. 12 simply supported square plate under uniform load. Extending of the yield-domain during increasing load.

7. Conclusions

The incremental numerical concept for the calculation of elastic-plastic deformations of plates in bending and compression given in this paper has been proved to be problem-adequate and effective. The assumption of large displacements and rotations of the middle surface of the plate but of small strains is valid for elasto-plastic deformations up to ultimate loads.

In two examples correspondence with known results can be shown. In the third example which treats the extension of yielding zones only the tendency of agreement with the cited paper is possible because no comparable results were available where Prandtl-Reuss equations are fulfilled exactly for v. Mises material over the whole thickness of the plate.

An optimization of the number of layer- and Gauss-points for some important elements has still to be done. This needs further practical calculations with the method.

References

1 Lehmann, Th.: Einige Bemerkungen zu einer allgemeinen Klasse von Stoffgesetzen für große elasto-plastische Formänderungen, Ing. Arch. 41, 297 - 31o (1972).

2 Truesdell, C.: Hypo-elasticity, J. Rat. Mech. Anal. 4, 83 - 133 (1955).

3 Bathe, K.-J.; Ramm, E.; Wilson, E. L.: Finite element formulation for large deformation dynamic analysis, Int. J. Num. Meth. Eng., 9, 353 - 386 (1975).

4 Bathe, K.-J.; Borlourchi, S.: A geometric and material nonlinear plate and shell element, Int. J. Comp. Struct., 11, 23 - 48 (198o).

5 Washizu, K.: Variational Methods in Elasticity and Plasticity, 2nd ed., Pergamon Press (1975).

6 Atluri, S.: On the hybrid stress finite element model for incremental analysis of large deflection problems, Int. J. Sol. Struct. 9, 1177 - 1191 (1973).

7 Larsen, P. K.; Popov, E. P.: Large displacement analysis of viscoelastic shells of revolution, Comp. Meth. Appl. Eng. 3, 237 - 253 (1974).

8 Paulun, J.: Plastische Inkompressibilität bei großen Formänderungen, Acta Mechanica 37, 43 - 51 (198o).

9 Klee, K.-D.; Paulun, J.: On numerical treatment of large elastic-viscoplastic deformations, Archives of Mechanics 3, 333 - 345, (198o).

1o Riks, E.: The application of Newton's method to the problem of elastic stability, J. Appl. Mech. 39, 1o6o - 1o65 (1972).

11 Riks, E.: The incremental Solution of some Problems in Elastic Stability, Nat. Aerospace Lab. NRL TR 74oo5 U, The Netherlands (1973).

12 Wessels, M.: Das statische und dynamische Durchschlagproblem der imperfekten flachen Kugelschale bei elastischer rotationssymmetrischer Verformung, Mitt. Inst. f. Statik Nr. 23, Universität Hannover (1977).

13 Bäcklund, J.: Mixed finite element analysis of elastoplastic plates in bending, Archives of Mechanics, 24, 319 - 335 (1972).

14 Eggers, H.; Kröplin, B.: Yielding of plates with hardening and large deformations, Int. J. Num. Meth. Eng., 12, 739 - 75o (1978).

15 Pflüger, A.: Zur plastischen Beulung von Flächenträgern, ZAMM 47, T 2o9 - T 211 (1967).

16 Pflüger, A.: Zur plastischen Beulung der Rechteckplatte, Ing. Arch. 41, 258 - 269 (1972).

17 Lehmann, Th.: Zur Beschreibung großer plastischer Formänderungen unter Berücksichtigung der Werkstoffverfestigung, Rheologica Acta 2/3, 247 - 254 (1962).

18 Paulun, J.: Zur numerischen Berechnung elastoplastischer Plattenbiegung, ZAMM 6o, T 144 - 146 (198o).

19 Stein, E.; Wunderlich, W.: Finite-Element Methode als direkte Variationsverfahren, in Finite Elemente in der Statik, W. Ernst & Sohn, Berlin (1973).

2o Laudiero, F.; Tralli, A.: Finite element incremental analysis of elastoplastic plate bending, Meccanica 8, 19o - 2o2 (1973).

21 Girkmann, K.: Traglasten gedrückter und zugleich querbelasteter Stäbe und Platten, Der Stahlbau 15, 57 - 61 (1942).

A Study of Some Generalized Constitutive Models for Elasto-Plastic Shells

O. M. EIDSHEIM, P. K. LARSEN
The Norwegian Institute of Technology, Trondheim, Norway

Abstract

Constitutive laws for elasto-plastic shells based on strains and curvatures of the middle surface and the corresponding stress resultants and stress couples are investigated. The stress-strain relationships are obtained using the classical theory of plasticity for work-hardening material, expressed in generalized variables. Two expressions for the consequtive yield - condition are studied, and modified to improve the behaviour. The models are verified against results obtained by numerical integration through the thickness for the special case of bending and stretching of beams.

Introduction

During the last decade our capability of predicting the load-deformation behaviour of large engineering structures has been significantly improved. Computer codes based on the finite element method and the finite difference method, incorporating both the effects of large displacements/rotations and material nonlinearities, are available for use in the analysis. However, in practical design work the computational cost and computer storage requirement may render these codes intractable. Extensive research efforts are hence directed towards the improvement of the codes with respect to computational efficiency.

Two alternative methods are available when describing the nonlinear material behaviour in shells. In the first approach the constitutive equations in terms of stresses and strains are applied to shell layers and integrated through the thickness. In order to reduce the computer storage requirement and reduce the computational effort the second approach formulates the constitutive law in terms of middle surface strains and curvatures and their conjugated stress resultants. Such constitutive laws have been proposed by Crisfield /1/, Bieniek and Funaro /2/, and Eggers and Kröplin /3/.

While reducing the computational effort, the latter approach requires additional assumptions regarding the stress distribution and the propagation of the plastic zone over the shell thickness. The subject of the present discussion is the accuracy and the efficiency of such generalized constitutive models. For the sake of simplicity the discussion is restricted to the case of infinitesimal displacements and rotations.

Variational formulation for elasto-plastic shells

In the context of a hybrid stress finite element model the governing equations of the discretized system are derived from an incremental variational principle of the Hellinger-Reissner type /4/. Introducing the incremental complementary energy function ΔU_c and relaxing the interelement continuity requirements the following functional is defined /5/

$$\Delta\pi = \sum_n \{-\int_{A_n} \Delta U_c(N_{\alpha\beta},M_{\alpha\beta})dA - \int_{A_n} \Delta u^T \Delta\bar{F} dA - \int_{S_{\sigma_n}} \Delta u^T \Delta\bar{T} ds$$
$$+ \int_{\partial A_n} \Delta T^T \Delta\tilde{u} ds\} \tag{1}$$

$N_{\alpha\beta}$ and $M_{\alpha\beta}$ are here the generalized stresses (stress resultants), and $\tilde{u}$ and T are the vectors of boundary displacements and tractions respectively. Prescribed values are "bared", and the summation is extended over all elements.

For a discussion of the actual element model reference is given to /4,5/. With respect to the material behaviour only the complementary energy function is of further interest, and for the present purpose ΔU_c is taken on the form

$$\Delta U_c = \tfrac{1}{2}(\Delta M^T \mathbf{D}_{mm} \Delta M + \Delta M^T \mathbf{D}_{mn} \Delta N + \Delta N^T \mathbf{D}_{nm} \Delta M + \Delta N^T \mathbf{D}_{nn} \Delta N) \tag{2}$$

where

$$\Delta M^T = (\Delta M_{11}, \Delta M_{22}, \Delta M_{12}) \; ; \qquad \Delta N^T = (\Delta N_{11}, \Delta N_{22}, \Delta N_{12})$$

The elasto-plastic compliance matrix $\mathbf{D}$ is given by

$$\mathbf{D} = \begin{bmatrix} \mathbf{D}_{nn} & \mathbf{D}_{nm} \\ \mathbf{D}_{mm} & \mathbf{D}_{mm} \end{bmatrix}$$

Note that Eqs. (1) and (2) are based on Love's first order shell theory, and that no explicite dependence on the transverse shear forces is indicated in Eq. (2).

Generalized constitutive models

Beyond the assumption of the existence of a compliance relationship the preceding variational functional was derived without further knowledge of the constitutive law of the material at hand. When applied to elasto-plastic shells the compliance matrix $\mathbf{D}$ relating increments of membrane forces and moments to increments of strains and curvatures is needed.

Elastic materials

For elastic materials the compliance relationship is given by Hooke's law, which on matrix form is written as

$$\begin{Bmatrix} d\varepsilon^e \\ d\kappa^e \end{Bmatrix} = \begin{bmatrix} \mathbf{D}^e_{nn} & 0 \\ 0 & \mathbf{D}^e_{mm} \end{bmatrix} \cdot \begin{Bmatrix} d\mathbf{N} \\ d\mathbf{M} \end{Bmatrix} = \begin{bmatrix} \mathbf{D}^e \end{bmatrix} \begin{Bmatrix} d\mathbf{N} \\ d\mathbf{M} \end{Bmatrix} \tag{3}$$

with

$$D^e_{nn} = \frac{1}{h} D_o \; ; \qquad D^e_{mm} = \frac{12}{h^3} D_o$$

and

$$D_o = \frac{1}{E} \begin{bmatrix} 1 & -\nu & o \\ -\nu & 1 & o \\ o & o & 2(1+\nu) \end{bmatrix}$$

Flow theory of plasticity

The plastic part of the compliance relationship is derived from the flow theory of plasticity in terms of generalized stresses. The development of the theory parallels that of the classical infinitesimal theory, incorporating the concept of additive decomposition of elastic and plastic deformations. The yield criterion takes the form

$$f(N_{\alpha\beta}, M_{\alpha\beta}, \varepsilon^P_{\alpha\beta}, \kappa^P_{\alpha\beta}, \kappa) = 0 \qquad (\alpha = 1,2)$$

where κ is a hardening parameter, the flow rule becomes

$$\begin{Bmatrix} d\varepsilon^P_{\alpha\beta} \\ d\kappa^P_{\alpha\beta} \end{Bmatrix} = d\lambda \begin{Bmatrix} \dfrac{\partial f}{\partial N_{\alpha\beta}} \\ \dfrac{\partial f}{\partial M_{\alpha\beta}} \end{Bmatrix} \qquad (4\ a,b)$$

The consistency condition and the loading/unloading criterion are derived from the yield criterion as in the classical theory.

Analogous to the classical theory it is necessary to develope an initial yield criterion describing the onset of plastic deformation in an infinitesimal shell element. The hardening rule now describes the subsequent yield surfaces as the plastic zone propagates throughout the element. Note that this hardening is related to the propagation of the plastic zone, and occurs even for elastic-perfectly plastic material. Finally, for materials of the latter type and for materials where the stress-strain curve goes asymptotically toward a limiting value a limit yield surface is defined, at which the entire section is fully plastified.

Initial and limit yield surface

For elastic shells the stress distribution over the thickness is linear, and the extreme fiber stress is

$$\sigma_{\alpha\beta} = \frac{N_{\alpha\beta}}{h} \pm \frac{6M_{\alpha\beta}}{h^2} \tag{5}$$

Introduction of Eq. (5) into the von Mises yield condition gives the following condition for initial yielding in the extreme fiber

$$\left(\frac{\bar{N}^2}{N_0^2} + \frac{\bar{M}^2}{M_0^2} \pm 2\,\frac{\overline{NM}}{N_0 M_0}\right)^{\frac{1}{2}} = 1 \tag{6}$$

The invariants of the generalized stresses are here defined by

$$\bar{N}^2 = N_{11}^2 + N_{22}^2 - N_{11}N_{22} + 3N_{12}^2 \tag{7a}$$

$$\bar{M}^2 = M_{11}^2 + M_{22}^2 - M_{11}M_{22} + 3M_{12}^2 \tag{7b}$$

$$\overline{NM} = N_{11}M_{11} + N_{22}M_{22} - \tfrac{1}{2}(N_{11}M_{22} + N_{22}M_{11}) + 3N_{12}M_{12} \tag{7c}$$

and

$$N_0 = h\sigma_0 \; ; \quad M_0 = \frac{1}{6}\,h^2\sigma_0 \tag{8}$$

σ_0 is here the uniaxial yield stress. Note that the effect of transverse shear stresses and the normal stress σ_{33} is neglected here. This is in accordance with the plane stress assumption of Love's first order theory.

The initial yield condition is given by Eq. (6) when the positive value of $\pm \overline{NM}$ is chosen

$$f_0 = F_0 - 1 = \left(\frac{\bar{N}}{N_0^2} + \frac{\bar{M}}{M_0^2} + 2\frac{|\overline{NM}|}{N_0 M_0}\right)^{\frac{1}{2}} - 1 = 0 \tag{9}$$

The most common of the limit yield conditions is that developed by Ilyushin /6,7/ for elastic-perfectly plastic materials

$$F_L = \left(\frac{\bar{N}^2}{N_L^2} + \frac{\bar{M}^2}{M_L^2} + \frac{1}{\sqrt{3}}\frac{|\overline{NM}|}{N_L M_L}\right)^{\frac{1}{2}} = 1 \tag{10}$$

where

$$N_L = h\sigma_0 \;; \qquad M_L = \frac{3}{2} M_0 = \frac{1}{4} h^2\sigma_0 \tag{11}$$

This criterion is based on von Mises' yield condition and the Kirchhoff-Love assumption, and neglects the effect of transverse shear stresses. It should be noted that Eq. (10) is an approximation, and is as such strictly valid only for bending dominant situations /7/.

Crisfield's subsequent yield surface

Crisfield's subsequent yield surface is derived for elastic-perfectly plastic materials, and has the form /1/

$$f_C = F_C - 1 = \left(\frac{\bar{N}^2}{N_L^2} + \frac{\bar{M}^2}{M^2} + \frac{1}{\sqrt{3}}\frac{|\overline{NM}|}{N_L M}\right)^{\frac{1}{2}} - 1 = 0 \tag{12}$$

where M models the moment-curvature relationship. For the uniaxial case the exact relationship is given by

$$M = \left(1 - \frac{1}{3}\left(\frac{\kappa_0}{\kappa}\right)^2\right) M_L \tag{13}$$

where $\kappa_0 = 2\sigma_0/(Eh)$. For a shell element in a multiaxial state of stress Crisfield gives M as a function of the equivalent plastic curvature $\bar{\kappa}^P$

$$M = a(\bar{\kappa}^P)M_L = \left(1 - \frac{1}{3}\cdot\exp\left(-\frac{8}{3}\frac{\bar{\kappa}^P}{\kappa_0}\right)\right)\cdot M_L \tag{14}$$

with

$$\bar{\kappa}_P = \int_0^{\bar{\kappa}^P} d\kappa^P = \int_0^{\bar{\kappa}^P} \frac{2}{\sqrt{3}} (d\kappa^{P^2}_{11} + d\kappa^{P^2}_{22} + d\kappa^P_{11} d\kappa^P_{22} + d\kappa^{P^2}_{12})^{\frac{1}{2}}$$

F_c approaches F_L in the limit when $\bar{\kappa}^P \to \infty$, while it is in error at initial yielding due to the coefficient $1/\sqrt{3}$ for the $|\overline{NM}|$ term. In order to improve the behaviour at initial yielding a modification of Eq. (12) on the form

$$F_c^* = (\frac{\bar{N}^2}{N_L^2} + \frac{\bar{M}^2}{a^2 M_L^2} + \frac{1}{\sqrt{3}} \frac{|\overline{NM}|}{bN_L M_L})^{\frac{1}{2}} = 1 \qquad (16)$$

is now introduced where the parameter b is taken as

$$b(\bar{\kappa}^P) = 1 - (3 - \frac{1}{\sqrt{3}})(1 - a(\bar{\kappa}^P)) \qquad (17)$$

In the further discussion the original model is denoted Model 1 and the modified one Model 1*.

Bieniek's subsequent yield surface

Based on the concept of translation of the yield surface in the moment hyperspace Bieniek and Funaro give the following yield surface /2/

$$f_B = F_B - 1 = (\frac{\bar{N}^2}{N_L^2} + \frac{\bar{M}^{*2}}{M_0^2} + \frac{1}{\sqrt{3}} \frac{|\overline{NM}|}{N_L M_L})^{\frac{1}{2}} - 1 = 0 \qquad (18)$$

Here, $\bar{M}^{*2}$ is the invariant of the components $(M_{\alpha\beta} - M^*_{\alpha\beta})$ as given by Eq. (7b), with translations $M^*_{\alpha\beta}$ determined by

$$dM^*_{\alpha\beta} = \begin{cases} B\, d\kappa^P_{\alpha\beta} & \text{for loading} \\ o & \text{for unloading/neutral loading} \end{cases}$$

The proportionality factor B is given by

$$B = 2(1 - F_L) \frac{M_0}{\kappa_0} \frac{F_s^2}{F_m^2} \qquad (19)$$

where F_s is the absolute value of the gradient of the yield surface in a dimensionless formulation, while F_m is the part of F_s contributed by bending.

The yield surface defined by $F_B = 1$ reproduces neither the correct initial yield surface nor the limit surface. In the former case agreement may be obtained by the introduction of the parameter $b(\bar{\kappa}^P)$, as in the modified Crisfield model, Eq. (16). In the numerical studies the original Bieniek model is denoted Model 2.

Fig. 1 depicts the various yield surfaces for the degenerate case of a beam-column subjected to moment M and axial force N. At initial yielding both Crisfield's surface $F_c(\bar{\kappa}^P{=}o)=1$ and Bieniek's surface $F_B(M^*_{\alpha\beta} = o) = 1$ deviate from the exact surface $F_o = 1$, causing delayed onset of plastic deformations. For this simple case it is also observed that Ilyushin's surface is in error compared with the exact one.

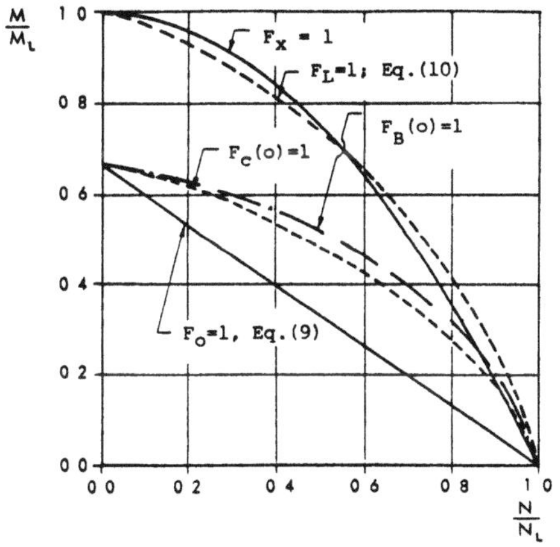

Fig.1 - Initial and limit yield surfaces for beam-column

Subsequent surfaces for hardening materials

The formulations of Crisfield and Bieniek are both derived for elastic-perfectly plastic materials. It is obvious that an initial yield surface exists also when the stress-strain curve has a hardening regime. The limit yield surface, however, is only defined when the stress-strain curve reaches an asymptotic value σ_L, Fig. 2. In the following the yield surfaces $F_C = 1$ and $F_B = 1$ are modified to take account of material hardening, using the concept of isotropic hardening.

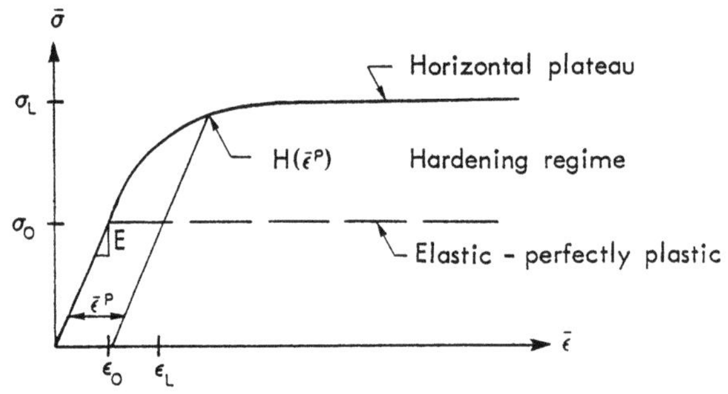

Fig.2 - Uniaxial stress-strain curve

Model 1

Crisfield's yield surface, Eq. (12), may be rewritten as

$$\left(\frac{\bar{N}^2}{h^2} + 16\,\frac{\bar{M}^2}{a^2h^4} + \frac{4s}{\sqrt{3}}\,\frac{\overline{NM}}{ah^3}\right)^{\frac{1}{2}} = \sigma_0 \tag{20}$$

where

$$s = \frac{|\overline{NM}|}{\overline{NM}}$$

Introducing the equivalent stress $\bar{\sigma}_C$ defined by the left hand side of Eq. (20) and a hardening function $H(\varepsilon^P_{\alpha\beta}, \kappa^P_{\alpha\beta})$ the yield function takes the form

$$f_C = \bar{\sigma}_C\,(N_{\alpha\beta}, M_{\alpha\beta}, a) - H(\varepsilon^P_{\alpha\beta}, \kappa^P_{\alpha\beta}) = 0 \tag{21}$$

The plastic compliance matrix D^P is easily determined from the flow rule, Eq. (4), using the relationship

$$d\lambda = \frac{\dfrac{\partial\bar{\sigma}_C}{\partial N_{\alpha\beta}}\,dN_{\alpha\beta} + \dfrac{\partial\bar{\sigma}_C}{\partial M_{\alpha\beta}}\,dM_{\alpha\beta}}{\dfrac{\partial H}{\partial\varepsilon^P_{\alpha\beta}}\,\dfrac{\partial\bar{\sigma}_C}{\partial N_{\alpha\beta}} + \dfrac{\partial H}{\partial\kappa^P_{\alpha\beta}}\,\dfrac{\partial\bar{\sigma}_C}{\partial M_{\alpha\beta}} - \dfrac{\partial\bar{\sigma}_C}{\partial a}\,\dfrac{\partial a}{\partial\bar{\kappa}^P}\,\dfrac{\partial\bar{\kappa}^P}{\partial\lambda}} \tag{22}$$

For futher development a plastic work function is defined by

$$dW^P = h\bar{\sigma}_c d\bar{\varepsilon}^P = N_{\alpha\beta} d\varepsilon^P_{\alpha\beta} + M_{\alpha\beta} d\kappa^P_{\alpha\beta}$$

which yields

$$\frac{\partial H}{\partial \varepsilon^P_{\alpha\beta}} = \frac{\partial H}{\partial \bar{\varepsilon}^P} \frac{\partial \bar{\varepsilon}^P}{\partial W^P} \frac{\partial W^P}{\partial \varepsilon^P_{\alpha\beta}} = H' \cdot \frac{1}{h\bar{\sigma}_c} N_{\alpha\beta} \tag{23a}$$

$$\frac{\partial H}{\partial \kappa^P_{\alpha\beta}} = \frac{\partial H}{\partial \bar{\varepsilon}^P} \frac{\partial \bar{\varepsilon}^P}{\partial W^P} \frac{\partial W^P}{\partial \kappa^P_{\alpha\beta}} = H' \cdot \frac{1}{h\bar{\sigma}_c} M_{\alpha\beta} \tag{23b}$$

Furthermore, one has

$$da = \frac{\partial a}{\partial \bar{\kappa}^P} d\bar{\kappa}^P = \frac{\partial a}{\partial \bar{\kappa}^P} \frac{\partial \bar{\kappa}^P}{\partial \lambda} d\lambda$$

The equivalent plastic strain increment $d\bar{\varepsilon}^P$ may be expressed in terms of the invariants of the plastic strain and curvature increments. Alternatively, the following expressions are applicable /5/

$$d\bar{\varepsilon}^P = \frac{1}{h} d\lambda \tag{24}$$

or

$$d\bar{\varepsilon}^P = \frac{1}{h\bar{\sigma}_c} (N_{\alpha\beta} d\varepsilon^P_{\alpha\beta} + M_{\alpha\beta} d\kappa^P_{\alpha\beta}) \tag{25}$$

Model 2

Following the previous development Bieniek's yield surface for a hardening material is chosen as

$$f_B = \bar{\sigma}_B (N_{\alpha\beta}, M_{\alpha\beta}, M^*_{\alpha\beta}) - H(\varepsilon^P_{\alpha\beta}, \kappa^P_{\alpha\beta}) = 0 \tag{26}$$

where

$$\bar{\sigma}_B = (\frac{\bar{N}^2}{h^2} + 36 \frac{\bar{M}^{*2}}{h^4} + \frac{4s}{\sqrt{3}} \frac{\overline{NM}}{h^3})^{\frac{1}{2}} \tag{27}$$

Again, the compliance relationship is determined from the flowrate, using the following expression for the proportionality constant

$$d\lambda = \frac{\dfrac{\partial\bar{\sigma}_B}{N_{\alpha\beta}} dN_{\alpha\beta} + \dfrac{\partial\bar{\sigma}_B}{M_{\alpha\beta}} dM_{\alpha\beta}}{\dfrac{\partial H}{\partial\varepsilon^P_{\alpha\beta}} \dfrac{\partial\bar{\sigma}_B}{\partial N_{\alpha\beta}} + \dfrac{\partial H}{\partial\kappa^P_{\alpha\beta}} \dfrac{\partial\bar{\sigma}_B}{\partial M_{\alpha\beta}} - \dfrac{\partial\bar{\sigma}_B}{\partial M^*_{\alpha\beta}} \dfrac{\partial\bar{\sigma}_B}{\partial M_{\alpha\beta}} \cdot B} \qquad (28)$$

The equivalent plastic strain increment is again given by Eq. (25). $\partial H/\partial\varepsilon^P_{\alpha\beta}$ and $\partial H/\partial\kappa^P_{\alpha\beta}$ are given by Eqs. (23), replacing $\bar{\sigma}_C$ by $\bar{\sigma}_B$.

Numerical studies show that the original Bieniek model in certain cases violates Drucker's postulate, leading to a nonpositive definite compliance matrix. This is noted by the fact that the parameter B, Eq. (19), becomes negative. The reason for this deficiency is that Bieniek's subsequent yield surface exceeds the limit yield surface $F_L = 1$ in certain regions of the stress space. In the present investigation this problem is avoided by using $F_L = 1$ as plastic potential instead of $\bar{\sigma}_B$ whenever this occur.

A further modification of the model is investigated, where Eq. (19) is replaced by

$$B = 2(1 - F_L) \frac{M_0}{\kappa_0} \frac{\sigma_0}{\sigma_L} \qquad (29)$$

Here, the effect of the scaling factor σ_0/σ_L depends on the shape of the stress-strain curve. In the following numerical studies the model based on Eq. (29) and including the parameter b as given by Eq. (17) is denoted Model 2*.

Numerical studies

In the present investigation the material models are verified by numerical simulations for the degenerate case of uniaxial bending and stretching. Comparison is made with the results obtained from the classical flow theory of plasticity using numerical integration through the thickness.

Elastic-perfectly plastic materials

In the following the results are presented on a nondimensional form, with the scaling factors N_L and M_L, Eq. (11), and

$$\varepsilon_L = \sigma_L/E \; ; \qquad \kappa_L = 3\,\sigma_L/Eh$$

Fig. 3 shows the moment-curvature relationship for a beam subjected to pure bending, where both Crisfield's (Model 1) and Bieniek's (Model 2) models show good agreement with the "exact" solution. In Figs. 4 and 5 the beam is subjected to a "pre-strain" $\varepsilon = 0{,}75\ \varepsilon_L$, which is kept constant while the curvature is varied. The delay in initial yielding is here noticeable for both models 1 and 2, while models 1* and 2* show improved behavior for $\kappa < 1{,}2\ \kappa_L$. The interaction (M-N) curves for the models are given in Fig. 6, where again the deficiency of the original models are observed. Simulation studies with other values of the pre-strain show similar results with those given here.

The results obtained when an initial curvature $\kappa = 1{,}4\ \kappa_L$ is imposed before the beam is stretched are depicted in Fig. 7. Here, the limit surface $F_L = 1$ was exceeded for $\varepsilon = 0{,}33\ \varepsilon_L$, resulting in a non-positive definite stress-strain relationship. For the remaining part of the deformation $f_B = 0$ was replaced by $F_L = 1$ as the plastic potential. For this case models 1* and 2* show only minor improvement over the original models, and the results are not presented here.

The pure bending and pure stretching cases presented so far are not indicative for the load history in a real structure. A more realistic test on the accuracy of the material models

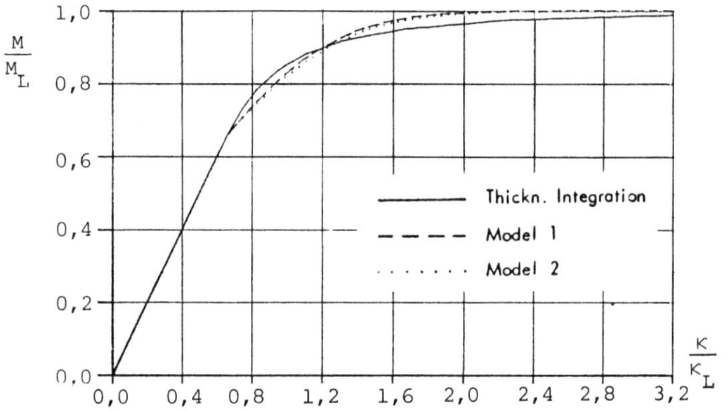

Fig.3 - Pure bending of beam section

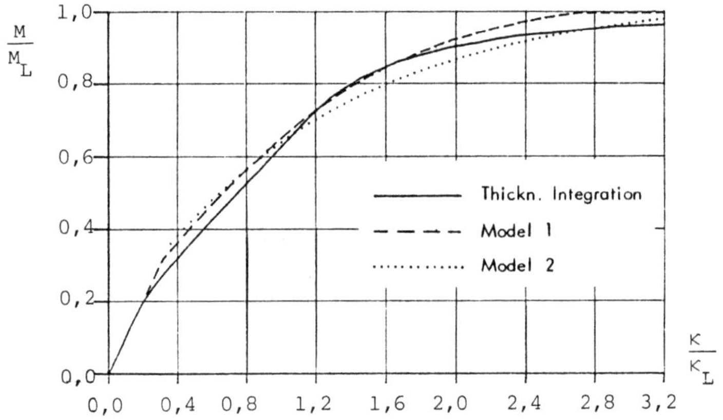

Fig.4 - Bending of beam section subjected to constant axial strain $\varepsilon = 0.75 \cdot \varepsilon_L$

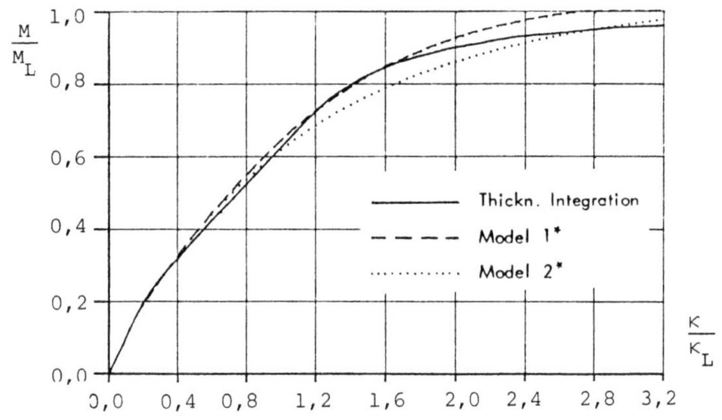

Fig.5 - Bending of beam section subjected to constant axial strain $\varepsilon = 0.75 \cdot \varepsilon_L$

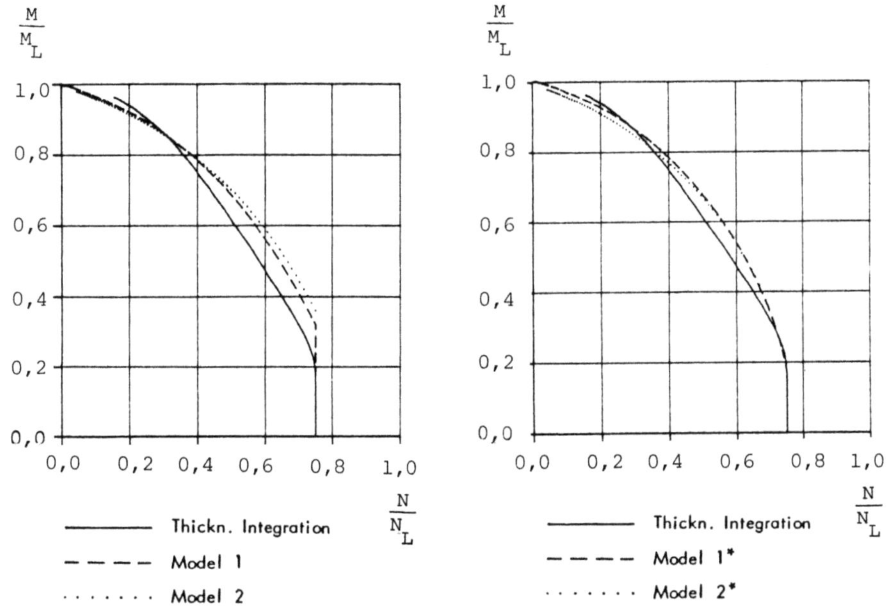

Fig.6 - Interaction diagrams for bending of beam section subjected to constant axial strain $\varepsilon = 0.75\ \varepsilon_L$

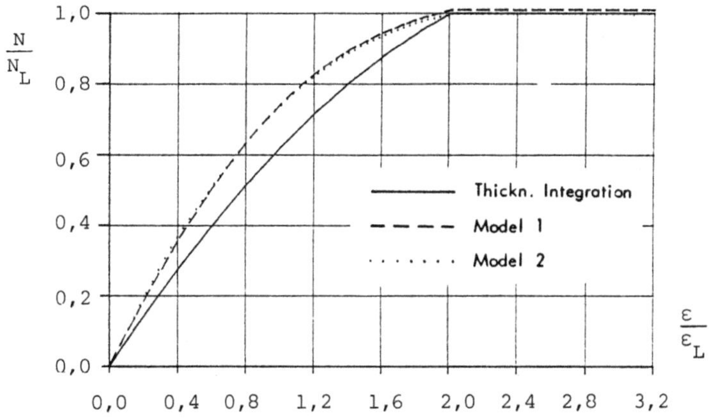

Fig.7 - Stretching of beam section subjected to constant curvature $\kappa = 1.4\ \kappa_L$

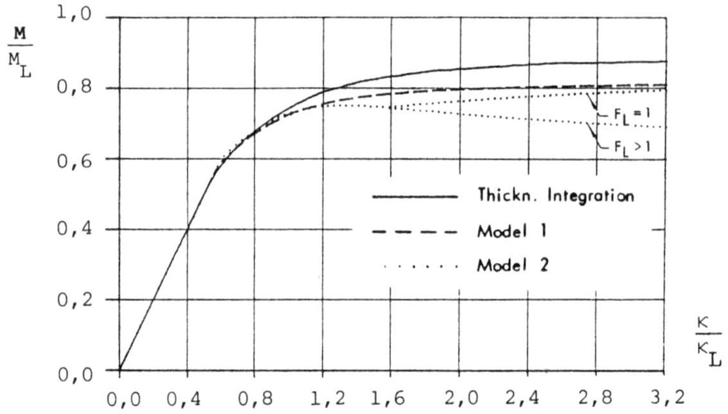

Fig.8 - Moment vs. curvature for proportional bending and stretching of beam section ($\frac{\kappa}{\kappa_L} = 2 \frac{\varepsilon}{\varepsilon_L}$)

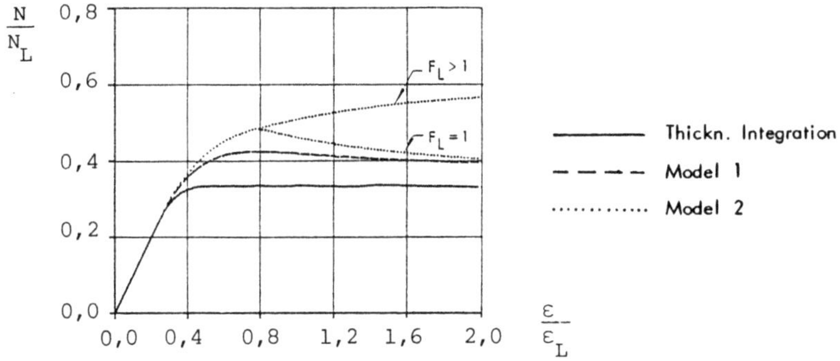

Fig.9 - Axial force vs. axial strain for proportional bending and stretching of beam section ($\frac{\kappa}{\kappa_L} = 2 \frac{\varepsilon}{\varepsilon_L}$)

is the case when ε and κ are varied proportionally. Two such strain paths, $\kappa/\kappa_L = 2\varepsilon/\varepsilon_L$ and $\kappa/\kappa_L = 0.5 \cdot \varepsilon/\varepsilon_L$, are investigated. The former corresponds to a neutral axis at $\zeta_3 = -h/6$ with limiting values $M = 0.89\ M_L$ and $N = 0.33\ N_L$, while the latter corresponds to a position $\zeta_3 = -2h/3$ (outside the beam) with limit values $M = 0$ and $N = N_L$. The results are presented in Figs. 8 - 11. Here, both the results of Bieniek's original model with negative definite strain-stress relationship when $F_L > 1$, and the model obtained when using $F_L = 1$ as a plastic potential, are included. Again the modified models 1* and 2* show only minor improvements compared with

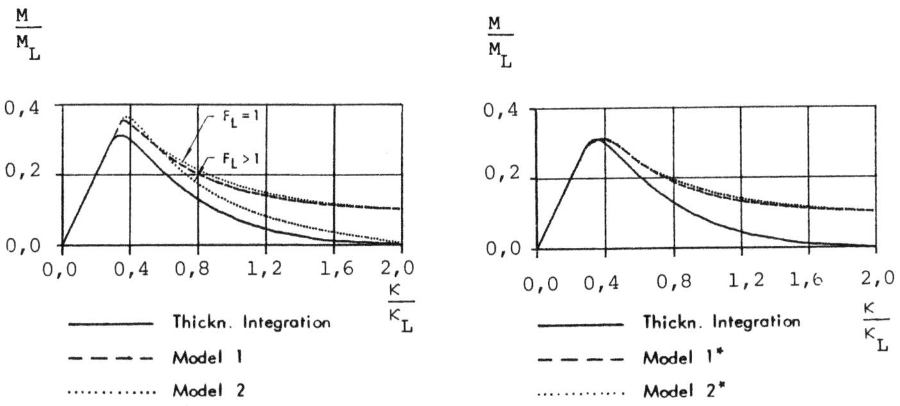

Fig.10 - Moment vs. curvature for proportional bending and stretching of beam section ($\frac{\kappa}{\kappa_L} = 0.5 \frac{\varepsilon}{\varepsilon_L}$)

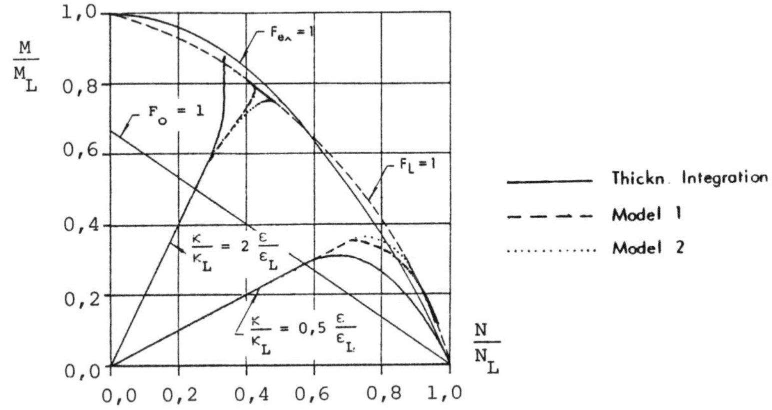

Fig.11 - Loading paths for proportional bending and stretching of beam section

models 1 and 2. The largest improvement is shown in Fig. 10 where the peak value of the moment is correctly represented. The loading paths in the stress space for the two proportional straining histories are given in Fig. 11.

When subjecting the beam segment to cyclic bending the moment-curvature relationships in Fig. 12 are obtained. Note that the lowered yield point obtained when the bending moment is reversed has nothing to do with hardening of the material, as an elastic-perfectly plastic material displays no Bauschinger effect. As expected, model 2 reproduces fairly closely the actual behaviour under cyclic bending. This is due to the translational quantity M* defining the subsequent yield surface. On the other hand, model 1 exhibits an "isotropic hardening" effect, and is hence less suited to predict the behaviour under reversed yielding. The results obtained when subjecting the section to a pre-strain before the cyclic curvature is imposed are not significantly different from those presented in Fig. 12.

Work-hardening materials

For the case of work-hardening materials the uniaxial stress-strain curve is taken as

$$\sigma = \sigma_0 + (\sigma_L - \sigma_0)(1 - e^{-(E\varepsilon - \sigma_0)/(\sigma_L - \sigma_0)}); \quad \varepsilon > \frac{\sigma_0}{E}$$

with $\sigma_L = 2\sigma_0$.

For the case of pure bending the accuracy of the two models is comparable with that of the non-hardening case. The results of bending or stretching of a prestrained section are given in Figs. 13 and 14. From Fig. 13 it is clear that the correct limit case M = 0 and $N = N_L$ is not reached for model 2*, which is due to the fact that the quantity M* in the expression for the equivalent stress goes asymptotically towards the value M* = 0.25 M_L instead of the correct value M* = 0. For the proportional loading path $\kappa/\kappa_L = 2\ \varepsilon/\varepsilon_L$ both models reproduce the correct moment-curvature relationship, while the relationship between N and ε shows greater deviations, Fig. 15. For the case $\kappa/\kappa_L = 0.5\ \varepsilon/\varepsilon_L$, however, the opposite is the case, and the moment-curvature relationships display significant deviation from the correct one, Fig. 16. Finally, the interaction diagram for the two cases are depicted in Fig. 17. The limit surface $F_L = 1$ is here exceeded for both models. This is due to the fact that finite strain increments are used in the solution, without proper scaling back to the yield surface.

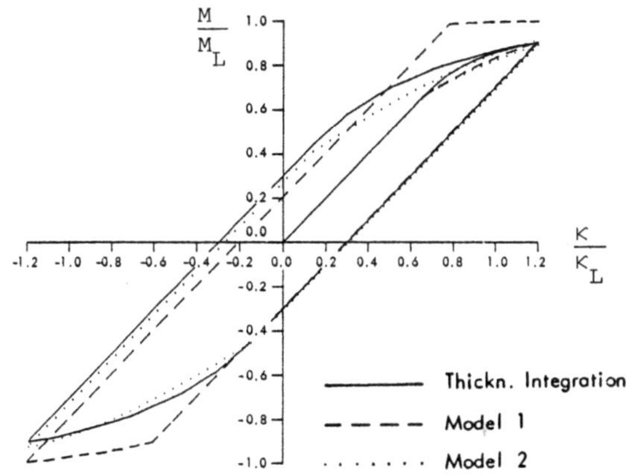

Fig.12 - Cyclic bending of beam section (ε = o)

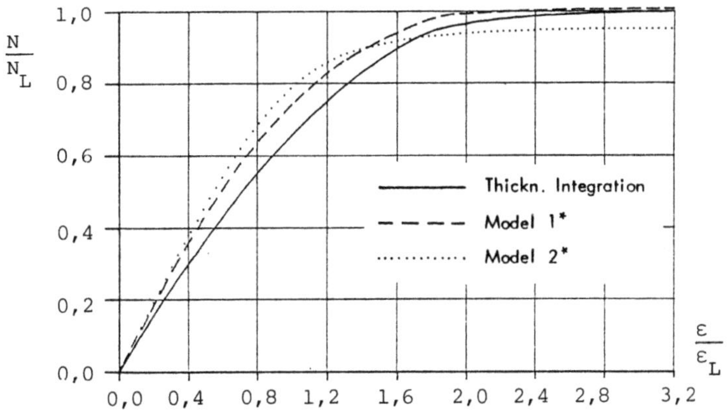

Fig.13 - Stretching of beam section of hardening material subjected to constant curvature $\kappa = \kappa_L$

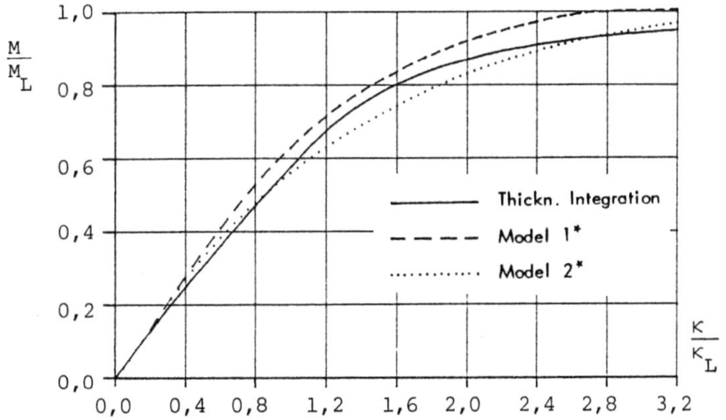

Fig.14 - Bending of beam section of hardening material subjected to constant axial strain $\varepsilon = \varepsilon_L$

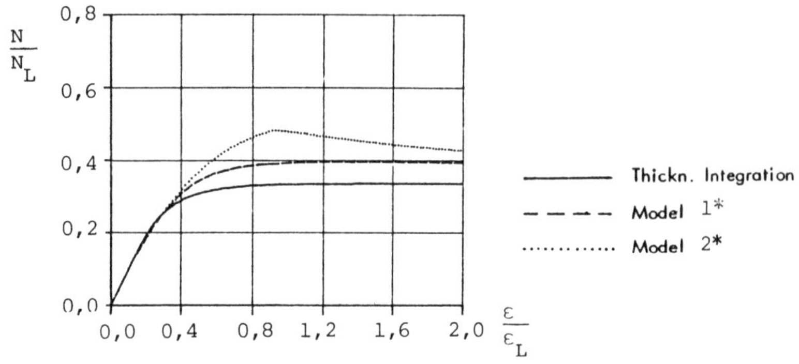

Fig.15 - Proportional stretching and bending of beam section of hardening material ($\frac{\kappa}{\kappa_L} = 2 \frac{\varepsilon}{\varepsilon_L}$)

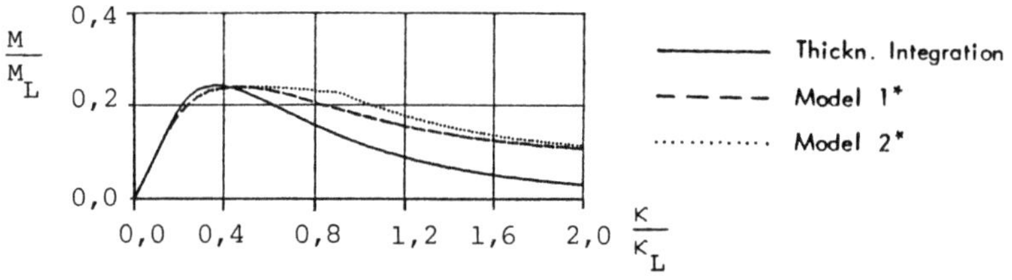

Fig.16 - Proportional stretching and bending of beam section of hardening material ($\frac{\kappa}{\kappa_L} = 0.5 \frac{\varepsilon}{\varepsilon_L}$)

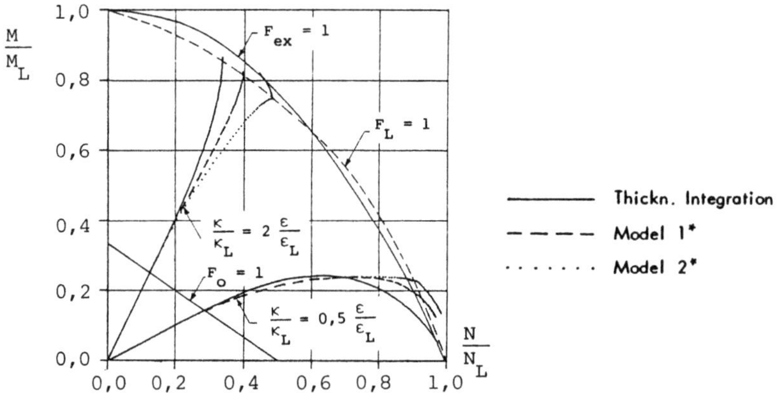

Fig.17 - Loading paths for proportional stretching and bending of beam section of hardening material

Summary and conclusions

Based on numerical studies the behaviour of the elasto-plastic material models of Crisfield and Bieniek has been compared with the classical flow theory of plasticity. For the degenerate case of a beam section subjected to combinations of bending and stretching both models give results that agree well with the classical theory for monotonic loading. However, negative definiteness of the stress-strain matrix was detected in some of the examples, necessitating modifications in the flow potential. For cyclic loading Bieniek's model give the best description of the behaviour.

The adjustment of the coefficient of the $\overline{NM}$ term in the yield conditions (model 1* and model 2*) seems to have a favourable influence on the behaviour of the models, although the improvement in some cases is not large.

For work-hardening materials only models 1* and 2* have been considered. The results indicate that the behaviour of these models is quite similar to that observed for the elastic-perfectly plastic material. The effect of varying the shape of the uniaxial stress-strain curve has not been studied in the present investigation.

The test cases presented here give information regarding the general behaviour of the models in the two-dimensional stress space. The importance of some of the deficiencies of the models in a more general stress space can only be ascertained by the analyses of some typical shell structures.

References

/1/ Crisfield, M.A.: "On an Approximate Yield Criterion for Thin Steel Shells", TRRL Report LR658, Transport and Road Research Laboratory, Crowthorne, Berkshire, 1974.

/2/ Bieniek, M.P. and Funaro, J.R.: "Elasto-Plastic Behaviour of Plates and Shells", Techn. Rep. DNA 3954T, Defense Nuclear Agency, Weidlinger Associates, New York, 1976.

/3/ Eggers, H. and Kröplin, B.: "Yielding of Plates with Hardening and Large Deformations", International Journal for Numerical Methods in Engineering, Vol. 12, pp. 739-750, 1978.

/4/ Horrigmoe, G.: "Nonlinear Finite Element Models in Solid Mechanics", Report No. 76-2, Division of Structural Mechanics, The Norwegian Institute of Technology, The University of Trondheim, 1976.

/5/ Eidsheim, O.M.: "Nonlinear Analysis of Elastic-Plastic Shells Using Hybrid Stress Finite Elements", Dr.ing. Thesis, Division of Structural Mechanics, The Norwegian Institute of Technology, The University of Trondheim, 1980.

/6/ Ilyushin, A.A.: "Plasticité", Editions Eyrolles, 1956.

/7/ Crisfield, M.A.: "Some Approximations in the Nonlinear Analysis of Rectangular Plates Using Finite Elements", TRRL Supplementary Report 51 UC, Transport and Road Research Laboratory, Crowthorne, Berkshire, 1974.

An Efficient Finite Element Method for Elastic-Plastic Analysis of Plane Frames

J BANOVEC
University Edvard Kardelj of Ljubljana, Yugoslavia

Summary

A general method for the elastic - plastic analysis of plane frames including nonlinear geometric effects (large displacements, large /quadratic/ strains) is presented. A mixed type element with three degrees of freedom d. o. f. at each end, and with n (n = 1,2,3,...) internal d.o.f. along the element is proposed. The mixed approach with the approximation of the axial force and the curvature presents better the elastic - plastic state than the well known displacement approach.

Introduction

Plane frames are almost the most frequent structures in the civil engineering and the nonlinear analysis of these structures has attained considerable interest. Many publications treat only material nonlinearity (elasto - plastic frame) or only geometrical nonlinearity (large displacements and quadratic strains; large displacements and small strains; the second order effects). Only a few papers analyse both nonlinearities, e. g. [Ref.3] to [5]. Most of these methods separate the frame columns and beams into large number of elements. Usually element uniform loads must be replaced by adequate modal forces and moments.

The aim of the presented paper is to develop a more effective method, which could decrease the necessary number of elements and will at the same time give sufficient accurate results.

Description of the problem

A typical beam element is shown in Fig. 1, where notation for element loads, end actions, displacements, and orthonormal coordinate systems is also defined.

It is assumed that the element follows Bernoulli - Navier hypothesis, thus shear deformation is neglected and the cross - sectional area does not change during the deformation. The element loads do not change their

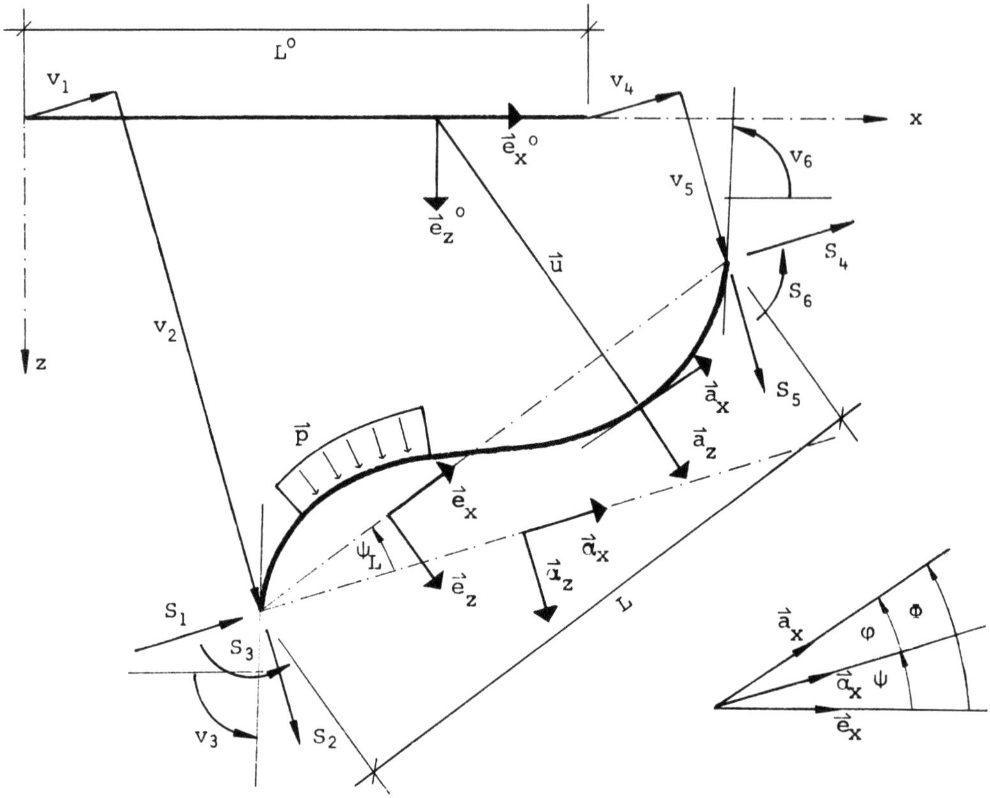

Fig. 1. Beam element

direction and the intensity of element load resultant forces is assumed constant. The element material follows some nonlinear stress - strain diagram.

Kinematics

The derivation starts with the expression for the extensional strain ε specialised for negligible transverse shear deformation [1]:

$$\varepsilon = D_N + z\, D_M$$

$$D_M = \frac{d\Phi}{dx} = \frac{d\varphi}{dx} \tag{1}$$

$$(1 + D_N)\vec{a}_x = \vec{e}_x^{\,o} + \frac{d\vec{u}}{dx}$$

or

$$D_N = \sqrt{1.0 + 2\,\vec{e}_x\,\frac{d\vec{u}}{dx} + \frac{d\vec{u}}{dx}\frac{d\vec{u}}{dx}} - 1.0 \qquad (2)$$

or in the component form

$$\begin{aligned}(1 + D_N)\cos\varphi &= \cos\psi + \frac{du}{dx}\\ (1 + D_N)\sin\varphi &= -\sin\psi - \frac{dw}{dx}\end{aligned} \qquad (3)$$

$$\frac{d\vec{u}}{dx} = \frac{du}{dx}\vec{\alpha}_x + \frac{dw}{dx}\vec{\alpha}_z \qquad (4)$$

where z is the normal coordinate; D_N, D_M are one dimensional stretching and bending deformations; u, w are translation components with respect to the reference coordinate system ($\vec{\alpha}_x$, $\vec{\alpha}_z$), which is defined by the rotation ψ (Fig. 1). The Eqs. (3) for D_N and φ are usually too complicated for practical use and they will be simplified. Rearranging them:

$$\begin{aligned}(1 + D_N)(\cos\varphi - 1) &= \cos\psi + \frac{du}{dx} - 1 - D_N\\ (1 + D_N)\sin\varphi &= -\sin\psi - \frac{dw}{dx}\end{aligned} \qquad (3^*)$$

and assuming that the reference system is chosen so that the local rotation φ is a small order quantity (Fig. 1):

$$\Phi = \psi + \varphi \quad ; \quad |\varphi| \ll 1.0 \qquad (5)$$

the left sides of (3^*) are expanded into Taylor's series about the point $\varphi = 0.0$:

$$F = (1 + D_N)(\cos\varphi - 1) \simeq F\Big|_{\varphi=0.0} + \frac{\partial F}{\partial\varphi}\Big|_{\varphi=0.0}\varphi + \frac{\partial^2 F}{\partial\varphi^2}\Big|_{\varphi=0.0}\frac{\varphi^2}{2}\Big|_{\varphi=0.0} \qquad (6)$$

$$G = (1 + D_N)\sin\varphi \simeq G\Big|_{\varphi=0.0} + \frac{\partial G}{\partial\varphi}\Big|_{\varphi=0.0}\varphi \qquad (7)$$

Considering (3^*), (6) and (7), simplified equations are obtained:

$$D_N = \frac{du}{dx} + \cos\psi - 1 + \frac{c_N}{2}\varphi^2$$
$$c_N\,\varphi = -\frac{dw}{dx} - \sin\psi \tag{8}$$

$$c_N = (1 + D_N)\Big|_{\varphi=0.0} \tag{9}$$

where higher terms are neglected. Considering (3), c_N becomes:

$$c_N = \frac{du}{dx} + \cos\psi \tag{10}$$

Kinematic relations (8) are still too complicated, so it is supposed that c_N is constant along the element:

$$c_N \cong \frac{1}{L^o}\int_o^{L^o} \left(\frac{du}{dx} + \cos\psi\right)dx = \frac{v_4 - v_1}{L^o} + \cos\psi \tag{11}$$

If φ is small (5), the rigid local rotation ψ_L is also a small order quantity (Fig. 1):

$$\frac{L}{L^o}\cos\psi_L = \cos\psi + \frac{v_4 - v_1}{L^o} \tag{12}$$

Assuming that:

$$\frac{L}{L^o}\cos\psi_L \cong \frac{L}{L^o} \tag{13}$$

c_N is given by:

$$c_N = \frac{L}{L^o} \tag{14}$$

Collecting results from (1), (5), (8) and (14):

$$D_N = \frac{du}{dx} + \cos\psi - 1 + \frac{c_N}{2}\varphi^2$$
$$c_N\,\varphi = -\frac{dw}{dx} - \sin\psi$$
$$\Phi = \varphi + \psi \quad ; \quad D_M = \frac{d\Phi}{dx} = \frac{d\varphi}{dx} \tag{15}$$
$$c_N = \frac{L}{L^o}$$

kinematic relations are obtained. They are taken into account in derivation of equilibrium equations by the energy principle. The influence of c_N will be discussed later.

Almost in all publications, which analyse plane frames, the reference coordinate system is identified with the rotated element system $(\vec{e}_x, \vec{e}_z)$, e. g. $\psi_L = 0.0$ and it is assumed that the deformed length of the element L is equal to L^o ($c_N = 1.0$).

Stress resultants

An element cross section is considered in Fig. 2.

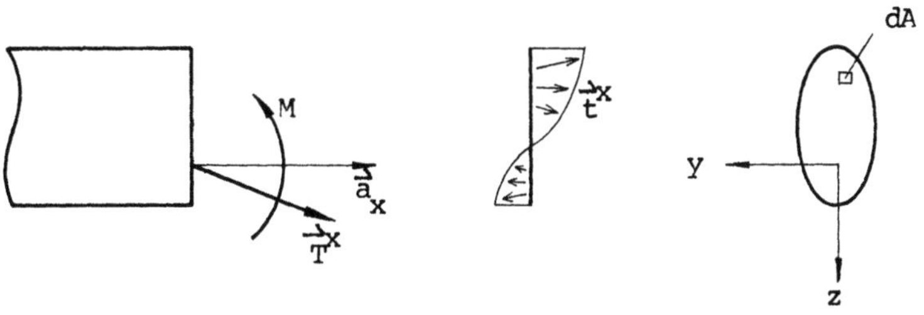

Fig. 2. Stress resultants

The stress vector $\vec{t}^x$ is written in the component form:

$$\vec{t}^x = \sigma \, \vec{a}_x + \tau \, \vec{a}_z \tag{16}$$

where due to Bernoulli - Navier hypothesis σ and τ present the physical components of the stress tensor. The resultant force:

$$\vec{T}^x = \int_A \vec{t}^x \, dA = N \, \vec{a}_x + Q \, \vec{a}_z \tag{17}$$

where:

$N = \int_A \sigma \, dA$ ... normal force

$Q = \int_A \tau \, dA$ ... shear force ,

and the bending moment:

$$M = \int_A \vec{t}^{\,x} \vec{a}_x \, z \, dA = \int_A \sigma \, z \, dA \tag{18}$$

are defined by the integration of the stress vector.

Constitutive equations

The material of the element is considered to be nonlinear (Fig. 3):

$$\sigma = \sigma \, (\varepsilon) \tag{19}$$

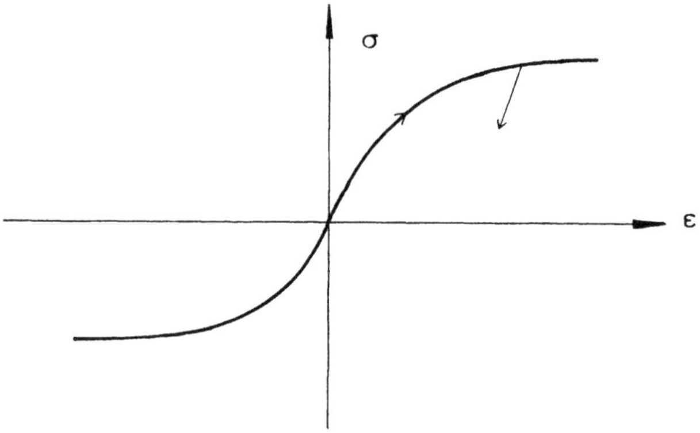

Fig. 3. Stress - strain diagram

All necessary equations will be derived for nonlinear hyperelastic material [6]:

$$\sigma = \frac{d\bar{U}}{d\varepsilon} \tag{20}$$

where $\bar{U}$ is element deformation energy per unit initial volume. Defining the energy per unit initial length as

$$U = \int_A \bar{U} \, dA \tag{21}$$

it can be concluded from (1) and (16) to (21) that the energy becomes function of stretching and bending deformations D_N and D_M:

$$U = U \, (D_N, D_M)$$

$$N = \frac{\partial U}{\partial D_N} \quad ; \quad M = \frac{\partial U}{\partial D_M} \tag{22}$$

In the case of real elasto - plastic material , when an irreversible process may occur, the incremental equations for reversible process can still be used if the history of previous load steps is taken into account [2]. The Eqs. (17) and (18):

$$N = \int_A \sigma \, dA = N\,(D_N, D_M)$$
$$M = \int_A \sigma \, z \, dA = M\,(D_N, D_M) \tag{23}$$

can be observed as two functions relating four variables N, M, D_N and D_M. Choosing deformations as independant variables the incremental equations are obtained, Eqs. (1), (19) and (23):

$$\begin{Bmatrix} \Delta N \\ \Delta M \end{Bmatrix} = \begin{bmatrix} A_t & S_t \\ S_t & I_t \end{bmatrix} \begin{Bmatrix} \Delta D_N \\ \Delta D_M \end{Bmatrix}$$
$$A_t = \int_A \frac{d\sigma}{d\varepsilon} \, dA \quad ; \quad S_t = \int_A \frac{d\sigma}{d\varepsilon} \, z \, dA \tag{24}$$
$$I_t = \int_A \frac{d\sigma}{d\varepsilon} \, z^2 \, dA$$

where e. g. ΔN presents the normal force increment. The Eqs. (24) are used in the displacement approach.

Selecting the normal force N and the bending deformation D_M for independant variables a modified form of (24) is obtained:

$$\begin{Bmatrix} \Delta D_N \\ \Delta M \end{Bmatrix} = \begin{bmatrix} \dfrac{1}{A_t} & -\dfrac{S_t}{A_t} \\ -\dfrac{S_t}{A_t} & I_t - \dfrac{S_t^2}{A_t} \end{bmatrix} \begin{Bmatrix} \Delta N \\ \Delta D_M \end{Bmatrix} \tag{25}$$

The Eq. (24) will be used in the mixed energy principle. It is convenient to introduce mixed deformation energy per unit initial length, (22):

$$U^*(N, D_M) = U\,(D_N, D_M) - N\,D_N$$
$$\frac{\partial U^*}{\partial N} = -D_N \quad ; \quad \frac{\partial U^*}{\partial D_M} = M \tag{26}$$

It follows from (25):

$$\frac{\partial^2 U^*}{\partial N\,\partial D_M} = -\frac{\partial D_N}{\partial D_M} = \frac{\partial^2 U^*}{\partial D_M\,\partial N} = \frac{\partial M}{\partial N} \tag{27}$$

than the second derivates of U^* are commutative.

Equilibrium equations

It is well known that equilibrium differential equations and statical conditions for the end actions can be derived using the stationary value potential energy principle. When using the "exact" kinematic relations (1), which means exact according to Bernoulli - Navier hypothesis, the "exact" equilibrium conditions together with the "exact" statical conditions can be obtained; while introducing the simplified kinematical relations (15) into potentional energy functional the approximate equilibrium conditions will be derived.

The element potential energy Π consist of two parts i. e. of the deformation energy and of potential of external forces:

$$\Pi\ (u,w) = \int_0^{L^o} U\ (D_N,D_M) + \int_0^{L^o} \bar{\Omega}\ (u,w)dx + \Omega\ (v_k) \tag{28}$$

where $\bar{\Omega}$ is the potential of the element loads and Ω presents the potential of the end actions (Fig. 1):

$$\begin{aligned} \frac{\partial\bar{\Omega}}{\partial u} &= -\vec{p}\,\vec{\alpha}_x = -p_x \\ \frac{\partial\bar{\Omega}}{\partial w} &= -\vec{p}\,\vec{\alpha}_z = -p_z \\ \frac{\partial\bar{\Omega}}{\partial v_k} &= -S_k\ ;\quad k = 1,2,\ldots 6 \end{aligned} \tag{29}$$

The admissible variations of the displacements δu and δw cause deformation changes (15):

$$\begin{aligned} \delta D_N &= \delta\left(\frac{du}{dx}\right) + c_N\,\varphi\,\delta\varphi \\ c_N\,\delta\varphi &= -\delta\left(\frac{dw}{dx}\right) \\ \delta D_M &= \delta\left(\frac{d\Phi}{dx}\right) = \delta\left(\frac{d\varphi}{dx}\right) \end{aligned} \tag{30}$$

where neglecting the terms which contain δc_N, additional assumption is made. The variations of end displacements become:

$$\begin{aligned} &\delta v_1 = \delta u(0) && \delta v_3 = \delta\Phi(0) && \delta v_5 = \delta w(L^o) \\ &\delta v_2 = \delta w(0) && \delta v_4 = \delta u(L^o) && \delta v_6 = \delta\Phi(L^o) \end{aligned} \tag{31}$$

Using (22), (28) and (29) the first variation of Π can be stated in the form:

$$\delta\Pi = \int_0^{L^o} (N\ \delta D_N + M\ \delta D_M)dx - \int_0^{L^o} (p_x\ \delta u + p_z\ \delta w)dx - \\ - \sum_{k=1}^{6} S_k\ \delta v_k = 0 \tag{32}$$

Considering (30) and (31), and using the integration by parts, approximate differential equations:

$$\begin{aligned} &\frac{dN}{dx} + p_x = 0 \\ &\frac{d}{dx}\left(\frac{1}{c_N}\frac{dM}{dx} - N\ \varphi\right) + p_z = 0 \end{aligned} \tag{33}$$

and statical boundary conditions

$$\begin{aligned} &S_1 = -N(0) \qquad\qquad S_4 = N(L^o) \\ &S_3 = -M(0) \qquad\qquad S_6 = M(L^o) \\ &S_2 = -\left(\frac{1}{c_N}\frac{dM}{dx} - N\ \varphi\right)_{x=0} \\ &S_5 = \left(\frac{1}{c_N}\frac{dM}{dx} - N\ \varphi\right)_{x=L^o} \end{aligned} \tag{34}$$

are obtained. The Eqs. (33) and (34) are quite simple in comparison with the "exact" ones [1]. It will be shown that relatively good results can be obtained.

In the case of small (linear) strains, the quadratic term $0.5\ c_N\ \varphi^2$ in (15) is neglected and corresponding equilibrium equations and statical conditions are recognised by omitting the terms $N\ \varphi$ in (33) and (34). The results strongly depend on the choise of the reference system $(\vec{\alpha}_x, \vec{\alpha}_z)$, (Fig. 1). If the reference system coincides with the rotated element system $(\vec{e}_x, \vec{e}_z)$, the large displacement analysis is considered. In the case when the rotation ψ is zero, the small displacement analysis is treated. Combining the mentioned possibilities four different analyses can be used:

- large displacement - large (quadratic) strains (*l.* - *l.*)
- large displacement - small strains (*l.* - *s.*)
- small displacement - large (quadratic) strains (s. 0. t.); usually it is named the second order theory
- small displacement - small strains (f. o. t.); it is the well known first order theory.

In the case of constant linear material along the element simplified equations can be solved analytically.

Fig. 4 shows the analytical results obtained in the analysis of cantilever subjected to an end moment. The comparison between two different methods, i. e. l. - l. and l. - s. shows that using the integration interval L^o, very good results are obtained by the l. - l. method for the rotation up to 90 degrees. Using two equal integration intervals the relative error for the end displacement w is negligible even in the case when the cantilever deforms into a half circle. Both methods, i.e. l. - l. and l. - s. give exact value for end rotation Φ. The error in results

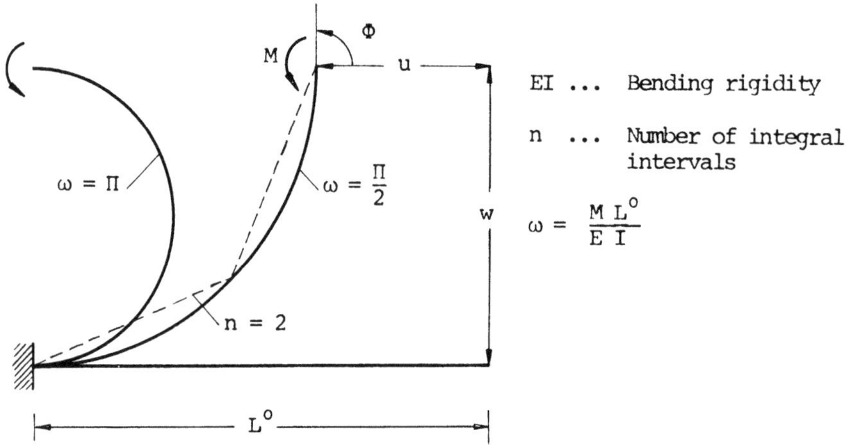

One integral interval							
"Exact" method				Simplified methods R_u, R_w = % error for u,w			
ω	u/L^o	w/L^o	Φ	l. - l.		l. - s.	
				R_u	R_w	R_u	R_w
0.1	0.0017	0.0500	0.1	- 0.01	0.00	- 24.98	0.04
0.5	0.0411	0.2448	0.5	- 0.18	0.01	- 24.45	1.05
1.0	0.1585	0.4597	1.0	- 0.64	0.12	- 22.78	4.29
Π/2	0.3634	0.6366	Π/2	- 1.26	1.26	- 19.40	11.07
Π	1.0000	0.6366	Π	0.00	11.31	0.00	57.08

Several integral intervals							
ω	l. - l. method		l. - s. method				
	n = 2		n = 2		n = 4		n = 8
	R_u	R_w	R_u	R_w	R_u	R_w	R_w
0.1	0.00	0.00	- 6.24	0.01	- 1.56	0.00	0.00
0.5	- 0.01	0.00	- 6.08	0.26	- 1.53	0.06	0.02
1.0	- 0.04	0.01	- 5.57	1.05	- 1.38	0.26	0.07
Π/2	- 0.08	0.05	- 4.56	2.63	- 1.11	0.65	0.16
Π	0.00	0.07	0.00	11.06	0.00	2.60	0.65

Fig. 4. Comparison of analytical results

obtained by the *l. - s.* method using 8 intervals is greater than the error in results obtained by *l. - l.* method with two intervals. The given example and other tests, which are not presented here, demonstrate that *l.* - *l.* method is very efficient and suitable numerical method should be chosen in the case of the elastoplastic beam elements.

Mixed variational principle

In defining the element potential functional (28) it was assumed that deformations were related to the displacements (15). The expression for D_N seems to be quite a complex constraint. To relax it a new variational principle is stated in the form:

$$\Pi^I = \Pi - \int_0^{L^o} \lambda(D_N - \frac{du}{dx} - \cos\psi + 1 - c_N \frac{\varphi^2}{2})dx \tag{35}$$

by introducing one independant Lagrangian multiplier λ. Thus u, w, λ and D_N can be varied independently. The variation of D_N gives:

$$\delta\Pi^I = 0 = \int_0^{L^o} (N - \lambda)\delta D_N \, dx \tag{36}$$

where U - N relation is given in (22). The Lagrangian multiplier can be identified as the normal force N. Considering Eqs. (26), (35) and (36) the modified functional is defined as

$$\Pi^{II} = \int_0^{L^o} U^* \, dx + \int_0^{L^o} \overline{\Omega} \, dx + \Omega + \int_0^{L^o} N(\frac{du}{dx} + \cos\psi - 1 + \frac{c_N}{2}\varphi^2)dx \tag{37}$$

Taking into account (29) and the variation of u:

$$\delta\Pi^{II} = 0 = \int_0^{L^o} N\,\delta(\frac{du}{dx})dx - \int_0^{L^o} p_x \,\delta u \, dx - S_1 \,\delta u(0) - S_4 \,\delta u(L^o) \tag{38}$$

and using integration by parts, equilibrium equations in the direction ($\vec{\alpha}_x$) are obtained:

$$\frac{dN}{dx} + p_x = 0$$
$$S_4 = N(L^o) \quad ; \quad S_1 = -N(0) \tag{39}$$

Solving them:

$$N = -\int_0^x p_x\, dx^* + \frac{1}{2}\int_0^{L^o} p_x\, dx + N_D$$
$$S_4 = -\frac{1}{2}\int_0^{L^o} p_x\, dx + N_D\ , \quad S_1 = -\frac{1}{2}\int_0^{L^o} p_x\, dx - N_D \tag{40}$$

the above equations can be used as the constraints; thus the element mixed deformation energy functional is defined:

$$\Pi^* (w, N_D) = \int_0^{L^o} [U^*(N,D_M) + \overline{\Omega}(u,w)]dx + \Omega(v_k) + + \int_0^{L^o} N(N_D)\left(\frac{du}{dx} + \cos\psi - 1 + \frac{c_N}{2}\varphi^2\right)dx \tag{41}$$

Dependent variables D_M, φ, and N are expressed in (15) and (40). It is suitable that new independent variable N_D is not the function of x. Π^* will be treated in derivation of the numerical solution of the problem.

Finite element approximation

It is now a simple matter to obtain general equations for a typical finite element.

The displacement w approximation. Returning to (41) it is observed that only the approximation for w can be chosen. Considering the element as the part of the frame the behaviour of w is described with the shape function w_s, which rigorously satisfies nodal conditions and with the bubble function w_b, which defines the deformation of rotated clamped - clamped element. Considering kinematic relations (15) and introducing the non dimensional coordinate ξ the following functions are proposed:

$$w = w_s + w_b$$
$$w_s = I_2\, v_2 + I_5\, v_5 + c_N L^o(I_3\, v_3 + I_6\, v_6) - - 0.25\, L^o(c_N\, \psi - \sin\psi)(\xi - \xi^3) \tag{42}$$
$$w_b = \sum_{k=7}^{n} I_k\, v_k \quad ; \quad n \geqslant 7$$
$$c_N\, \varphi = -\frac{dw}{dx} - \sin\psi\ ; \quad D_M = \frac{d\varphi}{dx}$$

where

$$x = 0.5\, L^o(1 + \xi)$$
$$I_2 = 0.25\ (2 - 3\,\xi + \xi^3)$$
$$I_5 = 0.25\ (2 + 3\,\xi - \xi^3) \tag{43}$$

$$I_3 = 0.125\ (-1 + \xi + \xi^2 - \xi^3)$$
$$I_6 = 0.125\ (1 + \xi - \xi^2 - \xi^3) \qquad (43)$$

and for $k \geqslant 7$:

$$I_k = (1 - \xi^2)\xi^{(7-k)}$$
$$I_k(-1) = I_k(1) = \frac{dI_k}{d\xi}(-1) = \frac{dI_k}{d\xi}(1) = 0 \qquad (44)$$

So w, φ, D_M depend on v_k $(k = 2,3,5,6,\ldots)$. The third order polynomials (43) are recognised almost in all publications, which analyse beam element. The above polynomials satisfy the condition for the shape function. *The element equations.* Π^* is now dependent on v_k $(k \neq 1,4)$ and on N_D. Equating to zero the first variation of Π^*, considering Eqs. (26), (29), (40), (41) and (42) following expressions are obtained:

$$S_k = \int_0^{L^o} (M \frac{\partial D_M}{\partial v_k} dx + c_N N \varphi \frac{\partial \varphi}{\partial v_k} - p_z \frac{\partial w}{\partial v_k})dx$$
$$k = 2,3,5,6,7,\ldots \qquad (45)$$

$$S_1 = -0.5 \int_0^{L^o} p_x\, dx - N_D \quad ; \quad S_4 = -\int_0^{L^o} p_x\, dx + N_D$$

$$\left.\begin{aligned} R_D &= \frac{1}{L^o} \int_0^{L^o} (-D_N + \frac{du}{dx} + \cos\psi - 1 + \frac{c_N}{2}\varphi^2)dx \\ &= \frac{1}{L^o} \int_0^{L^o} (-D_N + D_G)dx \\ \frac{1}{L^o} \int_0^{L^o} D_G\, dx &= \frac{v_4 - v_1}{L^o} + \cos\psi - 1 + \frac{c_N}{2L^o} \int_0^{L^o} \varphi^2\, dx \end{aligned}\right\} \qquad (46)$$

where:

$$R_D = 0$$
$$S_k = 0 \qquad \text{for } 7 \leqslant k \leqslant n \qquad (47)$$

Varying the shape function w_s the expressions for end actions S_2, S_3, S_5, and S_6 are obtained. The bubble function w_b does not displace the ends of the element and the corresponding generalized force S_k must be zero (47). Similarly N_D, Eq. (40), does not influence the global element equilibrium and the corresponding generalized stretching deformation R_D must be zero (47). The integrand in (46) can be treated as a difference between two stretching deformations, the first one D_N is defined by constitutive relations and the second one D_G by geometrical relations (46).

Incremental equations. The increments of S_k and R_D are caused by increments Δv_k and ΔN_D, Eqs. (45), (46):

$$\Delta S_k = \sum_{\substack{i=2 \\ i\neq 4}}^{n} \Delta v_i \int_0^{L^o} \left(\frac{\partial M}{\partial D_M} \frac{\partial D_M}{\partial v_i} + N\, c_N \frac{\partial \varphi}{\partial v_i} \frac{\partial \varphi}{\partial v_k} \right) dx$$

$$+ \Delta N_D \int_0^{L^o} \left(\frac{\partial M}{\partial N} \frac{\partial D_M}{\partial v_k} + c_N\, \varphi \frac{\partial \varphi}{\partial v_k} \right) dx$$

$$\Delta S_4 = \Delta N_D \quad ; \quad \Delta S_1 = - \Delta N_D \tag{48}$$

$$L^o\, \Delta R_D = \sum_{\substack{i=2 \\ i\neq 4}}^{n} \Delta v_i \int_0^{L^o} \left(- \frac{\partial D_N}{\partial D_M} \frac{\partial D_M}{\partial v_i} + c_N\, \varphi \frac{\partial \varphi}{\partial v_i} \right) dx \; +$$

$$+ \Delta v_4 - \Delta v_1 - \Delta N_D \int_0^{L^o} \frac{\partial D_N}{\partial N}\, dx$$

The increments (48) can be expressed in more compact form, if (24) and (26) are taken into account:

$$\Delta S_k = K_{ki}\, \Delta v_i + G_k\, \Delta N_D$$

$$L^o\, \Delta R_D = G_i\, \Delta v_i - \int_0^{L^o} \frac{dx}{A_t}\, \Delta N_D \tag{49}$$

$$i,k = 1,2,3,\ldots,n$$

$$K_{ik} = K_{ki}$$

and the summation convention is introduced. Using Newton - Raphson iteration, Eqs. (47) take the form:

$$\Delta R_D^{(s)} = - R_D^{(s)} \tag{50}$$

$$\Delta S_k^{(s)} = - S_k^{(s)} \quad \text{for } 7 \leqslant k \leqslant n \tag{51}$$

where (s) is the iteration number, $\Delta R_D^{(s)}$ and $\Delta S_k^{(s)}$ are corrections to $R_D^{(s)}$ and to $S_k^{(s)}$. In comparison with the "classical" beam element ($w_b = 0$) additional unknowns (v_k, $k \geqslant 7$) and corresponding equations (51) are introduced. Assembling the elements the nodal point equations are obtained using the known procedure [7], which is not presented here. To get symmetric tangent matrix nodal forces are supposed to be conservative. The Eqs. (50) and (51) are used for the statical condensation of corresponding increments N_D and v_k $(k \geq 7)$; thus the total frame matrix presents some modified stiffness matrix and it is of the same order as in the case of the classical beam analysis.

Comparison of different order bubble functions. The problem can be solved analytically only in some simple cases which can be used for testing the elements with different order bubble functions.

In Table 1 the symbol P_j $(j = 1,2,3,\ldots)$ is used to present the element in which the bending deformation D_M is the j-th order polynomial, it means that the corresponding bubble function is of order j+2, so that the notation for the classical beam element is P_1. The classical buckling of the clamped - clamped beam with linear constant rigidity is very suitable for testing the elements.

Element / Number of elements	P_1	P_3	P_4
	% Error	% Error	% Error
1		6.40	0.06
2	1.32	0.	0.
4	0.75		
6	0.16		
8	0.05		

Table 1. Classical buckling of the clamped - clamped beam

The results show that the relative error for critical normal force is negligible even in the case when one P_4 element presents the beam. To get nearly the same results 8 equal length P_1 elements have to be used.

Figure 5 shows the elasto - plastic cantilever which was analised using f. o. t. Numerical results are calculated by numerical integration along the element. The j-th order polynomial can be represented by j+1 values of D_M, which is approximated as the j-th order polynomial; thus the most suitable number of subdivision points for P_j element is j+1. Using greater number of points a strong divergence of results may occur. The results in Fig. 5 show that one element P_4 gives a quite good approximation of cantilever up to 97 % full plastic force P_{fp}.

By different position of the points different results are obtained. Using Lobatto integration [10] the P_{fp} is calculated exactly, while considering Legendre integration the results for w are more accurate, but the relative error for P_{fp} is 5 %. The comparison between P_1 and P_4 elements shows that one P_4 element gives better results than 4 P_1 elements.

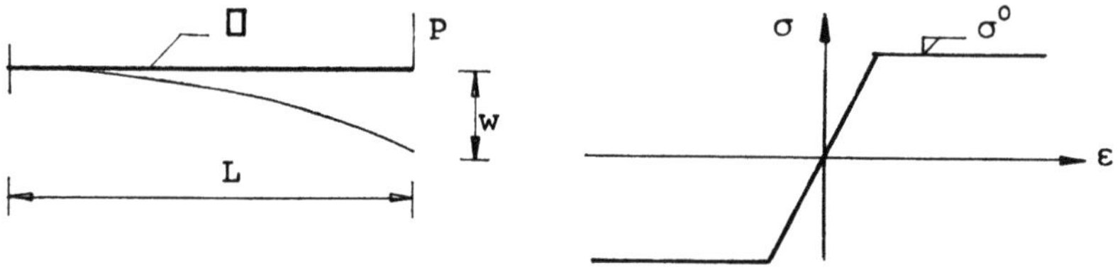

$$\omega = \frac{P\,L}{\sigma^o\,W} \quad ; \quad \omega_{fp} = \frac{P_{fp}\,L}{\sigma^o\,W} = 1.5$$

P_{fp} ... full plastic force

W ... section moduli of the cross section

First order theory						
ω	% Error for w					
	Number of P_1 elements				Different numerical integration one element P_4	
	2	3	4	8	Lobatto	Legendre
1.1	- 0.1	- 0.1	- 0.1	0.0	0.1	- 0.0
1.2	- 0.6	- 0.1	- 0.0	- 0.0	0.1	0.1
1.3	- 0.9	- 0.2	- 0.1	- 0.0	- 0.0	- 0.1
1.4	- 2.2	- 0.9	- 0.5	- 0.1	0.8	- 0.3
1.45	- 4.5	- 2.3	- 1.4	- 0.3	3.5	- 0.9
1.47	- 6.5	- 3.8	- 2.5	- 0.7	7.2	- 1.8
% Error for ω_{fp}	11.2	7.6	5.6	2.7	0.0	4.9

ω	% Error for w			
	One element - Lobatto integration			
	P_4	P_5	P_6	P_8
1.1	0.104	0.025	- 0.019	0.
1.2	0.055	0.081	0.008	- 0.009
1.3	- 0.043	0.034	0.072	- 0.005
1.4	0.841	0.414	- 0.051	0.042
1.45	3.465	1.524	0.497	0.120
1.47	7.200	3.377	1.464	0.382

Fig. 5. Elasto - plastic cantilever analysis (f. o. t.)

Discussion on c_N. Using simplified kinematic relations (15) the factor c_N was introduced. Its meaning can be seen if global element equilibrium is considered, Fig. 1 ($\vec{p} = \vec{0}$, $\vec{\alpha}_x = \vec{e}_x$):

$$\vec{F}^1 = S_1 \, \vec{e}_x + S_2 \, \vec{e}_z \; ; \; \vec{F}^2 = S_4 \, \vec{e}_x + S_5 \, \vec{e}_z \tag{52}$$

$$\vec{F}^1 + \vec{F}^2 = \vec{0} \; ; \; L \, (\vec{F}^2 \, \vec{e}_z) = S_3 + S_6 \tag{53}$$

Transformed form ($\vec{e}_x = \vec{\alpha}_x$) of the simplified equations (33), (34):

$$\vec{F}^1 + \vec{F}^1 = \vec{0} \; ; \quad c_N \, L^o (\vec{F}^2 \, \vec{e}_z) = S_3 + S_6 \tag{54}$$

is identical to (53) if c_N is equal to L/L^o (15). So the simplified equations exactly satisfy global element equilibrium. Using the incremental form of (53):

$$\Delta\vec{F}^1 + \Delta\vec{F}^2 = \vec{0} \tag{55}$$

$$\Delta L \, (\vec{F}^2 \, \vec{e}_z) + L \, (\Delta\vec{F}^2 \, \vec{e}_z) + L \, (\vec{F}^2 \, \Delta\vec{e}_z) = \Delta S_3 + \Delta S_6$$

where (Fig. 1)

$$\begin{aligned} (\Delta\vec{F}^2 \, \vec{e}_x) &= \Delta(\vec{F}^2 \, \vec{e}_x) - (\vec{F}^2 \, \Delta\vec{e}_z) \\ \Delta\vec{e}_x &= - \, \vec{e}_z \, \Delta\psi \; ; \; \Delta\vec{e}_z = \vec{e}_x \, \Delta\psi \end{aligned} \tag{56}$$

the expressions for the end action increments can be stated in the form:

$$\Delta\vec{F}^1 = - \, \Delta\vec{F}^2 \tag{57}$$

$$\begin{aligned} \Delta\vec{F}^2 \, \vec{e}_x &= S_5 \, \Delta\psi + \Delta \, S_4 \\ \Delta\vec{F}^2 \, \vec{e}_z &= - \frac{\Delta L}{L} S_5 - \Delta\psi \, S_4 + \frac{1}{L} (\Delta S_3 + \Delta S_6) . \end{aligned} \tag{58}$$

When using (57) and (58) the correct tangent matrix is obtained. Many authors use modified Eqs. (54) and (58) with c_N equal to 1 ; this way a nonsymmetric tangent matrix is derived and usually the term $S_5 \, \Delta\psi$ is neglected. Another way of obtaining an approximate tangent matrix is to introduce $L = L^O$ into Eqs. (53) and (58). In the method proposed here the global element equilibrium is satisfied exactly, while neglecting Δc_N leads to an approximate tangent matrix. In the case of zero bending deformation (truss) this matrix becomes the rigorous tangent matrix.

C o n c l u s i o n

A consistent formulation for the geometrically nonlinear behaviour of elasto - plastic plane frames has been presented. The numerical examples show that in most cases the whole beam (column) can be treated as a single P_4 element, when it is loaded only at the ends. The computer program NONFRAN was developped using the P_4 element. Thanks to the presented formulation it works with all four methods mentioned above, i.e. *l. - l.* , *l. - s.* , *s. o. t.* and *f. o. t.* The program has two versions; the first is based on the displacement method [11], while the second uses the mixed approach. It is often used in structural analysis of frames. This was the reason why the computer costs had to be lowered. It was done by using the P_4 element which is sufficiently accurate for practical use, but less time consuming than P_6 , though the latter gives better results. Tests show that mixed approach requires less iterations than the displacement approach.

R e f e r e n c e s

1 Pflüger, A.: Stabilitätsprobleme der Elastostatic. Springer 1964.

2 Kachanov, L. M.: Foundations of the Theory of Plasticity. North - Holland - Amsterdam - London 1971.

3 Frey, F.; Lemaire, E.: PAA - PBB element. FINEL G - User's manual, Université de Liège, Laboratoire de mecanique des materiaux et statique des constructions 1978.

4 Vinnakota, S.: The Elasto - Plastic Stability of Frames. I. C. E. Monthly / Vol. III No. 1. 1974.

5 Bäcklund, J.: Large Deflection Analysis of Elasto - Plastic Beams and Frames. I N T. J. Mechanical Science, 1976, Vol. 1, 8.

6. Oden, J. T.: Finite Element of Nonlinear Continua. McGraw - Hill 1972.

7 Zienkiewicz, O. C.: The Finite Element Method. McGraw - Hill 1977.

8 Argyris, J. H.; Dunne, P. C.; Scharpf, D. W.: On Large Displacement-Small Strain Analysis of Structures with Rotational Degrees of Freedom. Comp. Meths. Appl. Mech. Eng. 14 (1978), 401 - 451.

9 Flügge, W.: Handbook of Engineering Mechanichs. McGraw - Hill 1962.

10 Abramowitz, M.; Stegun, I. A.: Handbook of Mathematical Functions. Dover Publ. 1966.

11 Banovec, J.; Marinček, M.: Some computer programs for inelastic buckling and for instability of planar frames. Liège Colloquium on Stability of steel structures (1977), Preliminary Report.

Elasto-Plastic Boundary Element Analysis

J. C. F. TELLES,
Southampton University, U. K.,

C. A. BREBBIA
University of California, Irvine, USA

Summary

This paper presents the complete and correct formulation for the boundary element method applied to plasticity. Three alternative formulations - initial strain, initial stress and fictitious traction and body forces approach - are discussed. The paper explains how different yield criteria can be implemented and gives, in matrix form, the required expressions for the stepwise plasticity solution. This formulation relates stresses to the initial elastic solution and the plastic strains, being very simple to apply. The procedure is only possible due to the reduced number of unknowns originated by the boundary element method.

Several examples are presented to illustrate the advantages of using boundary elements to solve plasticity problems in preference to finite elements.

1. Introduction

In recent years researchers have become increasingly interested in solving complex stress analysis problems using boundary methods rather than domain techniques. This has resulted in a reassessment of classical boundary integral equation techniques and in their interpretation as an approximate method of analysis [1]. The concept of boundary elements - i.e. elements only on the surface of the domain but with interpolation functions similar to those used in finite elements - was introduced at the beginning of the 1970's, the combination of integral equations accuracy with finite elements versatility resulted in what is now called the boundary element method. This technique offers several advantages over finite elements such as, the possibility of working with boundaries extending to infinity; having smaller systems of equations and more accurate results for stress concentration regions and requiring much less data than finite elements to run the same problem. This last characteristic is, without doubt, the most interesting feature of boundary elements for the practicing engineer or user.

Several important papers have been published on the application of integral equations for solving plasticity problems [2, 3, 4] but none of them presented the complete expressions for strains or stresses at internal points. These expressions were given by Bui in 1978 [5] for three dimensional bodies and then Telles and Brebbia [6] in 1979 extended them to two dimensional problems. The latter authors presented the complete formulation for the boundary element method applied to two and three dimensional plasticity problems.

In this paper, the application of boundary elements to plasticity is presented, employing linear interpolation functions for the boundary elements as well as for the internal cells. The problem of accurately integrating the domain integrals has been solved using a semi-analytical approach, which can also be applied to compute principal values. Three alternative boundary element formulations are discussed; i) the initial strain, ii) the initial stress and iii) the fictitious traction and body forces approach. The paper explains how different yield criteria can be introduced into the formulation in conjunction with the normality principle.

The required stepwise solution procedure is formulated in matrix form and a simple recursive relation is presented relating the stresses to the initial elastic solution and the plastic strains. This approach is only possible due to the reduced number of unknowns originated by the boundary element method, but would not be suitable for finite element programs for instance.

The results found using boundary elements for a series of

examples are compared with analytical s...
element results. Some of the applicatio...
advantages of using boundary elements ove...
techniques to solve infinite medium proble...
show how one can take advantage of symme...
significantly reduce the required number ... I
all the cases the agreement between bou...
other solutions is satisfactory. The ...
the considerable reductions achieved in ...
required to run a problem using boundary ...
cells, for instance, are only needed in ...
plasticity, but they do not increase the ...
which are still those defined on the bounda...

2. Governing Equations

The total strain rate is assumed to be ...

$$\dot{\varepsilon}_{ij} = \tfrac{1}{2}\left(\dot{u}_{i,j} + \dot{u}_{j,i}\right) = \dot{\varepsilon}^{e}_{ij} + \dot{\varepsilon}^{p}_{ij}$$

where $\dot{u}_i$ are the displacement rate comp... $\dot{\varepsilon}^{p}_{ij}$ are respectively the elastic and plas... the strain rate tensor. Time derivati... a dot and space derivatives by a comma.

The equilibrium conditions are,

$$\dot{\sigma}_{ij,i} + \dot{b}_j = 0$$

in the interior Ω of the body, and

$$\dot{p}_i - \dot{\sigma}_{ij} n_j = 0$$

on the boundary Γ, being $\dot{b}_j$ the body fo... per unit volume, $\dot{\sigma}_{ij}$ the stress rate compone... traction rate components per unit area a... ward normal to the boundary of the body.

If we consider the plastic strain rates ... and apply Hooke's law to the elastic par... dimensional strain rate tensor, the follo... arises,

$$\dot{\sigma}_{ij} = 2G\left(\dot{\varepsilon}_{ij} - \dot{\varepsilon}^{p}_{ij}\right) + \frac{2G\nu}{1-2\nu}\left(\dot{\varepsilon}_{\ell\ell} \ldots\right.$$

G: shear modulus and ν: Poisson's ratio

The above expression can be rewritten in terms of initial stresses,

$$\dot{\sigma}_{ij} = 2G\dot{\varepsilon}_{ij} + \frac{2G\nu}{1-2\nu} \varepsilon_{\ell\ell} \delta_{ij} - \dot{\sigma}^p_{ij} \tag{5}$$

where $\dot{\sigma}^p_{ij}$ represents the components of the "initial stresses" given by,

$$\dot{\sigma}^p_{ij} = 2G \dot{\varepsilon}^p_{ij} + \frac{2G\nu}{1-2\nu} \dot{\varepsilon}^p_{kk} \delta_{ij} \tag{6}$$

The substitution of equation (4) into (2) and (3), together with equation (1) gives [7],

$$\dot{u}_{j,\ell\ell} + \frac{1}{1-2\nu} \dot{u}_{\ell,\ell j} = 2(\dot{\varepsilon}^p_{ij,i} + \frac{\nu}{1-2\nu} \dot{\varepsilon}^p_{kk,j}) - \frac{\dot{b}_j}{G} \tag{7}$$

and

$$\dot{p}_i + 2G(\dot{\varepsilon}^p_{ij} n_j + \frac{\nu}{1-2\nu} \dot{\varepsilon}^p_{kk} n_i) = \frac{2G\nu}{1-2\nu} \dot{u}_{\ell,\ell} n_i + G(\dot{u}_{i,j} + \dot{u}_{j,i}) n_j \tag{8}$$

Equation (7) is an extended form of Navier's equation and (8) represents its boundary conditions. The above expressions can be written in the following form,

$$\dot{u}_{j,\ell\ell} + \frac{1}{1-2\nu} \dot{u}_{\ell,\ell j} = - \dot{\bar{b}}_j/G \tag{9}$$

and

$$\dot{\bar{p}}_i = \frac{2G\nu}{1-2\nu} \dot{u}_{\ell,\ell} n_i + G(\dot{u}_{i,j} + \dot{u}_{j,i}) n_j \tag{10}$$

where $\dot{\bar{b}}_j$ and $\dot{\bar{p}}_i$ are pseudobody forces and pseudotractions given by,

$$\dot{\bar{b}}_j = \dot{b}_j - 2G(\dot{\varepsilon}^p_{ij,i} + \frac{\nu}{1-2\nu} \dot{\varepsilon}^p_{kk,j}) \tag{11}$$

and

$$\dot{\bar{p}}_i = \dot{p}_i + 2G(\dot{\varepsilon}^p_{ij} n_j + \frac{\nu}{1-2\nu} \dot{\varepsilon}^p_{kk} n_i) \tag{12}$$

One can notice that equation (9) represents a set of three quasi-linear partial differential equations for the displacements (plasticity terms appear on the right hand side). This enables us to apply as a fundamental solution for the boundary element formulation the singular solution of the Navier equations due to Kelvin [1].

Expressions (1) to (12) are valid for plane strain ($i, j, \ell = 1, 2$; $k = 1, 2, 3$) and also for plane stress ($i, j, k, \ell = 1, 2$) if ν is replaced by $\bar{\nu} = \nu/(1+\nu)$.

Fundamental Solutions The singular fundamental solutions corresponding to unit point loads can be written as,

$$u^*_{ij} = \frac{1}{16\pi(1-\nu)Gr}\{(3-4\nu)\,\delta_{ij} + r_{,i}\,r_{,j}\} \tag{13}$$

for 3-D problems

$$u^*_{ij} = \frac{-1}{8\pi(1-\nu)G}\{(3-4\nu)\,\ell n(r)\,\delta_{ij} - r_{,i}\,r_{,j}\} \tag{14}$$

for 2-D plane strain problems

$$p^*_{ij} = \frac{-1}{4\alpha\pi(1-\nu)r^\alpha}\{[(1-2\nu)\,\delta_{ij} + \beta r_{,i}\,r_{,j}]\frac{\partial r}{\partial n} - (1-2\nu)(r_{,i}\,n_j - r_{,j}\,n_i)\} \tag{15}$$

$\alpha = 2, 1$; $\beta = 3, 2$ for 3-D and 2-D plane strain respectively.

u^*_{ij} and p^*_{ij} represent the displacements and tractions in the j direction due to a unit force acting in i direction in which r is the distance from the point of application of the load to the point under consideration. The derivatives of r are taken with reference to the coordinates of the latter point and δ_{ij} is the Kronecker delta.

The strains (ε_{jk}) at any point within the body are given by,

$$\varepsilon^*_{jki} = \frac{-1}{8\alpha\pi(1-\nu)Gr^\alpha}\{(1-2\nu)\,(r_{,k}\,\delta_{ij} + r_{,j}\,\delta_{ik}) - r_{,i}\,\delta_{jk} + \beta r_{,i}\,r_{,j}\,r_{,k}\} \tag{16}$$

and the stresses,

$$\sigma^*_{jki} = C_{jkrs}\,\varepsilon^*_{rsi} = \frac{-1}{4\alpha\pi(1-\nu)r^\alpha}\{(1-2\nu)\,(r_{,k}\,\delta_{ij} + r_{,j}\,\delta_{ki} - r_{,i}\,\delta_{jk}) + \beta\,r_{,i}\,r_{,j}\,r_{,k}\} \tag{17}$$

... sotropic tensor of elastic constants ... ly written as,

$$\ldots \delta_{ij}\,\delta_{rs} + G(\delta_{ir}\,\delta_{js} + \delta_{is}\,\delta_{jr}) \tag{18}$$

... expressions are valid for plane stress if

... ve Boundary Element Formulations

... We start this section by introducing an ... ulation which is particularly simplified ... plastic strains to be incompressible. ... ctive but is justified by the fact that ... of the formulation is only intended for ... of the von Mises yield criterion.

... initial strain formulation for three ... ity problems leads to the following

$$\ldots u^{*}_{ij}\,\dot{p}_j\,d\Gamma - \int_{\Gamma} p^{*}_{ij}\,\dot{u}_j\,d\Gamma + \int_{\Omega} u^{*}_{ij}\,\dot{b}_j\,d\Omega + \int_{\Omega} \sigma^{*}_{jki}\,\dot{\varepsilon}^{p}_{jk}\,d\Omega \tag{19}$$

... valid for boundary points ($C_{ij} = \delta_{ij}/2$ ... es, see [19] otherwise) and also for ... with $C_{ij} = \delta_{ij}$. If metal plasticity is ... , plastic strains are deviatoric, i.e.

$$\dot{\varepsilon}^{p}_{11} + \dot{\varepsilon}^{p}_{22} + \dot{\varepsilon}^{p}_{33} = 0 \tag{20}$$

... stress rates at interior points ... by use of expressions (1) and (4). The ... of equation (19) yields,

$$\ldots \frac{\partial u^{*}_{ij}}{\partial x_m}\,\dot{p}_j\,d\Gamma - \int_{\Gamma} \frac{\partial p^{*}_{ij}}{\partial x_m}\,\dot{u}_j\,d\Gamma + \int_{\Omega} \frac{\partial u^{*}_{ij}}{\partial x_m}\,\dot{b}_j\,d\Omega + \int_{\Omega} \sigma^{*}_{jki}\,\dot{\varepsilon}^{p}_{jk}\,d\Omega\} \tag{21}$$

Where the derivatives are taken with reference to the load point (these derivatives are written explicitly to differentiate from the previous ones indicated by comma). Notice that we need to perform the derivative of the plastic strain rate integral, which is strongly singular. The proper expression for this derivative has been presented elsewhere [6] and is repeated here for completeness,

$$\frac{\partial}{\partial x_m} \left\{ \int_\Omega \sigma^*_{jki} \, \dot{\varepsilon}^p_{jk} \, d\Omega \right\} = \int_\Omega \frac{\partial \sigma^*_{jki}}{\partial x_m} \, \dot{\varepsilon}^p_{jk} \, d\Omega + \frac{(8-10\nu)}{15(1-\nu)} \, \dot{\varepsilon}^p_{im} \tag{22}$$

where the integral on the right hand side is to be interpreted in the Cauchy principal value sense and the last term comes from a boundary integral over the surface of a sphere of unit radius centred at the singular point.

From now on the reader is referred to appendix A for all components of the tensors related to the fundamental solutions.

The above expressions together with (1) and (4) allow for the determination of the internal stresses,

$$\dot{\sigma}_{ij} = \int_\Gamma (-\sigma^*_{ijk}) \, \dot{p}_k \, d\Gamma - \int_\Gamma p^*_{ijk} \, \dot{u}_k \, d\Gamma + \int_\Omega (-\sigma^*_{ijk}) \, \dot{b}_k \, d\Omega + \int_\Omega \sigma^*_{ijk\ell} \, \dot{\varepsilon}^p_{k\ell} \, d\Omega - \frac{2G(7-5\nu)}{15(1-\nu)} \, \dot{\varepsilon}^p_{ij} \tag{23}$$

where the last two terms represent the influence of the plastic strains.

For plane strains the procedure is analogous, the only difference being the fact that the plastic strain rate integrals still have to take into consideration the work performed in the third direction ($\sigma^*_{33i} \, \dot{\varepsilon}^p_{33}$). This effect is easily incorporated through the assumption of incompressible plastic strains [4] leading to what follows,

$$C_{ij}\, u_j = \int_\Gamma u^*_{ij}\, \dot{p}_j\, d\Gamma - \int_\Gamma p^*_{ij}\, \dot{u}_j\, d\Gamma + \int_\Omega u^*_{ij}\, \dot{b}_j\, d\Omega + \int_\Omega \hat{\sigma}^*_{jki}\, \dot{\varepsilon}^p_{jk}\, d\Omega \tag{24}$$

where

$$\hat{\sigma}^*_{jki} = \sigma^*_{jki} + \frac{2\nu\delta_{jk} r_{,i}}{4\pi(1-\nu) r} \tag{25}$$

and the internal stress rates,

$$\dot{\sigma}_{ij} = \int_\Gamma (-\sigma^*_{ijk})\, \dot{p}_k\, d\Gamma - \int_\Gamma p^*_{ijk}\, \dot{u}_k\, d\Gamma + \int_\Omega (-\sigma^*_{ijk})\, \dot{b}_k\, d\Omega + \int_\Omega \hat{\sigma}^*_{ijk\ell}\, \dot{\varepsilon}^p_{k\ell}\, d\Omega - \frac{G}{4(1-\nu)} \left[2\dot{\varepsilon}^p_{ij} + (1-4\nu)\, \varepsilon_{\ell\ell}\, \delta_{ij}\right] \tag{26}$$

in which

$$\hat{\sigma}^*_{ijk\ell} = \sigma^*_{ijk\ell} + \frac{G}{2\pi(1-\nu) r^2} \left[4\nu\, r_{,i}\, r_{,j}\, \delta_{k\ell} - 2\nu\, \delta_{ij}\, \delta_{k\ell}\right] \tag{27}$$

and the last integral is to be taken in the Cauchy principal value sense.

Plane stress problems can also be solved by using equations (24) and (26) with $\hat{\sigma}^*_{jki} = \sigma^*_{jki}$, $\hat{\sigma}^*_{ijk\ell} = \sigma^*_{ijk\ell}$, ν replaced by $\bar{\nu}$ in all $(\ )^*$ tensors and the last term in (26) being substituted by,

$$-\frac{G}{4(1-\bar{\nu})} \left[2\dot{\varepsilon}^p_{ij} + \dot{\varepsilon}^p_{\ell\ell}\, \delta_{ij}\right] \tag{28}$$

Initial Stress: In order to introduce the initial stress formulation, let us merely study the plastic strain rate integral presented in (19). Recalling expression (17) we see that,

$$\int_{\Omega} \sigma^{*}_{jki} \, \dot{\varepsilon}^{p}_{jk} \, d\Omega = \int_{\Omega} C_{jkrs} \, \varepsilon^{*}_{rsi} \, \dot{\varepsilon}^{p}_{jk} \, d\Omega \tag{29}$$

by simple inspection of (18) we can make,

$$C_{jkrs} = C_{rsjk} \tag{30}$$

moreover

$$C_{rsjk} \, \dot{\varepsilon}^{p}_{jk} = \dot{\sigma}^{p}_{rs} \tag{31}$$

where $\dot{\sigma}^{p}_{rs}$ was given in (6).

Thus,

$$\int_{\Omega} \sigma^{*}_{jki} \, \dot{\varepsilon}^{p}_{jk} \, d\Omega = \int_{\Omega} \varepsilon^{*}_{jki} \, \dot{\sigma}^{p}_{jk} \, d\Omega \tag{32}$$

Hence the initial stress formulation can be applied with equal ease. For plane strain problems the formulation follows the same pattern and the equivalent expression is,

$$\int_{\Omega} \hat{\sigma}^{*}_{jki} \, \dot{\varepsilon}^{p}_{jk} \, d\Omega = \int_{\Omega} \varepsilon^{*}_{jki} \, \dot{\sigma}^{p}_{jk} \, d\Omega \tag{33}$$

and internal stresses can be computed by,

$$\begin{aligned} \dot{\sigma}_{ij} = & \int_{\Gamma} (-\sigma^{*}_{ijk}) \, \dot{p}_{k} \, d\Gamma - \int_{\Gamma} p^{*}_{ijk} \, \dot{u}_{k} \, d\Gamma \\ & + \int_{\Omega} (-\sigma^{*}_{ijk}) \, \dot{b}_{k} \, d\Omega + \int_{\Omega} \varepsilon^{*}_{ijk\ell} \, \dot{\sigma}^{p}_{k\ell} \, d\Omega \\ & - \frac{1}{8(1-\nu)} \left[2\dot{\sigma}^{p}_{ij} + (1-4\nu) \, \dot{\sigma}^{p}_{\ell\ell} \, \delta_{ij} \right] \end{aligned} \tag{34}$$

where the initial stress integral is to be interpreted in the principal value sense.

It is worth noting that here the initial stress integrals do not require the contribution of the work performed in the third direction, nor the assumption of incompressibility of the plastic strains needs to be made. This is because $\varepsilon^{*}_{33i} = 0$ and the effect of $\dot{\varepsilon}^{p}_{33}$ is already included into the components of $\dot{\sigma}^p_{ij}$. As a consequence, plane stress problems can be handled by the above expressions with the replacement of ν by $\bar{\nu}$ being the only modification.

Fictitious tractions and body forces: The last integral presented in equation (19) can be written in terms of the derivatives of u^*_{ij} as follows,

$$\int_\Omega \sigma^*_{jki}\ \dot{\varepsilon}^p_{jk}\ d\Omega = \int_\Omega \{G(u^*_{ij,k} + u^*_{ik,j}) + \frac{2G\nu}{1-2\nu}\ u^*_{i\ell,\ell}\ \delta_{jk}\}\ \dot{\varepsilon}^p_{jk}\ d\Omega \tag{35}$$

which after integrating by parts gives the identity,

$$\int_\Omega \sigma^*_{jki}\ \dot{\varepsilon}^p_{jk}\ d\Omega = \int_\Gamma u^*_{ij}\ 2G\ (\dot{\varepsilon}^p_{jk}\ n_k + \frac{\nu}{1-2\nu}\ \dot{\varepsilon}^p_{\ell\ell}\ n_j)\ d\Gamma - \int_\Omega u^*_{ij}\ 2G\ (\dot{\varepsilon}^p_{jk,k} + \frac{\nu}{1-2\nu}\ \dot{\varepsilon}^p_{\ell\ell,j})\ d\Omega \tag{36}$$

The substitution of (36) in (19) gives as a result,

$$c_{ij}\ \dot{u}_j = \int_\Gamma u^*_{ij}\ \dot{\bar{p}}_j\ d\Gamma - \int_\Gamma p^*_{ij}\ \dot{u}_j\ d\Gamma + \int_\Omega u^*_{ij}\ \dot{\bar{b}}_j\ d\Omega \tag{37}$$

where $\dot{\bar{b}}_j$ and $\dot{\bar{p}}_j$ were given in (11) and (12) respectively. Therefore, we have arrived at a plasticity formulation in which tractions and body force rates are fictitious (depend on the plastic strains), but the displacements are correct. In order to apply equation (37) one has to be aware that although it looks like the elastic application of the boundary element method, the internal stresses still have to be computed by use of equations (1) and (4), i.e.

$$\dot{\sigma}_{ij} = \int_\Gamma (-\ \sigma^*_{ijk})\ \dot{\bar{p}}_k\ d\Gamma - \int_\Gamma p^*_{ijk}\ \dot{u}_k\ d\Gamma + \int_\Omega (-\ \sigma_{ijk})\ \dot{\bar{b}}_k\ d\Omega - C_{ijk\ell}\ \dot{\varepsilon}^p_{k\ell} \tag{38}$$

Another feature of this formulation is that in contrast with the two previous approaches, it needs computation of space derivatives of the plastic strains (see eq. (11)). This may be considered as a disadvantage for numerical implementation, but, nevertheless, it is a valid procedure for formulating the B.E.M. to plasticity and still remains to be properly attempted.

4. Boundary Element Discretization

The numerical implementation of the initial strain formulation presented in (24) and (26) was given in [9]. It is the aim of this section to illustrate the initial stress formulation by an entirely similar procedure.

Let us suppose that the boundary of the body is represented by surface elements while the part (or parts) of its interior where plasticity is likely to occur is discretized into a number of internal cells. Under these assumptions, linear piecewise functions are chosen to interpolate tractions, displacements and "initial stresses", the two former over boundary elements and the latter over internal cells (body forces are not considered in this text). The unknowns produced by this discretization are interrelated by a series of coefficients found by integrating over boundary elements and internal cells.

From the substitution of (33) into (24) comes the following matrix relationship,

$$\underset{\sim}{H}\,\dot{\underset{\sim}{u}} = \underset{\sim}{G}\,\dot{\underset{\sim}{p}} + \underset{\sim}{Q}\,\dot{\underset{\sim}{\sigma}}^{p} \tag{39}$$

where matrices $\underset{\sim}{H}$ and $\underset{\sim}{G}$ are the same as obtained for elastic analysis and matrix $\underset{\sim}{Q}$ is due to the initial stress integral.

Similarly, equation (34) gives,

$$\dot{\underset{\sim}{\sigma}} = \underset{\sim}{G}'\,\dot{\underset{\sim}{p}} - \underset{\sim}{H}'\,\dot{\underset{\sim}{u}} + (\underset{\sim}{Q}' + \underset{\sim}{E}')\dot{\underset{\sim}{\sigma}}^{p} \tag{40}$$

in which $\underset{\sim}{E}'$ is a well defined matrix that represents the independent terms and $\underset{\sim}{Q}'$ stands for the initial stress rate integral. Matrices $\underset{\sim}{H}'$ and $\underset{\sim}{G}'$ are the same as those obtained for elastic analysis.

In order to extend the validity of equation (40) to stresses at boundary nodes, expressions computed by means of strain displacement relationships and traction rate values along each boundary element have to be employed. These expressions do not require any integration and in the local coordinate system $(\eta_1;\eta_2)$ of a linear boundary element located between nodes i and j are given by,

$$\dot{\sigma}_{11} = \frac{1}{1-\nu}\left\{\frac{2G}{\ell}\left({}^{j}\dot{u}_1 - {}^{i}\dot{u}_1\right) + \nu\dot{p}_2\right\} + \frac{\nu}{1-\nu}\dot{\sigma}^{p}_{22} - \dot{\sigma}^{p}_{11} \tag{41}$$

$$\dot{\sigma}_{12} = \dot{p}_1 \tag{42}$$

$$\dot{\sigma}_{22} = \dot{p}_2 \tag{43}$$

where ${}^{i}\dot{u}_1$ and ${}^{j}\dot{u}_1$ are the displacement rate values at nodes i and j and ℓ is the element size. For plane stress ν is replaced by $\bar{\nu}$

5. Numerical Integration

The process of integration over boundary elements is well-known and the procedure presented in [1] was followed.

For the initial stress rate integrals the semi-analytical integration scheme fully described in [9] was adopted. The technique is formulated with reference to a cylindrical coordinate system (r,ϕ) based at the singular point. Due to the very nature of the fundamental solution, analytical integration with respect to r turns to be very simple and removes the singularities. Integration with respect to the angle ϕ is then performed using one dimensional Gaussian quadrature formulae, five integration points being sufficient.

Notice that the procedure is also applied to compute the principal value of the domain integral presented in (34). But if higher order interpolation functions as well as different cell shapes were chosen, the principal value could still be computed by applying equation (40) to represent a state of constant initial stresses, in much the same way as it was shown in [6].

6. Elasto-plastic Stress Strain Relations

For the formulation of a theory which models elasto-plastic material deformation, three requirements have to be met, these are:

a) Explicit elastic relationship between stress and strain before the onset of plastic deformation.
b) A yield criterion indicating the stress level at which plastic flow commences.
c) Relationship between stress and strain for post yield behaviour.

The yield criterion for isotropic hardening can be written in general form as,

$$F(\sigma_{ij}, k) = 0 \tag{44}$$

where k is a hardening parameter that gives the instantaneous position of the yield surface in the n-dimensional stress space. Here, the work hardening hypothesis has been adopted, hence

$$k = W^p = \int \sigma_{ij} \, d\varepsilon^p_{ij} \tag{45}$$

On physical grounds, one can notice that the yield criterion to be independent of the orientation of the coordinate system employed, should be a function of the three stress invariants only. It is common to represent two of these invariants as functions of the deviatonic stresses,

$$\begin{aligned} J_1 &= \sigma_{ii} \\ J_2 &= \frac{1}{2} S_{ij} S_{ij} \\ J_3 &= \frac{1}{3} S_{ij} S_{jk} S_{ki} \end{aligned} \tag{46}$$

where

$$S_{ij} = \sigma_{ij} - \frac{1}{3} J_1 \delta_{ij} \tag{47}$$

In the present work, instead of J_3 the alternative stress invariant θ, known as the Lode angle [10] was used.

$$-\frac{\pi}{6} \leq \theta = \frac{1}{3} \sin^{-1} \left(-\frac{3\sqrt{3}}{2} \frac{J_3}{J_2^{3/2}} \right) \leq \frac{\pi}{6} \tag{48}$$

By using these stress invariants, different yield criteria can be applied, such as

Tresca:

$$2 \sqrt{J_2} \cos\theta - \sigma_o(k) = 0 \tag{49}$$

where $\sigma_o(k)$ is the uniaxial yield stress.

Von Mises:

$$\sqrt{3J_2} - \sigma_o(k) = 0 \tag{50}$$

Mohr-Coulomb:

$$\frac{J_1}{3} \sin\phi + \sqrt{J_2}(\cos\theta - \frac{1}{\sqrt{3}} \sin\theta \sin\phi) - c \cos\phi = 0 \tag{51}$$

in which ϕ is the angle of internal friction and c is the cohesion of the material.

Drucker-Prager:

$$\alpha J_1 + \sqrt{J_2} - K = 0 \tag{52}$$

where,

$$\alpha = \frac{2 \sin\phi}{\sqrt{3}(3-\sin\phi)} \quad ; \quad K = \frac{6c \cos\phi}{\sqrt{3}(3-\sin\phi)} \tag{53}$$

Mohr-Coulomb hypothesis can be simulated by the Drucker-Prager criterion in plane strains if α and K are written as [11],

$$\alpha = \frac{\tan\phi}{(9+12\tan^2\phi)^{\frac{1}{2}}} \quad ; \quad K = \frac{3c}{(9+12\tan^2\phi)^{\frac{1}{2}}} \tag{54}$$

For our practical purposes, equation (44) can then be written as,

$$F(\sigma_{ij},k) = f(\sigma_{ij}) - \psi(k) = 0 \tag{55}$$

where one can notice that $f(\sigma_{ij})$ is a scalar function of σ_{ij} which plays the role of an equivalent stress here designated by σ_e. As a consequence, we can define an equivalent plastic strain ε_e^p whose increment produces an increment in the plastic strain energy as follows,

$$\sigma_e \, d\varepsilon_e^p = \sigma_{ij} \, d\varepsilon_{ij}^p = dk \tag{56}$$

In order to obtain the stress-strain relations for post yield behaviour, let us first rewrite equation (4),

$$\dot{\sigma}_{ij} = C_{ijk\ell}(\dot{\varepsilon}_{k\ell} - \dot{\varepsilon}_{k\ell}^p) \tag{57}$$

Within the context of associated plasticity, the flow rule also known as normality principle, can be described by,

$$\dot{\varepsilon}_{ij}^p = d\lambda \, \frac{\partial F}{\partial \sigma_{ij}} \tag{58}$$

where $d\lambda$ is a proportionality factor, termed the plastic multiplier.

The substitution of (58) into (57) gives,

$$\dot{\sigma}_{ij} = C_{ijk\ell}(\dot{\varepsilon}_{k\ell} - a_{k\ell} \, d\lambda) \tag{59}$$

in which $a_{k\ell} = \frac{\partial F}{\partial \sigma_{k\ell}} = \frac{\partial f}{\partial \sigma_{k\ell}}$ (60)

When plastic yielding is occurring, the stresses satisfy equation (55) which by differentiating gives,

$$\dot{F} = a_{ij}\,\dot{\sigma}_{ij} - \frac{d\psi}{dk}\,\dot{k} = 0 \tag{61}$$

Or, according to (45),

$$a_{ij}\,\dot{\sigma}_{ij} - \frac{d\psi}{dk}\,\sigma_{ij}\,\dot{\varepsilon}^{p}_{ij} = 0 \tag{62}$$

From the application of the normality principle to equation (62) results

$$a_{ij}\,\dot{\sigma}_{ij} - \frac{d\psi}{dk}\,\sigma_{ij}\,a_{ij}\,d\lambda = 0 \tag{63}$$

If we substitute (59) into (63) and solve for $d\lambda$ comes,

$$d\lambda = \frac{1}{\gamma}\,a_{ij}\,C_{ijk\ell}\,\dot{\varepsilon}_{k\ell} \tag{64}$$

where

$$\gamma = a_{ij}\,C_{ijk\ell}\,a_{k\ell} + \frac{d\psi}{dk}\,\sigma_{ij}\,a_{ij} \tag{65}$$

Before we go further the last term in (65) can be examined. It can be shown that $f(\sigma_{ij})$ is homogeneous of degree one and this allows for the application of Euler's theorem as follows,

$$\sigma_{ij}\,\frac{\partial f}{\partial \sigma_{ij}} = f(\sigma_{ij}) = \sigma_e \tag{66}$$

The substitution of (66) and (56) into (65) gives,

$$\gamma = a_{ij}\,C_{ijk\ell}\,a_{k\ell} + H' \tag{67}$$

where $H' = d\psi/d\varepsilon^{p}_{e}$ can be interpreted as the slope of the uni-axial curve plotted as true stress versus true plastic strain.

Equation (64) can now be used to substitute $d\lambda$ in (59) providing the required incremental stress-strain relations

$$\dot{\sigma}_{ij} = C^{ep}_{ijk\ell}\,\dot{\varepsilon}_{k\ell} \tag{68}$$

in which

$$C^{ep}_{ijk\ell} = C_{ijk\ell} - \frac{1}{\gamma} C_{ijmn}\, a_{mn}\, a_{op}\, C_{opk\ell} \qquad (69)$$

For the application of the above relations to the initial stress formulation, a further modification has proved to be convenient. Let us adopt the following notation,

$$\dot{\sigma}^{e}_{ij} = C_{ijk\ell}\, \dot{\varepsilon}_{k\ell} \qquad (70)$$

where $\dot{\sigma}^{e}_{ij}$ stands for the components of the elastic stress rate tensor (i.e. these represent the stress values as if a pure elastic problem were being solved).

Equation (68) can then be rewritten in the following form

$$\dot{\sigma}_{ij} = \dot{\sigma}^{e}_{ij} - \frac{1}{\gamma} C_{ijmn}\, a_{mn}\, a_{k\ell}\, \dot{\sigma}^{e}_{k\ell} \qquad (71)$$

which means that the true stresses can be computed from the corresponding elastic stresses. In addition to this, the initial stress rates presented in (6) can also be calculated by the relation,

$$\dot{\sigma}^{p}_{ij} = \dot{\sigma}^{e}_{ij} - \dot{\sigma}_{ij} = \frac{1}{\gamma} C_{ijmn}\, a_{mn}\, a_{k\ell}\, \dot{\sigma}^{e}_{k\ell} \qquad (72)$$

By simply examining equation (5) we notice that equation (40) can be applied for the computation of $\dot{\sigma}^{e}_{ij}$ if matrix $\underset{\sim}{E}'$ is replaced by

$$\underset{\sim}{\bar{E}} = \underset{\sim}{E}' + \underset{\sim}{I} \qquad (73)$$

where $\underset{\sim}{I}$ is the identity matrix.

This gives,

$$\underset{\sim}{\dot{\sigma}}^{e} = \underset{\sim}{G}'\, \underset{\sim}{\dot{p}} - \underset{\sim}{H}'\, \underset{\sim}{\dot{u}} + \underset{\sim}{Q}^{*}\, \underset{\sim}{\dot{\sigma}}^{p} \qquad (74)$$

where,

$$\underset{\sim}{Q}^{*} = \underset{\sim}{Q}' + \underset{\sim}{\bar{E}} \qquad (75)$$

Before the application of the above expressions to solve plasticity problems, it should be pointed out that although solution algorithms are incremental, always finite sized load increments are prescribed and this may create some drifts of the stress level beyond the yield surface. If load increments are kept sufficiently small this problem is practically eliminated, but if relatively large load increment sizes are to be permitted, special techniques of the type presented in [12] have been found necessary to maintain the stresses on the yield surface.

7. Outline of Solution Technique

In order to minimise the computer effort for the initial stress formulation, let us reexamine equations (39) and (74). For a well-posed problem, a sufficient number of tractions and boundary displacements needs to be prescribed. The unknowns are then reordered leading to

$$\underset{\sim}{A}\,\dot{\underset{\sim}{x}} = \dot{\underset{\sim}{f}} + \underset{\sim}{Q}\,\dot{\underset{\sim}{\sigma}}^{p} \tag{76}$$

and similarly,

$$\dot{\underset{\sim}{\sigma}}^{e} = -\,\underset{\sim}{A}'\,\dot{\underset{\sim}{x}} + \dot{\underset{\sim}{f}}' + \underset{\sim}{Q}^{*}\,\dot{\underset{\sim}{\sigma}}^{p} \tag{77}$$

where vector $\dot{\underset{\sim}{x}}$ contain the unknown tractions and boundary displacements.

From the multiplication of (76) by $\underset{\sim}{A}^{-1}$ we get

$$\dot{\underset{\sim}{x}} = \underset{\sim}{R}\,\dot{\underset{\sim}{\sigma}}^{p} + \dot{\underset{\sim}{m}} \tag{78}$$

where

$$\underset{\sim}{R} = \underset{\sim}{A}^{-1}\,\underset{\sim}{Q} \tag{79}$$

and

$$\dot{\underset{\sim}{m}} = \underset{\sim}{A}^{-1}\,\dot{\underset{\sim}{f}} \tag{80}$$

Substituting (78) into (77) gives,

$$\dot{\underset{\sim}{\sigma}}^{e} = \underset{\sim}{S}\,\dot{\underset{\sim}{\sigma}}^{p} + \dot{\underset{\sim}{n}} \tag{81}$$

in which

$$\underset{\sim}{S} = \underset{\sim}{Q}^* - \underset{\sim}{A}' \underset{\sim}{R} \tag{82}$$

and

$$\dot{\underset{\sim}{n}} = \dot{\underset{\sim}{f}}' - \underset{\sim}{A}' \dot{\underset{\sim}{m}} \tag{83}$$

Note that vectors $\underset{\sim}{m}$ and $\underset{\sim}{n}$ represent the elastic solution to the incremental problem (actual solution in absence of plasticity). Furthermore, equations (78) and (81) remain valid if instead of incremental loading, the total load is applied. The only reason to proceed incrementally being the constitutive relations presented in (68). This enables us to compute load at first yield by simply scaling down the total elastic solution by a load factor λ_o. The incremental process starts at this load level and further values of the load factor are given by,

$$\lambda_i = \lambda_{i-1} + \beta \tag{84}$$

where $\beta = \lambda_o \omega$; ω is the given value of the load increment with reference to load at first yield.

For elasto-plastic solutions, equations (78) and (81) can therefore be applied as,

$$\underset{\sim}{x} = \underset{\sim}{R}(\underset{\sim}{\sigma}^p + \dot{\underset{\sim}{\sigma}}^p) + \lambda_i \underset{\sim}{m} \tag{85}$$

and

$$\underset{\sim}{\sigma}^e = \underset{\sim}{S}(\underset{\sim}{\sigma}^p + \dot{\underset{\sim}{\sigma}}^p) + \lambda_i \underset{\sim}{n} \tag{86}$$

or alternatively for pure incremental relations,

$$\dot{\underset{\sim}{x}} = \underset{\sim}{R}\, \dot{\underset{\sim}{\sigma}}^p + \beta \underset{\sim}{m} \tag{87}$$

and

$$\dot{\underset{\sim}{\sigma}}^e = \underset{\sim}{S}\, \dot{\underset{\sim}{\sigma}}^p + \beta \underset{\sim}{n} \tag{88}$$

where in both cases vectors $\underset{\sim}{m}$ and $\underset{\sim}{n}$ correspond to the application of the total load and $\dot{\underset{\sim}{\sigma}}^p$ stands for the current initial stress increment.

For a typical load increment (i.e., a given value of λ_i), the initial stress increment can be determined iteratively at each boundary node and internal point exhibiting elasto-plastic behaviour by two different processes. The former is in fact a pure incremental procedure. Once the load increment $\beta \underset{\sim}{n}$ is applied, the initial stress increment corresponding to the solution of the elastic problem is computed and has to be applied back into the body, providing an elastic stress redistribution. This operation, again generates a new initial stress field to be redistributed elastically and so on. Iteration is halted when the contribution of the last initial stress increment can be neglected.

The process is in essence comparable to what was presented in [13] for the finite element method and is summarized as follows,

a) compute elastic stress increment by,
   eq. (88) if first iteration is being performed or
   $\dot{\underset{\sim}{\sigma}}^e = \underset{\sim}{S}\ \dot{\underset{\sim}{\sigma}}^p$ otherwise
b) find true stress increment $\dot{\sigma}_{ij}$ (eq. (71))
c) verify convergence, i.e.
   compare $d\varepsilon_e^p$ calculated with its accumulated value obtained during the current load increment to see if it can be neglected.
d) calculate initial stress increment by,
   $\dot{\sigma}^p_{ij} = \dot{\sigma}^e_{ij} - \dot{\sigma}_{ij}$
e) accumulate values of initial stress and true stress,
   $\sigma^p_{ij} = \sigma^p_{ij} + \dot{\sigma}^p_{ij}$
   $\sigma_{ij} = \sigma_{ij} + \dot{\sigma}_{ij}$
f) continue with next node or point and start with b) until all nodes and points have been considered.
g) go to a) for a new iteration.

Iterations are performed until convergence is obtained (within prescribed tolerance) at every node or point.

It is interesting to note that in order to avoid cumulative

errors, $\dot{\underset{\sim}{\sigma}}^p$ obtained at the end of iterations is applied together with $\beta \underset{\sim}{n}$ in eq. (88) for the first iteration of the next load increment.

The second process, which proved to be less dependent on the load increment size but not always more economical, deals with accumulated values of the elastic stress in a similar fashion to the procedure adopted for the initial strain implementation [9].

The initial stress increment is kept separate from its accumulated values until convergence is obtained as follows,

a) compute elastic stress (eq.(86))

b) calculate elastic stress increment by,
$$\dot{\sigma}^e_{ij} = \sigma^e_{ij} - \sigma_{ij} - \sigma^p_{ij}$$

c) find true stress increment $\dot{\sigma}_{ij}$ (eq. (71))

d) verify convergence i.e.
compare $d\varepsilon^p_e$ calculated with its previous value.

e) calculate new estimate of initial stress increment by,
$$\dot{\sigma}^p_{ij} = \dot{\sigma}^e_{ij} - \dot{\sigma}_{ij}$$

f) continue with next node or point and start with b) until all nodes and points have been considered.

g) go to a) for a new iteration.

Once that convergence is obtained for all nodes and points, the initial stress increment is accumulated and its value is also used as an initial guess for the next load increment.

Note that neither procedure requires computation of the boundary unknowns. Consequently equation (85) need only be used to print the boundary unknowns once convergence is achieved. In addition we should emphasize that matrices $\underset{\sim}{R}$ and $\underset{\sim}{S}$ as well as vectors $\underset{\sim}{m}$ and $\underset{\sim}{n}$ remain unchanged throughout the incremental process. These features, also common with the initial strain procedure, represent a great saving in computer time.

Finally a question is posed,

which procedure is more efficient?

The answer can only be obtained in the light of experience. Regarding precision, the examples solved show that if load increment is kept small, the initial strain and both initial stress processes lead to entirely equivalent results (difference within the prescribed tolerance). Nevertheless with relation to computer time, the initial strain version has proved to be faster than the initial stress. This difference however, seems to be due to the fact that in contrast with the initial strain program which is restricted to von Mises yield criterion, both versions of the initial stress routine are more general and can handle four different yield criteria.

As load increment was increased, the initial strain and the second initial stress versions exhibited more stable results than the former initial stress procedure, despite special techniques to avoid stress drifts beyond the yield surface. But in some of the examples run with relatively large load increments, the pure incremental form of the initial stress program produced equally reliable results, requiring less computer time than its second version.

In conclusion, the above discussion demonstrates the equivalence of the three processes and furthermore shows that the efficiency of the procedure is problem dependent. Ideally, all three options should be available for a general case.

8. Examples

Following the discussion presented in the end of the last section, the results for a series of examples solved by the boundary element technique are now compared with analytical solutions where such solutions are available and with finite element results.

Notched Tensile Specimen:

This first example is one of the very early plasticity problems solved by using the finite element technique. Plane stress and plane strain results have been presented in several papers, creating a good opportunity to compare the boundary element computations.

Material parameters are as follows,

$E = 7000\ kg/mm^2$
$\sigma_o = 24.3\ kg/mm^2$
$\nu = 0.2$
$H' = 0$ (von Mises yield criterion)

Plane stress analysis was carried out using the discretization shown in figure 1. Note that symmetry was taken into consideration without boundary discretization of the symmetry axes, this is due to a direct condensation process which integrates automatically over reflected elements and cells in such a way that the size of the final matrices correspond to the reduced number of boundary nodes and internal points presented.

Figure 2 gives the load displacement curve for this case. It is seen that the curve remains nearly straight until very close to the limit load, when a sharp bend then occurs. Such behaviour was also observed by Yamada et al [14] in an entirely similar problem. The limit load achieved by the boundary elements ($2\sigma_a/\sigma_o = 1.21$) coincides with the results presented by Nayak and Zienkiewicz [12] using four different finite elements to analyse the same problem. Their limit load was found to vary between $2\sigma_a/\sigma_o = 1.19 - 1.23$ and simple triangular, isoparametric linear, quadratic and cubic elements were used, all four meshes with approximately 97 nodes.

For the plane strain case, because of a large spread of plastic zones before limit load is achieved, the number of internal points and cells was increased from 33 and 51 to 59 and 97 respectively. Load displacement curve is shown in figure 3 where equivalent finite element results presented by Chen [15] are also given. The limit load obtained by BEM ($2\sigma_a/\sigma_o = 1.64$) is below the value given by the finite element method ($2\sigma_a/\sigma_o = 1.85$). But, as stated by Chen, bound theorems demonstrate that the maximum load should lie between 1.52 and 1.73, which supports the boundary element results.

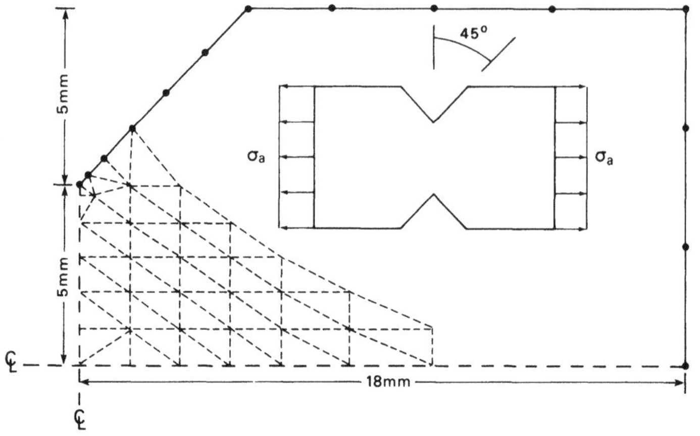

Fig 1 Notched tensile specimen Boundary element and internal cell discretization (plane stress case)

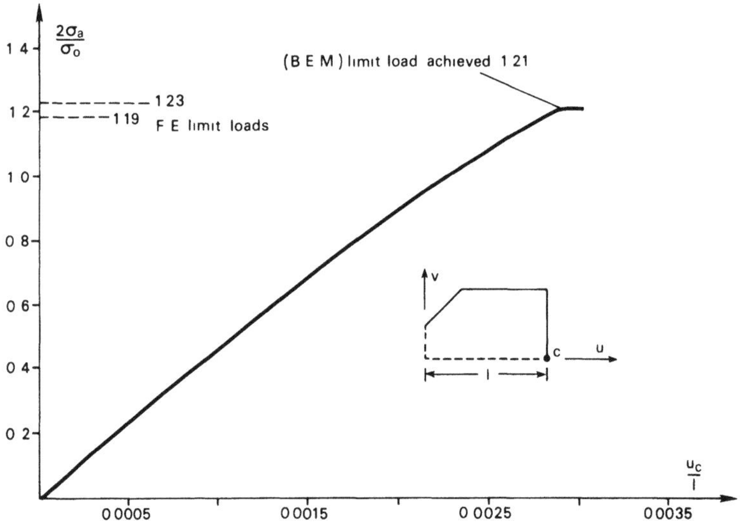

Fig 2 Load – displacement curve for notched specimen in plane stress

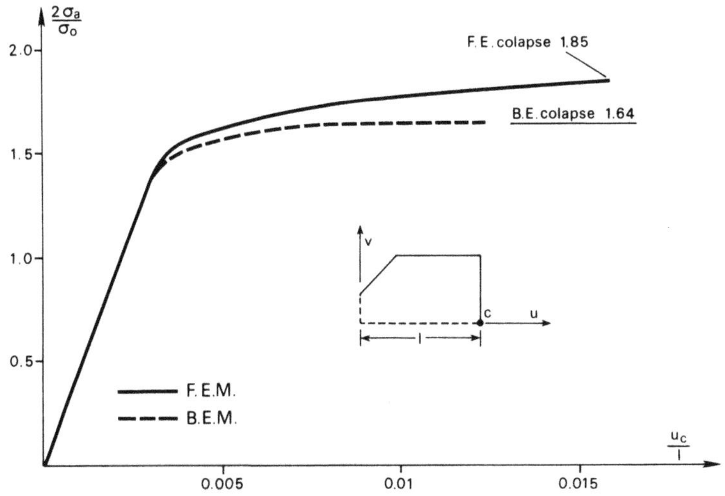

Fig 3 Load - displacement curves for notched specimen in plane strain

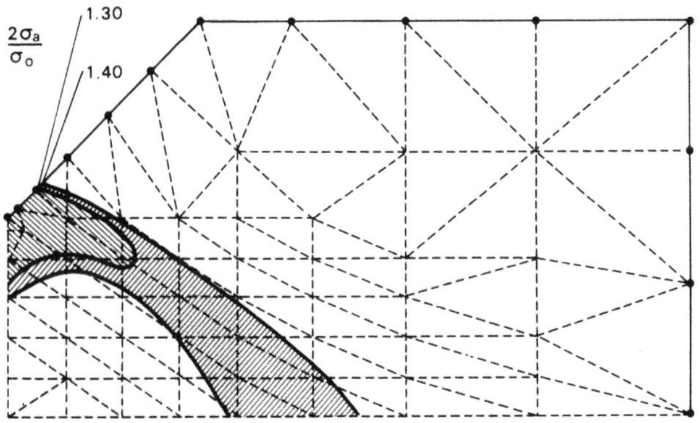

Fig 4 Plastic zones obtained by B E M for different load levels (plane strain)

Spread of plastic zones at lower load levels presented in figure 4 exhibits good agreement with finite element computations [13],[15] for the same problem.

Deep Circular Tunnel:
The second example was selected to emphasize the advantages of boundary elements over "domain" type techniques to solve infinite medium problems.

A circular excavation studied by Reyes [16] and later by Baker et al [17] with linear displacement triangular and quadrilateral finite elements respectively is here compared with boundary element results.

The plane strain problem was analysed under the Drucker-Prager simulation of Mohr-Coulomb yield criterion (α and K given by eq. (54)) and by assuming the infinite domain to be initially subjected to a uniform stress field of 1 ksi vertical and 0.4 ksi horizontal ($K_o = 0.4$). For the present study, external loads corresponding to the relaxation of this free stress field were applied along the surface of the opening.

The material (Rock) was assumed to be perfectly plastic with,

$$E = 500 \text{ ksi}$$
$$c = 0.28 \text{ ksi}$$
$$\nu = 0.2$$
$$\phi = 30^o$$

Boundary element and internal cell discretization is presented in figure 5 where the plastic zone on complete removal of the free stress from the boundary of the cavity is also given.

Stresses along the horizontal section computed at the end of the relaxation process are compared with the corresponding results presented by Reyes and Baker in figure 6. Here, internal stresses outside the discretized region were calculated at simple internal points not connected to any internal cells.

It is important to note that the refinement of the two finite

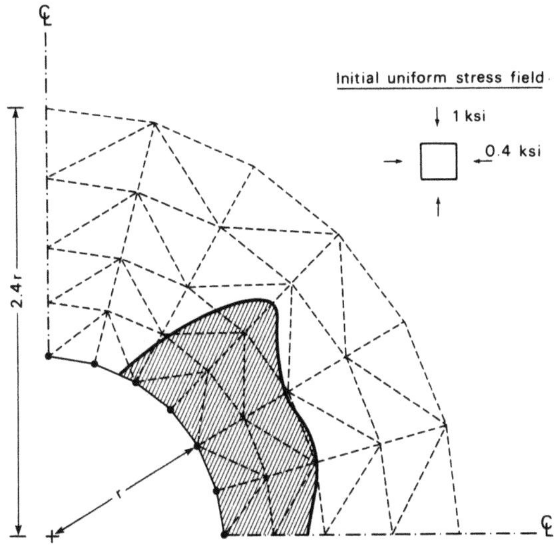

Fig 5 Deep circular tunnel problem Discretization used for B E results and spread of plastic zone at the end of relaxation process

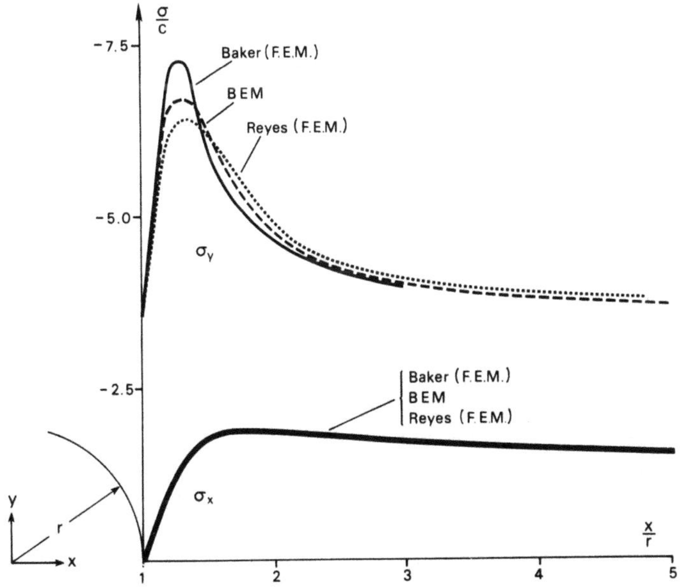

Fig. 6 Final stresses along the horizontal section through the infinite medium

element meshes (about 253 nodes) should not lead to the differences in the σ_y values in fig. 6. Although no reference was made by the authors, this discrepancy is probably due to the outer boundary conditions considered in the two analyses. The boundary element technique does not require any outer boundary discretization, but in order to study its influence in the results a quarter of a circle with radius equal to nine times the radius of the cavity was discretized into six boundary elements, this is approximately the extent of the finite element meshes. The outer circle was then considered to be free to displace, giving as a result a better agreement with Baker computations. The second alternative was carried out by prescribing zero displacements over the outer boundary, leading now to improved agreement with Reyes results.

Rough Punch:
In the third example the elastic-plastic behaviour of a square block compressed by two opposite perfectly rough rigid punches is studied. The problem is analysed under plane strain condition and material is considered to be perfectly plastic obeying von Mises yield criterion.

By using a very refined mesh of 274 linear displacement triangular finite elements and 173 nodal points (fig. 7a), results to this problem were presented by Chen [15] [18]. The boundary element analysis was performed with the discretization shown in figure 7b, requiring less than one third of the FE data to run the problem.

The indentation process was developed by prescribing the flat punch displacements leading to the average pressure-applied displacement curve presented in figure 8. As can be seen, agreement between the two analyses has been thoroughly obtained, both methods slightly exceeding (4%) the theoretical limit load $\sqrt{3}\ p/2\sigma_o = 2.5$.

9. Conclusions

In this paper the correct and complete formulation for the boundary element method applied to plasticity has been pres-

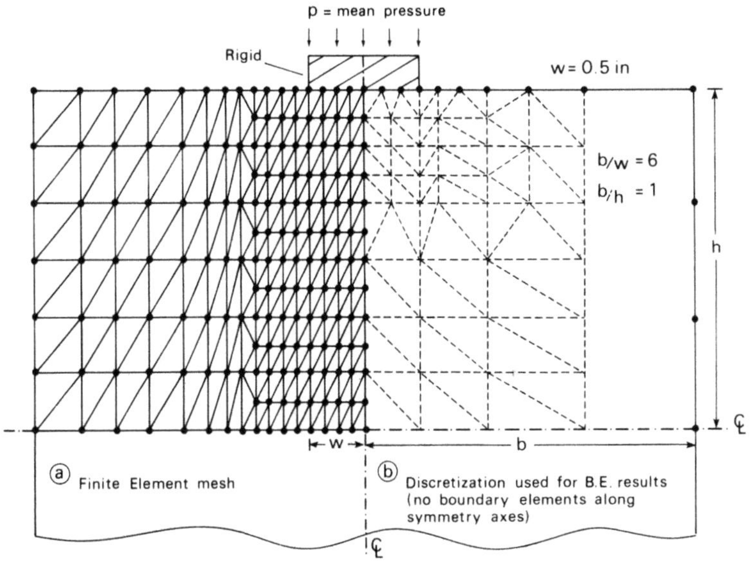

Fig 7 Geometry of rough punch problem

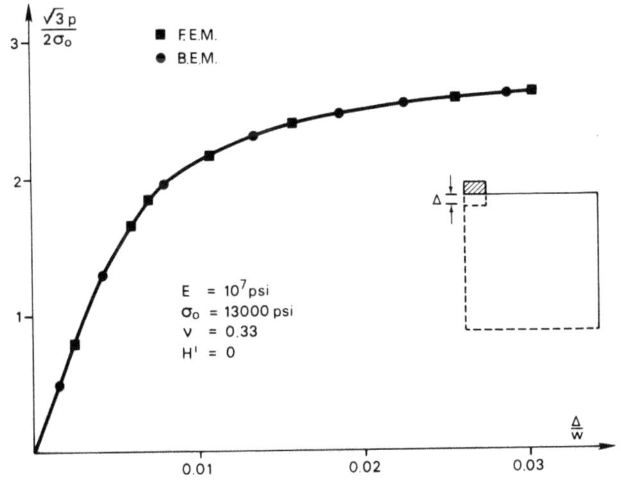

Fig 8 Mean pressure – applied displacement curve for rough punch problem

ented. Three alternative procedures - initial stress, initial strain and fictitious forces - were discussed together with the possibility of applying different yield criteria. The required stepwise procedure has been formulated in matrix form using a simple recursive relation relating stresses to initial elastic solution and plastic strains.

The examples presented show the advantages of using boundary elements in plasticity in preference to finite elements. Considerable reductions in the data required to run a problem can be achieved for the same degree of accuracy. In addition, problems with boundaries at infinity can be properly modelled using boundary elements and the resulting matrices are smaller than those corresponding to finite element or finite difference techniques.

References

1 Brebbia, C.A.; Walker, S.: The Boundary Element Techniques in Engineering. Butterworths, London (1979).

2 Swedlow, J.L.; Cruse, T.A.: Formulation of Boundary Integral Equations for Three-Dimensional Elasto-Plastic Flow. Int. J. Solids Structures 7 (1971) 1673-1683.

3 Mendelson, A.: Boundary Integral Methods in Elasticity and Plasticity. Report No. NASA-TN-D-7418 (1973).

4 Mukherjee, S.: Corrected Boundary Integral Equation in Planar Thermoelastoplasticity. Int. J. Solids Structures 13 (1977) 331-335.

5 Bui, H.D.: Some Remarks about the Formulation of Three-Dimensional Thermoelastoplastic Problems by Integral Equations. Int. J. Solids Structures 14 (1978) 935-939.

6 Telles, J.C.F.; Brebbia, C.A.: On the Application of the Boundary Element Method to Plasticity. Appl. Math. Modelling 3 (1979) 466-470,

7 Lin, T.H.: Theory of Inelastic Structures. Wiley, New York (1968).

8 Riccardella, P.C.: An Implementation of the Boundary-Integral Technique for Planar Problems in Elasticity and Elasto-Plasticity. Report No. SM-73-10, Dept. Mech. Eng. Carnegie Mellon University, Pittsburgh (1973).

9 Telles, J.C.F.; Brebbia, C.A.: The Boundary Element Method in Plasticity. Proc. 2nd Int. Seminar on Recent Advances in Boundary Element Methods, (edited by C.A. Brebbia), University of Southampton (1980) 295-317.

10 Nayak, G.C.; Zienkiewicz, O.C.: Convenient Form of Stress Invariants for Plasticity. Proc. Am. Soc. Civ. Engrs., J. Struc. div. 98 (1972) 949-954.

11 Drucker, D.C.; Prager, W.: Soil Mechanics and Plastic Analysis or Limit Design. Q. Appl. Math. 10 (1952) 157-165.

12 Nayak, G.C.; Zienkiewicz, O.C.: Elasto-plastic Stress Analysis. A Generalization for Various Constitutive Relations including Strain Softening. Int. J. Num. Meth. Engng. 5 (1972) 113-135.

13 Zienkiewicz, O.C.; Valliappan, S.; King, I.P.: Elasto-plastic Solutions of Engineering Problems; Initial Stress Finite Element Approach. Int. J. Num. Meth. Engng. 1 (1969) 75-100.

14 Yamada, Y.; Yoshimura, N.; Sakurai, T.: Plastic Stress-strain Matrix and its Application for the Solution of Elastic-plastic Problems by the Finite Element Method. Int. J. Mech. Sci. 10 (1968) 343-354.

15 Chen, W.F.: Limit Analysis and Soil Plasticity. Elsevier Scientific Publishing Co., Amsterdam, The Netherlands (1975).

16 Reyes, S.F.; Deere, D.U.: Elastic-plastic Analysis of Underground Openings by the Finite Element Method. Proc. 1st. Int. Congr. Rock Mechanics, Lisbon (1966) 477-483.

17 Baker, L.E.; Sandhu, R.S.; Shieh, W.Y.: Application of Elasto-plastic Analysis in Rock Mechanics by Finite Element Method. Proc. 11th Symp. Rock Mechanics, (edited by W.H. Somerton), University of California, Berkeley (1969) 237-251.

18 Chen, A.C.T.; Chen, W.F.: Constitutive Equations and Punch-indentation of Concrete. Proc. Am. Soc. Civ. Engrs., J. Engng. Mech. div. 101 (1975) 889-906.

19 Lachat, J.C.; Watson, J.O.: Effective Numerical Treatment of Boundary Integral Equations, a Formulation for Three-dimensional Elastostatics. Int. J. Num. Meth. Engng. 10 (1976) 991-1005.

Appendix A

Tensors related to the fundamental solutions that appear in the text are of the following form,

$$p^*_{ijk} = \frac{G}{2\alpha\pi(1-\nu)r^\beta}\Big\{\beta\frac{\partial r}{\partial n}\big[(1-2\nu)\delta_{ij}r_{,k} +$$

$$+ \nu(\delta_{ik}r_{,j} + \delta_{jk}r_{,i}) - \gamma r_{,i}r_{,j}r_{,k}\big]$$

$$+ \beta\nu(n_i r_{,j}r_{,k} + n_j r_{,i}r_{,k}) +$$

$$(1-2\nu)(\beta n_k r_{,i}r_{,j} + n_j\delta_{ik} + n_i\delta_{jk})$$

$$- (1-4\nu)n_k\delta_{ij}\Big\} \tag{A1}$$

$$\sigma^*_{ijk\ell} = \frac{G}{2\alpha\pi(1-\nu)r^\beta}\Big\{\beta(1-2\nu)(\delta_{ij}r_{,k}r_{,\ell} + \delta_{k\ell}r_{,i}r_{,j})$$

$$+ \beta\nu(\delta_{\ell i}r_{,j}r_{,k} + \delta_{jk}r_{,\ell}r_{,i} + \delta_{ik}r_{,\ell}r_{,j}$$

$$+ \delta_{j\ell}r_{,i}r_{,k}) - \beta\gamma r_{,i}r_{,j}r_{,k}r_{,\ell} +$$

$$+ (1-2\nu)(\delta_{ik}\delta_{\ell j} + \delta_{jk}\delta_{\ell i}) - (1-4\nu)\delta_{ij}\delta_{k\ell}\Big\} \tag{A2}$$

where $\alpha = 2,1$; $\beta = 3,2$; $\gamma = 5,4$ for 3-D and plane strain respectively.

Also,

$$\varepsilon^*_{ijk\ell} = \frac{1}{4\pi(1-\nu)r^2}\Big\{(1-2\nu)\big[\delta_{ik}\delta_{\ell j} + \delta_{jk}\delta_{\ell i} - \delta_{ij}\delta_{k\ell}$$

$$+ 2\delta_{ij}r_{,k}r_{,\ell}\big] + 2\nu\big[\delta_{\ell i}r_{,j}r_{,k} + \delta_{jk}r_{,\ell}r_{,i}$$

$$+ \delta_{ik}r_{,\ell}r_{,j} + \delta_{j\ell}r_{,i}r_{,k}\big]$$

$$+ 2\delta_{k\ell}r_{,i}r_{,j} - 8r_{,i}r_{,j}r_{,k}r_{,\ell}\Big\} \tag{A3}$$

for the plane strain initial stress formulation.

Simplified Calculation Models Applied to Postbuckling Analysis of Thin Plates

B. H. KRÖPLIN, H. DUDDECK
Technische Universität Braunschweig, Germany

Summary
A yield function, which describes the plastic flow of a plane cross section of a thin plate approximately is discussed and compared to the layer approach. Further an artificial (viscous) damping is suggested to overcome instabilities due to snap throughs with or without bifurcations.

Introduction

In the first part of this paper some aspects of a yield function model for thin plates are discussed and compared to the well known layer approach, /1,2,3/.
In elastic-plastic analysis the theory of thin members loses much of its advantages for the stress resultants, axial forces and bending moments can only be determined by a discretisation over the depth of the cross section or by a layer approach. To simplify the calculation process, in order to safe computing time, yield function models are developed, in which yielding of a cross section is described as a function of quantities of the reference surface only. The yielding process is reduced to its most important features. The solution obtained is not generally valid, but often sufficient accurate, if the structural behaviour is sought, e.g. the limit load.
The considered yield function is developed by Eggers in /4/. Yielding depends on the axial forces, the bending moments and a plastic strain parameter.
The yield function is applied to the post buckling analysis of thin steel plates under axial compression and compared to results obtained by a layer approach.

The second part deals with a method to overcome snap through problems with or without bifurcations. The structure is treated as a viscous one with an artificial damping, such that equilibrium states before and behind the snap through can be reached in a stable manner. Examples are carried out on a plate strip.

1. Yield function and the layer model

Thin plate theory

One of the most advantageous features of the thin plate theory is that the internal work of the structure can be expressed in quantities of the reference surface only. With the Bernoulli hypotheses taken to be valid, the strains are expressed in strains and curvatures of the middle surface,

$$\underline{\varepsilon} = \bar{\varphi}\,\underline{\varepsilon}_M + \bar{\psi}\,\underline{\chi}_M \quad . \tag{1}$$

$\bar{\varphi}$ and $\bar{\psi}$ are symmetric and antimetric shape functions, fig. 1. The subscript M indicates quantities of the middle surface.

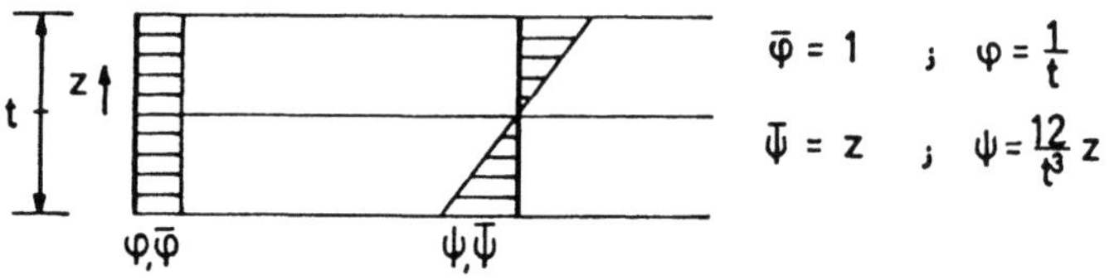

fig. 1: Shape functions for strains and elastic stresses

Analogue to (1) the stress can be expressed as function of the axial forces and bending moments

$$\underline{\sigma} = \varphi\,\underline{n} + \psi\,\underline{m} \quad , \tag{2}$$

if for $\underline{n}$ and $\underline{m}$ the following integral conditions hold /5/,

$$\int_z \varphi\, dz = 1 \quad , \quad \int_z \varphi\, z\, dz = 0 \quad , \tag{3}$$

$$\int_z \psi\, dz = 0 \quad , \quad \int_z \psi\, z\, dz = 1 \; . \tag{4}$$

According to the orthogonality condition (3,4) $\underline{n}$ and $\underline{m}$ are uncoupled.

The layer approach

In the common layer approach the stress increments are calculated from the strains of every layer (5) and an incremental constitutive law (6).

$$\dot{\underline{\varepsilon}} = \bar{\varphi}\,\dot{\underline{\varepsilon}}_M + \bar{\psi}\,\dot{\underline{\varkappa}}_M \tag{5}$$

$$\dot{\underline{\sigma}} = \underline{C}_{(\sigma,\varepsilon^P)}\,\dot{\underline{\varepsilon}} \tag{6}$$

The compatibility is satisfied a priori by (1). An iterative correction is necessary to satisfy the constitutive law (6) for finite increments. The equilibrium of the stresses for a single particle in the cross section is not considered. To avoid undue complexity σ^{13} is often neglected and every layer is treated independently /1,2/. Integration of the layer stresses over the depth of the cross section leads to axial forces and moments (7). Equation (2) is not needed.

$$\underline{m} = \int_z \underline{\sigma}\, z\, dz \qquad \underline{n} = \int_z \underline{\sigma}\, dz \tag{7}$$

Improved strains $\underline{\varepsilon}_M$, $\underline{\varkappa}_M$ are calculated by an iteration with unbalanced forces on structural level.

The yield function model

The derivation of the yield function model starts from the alternative assumption that yielding of the cross section can be described by an elastic limit state, an intermediate state and a plastic limit state. The elastic limit state is defined by the first fiber yielding; the plastic limit state is reached, if all fibers yield.

With $\underline{\sigma}$ (2) substituted into the yield condition for the continuum

$$Y = \sigma^{\alpha\beta} J_{\alpha\beta\rho\lambda}\, \sigma^{\rho\lambda} - \sigma_y^2 = 0 \qquad \text{(Hill /5/)} \tag{8}$$

a yield condition for the cross section (9) is obtained,

$$Y_i = A_{11} t^2 \varphi^2 + \frac{1}{2} A_{12} t^3 \varphi \psi + \frac{1}{16} A_{22} t^4 \psi^2 - S_{yi}^2 = 0 \,, \tag{9}$$

where

$$A_{11} = n^{\alpha\beta} J_{\alpha\beta\rho\lambda}\, n^{\rho\lambda} \,, \qquad A_{22} = \frac{4}{t^2}\, m^{\alpha\beta} J_{\alpha\beta\rho\lambda}\, m^{\rho\lambda} \,,$$

$$A_{12} = \frac{4}{t}\, n^{\alpha\beta} J_{\alpha\beta\rho\lambda}\, m^{\rho\lambda}, \qquad S_{yi} = \sigma_{yi}\, t \,.$$

In general φ and ψ depend on the constitutive relationship. Only for the limit states simple form occur.

The limit states

For the elastic limit state the elastic stress distribution is given by $\varphi = 1/t$ and $\psi = (12/t^3)\ z$, fig. 1. Substitution of (2) into (8) leads to the elastic limit state

$$Y_0 = A_{11} + 3 \operatorname{sign}(A_{12})\, A_{12} + \frac{9}{4} A_{22} - S_{y0}^2 = 0 \,. \tag{10a}$$

Sign (A_{12}) provides yielding of the top and the bottom fiber.

For the plastic limit state biconstant functions φ and ψ fig. 2, are chosen, which are developed from a stress

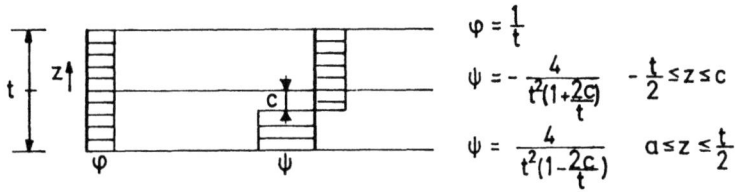

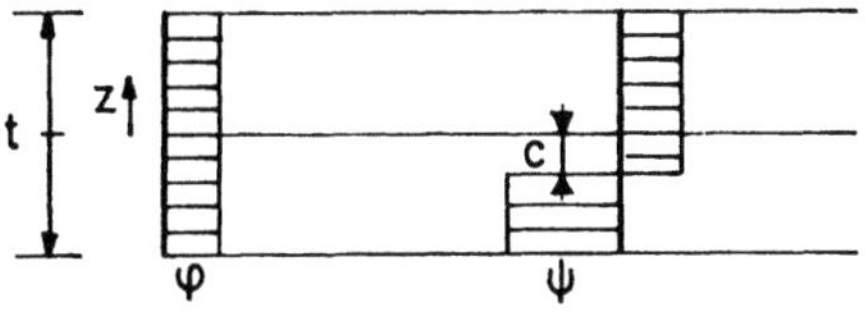

$$\varphi = \frac{1}{t}$$

$$\psi = -\frac{4}{t^2(1+\frac{2c}{t})} \qquad -\frac{t}{2} \le z \le c$$

$$\psi = \frac{4}{t^2(1-\frac{2c}{t})} \qquad a \le z \le \frac{t}{2}$$

fig. 2: Shape functions for stresses in the plastic limit state

function and satisfy the orthogonality condition (3) in advance. After substituting (2) in (8) and eliminating c the plastic limit state Y_1 (10b) is given by

$$Y_1 = A_{11} + \frac{1}{2} A_{22} + \frac{1}{2} \sqrt{4A_{12}^2 + A_{22}^2} - S_{y1}^2 = 0 . \qquad (10b)$$

The intermediate state

In order to avoid complicated functions in the partial plastic state the yielding is considered as gradual change from the elastic to the plastic limit state controlled by means of a plastic strain parameter $\bar{\varepsilon}^P$. As yield function for the cross section

$$Y = (Y_1 (\bar{\varepsilon}^P)^a + Y_0\, b) / ((\bar{\varepsilon}^P)^a + b) \le 0 \qquad \begin{array}{l} 0 < a < 1 \\ 0 < b \end{array} \qquad (11)$$

is chosen. Eq. (11) satisfies the boundary conditions for the elastic and the plastic limit state, /4/. Due to the choice of (11) the yield history of the intermediate state has to be expressed by the parameters a and b only. They are identified such that $\bar{\varepsilon}^P$ calculated from the yield function (11) fits as close as possible to the plastic strains and curvatures of a plane reference cross section under monotonous yielding, fig. 3.

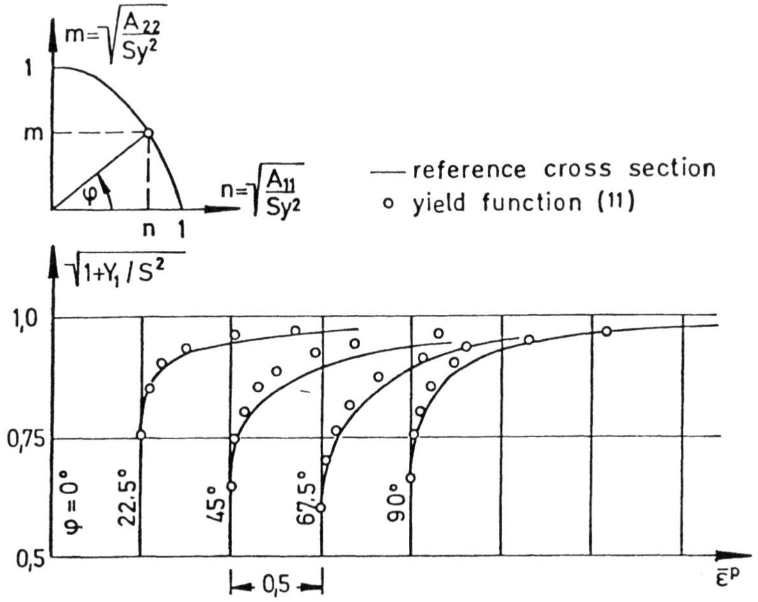

fig. 3: Accuracy of the approximation for a = .9, b = .0666, $A_{12}/\sqrt{A_{11}\cdot A_{22}} = 1$.

Clearly partial unloadings of the cross section are neglected, which are included in the layer approach. Rotations of the axial forces and bending moments are possible for the path on the yield function is not predetermined. Depending on A_{12} the yield function changes its shape, fig. 4. The neighbourhood of a actual stress state is described by the tangential plane of the yield function in stress and strain space,

$$\dot{Y} = \frac{1}{2S}\left(\frac{\partial Y}{\partial n^{\alpha\beta}}\dot{n}^{\alpha\beta} + \frac{\partial Y}{\partial m^{\alpha\beta}}\dot{m}^{\alpha\beta} + \frac{\partial Y}{\partial \varepsilon^{P}}\dot{\bar{\varepsilon}}^{P}\right) = 0. \tag{12}$$

In opposite to other yield function approaches /6/ a S-$\bar{\varepsilon}^P$ diagram is not needed. The fictious force S depends on the integration path,

$$S = S_0 + \int_{\bar{\varepsilon}^P} H\, d\dot{\bar{\varepsilon}}^{P}. \tag{13}$$

S is coupled with $\bar{\varepsilon}^p$ by the relationship

$$\frac{\partial Y}{\partial \bar{\varepsilon}^p} \bar{\varepsilon}^p = 2\, S\dot{S}. \qquad (14)$$

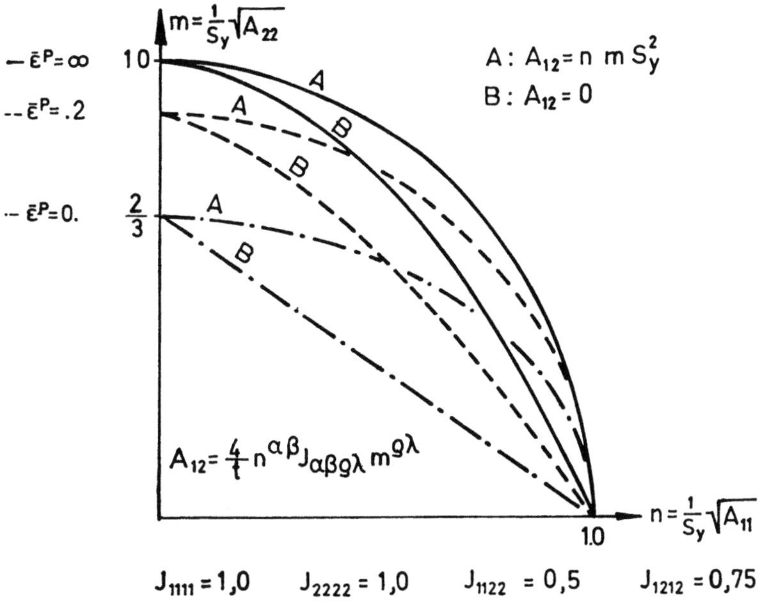

fig. 4: Change of the yield surface with A_{12} and $\bar{\varepsilon}^p$

In the incremental functional the forces, n, m are constrained to the tangential plane by a Lagrangian multiplier λ , which is interpreted from the Euler equations as the plastic strain parameter $\bar{\varepsilon}^p$ / 4,7/. The constrained funtional then reads

$$\dot{W}_M = \dot{W}_M^e - \frac{\dot{\bar{\varepsilon}}^p}{2S} \left(\frac{\partial Y}{\partial S} \dot{S} + \frac{\partial Y}{2\partial \dot{\bar{\varepsilon}}^p} \dot{\bar{\varepsilon}}^p \right) = \text{stat.}. \qquad (15)$$

Layer approach versus yield function model

The derivation of the yield function model restricts the application clearly to problems, where

- the unloadings in parts of the cross sections and
- the rotations of the forces and moments

are of small influence.
The ultimate state Y_1 is sufficient exact (about 4.5 %) /8/, the plastic strains and curvatures are components of the plastic strain parameter. They depend on the integration path, while the plastic strain parameter itself is determined by the yield condition based on monotonous yielding. Thus errors in compatibility may occur, which affect the distribution of the stress resultants in the intermediate state.
Since the layer model satisfies compatibility first (1) there such errors don't occur, but the equilibrium of the stresses is only controlled in an integrated manner (7).
In engineering applications the loss in generality of the yield function has to be often considered against the gain of numerical effectivity.
In the layer model the stresses have to be calculated in every layer for every strains $\underline{\varepsilon}_M$, $\underline{\varkappa}_M$ and have to be stored. In the yield function model the yield function is determined in advance. Only the iteration on structural level is needed. This means correction of the deviation from a yield surface one time while it is necessary for every layer in the layer approach. For the same reason the element matrix can be calculated in the yield function approach in one step while it is calculated in the layer approach by the sum of the layer matrices.
After own estimations the calculation is about ten times as fast as the layer model with five layers.

Application

The yield function model was applied to post buckling

analysis of thin steel plates under axial compression. Since the rotations are small, a total Lagrangian description was used with the strain in the middle surface

$$a_{\alpha\beta} = \varepsilon^{L}_{\alpha\beta} + \frac{1}{2} w_{,\alpha} w_{,\beta}$$

The incremental functional is given in (17)

$$\dot{W} = \int_A \left[\dot{\underline{S}}^T \dot{\underline{\varepsilon}}^L - \frac{1}{2} \dot{\underline{S}}^T \underline{F} \dot{\underline{S}} + \sum_2 \dot{\underline{\eta}}_1^T \dot{\underline{\varepsilon}}_1^N \right] dA - \int_A \dot{\underline{p}} \dot{\underline{u}} \, dA + b = \text{stat.}$$

where for moderate deflections

$$\dot{\underline{\eta}}_1 = \begin{bmatrix} \dot{n}^{\alpha\beta} \\ n^{\alpha\beta} \end{bmatrix} \qquad \dot{\underline{\varepsilon}}_1^N = \frac{1}{2} \begin{bmatrix} \dot{w}_{,\alpha} w_{,\beta} + w_{,\alpha} \dot{w}_{,\beta} \\ \dot{w}_{,\alpha} \dot{w}_{,\beta} \end{bmatrix}$$

and for plasticity

$$\dot{\underline{\eta}}_2 = \begin{bmatrix} \dot{s}^{\alpha\beta} \\ \frac{\partial Y}{\partial \bar{\varepsilon}^P} \frac{\dot{\varepsilon}^P}{2} \end{bmatrix} \qquad \dot{\underline{\varepsilon}}_2^N = \frac{1}{2S} \begin{bmatrix} \frac{\partial Y}{\partial s^{\alpha\beta}} \\ \dot{\bar{\varepsilon}}^P \end{bmatrix}$$

$\underline{S}$ $= [\underline{n}, \underline{m}]$

$\dot{\underline{\varepsilon}}^L$ $= [\underline{\varepsilon}^L, \underline{\chi}]$

$\underline{F}$ = Flexibility of the cross section

b = boundary integrals.

Examples

The load deflection curve of a slender plate, b/t = 80 calculated by various approaches are compared in fig. 5.

The approaches are:

1. A layer model with five layers and 36 Gauß integration points per element for the material properties, i.e. 144 integration points in the middle surface of a square plate. An element with a complete cubic shape function was used /9/.

2. The yield function model applied in a displacement

approach with a triangular element. Cubic shape functions out of plane and quadratic shape functions in plane are used /10/. The element variables are displacements and midside rotations.

3. The yield function model was applied in a mixed FEM formulation with linear shape functions for all unknowns. /4/.

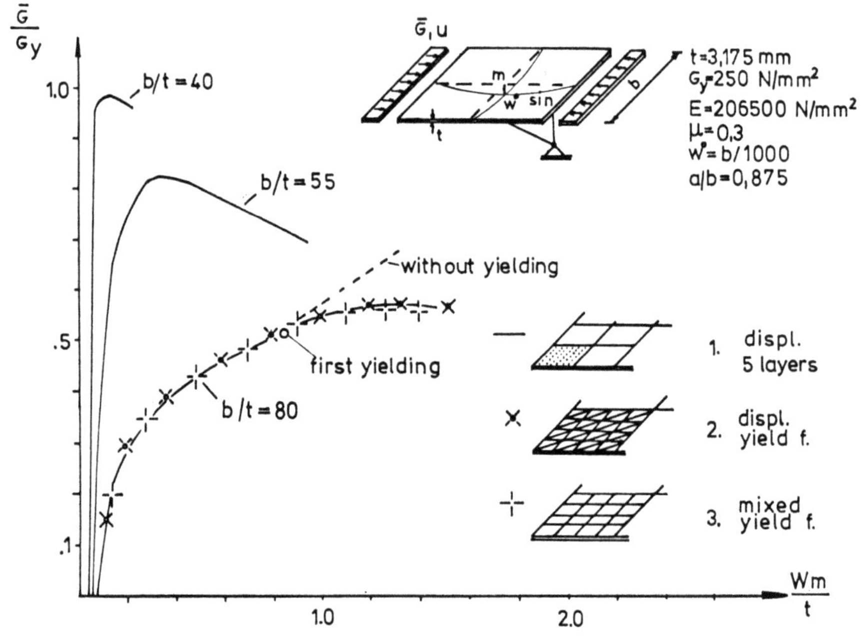

fig. 5: Load deflection curve of axially compressed plates - Comparison of different approaches -

The load deflection curves of the layer approach (1.) and the yield function model (2.) show excellent agreement. With the mixed model the limit load is predicted a little less for the linear shape functions overestimate the plastic zones of the plate.

In comparison to the mixed formulation a refined mesh was necessary to arrive at the same accuracy in axial forces and bending moments with the displacement approach. Due to the simpler structured nonlinear element matrix in the mixed approach the computation was considerably faster than in the displacement approach, despite indefinite finite matrizes and more unknowns.

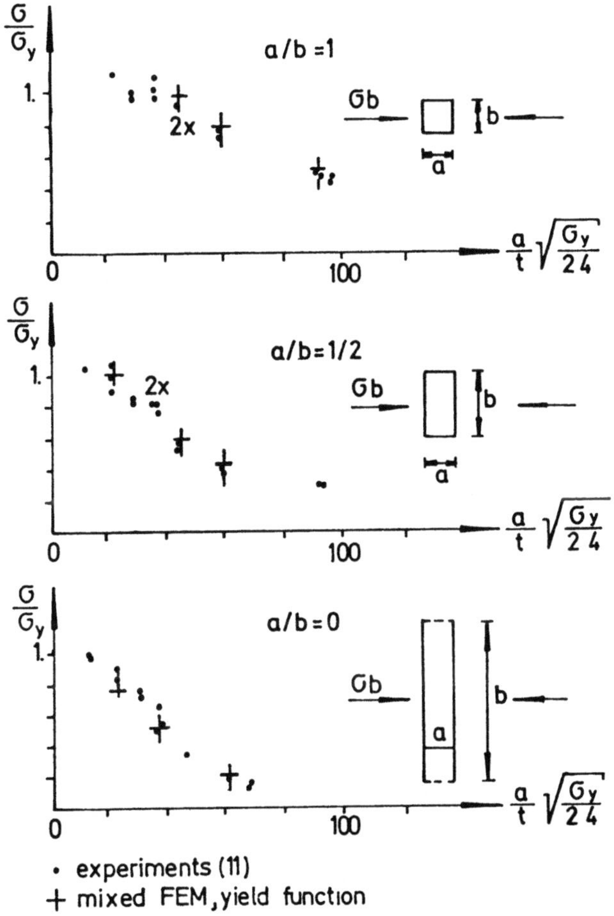

fig. 6: Comparison of experimental and FEM results

In fig. 6 the results calculated by the mixed formulation and the yield function are compared to the results of experiments /11/. The influence of the shape, a/b was investigated for various slenderness ratios. The calculated and experimental results coincide in the full range of the investigation.

Conclusion 1:

A comprehensive comparison of the results leads to the engineering judgement, that the accurate description of the structural behaviour, i.e. various yield zones and plastic mechanism is more important for predicting a correct limit load in the postbuckling range of thin plates than a complete consideration of the cross section yielding with an improved description of the stress history in the intermediate state. Especially in problems, which require consideration of geometric and material nonlinearity the yield function model reduces the numerical effort considerably.
However further investigations are needed to define the application range of the yield function model in general.

2. A viscous approach to structural instabilities

In post buckling of plates snap throughs may occur, even when imperfect structures are considered. At the point of instability the Newton Raphson iteration diverges. An alternative to various techniques suggested, /12, 13, 14/ is to treat the structure as a viscous one. The static equations are augmented by damping force $\underline{D}\dot{\underline{x}}$, which carries a part of the external load or the unbalanced forces until a steady state is reached /15/,

$$\underline{K}_{(x)}\underline{x} + \underline{D}\underline{x} = \underline{p} \ . \tag{18}$$

$\underline{x}$ means the vector of unknowns, $\underline{K}_{(x)}$ describes the non-linear structural matrix, $\underline{p}$ contains the applied loads, $\underline{D}$ is an artificial damping and $\dot{\underline{x}}$ the derivation of $\underline{x}$ with respect to the time. With a backward time integration scheme and

$$\Delta x_n = \underline{x}_n - \underline{x}_{n-1} \tag{19}$$

(18) can be written as

$$(D_n^* + K_{(\underline{x}_n)})\Delta \underline{x}_n = p_n^u \qquad \underline{p}_n^u = \underline{p} - \underline{p}'_n \qquad \underline{D}_n^* = \frac{2}{\Delta t}\underline{D} \tag{20}$$

$\underline{p}_n^u$ are the unbalanced forces. Eq. (20) represents a Newton Raphson type iteration, where $\underline{D}^*$ is chosen as a function of the unbalanced forces, thus speeding convergence,

$$\ulcorner D_n^* \lrcorner = \left| \frac{\alpha \underline{p} - \beta \underline{p}_1}{\gamma x_{n-1}} \right| \qquad \alpha = 1.0 \qquad \beta = 0.8 \qquad \gamma = 0.5. \tag{21}$$

Examples

The method is demonstrated on a snap through of a simply supported beam, calculated as plate strip with and without geometrical imperfections, which is loaded by the displacement u_1, fig. 7.

With an imperfection, composed of the first and the second buckling mode $\overset{\circ}{w} = a_1 \sin \pi x/l + a_2 \sin 2\pi x/l$ and the damping a stable equilibrium state after the snap through was reached within 12 steps, (curve B^*). The same problem was calculated with the modified Newton Raphson procedure. After careful adaption of the increments the same equilibrium state was reached after 11 increments and in the whole 86 iterations.

With a true antimetric imperfection, $\overset{\circ}{w} = a_2 \sin 2\pi x/l$, curve C, no instability occured at the critical load of the first buckling mode. With the mixed method the first buckling mode with a lower level of energy was missed.

To avoid an accompanying eigen value calculation in every step a disturbed unbalanced force

$$\Delta p = \frac{\Delta \underline{p} + \alpha |\Delta \underline{p}|}{1+a} \quad , \quad \alpha = 0\ 2 \tag{22}$$

was applied together with the damping D^{*}. $\Delta \underline{p}$ indicated a snap through from the second to the first buckling mode curve C^{*}. Curve D shows the iteration without damping.

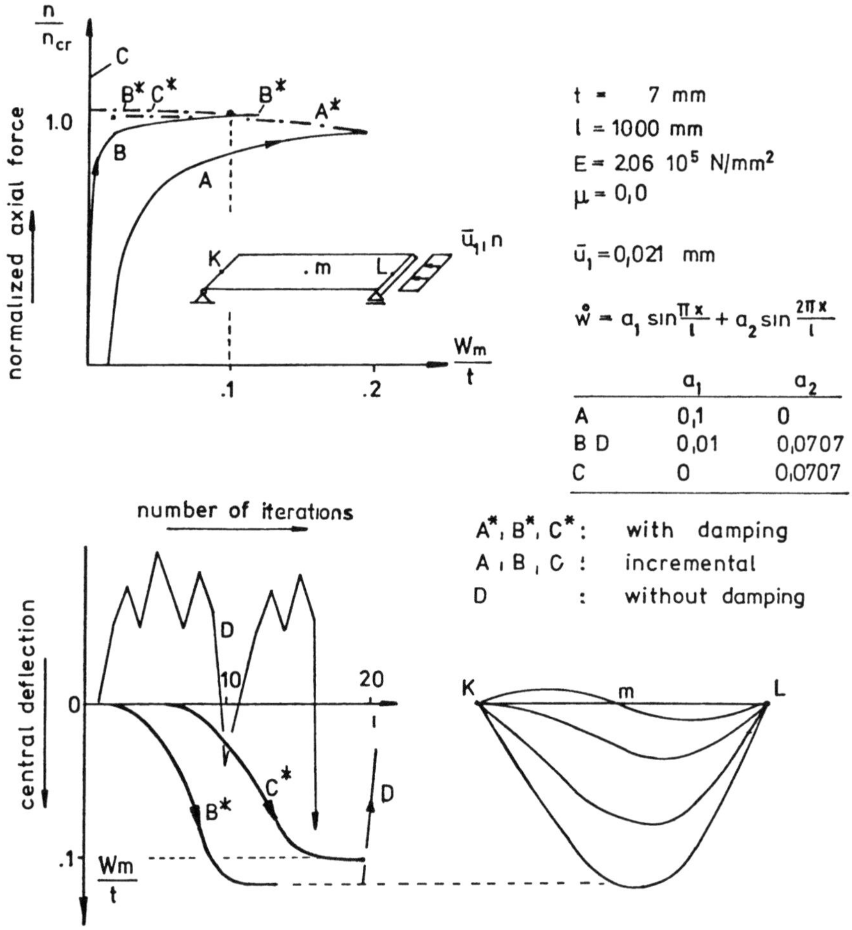

fig. 7: Snap through with and without damping

Conclusion 2

With an artificial damping elastic snap throughs can be calculated in a stable manner. The increments can be enlarged or the load can be applied in one step. No adulterations of the results occur.

The stability of the reached equilibrium state can be tested by a disturbance of the unbalanced forces or the damping orthogonal to the reached deflection mode.

Optimisation of the damping with respect to the efficiency of the calculation are subjects for further investigations as well as the definition of a significant disturbance for the stability test.

Acknowledgement

The author would like to acknowledge the support provided by the Heisenberg Stipendium der Deutschen Forschungsgemeinschaft and the work of Mr. D. Dinkler on the numerical examples.

Literatur:

1 Wegmuller, A.W.: Full Range Analysis of Eccentrically Stiffened Plates, ASCE J. Struct. Div. 100, 143-159 (1974)

2 Baums, B.: Über das Tragverhalten längsgedrückter quadratischer elastisch-plastischer Platten bei großen Deformationen. Diss. TU Braunschweig 1977

3 Eggers, H.: Stiffness and Force-Moment Interaction for Plates with Plastic Flow, J. Struct. Mech. 4, 123-139 (1976)

4 Eggers, H., Kröplin, B.: Yielding of Plates with Hardening and Large Deformations. Int. J. num. Meth. Engng. 12, 1978 (737-750)

5 Hill, R.: The Mathematical Theory of Plasticity. Oxford University Press London 1950, 318 ff.

6 Crisfield, M.A.: Full Range Analysis of Steel Plates and Stiffened Plating under Uniaxial Compression, Proc. Instn. Civ. Engrs. Part 2, 59 (1975) 595-624

7 Eggers, H.: Variational Principles for Elastiplastic Continua, J. Struct. Mech. 3, (1974-1975) 345-358

8 Robinson, M.: A Comparison for Yield Surfaces for Thin Shells, Int. J. mech. Sci. 13, ;(1971) 345-354

9 Dumont, N.,A.: Traglastberechnung beulgefährdeter Rechteckplatten im elastisch-plastischen Bereich nach der geometrisch nichtlinearen Theorie unter Berücksichtigung geometrischer und werkstofflicher Imperfektionen. Dissertation D17, Darmstadt 1978

10 Meier, F.: Grenzlastberechnung ausgesteifter Platten mit Imperfektionen. Bericht Nr. 80-33 aus dem Institut für Statik, TU Braunschweig

11 Fischer, M., Harre, W.: Ermittlung der Traglastkurven von einachsig gedrückten Rechteckplatten aus Baustahl der Seitenverhältnisse $\alpha \leq 1$ von Versuchen, Der Stahlbau 8, 1978 (239-247

12 Haisler, W.E., Stricklin, J.A., Key, J.E.: Displacement Incrementation in Non-Linear Structural Analysis by the Self Correcting Method, Int. J. num. Meth. Engng. 11, (1977) 3-10

13 Bergan, P.G.: Solution Algorithms for Nonlinear Structural Problems, Proceedings Int. Conf. on engineering Application of the Finite Element Method, A.S. Computas, Hovik, 9-11 May 1979

14 Argyris, I.H., Hilpert, O., Hindenlang, K., Malejannakis, G.A., Schelke, E.: Flächentragwerke im Konstruktiven Ingenieurbau, ISD-Bericht Nr. 263 (1979)

15 Kröplin, B.-H.: A Viscous Approach to Post Buckling Analysis, to appear in: Engng. Struct.

Finite Element Analysis of Reinforced Concrete Slabs and Panels

G. MEHLHORN, D. KLEIN
Technische Hochschule Darmstadt, Germany

1. Introduction

The real behaviour of reinforced concrete structures differs widely from the results of a linear elastic computation. Because of cracking of concrete, the nonlinear stress-strain relations of concrete and reinforcement, and the bond-slip between concrete and reinforcement, the deformation behaviour of reinforced concrete structures is extremely nonlinear even under working loads.

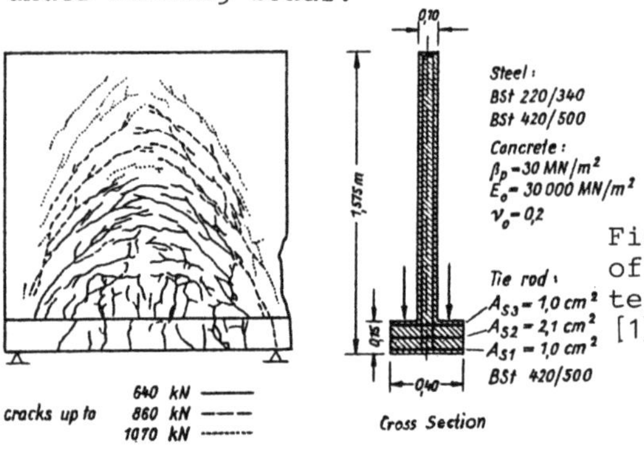

Fig. 1 Crack distribution of Shear Wall WT7 of the tests by Leonhardt/Walther [1]

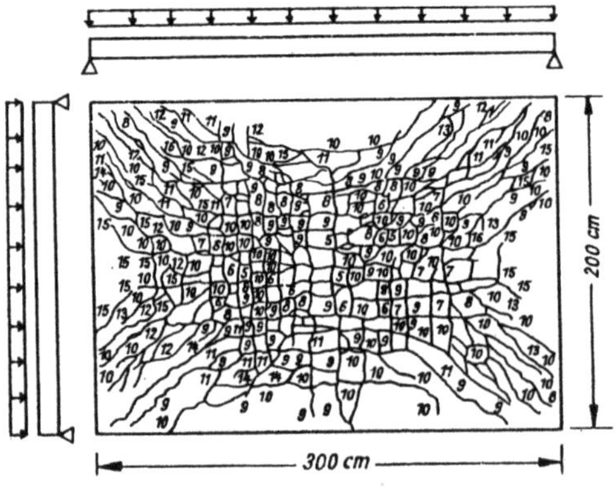

Fig. 2 Crack distribution of a simply supported rectangular slab tested by Franz [2]

In Figs. 1 and 2 the crack distribution of two characteristic planar structures is shown. They are the shear wall WT7 of the tests of Leonhardt/Walther [1] and a simply supported, rectangular slab that was tested by Franz [2] . Through the load-midpoint deflection curve of the slab in Fig. 3 the characteristic behaviour of a reinforced concrete structure is demonstrated. The curve can be divided into three parts. After nearly linear elastic behaviour, the first change of the stiffness is caused by the beginning of cracking. The further softening of the curve depends first of all upon the spreading of the cracks and the nonlinear deformation behaviour of concrete. When the reinforcement begins to yield, a second reduction of the stiffness is evident, which finally leads to the failure by crushing of concrete. The test results were measured only up to a load of about 52 kN/m^2 in order to save the test equipment in case of sudden failure.

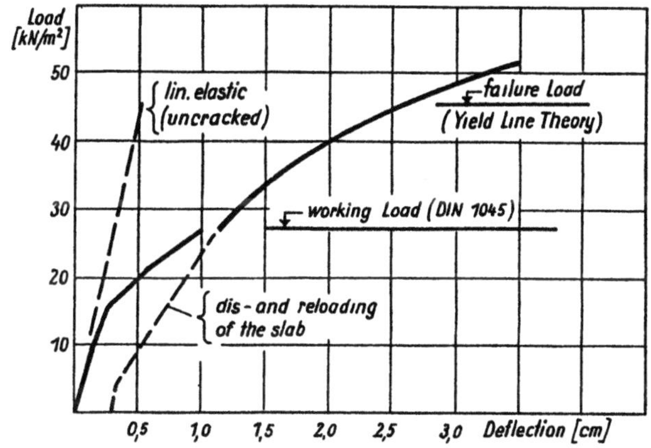

Fig. 3. Load-midpoint deflection curve of a simply supported rectangular slab tested by Franz [2]

Of all the calculation methods that were developed to consider the deviation of the behaviour of reinforced concrete structures from the linear elastic range, the Finite Element Method proved to be one of the most powerful tools. This method was used by many authors, for example [10] to [29] .

The purpose of this report is to present the main problems of computation of the real behaviour of planar reinforced concrete structures using the Finite Element Method and to present solutions obtained by the Darmstadt Group. More general reviews are given by Eibl/Ivanyi [3] , Wegner [4] , Schnobrich [5] and others.

2. Computation of Planar Reinforced Concrete Structures by the Finite Element Method

2.1 Load Carrying Behaviour of Reinforced Concrete Structures

Reinforced concrete is a composite material consisting of the two components, concrete and steel. Its load carrying behaviour is characterized by the fact that the concrete is able only to resist small tensile forces. After cracking of the concrete, the tensile forces are transmitted by bond to the reinforcement consisting of steel bars. The direction and the diameter of the reinforcement bars are determined by the designer, usually following the results of an elastic calculation of the structure, assuming homogeneous behaviour.

After cracking, however, reinforced concrete is inhomogeneous and nonlinear. Its deformation behaviour depends on the direction and history of the loading and is represented by the material properties of the components concrete and steel and the quality of the bond between concrete and reinforcement.

2.1.1 Material Behaviour of Concrete in the Biaxial Stress State

For the computation of planar structures, the behaviour of the concrete must be considered in the biaxial domain. The material properties are influenced by many different factors, among them are the composition of the concrete (kind of aggregate, cement and water matrix), rate of hardening, consolidation, temperature, loading history (age of concrete at loading, loading rate, loading duration, cyclic loading, direction and size of the previous load by reloading) and long time behaviour (shrinkage, creep).

Because of the different influences, the form of the failure criterion and the stress-strain-relation of concrete in the biaxial stress state depends on the results of the tests that are performed to obtain these relations. For short time loading and proportionally increasing load, the tests of Kupfer [6] proved to be the most reliable. Different authors used these test results to develop analytical formulations of the failure and deformation behaviour of concrete for application in numerical calculations such as the finite element method.

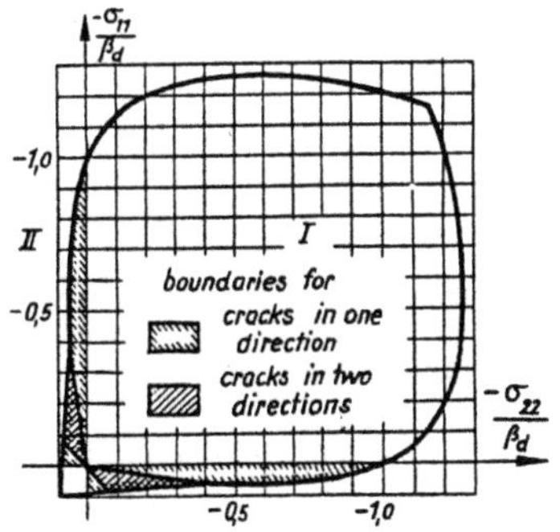

Fig. 4. Failure criterion of concrete in the biaxial state of stresses by the formulation of Link [7] based on Kupfers tests [6]

In Fig. 4 the failure criterion by the formulation of Link [7] is shown. The stresses are normalized in terms of the uniaxial cylinder strength, so that the formulation can be used for different grades of concrete. The whole stress region is described by one relation without considering the boundaries between the regions of compression-compression, compression-tension or tension-tension. Failure criteria such as that of Link [7] cannot be used as plasticity condition. They only describe a boundary for the maximum stresses and do not allow any statements about the plastic deformations.

The stress-strain-relation of concrete is nonlinear from the beginning of loading. Because of the nonlinear deformation behaviour, the concrete ceases to be isotropic in the biaxial stress field. The material properties depend on the existing state of stresses and strains. Different ways are possible for the computational adaptation of the nonlinear stress-strain relation. One way is the finite formulation of the stress strain relation by analytical functions. In the biaxial stress state, however, differential formulations are more suitable for the description of redistribution of the stresses. Link [7] developed an incremental formulation for the tangent stiffness of the concrete by considering the anisotropy in the biaxial stress state. In Fig. 5 some stress-strain-relations that follow the formulation of Link [7] are compared with the test results of Kupfer [6] .

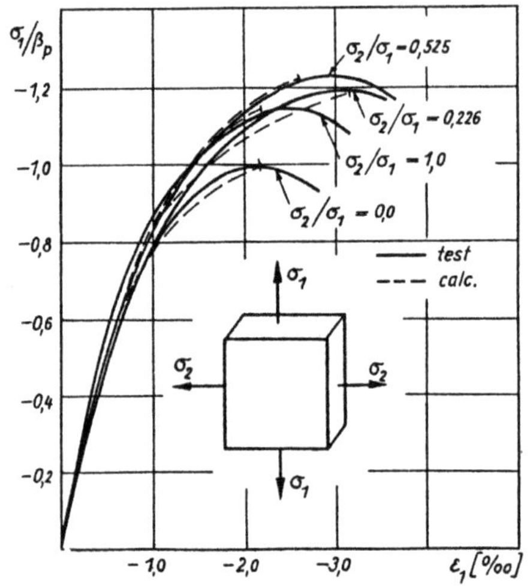

Fig. 5. Stress-strain relations of concrete. Comparison between test results of Kupfer [6] and analytical formulation of Link [7]

2.1.2 Material Behaviour of Reinforcement

As in reinforced concrete only steel bars are used it is sufficient to study the material properties of the steel under uniaxial stresses. In Fig. 6 the stress-strain-relations of the preferably used two kinds of steel are shown. After a nearly perfect-elastic behaviour, the mild steel shows a strongly marked yield strength followed by a more or less constant stress after yielding. At large strains a hardening effect occurs. The cold-worked steel does not have a clearly defined yield strength. The end of the elastic region is marked by a loss of stiffness with a further increase of the stress in the plastic range. For computation the mild steel may be idealized as an elastic-perfectly plastic material with consideration of the hardening at large strains while the stress-strain relation of the cold-worked steel can be approached by a bilinear relation.

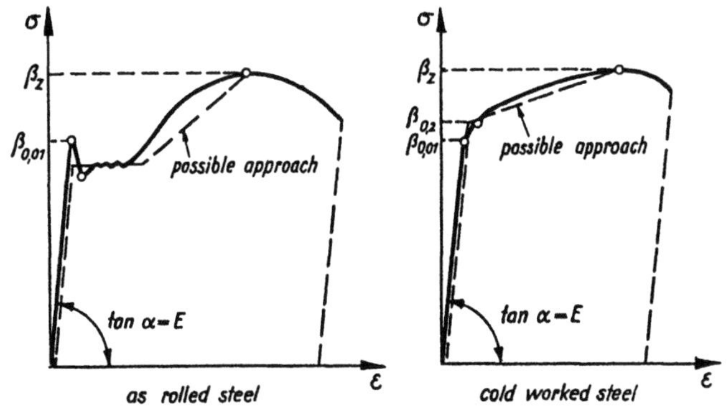

Fig. 6. Stress-strain relations of reinforcement

Under unloading the steel obeys the same law as under first loading. At cyclic loading with many cycles, however, the strength of the steel is diminished. Temperature does influence the steel properties very much. So the strength is reduced at high temperature.

2.1.3 Bond Between Concrete and Reinforcement

The interaction of concrete and reinforcement is first of all influenced by the quality of bond between the two components. The tensile forces must be transmitted by bond after cracking from the concrete by shear stresses into the steel bars. The properties of the bond should guarantee a good distribution of the cracks so that the crack width is small enough to protect the reinforcement against corrosion.

The quality of bond depends on the condition of the surface of reinforcement, the diameter of reinforcement bars, the strength of concrete, the load history and load direction and the state of stresses. So the transferable shear stresses are increased by transverse compression. In Fig. 7 the mechanism of bond between concrete and ribbed steel bars is demonstrated. The Fig. is taken from the tests of Goto [8]. Before cracking, the strain under tensile forces will be the same in concrete and in steel. When the limit of strain in concrete is reached,

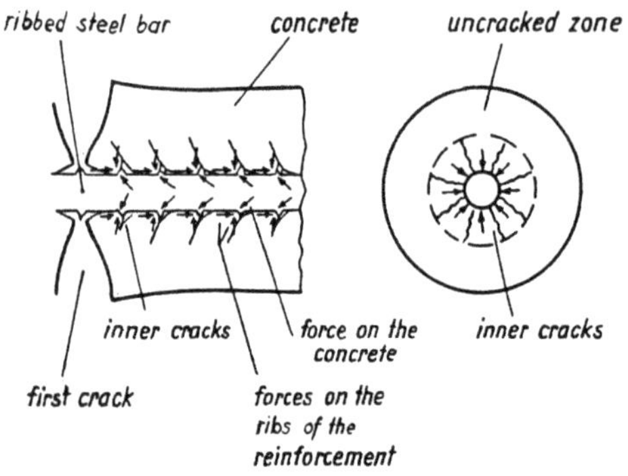

Fig. 7. Crack mechanism (tests of Goto [8])

the concrete will crack throughout and by contracting will be propped at the ribs of the reinforcement so that all tensile forces now will be transmitted into the steel bars. At the ribs, smaller, inner cracks will develop.

Different ways have been developed to test the bond behaviour of reinforcement. In Fig. 8 some of the generally used test specimens are shown. The concentric pull-out specimen is used to test the bond strength under tension while in the excentric pull-out specimen the conditions in the end of a cantilever beam is simulated with lateral compression forces in the anchorage zone. With the lapped-splice specimen the overlapping of reinforcement is tested. In the tension specimen the bond behaviour in the tensile zone of the construction is simulated. This specimen has preferably been used for the derivation of shear stress-slip relations as representation of the stiffness of bond. In Fig. 9 some characteristic bond-slip relations of the tests of Dörr [9] under different transverse compression are shown. The slip is defined as relative displacement between reinforcement bar and surrounding concrete.

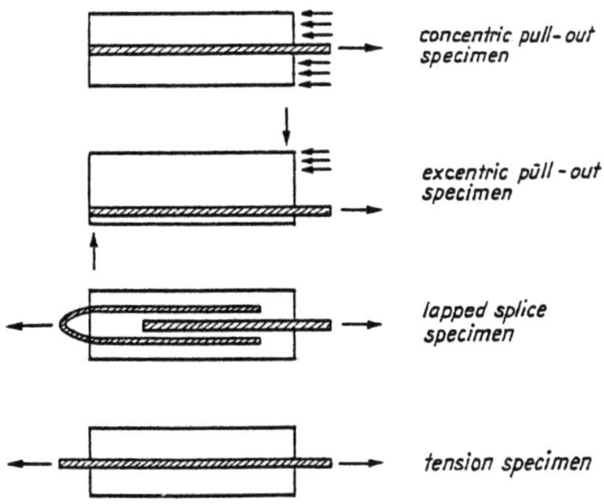

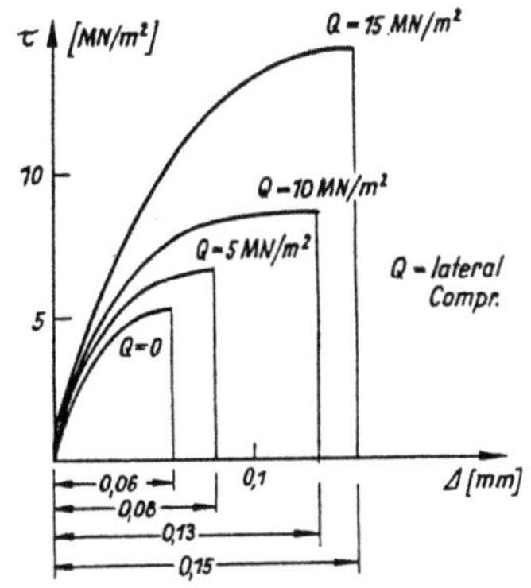

Fig. 8. Some test specimens for the investigation of bond between reinforcement and concrete

Fig. 9. Comparison of the τ- Δ -relationships with lateral pressure by the tests of Dörr [9]

The bond is usually taken to be perfect, that means the displacement of reinforcement and concrete is the same, the slip is zero. Only the shear strength may be limited. In structures in which bond is specially important, however, for example if the load is induced into the structure first of all by the reinforcement, the slip between concrete and reinforcement may not be neglected for a realistic computation of the deformation behaviour of the structure.

2.1.4 Reinforced Concrete Under In-Plane Forces

Fig. 10 shows a planar element of cracked reinforced concrete under in-plane forces. The situation of the element is defined by three geometrical parameters: the direction of the reinforcement bars, the direction of the cracks and the direction of the principal stresses. The first cracks will develop perpendicular to the principal tensile stress. Under further increase of the load, the direction of the principal stresses can change because of the transmission of shear stresses across the crack, so that a second crack may occur at any angle to the first crack. The shear stresses are transmitted by aggregate interlock and by dowel action of the

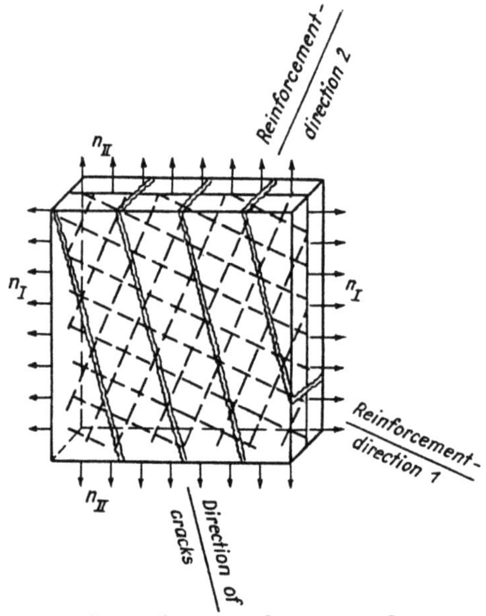

Fig. 10. Planar element of cracked reinforced concrete

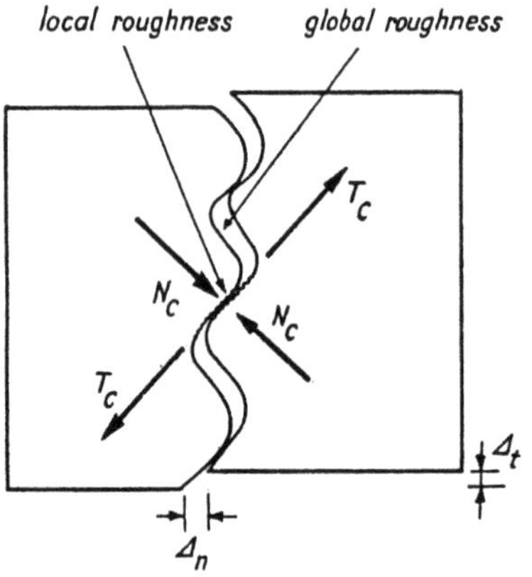

Fig. 11. Transmission of shear forces across a crack by aggregate interlock

reinforcement. Fig. 11 shows a model for the aggregate interlock at the crack. Shear forces are transmitted by global roughness that results from the meshing of the grains in the aggregate so that a tangential displacement Δ_t of the two edges of the crack is possible only by a widening Δ_n of the crack. The inclination of the direction of the forces in the crack is caused by local roughness of the mortar. The dowel action of the reinforcement is demonstrated in Fig. 12. Any relative displacement of the edges of the crack causes a deformation of the reinforcement bar and a transmission of shear forces across the crack.

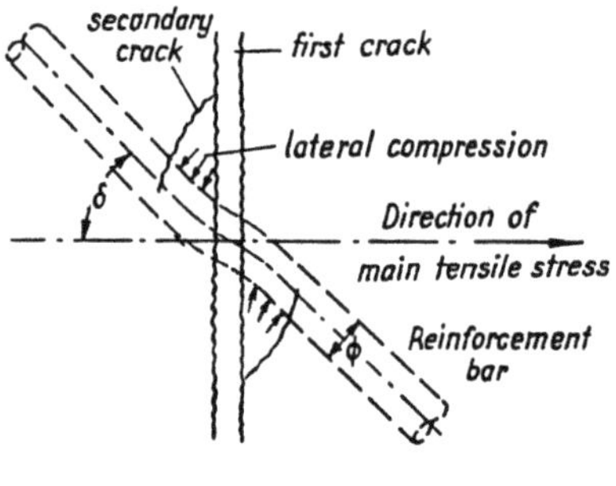

Fig. 12. Dowel action of reinforcement

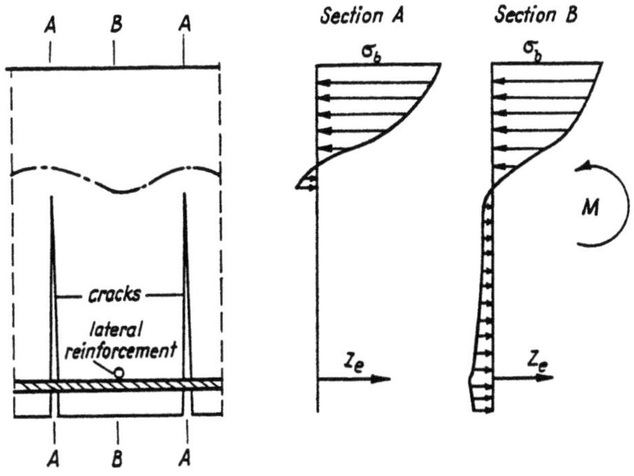

Fig. 13. Cross-section through a reinforced concrete element under bending moment

2.1.5 Flexural Behaviour of Reinforced Concrete

In Fig. 13 a section through a cracked element of reinforced concrete, loaded by bending moments, and the distribution of stresses in the cross-section is represented. In section A-A, the tensile forces are transmitted only by the reinforcement, besides a small resultant of possible tensile stresses in the concrete. Between the cracks, however, section B-B, the concrete is not free of stresses in the tensile zone as well as in the compression zone. The tensile stresses are caused by shear forces (bond) between the reinforcement and the concrete and by the effect of the compression stresses into the concrete tooth between the cracks. The neutral axis is not plane anymore after cracking.

2.2 Modelling of Planar Reinforced Concrete Structures by the Finite Element Method

By adopting the finite element method to describe the load-deformation behaviour of structures, the system is subdivided into finite elements that are interconnected at selected nodes with equal deformations and forces. Linear stiffness relations between the nodal forces and deformations of the elements are derived by the use of the displacement method, the force method, or mixed methods. The assembly of all elements of the complete structure yields a system of linear equations between the nodal forces and nodal deformations that can be solved for different load stages or given deformations.

2.2.1 Types of Finite Elements

The most versatile but also most expensive method for the modelling of reinforced concrete is the separate idealization of concrete and reinforcement by elements (see Fig. 14).

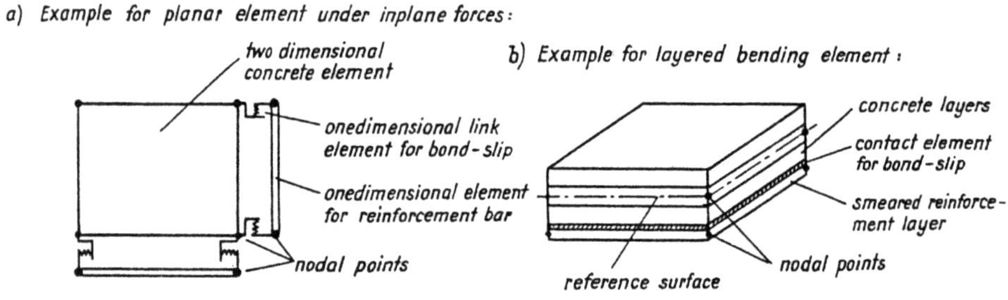

Fig. 14. Idealization of concrete and reinforcement by finite elements

The concrete is described by elements for bending, in-plane forces, the combination of bending and in-plane forces, or for shallow shells, dependent upon the problem. The possible elements differ in their shape, the interpolation functions for the fields of deformations or stresses, the number of the unknowns and the method of integration about the area of the element. The material properties of the concrete may be described by one constitutive law that is valid over the whole thickness of the planar element. If the material properties vary over the thickness of the element, for example after cracking under bending, a subdivision of the element into layers with different thicknesses or partitioning of the thickness by integration points at different distances is more suitable. In each element layer or integration point the actual stiffness relations can be formulated.

The reinforcement is idealized by one-dimensional elements for each steel bar or by planar elements with smeared material properties if reinforcement meshes are used.

Because of compatibility, the shape functions of the elements for the reinforcement depend on the kind of the concrete elements used.

The bond between concrete and reinforcement is realized by link elements with one-dimensional or orthogonal springs between the nodes of concrete elements and reinforcement elements or by contact elements. The contact elements are fictitious elements with one or two dimensions and no expansion in the third direction. They transmit normal stresses and shear stresses but allow only a relative tangential movement of the adjacent surfaces. As compared to the link elements the contact elements offer not only a more distinct idealization of bond but allow also a better match of the concrete and reinforcement elements. The link elements may result in incompatibilities if higher order finite elements are used.

If perfect bond is assumed, the elements of reinforcement can be combined with the concrete elements by the assumption of equal deformations in the nodes of reinforcement elements and concrete elements into one macro-element. The deformations in any plane of the element then are related to the deformations in one reference plane. This yields a reduction of the number of unknowns and a shortening of the computation time. In reinforced concrete slabs the deformations within the reference plane must be considered because the position of the neutral axis is not plane and its height differs with changing load while cracking.

The most economical method is the integrated modelling of reinforced concrete in one element. The material properties of the components concrete, steel, and bond are combined to a fictitious material whose deformation behaviour is described by in-plane force-strain relations or moment-curvature relations. Because of its economy of computation this method is suitable for the investigation of the behaviour of a structure by a series of iterations.

2.2.2 Propagation of the Crack Development

The development of cracks is considered as single cracks along the element boundaries, Fig. 15a, or as smeared cracks (crack fields) within an element, Fig. 15b. The disadvantage of the first method is that the direction of the cracks is assumed and that a renumbering of the element mesh is necessary after the development of the cracks. Within the element the crack is taken into account by modifying the material properties without changing the topology. The cracked concrete element is described by an orthotropic stiffness matrix, usually derived by the assumption that tension perpendicular to the crack cannot occur in the

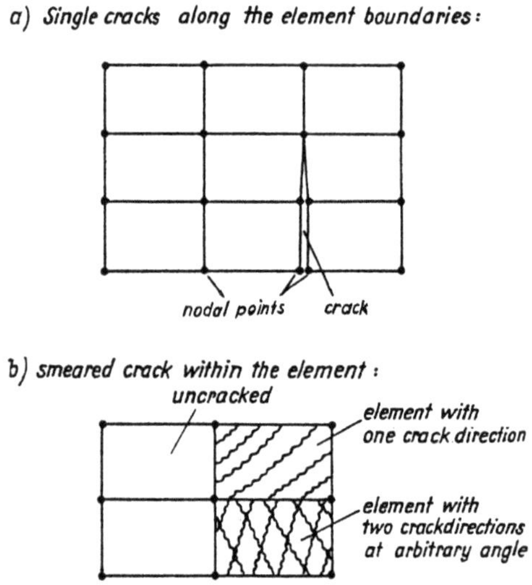

Fig. 15. Description of the crack devolopment with finite elements

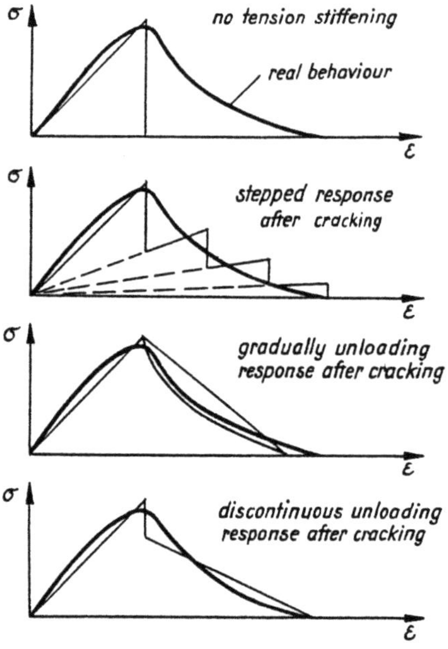

Fig. 16. Some possible assumptions for strain-softening of concrete after cracking

concrete. The shear forces parallel to the crack, which are transfered by aggregate interlock and dowel action of the reinforcement, are taken into consideration by reduction of the shear stiffness of concrete.

The tension-stiffening of the concrete, that is the carrying capacity of the concrete between the cracks, may be represented by an average additional charge to the steel stresses or by an unloading section in the stress-strain relation of concrete, after the maximum tensile stress is reached. In Fig. 16 some assumptions for the strain-softening of concrete are compared.

2.2.3 Numerical Treatment of the Nonlinearities

By consideration of real material behaviour, the system of equations describing the relations between the nodal deformations and nodal forces is nonlinear and cannot be solved directly. For the iterative treatment different solution procedures have been developed. Fig. 17 shows some of them. In the pure incremental method the load is increased stepwise

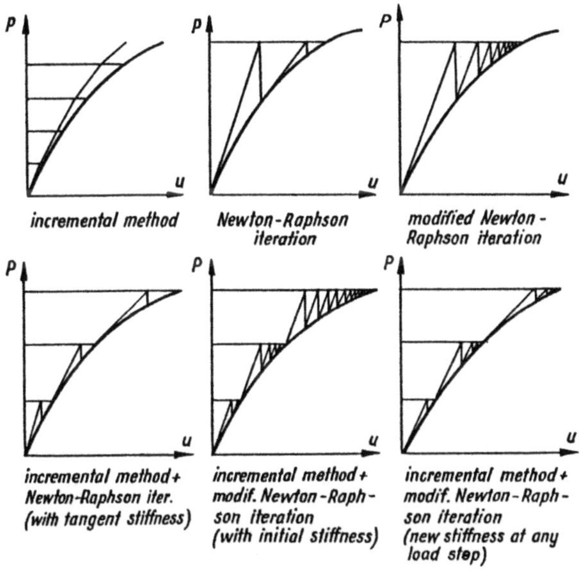

Fig. 17. Some solution procedures for the numerical treatment of the nonlinearities

and the stiffness is recomputed in each load step. However, the solution drifts away from the real load-deformation relation, if no equilibrium correction is performed. In the

2.2.4 Some Remarks About the Idealization of a Structure by Finite Elements

A conscientious choice of the right type of element and a careful discretization of the structure into elements is necessary to obtain satisfactory results. The best type of element that should be used depends on the problem to be studied. The appropriate modelling of the load-carrying behaviour of the reinforced concrete structure is more important than the accuracy of the numerical calculations. The question wether simple elements or higher-order elements should be preferred cannot be answered satisfactorily. Simple elements may produce convergence difficulties, if they are not compatible. On the other hand, by the use of higher-order elements, the system can be divided only into a few large elements to reduce the numerical efforts so that the change of stiffness, for example by cracking, is considered in too large regions if the element itself is not subdivided into zones with different material properties. With simple elements it is easier to refine the discretization of the structure in regions with a sharp change of the stress or strain gradient, for example at a concave corner or under single loads, or at sudden variation of the stiffness.

2.3 Synopsis of the Solution Methods Used by the Working Group of Darmstadt

In Table 1 the different methods are summarized that are used by the investigators of the working group of the Institute of Massivbau, TH Darmstadt, and the considered nonlinearities are stated. All the investigators use the displacement method. Only the short time behaviour of concrete is regarded except by Hoshino [11], who considers the shrinkage of concrete, and by Zeitler/Dietrich/Mehlhorn [17], who investigate the influence of creep and relaxation. Geometrical nonlinearities are included by Kristjansson [15] into the computation of slabs.

3. Numerical Examples

3.1 Shear Wall WT7 of the Tests of Leonhardt/Walther [1]

The system and the discretisation into finite elements used by Dörr [19] is shown in Fig. 18. The wall is characterized by a concentrated tie rod at the lower side and strong vertical reinforcement bars. After cracking of the concrete in the tension zone, the vertical reinforcement must transmit the load applied to the lower side of the panel upward into the uncracked compression zone, where it is transmitted by bond to the concrete and carried by arching action to the supports. To study the effect of bond-slip between the reinforcement and the concrete, a matter of main importance in this example, Dörr [19] idealized the reinforcement by separated bar elements and represented the bond-slip by realistic shear stress-slip relations including the influence of lateral compression. Fig. 19 shows the crack-development of the solution of Dörr [19].

In Fig. 20 the load-midspan deflection relation of Dörr according to two different assumptions about the bond-slip behaviour is shown and compared with the test results and the results of other authors.

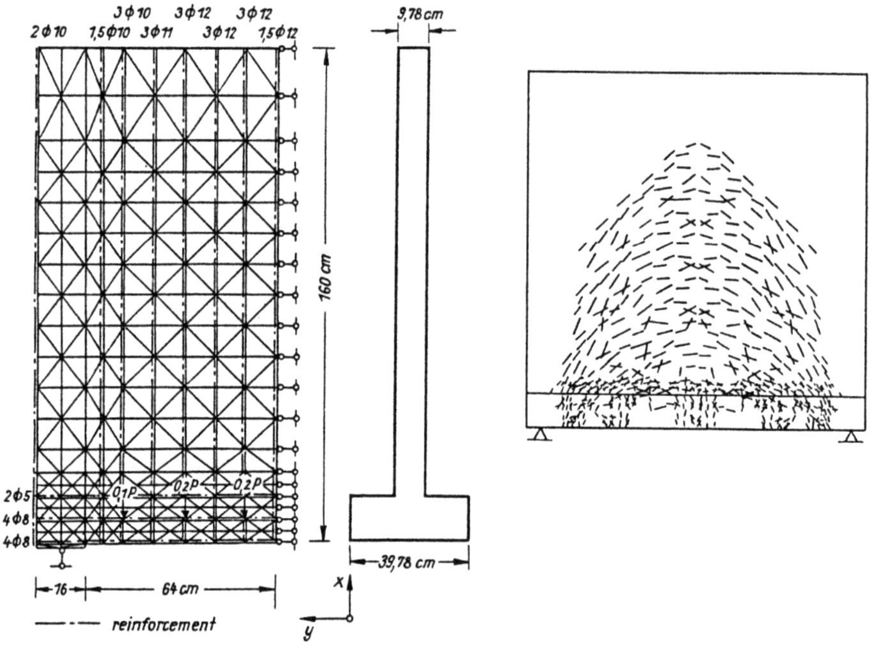

Fig. 18. Finite element idealization of Dörr [19] for the recalculation of the Shear Wall WT7 (see Fig. 1)

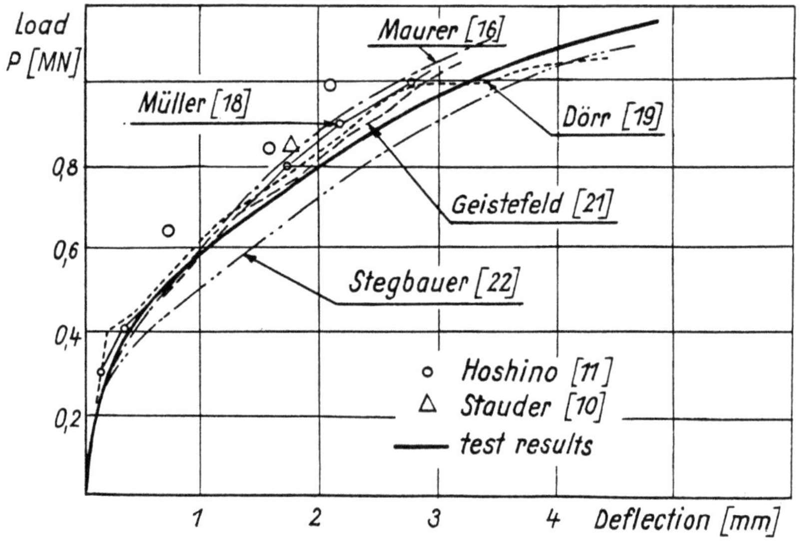

Fig. 20. Comparison of calculated load-midspan deflection relations with the test result of Leonhardt/Walther [1]

3.2 Simply Supported, Rectangular Slab Under Uniform Load of the Tests of Franz [2]

Fig. 21 shows the reinforcement of the slab and the discretisation into elements and layers of the calculation of Klein/Kristjansson [23] . The finite element idealization depends on the complicated distribution of the upper and lower reinforcement.

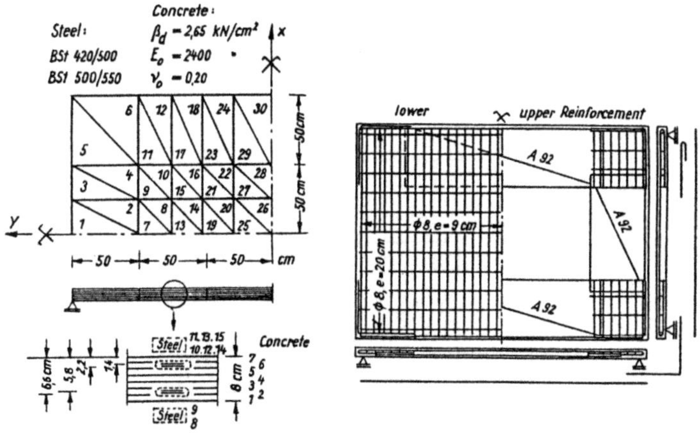

Fig. 21. Finite element idealization of Klein et al. [23] for the recalculation of the rectangular slab of the tests of Franz [2]

In Fig. 22 the load-midpoint deflection relation of the slab is compared to the results of various calculations. The calculations differ, besides the use of different element types and element meshes, first of all in the consideration of the tension stiffening of concrete between the cracks. At the load stages beyond the working load, all the calculated curves are close together and agree quite well with the test result. After primary cracking, however, tension stiffening is of significant importance, so that in the calculations without consideration of tension stiffening effects the stiffness of the slab is underestimated.

The distribution of the cracks at the lower and upper side of the slab by the calculation of Klein/Kristjansson [23] is shown in Fig. 23 together with crack distribution of the test.

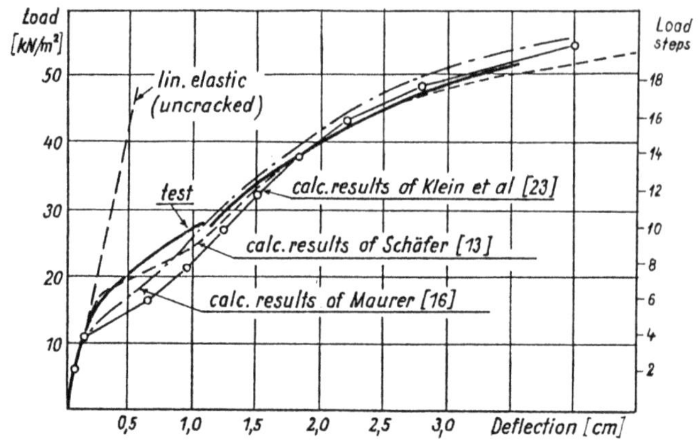

Fig. 22. Comparison of calculated load-midpoint deflection relations with the test result of Franz [2]

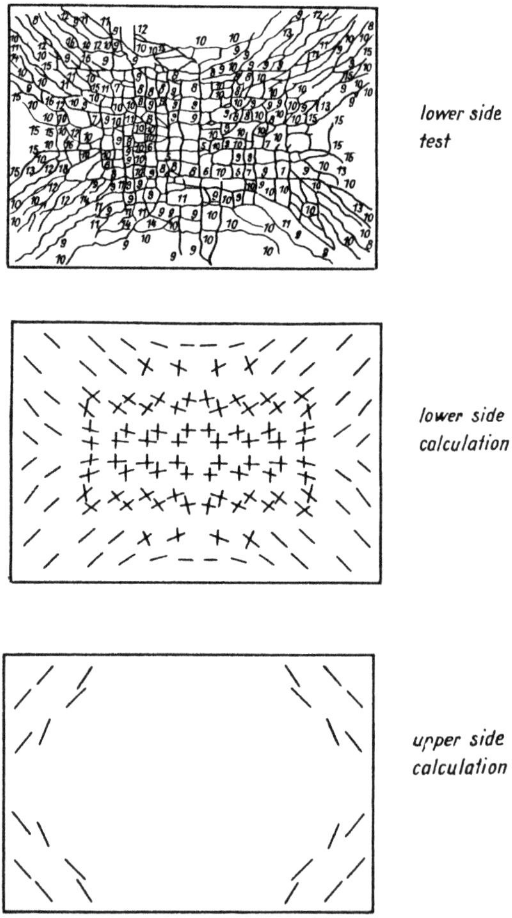

Fig. 23. Comparison of the crack development of the test-slab with the calculation result of Klein et al. [23]

3.3 Square, Corner-Supported Two-Way Slab Under Central Point Load

In this slab tested by McNeice [24] the influence of in-plane forces on the load bearing capacity of reinforced concrete slabs is demonstrated. For this purpose, the slab was computed by Mehlhorn/Kristjansson/Klein [14] with and without prevention of the horizontal movement of the corners in the midplane of the slab. The slab was calculated for both support conditions by including as well as excluding the nonlinear terms of the strain-deformation relations.

In Fig. 24 the calculated load-midpoint deflection relations are shown. After cracking of concrete the neutral plane of the slab moves upwards, so that the midplane is stretched and the corners move outwards. If the expansion of the slab is prevented by fixed corners, the spreading of the cracks is reduced. An in-plane compressive force is induced which leads to a stiffening of the slab and provides a higher load capacity. By the assumption of geometric linearity, this influence is overestimated.

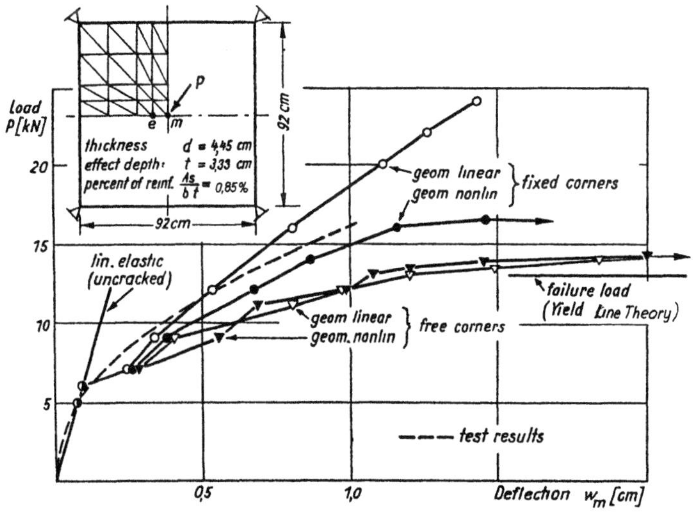

Fig. 24. Load-midpoint deflection relations of the square, corner supported slab by the tests of McNeice [24]. Calculation results of Mehlhorn/Kristjansson/Klein [14]

A comparison of the results of Mehlhorn/Kristjansson/Klein [14] with the results of other authors is given in Fig. 25. Only calculations without consideration of tension stiffening of concrete as in the calculation of Mehlhorn/Kristjansson/Klein [14] are chosen. For comparison the geometrically linear calculations of Mehlhorn/Kristjansson/Klein are used because the other mentioned authors neglect the geometric nonlinear terms. Scanlon [25], Hand/Schnobrich/Pecknold [26], Lin [27] and Gilbert/Warner [28] use the layered Finite Element Model. Hand/Schnobrich/Pecknold [26] and Lin [27] allow for the influence of in-plane forces. For the higher load levels, the agreement of the results by the different calculations is quite sufficient. The differences after primary cracking depend on the assumed tensile strength of concrete. Bashur/Darwin [29] obtain very good agreement with the test results. They use the integrated model for the description of the deformation behaviour of the reinforced concrete. In their moment-curvature relations, cracking of concrete and yielding of reinforcement are considered.

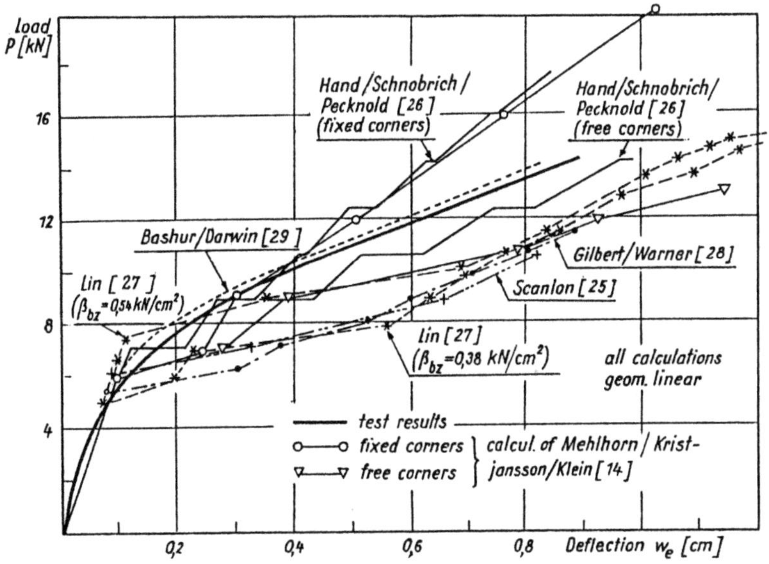

Fig. 25. Comparison of the calculation results of Mehlhorn/ Kristjansson/Klein [14] with the results of other authors

Tab. 1: Solution Methods Used by the Darmstadt Group

Author	Investig. Structures	Idealisation by Finite Elements	
		Kind of Model	Type of Element
Stauder [10]	panels under in-plane forces	sep. ideal. of concrete and reinforcement	concrete: 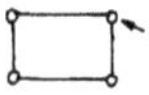 u; v reinf.:  u bond: onedim. link element
Hoshino [11]	panels under in-plane forces	sep. ideal. of concrete and reinforcement	concrete: 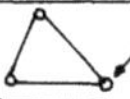 u; v reinf.: 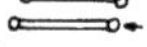 u bond: onedim. contact element
Schwing [12]	prefabricated panels under in-plane forces	sep. ideal. of concrete and reinforcement	concrete, reinf. and bond slip like Stauder bond between prefabr. panels: two-dim. link element
Schafer [13]	slabs, including in-plane forces	comb. plate panel element (int. points)	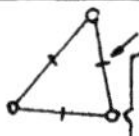 $u, u_{,n}, v; v_{,n}; w_{,n}$ $u; u_{,x}; u_{,y}, v; v_{,x};$ $v_{,y}; w; w_{,x}; w_{,y};$ $w_{,xx}; w_{,xy}; w_{,yy}$
Mehlhorn/Krist-jansson/Klein [14], [15], [23]	slabs, including in-plane forces	comb. layered plate panel element	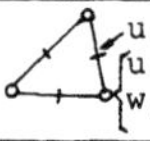 $u, v; w_{,n}$ $u; v; w; w_{,x}; w_{,y}$ $w_{,xx}, w_{,xy}; w_{,yy}$
Maurer [16]	folded plate structures	comb. layered plate panel element + reinf. bars	concrete comb. with reinf. meshs 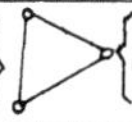 $u; v, w, \theta_x; \theta_y; \theta_z$ reinf. bars.  u
Zeitler/Dietrich/ Mehlhorn [17]	slabs, including in-plane forces	comb. layered plate panel element	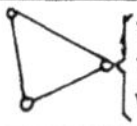 $u, u_{,x}; u_{,y};$ $v, v_{,x}, v_{,y};$ $w; w_{,x}; w_{,y}$
Muller [18]	panels under in-plane forces	sep. ideal. of concrete and reinf.	concrete + reinf. mesh 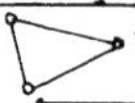 u; v reinf. bars:  u bond: contact element
Dorr [19]	panels under in-plane forces	sep. ideal. of concrete and reinf. bars	concrete comb. with reinf. mesh 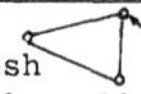 u; v reinf. bars:  u bond: onedim. cont. element

Formulation of the Material Properties				Description of the Cracks			
σ - ε-Rel. of Concrete	Fail. Crit. of Concrete	σ - ε -Rel. of Reinf.	Bond-Slip	Develop. of Cracks	Transm. of Shear	Tens. Stiff. of Concrete	Iteration Procedure
uniaxial nonlinear (DIN 1045)	hyp. of Mises in compr., max σ in tension	elastic-ideal-plastic	elastic-ideal-plastic	along element boundar	fully	stiffening of bond-links	pure iterat. method with secant stiffness
uniaxial linear elastic	unrestrained in compr., max σ in tension	linear elastic	linear-elastic	crack-field within the element	fully	no	gradual cons of cracks by initial stresses
uniaxial elastic-ideal plastic	hyp. of Mises (no reduct. of tensile strength)	elastic-ideal plastic	elastic-ideal-plastic	along the joints of prefabr. panels	nonlinear normal shear force interact.	---	pure iterat. method with secant stiffness
biaxial nonlinear (Link [7])	biaxial (Link [7])	nonlinear	perfect	crack field within the element	reduct of shear stiffness	strain soft. of concrete	incremental+ mod. Newton/ Raphson iter
biaxial nonlinear (Link [7])	biaxial (Link [7])	nonlinear or elast. ideal pl.	perfect	crack field within the element	reduction of shear stiffness	no	incremental+ mod. Newton/ Raphson iter
biaxial nonlinear (Link [7])	biaxial (Link [7])	nonlinear or bilinear	perfect	crack field within the element	reduct. of shear stiffness	strain soft. of concrete	incremental+ mod. Newton/ Raphson iter
uniaxial linear elastic	unrestr. comp., max σ in tension	linear elastic	perfect	crack field within the element	fully	no	gradual cons of cracks by new stiffness
biaxial nonlinear (Link [7])	biaxial (Link [7])	bilinear	non-linear	crack field within the element	reduct., depends on the crack-width	no	incremental+ mod. Newton/ Raphson iter
biaxial nonlinear (Link [7])	biaxial (Link [7])	bilinear	nonlinear, depends on lateral compr.	crack field within the element	reduct., depends on the crack-width	stiffening of reinf. bars	incremental+ Newton/ Raphson iter

References

1 Leonhardt, F. und R. Walther: Wandartige Träger. Deutscher Ausschuß für Stahlbeton, Heft 178. Verlag Wilhelm Ernst & Sohn, Berlin 1966.

2 Franz, G.: Über die Beanspruchung der Bewehrung von kreuzweise bewehrten vierseitig frei drehbar gelagerten Rechteckplatten aus Stahlbeton. Untersuchungsbericht des Instituts für Beton und Stahlbeton, Karlsruhe, 1970.

3 Eibl, J., G. Ivânyi: Studie zum Trag- und Verformungsverhalten von Stahlbeton. Deutscher Ausschuß für Stahlbeton, Heft 260, Verlag Wilhelm Ernst & Sohn, Berlin 1976.

4 Wegner, R.: Finite Element-Models for Reinforced Concrete. US-German Symposium on "Formulation and Computational Algorithms in Finite Element Analysis", M.I.T. Boston, Aug. 1976.

5 Schnobrich, W. C.: Behavior of Reinforced Concrete Structures, Predicted by the Finite Element Method. Computers and Structures, Vol. 7, pp. 365 - 376, Pergamon Press 1977, Great Britain.

6 Kupfer, H.: Das Verhalten des Betons unter mehrachsiger Kurzzeitbelastung unter besonderer Berücksichtigung der zweiachsigen Beanspruchung. Deutscher Ausschuß für Stahlbeton, Heft 229, Verlag Wilhelm Ernst & Sohn, Berlin 1973.

7 Link, J.: Eine Formulierung des zweiaxialen Verformungs- und Bruchverhaltens von Beton und deren Anwendung auf die wirklichkeitsnahe Berechnung von Stahlbetonplatten. Deutscher Ausschuß für Stahlbeton, Heft 270, Verlag Wilhelm Ernst & Sohn, Berlin 1976.

8 Goto, Y.: Cracks formed in Concrete Around Deformed Tension Bars. Journal of the American Concrete Inst., April 1971.

9 Dörr, K., G. Mehlhorn: Berechnung von Stahlbetonscheiben im Zustand II bei Annahme eines wirklichkeitsnahen Werkstoffverhaltens. Forschungsberichte aus dem Institut für Massivbau der Technischen Hochschule Darmstadt, Heft Nr. 39, 1979.

10 Stauder, W.: Ein Beitrag zur Untersuchung von Stahlbetonscheiben mit Hilfe finiter Elemente unter Berücksichtigung eines wirklichkeitsnahen Werkstoffverhaltens. Dissertation, TH Darmstadt, 1973.

11 Hoshino, M.: Ein Beitrag zur Untersuchung des Spannungszustandes an Arbeitsfugen mit Spanngliedkoppelungen von abschnittsweise in Ortbeton hergestellten Spannbetonbrücken. Dissertation, TH Darmstadt, 1974.

12 Schwing, H.: Zur wirklichkeitsnahen Berechnung von Wandscheiben aus Fertigteilen. Dissertation, TH Darmstadt, 1975.

13 Schäfer, H.: Zur Berechnung von Stahlbetonplatten. Dissertation, TH Darmstadt, 1976.

14 Mehlhorn, G., R. Kristjansson, D. Klein: Berechnung von dünnen Stahlbetonplatten bei Berücksichtigung eines wirklichkeitsnahen Werkstoffverhaltens. Forschungsbericht aus dem Institut für Massivbau der TH Darmstadt, Heft Nr. 44, 1980.

15 Kristjansson, R.: Physikalisch und geometrisch nichtlineare Berechnung von Stahlbetonplatten mit Hilfe Finiter Elemente. Dissertation, TH Darmstadt, 1977.

16 Maurer, G.: Zum Tragverhalten des einzelligen Spannbetonkastenträgers. Dissertation, TH Darmstadt, 1979.

17 Zeitler, W., R. Dietrich, G. Mehlhorn: Erfassung des räumlichen Spannungszustandes in Querschnitten von Massivbrücken unter besonderer Berücksichtigung des Spannungszustandes an Bauabschnittsgrenzen mit Spanngliedkopplungen. Forschungsbericht aus dem Institut für Massivbau der TH Darmstadt, Heft 42, 1978.

18 Müller, T.: Stahlbetonscheiben mit Öffnungen. Dissertation, TH Darmstadt, 1979.

19 Dörr, K.: Ein Beitrag zur Berechnung von Stahlbetonscheiben unter besonderer Berücksichtigung des Verbundverhaltens. Dissertation, TH Darmstadt, 1980.

20 Cornelius, V.: Zum Einfluß der Umlenkkräfte aus Vorspannung auf die Tragfähigkeit von Spannbetonbauteilen. Dem Fachbereich Konstruktiver Ingenieurbau der TH Darmstadt vorgelegte Dissertation, 1980.

21 Geistefeld, H.: Stahlbetonscheiben im gerissenen Zustand - Berechnung mit Berücksichtigung der rißabhängigen Schubsteifigkeit im Materialgesetz. Dissertation, TU Braunschweig, 1976.

22 Stegbauer, A.: Beitrag zur Berechnung von Stahlbetonflächentragwerken mit wirklichkeitsnahen Materialgesetzen unter besonderer Berücksichtigung des Plattenbalkens. Dissertation TU München, 1975.

23 Klein, D., R. Kristjansson, J. Link, G. Mehlhorn, H. Schäfer, K. Schneider: Zur Berechnung von dünnen Stahlbetonplatten bei Berücksichtigung eines wirklichkeitsnahen Werkstoffverhaltens. Forschungsbericht Nr. 32 aus dem Institut für Massivbau der TH Darmstadt, 1976.

24 Jofriet, J. C., G. M. McNeice: Finite element analysis of reinforced concrete slabs. Journal of the Struct. Division, Proc. ASCE, Vol. 97, No. ST3, 1971.

25 Scanlon, A.: Time dependent deflections of reinforced concrete slabs. Ph. D. Thesis, Department of Civil Engineering, University of Alberta, Edmonton, Canada, 1971.

26 Hand, F. R., W. C. Schnobrich, D. A. W. Pecknold: A Layered Finite Element Nonlinear Analysis of Reinforced Concrete Plates and Shells. University of Illinois, Urbana, Illinois, SRS No. 389, Civil Engineering Studies, Aug. 1972.

27 Lin, C. S.: Nonlinear Analysis of Reinforced Concrete Slabs and Shells. Ph. D. Thesis, Div. of Structural Engineering and Structural Mechanics, University of California, Berkeley, April 1973.

28 Gilbert, R. J. und R. F. Warner: Tension Stiffening in Reinforced Concrete Slabs. Journal of the Structural Division, Proceedings of ASCE, Vol. 104, ST 12, Dez. 1978.

29 Bashur, F. K. und D. Darwin: Nonlinear Model for Reinforced Concrete Slabs. Journal of the Structural Division, Proceedings of ASCE, Vol. 104, ST 1, Jan. 1978.

Part IV:
Nonlinear Dynamics

Finite Element Analysis of Transient Nonlinear Response of Reinforced Concrete Structures

M. J. MIKKOLA, H. S. SINISALO
Helsinki University of Technology, Finland

Abstract

Material nonlinearities of reinforced concrete are described as follows: tensile cracking by a criterion defined in terms of stresses, inelastic response in compression by elastic plastic relationships, and effects of aggregate interlocking, dowel action and bond failure by approximate means. Governing equations are derived using the total Lagrangian approach. In time integration, the central difference method is employed. Numerical applications include quasistatic, cyclic and impulsive loadings of slabs and beams. Computed results provided by the finite element procedure are compared with available experimental data.

Introduction

Major sources of nonlinearity in reinforced concrete structures are the progressive cracking of concrete in tension zones and the inelastic response in compression, the inelastic deformation of reinforcing steel, and other nonlinearities related to reinforcement and its interaction with concrete. In long term loading, the time dependency of deformation of concrete becomes a characteristic part of the response. In transient loading, the effect of high strain rate on the behaviour of concrete and steel should also be taken into consideration.

The finite element method has enabled the solution of the complicated problems represented by the nonlinear behaviour of reinforced concrete structures. Several studies on reinforced concrete under static loads have been published in recent years, see the excellent review [1] and the references given there. Dynamic behaviour of reinforced concrete structures has been investigated in earthquake engineering. However, relatively few reports deal with the nonlinear transient response of reinforced concrete structures [2]-[9]. Various aspects of impact and impulsive loadings on buildings are also discussed in the extensive review by a RILEM committee [10]. It is obvious that in many problems of structural design, particularly in new fields like nuclear and off-shore engineering, the application and development of powerful and accurate tools, such as the finite element method, are required for the assessment of safety and deformational characteristics of structures.

This paper reports on a study of reinforced concrete beams and slabs subjected to transient impulsive loading. In the initial state, the composite material formed by concrete and reinforcement is assumed to behave elastically. Tensile cracking of concrete is determined by a stress criterion. Inelastic response in compression is described by the theory of plasticity, which is extended also to the softening region. As a second model, a yield function and cracking criterion depending on the strain rate are applied. The purpose is to allow for the rate effect in tensile and compressive regions. The reinforcement is described as smeared and orientated steel layers. The cracks in concrete are treated as smeared, and the nonlinearity effects in cracks, such as the aggregate interlocking, the dowel action and the tension stiffening are given by an approximate formula.

The equation of motion is derived using the principle of virtual displacements in total Lagrangean formulation. The kinematic equations of plates are taken in accordance with the theories of von Kármán and Mindlin. The problem is discretized spatially and temporally by use of the finite element and finite difference methods, respectively. Some problems of beams and slabs subjected to static, cyclic and dynamic loads are analyzed and the computed results compared with available experimental and numerical values.

Modelling of reinforced concrete

The behaviour of plain concrete in multiaxial states of stress has been the objective of intensive research for many years. Constitutive modelling of inelastic short-term behaviour of concrete has been performed on the basis of nonlinear elastic, hypoelastic, elasticplastic, endochronic and fracture models [11]-[15]. From the experimental investigations the references [16] and [17] can be mentioned here. A considerable variation of experimental results exists. The primary reason seams to be in test arrangements. A review of the proposed plastic yield functions and their interrelations with experiments is given by Wastiels [18]. Theory of plasticity provides a satisfactory model for the behaviour of concrete. However, it is apparent that the models which are capable to describe changes in microstructure of concrete due to straining (microcracking) and the corresponding changes in response are more realistic. Such are the endochronic and the plastic-fracturing models.

The properties of concrete at high strain rates are much less known than in quasistatic loading. Some experimental results for uniaxial loading exist [10], [19]-[22]. They indicate that the compressive strength is increased by 0-80 %, the tensile strength by 0-50 %, and the initial modulus of elasticity by 0-40 %, when the strain rate varies in the range 10^{-4}...10 1/s. The effect of two- or three-dimensional state of stress on the dynamic response corresponds to that of quasistatic case [22]. The strain rate effect on strength can be incorporated into viscoplastic model or elastic plastic model with a yield function depending on strain rate.

In this study, the response of concrete is described by the elastic plastic model. In compressive region, the strain hardening and strain softening are included. In the tensile region, the cracking is determined by a stress criterion. The strain rate effect is included by rate dependency of the yield function.

Elastic plastic model

The behaviour in one dimensional fashion is illustrated in Fig. 1. After tensile fracture, the stress should drop to zero, but in this model it decreases linearly, which is intended to simulate the tension stiffening effect observed in reinforced concrete. In compression, the behaviour is linear up to stress value $-0.4\ f_c$. Beyond this point, the equation proposed by Sargin [23]

$$\sigma/f_c = (AX+BX^2)/(1+CX+DX^2),\ X = \varepsilon/\varepsilon_o \qquad (1)$$

is employed. f_c is the maximum compressive stress and ε_o the corresponding strain. A, B, C, and D are experimental parameters. This equation also describes strain softening. Strain softening does not comply with the stability requirements of the theory of plasticity, but it is accepted here as well as the normality rule for plastic strain rates in softening region (see [24] and [25]).

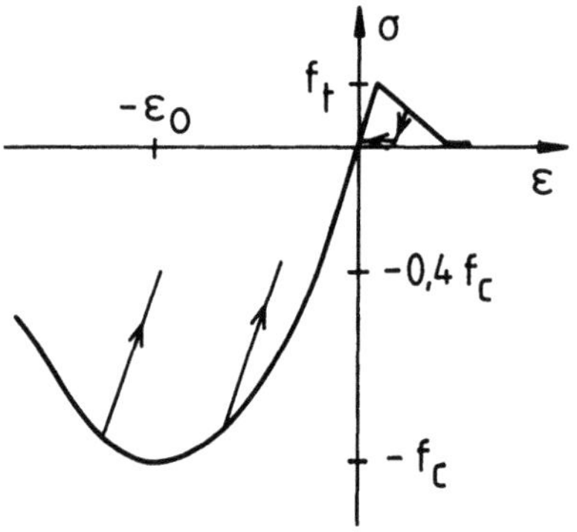

Fig. 1. Uniaxial stress strain diagram of concrete

In unloading, the stress reversal obeys the initial modulus of elasticity. In tensile region, the stress is first relieved to zero and the cracks are closed before the material can take any compressive stress (Fig. 1).

In general three- or two-dimensional case, the response of the the elastic plastic material is

$$\dot{\sigma} = C\dot{\varepsilon} - C\lambda n \qquad (2)$$

where C is the elasticity matrix, $n \equiv \partial f/\partial\sigma$ the gradient of the yield surface $f = 0$ and λ the plasticity parameter

$$\lambda = H^{-1}\langle n^T\dot{\sigma}\rangle = \langle n^T C\dot{\varepsilon}\rangle/(H+n^T Cn) \tag{3}$$

The notation $\langle x\rangle$ means $\langle x\rangle = x$ if $x \geq 0$, $\langle x\rangle = 0$ if $x < 0$. The modulus of plasticity $H(\varepsilon^p) \equiv d\bar{\sigma}/d\varepsilon^p$ can be found on the basis of the uniaxial stress-strain relationship. The actual yield condition, which was applied in numerical calculations, is

$$f(\sigma,\varepsilon P) = a\sqrt{3gJ_2} + bJ_1 - \bar{\sigma} \tag{4}$$

The function $g(\theta)$

$$g(\theta) = \frac{2(r^2-1)\cos\theta+(2-r)[4(r^2-1)\cos^2\theta+5-4r]^{1/2}}{4(r^2-1)\cos^2\theta+(r-2)^2}, \tag{5}$$

taken in accordance with Ref. [26], defines an elliptic form for the sixth part of the yield locus in the deviatoric plane. The angle θ, varying between 0° and 60°, is

$$\theta = (1/3)\ \mathrm{arc}\ \cos(3\sqrt{3}/2J_3{}^{3/2}) \tag{6}$$

The factor $r = g(60^\circ)/g(0^\circ)$ determines the ratio of the radius at the "compressive" meridian to the radius at the "tensile" meridian. The value $r = 1.5$ was found to be appropriate [17]. J_1, J_2, and J_3 are the invariants of the stress tensor σ_{ij} and deviator s_{ij}

$$J_1 = \sigma_{kk},\ J_2 = s_{ij}s_{ij}/2,\ J_3 = s_{ij}s_{jk}s_{ki}/3 \tag{7}$$

ε^p is the effective plastic strain

$$\varepsilon^p = \int d\varepsilon^p = \int\sqrt{2d\varepsilon^p_{ij}\ d\varepsilon^p_{ij}/3} \tag{8}$$

a and b are parameters to be determined on the basis of experimental data [16]. Tensile strength $f_t = 0.1\ f_c$ and biaxial strength $f_{c2} = 1.16\ f_c$ yield $a = 1.138$, $b = 0.138$

for compressive region and a = 5.5, b = 4.5 for tensile region. The biaxial behaviour of the material is illustrated by Fig. 2. In Fig. 3, the predicted diagrams of the axial and lateral strains and the dilatation in uniaxial compression are shown together with experimental curves [16].

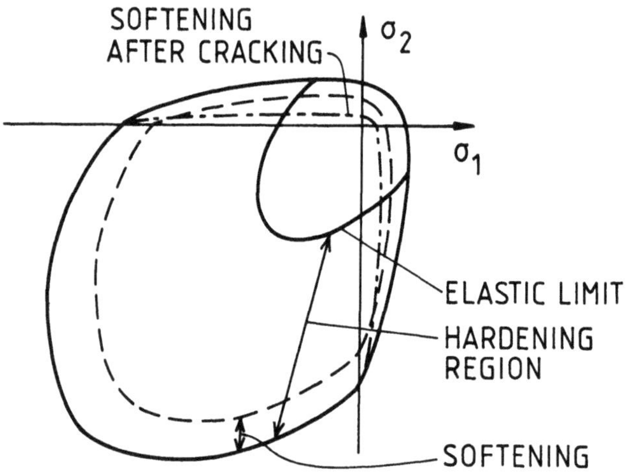

Fig. 2. Biaxial behaviour of concrete

Elastic viscoplastic model

The strain rate dependency of the strength of concrete could be described by Perzyna's viscoplastic model

$$\dot{\varepsilon}^{vp} = \gamma\langle\Phi(f)\rangle \frac{\partial f}{\partial \sigma} \tag{9}$$

which leads to the dynamic yield condition

$$f - \Phi^{-1}[\gamma^{-1}(\frac{1}{2}\frac{\partial f}{\partial \sigma}\frac{\partial f}{\partial \sigma})^{-\frac{1}{2}}\dot{\bar{\varepsilon}}^{vp}] = 0 \tag{10}$$

where $\dot{\bar{\varepsilon}}^{vp} = \sqrt{2\dot{\varepsilon}_{ij}^{vp}\dot{\varepsilon}_{ij}^{vp}/3}$ is the effective viscoplastic strain rate. Instead of the common viscoplasticity theory, another approach is chosen here. Following the suggestion by Nilsson [7], a yield limit, which is dependent on the strain rate, is inserted into the yield function (4)

$$a\sqrt{3gJ_2}+bJ_1-\bar{\sigma}[c_1+c_2\ln\dot{\varepsilon}+c_3(\ln\dot{\varepsilon})^2] = 0 \tag{11}$$

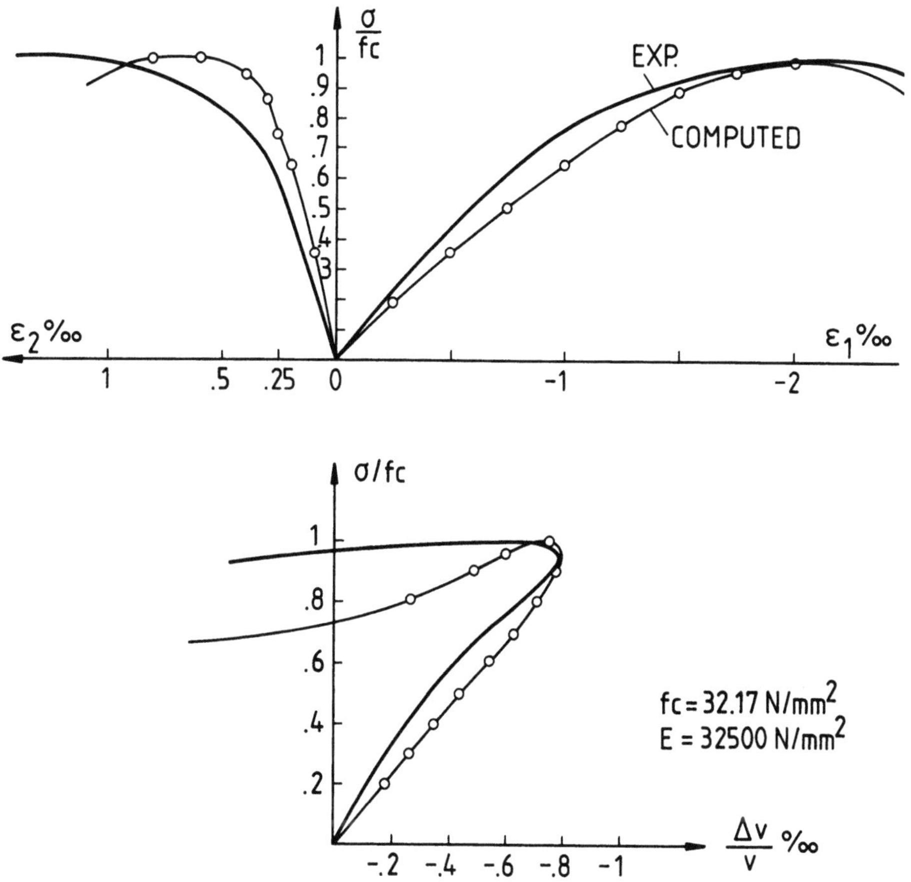

Fig 3. Axial and lateral strains and dilatation in uniaxial compression

The effective strain rate is $\dot{\varepsilon} = \sqrt{2\dot{\varepsilon}_{ij}\dot{\varepsilon}_{ij}/3}$ and the coefficients are $c_1 = 1.6$, $c_2 = 0.104$, $c_3 = 0.0045$, obtained by curve fitting on experimental data. Yield of concrete is checked with respect to the dynamic yield condition (11). The cracking criterion is changed in similar fashion due to the strain rate effect.

Cracking

Cracking of concrete occurs when the cracking criterion of the form (4) is satisfied. The direction of the cracking plane is perpendicular to the largest principal stress. After cracking, the three-dimensional state of stress changes to two-dimensional, etc. The crack is closed, when the strain perpendicular to the cracking plane becomes zero or compressive.

Reinforcement

For reinforcing steel, the uniaxial bilinear elastic plastic material model is used.

Reinforced concrete

Complete compatibility between concrete and reinforcement is assumed, so that the deformation is continuous over the adjoining boundaries of concrete and steel. In slabs and panels, the reinforcement net is treated as smeared layers of thin wires. The effect of separate reinforcing bars is transferred to the nodes of the finite element mesh through the shape functions. The shear resistance, which still exists in the cracked region due to aggregate interlocking and dowel action, is taken into account in reduced shear modulus of concrete [27]

$$G_{red} = 0.1\ G \tag{12}$$

Equations of equilibrium

In total Lagrangian approach, the principle of virtual work can be presented in the form

$$\int_{V_o} S\delta E dV_o - \int_{V_o} f\delta u dV_o - \int_{S_{ot}} t\delta u dS_o + \int_{V_o} \rho_o \ddot{u}\delta u dV_o = 0 \tag{13}$$

where S is the 2nd Piola-Kirchhoff stress, E the Green-Lagrange strain $E_{ij} = (\partial u_i/\partial x_j + \partial u_j/\partial x_i + \partial u_k/\partial x_i \partial u_k/\partial x_j)/2$, f the body force, t the surface traction, ρ_o the mass density, u the displacement, $\ddot{u}$ the acceleration, and V_o and S_o the volume and surface in the reference configuration, respectively. Use of the finite element approximation $u = Nq$ results in a system of nonlinear differential equations with respect to time

$$R(q) + M\ddot{q} = Q \tag{14}$$

where

$$R = \int_{V_o} B^T S \, dV_o,$$
$$M = \int_{V_o} N^T \rho_o N \, dV_o, \qquad (15)$$
$$Q = \int_{V_o} N^T f dV_o + \int_{S_{ot}} N^T t dS_o,$$

are the internal force vector, the mass matrix, and the external force vector, respectively. The incremental form of the equation (14) is

$$^1K_t \Delta q + M\,^2\ddot{q} = {}^2Q - {}^1R \qquad (16)$$

where the tangent stiffness $^1K_t = {}^1K_L + {}^1K_G + {}^1K_S$ consists of linear, geometric and stress stiffnesses. The superscripts 1 and 2 refer to two adjacent configurations of the structure on its deformation path. (For details, see Ref. [28] Ch. 19). The increment of stress $\Delta S = {}^2S - {}^1S \cong \dot{S}\Delta t$ is determined by the constitutive equation

$$\overset{*}{S} = C\dot{E} \qquad (17)$$

The Jaumann rate of the Kirchhoff stress (the components of which are equal with those of the 2nd Piola-Kirchhoff stress) is related to the convective rate as follows

$$\overset{*}{S}{}^{ij} = \dot{S}^{ij} + G^{ik}S^{j\ell}\dot{E}_{k\ell} + G^{jk}S^{i\ell}\dot{E}_{k\ell} \qquad (18)$$

G_{ij} is the metric tensor in convective coordinates.

Here the equations of equilibrium have been presented in a form valid for large deformations and rotations. In application to reinforced concrete structures, however, the observation was made that only in some cases of flexible beams and slabs the effect of large rotations was necessary to be taken into account. In all other cases, the accuracy provided by the geometrically linear theory was satisfactory.

Solution techniques

Several direct time integration schemes have been introduced and discussed in papers on structural dynamics. Comparative computations have shown [29] that the central difference (CD) method and the Newmark β-method perform accurately and effectively in various problems of transient loading. In the solution of transient response of reinforced concrete structures, the explicit CD-scheme and the implicit trapezoidal rule or the Newmark β-method with parameters $\beta = 0.25$ and $\gamma = 0.5$ were applied.

When the CD-scheme is used, the solution q_{n+1} of the equation (14) at the instant t_{n+1} is computed from formula

$$q_{n+1} = h^2 M^{-1}(Q_n - R_n) + 2q_n - q_{n-1} \tag{19}$$

where $h = t_{n+1} - t_n$ is the step length. The strain and stress increments are

$$\Delta E = B_n(q_{n+1} - q_n), \quad \Delta S = C_n \Delta E \tag{20}$$

The internal force vector at the instant t_{n+1} is evaluated in accordance with equation (15), using B_{n+1} and $S_{n+1} = S_n + \Delta S$. The initial condition $\dot{q}_o = (q_1 - q_{-1})/2\Delta t$ is used for elimination of q_{-1} in the first step. The CD-scheme with diagonal mass matrix is accurate and simple. As an explicit difference method its step length is limited by the largest natural frequency of the finite element equation system, $h < 2/\omega_{max}$.

The incremental equation (20) is solved applying the trapezoidal rule

$$\begin{aligned} \dot{q}_{n+1} &= \dot{q}_n + h\ddot{q}_n/2 + h\ddot{q}_{n+1}/2 \\ q_{n+1} &= q_n + h\dot{q}_n + h^2\ddot{q}_n/4 + h^2\ddot{q}_{n+1}/4 \end{aligned} \tag{21}$$

This is an implicit scheme and therefore iteration has to carried out at each time step. The displacement vectors q^i_{n+1} and q^{i+1}_{n+1} in the ith and (i+1)th iteration cycles correspond to configurations 1 and 2 in equation (16). Use of equations (16) and $(21)_2$ yields

$$(K^{i}_{t,n+1}+4M/h^{2})\Delta q^{i+1} = Q_{n+1}-R^{i}_{n+1}+M[-4(q^{i}_{n+1}-q_{n})/ /h^{2}+4\dot{q}_{n}/h+\ddot{q}_{n}] \tag{22}$$

where $\Delta q^{i+1} = q^{i+1}_{n+1} - q^{i}_{n+1}$. For the first iteration cycle, $q^{1}_{n+1} = q_{n}$ is taken. The iteration is continued until $\| \Delta q^{i+1} \| < \varepsilon \| q^{i+1}_{n+1} - q_{n} \|$ where ε is a tolerance parameter. Consistent mass matrix is used. In order to avoid too drastic or artificial changes due to cracking, the tangent stiffness was updated in the first iteration cycle but held constant thereafter. This also reduces computing time.

The stability of the solution was followed by an energy balance check [30], [29]. If T, U and W denote the kinetic, internal and external energies, respectively, the energy balance criterion is

$$\Delta E = \Delta U + \Delta T - \Delta W < \varepsilon(U+T) \tag{23}$$

where ε is a given tolerance. The energy increments can be calculated using the trapezoidal integration

$$\begin{aligned} \Delta U &= \Delta q^{T}(R_{n} + R_{n+1})/2, \\ \Delta T &= \Delta\dot{q}^{T}M(\dot{q}_{n} + \dot{q}_{n+1})/2, \\ \Delta W &= \Delta q^{T}(Q_{n} + Q_{n+1})/2. \end{aligned} \tag{24}$$

In applications, it was found that the CD-method was much more economical than the trapezoidal rule. Therefore most of the results were calculated using the CD-procedure.

In quasistatic cases, solutions were achieved by the incremental equation (16) using similar iteration as in the implicit scheme.

Applications

Statically loaded slab

The corner supported square slab tested under central point load by Jofriet & McNeice [31] was analyzed in order to check the performance of the computing procedure. The plate theories by von Kármán and Mindlin were applied. The non-linear strain-displacement relationships of Mindlin's theory are

$$
\begin{aligned}
\varepsilon_x &= \partial u/\partial x + (\partial w/\partial x)^2/2 + z\partial\varphi/\partial x \\
\gamma_{xy} &= \partial u/\partial y + \partial v/\partial x + (\partial w/\partial x)(\partial w/\partial y) + z(\partial\varphi/ \\
&\quad /\partial y + \partial\Psi/\partial x) \\
\gamma_{xz} &= \partial w/\partial x + \varphi
\end{aligned}
\tag{25}
$$

etc. where φ and Ψ represent the rotations of the normal with respect to x- and y-axes. The expressions of von Kármán follow when the constraints $\gamma_{xz} = \gamma_{yz} = 0$ are imposed.

In von Kármán's plate theory, a 24 DOF rectangular element was used with 4 nodes and displacement parameters u, v, w, $\partial w/\partial x$, $\partial w/\partial y$, $\partial^2 w/\partial x\partial y$ at each node. Integration was carried out by 2x2 Gaussian formula. In Mindlin's plate theory, a 20 DOF isoparametric element with 4 nodes and displacement parameters u, v, w, φ, Ψ at each node was employed. 2x2 Gaussian integration rule was used except for the shear deformation only one integration point. Integration through thickness was performed by Simpson's rule with 7 integration points.

Load-deflection curves of the point A are shown in Fig. 4. The influence of tension stiffening was investigated using three different assumptions for the tensile stress-strain behaviour. Assumption 2 provides the best fit with the experimental curve. Von Kármán element with a 3 by 3 mesh for a plate quadrant was used. The mesh for Mindlin element was 2 by 2.

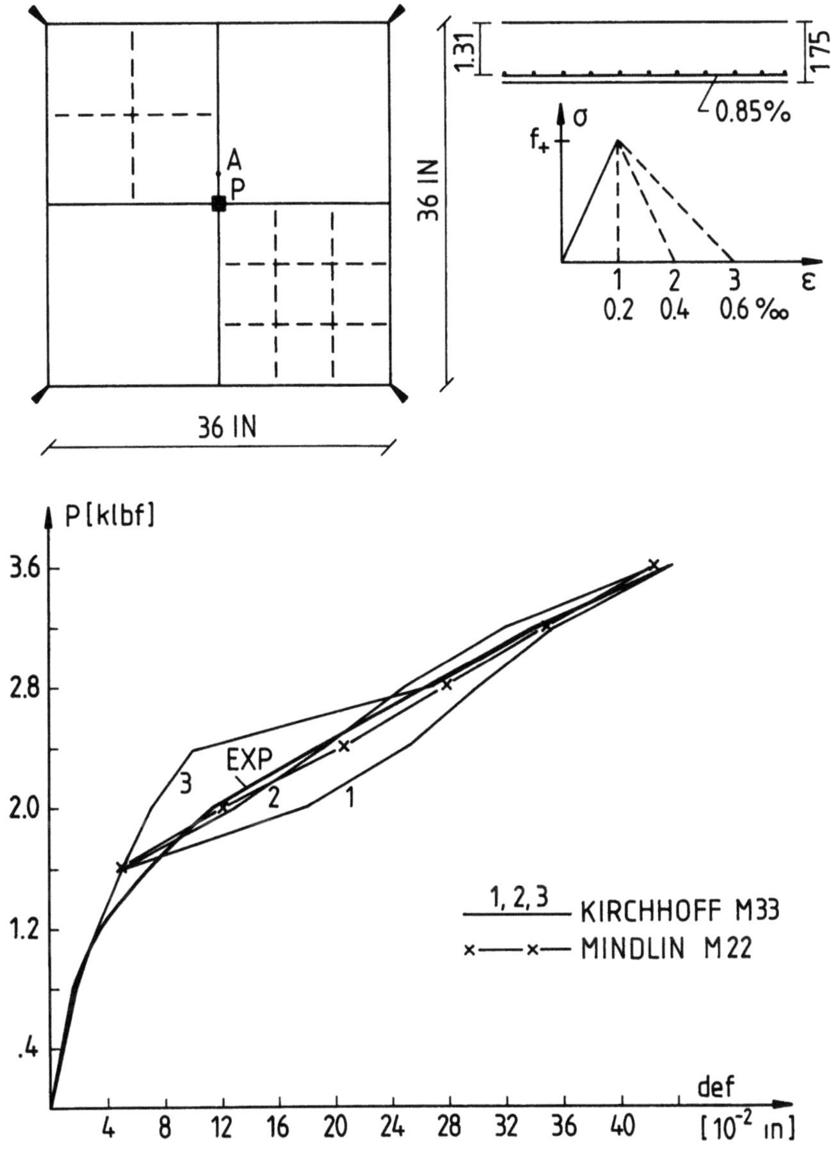

Fig. 4. Load-deflection curves of two-way slab tested by McNeice [31]. The data are $f_c = 5500$ psi, $f_t = 550$ psi, $E_c = 4.15\ 10^6$ psi, $\nu_c = 0.15$, $f_s = 60\ 000$ psi, $E_s = 29\ 10^6$ psi.

Cyclically loaded beam

Burns & Siess [32] made tests on reinforced concrete beams subjected to cyclically varying point load. The computation was based on a nonlinear strain-displacement relationship corresponding to the von Kármán plate theory. Six 8-DOF beam elements with displacement parameters u, du/dx, w, dw/dx at each node were used for a beam half. Load-deflection curves for 4 cycles are shown in Fig. 5. The loading criterion for the 1st cycle was to obtain a crack through the thickness in the middle of the beam, for the 2nd cycle to apply a load which is half of the load inducing first yield of the reinforcement, for the 3rd and 4th cycles to obtain the same permanent deflection at zero load as in the test.

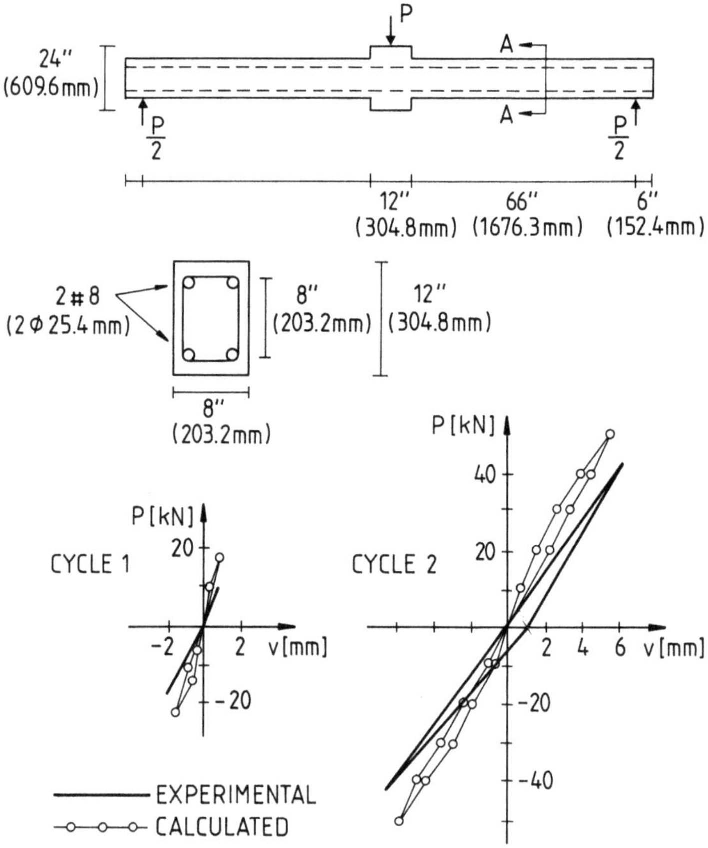

Fig. 5. Load-deflection curves of cyclically loaded beam tested by Burns & Siess [32].

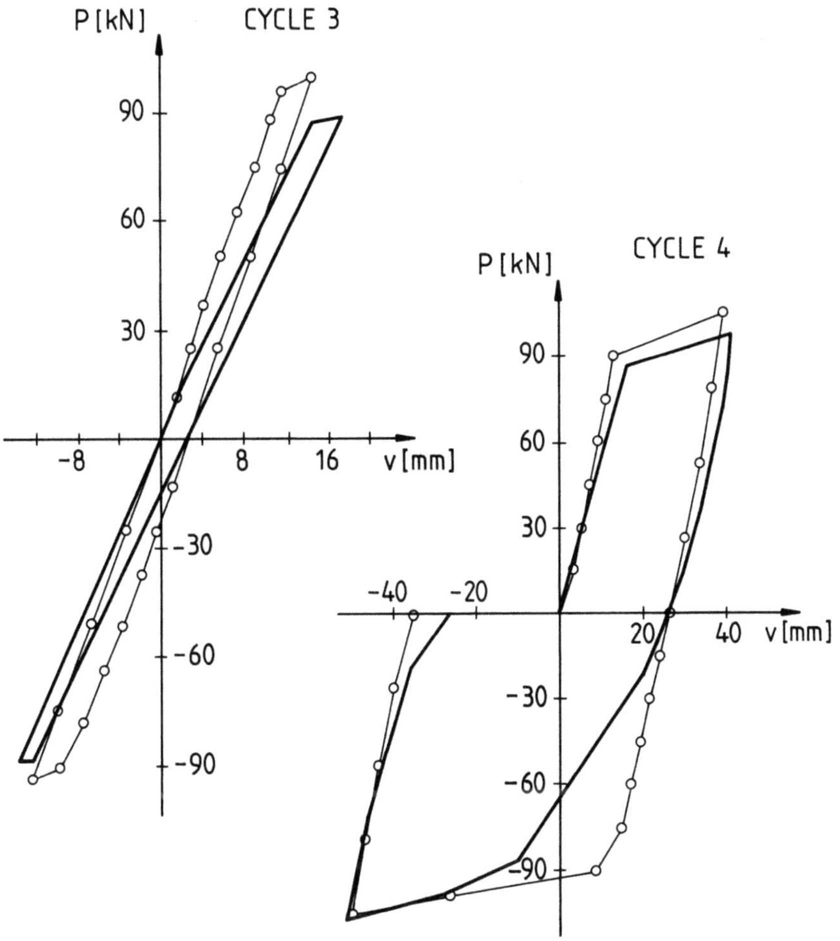

Fig. 5. Continued

Dynamically loaded beam

Nosseir [2] made tests on simply supported reinforced beams loaded by a freely falling mass. Based on some measurements, he estimated the contact force between the beam and mass by a bilinear expression with a peak value of 123.9 kN. Midpoint deflection of the beam versus time is presented in Fig. 6. The computed curve agrees qualitatively with the experimental one, but the computed deflection is somewhat smaller and the phase is lagging behind the experimental. Constant time step h = 0,6 ms was used.

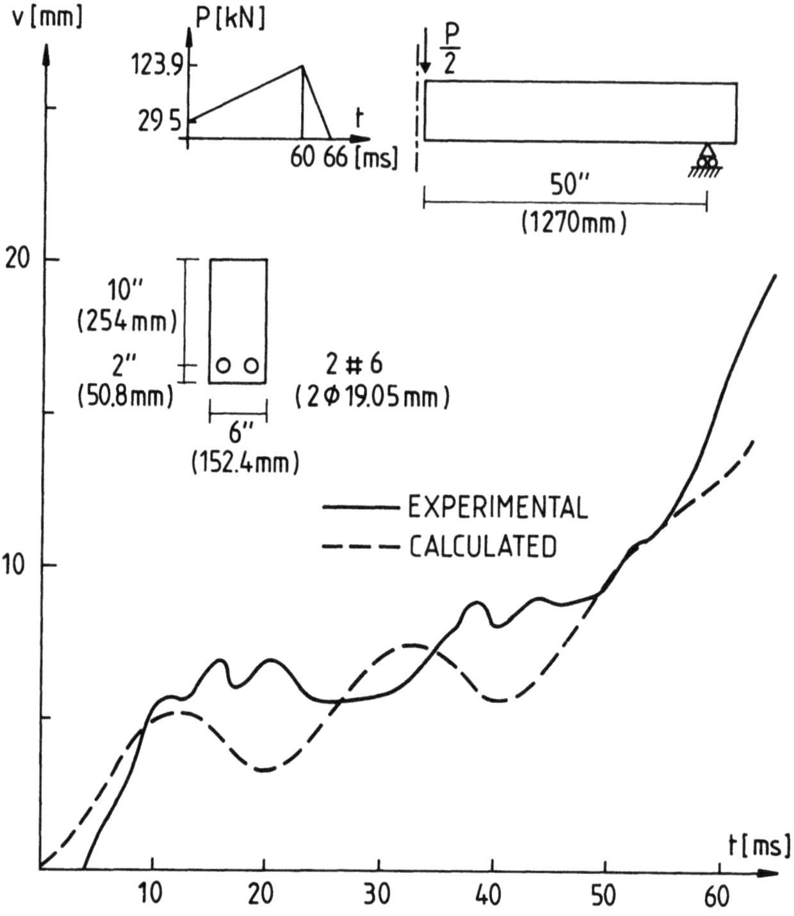

Fig. 6. Midpoint deflection vs. time of beam under dynamic load

Dynamically loaded slabs

A clamped rectangular slab subjected to a jet force at the center was analyzed by Stangenberg [3] using a difference method. The deflection histories of the midpoint are shown in Fig. 7. The finite element results based on plasticity and thin plate theory compare favourably with the finite difference solution, but a considerable deviation occurs for the results of the thick plate theory. The rate dependent model predicts smaller deflections than the elastic plastic model.

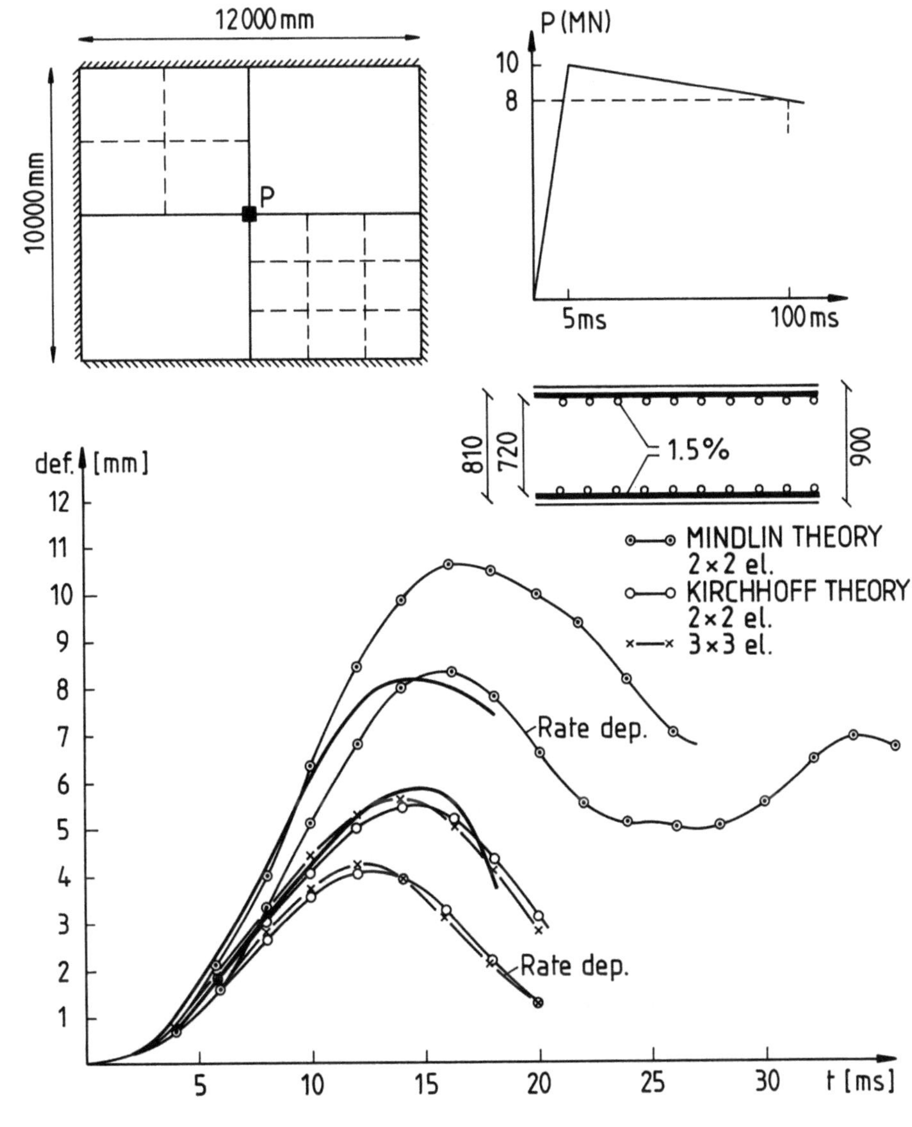

Fig. 7. Midpoint deflection vs. time of clamped slab under central jet force. Data f_c = 22.7 MPa, f_t = 2.27 MPa, E_c = 33 GPa, ν_c = 0.15, f_s = 210 GPa, $\rho \cong$ 2400 kg/m^3

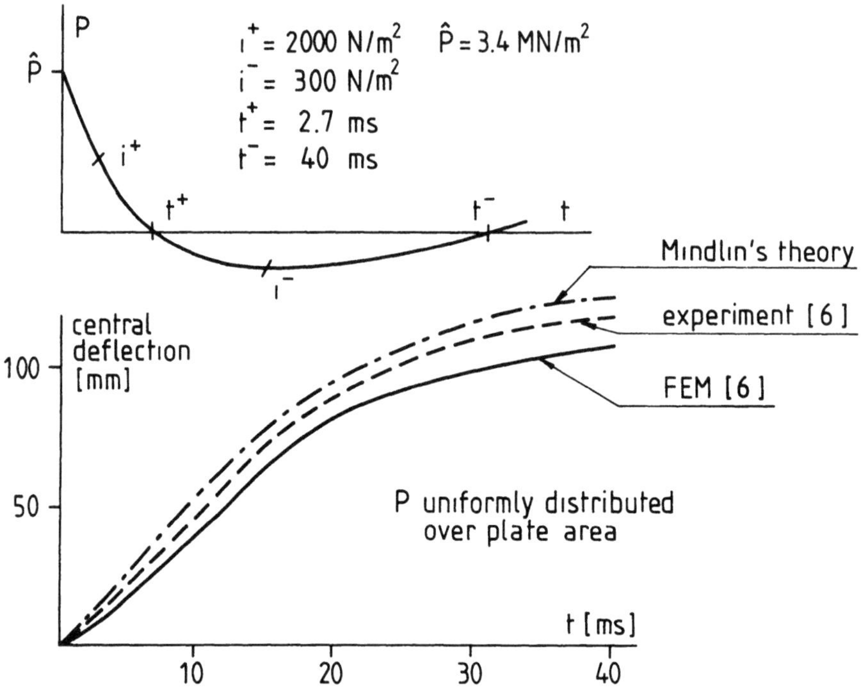

Fig. 8. Central deflection vs. time of one-way slab under uniform pressure. Data. reinforcement 0.17 % and 0.085 % f_c = 36.6 MPa, f_t = 4.90 MPa, E_c = 28.4 GPa, ν_o = 0.15, f_s = 450 MPa, E_s = 210 GPa, span 2280 mm, width 1230 mm depth 130 mm.

A rectangular slab, simply supported on two opposite edges and free on the remaining edges, was considered by Nilsson & Johansson [6] . The slab was subjected to a uniformly distributed pressure load varying with time. In addition to the test, the authors made finite element calculations which included cracking and plastic deformations. The central deflection-time history of this study was computed using four elements for a slab quadrant (Fig. 8). Varying time step h = 0.025...0.5 ms was used for the Kirchhoff theory with 2 by 2 mesh and h = 0.01...0.2 ms for the Kirchhoff theory with 3 by 3 mesh and for the Mindlin theory.

Discussion

The numerical results obtained for the problems of beams and slabs indicate that the present method is capable to predict satisfactorily the response of reinforced concrete structures subjected to transient loading. More numerical results and experimental data are still needed. Modelling of rate dependent behaviour should be developed. More realistic description of aggregate interlocking, dowel action and bond slip is also necessary.

References

1 Bergan, P.G.; Holand, I.: Nonlinear finite element analysis of concrete structures. Computer Methods in Applied Mechanics and Engineering 17/18 (1979) 443-467.

2 Nosseir, Mohamed S.E.B.: Static and dynamic analysis behaviour of concrete beams failing in shear. Ph. D. Thesis, University of Texas, 1966.

3 Stangenberg, F.: Nonlinear dynamic analysis of reinforced concrete structures. Nucl. Eng. and Design 29 (1974) 71-88.

4 Rebora, B.; Zimmermann, Th.; Wolf, J.P.: Dynamic rupture analysis of reinforced concrete shells. Nucl. Eng. and Design 37 (1976) 269-

5 Buyukozturk, O.; Connor, J.J.: Nonlinear dynamic response of reinforced concrete under impulsive loading: Research status and needs. Nucl. Eng. and Design 50 (1978) 83-92.

6 Nilsson, L.; Johansson, I.: Analys med finit elementmetod av luftstötvågsbelastade enkelspända betongplattor. FOA Rapport C20204-D4, Stockholm, November 1977.

7 Nilsson, L.: Impact loading of concrete structures. Publication 79:1, Department of Structural Mechanics, Chalmers University of Technology, 1979.

8 Sinisalo, H.S., Tuomala, M.T.E.; Mikkola, M.J.: Nonlinear finite element analysis of reinforced concrete slabs subjected to transient impulsive loading. Int. Conference on Concrete Slabs, 3-5 April 1979, Dundee, Scotland. Advances in Concrete Slab Technology, Proceedings published by Pergamon Press, 1980.

9 Sinisalo, H.S.; Tuomala, M.T.E.; Mikkola, M.J.: Nonlinear dynamic analysis of reinforced concrete slabs. Plasticity in Reinforced Concrete, Final Report, IABSE Colloquium, Copenhagen, 1979, pp. 239-246.

10 The effect of impact loading on buildings. State-of-the-art reports. Matériaux et Constructions (RILEM) 8 (1975) 44, 77-129.

11 Cedolin, L.; Crutzen, Y.R.J.; Dei Poli, S.: Triaxial stress-strain relationship for concrete. J. Eng. Mech. Div. ASCE 103 (1977) EM3, 423-439.

12 Coon, M.D.; Evans, R.J.: Incremental constitutive laws and their associated failure criteria with application to plain concrete. Int. J. Solids Struct. 8 (1972) 1169-1183.

13 Chen, W.F.: Constitutive equations for concrete. Plasticity in Reinforced Concrete, Introductory Report, IABSE Colloquium, Copenhagen, 1979, pp. 11-34.

14 Bažant, Z.P.; Bhat, P.D.: Endochronic theory of inelasticity and failure of concrete. J. Eng. Mech. Div. ASCE 102 (1976) EM4, 701-722.

15 Bažant, Z.P.; Kim, S.-S.: Plastic-fracturing theory for concrete. J. Eng. Mech. Div. ASCE 105 (1979) EM3, 407-428.

16 Kupfer, H.B.; Hilsdorf, H.K.; Rüsch, H.: Behaviour of concrete under biaxial stresses. J. of ACI 66 (1969) 8, 656-666.

17 Gerstle, K.H. et al.: Behaviour of concrete under multiaxial stress states. ASCE Annual Convention, Chigao, Illinois, October 1978.

18 Wastiels, J.: Failure criteria for concrete under multiaxial stress states. Plasticity in Reinforced Concrete, Final Report, IABSE Colloquium, Copenhagen 1979, pp. 3-10.

19 Read, H.E.; Maiden, C.J.: The dynamic behaviour of concrete. Systems, Science and Software, Report 3 SR-707, 1971.

20 Goldsmith, W.; Polivka, M.; Yang, T.: Dynamic behaviour of concrete. Experimental Mechanics 6 (1966) 2, 65-79.

21 Griner, G.R.; Sierakowski, R.L.; Ross, C.A.: Dynamic properties of concrete under impact loading. Shock and Vibration Bulletin 45 (1974-75) 131-142.

22 Takeda, J.; Tachikawa, H.; Fujimoto, K.: Mechanical behaviour of concrete under higher rate loading than in static test. Mechanical Behaviour of Materials, Symposium Procedure, Kyoto, Japan, 2 (1974) 479-486.

23 Wang, P.T.; Shah, S.P.; Naaman, A.E.: Stress strain curves of normal and lightweight concrete in compression. ACI Journal 75 (1978) November, 603-611.

24 Palmer, A.C; Maier, G.; Drucker, D.C.: Normality relations and convexity of yield surfaces for unstable materials or structural elements. J. Appl. Mech., Trans. ASME 34 (1967) June, 464-470.

25 Nguyen, Q.S.; Bui, H.D.: Sur les matériaux elastoplastiques à écrouissage positif ou negatif. J. de Mécanique 13 (1974) 2, 321-342.

26 Willam, K.J.; Warnke, E.P.: Constitutive model for the triaxial behaviour of concrete. IABSE Seminar on "Concrete structures subjected to triaxial stresses" May 17-19, 1974, ISMES, Bergamo, Italy.

27 Lin, C.S.; Scordelis, A.C.: Nonlinear analysis of RC shells of general form. J. Struct. Div. ASCE 101 (1975) ST3, 523-538.

28 Zienkiewicz, O.C.: The finite element method (3rd ed.). McGraw-Hill, London 1977.

29 Mikkola, M.J.; Tuomala, M.T.E.; Sinisalo, H.S.: Comparison of numerical integration methods in the analysis of impulsively loaded elasto-plastic and viscoplastic structures (Submitted for publication).

30 Belytschko, T.; Schoeberle, D.F.: On the unconditional stability of an implicit algorithm for nonlinear structural dynamics. J. Appl. Mech. ASME 42 (1975) 865-869.

31 Jofriet, J.C.; McNeice, G.M.: Finite element analysis of reinforced concrete slabs. J. Struct. Div. ASCE 97 (1971) ST3, 785-806.

32 Burns, N.H.; Siess, C.P.: Repeated and reversed loading in reinforced concrete. J. Struct. Div. ASCE 92 (1966) ST5

An Analysis of the Stability and Convergence Properties of a Crank-Nicholson Algorithm for Nonlinear Elasto-Dynamics Problems

L. C. WELLFORD, JR.,
University of Southern California,
Los Angeles, USA

S. M. HAMDAN
Drexel University,
Philadelphia, USA

Summary

The stability of nonlinear step by step computing schemes for solid mechanics problems can be insured by either requiring that the energy of the approximation be conserved or by insisting that the energy be bounded for all time. While techniques exist for analyzing conserving procedures, few analysis methods exist which are capable of establishing the boundedness of the energy of nonconserving computing schemes. In this paper an analysis procedure for demonstrating the stability and convergence of nonconserving step by step solution algorithms is introduced. As a typical example this technique is used to analyze a Crank-Nicholson-Galerkin algorithm for nonlinear elastodynamics problems.

Introduction

Energy analysis procedures have been used by several investigators [1-4] to study the stability of step by step solution algorithms for nonlinear structural dynamics problems. This previous work has been principally directed to developing algorithms which, either implicitly or by constraint, conserve some scalar or energy like quantity from time step to time step. Various unconditionally stable algorithms have been constructed using these procedures. In this paper a method of analysis for nonconserving algorithms is developed. This analysis is based on the premise that, in the approximation, energy should remain bounded for all time. This method of analysis should be useful in analyzing various nonconserving implicit algorithms, load extrapolation algorithms, and explicit algorithms.

The Physical Problem

In the Lagrangian formulation of nonlinear elasticity the motion of a solid body is defined by the following equation:

$$\rho \ddot{u}_i - [t^{kj}(\delta_{ij} + u_{i,j})]_{,k} = f_i \tag{1}$$

where ρ is the density, u_i is the displacement vector, t^{kj} is the stress tensor, and f_i is the body force vector. The techniques to be introduced here can be used in analyzing discrete algorithms for (1). However, in order to concisely present the method, it is useful to consider an alternate problem. This problem is a one dimensional analog of (1). In the one dimensional case the motion of a flat slab of material, I, is considered. The function u represents the displacement component, perpendicular to the slab, of planes in the slab. The displacement component u is governed by the following equation of motion:

$$\rho \ddot{u} - D[\bar{t}(1 + Du)] = f \tag{2}$$

where $D \equiv \partial/\partial X$, $\bar{t}$ is the stress, and f is the body force.

The analysis procedure to be introduced here is designed for the case in which the stress $\bar{t}$ is known in terms of powers of the strain. Thus it would be particularly useful in considering hyperelastic materials and the case in which the strain energy is known in terms of a power series in the strains. However, in order to analyze a problem for which existence and regularity properties are available, a particular case is considered. This is the geometrically nonlinear problem. In this case $\bar{t} = E\varepsilon_x$ and $\varepsilon_x = Du + 1/2(Du)^2$ where E is the modulus of elasticity and ε_x is the strain. In this problem it can be shown, using energy arguments that, in the case when homogeneous boundary conditions are applied, if $f \in L_2(0,\infty;L_2(I))$, a solution exists such that

$$u \in L_\infty(0,\infty;\overset{o}{W}{}^1_2(I)) \qquad \dot{u} \in L_\infty(0,\infty;L_2(I)) .$$

This is the regularity that would be expected in a shock wave propagation problem. In the transient vibration problems considered here the solution is much more regular. Thus it is assumed that for some integer $m \geq 2$

$$u \in L_\infty(0,\infty;\overset{o}{W}{}_2^m(I)) \qquad \dot{u} \in L_\infty(0,\infty;\overset{o}{W}{}_2^{m-1}(I)) \ .$$

Similar regularity assumptions will be introduced for higher time derivatives when needed subsequently.

Regularization Procedures

Stability and convergence results, of the type discussed in this paper, are extremely difficult to derive using the second order equation (2). Thus (2) is represented as a first order system and certain regularization (artificial viscosity) terms are added to obtain the following form:

$$\rho\dot{v} - D(\bar{t}(1 + Du)) + \varepsilon v = f$$

$$\dot{u} - v + \alpha u = 0 \qquad\qquad (\varepsilon,\alpha > 0) \qquad\qquad (3)$$

The analysis is limited by considering the case in which homogeneous boundary conditions on u are applied. In addition, the nonlinearity in the equation is simplified by neglecting the third degree term in the displacement gradients. This restriction could be eliminated but at the expense of an increase in the complexity of the arguments. Then the equation $(3)_1$ is satisfied in its weak form and $(3)_2$ is satisfied in its strong form to obtain

$$(\rho\dot{v},w) + (E(Du + \tfrac{3}{2}\,Du^2),Dw) + \varepsilon(v,w) = (f,w)$$

$$\dot{u} - v + \alpha u = 0 \qquad\qquad (4)$$

for all w in the original solution space of velocities.

The Semi-Discrete Solution

In order to form an approximate model which is discretized in the spatial variable, a set of global finite element basis functions is introduced. These basis functions define a finite dimensional subspace $S_h(I) \times S_h(I)$ of the original solution space. The elements of $S_h(I)$ are assumed to satisfy an inverse hypothesis [5] of the following form (for positive constant C^*):

$$||DU||_{L_2(I)} \leq \frac{C^*}{h}||U||_{L_2(I)} \qquad (U \in S_h(I)) \ . \qquad (5)$$

It is assumed that an imbedding relationship [6] of the following form is satisfied by the elements of $S_h(I)$:

$$||DU||_{L_{p+1}(I)} \leq ch^{-\frac{1}{p(p+1)}} ||DU||_{L_p(I)} \ . \qquad (6)$$

In addition, if k denotes the degree of the piecewise polynomials in $S_h(I)$, it is assumed that there exists a constant K independent of h such that

$$\underset{U \in S_h(I)}{\text{INF}} ||u-U||_{\overset{o}{W}{}_2^m(I)} \leq Kh^{k+1-m}||u||_{\overset{o}{W}{}_2^{k+1}(I)} \qquad (7)$$

Let $(U,V) \in S_h(I) \times S_h(I)$ be the approximation for the displacement u and the velocity v. Then the approximate version of (4) is defined as follows:

$$(\rho\dot{V},W) + (E(DU + \tfrac{3}{2} DU^2),DW) + \varepsilon(V,W) = (f,W)$$

$$\dot{U} - V + \alpha U = 0 \qquad (8)$$

The equation $(8)_1$ is satisfied for all $W \in S_h(I)$ while $(8)_2$ is satisfied pointwise at the nodes.

Stability and Convergence of the Semidiscrete Approximation

Let $\bar{U}$ and $\bar{V}$ denoted weighted projections of u and v into the subspace $S_h(I)$ defined by

$$(E[1+3Du]D(u-\bar{U}),DW) = 0 \qquad \forall W \in S_h(I)$$

$$(E\nu D(v-\bar{V}),DW) = 0 \qquad (9)$$

where $\nu = 1 + 3/2(Du + D\bar{U})$. It is assumed that the projection is optimal in the sense of (7). Let the approximation errors be defined by $e_u = u - U$ and $e_v = v - V$. These errors can be decomposed as follows:

$$e_u = E_u + \mathcal{E}_u \qquad E_u = u - \bar{U} \qquad \mathcal{E}_u = \bar{U} - U$$

$$e_v = E_v + \mathcal{E}_v \qquad E_v = v - \bar{V} \qquad \mathcal{E}_v = \bar{V} - V \qquad (10)$$

The energy error can be defined in certain weighted norms

$$U(W) = \tfrac{1}{2}(E\nu DW,DW) \text{ - Strain Energy Error}$$

$$\Omega(W) = \tfrac{1}{2}\rho||W||_{L_2(I)}^2 \text{ - Kinetic Energy Error .}$$

It is necessary, in the nonlinear problem, to scale these

energy errors by defining auxiliary energy error measures

$$U^*(W) = U(W)e^{-\chi t} \qquad \Omega^*(W) = \Omega(W)e^{-\chi t} \tag{11}$$

where

$$\chi = \sup_{\substack{t \in [0,\infty] \\ X \in I}} \left[\frac{3}{2\nu}(D\dot{u}(t) + D\dot{\bar{U}}(t))\right] .$$

In addition total energy errors can be defined by $F(w,r) = \Omega(w) + U(r)$ and $F^*(w,r) = \Omega^*(w) + U^*(r)$. The errors $\mathscr{E}_u$ and $\mathscr{E}_v$ can be established through the following Theorem:

Thm. 1. If the exact solutions u and v are sufficiently regular and the mesh discretization parameter h is sufficiently restricted, there exist positive constants g_1, g_2, and g_3, independent of h and t, such that

$$\frac{d}{dt} F^*(\mathscr{E}_v,\mathscr{E}_u) \leq g_1 e^{-\chi t} h^{2k-1} - g_2 F^*(\mathscr{E}_v,\mathscr{E}_u) + g_3 e^{\chi t} h^{-3} F^{*2}(\mathscr{E}_v,\mathscr{E}_u) \tag{12}$$

Proof: Setting w = W in $(4)_1$, subtracting $(8)_1$ from $(4)_1$, making use of (10), and setting W = $\mathscr{E}_v$, the following equation is obtained:

$$\begin{aligned} \frac{d}{dt}\Omega(\mathscr{E}_v) + (E\nu D\mathscr{E}_u, D\mathscr{E}_v) + \frac{2\varepsilon}{\rho}\Omega(\mathscr{E}_v) &= (\tfrac{3}{2} ED\mathscr{E}_u^{\,2}, D\mathscr{E}_v) - (\rho\dot{E}_v, \mathscr{E}_v) \\ &- (E(\nu DE_u - \tfrac{3}{2} D\mathscr{E}_u DE_u, D\mathscr{E}_v) \\ &- \varepsilon(E_v, \mathscr{E}_v) \end{aligned} \tag{13}$$

Similarly differentiating $(4)_2$ and $(8)_2$ with respect to the spatial coordinate, subtracting the two equations, multiplying by $E\nu D\mathscr{E}_u$, and integrating over the spatial domain, the following equation is obtained:

$$\begin{aligned} \frac{d}{dt} U(\mathscr{E}_u) - (E\nu D\mathscr{E}_v, D\mathscr{E}_u) + 2\alpha U(\mathscr{E}_u) &= (\tfrac{3E}{4}[D\dot{u} + D\dot{\bar{U}}]D\mathscr{E}_u, D\mathscr{E}_u) \\ &- (E\nu[D\dot{E}_u - DE_v + \alpha DE_u], D\mathscr{E}_u) \end{aligned} \tag{14}$$

Adding (13) and (14), it can be seen that

$$\frac{d}{dt} F(\mathscr{E}_v,\mathscr{E}_u) + \frac{2\varepsilon}{\rho}\Omega(\mathscr{E}_v) + 2\alpha U(\mathscr{E}_u)$$

$$= (\frac{3E}{4} (D\dot{u} + D\dot{\bar{U}}) D\mathcal{E}_u, D\mathcal{E}_u)$$

$$+ (\frac{3}{2} ED\mathcal{E}_u^{\ 2}, D\mathcal{E}_v) + \phi \qquad (15)$$

where ϕ is a truncation error term

$$\phi = -(\rho\dot{E}_v + \varepsilon E_v, \mathcal{E}_v) - (E[\nu DE_u - \frac{3}{2} D\mathcal{E}_u DE_u], D\mathcal{E}_v)$$

$$- (E\nu[D\dot{E}_u - DE_v + \alpha DE_u], D\mathcal{E}_u)$$

Using (9) certain troublesome terms can be omitted from this expression to obtain

$$\phi = -(\rho\dot{E}_v + \varepsilon E_v, \mathcal{E}_v) + (\frac{3}{2} E[DE_u^{\ 2} + D\mathcal{E}_u DE_u], D\mathcal{E}_v)$$

$$- (E\nu[D\dot{E}_u + \alpha DE_u], D\mathcal{E}_u)$$

The terms on the right in (15) can be estimated using the Hölder inequality, (5), and the elementary inequality $ab \leq \frac{1}{2\gamma} a^2 + \frac{\gamma}{2} b^2 (\gamma > 0)$. In addition to estimate the nonlinear term the following relationship, obtained from (6) is required:

$$||D\mathcal{E}_u||^4_{L_4(I)} \leq \frac{C_1}{h} ||D\mathcal{E}_u||^4_{L_2(I)}$$

Typically, the first two terms on the right of (15) take the following form:

$$(\frac{3E}{4}(D\dot{u} + D\dot{\bar{U}}) D\mathcal{E}_u, D\mathcal{E}_u) \leq \chi U(\mathcal{E}_u)$$

$$\begin{aligned}(\frac{3}{2} ED\mathcal{E}_u^{\ 2}, D\mathcal{E}_v) &\leq \frac{3}{2} E||D\mathcal{E}_u||^2_{L_4(I)} ||D\mathcal{E}_v||_{L_2(I)} \\ &\leq \frac{3E}{4h^2\gamma_1} ||D\mathcal{E}_u||^4_{L_4(I)} + \frac{3Eh^2\gamma_1}{4} ||D\mathcal{E}_v||^2_{L_2(I)} \\ &\leq \frac{3EC_1}{4h^3\gamma_1} ||D\mathcal{E}_u||^4_{L_2(I)} + \frac{3EC^{*2}\gamma_1}{4} ||\mathcal{E}_v||^2_{L_2(I)} \\ &\leq \frac{3C_1}{\nu^{*2}Eh^3\gamma_1} U(\mathcal{E}_u)^2 + \frac{3EC^{*2}\gamma_1}{2\rho} \Omega(\mathcal{E}_v)\end{aligned}$$

where $\nu^* = 1 - \frac{3}{2}||Du + D\bar{U}||_{L_\infty(I)}$. Using similar methods to bound the terms in ϕ, it can be shown that, for certain parameters $\gamma_1, \ldots, \gamma_6$ (with dependence on h indicated), the right hand side of (15) denoted R is

$$R \leq \chi U(\mathcal{E}_u) + d_1 U(\mathcal{E}_u)^2 + d_2\Omega(\mathcal{E}_v) + d_3 U(\mathcal{E}_u) + d_4 \qquad (16)$$

where

$$d_1 = \frac{3C_1}{\nu^{*2} E h^3 \gamma_1}$$

$$d_2 = \frac{3EC^{*2}\gamma_1}{2\rho} + \gamma_2 + \frac{3E\gamma_3(\frac{1}{h})\,||DE_u||_{L_\infty(I)}\,C^{*2}}{2\rho} + \frac{3E\,||DE_u||_{L_\infty(I)}\,C^{*2}\gamma_4(\frac{1}{h})}{2\rho h} + \frac{\varepsilon\gamma_5}{\rho}$$

$$d_3 = \frac{3\,||DE_u||_{L_\infty(I)}}{2\nu^*\gamma_4(\frac{1}{h})\,h} + \frac{\gamma_6(h)\,\overline{\nu}(1+\alpha)}{\nu^*}$$

$$d_4 = \frac{\rho}{2\gamma_2}\,||\dot{E}_v||^2_{L_2(I)} + \frac{3E\,||DE_u||_{L_\infty(I)}\,||DE_u||^2_{L_2(I)}}{4\gamma_3(\frac{1}{h})\,h^2} + \frac{\varepsilon}{2\gamma_5}\,||E_v||^2_{L_2(I)} + \frac{1}{2}\,\frac{E\overline{\nu}}{\gamma_6(h)}\,(||D\dot{E}_u||^2_{L_2(I)} + \alpha\,||DE_u||^2_{L_2(I)})$$

where $\overline{\nu} = 1 + \frac{3}{2}||Du + D\overline{U}||_{L_\infty(I)}$.

The constants $\gamma_1,\ldots,\gamma_6$ are adjustable. They can be chosen such that $d_1 = d_5/h^3$, $d_2 = \chi$, and $d_3 = d_6 h$ where d_5 and d_6 are independent of h. Then by making minimal restrictions on the regularity of the solutions, it is possible to show that, because of (7), $d_4 \leq d_7 h^{2k-1}$. Then from (15) and (16)

$$\frac{d}{dt} F(\mathscr{E}_v,\mathscr{E}_u) + \frac{2\varepsilon}{\rho}\,\Omega(\mathscr{E}_v) + (2\alpha - d_6 h)\,U(\mathscr{E}_u) \leq \chi F(\mathscr{E}_v,\mathscr{E}_u) + d_5 h^{-3} U(\mathscr{E}_u)^2 + d_7 h^{2k-1}$$

Introducing the expressions for the scaled energy error from (11), it can be seen that

$$\frac{d}{dt} F^*(\mathscr{E}_v,\mathscr{E}_u) + \frac{2\varepsilon}{\rho}\,\Omega^*(\mathscr{E}_v) + (2\alpha - d_6 h)\,U^*(\mathscr{E}_u) \leq d_5 e^{\chi t} h^{-3} U^{*2}(\mathscr{E}_u) + d_7 e^{-\chi t} h^{2k-1}\ .$$

It is clear from this equation that, if h is sufficiently restricted, the algorithm has positive damping and the result (12) is obtained. ■

If the initial values of velocity and displacement are defined by weighted energy projections of the type introduced in (9), the value of $F^*(\mathscr{E}_v,\mathscr{E}_u)$ at $t=0$ is zero. This is the case which will be considered in this analysis. The case, when $F^*(\mathscr{E}_v,\mathscr{E}_u)$ at $t=0$ is not zero, could be handled by a straightforward modification of the techniques presented here. The inequality (12) has the following form:

$$\frac{d}{dt} F^* \le f_1(t) - f_2 F^* + f_3(t) F^{*2} \qquad (17)$$

Suppose, the calculations are to be carried out until time $t=T$. Then for any $t\in[0,T]$, the inequality (17) implies that

$$\frac{d}{dt} F^* \le f_1(0) - f_2 F^* + f_3(T) F^{*2}$$

When equality occurs in this equation, it can be transformed to a Bernoulli equation. The resulting equation can be solved, and the solution can be used to establish the following bound:

$$F^*(t) \le f_5 \left(\frac{1 - e^{-f_4 t}}{1 + \left[\frac{f_4 - f_2}{f_4 + f_2}\right] e^{-f_4 t}} \right)$$

where $f_4 = \sqrt{f_2^2 - 4f_1 f_3}$ and $f_5 = (f_2 - f_4)/2f_3$. The algorithm is stable if the numerical damping constant f_4 is real and the term $f_4 - f_2$ is small. The algorithm converges (for a fixed amount of damping) only if f_5 decreases with h. The algorithm converges to the undamped solution only if the numerical damping constant f_4 decreases with h. From (12) it can be seen that

$$f_4 = \sqrt{g_2^2 - 4 g_1 g_3 e^{\chi T} h^{2k-4}} \; .$$

Clearly if $k \ge 3$, there always exists a value for h such that f_4 is real and arbitrarily close to g_2. Thus under the condition $k \ge 3$, the algorithm can be stabilized by simply refining the mesh. In addition if $k \ge 3$, the artificial damping constant can be made proportional to h^μ where $\mu < 2$ while still retaining the stability property. Thus, without causing instability, the numerical damping constant can be made proportional to h, resulting in convergence to the undamped solution. The size of f_5 can be established by employing the binomial theorem to define f_4. Asymptotically as $h \to 0$, $f_5 = 0(h^{2k-1})$. Thus

$$F^*(\mathcal{E}_v,\mathcal{E}_u) \leq ch^{2k-1}$$

The total energy error of the approximation can be estimated using the triangle inequality

$$F^*(e_v,e_u) \leq c\left[F^*(E_v,E_u) + F^*(\mathcal{E}_v,\mathcal{E}_u)\right] \tag{18}$$

The first term on the right in (18) is a truncation error term which can be estimated from (7) and (9). The second term on the right hand side of (18) is dominant, and the following result is obtained:

Thm. 2. If the exact solution to (4) is sufficiently regular, if the subspace $S_h(I)$ is characterized by $k \geq 3$, and if h is sufficiently restricted, the approximation (8) is stable and convergent, and

$$F(e_v,e_u) \leq ch^{2k-1}$$

■

The following comments about the results of this analysis seem appropriate:

1. The energy growth equation (12) contains, on the right hand side, two competing forces. The interpolation error term, characterized by g_1, acts to cause convergence. The nonlinear term characterized by g_3 acts to cause divergence. The damping term, characterized by g_2, is, in a way, an innocent bystander. Only when the truncation error term is made more powerful (by increasing k) can the truncation error overcome the nonlinear effects to produce stability.

2. The restriction that higher degree polynomials ($k \geq 3$) be used in the finite element model can be relaxed if we give up some of the results of the analysis. For lower degree interpolation procedures short time stability, with damping constants depending inversely on h, can be obtained.

3. From Thm. 2 standard W_2^1 error estimates can be obtained by taking the square root of both sides. The result shows that there is a loss of accuracy of ½ power in h, as compared to the error of the linear problem.

The Crank-Nicholson-Galerkin Procedure

The semi-discrete algorithm, discussed in the previous section, satisfactorily incorporates a spatial discretization which leads to stability. The objective of this section is to define a temporal discretization procedure, motivated by the semi-discrete model, which has a stable temporal discretization and inherits the stable spatial discretization of the semi-discrete model. Let, typically, $u^n = u(n\Delta t)$ where n is the time step number in a step by step solution algorithm with time step length Δt. Let $u^{n+\frac{1}{2}} = (u^{n+1} + u^n)/2$. Finally, let a Crank-Nicholson-Galerkin version of (8) be defined as follows:

$$\left(\rho \frac{V^{n+1} - V^n}{\Delta t}, W\right) + (EDU^{n+\frac{1}{2}}, DW) + \frac{3}{2}(E(DU^{n+\frac{1}{2}})^2, DW)$$
$$+ \varepsilon(V^{n+\frac{1}{2}}, W) = (f^{n+\frac{1}{2}}, W)$$
$$\frac{U^{n+1} - U^n}{\Delta t} - V^{n+\frac{1}{2}} + \alpha U^{n+\frac{1}{2}} = 0 \tag{19}$$

for $\forall W \in S_h(I)$. The equation $(19)_2$ is satisfied at each node and everywhere in I.

Weighted projections $\bar{U}^n$ and $\bar{V}^n$ of the exact solution u^n and v^n are defined through the semi-discrete methods (defined in (9)) with all time functions being evaluated at time step n. Total errors and decomposed errors are defined as in (10). Typically, $e_u^n = u^n - U^n$. The energy error at discrete time points is defined in the following weighted measures:

$$\mathcal{U}^n(W) = \frac{1}{2}(E\nu^n DW^n, DW^n)$$
$$\Omega^n(W) = \frac{1}{2}\rho||W^n||^2_{L_2(I)}$$

Again, it is necessary to scale the energy error measures by defining temporally discrete auxiliary energy error measures

$$\mathcal{U}^{*n}(W) = \mathcal{U}^n(W)\lambda^n \qquad \Omega^{*n}(W) = \Omega^n(W)\lambda^n$$

where

$$\lambda^n = \left[\frac{1 - \frac{\bar{\chi}\Delta t}{2}}{1 + \frac{\bar{\chi}\Delta t}{2}}\right]^n \tag{20}$$

and

$$\bar{\chi} = \sup_{\substack{n=0,\ldots,\infty \\ X\in I}} \left[\frac{3}{2\nu^{n+1}} \left(\frac{D\bar{U}^{n+1} - D\bar{U}^{n+\frac{1}{2}}}{\frac{\Delta t}{2}} + \frac{Du^{n+1} - Du^{n+\frac{1}{2}}}{\frac{\Delta t}{2}} \right) + \frac{3}{2\nu^{n}} \left(\frac{D\bar{U}^{n+\frac{1}{2}} - D\bar{U}^{n}}{\frac{\Delta t}{2}} + \frac{Du^{n+\frac{1}{2}} - Du^{n}}{\frac{\Delta t}{2}} \right) \right]$$

As before fully discrete total energy errors can be defined. For example, $F^{*n}(w,r) = \Omega^{*n}(w) + U^{*n}(r)$. The fully discretized errors $\mathscr{E}_v^n$ and $\mathscr{E}_u^n$ can be estimated through the following theorem:

Thm. 3. If the exact solutions u and v are sufficiently regular and if the discretization parameters h and Δt are sufficiently restricted, there exist positive constant $\bar{g}_1$, $\bar{g}_2$, and $\bar{g}_3$ independent of h and Δt, such that

$$\frac{1}{\Delta t}\{F^{*n+1}(\mathscr{E}_v,\mathscr{E}_u) - F^{*n}(\mathscr{E}_v,\mathscr{E}_u)\} \leq \bar{g}_1 e^{-\bar{\chi}n\Delta t}(h^{2k-1} + \Delta t^4) - \bar{g}_2 F^{*n+\frac{1}{2}}(\mathscr{E}_v,\mathscr{E}_u) + \bar{g}_3 e^{\bar{\chi}(n+1)\Delta t} h^{-3}(F^{*n+\frac{1}{2}}(\mathscr{E}_v,\mathscr{E}_u))^2 \quad \blacksquare \quad (21)$$

Theorem 3 can be proved by a straightforward adaptation of the methods presented in Theorem 1. In order to establish the stability and convergence properties of the algorithm (19), the fully discretized energy growth equation (21) must be utilized. The size of the energy error can be estimated through the following Lemma:

Lemma 1. Let $\omega(t)$ be a non-negative function. Let $\omega(t)$ be defined on the discrete set $P = \{0,\Delta t,2\Delta t,\ldots,N\Delta t\}$ by $\omega^n = \omega(n\Delta t)$. Suppose that there exist positive constants a_1, a_2, and a_3 such that

$$\omega^{n+1} - \omega^n \leq \Delta t a_1 - \Delta t a_2\left(\frac{\omega^{n+1}+\omega^n}{2}\right) + \Delta t a_3\left(\frac{\omega^{n+1}+\omega^n}{2}\right)^2 \quad (22)$$

for $n = 0,1,2,\ldots,\infty$. In addition, suppose that $\omega^{\circ} = 0$ and that $a_4 = \sqrt{a_2^2 - 4a_1a_3} > 0$. Then there exists a constant c such that if

$$\Delta t \leq c$$

the discrete function ω^{n+1} satisfies

$$\omega^{n+1} \leq a_5(1-e^{-2(a_4+a_3a_5)(n+1)\Delta t}) \qquad (23)$$

where

$$a_5 = (a_2 - a_4)/2a_3 \ .$$

Proof: Letting $\omega^{n+1} = \omega^{*n+1} + a_5$ and $\omega^n = \omega^{*n} + a_5$ in (22), a simplified formula can be obtained to define a bound for the solution

$$\omega^{*n+1} - \omega^{*n} \leq -\Delta t a_4 (\frac{\omega^{*n+1} + \omega^{*n}}{2}) + \Delta t a_3 (\frac{\omega^{*n+1} + \omega^{*n}}{2})^2$$

Equivalently, a bound for the solution can be defined by $\omega^{*n+1} \leq \tilde{\omega}^{*n+1}$ where $\tilde{\omega}^{*n+1}$ is the smallest root of the following equation:

$$\tilde{\omega}^{*n+1} - \omega^{*n} = -\Delta t a_4 (\frac{\tilde{\omega}^{*n+1} + \omega^{*n}}{2}) + \Delta t a_3 (\frac{\tilde{\omega}^{*n+1} + \omega^{*n}}{2})^2 \ .$$

It can be easily shown that

$$\tilde{\omega}^{*n+1} = \frac{-B(1 - \sqrt{1 - \frac{4AC}{B^2}})}{2A} \qquad (24)$$

where

$$A = -\frac{\Delta t a_3}{4}$$

$$B = 1 + \frac{\Delta t}{2}(a_4 - a_3\omega^{*n})$$

$$C = -(1 - \frac{\Delta t}{2}(a_4 - \frac{a_3\omega^{*n}}{2}))\omega^{*n}.$$

By reducing Δt, it is possible to make the term $4AC/B^2$ in (24) small. Thus the square root term in (24) can be evaluated using a binomial series. The following expression is obtained:

$$\tilde{\omega}^{*n+1} = -\frac{C}{B} - \frac{AC^2}{B^3}(1 + \frac{2AC}{B^2} + \alpha). \qquad (25)$$

The term α represents the summation of the higher order terms *in the binomial series.* These terms are proportional to powers of Δt (in particular powers > 1). Thus as $t \to 0$, $\alpha \to 0$. Simplifying (25) using (24), it can be seen that

$$\omega^{*n+1} \leq \frac{(1-\frac{\Delta t}{2}(a_4 - a_3\omega^{*n}))\omega^{*n}}{1+\frac{\Delta t}{2}(a_4 - a_3\omega^{*n})} + \beta$$

where β is a term which approaches zero as $\Delta t \to 0$. The corresponding expression in terms of ω^{n+1} and ω^n is defined as follows:

$$\omega^{n+1} \leq a_5(1-\mu_n) + \omega^n\mu_n + \beta \tag{26}$$

where

$$\mu_n = \frac{1-\frac{\Delta t}{2}(a_4 + a_3a_5 - a_3\omega^n)}{1+\frac{\Delta t}{2}(a_4 + a_3a_5 - a_3\omega^n)} .$$

Let us investigate the case when $\omega^n \leq a_5$ (or correspondingly $-a_5 \leq \omega^{*n} \leq 0$). This condition is satisfied when $n = 0$ because of the hypothesis $\omega^o = 0$. Subsequently in this proof it will be shown that, in fact, this condition is satisfied for all n. Because of this limitation on the size of ω^n, the first two terms on the right in (26) involve exponentially decaying terms. These terms behave nicely and can not cause instability. The third term on the right is an error term caused by the nonlinearity in the equation. At any particular time step, by shrinking Δt to zero, this term can be eliminated. However, small errors, in particular time steps, can build up over many time steps to produce instability. To insure stability it is, thus, necessary to show that the error term β, defined as follows, is negative:

$$\beta = \frac{\Delta t a_3(\omega^{*n})^2}{4\tau}(\psi^2\gamma - 1)$$

where

$$\psi = 1 - \frac{\Delta t}{2}(a_4 - \frac{a_3}{2}\omega^{*n})$$

$$\tau = 1 + \frac{\Delta t}{2}(a_4 - a_3\omega^{*n})$$

$$\gamma = \frac{1}{\tau 2}\left\{1 + \frac{\Delta t a_3\psi\omega^{*n}}{2\tau^2} + \alpha\right\}$$

Because of the condition that $\omega^{*n} \leq 0$, the fact that α is proportional to powers of $\Delta t \geq 2$, and the fact that $\alpha = 0$ when

$\omega^{*n} = 0$, there always exists a time step Δt which is small enough so that $\gamma < 1$. Thus if Δt is sufficiently restricted in size, $\beta < 0$ and from (26)

$$\omega^{n+1} \leq a_5(1 - \mu_n) + \omega^n \mu_n \tag{27}$$

By employing the inequality (27) and the condition that $\omega^o = 0$, it can be shown that as long as Δt is sufficiently restricted and $\omega^n \leq a_5$ for $n = 0,\ldots,N$

$$\omega^{N+1} \leq a_5(1 - \prod_{i=0}^{N} \mu_i) \tag{28}$$

for $N = 0,1,\ldots,\infty$. The size of the term on the right hand side of (28) can be fairly easily estimated. In fact

$$1 - \prod_{i=0}^{N} \mu_i \leq 1 - \mu^{N+1} \tag{29}$$

where

$$\mu = \frac{1 - \frac{\Delta t}{2}(a_4 + a_3 a_5)}{1 + \frac{\Delta t}{2}(a_4 + a_3 a_5)} \tag{30}$$

In addition, for $x \geq 0$ small enough

$$e^{-2x} \leq \frac{1 - \frac{x}{2}}{1 + \frac{x}{2}} \leq e^{-x} \tag{31}$$

From (30) and (31), it can be seen that for Δt small enough

$$e^{-2(a_4+a_3a_5)(N+1)\Delta t} \leq \mu^{N+1} \leq e^{-(a_4+a_3a_5)(N+1)\Delta t} \tag{32}$$

Finally from (28), (29), and (32), it can be seen that if Δt is sufficiently restricted and if $\omega^n \leq a_5$ for $n = 0,\ldots,N$

$$\omega^{N+1} \leq a_5(1 - e^{-2(a_4+a_3a_5)(N+1)\Delta t}) \tag{33}$$

The result (23) can now be obtained by an induction proof. Let $N = 0$ and note that by hypothesis $\omega^o = 0$. Thus, from (33)

$$\omega^1 \leq a_5(1 - e^{-2(a_4+a_3a_5)\Delta t}) < a_5$$

Similarly letting $N = 1$ and noting that $\omega^o < a_5$ and $\omega^1 < a_5$, equation (33) gives

$$\omega^2 \leq a_5(1 - e^{-4(a_4+a_3a_5)\Delta t}) < a_5$$

and so on. ■

From the results (21) and (23) the energy of $\mathscr{E}_v$ and $\mathscr{E}_u$ can be estimated. Initially it is convenient to assume that, as in the semi-discrete case, $F^{*n}(\mathscr{E}_v,\mathscr{E}_u)=0$ when n=0. The energy error growth condition (21) contains certain time dependent functions. To make the equations agree, certain restrictions must be employed. Suppose that the calculations are to be carried out until $t=(N+1)\Delta t$. Then for any time step $n=0,\cdots,N$, equation (21) gives

$$\begin{aligned} F^{*n+1}(\mathscr{E}_v,\mathscr{E}_u) - F^{*n}(\mathscr{E}_v,\mathscr{E}_u) \leq\ & \Delta t\bar{g}_1(h^{2k-1}+\Delta t^4) \\ & - \Delta t\bar{g}_2 F^{*n+\frac{1}{2}}(\mathscr{E}_v,\mathscr{E}_u) \\ & + \Delta t\bar{g}_3 e^{\bar{\chi}(N+1)\Delta t} h^{-3}(F^{*n+\frac{1}{2}}(\mathscr{E}_v,\mathscr{E}_u))^2 \end{aligned} \qquad (34)$$

Comparing (34) and (22), it is clear that as long as $k\geq 3$, $\Delta t=ch^{\ell}$ $(\ell\geq 2)$, and h is sufficiently restricted, there exists a positive numerical damping constant

$$q_4 = \sqrt{\bar{g}_2^{\,2} - 4\bar{g}_1\bar{g}_3 e^{\bar{\chi}(N+1)\Delta t} h^{-3}(h^{2k-1}+\Delta t^4)} \qquad (35)$$

such that for all time steps $n\leq N$

$$F^{*n+1}(\mathscr{E}_v,\mathscr{E}_u) \leq q_5(1 - e^{-2(q_4+q_3q_5)(n+1)\Delta t})$$

where

$$q_3 = \bar{g}_3 e^{\bar{\chi}(N+1)\Delta t} h^{-3}$$

$$q_5 = \frac{\bar{g}_2 - q_4}{2q_3}$$

Estimating q_5 using the binomial Theorem (in the way introduced in the semi-discrete analysis), the following estimate is obtained:

$$\mathcal{F}^{*n+1}(\mathscr{E}_v,\mathscr{E}_u) \leq c(h^{2k-1} + \Delta t^4)$$

Finally, employing the triangle inequality as in (18), it can be shown that the following result holds:

Thm. 4 If the exact solution to (4) is sufficiently regular, if the subspace $\mathcal{S}_h(I)$ is characterized by $k\geq 3$, if $\Delta t=ch^\ell$ for $\ell\geq 2$, and if h is sufficiently restricted, the fully discretized algorithm (19) is stable and convergent, and

$$\mathcal{F}^{n+1}(e_v,e_u) \leq c(h^{2k-1} + \Delta t^4) \qquad (36)$$

■

The following comments, concerning these results, seem appropriate:

1. The stability property of the algorithm, inherent in the numerical damping term (q_4 in (35)), depends on the temporal truncation error. From (35) it can be seen that, if there is real physical damping in the problem (characterized by positive values of α,ε, and $\bar{g}_2$), positive numerical damping can be obtained by simply making $\Delta t=ch$. However, if there is no physical damping in the problem, convergence to an undamped solution can be obtained only if $\bar{g}_2=f(h)$. Then, in order to obtain positive numerical damping, it is necessary to make $\Delta t=ch^\ell$ for $\ell\geq 2$.

2. Other temporal discretization schemes could be introduced to integrate the equations of motion forward in time. However, it should be noted that any scheme, with a temporal truncation greater than the Crank-Nicholson method, may be very difficult to stabilize. In these schemes the loss of temporal convergence rate may make it very difficult to produce a positive value for the numerical damping parameter q_4.

3. If the square root is taken on both sides of (36), standard L_2 type error estimates can be obtained.

4. From the error estimate (36) it can be seen that the nonlinear problem has no loss of convergence rate, in the temporal discretization, as compared to the linear problem.

References

1. Belytschko, T., and Schoeberle, D.F., "On the Unconditional Stability of an Implicit Algorithm for Nonlinear Structural Dynamics", Journal of Applied Mechanics,42,1975, pp. 865-869.

2. Hughes, T.J.R., "Stability, Convergence and Growth and Decay of Energy of the Average Acceleration Method in Nonlinear Structural Dynamics", Computers and Structures, 6, 1976, pp. 313-324.

3. Hughes, T.J.R., Caughey, T.K., and Liu, W.K., "Finite Element Methods for Elastodynamics which Conserve Energy", Journal of Applied Mechanics,45,1978, pp. 366-370.

4. Wellford, L.C., and Hamdan, S.M., "An Analysis of an Implicit Finite Element Algorithm for Geometrically Nonlinear Problems of Structural Dynamics, Part 1: Stability", Computer Methods in Applied Mechanics and Engineering,14, 1978, pp. 377-390.

5. Oden, J.T., and Reddy, J.N., Mathematical Theory of Finite Elements, Wiley Interscience, New York, 1976.

6. Wellford, L.C., "On the Theoretical Basis of Finite Element Methods for Geometrically Nonlinear Problems of Structural Analysis", Formulations and Computational Algorithms in Finite Element Analysis, Ed.,K.J. Bathe, et. al., MIT Press, Cambridge, Massachusetts, 1977.

Part V:
Solution Methods

Direct Solution of Equations by Frontal and Variable Band, Active Column Methods

R. L. TAYLOR, E. L. WILSON,
University of California, Berkeley, USA

S. J. SACKETT
University of California, Livermore, USA

ABSTRACT

The variable band, active column and the frontal methods for the solution of linear algebraic equations are compared. Areas considered in the comparison of the two methods are:

1. The number of numerical operations required for the triangular decomposition and resolution for different load vectors.

2. Logical operations and data transfer requirements associated with the formation of equations, decomposition and resolution.

3. Use of the methods for nonlinear problems, dynamics and substructure analyses.

4. Ease of computer implementation on mini-computers, micro-computers and vectorized computers.

A major contribution of the paper is the demonstration that the blocked, variable band, active column method requires exactly the same number of numerical operations as the frontal method. Based on advantages in other areas, it is concluded that the variable band, active column method is preferable for the solution of a broad range of problems in computational mechanics.

INTRODUCTION

In this paper we consider two techniques which are commonly employed to directly solve the algebraic equations resulting from a structural or a finite element analysis. The two methods compared are the variable band, active column method (also commonly referred to as profile or skyline method) and the frontal method. Each of these methods has been extensively described in the literature: for the frontal method see [1]-[8], and for variable band, active column methods see [9]-[14]. Proponents for each of these methods make claims that one method is superior to the other, usually with little or no analysis to substantiate the claims. In our paper we examine several of the factors which influence the cost of solving large sets of algebraic equations, e.g. those resulting in three dimensional analyses. The factors considered include number of operations to effect triangular decomposition, ordering of the elimination process, resolution vs. decomposition effort, vectorization of algorithms, etc. Based upon our evaluation and experiences we conclude that variable band, active column methods are generally preferable to frontal methods for a broad range of problems in computational mechanics.

DIRECT SOLUTION OF FINITE ELEMENT EQUATIONS

In many structural or finite element methods it is necessary to solve a large set of algebraic equations of the form

$$\underline{K}\,\underline{u} = \underline{f} \qquad (1)$$

where $\underline{f}$ is a specified nodal vector, $\underline{u}$ is the nodal solution vector to be computed, and $\underline{K}$ is a symmetric, positive definite, sparse matrix. In linear static analysis (1) occurs naturally, in linear dynamic analyses (1) occurs when using step-by-step integration methods, and in nonlinear analyses (1) occurs after linearization by, for example, Newton's method. The sparsity of $\underline{K}$ results from the basic structure of finite element approximation by locally based functions. The arrays are normally formed by direct sums over elements (direct stiffness) and with $\underline{k}_e$ and $\underline{f}_e$ as the appropriate element arrays for the problem class being considered we may write

$$\underline{K} = \sum_e \underline{k}_e \qquad (2)$$

and

$$\underline{f} = \sum_e \underline{f}_e \qquad (3)$$

The connection properties of each element will define the non-zero structure of the sparse matrix $\underline{K}$. In variable band methods the non-zero envelope of $\underline{K}$ is called the profile of the matrix. The profile will be defined in terms of equation numbers assigned to each degree-of-freedom at each node and the list of nodal numbers attached to each element. In most implementations of the variable band method the equation numbers are assigned independently of the nodal numbers. Often profile optimization schemes are employed to obtain "good" ordering of the equation numbers. We shall discuss this further in a subsequent section of the paper.

In the frontal method the connection properties of the elements and the element ordering will define the maximum front width. For either approach a different ordering of equations or elements will create a different sparsity of the $\underline{K}$ matrix. This fact is crucial to the efficient solution of (1) by direct methods. What we seek always is the ordering to effect a solution with a minimum effort.

In the next section we will summarize the steps required to solve (1) by variable band and frontal methods.

SOLUTION STEPS FOR VARIABLE BAND AND FRONTAL METHODS

All practical direct solution methods are versions of elimination or factorization algorithms. The direct solution to (1) is obtained by constructing the triangular decomposition

$$\underline{K} = \underline{U}^t \underline{D} \underline{U} \tag{4}$$

where $\underline{D}$ is a diagonal matrix, $\underline{U}$ is an upper triangular matrix with unit diagonals and $\underline{U}^t$ denotes matrix transpose of $\underline{U}$. The solution to (1) may now be constructed by solving

$$\underline{U}^t \underline{x} = \underline{f} \tag{5}$$

$$\underline{D} \underline{y} = \underline{x} \tag{6}$$

and

$$\underline{U} \underline{u} = \underline{y} \tag{7}$$

The solutions of (5) to (7) are relatively simple. The solution to (5) is called a forward solution since we consider the equations in forward order. Equation (6) is a set of uncoupled equations which is solved sequentially by simple divisions. The solution to (7) is obtained by starting with the last equation and working to the first - - this step is usually called backsubstitution.

It is clear that once the triangular decomposition is constructed the solution can be obtained for any number of specified right-hand sides $\underline{f}$, either simultaneously or sequentially. In the sequel we shall refer to equations (5) to (7) as the resolution step. In iterative or step-by-step time integration methods the same factors of $\underline{K}$ are often used many times, consequently, the efficient resolution of equations can be as important as efficient triangular decomposition of $\underline{K}$ (which may be performed very few times in an analysis compared to the number of resolutions).

In large analyses the number of nonzero entries in the triangular factors greatly exceed the core capacities of modern computers. In fact, the capacity of the backing storage may be severly taxed by this class of problems. For example a problem with 20,000 equations may have more than 10,000,000 terms in the triangular factor for $\underline{U}$. Therefore, we cannot consider in-core methods as a general approach to solving (1)

In the variable band algorithm we will assume that $\underline{U}$ is blocked into groups of columns approximately equal to one-half of the available high-speed storage and all blocks are placed in secondary storage*. The factorization is performed in sequential block order by reading previously factored blocks into high-speed storage for each block to be factored. The limiting size on the problem to be solved is that two blocks must reside in-core simultaneously. All accesses to secondary storage are for large blocks of data. The number of numerical operations between reads is significant since one large block of data is operating on another large block of data. In this way, the access time is small compared to in-core numerical effort for most structural and finite element problems.

The main steps to solving (1) by a blocked, variable band algorithm may be summarized as:

(1) Determine equation numbers for nodes, compute profile of matrix $\underline{K}$ and determine block structure for triangular factors.

(2) Compute element arrays $\underline{k}_e$ and $\underline{f}_e$, equation numbers for each node on the element and store results on low-speed storage.

(3) Assemble arrays $\underline{K}$ and $\underline{f}$ using previously computed element arrays and blocking for the triangular factors. Store assembled blocks on low-speed storage.

(4) Construct triangular factors of $\underline{K}$ using assembled blocks. Store blocks of factors on low-speed storage.

---

* For the CDC-7600 the blocks may be stored in the large core memory and single vectors moved to the fast core for the triangular decomposition and resolution steps. See section on vectorization for further details.

(5) Perform resolution using blocks of factors and $\underline{f}$.

When a new solution is required only step 5 and those steps involving $\underline{f}_e$ and $\underline{f}$ are required.

The frontal approach solves the set of algebraic equations as they are assembled. As pointed out in the paper of Irons [1], the only requirement to perform a Gauss elimination for any equation is that the elements of the equation need to be completely assembled. In the sparse matrix defined by a finite element problem interaction with an equation being eliminated on the "front" of active nodes is defined by the elements processed to date. Furthermore, all equations already eliminated are removed from subsequent consideration until the resolution step. Thus, the number of nodes active in the front depends on the order elements are introduced. In the frontal method the total matrix $\underline{K}$ is never fully assembled prior to factorization, rather the two steps are interweaved. The only information required in high speed core during solution is the active part of the front. In addition, the equation(s) to be eliminated after coefficients of each element are assembled may be anywhere in the front, in contrast to other direct procedures where elimination proceeds in equation order. In the frontal procedure once an equation is eliminated it is no longer required for the remainder of the triangular decomposition and may be removed from the front. The removal of an equation leaves a zero row/column in the active frontal equations. When additional equations are eliminated the existence of a zero row/column will increase the apparent number of operations (even with zero checks in the outer loops). Furthermore, when the next element is added all zero row/columns may not be filled. To minimize this detrimental effect two procedures have been proposed:

(a) Longevity based frontal methods where equations which survive for longer durations are put in the lower equation numbers of the active front [4]. This method requires additional computations to determine the order of elimination and placing in the front - - a step called prefront by Irons.

(b) Shift methods to fill the vacated row/column positions [3]. An equation to be eliminated is first moved to a buffer and then used to modify the other terms. As the other terms are modified those in equation positions larger than the one eliminated are shifted to the left and/or up one position to fill the vacated space. The added indexing to effect the shift is minimal.

Other strategies have been proposed, however the two cited appear to be the most common. The shift strategy is optimal in that no zero row/columns can exist after elimination of an equation. In our construction of operation counts for the frontal method we assume that shifts are employed to fill zeros.

We can summarize the steps to effect a frontal elimination as follows:

(1) Perform a prefront to determine the order for elimination of equations [7]. Compute maximum front width.

(2) Compute element arrays $\underline{k}_e$ and $\underline{f}_e$, corresponding equation elimination information, and store results on low speed storage.

(3) For each element retrieve the element arrays, assemble into frontal arrays, and eliminate equations which are completely assembled at this step. Store eliminated equations in buffer and when buffer is full transfer to low-speed storage.

(4) Perform resolution steps using frontal equations in low-speed core.

For a single set of equations steps 2 to 4 can be combined. Figure 1 illustrates the steps in a frontal solution using a very simple example.

OPERATION COUNTS - FRONTAL METHODS

In order to assess relative efforts between frontal and variable band solution methods the operations required to perform a triangular decomposition and a resolution will be computed for a set of sample problems. The problems are simple three dimensional meshes constructed from 8-node Lagrangian elements, 20-node Serendipity elements, or 27-node Lagrangian elements. The meshes are described by a block of elements with m, m, and n elements in each direction (subsequently described as "mxmxn" meshes of elements). The maximum front widths (in nodes) are described in terms of the number of elements on an m-edge as:

- $m(m+1)+2$, for 8-node elements
- $(2m+1)(m+2)+(m+1)(m+1)+7$, for 20-node elements
- $(2m+1)(2m+3)+12$, for 27-node elements

The results are given in Table 1 for several values of the number of nodes, M, on an m-edge where

- $M = m+1$, for 8-node elements
- $M = 2m + 1$, for 20-node and 27-node elements.

Table 1 shows that for a given mesh of nodal layouts, use of 20 node elements with quadratic interpolation give smaller front-widths in three dimensions than 8-node elements which use only linear interpolations (this is primarily due to the fact that one node in the middle of each face is omitted). In addition the use of 8-node elements requires processing of 8 times the number of 20-node elements. To the authors knowledge this rather surprising result has not been noted previously, although it has been recognized that mid-side nodes are treated very efficiently in frontal methods [1]. Moreover, use of 27-node Lagrangian elements produces front-widths which are asymptotically $1 + 1/M$ times those of the 8-node element meshes. Thus, for a given number of nodes use of quadratic order elements should not, in general, significantly increase the cost of solving equations over that required for linear order elements (e.g., in Table 3 compare 8-node 8x8x32 mesh with a 27-node 4x4x16 mesh, since these two have the same number of nodes). The number of operations to perform a triangular decomposition by the frontal method may be estimated by summing the operations performed for each element added to the mesh. If after adding an element the resulting front width is I and after eliminating all the equations appearing for the last time the front width is J, an estimate for the number of operations performed for each element is

- I*(I+1)*(I+2)/6 - J*(J+1)*(J+2)/6 for triangular decomposition,
- I*(I+1)/2 - J*(J+1)/2 for a forward solution or backsubstitution.

For each of the meshes considered the total number of operations for triangular decomposition is given in Table 3 and the total number of operations for a forward solution or backsubstitution is given Table 6. The triangular decomposition is approximately proportional to maximum front width squared for the problems considered, whereas, a resolution is proportional to maximum front width (e.g., the 8-node 8x8x32 mesh requires only 25% less effort in the triangular decomposition than the 27-node 4x4x16 mesh, and from Table 6, resolution is only 10% less). Similarly we note in the tables that using 20 node elements requires 43% and 50% less effort to construct a triangular decomposition and resolution for the resulting algebraic equations than those of the 8-node element, respectively. The case for the 20-node element seems conclusive, however, one should recall the study in [15] where serious deterioration of results were observed for distorted Serendipity elements subjected to high gradient solutions. The quadratic Lagrangian elements do not degrade significantly when subjected to the same tests (recently Wachspress has shown that Lagrangian quadratic elements maintain complete quadratic polynomial solutions provided boundaries of the element are planar and nodes are uniformly spaced). For parallelepiped element meshing the

20-node element is the obvious choice.

In order to indicate possible limitations of the frontal method we have computed core demands for a three-dimensional problem with three degrees-of-freedom at each node This is typical of problems in continuum and fluid mechanics. For a mesh with equal number of nodes in the two directions defining the front surface only small meshes can be considered before the high speed core of most modern computers is exhausted (In most engineering applications M would normally be substantially greater than eleven). In addition to the core required for the front matrix, additional memory is required for front vectors and a buffer to store the eliminated equations. If each eliminated equation were written to low-speed storage as eliminated the peripheral processing costs on many computers would greatly exceed the numerical operation costs. Thus, the buffer should be substantial so that only large blocks of eliminated equations are processed. After the high-speed core demands for the frontal equations is exceeded, the front matrix must be partitioned in a manner similar to that used for the variable band algorithm. Once this occurs any advantages of frontal methods rapidly evaporate since multiple passes through the blocks are required to eliminate each equation [7]. In the next section we consider the operation counts for the variable band method. Guided by the order frontal methods process equations we are able to demonstrate that the variable band method is as efficient as frontal methods and for large problems undoubtably superior. Before proceeding to this section, however, we should note that frontal methods require extensive moving of information into and out of the frontal matrix and vectors. In addition searches for the location of equations in the front are required. In the resolution processes the moving of information is very detrimental to overall efficiency, whereas, in triangular decomposition it is necessary to keep the front as small as possible . Furthermore the cost of moving information is offset by the ease of assembling element arrays into the front, a step which is not trivial for variable band methods.

OPERATION COUNTS - VARIABLE BAND METHODS

The number of operations necessary to perform a triangular decomposition for equations solved by a variable band method depends on the ordering of the equations. Various methods have been proposed to automatically compute a "good" numbering order (e.g., see [16]). In Figure 2, we show schematically how we assume our equations to be partitioned and in Figure 3 we indicate how the columns interact during the triangular decomposition step. Figure 3 shows that a mixture of tall columns with short columns is very desirable since it is the intersection of the columns which defines the number of operations which will actually be performed during the triangular decomposition of $\underline{K}$. For a given mesh it is extremely difficult to estimate the number of operations required to effect a decomposition, however, the number of terms within the profile defines the number of operations required to perform a forward solution or a back-substitution. In order to obtain the operation counts for triangular decomposition we constructed software to simulate the solution process and sum the numerical operations that would be required to perform each reduction. In constructing operation counts we considered two numbering systems:

(a) Minimum bandwidth numbering. A numbering which starts numbering equations at one corner, increments along the shortest edge, moves to the next row of nodes in a continuous process until all rows of nodes on one level are numbered, and repeats for each level in the mesh will produce a minimum bandwidth numbering for the meshes considered (see Figure 4a).

(b) Frontal Numbering. In this method we merely number the equations in the order the frontal method would perform eliminations. This gives maximum column heights significantly larger than band numbering but produces a mix of tall and short columns such that the mean column height is significantly less than that for band numbering (see Figure 4b).

No operation counts were constructed using an automatic numbering scheme such as [16], nor do we infer that frontal numbering is actually optimal for the variable band method.

Simulations were performed for the same meshes considered for the frontal method and the results are given in Tables 4 and 5 for the 27-node Lagrangian element and the 20-node Serendipity element. The operation counts for the 8-node linear element were not constructed, however, we note that band and frontal numbering coincide in this case. Furthermore, based on results obtained for 20-node and 27-

node elements operation counts are expected to be similar to those of the frontal method.

Our results show that frontal numbering of equations for variable band solution produces substantially the same number of operations as for a frontal solution (in two dimensional problems set up and solved by hand we obtained identical number of operations for the two methods, consequently, any differences noted in the tables may be due to the estimators used for frontal operation counts). In contrast to this, using band numbering the operation counts were significantly increased, often double the frontal for the triangular decomposition. Comparing number of terms within the profile of the variable band with resolution operation estimates for the frontal method we again obtained agreement for frontal numbering but substantial differences for band numbering - - especially for the 27-node element meshes. In resolution using a variable band method very little indexing or manipulation exists compared to numerical operations (compared to fairly large amounts of indexing and manipulation in frontal programs we have seen). In calculations we have performed, actual timing on various computers for programs written in ANSI-FORTRAN always show variable band methods to execute faster than frontal methods provided equations are numbered in frontal elimination order for the variable band method.

The one area where variable band methods may be inferior to frontal methods (when the front matrix is completely contained in the high-speed core) is assembly of the blocks of equations. Front is nearly optimal since all the required coefficients of $\underline{K}$ are in the high-speed core, whereas, in variable band methods the coefficients required for each element can be in more than one block of $\underline{K}$. Use of frontal numbering of equations will generally give better ordering of element for assembly, but more care must be taken to produce efficient assembly by variable band methods than for frontal methods. Of course, once the frontal matrix cannot reside in the high-speed core frontal methods begin to suffer from the same difficulties. The simplicity of the triangular decomposition by variable band methods will more than compensate for the added assembly effort in most large problems, in resolution assembly is not a factor. The use of substructuring methods can also be used to minimize both assembly and solution costs by reducing the number of elements in each part to be assembled, as well as, avoiding operations on unfilled zeroes within the profile (or front!). We consider this aspect in a subsequent section. First we will discuss briefly our experiences with some notions of vectorization for the variable band method.

VECTORIZATIONS FOR THE VARIABLE BAND METHOD

The variable band method can make extensive use of parallel operations or vectorization features of modern computers. In the triangular decomposition and forward solution the basic operations in the inner loops of the algorithm are vector dot products. In the backsubstitution step the basic operation is a subtraction of a vector times a constant from another vector. For the blocked variable band algorithm (called FISSLE - a Fast Implicit Solver for Systems of Linear Equations) used in the NIKE-finite element programs at the Lawrence Livermore National Laboratories [17] vectorization notions have been programmed for the CDC-7600 and CRAY-1 computers. Most of the subroutines used in FISSLE are written in ANSI-FORTRAN, only three subprograms are vectorized:

(a). A dot product routine RECOL to compute the inner products for the triangular decomposition and forward solution steps.

(b). A special inner product routine RECOLD to compute the elements of the diagonal matrix $\underline{D}$ in the triangular decomposition.

(c). A routine COLBAC to compute the vector sum of a scalar with a vector in the backsubstitution step.

The solution algorithm has been implemented on the CDC-7600 and CRAY-1 computers at LLNL. The FTN compiler on the CDC-7600 optimizes the loops in the routines while the CFT compiler on the CRAY-1 automatically produces code to utilize the CRAY vector hardware for the computations.

Although the FTN and CFT compilers produce very efficient object code, it is still possible to obtain a significant increase in execution speed by coding RECOL, RECOLD, and COLBAC in assembly language (COMPASS for the CDC-7600 and CAL for the CRAY-1). On the CDC-7600 the blocks are all located in large core memory (LCM). Since the data is accessed sequentially it is possible to fetch operands considerably faster than if the data were located in the fast core (SCM). Achieving high access rate, however, requires that no bank conflicts occur. By coding in in assembly language it is possible to move a vector operand to SCM at very little cost. With one vector in LCM and one in SCM no bank conflicts will occur. In addition it is possible to utilize the machine registers more efficiently and by splitting the loop in half, to overlap memory access with computations. It is possible to take advantage of the machine features on the CRAY-1 to yield similar speed increases. The assembly language versions of RECOL, RECOLD and COLBAC are in fact an average of 2.7 times as fast as their Fortran counterparts on the CDC-7600 and an average of 2.13 times as fast as their Fortran counterparts on the CRAY-1 (see Table 8.).

Figure 5 shows the timing curves obtained with FISSLE on the CDC-7600 and the CRAY-1. These curves were constructed by solving problems similar to the model problems used in this paper. The advantages resulting from vectorization are readily apparent in the fact that the asymptotic slope of the factorization timing curve remains less than the theoretical limit of two. This is due principally to the operation overlap obtained by using assembly language and, on the CRAY-1, the exploitation of chaining in vector operations. It is also interesting to note that a resolution is over two orders of magnitude faster than a triangular decomposition for large average column heights. This has important implications for the solution of linear step-by-step dynamic problems and for the solution of non-linear problems by quasi-Newton or other iterative solution algorithms which re-use triangular factors many times.

SUBSTRUCTURING

In many problems, especially those with branches, zeroes within the profile may not fill during the triangular decomposition. This is quite detrimental to the overall efficiency of solving equations. When a frontal numbering order is used to solve equations these zeros will persist in both the variable band and the frontal methods - - neither is immune to the problem. A simple example of a problem in which zeroes persist within the profile is shown in Figure 6. If the equations are numbered as shown then a group of zeroes persists at the branch line after the triangular factors are computed. In order to eliminate operations with zero elements a check on every possible operation must be performed - - which is obviously more inefficient than performing operations with a small number of zero elements. It is possible to skip these operations by employing a substructure analysis on each of the branches first. The results of each substructure analysis can then be assembled with the remainder of the problem and the solution process completed.

If operation counts are constructed for the substructure analysis one again reaches the conclusion that the frontal ordering of the equations is the crucial aspect of the analysis -- both frontal and variable band methods will require about the same solution effort. Furthermore, both methods are equally adaptable to generating the substructure matrices with only minor coding changes to the algorithms.

The use of substructuring also will make the list of elements significantly shorter in each part. This will further enhance the assembly step in the variable band method by reducing the possible number of searches through the element file to assemble each block.

CLOSURE

In this paper we have demonstrated that both the frontal method and the variable band, active column method require the same number of operations to perform a triangular decomposition or a resolution. This is contrary to the prevailing opinion and even Irons has communicated directly to us his belief that variable band techniques are about midway between the band with full triangle and the front methods for numerical operations [18]. It is our belief that the case is now made in favor of variable band methods over frontal methods especially for very large problems. The algorithms are intrinsically simpler to code and in resolution require much less indexing and no relocation of information. In this study we have compared only the variable band and frontal methods for the solution of large problems. There is every reason to believe that neither of these methods is optimal for three-dimensional problems. Either sparse matrix methods or iterative methods may ultimately prevail. Nevertheless, many existing programs are using variable band methods and it should now be apparent that knowledge gained from the frontal method can enhance the efficiency of solving problems which contain elements with mid-side nodes.

REFERENCES

1. B. M. Irons, "A Frontal Solution Program for Finite Element Analysis," Int. J. Num. Meth. Engr., 2, 5-32, 1970.

2. R. J. Melosh and R. M. Bamford, "Efficient Solution of Load-Deflection Equations," J. Struct. Div. , ASCE, 95, 60-76, 1969.

3. E. Hinton and D. R. J. Owen, Finite Element Programming, Academic Press, London, 1977.

4. P. Hood, "Frontal Solution Program for Unsymmetric Matrices, Int. J. Num. Meth. Engr., 10, 379-399, 1976.

5. R. J. Collins, "Dynamic Destinations: A Method for Removing Redundant Operations in the Front Algorithm," Int. J. Num. Meth. Engr., 12, 1042-1043, 1977.

6. S. F. Abbas, "Some Novel Applications of the Frontal Concept," Int. J. Num. Meth. Engr., 15, 519-536, 1980.

7. G. Beer, "A Second Generation Frontal Solution Program," (to appear).

8. R. J. Collins, "A Modified Prefrontal Routine," Int. J. Num. Meth. Engr., 11, 765-766, 1977.

9. A. Jennings and A. D. Tuff, "A Direct Method for the Solution of Large Sparse Symmetric Simultaneous Equations," Large Sparse Sets of Linear Equations, Ed. J.K. Reid, Academic Press, London, 1971.

10. D. P. Mondkar and G. H. Powell, "Large Capacity Equation Solver for Structural Analysis," Comp. and Struct. , 4, 699-728, 1974.

11. C. A. Felippa, "Solution of Linear Equations with Skyline-Stored Symmetric Matrix," Comp. and Struct. , 5, 13-29, 1975.

12. E. L. Wilson and H. H. Dovey, "Solution or Reduction of Equilibrium Equations for Large Complex Structural Systems," Adv. Engr. Software , 1, 19-25, 1978.

13. E. L. Wilson, "Solution of Sparse Stiffness Matrices for Structural Systems," Sparse Matrices , Ed. I. S. Duff, SIAM, Philadelphia, 1979.

14. R. L. Taylor, "Computer Procedures for Finite Element Analysis," Ch.24, The Finite Element Method , 3rd. ed., by O.C.Zienkiewicz , McGraw-Hill, London, 1977.

15. J.A. Stricklin, W. S. Ho., E. Q. Richardson and W. E. Haisler, "On Isoparametric vs Linear Strain Triangular Elements," Int. J. Num. Meth. Engr., 11,1041-1043, 1977.

16. N. E. Gibbs, W. G. Poole, Jr., and P. K. Stockmeyer, "An Algorithm for Reducing the bandwidth an Profile of a Sparse Matrix," SIAM J. Num. Anal. ,13, 236-250, 1976.

17. J. O. Hallquist, "NIKE2D: An Implicit, Finite-Deformation, Finite-Element Code For Analyzing the Static and Dynamic Response of Two-Dimensional Solids," Lawrence Livermore Laboratory Report, UCRL-52678, March 1979.

18. B. M. Irons, personal communication.

TABLES AND FIGURES

Table 1.

Maximum Front Width for 3-d Meshes with M-nodes on Each Edge

M	Element Type 8-node	20-node	27-node
3	14	22	27
5	32	36	47
7	58	58	75
9	92	86	111
11	134	120	155

Table 2.

3-d Element Types with 3-dof/node

	8-node	(L)	20-node	(S)	27-node	(L)
M	Front	Core	Front	Core	Front	core
3	42	903	66	2211	81	3402
5	96	4656	108	5886	153	11781
7	174	15225	174	15225	237	28203
9	276	38226	258	33411	345	59685
11	402	81003	360	64980	477	114003

Table 3.

Triangular Decomposition Operations (in millions)

Mesh	Element Type 8-node	20-node	27-node
2x2x8	.01	.11	.29
3x3x12	.04	.91	2.52
4x4x16	.19	4.54	13.15
5x5x20	.65	16.83	50.21
6x6x24	1.86	50.76	154.84
8x8x32	10.50	304.94	959.65

Table 4.

27 node Lagrangian Element

		Band Numbering			Frontal Numbering		
Mesh Size	No. Eqs.	1000 Terms	Mean Column	Million Ops.	1000 Terms	Mean Column	Million Ops.
2x2x8	425	18	43	.42	15	36	.28
3x3x12	1225	100	82	4.32	77	63	2.44
4x4x16	2673	353	132	24.50	260	97	12.89
5x5x20	4961	964	194	98.20	694	140	49.52
6x6x24	8281	2225	267	312.61	1578	191	153.27
8x8x32	18785	8514	453	2010.	5937	316	953.72

Table 5.

20 node Serendipity Element

		Band Numbering			Frontal Numbering		
Mesh Size	No. Eqs.	1000 Terms	Mean Column	Million Ops.	1000 Terms	Mean Column	Million Ops.
2x2x8	261	8.2	31	.13	7.5	29	.11
3x3x12	712	40	56	1.12	35	50	.87
4x4x16	1505	132	88	5.82	115	77	4.43
5x5x20	2736	345	126	22.00	300	110	16.54
6x6x24	4501	773	172	67.08	669	149	50.10
8x8x32	10017	2842	284	408.00	2449	245	302.51

Table 6.

Forward Solution or Backsubstitution

Operations (in thousands)

	Element Type		
Mesh	8-node	20-node	27-node
2x2x8	.9	7.3	15.2
3x3x12	4.0	32.8	76.5
4x4x16	12.2	114.2	259.7
5x5x20	30.4	297.7	693.9
6x6x24	66.0	665.0	1578.3
8x8x32	232.9	2440.6	5937.0

Table 7.
Vectorization Timings in Fortran and Assembly Language
for CRAY-1 and CDC-7600 Computers
Time in nano-seconds to produce a single element for an n-vector.

	RECOL		RECOLD		COLBAC	
	Time	/CAL	Time	/CAL	Time	/CAL
n=1						
CAL	800.0	1.00	3000.0	1.00	600.0	1.00
CFT(ON=V)	2200.0	2.75	9000.0	3.00	1400.0	2.33
CFT(OFF=V)	2100.0	2.63	4000.0	1.33	1300.0	2.17
COMPASS	2300.0	2.88	2000.0	0.67	1000.0	1.67
FTN(OPT=2)	3500.0	4.38	6000.0	2.00	1300.0	2.17
n=8						
CAL	362.5	1.00	375.0	1.00	175.0	1.00
CFT(ON=V)	937.5	2.59	1125.0	3.00	337.5	1.93
CFT(OFF=V)	850.0	2.34	1000.0	2.67	662.5	3.79
COMPASS	1012.5	2.79	1000.0	2.67	1050.0	6.00
FTN(OPT=2)	1487.5	4.10	1500.0	4.00	1425.0	8.14
n=64						
CAL	70.3	1.00	109.4	1.00	56.3	1.00
CFT(ON=V)	167.2	2.38	203.2	1.86	75.0	1.33
CFT(OFF=V)	576.6	8.20	734.4	6.71	531.3	9.44
COMPASS	446.9	6.36	562.5	5.14	478.1	8.50
FTN(OPT=2)	806.3	11.47	5015.6*	45.86	817.2	14.53
n=1000						
CAL	30.7	1.00	57.0	1.00	46.6	1.00
CFT(ON=V)	69.8	2.27	97.0	1.70	49.2	1.06
CFT(OFF=V)	540.0	17.59	666.0	11.68	513.7	11.02
COMPASS	365.3	11.90	442.0	7.75	386.4	8.29
FTN(OPT=2)	1113.9*	36.28	872.0	15.30	1482.9*	31.82

* Due to large number of bank conflicts.

$$\left[\begin{array}{ccc:c:c} -4 & 5 & 2 & -7 & 8 \end{array}\right] = LG$$

$$\left[\begin{array}{ccc|c|c} K_{11} & K_{12} & K_{13} & K_{14} & K_{15} \\ & K_{22} & K_{23} & K_{24} & K_{25} \\ & & K_{33} & K_{34} & K_{35} \\ \hline & & & K_{44} & K_{45} \\ \hline & & & & K_{55} \end{array}\right]$$

FRONT Coefficient Array before Elimination.

$$\left[\begin{array}{ccccc} K_{14} & K_{24} & K_{34} & K_{44} & K_{45} \end{array}\right] = EQ$$

Equation being Eliminated.

K = 4 Pivot location

JJ= 7 Global equation number

$$\left[\begin{array}{ccc:c} -4 & 5 & 2 & 8 \end{array}\right] = LG$$

$$\left[\begin{array}{ccc|c} K_{11} & K_{12} & K_{13} & K_{15} \\ & K_{22} & K_{23} & K_{25} \\ & & K_{33} & K_{35} \\ \hline & & & K_{55} \end{array}\right]$$

FRONT Coefficient Array after Elimination. Note treatment of row/column 5. Shift prevents zeroes in frontal rows and columns.

Figure 1. Elimination of an Equation During Frontal Elimination

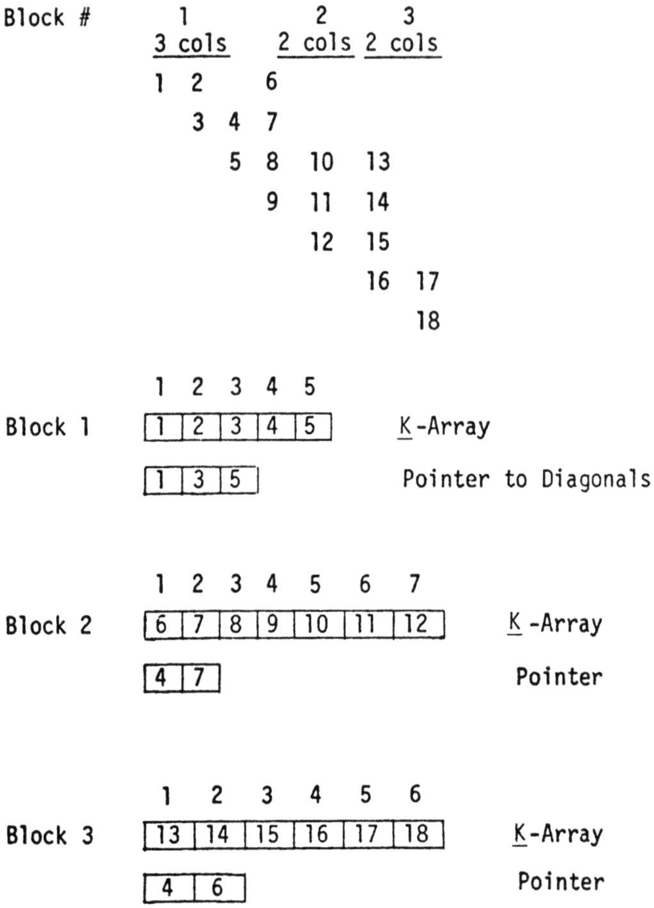

Figure 2. Blocking Format for Variable Band Solution.

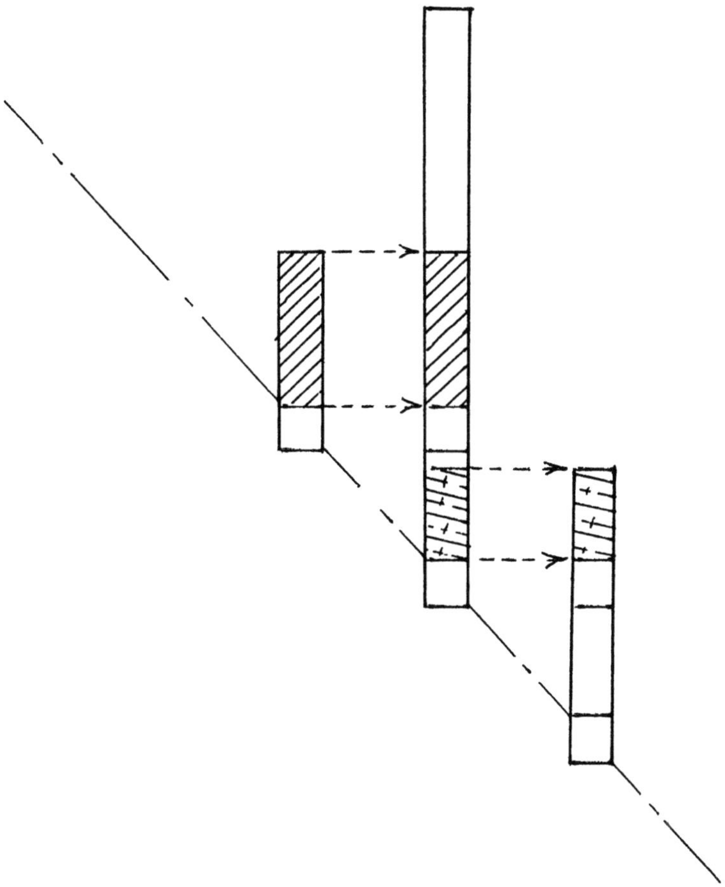

Figure 3. Interaction Between Tall and Short Columns for Variable Band Methods during Triangular Decomposition.

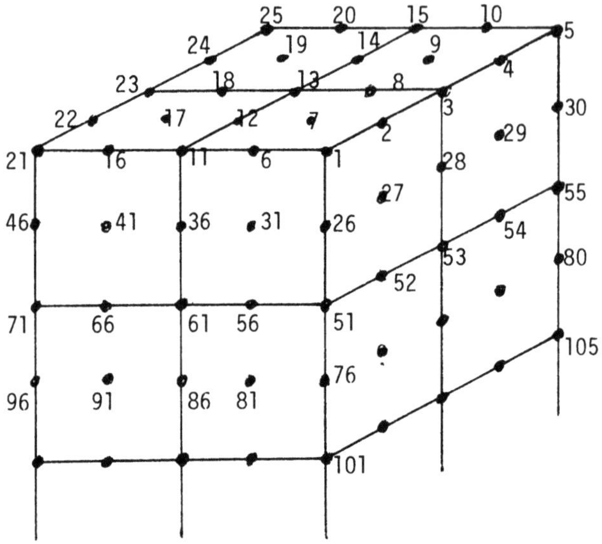

Figure 4a. Band Numbering for 27-node Lagrangian Element.
(2x2xn Mesh)

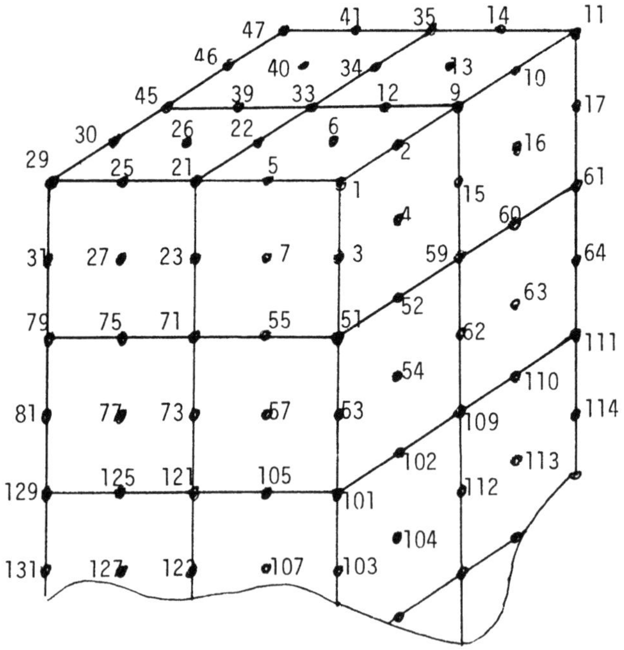

Figure 4b. Frontal Numbering for 27-node Element.
(2x2xn Mesh)

Figure 5. FISSLE CPU COST

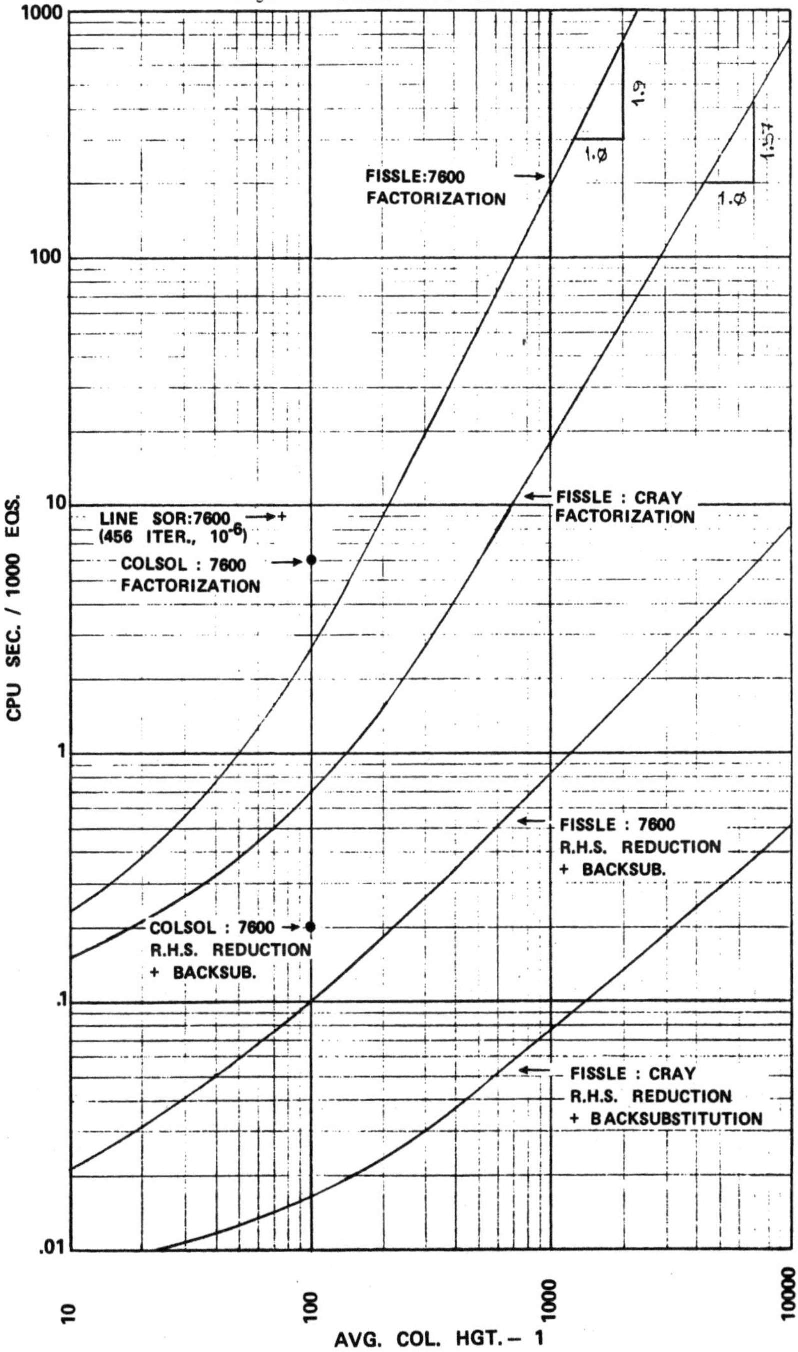

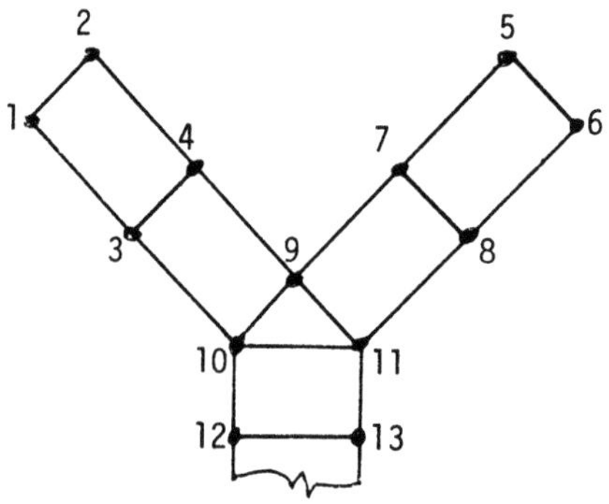


```
1 2 3 4 5 6 7 8 9--11--13
X X X X
  X X X
    X X         X X
      X         X X
        X X X X 0 0
          X X X 0 0
            X X X 0 X
              X X 0 X
                X X X
                  X X X X
                    X X X
                      X X
                        X
```

X - Non-zero in Triangular Factor.

0 - Zero in Triangular Factor.

⊗ - Substructure Coefficients.

Substructure 1.

```
1 2 3 4 10 9
X X X X
  X X X
    X X X X
      X X X
        ⊗ ⊗
          ⊗
```

Substructure 2.

```
5 6 7 8 9 11
X X X X
  X X X
    X X X X
      X X X
        ⊗ ⊗
          ⊗
```

Substructure 3.

```
9--11--13
⊗ ⊗ ⊗
  ⊗ X X X
    ⊗ X X
      X X
        X
```

Figure 6. Example with Persistent Zeroes within Profile of K.

Solution by Iteration in Displacement and Load Spaces

P. G. BERGAN
The Norwegian Institute of Technology, Trondheim, Norway

Summary

A method of adjusting the load intensity in order to minimize the residual (unbalanced) forces from the equilibrium equations is suggested. This technique may be used as a means of controlling the steps for load or displacement pattern incrementation schemes. It may also be used for obtaining improved convergence during equilibrium iterations of initial stress (strain), quasi-Newton or true Newton-Raphson type.

Introduction

The solution algorithms for analysis of nonlinear static analysis are normally based on stepwise incrementation of loads or of displacements. The load stepping schemes usually combines a simple Euler-Cauchy incrementation with equilibrium iterations of the Newton-Raphson type. For reasons of efficiency and reliability the size of the steps should be adapted to the degree of nonlinearity of the system during the course of solution. Special measures must be taken in order to deal with limit points, bifurcations and unstable branches. Solution algorithms of this type are discussed in Refs. /1/ - /4/.

Stepwise solution by incrementing a characteristic displacement component /5/ has also been a popular method. The load intensity is adjusted during the equilibrium iterations so that the force reaction at the preselected component vanishes. The main advantage by this method is that the modified stiffness matrix may remain positive definite even for unstable systems. Even though this approach may be very successful for small-scale and lucid problems it is often not suited for large-scale, practical problems.

This paper discusses a solution technique that has similarities with the two methods discussed above. It allows for incrementation of either loads or a suitable displacement pattern. Further, both the load intensity and the displacements are continuously corrected during the equilibrium iterations /6/.

The basic idea

In conventional incremental-iterative solution schemes the objective of the equilibrium iterations is to find the configuration of the structure for which the associated internal forces exactly balance the applied external forces. In other words, the point in load space is being kept unchanged while iterations are carried out in the displacement space. In most situations the analyst is not particularly interested in a solution for specific values of the loading vector $\mathbf{R}$; he rather wants to be able to define the over-all solution path by some typical discrete points. Intermediate values are easily obtained by interpolation. Since it is not essential to fix the loads at prescribed values it is tempting continuously to adjust the load intensity p along with the iterative adjustments of the displacements $\mathbf{r}$.

The principle for the method is shown in Fig. 1. Let points "a" in the two spaces denote a previously found equilibrium configuration for which the resulting internal forces R_{int} exactly balance the applied forces R_a. A new load increment ΔR_{ab} is now applied and the corresponding incremental displacements are found using the Euler-Cauchy method or some similar scheme.

$$\Delta \mathbf{r}_1 = K_I^{-1} \Delta R_{ab} \tag{1}$$

and

$$\mathbf{r}_1 = \mathbf{r}_a + \Delta \mathbf{r}_1 \tag{2}$$

where K_I is the incremental or tangential stiffness. The new configuration r_1 produces internal forces $R_{int,1}$, see points marked "1" in Fig. 1. The difference between the applied forces R_b and the internal forces $R_{int,1}$ is usually interpreted as unbalanced or residual forces and applied in a Newton-Raphson displacement correction. Instead of doing this the residual force is here minimized by adjusting the intensity of the applied forces to another point "b1" on the loading path. The corresponding minimized unbalanced force $R_{unbal,1}$ is then used in a Newton-Raphson correction of the displacements. In general, these corrections may be written

$$r_{j+1} = r_j + K_{I,j}^{-1} R_{unbal,j} \tag{3}$$

where

$$R_{unbal,j} = R - R_{int,j} \tag{4}$$

Index j indicates the number of the iteration cycle and $K_{I,j}$ is the stiffness matrix that may represent the true gradient or be an approximation to it.

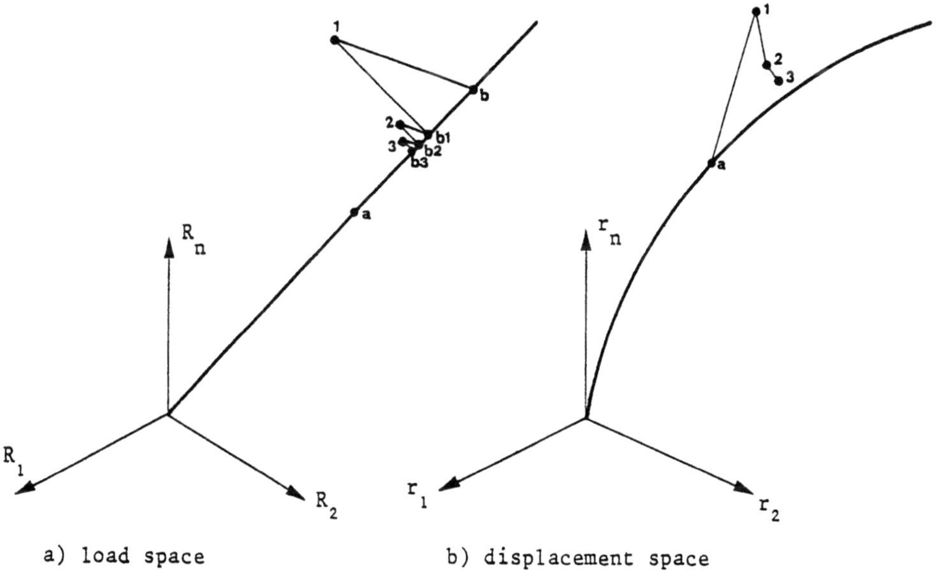

Fig. 1 - Iteration with adaptation of load intensity

The minimization of unbalanced forces and corresponding adjustment of load intensity should normally be carried out for every iteration cycle. Such a sequence of operations is illustrated in Fig. 1. It is also possible to carry out the force minimization only directly after a new increment has been applied and keep the loading level fixed during the following equilibrium iterations. Constant load increments could then be used throughout. The adjustments of load intensities would then produce an automatic loading algorithm in which the degree of nonlinearity governs the actual loading steps. This is illustrated in Fig. 2 for a one degree-of-freedom snap-through problem. The initial, elastic stiffness has been used in the incrementation, cf. Eq. (1) and the trial load increment is constant throughout. The minimization of unbalanced forces brings the loading level directly back on the curve in this case, and no further iterations have been necessary. It is seen that the actual load increments are nicely adjusted to the nonlinearity of the system. There has been no need for a special unloading criterion. A similar case with several degrees of freedom

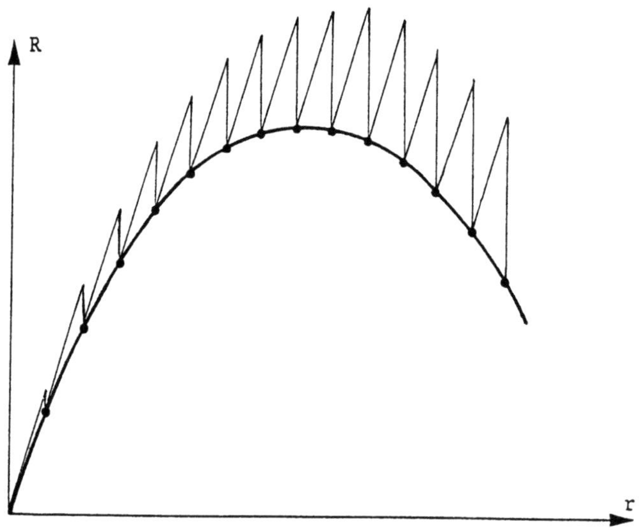

Fig. 2 - Solution of a one degree-of-freedom system with constant load steps and load intensity adaptation

would require equilibrium iterations, but the properties of adapted load steps and automatic unloading essentially remain. The special case shown in Fig. 2 corresponds to incrementation of the displacement by constant amounts.

Solution for proportional loading

In case of proportional loading applied to the structural system the loads may be related directly to a reference load vector $\mathbf{R}_{ref}$ in terms of a loading parameter p

$$\mathbf{R} = p\,\mathbf{R}_{ref} \tag{5}$$

This linear relationship implies that the load path is a straight line in load space, see also Fig. 3. Assuming that a new step has been taken from point "a" or that equilibrium iterations are being carried out, the condition of minimizing the unbalanced forces may be stated

$$\frac{\partial}{\partial p}\,||\mathbf{R}_{unbal}|| = \frac{\partial}{\partial p}||\,p\mathbf{R}_{ref} - \mathbf{R}_{int}||\;\; = 0 \tag{6}$$

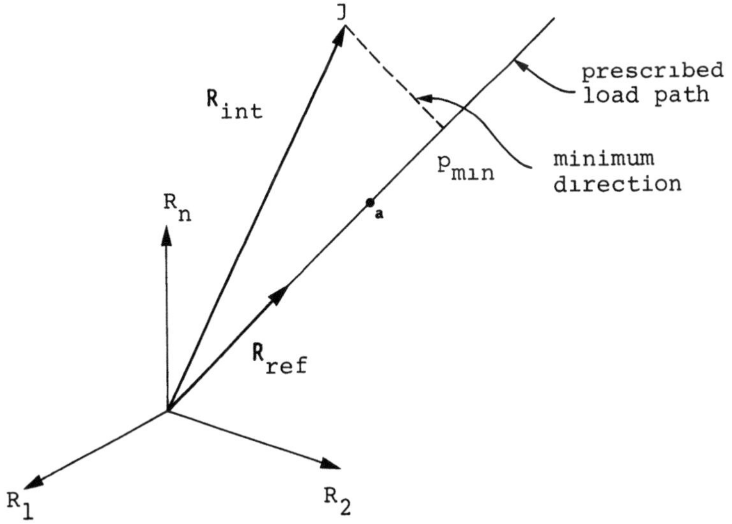

Fig. 3 - Minimization of unbalanced forces for proportional loading

This is the condition for finding the best suited load level p_{min}. The "size" of the unbalanced force should be measured in terms of some norm, e.g. a weighted Euclidean norm defined by

$$||\mathbf{x}|| = \left(\sum_{k=1}^{N} W_k^2 x_k^2\right)^{\frac{1}{2}} \tag{7}$$

where W_k are the weights and N is the dimension of the vector $\mathbf{x}$. Using this in Eq. (6) the best adapted load level is

$$p_{min} = \frac{\sum_{k=1}^{N} W_k^2 R_{ref,k} R_{int,k}}{\sum_{k=1}^{N} W_k^2 R_{ref,k}^2} \tag{8}$$

The direction of the minimized unbalanced force is perpendicular to the load path in the scaled load space, see Fig. 3. Note again that these operations of load adaption may be carried out after each new load step as well as after each cycle during equilibrium iterations.

The weights may in many cases be chosen as unity. However, when both translational and rotational degrees of freedom with corresponding loads are involved, the weights may be used to counteract that the forces have different units. As will be mentioned later, the possibility of using weight may also be advantageous when convergence problems occur.

Solution for non-proportional loading

Eq. (5) is not valid for a non-proportional load history, and the load intensity must be expressed in a more general form

$$\mathbf{R} = \mathbf{R}(p, \mathbf{r}) \tag{9}$$

It is implied that dependency of forces on the loading parameter must be known in advance or can be established on the bases of the deformed configuration (e.g. pressure loads).

The process of minimizing the unbalanced forces is now based on an auxiliary local linearization of the loading path. Initially, it is assumed that **R** depends only on p. The process starts by taking a load step from the equilibrium state "a" to a new load level $\mathbf{R}(p_b)$, see Fig. 4. The new displacements are found from Eqs. (1) and (2) and the internal forces $\mathbf{R}_{int}$ are established accordingly. It is then assumed that the load path may be approximated by a secant between "a" and "b", and the unbalanced force is minimized with respect to this line through the condition

$$\frac{\partial}{\partial p}||\mathbf{R}_{unbal}|| = \frac{\partial}{\partial p}||\mathbf{R}_a + \frac{p-p_a}{p_b-p_a}(\mathbf{R}_b-\mathbf{R}_a) - \mathbf{R}_{int}|| = 0 \quad (10)$$

Adopting the weighted norm in Eq. (7) the minimizing load level is

$$p_{min} = \frac{\sum_{k=1}^{N} W_k^2(R_{b,k}-R_{a,k})((p_b-p_a)R_{int,k} + p_a R_{b,k}-p_b R_{a,k})}{\sum_{k=1}^{N} W_k^2(R_{b,k} - R_{a,k})^2}$$

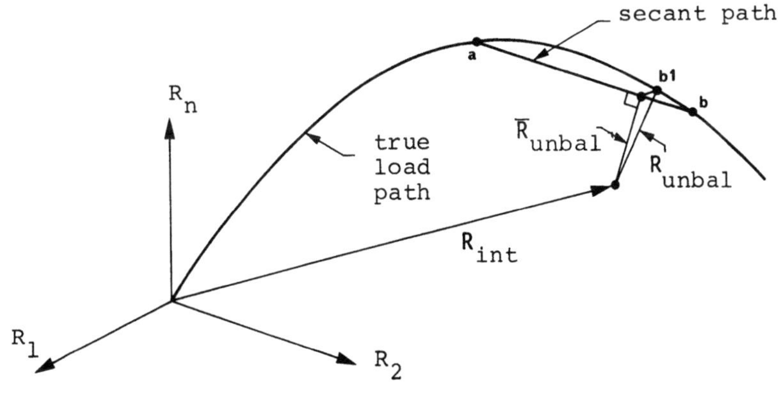

Fig. 4 - Minimization of unbalanced forces for non-proportional loading

The unbalanced force with respect to the corresponding point on the true load path (marked "bl" in Fig. 4) is now

$$\mathbf{R}_{unbal} = \mathbf{R}(p_{min}) - \mathbf{R}_{int} \tag{12}$$

These operations may now be repeated in connection with equilibrium iterations. $\mathbf{R}_b$ should then be taken as the adapted loading level from the last iteration cycle while $\mathbf{R}_a$ is kept constant.

It can be expected that the method of auxiliary linearizations works well for relatively smooth loading paths, however, it can also be applied in cases of "kinks" on the path. It may also be used without modifications for deformation-dependent loading. Calculation of $\mathbf{R}_b$ is then based on the last found (approximate) configuration with associated load intensity.

Incrementation of displacements

The fact that the present technique automatically finds a suitable load level opens the possibility for a simple way of incrementing displacements rather than loads. Of course, also the load incrementation of Eq. (1) produces an incremental displacement pattern. The quality of these displacements primarily depends on the gradient that is used; the true tangent stiffness usually gives a good pattern while approximate gradients or the elastic stiffness may give a very poor set of incremental displacements. For instance, a symmetric structure with symmetric loading may buckle into a asymmetric mode whereas use of the elastic stiffness would always give a symmetric displacement pattern.

A good guess for a new set of incremental displacements is that it should have the same shape as the incremental displacements between the last two equilibrium configurations

$$\Delta \mathbf{r}_{ab} = \gamma \Delta \mathbf{r}_i \tag{13}$$

The scaling factor γ assures that $\Delta \mathbf{r}_{ab}$ assumes a suitable size. For instance, the norm of the incremental step may be required to be the same as for some reference step $\Delta \mathbf{r}_{ref}$ by choosing

$$\gamma = \frac{||\Delta \mathbf{r}_{ref}||}{||\Delta \mathbf{r}_i||} \tag{14}$$

$\Delta \mathbf{r}_{ref}$ which should have an ideal size, may for instance be taken as the incremental displacements at the first linear step (the only step that is calculated from incremental loads).

Discussion of performance

As mentioned earlier, load level adaptation may be applied directly after each new increment as a means of step control. The method then implies automatic unloading after limit points, even without introducing any special unloading criterion. Experience so far indicates that the method is well suited for this purpose, particularly for systems with a smooth behaviour.

Even more interesting is the aspect of using the method as a way of improving convergence during equilibrium iterations. This is particularly true in connection with initial stress (strain) type iteration and iteration with approximate gradients; the prospective gains are greatest for these techniques. It can be expected that the convergence in connection with true gradient iteration (Newton-Raphson) will be at least as good with load level adaptation as without. In any case, the load level corrections do not cost much computationally.

In general terms, convergence is hardest to obtain when using only the elastic stiffness throughout the entire calculations; this case will now be discussed. The unbalanced force minimization always gives an unbalanced force that is perpendicular to the load path, which is a straight line for proportional loading. Thus, these unbalanced force components always have a linear dependency between them. When these forces are multiplied by K^{-1}, which is kept constant, a corresponding linear dependency also is introduced between the components of the displacement corrections. In an N-dimensional case the displacement corrections thus follow a hyperplane, which in an unfortunate case may have no intersection with the true solution path. In such a situation the iteration will not converge.

The question of convergence is illustrated for a system with only two degrees-of-freedom in Fig. 5. The path for proportional loading is a straight line in the R_1-R_2 plane and the minimized unbalanced forces are always perpendicular to this direction. When these forces are sequentially multiplied by the same inverse gradient they produce displacement corrections that all follow the same straight line. The final solution is simply the point of intersection between this line and the true solution path. Fig. 5 also shows that, if the iterations start from a point in the displacement plane that lies on the convex side of the path, it may happen that

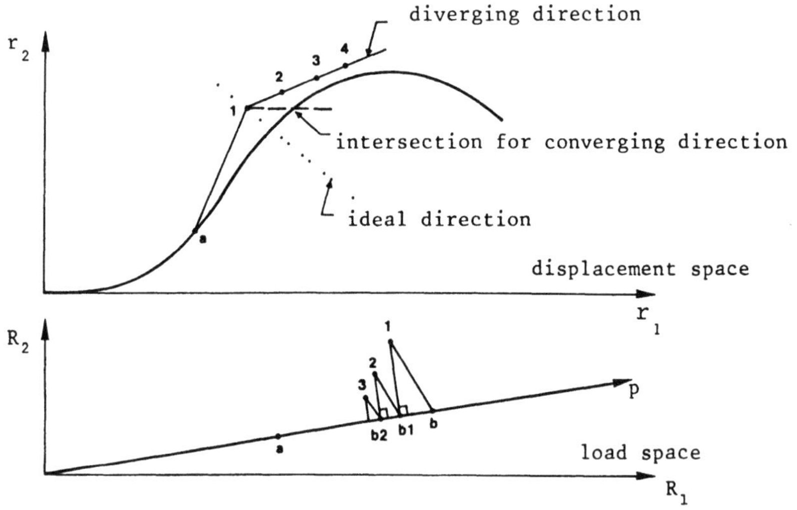

Fig. 5 - Convergence properties when using constant gradient

there is no intersection between the line of displacement corrections and the solution path. Thus, there will be no convergence. The figure also indicates that the best direction for the displacement corrections would be a line perpendicular to the true displacement path.

The simplest way of assuring converging iteration is to update the gradient at intervals. Another possibility is to start up the iterations from a different point in solution space whenever convergence is poor. This could be done by going back to the last equilibrium configuration and using only half of the previously used increment. In Fig. 5 this corresponds to iterating from the midpoint between points "a" and "1", which would assure convergence for that particular situation.

A third remedy that could be used is to rotate the direction of the displacement corrections. This may be done through matrix transformations or by use of acceleration factors. There is no point in using the same factor for all components since that would not change the direction of displacement corrections. One way of achieving acceleration is through use the following factor for component k in Eq. (3)

$$\alpha_k = \left| \frac{\Delta r_{k,j-1} + \Delta r_{k,j}}{\Delta r_{k,j-1}} \right| \tag{15}$$

where indices j-1 and j refer to the two previous iteration cycles. Eq. (15) should be overruled by requirements for minimum and maximum values, e.g. 0.5 and 2.0, respectively.

Scaling of the components in load space may have an effect similar to using acceleration factors.

Numerical studies

The algorithm is tested for a spring system suggested in Ref. /6/ and shown in Fig. 6. This case is a variation of the classical pinned bars snap-through problem with the extension that this system also may buckle out of the A-B-C plane. In this two degrees-of-freedom system v_1 is the snap-through component while v_2 is associated with bifurcation instability. The third displacement component in space is restrained. By varying spring constants, geometry, load combination and imperfections it is possible to simulate a wide range of coupled nonlinear phenomena involving limit points, instable branches, symmetric and asymmetric bifurcations etc.

Further, it is relatively simple to present the results on graphical form for a system with only two freedoms. The potential equilibrium and incremental forms for this problem are given in an Appendix.

The current stiffness parameter /3/,/4/,/6/ has been utilized extensively in the numerical studies. For a case of proportional loading and incremental solution with true gradient (tangent stiffness K_T) this quantity is defined by for step i

$$S_p = \left(\frac{\Delta p_i}{\Delta p_1}\right)^2 \frac{\Delta r_1^T K_{T1} \Delta r_1}{\Delta r_i^T K_{Ti} \Delta r_i} \tag{16}$$

where index 1 refers to the initial state and the first step.

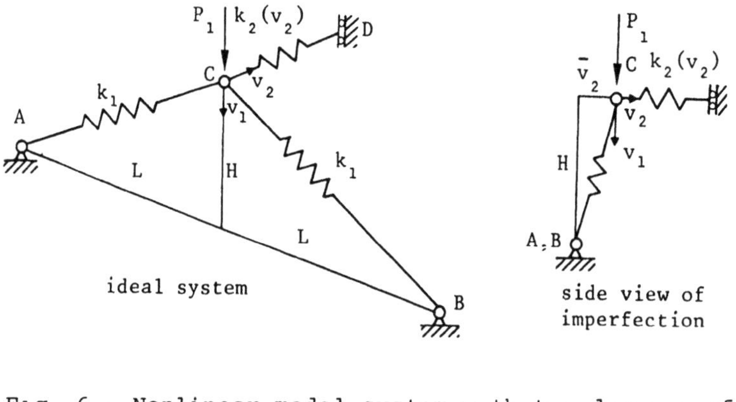

Fig. 6 - Nonlinear model system with two degress-of-freedom

When only an approximate gradient or the elastic stiffness is used S_p may rather be found from

$$S_p = \frac{\Delta p_i}{\Delta p_1} \frac{||\Delta r_1||}{||\Delta r_i||} \tag{17}$$

where $||\Delta r_i||$ is an Euclidean norm taken of the incremental displacements between the last two equilibrium configurations.

It should only be noted that the scalar quantity S_p gives a good characterization of the over-all stiffness of a structural system. S_p starts out with value 1 for the initial state, it is less than unity when the system becomes softer, and it is larger than unity when the system becomes stiffer than in the initial state. An important property of S_p is also that it is exactly zero when passing limit points and is negative for unstable branches of the solution path. Moreover, the current stiffness parameter may be used for automatic step size control be prescribing a desired change $\Delta\overline{S}_p$ for S_p during each new step. S_p may also be used to guide the algorithm at limit points. These properties are discussed in more depth in Refs. /3/, /4/ and /6/.

The first study of the model problem involves a case of slight geometric imperfection with initial displacement $\overline{v}_2 = 0.01$ L. Further data are given in Fig. 7. Curve A shows the results for pure Euler-Cauchy incrementation with $\Delta\overline{S}_p = 0.01$, curve is Euler-Cauchy combined with true Newton-Raphson iteration and $\Delta\overline{S}_p = 0.05$, and, finally curve C is the new method with only elastic stiffness, load incrementation and $\Delta\overline{S}_p = 0.05$ for the trial load steps. The results for the last two methods coincide in the figure while curve A for pure Euler-Cauchy deviates somewhat because of truncation errors. The jump of the stiffness parameter because of transition to a bifurcation path is shown in Fig. 7c.

The method of adjusting load intensity during increment and iterations is demonstrated for a case of a geometrically

a.

b.

c.

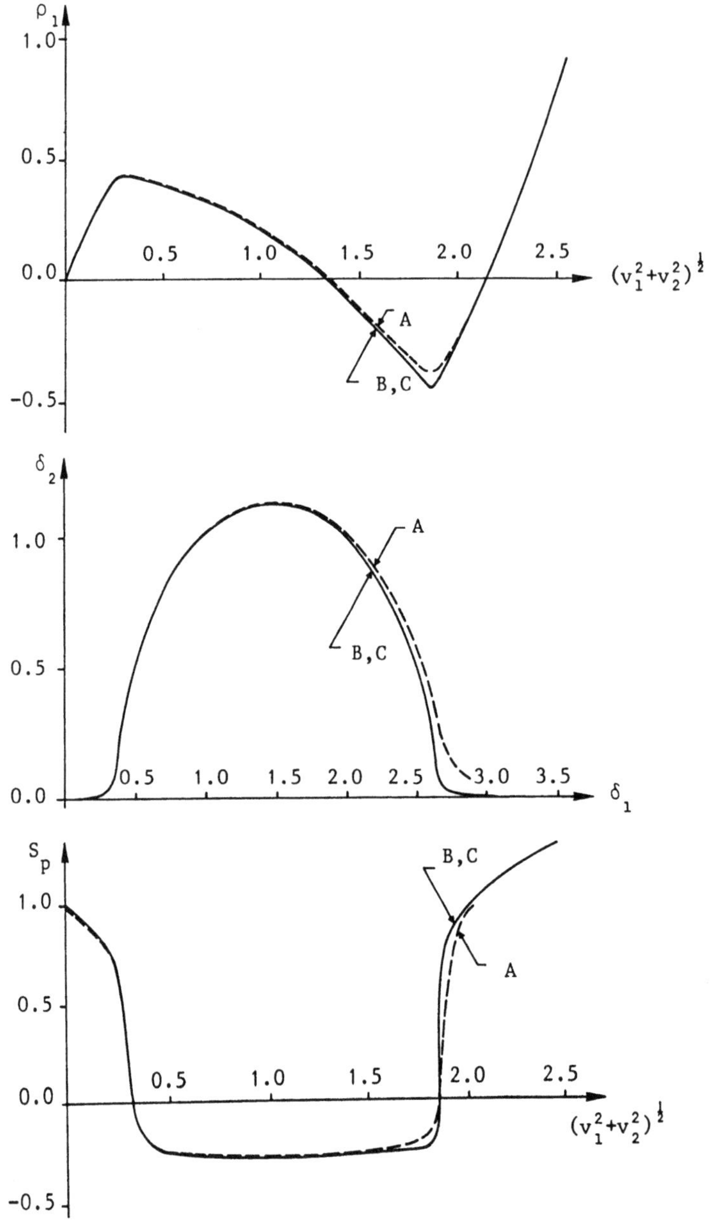

Fig. 7 - Study of model problem with geometric imperfection. $H_o=1,5$, $v_o=0.01$, $m_o=0.4$, $m_1=0.0$, $\rho_2=0.0$. Curve A: Euler-Cauchy with $\Delta\overline{S}_p=0.01$, Curve B: Incrementation with Newton-Raphson, Curve C: Elastic stiffness matrix and load adaptation.

perfect system ($\bar{v}_2 = 0$) with a small load perturbation $P_2 = 0.01\ P_1$ in Fig. 8. The very sharp deformation mode transition after the completion of the transverse bifurcation mode should be noted. Even with many attempts it was impossible to make the usual Euler-Cauchy incrementation with Newton-Raphson iteration pass this critical point of the path. Therefore, it was very satisfying that the new method with only linear elastic stiffness, but with acceleration factors, did work. The extreme singularity at the critical point is accentuated by the plot of the current stiffness parameter in Fig. 8.b.

a.

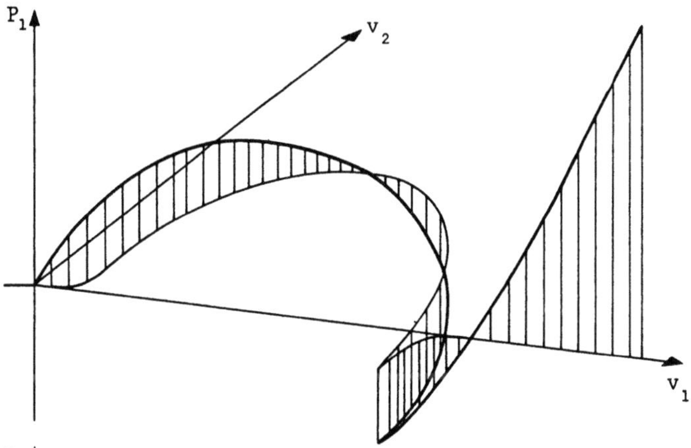

b.

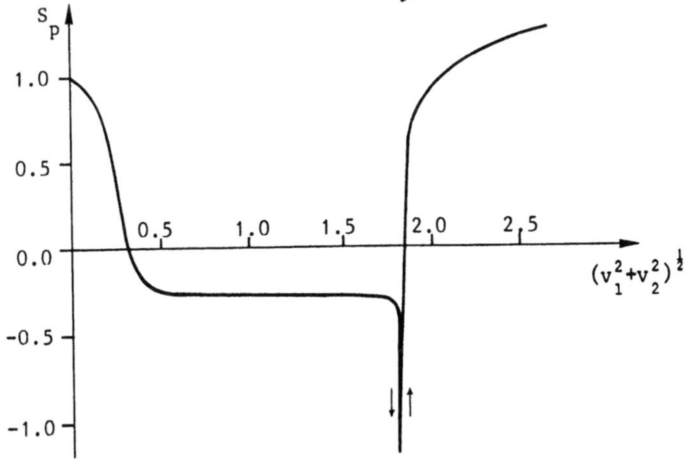

Fig. 8 - Study of model problem with load pertubation. $H_o=1.5$, $v_o=0.0$, $m_o=0.4$, $m_1=0.0$, $\rho_2=\rho_1/100$.

The model problem also lends itself to a study of asymmetric buckling. This is done by making the force-displacement relation of the transverse spring nonlinear, see Appendix.

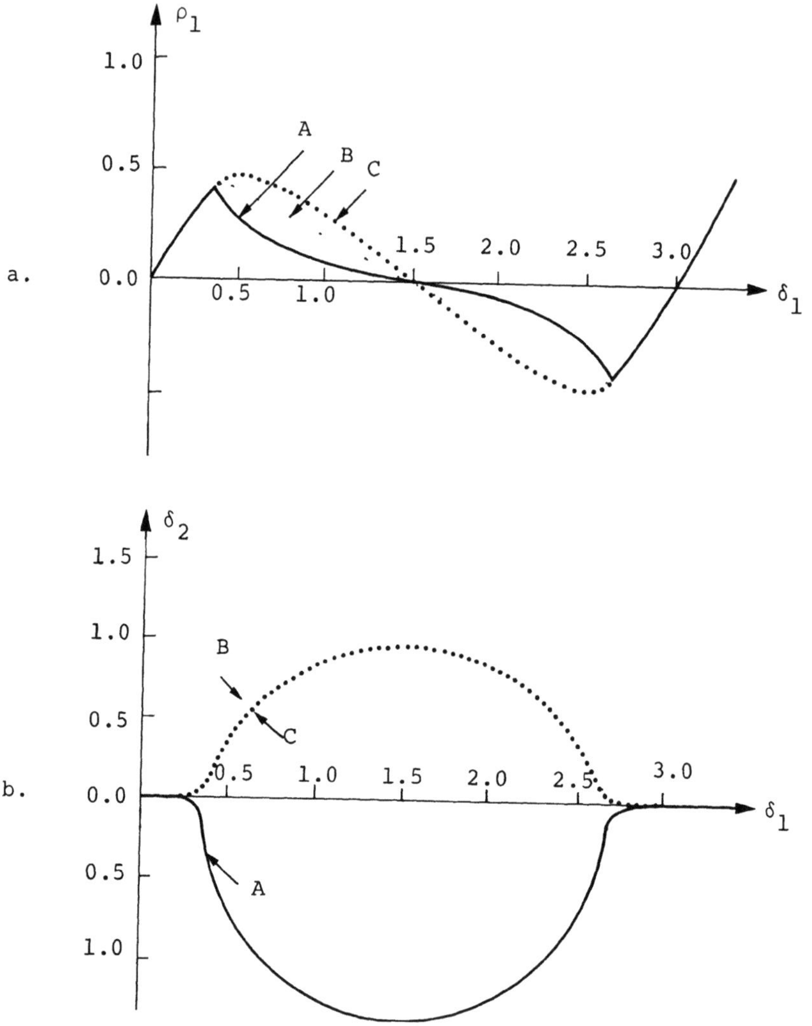

Fig. 9 - Study of model problem with nonlinear spring. H_o=1,5, v_o=0.0, m_o=0.4, ρ_2=0. Curve A: m_1=0.2, v_o=-0.01, Curve B: m_1=0.0, v_o=0.01, Curve C: m_1=0.2, v_o=0.01.

In this connection the cases of a small negative, and a positive geometrical imperfection were considered, see curves A and C in Fig. 9. For reference, also a constant spring coefficient was tested along with positive imperfection, see curve B. An interesting aspect here is that the lateral buckling follows the initially unstable assymetric branch in case of a negative imperfection while an imperfection in the opposite direction guides the solution into the initially stable, asymmetric branch. Curve B follows the lateral path for symmetric buckling. Also these results have been produced by the technique with load adaptation during iterations.

Results from some more test cases are described in Ref. /8/.

References

/1/ Oden, J.T.: Finite elements of nonlinear continua, McGraw-Hill (1972).

/2/ Tillerson, J.R.; Stricklin, J.A.; and Haisler, W.E.: Numerical methods for the solution of nonlinear problems in structural analysis, in Nonlinear structural problems, (editor Hartung, R.F.), AMD Vol. 6, (1973) 67-101.

/3/ Bergan, P.G.; Horrigmoe, G.; Kråkeland, B.; and Søreide, T.H.: Solution techniques for nonlinear finite element problems. Int. J. of Num. Meth. in Engng., Vol. 12 (1978) 1677-1696.

/4/ Bergan, P.G.: Automated incremental-iterative solution schemes. First Int. Conf. Numerical Methods for Nonlinear Problems, Swansea, U.K., Sept. 2-5, 1980.

/5/ Argyris, J.H.: Continua and discontinua, Proc. First Conf. Matrix Meth. in Struct. Mech., Wright-Patterson A.F.B. Ohio (1965).

/6/ Bergan, P.G.: Solution algorthms for nonlinear structural problems. Proc. Int. Conf. on Engineering Application of the Finite Element Method. Computas, Norway, (1979), also to be publ. in Computers and Structures, (1980).

/7/ Bergan, P.G. and Søreide, T.H.: Solution of Large Displacement and Instability Problems Using the Current Stiffness Parameter, Proc. Int. Conf. on Finite Elements in Nonlinear Solid and Structural Mechanics, Geilo, Norway, 1977. Also in Finite Elements in Nonlinear Mechanics, Vol. 2, Ed. Bergan P.G. et al., Tapir Publishers, Trondheim, (1978) 647-669.

/8/ Marum, S.E.: Testing of a New Method for Solution of Nonlinear Problems. (In Norwegian). Student Project. Division of Structural Mechanics, The Norwegian Institute of Technology, Trondheim, (1980).

Appendix - Equations for the model problem

The initial geometry is defined by

$$H_o = \frac{H}{L}, \quad v_o = \frac{\bar{v}_2}{L}$$

Spring coefficients

$$k_1 = \text{const.}, \quad k_2 = a+bv_2, \quad m_o = \frac{a}{k_1}, \quad m_1 = \frac{bL}{k_1}$$

Nondimensional forces and displacements

$$\rho_1 = \frac{P_1}{k_1 L}, \quad \rho_2 = \frac{P_2}{k_1 L}, \quad \delta_1 = \frac{V_1}{L}, \quad \delta_2 = \frac{V_2}{L}$$

Auxilliary quantities

$$f_o = 1 + H_o^2 + v_o^2 \qquad f_2 = \left(\frac{f_o}{f_1}\right)^{\frac{1}{2}} - 1$$

$$f_1 = 1 + (H_o - \delta_1)^2 + (v_o + \ _2)^2 \qquad f_3 = \frac{1+f_2}{f_1}$$

Nondimensional potential energy

$$\bar{\pi} = \frac{\pi}{k_1 L^2} = (\sqrt{f_o} - \sqrt{f_1})^2 + \tfrac{1}{2} m_o \delta_2^2 + \tfrac{1}{3} m_1 \delta_2^3 - \rho_1 \delta_1 - \rho_2 \delta_2$$

Equilibrium equations

$$\rho_1 = 2f_2(H_o - \delta_1)$$

$$\rho_2 = m_o \delta_2 + m_1 \delta_2^2 - 2f_2(v_o + \delta_2)$$

Incremental equilibrium equations

$$\begin{bmatrix} \Delta\rho_1 \\ \Delta\rho_2 \end{bmatrix} = \begin{bmatrix} 2f_3(H_o-\delta_1)-2f_2 & -2f_3(H_o-\delta_1)(v_o+\delta_2) \\ -2f_3(H_o-\delta_1)(v_o+\delta_2) & m_o+2m_1\delta_2+2f_3(v_o+\delta_2)^2-2f_2 \end{bmatrix} \begin{bmatrix} \Delta\delta_1 \\ \Delta\delta_2 \end{bmatrix}$$

Partitioned and Adaptive Algorithms for Explicit Time Integration

T. BELYTSCHKO
Northwestern University, Evanston, USA

Summary

Two computational algorithms for the explicit time integration of the equations of motion with different time steps within a mesh are presented. The first is a mesh partition, which has been generalized so that an arbitrary number of time steps can be used. The second method chooses the time step for each node and element, and this time step depends on the current material properties, so that the method is adaptive. Results are presented for some simple one dimensional nonlinear problems.

Introduction

In explicit time integration of transient problems, the time step is limited by numerical stability. This limit is inversely proportional to the maximum frequency of the mesh, which in turn, varies inversely with the traversal time for a wave across the element. Hence, if a mesh contains very short or stiff elements, the time step required for stability becomes very small.

In finite difference treatments of continua, this characteristic of explicit methods often imposes only small penalties, for finite difference grids are usually quite regular and only use a single type of "zone" or element. On the other hand, in finite element applications, large variations in element size are commonplace. Furthermore, different types of elements, such as beams, constant strain, and linear strain elements are often combined in a mesh, and the maximum frequency of these elements varies substantially. Thus the computational penalties incurred by single time step explicit integration of finite element meshes are often substantial.

Several remedies have been proposed for this difficulty. Belytschko and Mullen [1,2] have developed explicit-implicit methods where the mesh is partitioned into domains by nodes and the partitions are simultaneously integrated by explicit and implicit methods. They have shown that the maximum frequency in the explicit partition governs the time step, so that if the stiff portions are integrated implicitly, the time step can be much larger. Hughes and Liu [3] have proposed an explicit-implicit partition by elements with similar stability properties; the element basis of their partition simplifies the algorithm and enhances its compatibility with general purpose finite element software.

However, explicit-implicit methods suffer from the disadvantage that the addition of the implicit algorithms increases the size and complexity of the program. Moreover, in nonlinear applications, the reliability of implicit algorithms does not equal that of explicit algorithms.

For these reasons, in [4] and [5] partitioned algorithms were developed and examined in which explicit integration is used throughout the mesh but different time steps are used in the partitions. The algorithms used linear interpolations in time of the displacement at the interfaces between partitions, which in [5] were shown to yield stability for a model problem. Wright [6] has independently proposed a similar method.

In this paper, two further developments are reviewed. The first, for which a more detailed description is in preparation [7], is based on a constant velocity interpolation on interfaces. It increases the versatility of the partition immensely, and by using this approach, a single time step code has been modified to treat an arbitrary number of partitions within a mesh by a few FORTRAN statements. This algorithm is presented in the third section.

The second development, described in the fourth section, is a *generalization and* extension of this scheme, in which each node and element is associated with an independent "clock" and time step. The time steps are automatically computed and altered throughout the simulation. Thus the method is in many senses adaptive: as the elements change in sizes or change in stiffness because of the state of stress, the local time steps are changed so as to minimize computations. The scheme is not without flaws or shortcomings: it requires more storage for each element and node, and it does not equal the reliability of the partitioned technique. However, this is only a first, tentative step in this direction. This method may have significant potential, but its full realization will require further development and a better understanding of its stability characteristics and time step restrictions.

Governing Equations

We will here restrict our treatment to the semidiscretization of the momentum equations which will be written in the form

$$\underset{\sim}{M}\,\ddot{\underline{u}} = \underline{f} \qquad (1)$$

where

$$\underline{f} = \underline{f}^{ext} - \underline{f}^{int} \qquad (2)$$

Here a superposed dot denotes a time derivative, $\underset{\sim}{M}$ is the mass matrix, $\underline{u}$ a column matrix of nodal displacements, $\underline{f}^{ext}$ the external nodal forces and $\underline{f}^{int}$ the internal nodal forces. For a nonlinear problem, the internal nodal forces are obtained by

$$\underline{f}^{int} = \sum \underset{\sim}{L}_E^T \int_{V_E} \underset{\sim}{B}^T \underline{\sigma}\, dV \qquad (3)$$

where $\underset{\sim}{L}_E$ is the Boolean connectivity matrix, V_E the volume of element E, and $\underset{\sim}{B}$ the strain rate-displacement matrix which gives the element strain rates in terms of the nodal velocities by

$$\dot{\underline{\varepsilon}} = \underset{\sim}{B}\, \dot{\underline{u}}_E = \underset{\sim}{B}\, \underset{\sim}{L}_E\, \dot{\underline{u}} \tag{4}$$

For a linear, undamped structure

$$\underline{f}^{int} = \underset{\sim}{K}\, \underline{u} \tag{5}$$

so that Eq. 1 can be written as

$$\underset{\sim}{M}\, \ddot{\underline{u}} + \underset{\sim}{K}\, \underline{u} = \underline{f}^{ext} \tag{6}$$

The central difference formulas give

$$\dot{\underline{u}}^{n+\frac{1}{2}} = \dot{\underline{u}}^{n-\frac{1}{2}} + \Delta t^{n}\, \ddot{\underline{u}}^{n} \tag{7}$$

$$\underline{u}^{n+1} = \underline{u}^{n} + \Delta t^{n+\frac{1}{2}}\, \dot{\underline{u}}^{n+\frac{1}{2}} \tag{8}$$

$$\Delta t^{n} = \frac{1}{2}\left(\Delta t^{n-\frac{1}{2}} + \Delta t^{n+\frac{1}{2}}\right) \tag{9}$$

where superscripts designate the time step and fractional superscripts designate a midstep value.

Combining Eq. (1) with Eq.(7), we obtain

$$\dot{\underline{u}}^{n+\frac{1}{2}} = \dot{\underline{u}}^{n-\frac{1}{2}} + \Delta t^{n}\, \underset{\sim}{M}^{-1}\, \underline{f}^{n} \tag{10}$$

For the purpose of computing the new internal forces, we also need to update the stresses by

$$\underline{\sigma}^{n+1} = \underline{\sigma}^{n} + \boldsymbol{\sigma}'\left(\dot{\underline{\varepsilon}}^{n+\frac{1}{2}},\ \Delta t^{n+\frac{1}{2}}\right) \tag{11}$$

where $\boldsymbol{\sigma}'$ reflects any frame invariant stress rate that may be needed. Equations (8-10), in conjunction with (3), (4) and (11), provide the explicit algorithm.

The stability limit for the central difference method with a constant time step is given by

$$\Delta t \leq \frac{2}{\omega_{max}} \tag{12}$$

where ω_{max} is the maximum frequency of the mesh, which corresponds to the maximum eigenvalue of

$$\underset{\sim}{K}\,\underline{u} = \omega^2\,\underset{\sim}{M}\,\underline{u} \tag{13}$$

For a constant strain element, an upper bound on the maximum frequency is given by

$$\omega_{max} = \frac{2c}{\ell} \tag{14}$$

where c is the wavespeed and ℓ minimum length across an element. Hence

$$\Delta t \leq \frac{\ell}{c} \tag{15}$$

which is the smallest wave traversal time across an element in the mesh.

This stability condition has been proven only for linear problems; see, for example Fujii[8]. For nonlinear systems, no strict stability conditions are available but using a time step of 70% to 90% usually leads to stable computations. Stability in time integration of nonlinear systems should be checked post facto by an energy balance.

For this purpose, the following formulas can be used

$$W^{n+1,ext} = W^{n,ext} + \frac{1}{2}\,\Delta t(\dot{\underline{u}}^{n+\frac{1}{2}})^T(\underline{f}^{n,ext} + \underline{f}^{n+1,ext}) \tag{16}$$

$$W_E^{n+1,int} = W_E^{n,int} + \frac{1}{2}\,\Delta t(\dot{\underline{u}}_E^{n+\frac{1}{2}})(\underline{f}_E^{n,int}+\underline{f}_E^{n+1,int}) \tag{17a}$$

$$= W_E^{n,int} + \frac{1}{2} \Delta t \int_{V_E} \underline{\dot{\varepsilon}}^{n+\frac{1}{2}} (\underline{\sigma}^n + \underline{\sigma}^{n+1}) \, dV \qquad (17b)$$

$$W^{int} = \sum_E W_E^{int} \qquad (17c)$$

$$T^{n+\frac{1}{2}} = \frac{1}{2} (\underline{\dot{u}}^{n+\frac{1}{2}})^T \underset{\sim}{M} \, \underline{\dot{u}}^{n+\frac{1}{2}} \qquad (18)$$

where W^{int} and W^{ext} are the internal energy and external work, and T is the kinetic energy. Energy balance then requires that

$$W^{n,ext} + W^{n,int} + \frac{1}{2} (T^{n-\frac{1}{2}} + T^{n+\frac{1}{2}}) < TOL \, (|W^{n,ext}| + |W^{n,int}|) \qquad (19)$$

where TOL is usually less than 0.05.

If the time step varies during the computation, Eq.(12) does not hold. In a specialized analysis given in [9], where the ratio of successive time steps and $\omega\Delta t$ was considered constant, it was shown that varying the time step introduces an artificial viscosity which is positive when the time step decreases, negative when the time step increases. The latter is never stable according to a Neumann analysis. The analysis given in [9] implied a constantly increasing time step; numerical experiments have shown that limited increases in time step over restricted time intervals are not destabilizing. However, increasing time steps should be treated with care until this aspect is understood better.

Partitioned Algorithm

In this scheme we subdivide the mesh into NG groups of elements. The element group G is updated with a time step Δt_G

and the group time is denoted by t_G. In addition, we use a master time t_M which is incremented by a time step Δt_M. It is required that all Δt_G be integer multiples of Δt_M, and that for adjacent groups, the time steps be integer multiples of each other.

Only the element alignment within the groups needs to be specified by the data. The nodal time steps are automatically generated within the algorithm. For this purpose, each node is allocated two words of storage, the nodal time of node N, t_N, and its last time step value Δt_N.

The essence of the procedure is as follows. Whenever the master clock t_M is incremented, all element groups are checked. Any group which has fallen behind the master time is updated with its own time step Δt_G and the stresses and the clock time t_G are updated for this group. At the same time, any nodes connected to element group G are assigned the time step Δt_G if it is larger than the previously assigned time step. Therefore, a node at an interface will always be updated with the largest time step among the element groups connected to it.

After the element groups are completed, the nodes are checked. Any nodes for which the clock is not ahead of the master time are updated.

A flow chart of the scheme is given in Table 1. Other than the data for inputting element groups, only about 15 FORTRAN statements are required to implement this scheme in a single time step scheme.

The scheme is most appropriate for elements with a velocity strain formulation. Otherwise, the displacements at the element group time must be obtained by linear interpolate of the displacement prior to computing the strains.

Table 1

Partitioned Explicit Algorithm

0. initial conditions: $\underline{u}^o$, $\dot{\underline{u}}^{-\frac{1}{2}}$, $t_N=0$ for all nodes; $t_G=0$

for all groups; $t_M=0$

1. $t_M \leftarrow t_M + \Delta t_M$
2. compute $\underline{f}^{ext}$, set $\underline{f}^{int} = 0$, $\Delta t_N = 0$
3. loop over element groups: G = 1 to NG
4. if $t_G > t_M$, go to 7
5. update stresses in all elements in group G using time step Δt_G using nodal velocities $\dot{\underline{u}}$ and Eqs.(4) and (11); compute element internal forces and add into $\underline{f}^{int}$; increment group time $t_G \leftarrow t_G + \Delta t_G$
6. for all nodes connected to group G, if $\Delta t_G > \Delta t_N$ let $\Delta t_N = \Delta t_G$ (i.e. pick time step to be the largest of any element group connected to it)
7. end of loop over elements
8. loop over nodes: N + 1 to NNODE
9. if $t_N > t_M$, go to 11
10. update velocities and displacements of node N using Δt_N, $t_N \leftarrow t_N + \Delta t_N$
11. end of loop over nodes
12. go to 1

Adaptive Algorithm

The second algorithm to be discussed here is intended to take advantage of changes in local stable time steps that occur as the solution evolves. An example is the elastic-plastic response of structures, where the stable time step increases substantially whenever the material becomes plastic and continues to be loaded. The previous scheme is not well suited to taking advantage of such situations because the element groups are identified prior to the run as element data and because of the restrictions on the ratios of the time steps between partitions.

In this algorithm, each element is advanced by its own time step Δt_E, and each node is advanced by its own time step Δt_N. The nodal time step is the minimum of the element time steps of all elements connected to the node.

The basic feature of this scheme is that both the internal forces and the time derivatives of the internal forces, $\dot{\underline{f}}^{int}$, are stored for all nodes. Whenever a node is to be updated, we first update $\dot{\underline{f}}^{int}$ by

$$\underline{f}^{n+1,int} = \underline{f}^{n,int} + \Delta t^{n+\frac{1}{2}} \, \dot{\underline{f}}^{int} \tag{20}$$

where $\dot{\underline{f}}^{int}$ is the latest value of the time derivative.

Elements require the storage of the stresses at the previous to time steps. Whenever an element is updated, its new stresses are computed and the change in $\Delta\dot{\underline{f}}_E^{int}$ is computed by

$$\Delta\dot{\underline{f}}_E^{int} = \int_{V_E} \underset{\sim}{B}^T \left[\frac{1}{\Delta t_E^{n+\frac{1}{2}}} (\underline{\sigma}^{n+1} - \underline{\sigma}^n) + \frac{1}{\Delta t_E^{n-\frac{1}{2}}} (\underline{\sigma}^n - \underline{\sigma}^{n-1}) \right] dV \tag{21}$$

The purpose of the second term in the above equation is to eliminate the element's old contribution to $\dot{\underline{f}}^{int}$. The contribution of the element is then added to the nodal forces by

$$\dot{\underline{f}}^{int} \leftarrow \dot{\underline{f}}^{int} + \underset{\sim}{L}_E^T \, \Delta\dot{\underline{f}}_E^{int} \tag{22}$$

To drive the algorithm, a master time, t_M is needed. This master time is incremented by Δt_M, which is always the smallest increment Δt_E and Δt_N among all elements and nodes, respectively.

A flowchart of the algorithm is given in Table 2. Note that the time step for each element is recomputed continuously; however, in order to reduce truncation error and the destabilizing effects of increasing time steps, limits are set on the maximum increase.

The threshold in step 2 is essential to the method. If the element time lags the node time by only a small amount, failure to update the element can result in the development of rather large errors.

This adaptive scheme requires substantially more storage than a single time-step, explicit scheme and is also more susceptible to roundoff errors. However, on a computer with 12 digits, this roundoff error was not found to be significant over several thousand time steps in tests that have been made.

Table 2

Adaptive Explicit Algorithm

0. initial conditions: $\underline{u}^o$, $\dot{\underline{u}}^{-\frac{1}{2}}$, $\dot{\underline{f}}^{int} = 0$
   $t_E = t_N = 0$ for all elements and nodes, $t_M = 0$, $\Delta t_M = \infty$

1. loop over elements: E = 1 to NELE

2. if $t_E > t_M$ - threshold, go to 7

3. compute new element time step Δt_E while computing new stresses by Eqs.(4) and (11)

4. modify total nodal force derivative matrix by Eqs.(21) and (22)

5. if $\Delta t_M > \Delta t_E$, set $\Delta t_M = \Delta t_E$

6. update element time: $t_E \leftarrow t_E + \Delta t_E$; modify nodal time steps for nodes connected to element E

7. end of loop over elements

8. loop over nodes: N = 1 to NNODE

9. if $t_N > t_M$, go to 14

10. update $\underline{f}^{int}$ for node N by Eq.(20)

11. compute $\underline{f}^{ext}$ for node N

12. update nodal velocities and displacements for node N by Eqs.(7) and (8).

13. update nodal time $t_N \leftarrow t_N + \Delta t_N$

14. end of loop over nodes

15. $t_M \leftarrow t_M + \Delta t_M$

16. go to 1

The time step for adjacent nodes and elements cannot be completely arbitrary. One important restriction is that the element time step must be smaller than the nodal time steps of its nodes. This restriction detracts from the method substantially, for it would be desirable to be able to increment the element at the largest possible time step, since updating an element is far more costly than updating a node.

Example

An example to which the two methods were applied is shown in Figure 1. It is a bar in a uniaxial state of stress. A solution by an explicit-explicit partition with linear displacement interpolation was previously reported in [5].

A compressive stress of magnitude 0.1 was applied at the left-hand end as a step function in time. Figure 2 shows the stress in the fifth element from the left (x=4.5) obtained by the partioned technique described here. The results, as those in [5], show the usual dispersion and aliasing, which are exacerbated by the large discontinuity in element lengths. Figure 3 shows the stress in the same element for the adaptive solution, which is quite similar in character. In this problem, the adaptive method provides no advantage, since the stability limits do not change with time. Both computations are clearly stable, which was confirmed by an energy balance check.

The adaptive method was also tried for other problems of the same type. At times the solution was observed to drift from the correct solution. It was not a violent Neumann type of instability, although energy balance was violated. These appear to be due to certain combinations of time steps in adjacent nodes and elements and the choice of the element threshold (step 2 in Table 2); adequate restrictions and guidelines remain to be developed.

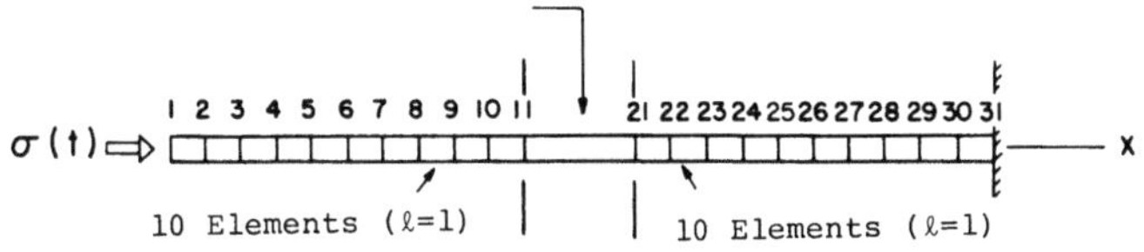

Figure 1. Mesh for sample problem; wavespeed c = 1.0

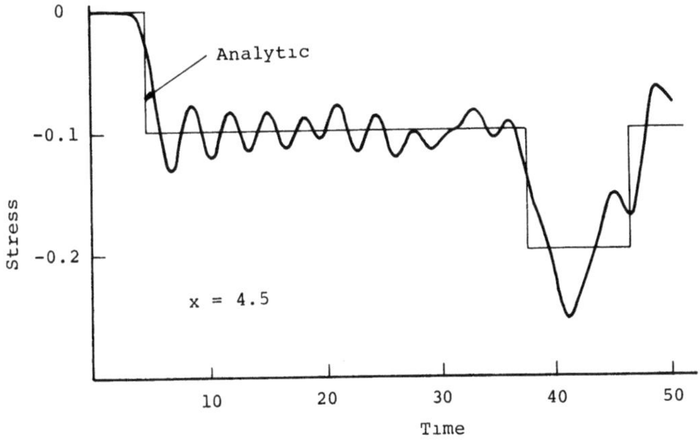

Figure 2. Comparison of solution by partitioned algorithm with analytic solution.

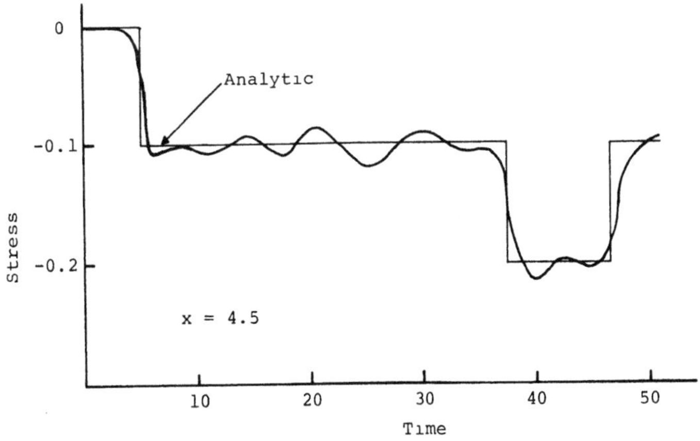

Figure 3. Comparison of solution by adaptive algorithm with analytic solution

Acknowledgement

We would gratefully like to ackowledge the support of the Electric Power Research Corporation and the National Science Foundation.

References

1. Belytschko, T.B. and Mullen, R.: Mesh Partitions of Explicit-Implicit Time Integration. Formulations and Computational Algorithms in Finite Element Analysis, edited by J. Bathe, et. al., M.I.T. Press, (1977) 673-690.

2. Belytschko, T. and Mullen, R.: Stability of Explicit-Implicit Mesh Partitions in Time Integration. International Journal for Numerical Methods in Engineering, Volume 12, (1978) 1575-1586.

3. Hughes, T.J.R. and Liu, W.K.: Implicit-Explicit Finite Elements in Transient Analysis: Stability Theory. ASME Journal of Applied Mechanics, Volume 45, (1978) 371-374.

4. Belytschko T. and Mullen, R.: "WHAMS: A Program for Transient Analysis of Structures and Continua. Structural Mechanics Software Series, Volume 2, edited by N. Perrone, et. al., (1978) 151-212.

5. Belytschko, T.; Yen, J. and Mullen, R.: Mixed Methods for Time Integration. Computer Methods and Applied Mechanics in Engineering, Volume 17/18, (1979) 259-275.

6. Wright, J.P.: Mixed Time Integration Schemes. Computers and Structures, Volume 10, (1979) 235-238.

7. Belytschko, T. and Robinson, R.: An Algorithm for Mixed Time Step Integration. Submitted for publication.

8. Fujii, F.: Finite Element Schemes: Stability and Convergence. Advances in Computational Methods in Structural Mechanics and Design. Edited by J.T. Oden, et. al., University of Alabama Press, Alabama (1972) 201-218.

9. Belytschko, T. and Mullen, R.: Explicit Integration of Structural Problems. Finite Elements in Nonlinear Mechanics. Edited by P. Bergan, et. al., Volume 2, (1977) 697-720.

Dynamic Relaxation Applied to the Quasi-Static, Large Deformation, Inelastic Response of Axisymmetric Solids *

S. W. KEY, C. M. STONE, R D KRIEG
Sandia Laboratories**, Albuquerque, NM, USA

ABSTRACT

The use of dynamic relaxation as a solution strategy for the quasi-static, large deformation, inelastic response of solids is examined. The underlying mechanics, the constitutive theories of interest, the incremental form of the equations, the spatial discretization, and the implementation of dynamic relaxation for path and/or time dependent material response are each discussed. The mechanics are carried out in the current configuration of the body described by a fixed spatial coordinate system and using the Cauchy stress. Finite strain constitutive theories for elastic, elastoplastic, and creep behavior are introduced. An incremental form of the problem allowing a sequence of equilibrium solutions to be found is presented. A constant bulk strain, bilinear displacement isoparametric quadrilateral finite element is employed for the spatial discretization. The solution strategy used to generate the sequence of equilibrium solutions is dynamic relaxation which in the form adopted is based on explicit central difference pseudo-time integration and artificial damping. It is used to find the next solution as a result of an increment in time and/or load. Each solution must satisfy equilibrium to within a prescribed tolerance before proceeding to the next increment. Several example calculations are presented.

*This article sponsored by the U. S. Department of Energy under Contract DE-AC04-76-DP00789.

**A U. S. Department of Energy Facility.

1. INTRODUCTION

The use of dynamic relaxation as a solution strategy for the quasi-static, large deformation, inelastic response of solids is examined. Dynamic relaxation as a means for generating solutions to statics problems goes back at least to Otter, Cassell and Hobbs [1]. The basic idea is to convert a static problem into a dynamic problem by adding acceleration and damping terms to the equations and then with "optimal" damping integrate the transient response out until the quiescent or static solution is obtained. The idea is eminently adaptable to the problem of large deformations and inelastic materials. The result is an explicit, iterative strategy which has a large radius of convergence as noted by Underwood [2]. A considerable literature concerned with the transient dynamic response of solids and structures can be brought to bear on the problem [3-11].

The purpose here is to present an analysis procedure where large deformations, finite strains, inelastic time independent and dependent constitutive modeling, and arbitrary load incrementation are all combined for problem solving.

Of particular significance is the disassociation of the incremental constitutive evaluation scheme from the load incrementation time step and related equilibrium iteration procedure. The approach is to integrate the constitutive equaticns accurately with subcycling if necessary, independent of the time step selected for load incrementation. This is an approach urged by Bushnell [12] and a practice of long standing in transient dynamic structural response calculations [13].

2. MECHANICS

While the purpose is to consider axisymmetric and planar problems, the treatment in this section is general for the sake of brevity and completeness. The treatment of continuum mechanics found in Truesdell and Toupin [14] is used.

Equations of Equilibrium. A body $\mathcal{V}$ is given which occupies a finite region of Euclidean space. Subjected to prescribed body forces and surface tractions, the body $\mathcal{V}$ undergoes the motion $x^i \equiv \chi^i(X^\alpha, t)$. The particles of the body are identified by the coordinates X^α. They are referred to as material coordinates, and the relation of the particles to the coordinates X^α does not change in time. The places in space which the particles occupy during the motion are identified by the coordinates x^i. The function χ^i describes the motion of the particles X^α through space as a function of time t. It is the motion χ^i which is sought.

The place occupied by the body at $t = 0$ is taken as the reference configuration. In this configuration the body is assumed to be strain free, though not necessarily stress free. Only material coordinates X^α which coincide with the spatial coordinates x^i in the reference configuration are considered. Thus, in the reference configuration, $\chi^i(X^\alpha, 0) \equiv X^\alpha$.

The problem is stated in terms of the principle of virtual work. The differential form

$$\delta\pi = \int_{\mathcal{V}} t^{km}\delta x_{k,m}\, dv - \int_{\mathcal{V}} \rho f^k \delta x_k\, dv - \oint_{\mathcal{S}^1} s^k \delta x_k\, da \qquad (1)$$

is to vanish at all points along the path of motion for all variations δx_k satisfying the displacement boundary conditions on $\mathcal{S}^2$. The integration is performed over the current configuration of the body $\mathcal{V}$, where ρ is the mass density in that configuration, t^{km} is the Cauchy stress--the stress in the current configuration, and s^k is the surface traction which is acting on $\mathscr{S}^1$. The comma in $x_{k,m}$ denotes covariant differentiation.

The divergence theorem is employed to display the equilibrium equations. In anticipation of using the finite element method to generate approximate solutions, the case where $\delta x_{k,m}$ is only piecewise continuous is considered. Interior surfaces where the discontinuities of $\delta x_{k,m}$ occur are denoted by $\mathscr{S}^o$. Only surfaces $\mathscr{S}^o$ which are stationary with respect to the material are considered. The situation is pictured in Figure 1 where n_k is the normal to $\mathscr{S}^o$ and the symbols + and - denote the respective sides of the surface. The result is

$$- \int_{\mathscr{V}} (t^{km}{}_{,m} + \rho f^k)\delta x_k \, dv + \oint_{\mathscr{S}^o} (t_+^{km} - t_-^{km}) n_m \delta x_k \, da$$

$$+ \oint_{\mathscr{S}^1} (t^{km} n_m - s^k)\delta x_k \, da = 0 \quad . \tag{2}$$

The differential form will vanish if and only if the respective integrands vanish. The resulting expressions are equilibrium

$$t^{km}{}_{,m} + \rho f^k = 0 \quad \text{in } \mathscr{V} \quad , \tag{3}$$

the jump condition at a contact discontinuity

$$(t_+^{km} - t_-^{km}) n_m = 0 \quad \text{on } \mathscr{S}^o \quad , \tag{4}$$

and the traction boundary conditions

$$t^{km} n_m = s^k(t) \quad \text{on } \mathscr{S}^1 \quad . \tag{5}$$

The displacement boundary conditions are

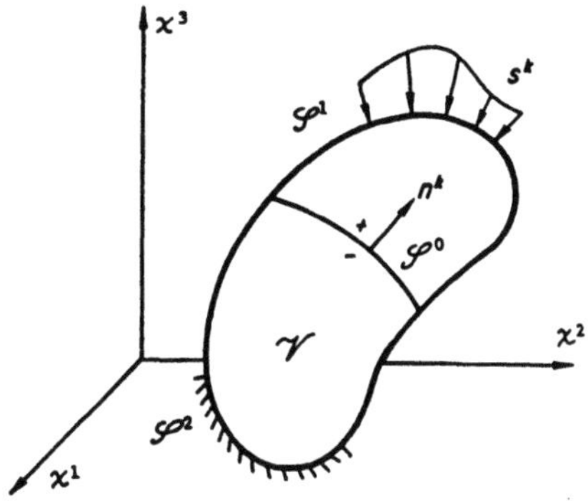

Fig 1 The body $\mathcal{V}$ with surface tractions s^k on the boundary $\mathcal{S}^1$ and a prescribed motion on the boundary $\mathcal{S}^2$ An interior boundary $\mathcal{S}^0$ with a unit normal vector n^k is pictured

$$\chi^i(X^\alpha, t) = \kappa^i(t) \quad \text{on } \mathcal{S}^2 \quad . \tag{6}$$

It is important to realize that these equations are completely general and applicable for arbitrarily large deformations.

Strain and Strain Rate. In finite deformations there are many strain measures which are useful. The majority of them can be computed from the deformation gradient F^k_α defined by

$$F^k_\alpha \equiv \frac{\partial \chi^k}{\partial X^\alpha}(X^\beta, t) \quad . \tag{7}$$

The left Cauchy-Green tensor is computed from the deformation gradient as

$$B^{km} = F^k_\alpha F^m_\beta G^{\alpha\beta} \quad . \tag{8}$$

The quantity $G^{\alpha\beta}$ is the metric tensor in the coordinates of the reference configuration. The velocity v^k is defined as

$$v^k(X^\alpha,t) \equiv \frac{\partial \chi^k}{\partial t}(X^\alpha,t) \quad . \tag{9}$$

The stretching is given by

$$d_{km} = \frac{1}{2}(v_{k,m} + v_{m,k}) \quad , \tag{10}$$

with the spin by

$$w_{km} = \frac{1}{2}(v_{k,m} - v_{m,k}) \quad , \tag{11}$$

In the constitutive models which follow, only the left Cauchy-Green tensor B^{km} and the stretching d_{km} are used as strain and strain rate measures, respectively. There are a wealth of additional strain and strain rate variables which could be introduced but they are not needed here. Further reference to them can be found in Truesdell and Toupin [14].

3. CONSTITUTIVE THEORIES

The stress relations considered are those of elementary character. They represent three classical forms of material behavior and are remarkably adaptable in performing useful engineering analyses. They are rubber elasticity which exhibits path independent and time independent behavior, classical plasticity which exhibits path dependent but time independent behavior, and classical creep which has both path dependent and time dependent behavior. These attributes have important consequences when incremental strategies are considered.

Rubber Elasticity. Rubber is a material used because of its ability to remain elastic when subjected to large deformations. Blatz and Ko [15] have made an extensive study of its stress-strain behavior when subjected to large strains. They consider the behavior of both foam and continuum rubber. Here only the continuum rubber is treated. The left Cauchy-Green strain (8) is used. Using the third invariant of the strain which is given by

$$J_3 = |\det(F^k_\alpha)| = \frac{\rho_R}{\rho} \tag{12}$$

the cauchy stress is computed as

$$t^{rs} = \mu\left(J_3^{-1}B^{rs} - J_3^{-\frac{1}{1-2\nu}} g^{rs}\right) . \tag{13}$$

Here, g^{rs} is metric tensor in the spatial coordinates, μ is the shear modulus and ν is their generalization of Poisson's ratio to finite strains. Based on data sited by them, ν is taken equal to 0.463. This gives J_3 an exponent of 13.3 and a marked hardening behavior under compression. This description for bulk compression matches the data to 700,000 pounds per square inch. This is a hyper-elastic relation.

Plasticity. Plasticity is the behavior characteristic of ductile metals. Figure 2 shows results which are typical of the behavior of a metal bar loaded first in uniaxial tension followed by uniaxial compression. The straight line representation in Figure 2 is an idealization of this behavior. This is the approximation which results from the plasticity relations employed here, Goel and Malvern [16]. It is based on the notion of a universal hardening curve from which general triaxial behavior is predicted. The finite strain treatment of Key, Krieg, and Biffle [17] is used. It is stated in a rate form where the invariant co-rotational or Jaumann stress rate is related through a tangent modulus to the stretching. The result is equation (14).

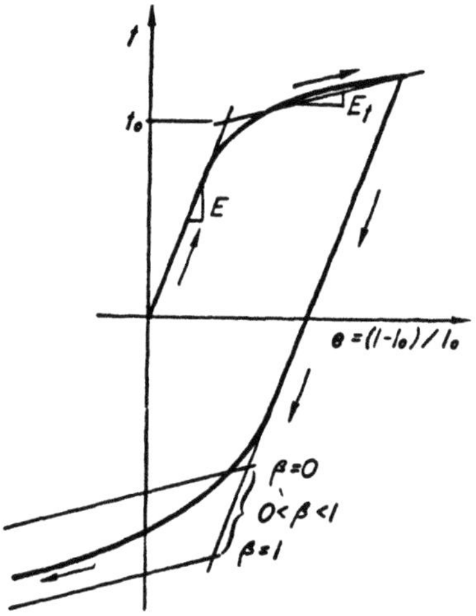

Fig. 2. The typical behavior of a ductile metal bar loaded first in uniaxial tension followed by uniaxial compression. The straight line approximation is characterized as an elastic modulus E, a yield stress t_o, a strain hardening modulus E_t, and a hardening parameter β, where kinematic hardening is obtained with $\beta = 0$, isotropic hardening is obtained with $\beta = 1$, and a linear combination of the two is obtained for β between zero and one.

$$\overset{\nabla}{t}{}^{rs} = \dot{t}^{rs} - w^r{}_m t^{ms} + t^{rm} w_m{}^s = C^{rsmn} d_{mn} \quad . \tag{14}$$

When no yielding is occurring, C^{rsmn} is the isotropic tensor $\lambda g^{rs} g^{mn} + 2\mu g^{rm} g^{sn}$, where λ and μ are the Lamé parameters. When plastic yielding is occurring, that is, when $\frac{1}{2}\xi'_{ij}\xi'^{ij} - k^2 = 0$ and $\xi'_{rs}\overset{\nabla}{\xi}{}'^{rs} > 0$, the moduli are replaced by

$$C^{rsmn} = \lambda g^{rs} g^{mn} + 2\mu (g^{rm} g^{sn} - n^{rs} n^{mn}) \tag{15}$$

where

$$n^{rs} = \xi'^{rs}/[2k^2(1 + H/3\mu)]^{1/2} \quad ,$$

$$\xi'_{rs} = t'_{rs} - \alpha'_{rs} \quad ,$$

$$\sqrt{2}\dot{k} = \beta \frac{2}{3} H|d^p_{rs}| \quad , \quad k(0) = (2/3)^{1/2} t_o \quad ,$$

$$\overset{\nabla}{\alpha}_{rs} = (1 - \beta) \frac{2}{3} H d^p_{rs} \quad , \quad \alpha_{rs}(0) = 0 \quad .$$

Isotropic hardening is described by $\beta = 1$. Kinematic hardening is included in a rather obvious way by letting $\beta = 0$ and prescribing the center of the yield surface α_{ij} to move according to $\overset{\nabla}{\alpha}_{ij} = (1 - \beta) \frac{2}{3} H d^p_{rs}$. The prime denotes deviatoric components, the superscript p denotes the plastic part of the stretching, and $H = EE_t/(E - E_t)$.

Creep. Creep occurs in man-made and natural materials, and in ductile materials becomes significant when the temperature on an absolute scale reaches one third to one half the material's melt temperature. It is a time dependent material mechanism in that the stress depends principally on the rate at which the material is deformed. Elastic terms are also necessary in this temperature range and are included here. In what follows the common Norton power law for secondary creep will be used, Penny and Marriott [18]. It is particularly descriptive of dislocation glide and climb deformation mechanisms in metals, Gittus [19].

It is stated in a rate form where the invariant co-rotational stress rate is related linearly to the difference between the total stretching minus the inelastic stretching d^c_{rs}. Thus

$$\overset{\nabla}{t}^{rs} = C^{rsmn}(d_{mn} - d^c_{mn}) \tag{16}$$

where

$$d^c_{mn} = \frac{3}{2} D\left(\frac{3}{2} t'_{ij} t'^{ij}\right)^{\frac{n-1}{2}} \exp(-Q/R\theta)\, t'_{mn}$$

D and n are constants determined from the material, θ is temperature, Q is the effective activation energy, and R is the universal gas constant.

4. INCREMENTAL FORM OF THE PROBLEM

The notion of an incremental solution is fundamental to the bulk of the methods for finding a motion $\chi^i(X^\alpha, t)$ which generates a stress history in equilibrium with the applied loads. It is assumed that at time t_n, the stress t^n_{rs} satisfies the equilibrium equations (3), (4), and (5); and the stress t^n_{rs} is the result of integrating the constitutive models with the strain histories derived from the known motion up to t_n. The prescribed loads are incremented to time t_{n+1} and a predictor/corrector method is introduced to find the new configuration $x^i_{n+1} = \chi^i(X^\alpha, t_{n+1})$ which has all the equilibrium properties which were deemed necessary at t_n and accuracies acceptable to the constitutive model evaluation. To indicate the incremental quantities a Δ is used. For example, $\Delta t = t_{n+1} - t_n$, $\Delta t_{rs} = t_{rs}^{n+1} - t_{rs}^n$, etc.

A basic assumption which underlies most incremental treatments, including the one here, but which is rarely stated is that the motion between x_n^i and x_{n+1}^i is linear. As a consequence, the incremental velocity given by $v_{n+1/2}^i = \Delta x^i/\Delta t$ is constant over the time increment. In what follows equilibrium is tested in the configuration at t_{n+1} which is a trial configuration until equilibrium is established. To do this the stresses at t_{n+1} must be evaluated. Following Hughes and Winget [20], the one parameter family of configurations is introduced

$$x_{n+\alpha}^i = (1 - \alpha)x_n^i + \alpha x_{n+1}^i \quad . \tag{17}$$

The gradient h_{ij} of $u_i = \Delta x^i$ with respect to $x_{n+\alpha}^j$ is given by

$$h_{ij} = u_{i,j} \tag{18}$$

From this gradient the strain increment e_{ij} is given by

$$e_{ij}(\alpha) = \frac{1}{2}\left[h_{ij} + h_{ji} + (1 - 2\alpha)h_{ki}h_{kj}\right] \tag{19}$$

Thus, $e_{ij}(0)$ is the Green-St. Venant strain increment, $e_{ij}(1)$ is the Signorini strain increment, and

$$e_{ij}\left(\frac{1}{2}\right) = \Delta t \; d_{ij}^{n+1/2} = \text{sym}\left(\frac{\partial u_i}{\partial x_{n+1/2}^j}\right) \quad . \tag{20}$$

Without the need for further linearization the configuration halfway between n and n+1 is selected for evaluating the stretching, spin, and computing Δt_{rs}. The midpoint configuration is optimal in the sense that no quadratic terms are needed to accurately evaluate $(dx^i dx^i)_{n+1} - (dx^i dx^i)_n$. The terms in the co-rotational derivative involving the spin w^{rk} used in the constitutive equations (14) and (16) are for the purposes of taking rigid body rotations of a material point relative to the spatial coordinates x^i into account. In incremental form they are an orthogonal rotation through an incremental angle. Hughes and Winget [20] have provided a modern account of this process and have provided a direct way to evaluate the orthogonal rotation matrix R_{ij} from the spin w_{ij}. Thus,

$$R_{ij} = (\delta^i_k - \Delta t \, \frac{1}{2} \, w^i{}_k)^{-1} (g_{kj} + \Delta t \, \frac{1}{2} \, w_{kj}) \quad . \tag{21}$$

Half angle trigonometric formulas are used to get the square root of [R]; $R_{ij} = R^{\frac{1}{2}k}_i \, R^{\frac{1}{2}}_{kj}$. With these constructions the constitutive models (14) and (16) can be integrated over the increment from n to n+1. First the stress t^n_{rs} and the applicable state variables α^n_{rs} are advanced to n + ½ by

$$\bar{t}^{n+\frac{1}{2}}_{rs} = R^{\frac{1}{2}i}_r \, R^{\frac{1}{2}j}_s \, t^n_{ij} \tag{22}$$

$$\bar{\alpha}^{n+\frac{1}{2}}_{rs} = R^{\frac{1}{2}i}_r \, R^{\frac{1}{2}j}_s \, \alpha^n_{ij} \tag{23}$$

Using $d^{n+1/2}_{rs}$ and Δt, the constitutive equations are integrated and new stresses $t^{n+1/2}_{rs}$ and state variables $\alpha^{n+1/2}_{rs}$ are obtained. These are then rotated from n + ½ to n + 1 by the same process as in equations (22) and (23). Mid-interval constitutive evaluation is used by Hallquist [21] and Biffle [22].

Of particular importance is the integration of the constitutive equations from time n to n + 1. The rubber elasticity is straightforward in that a new left Cauchy-Green strain is computed everytime from which the Cauchy stress is computed from (13). Since this is a hyperelastic model, it is independent of the path followed to get to the configuration at n + 1. The plasticity model is integrated from n to n + 1 assuming the stretching is constant which is consistent with (17). For a constant stretching path the integration is very nearly exact and is extremely reliable, and independent of the specific value for the time interval Δt, [23-25].

The creep model is also integrated from n to n + 1 assuming the stretching is constant. The creep equations, however, are not nearly as easy to numerically integrate. They are referred to in the mathematics community as "stiff". Only with considerable effort and great expense can they be numerically integrated with conventional methods from time n to n + 1. In this case, a semianalytic integration is used. Domains in stress and strain rate space are identified where various nearby differential equations with exact solutions are applicable. The solution path over a time step may remain within a single domain or may pass through two or more domains requiring the solutions to two or more of the differential equations to be applied, one after the other over the time step. In this way arbitrarily large strain or time intervals can be accurately and reliably used. An absolute maximum of seven subincrements are required so that computational time is not excessive. This approach, while conceptually straightforward is highly tailored to the constitutive model at hand.

This approach to constitutive equations uncouples stability and accuracy in their evaluation from the time step size used in load incrementing schemes with their attendant equilibrium iteration, cf., Bushnell [12]. This approach contrasts with that where Δt in the constitutive integration scheme corresponds to the Δt used in incrementing the load.

If this coupling is pursued, many numerical schemes can be developed. Argyris, Vaz and William [26] have documented a number of these and have provided the guidance necessary for their successful use.

While seemingly elaborate, these are in practice simple and quick calculations. This treatment has the property of being objective and uses a strain increment which while appearing "linear" is the same order of accuracy as using the nonlinear strain increments at either n or n + 1, respectively. It remains only to introduce the spatial discretization in order to discuss the dynamic relaxation solution strategy.

5. SPATIAL DISCRETIZATION

The treatment up to this point has been in terms of an arbitrary body undergoing an arbitrary motion. Of interest here are the axisymmetric and plane strain problems. Because the planar case may be obtained from the axisymmetric case by moving the symmetry axis arbitrarily far off, only the axisymmetric problem is discussed. Both of these problems are spatially two-dimensional. No attempt is made to do the axisymmetric problem in general but rather a specific element with specific interpolation functions is introduced from the outset.

If cylindrical coordinates, r, θ, z are introduced, an axisymmetric body can be characterized by a cross-section in the r-z plane which when revolved about the z-axis generates the body. Such a cross-section with a quadrilateral element is pictured in Fig. 3 (such an element is frequently called a "ring element" because when it is revolved about the z-axis a ring is generated). As indicated in Fig. 3, the element is a four node isoparametric element [27]. Thus, if y^{α} represents the value of the function y at the node with index

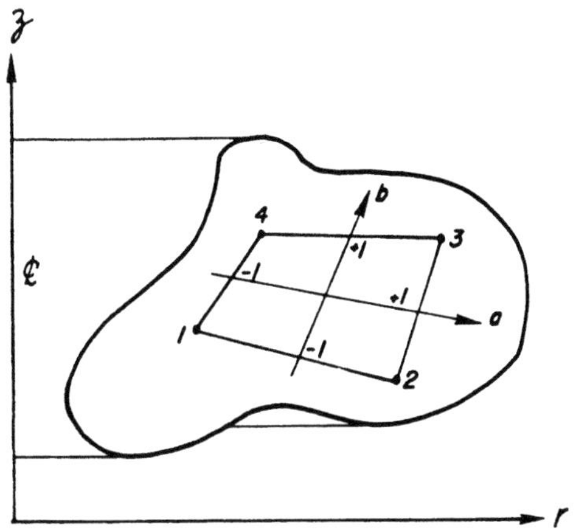

Fig. 3. The cross-section of an axisymmetric body with a four node isoparametric quadrilateral element. The isoparametric coordinates range between minus one and plus one.

α, and ϕ^α represents the interpolation function at the node with index α, the function y is described within the element by

$$y = \sum_{\alpha=1}^{4} y^\alpha \phi^\alpha(a,b) \tag{24}$$

where the bilinear isoparametric interpolation functions are given by

$$\begin{aligned} \phi^1 &= (1 - a)(1 - b)/4 \qquad & \phi^2 &= (1 + a)(1 - b)/4 \\ \phi^3 &= (1 + a)(1 + b)/4 \qquad & \phi^4 &= (1 - a)(1 + b)/4 \end{aligned} \tag{25}$$

When all the elements around node 1 are considered, the basis function obtained from the interpolation functions in each element can be pictured. Figure 4 shows the result for this case. As can be seen in Fig. 4, the basis functions obtained

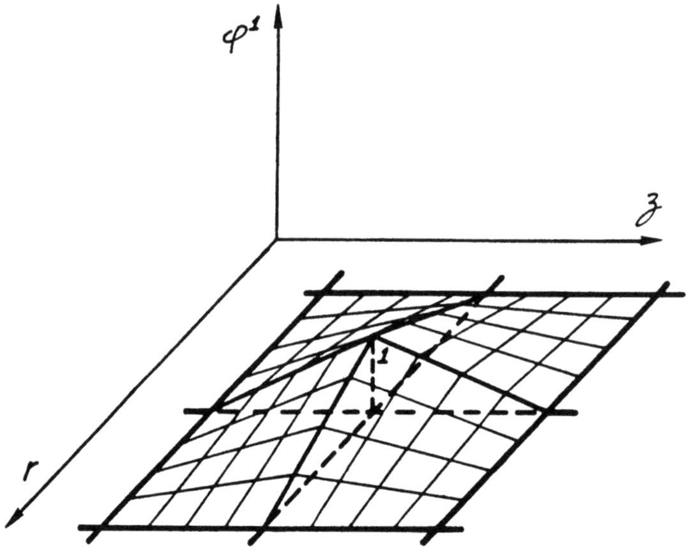

Fig. 4. The basis function φ^1 obtained from considering the contributions of the respective interpolation functions from each element surrounding nodal point 1.

in this manner have only piecewise continuous gradients along the element boundaries. As a result, implicit in the finite element equations which result from this choice of basis functions is an approximation to the jump conditions (4) along each element interface.

When the basis functions are introduced into the principle of virtual work and the variations taken, a discrete form of the differential equations (3), (4), and (5) is obtained as

$$[T] = [F] \tag{26}$$

where, with no sums on the index k,

$$[T] = \left[\int_{\mathcal{V}} t^{km}\phi^{\alpha}_{k,m}\, dv\right] \quad , \quad \text{a vector}$$

$$[F] = \left[\int_{\mathcal{V}} \rho f^k \phi_k^\alpha \, dv + \oint_{\mathcal{S}^1} s^k \phi_k^\alpha \, da \right] \quad , \quad \text{a vector.}$$

This is a discrete statement which requires the divergence of the stress field [T] to be in equilibrium with the prescribed body forces and surface tractions [F].

The integrals for [T] over each element are performed numerically using Gaussian quadrature rules. A 2x2 integration is used to evaluate the area integral. In the process a constant bulk strain is used. This may be viewed as a 1 point integration of the volumetric strain energy and a 2x2 integration of the deviatoric strain energy, [28-30].

6. DYNAMIC RELAXATION

As a solution strategy for statics problems, dynamic relaxation involves first converting the equilibrium equations into equations of motion by adding an acceleration term, second introducing an artificial damping, and finally integrating forward in time from initial conditions until the transient dynamic response has damped out to the static result with equilibrium satisfied. An early introduction of the idea is given by Otter, et al. [1], but a more recent work which summarizes all of the significant contributions on the topic since then is Underwood [2]. In reference [2] considerable detail is presented about dynamic relaxation and its use in nonlinear problems.

Dynamic relaxation is attractive for three reasons: it is vectorizable, it is versatile, and it is reliable. Because it can be made explicit, it is highly vectorizable for modern digital calculations. In an explicit form it is ideal for dealing with large deformations, finite strains and inelastic material behavior. It is incredibly reliabie and indefatigable in seeking an equilibrium state. The method does suffer from a certain lack of elegance.

To produce a dynamic problem, an acceleration term is added to the equilibrium equation (3). Thus,

$$t^{km}{}_{,m} + \rho f^k = r \frac{\partial^2 u^k}{\partial \tau^2} \quad , \tag{27}$$

where r is a spatially varying density selected to minimize the numer of steps needed to reach equilibrium and τ is a pseudo-time scale connected with the dynamic relaxation and separate from real time t. The acceleration term is discretized the same way that it would be in a true dynamics calculation, [31]. This leads to the discrete system

$$[M]\ddot{q} = [F] - [T] \tag{28}$$

where

$$\ddot{q} = [\ddot{u}^{k\alpha}(\tau)] \quad , \quad \text{a vector} \quad ,$$

$$[M] = \left[\int_{\mathcal{V}} r\phi_k^\alpha \, dv\right] \quad , \text{ a diagonal matrix} \quad .$$

At time t_n equilibrium is satisfied so that $[T]_n = [F]_n$. A new solution is initiated by incrementing the load to time t_{n+1}. In general, equilibrium will not be satisfied, so that

$$[M]\ddot{q} = [F]_{n+1} - [T]_{n+1} \tag{29}$$

Central difference expressions are introduced first for the acceleration in terms of a velocity p and then for the velocity in terms of the displacement q. Thus, the equations of motion

(29) become

$$p_{i+\frac{1}{2}} = \delta\left\{p_{i-\frac{1}{2}} + \Delta\tau[M]^{-1}\left[F_{n+1} - T^{i}_{n+1}\right]\right\} \quad , \tag{30}$$

$$q_{i+1} = q_i + \Delta\tau \; p_{i+\frac{1}{2}} \quad .$$

Here, δ is the damping, the optimum value of which leads to the minimum number of steps i to reach equilibrium.

Every step i leads to a new trial configuration $\left(x^{r}_{n+1}\right)^{i}$ and trial stress $\left(t^{n+1}_{rs}\right)^{i}$. The path in space traced out by the steps is artificial; it is a by-product of the dynamic relaxation, as is the advance in time τ. The trial states i represent equilibrium iterations. Figure 5 depicts the process in a multidimensional configuration space of the nodal point positions. The point n is an equilibrium solution and the point n + 1 is the equilibrium state being sought. The curved path between n and n + 1 traces out the true solution. The spiral path marked with tics and parameterized by steps in τ is the sequence of trial states generated by the dynamic relaxation method. The straight line from n to the last step calculated from dynamic relaxation is the interval over which the stress t^{n+1}_{rs} is evaluated using the real time step Δt. The integrals in (26) are reevaluated at each step i using the trial geometry and when equilibrium is achieved a straight line approximation to the true path between n and n + 1 is obtained. This scheme uncouples the path dependence and real-time dependence of the constitutive behavior from the arbitrary sequence of trial states generated by the dynamic relaxation method.

Convergence is based on achieving an acceptably small equilibrium unbalance. Because the converged solution is a straight line approximation, the true state n + 1 will not be found, but a nearby equilibrium state will be found nonetheless.

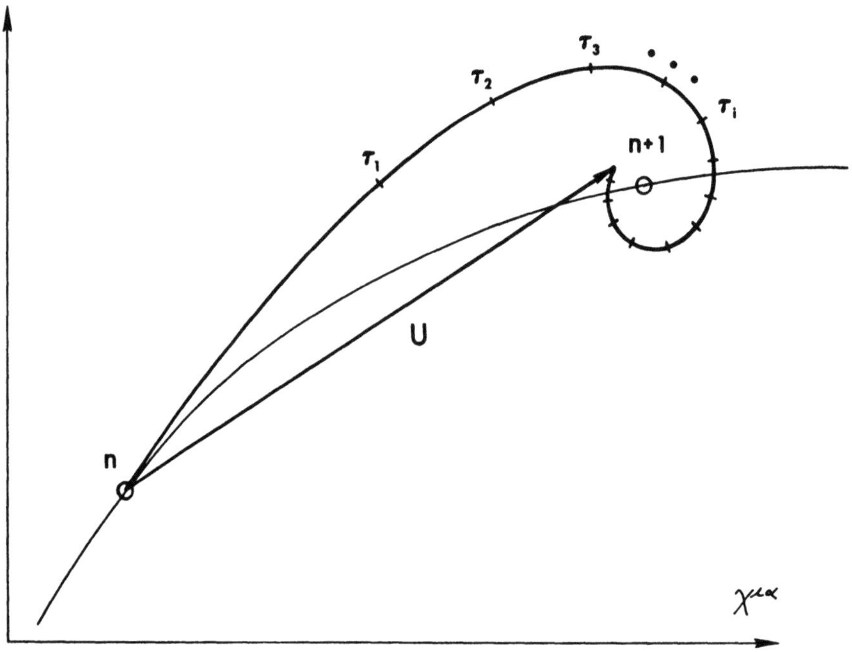

Fig. 5. A model equilibrium iteration sequence in a multidimensional configuration space of nodal point positions developed with dynamic relaxation showing convergence at load step n+1. The straight line path from n to the last step calculated from dynamic relaxation is the interval over which the stress t_{rs}^{n+1} is evaluated using the real time step Δt.

The only question remaining is how to select the variable density r, the pseudo-time step $\Delta\tau$, and the damping δ in order to find a converged solution in the minimum number of steps i. The performance of dynamic relaxation is tied to the minimum natural frequency ω_o and maximum natural frequency ω_1 of the discrete equations.

The damping per cycle which is demonstrated by Eqn. (30) is frequency dependent. For a given damping factor δ the decrease in amplitude per cycle is greatest for the lowest frequency component. The damping is then chosen to provide critical damping for the lowest frequency. This expression is

$$\delta = 1 - 4\omega_o\omega_1/(\omega_o + \omega_1)^2 \quad . \tag{31}$$

Note that the range on δ is (0,1). A stability analysis on this set of explicit equations produces a critical pseudo-time step given by

$$\Delta\tau_c = 2/[\delta^{\frac{1}{2}}(\omega_o + \omega_1)] \quad . \tag{32}$$

The damping is stabilizing although in a practical sense this is negligible. If the problem were linear so ω_o and ω_1 were fixed, then the numer of time steps required to reduce the amplitude by a factor of ten would be

$$N \approx 1.15 \; \omega_1/\omega_o \tag{33}$$

From Equation (33) it is seen that any action to reduce the ratio ω_1/ω_o speeds convergence.

For the linear problem and a uniform mesh of dimension h, the maximum frequency ω_1 is given by

$$\omega_1 = 2 \; c/h \quad . \tag{34}$$

Here c is the uniaxial strain sound speed given by $[(\lambda + 2\mu)/r]^{\frac{1}{2}}$. While the problem is nonlinear in the large, as equilibrium is approached at the end of each load step the behavior is linear. With (34) the condition on $\Delta\tau$ is essentially the well known Courant-Friedrichs-Levy condition which says that $\Delta\tau$ can be no larger than the time it takes for information to travel between the "closest" two nodes in the mesh.

On the basis of these observations, the concept of <u>mesh homogenization</u> is introduced. The densities r are selected element-by-element to give the same transit time across each element, regardless of the size of the element. Thus, information about equilibrium unbalance is transmitted uniformly over the mesh. This process gives the lowest possible value to the maximum frequency ω_1, the goal being to minimize the ratio ω_1/ω_o.

Even with changing sound speed and mesh distortions $\Delta\tau$ is easily controlled. If necessary, the density r can be redefined between load steps to recover a uniform element transit time.

The fundamental frequency ω_o is continuously estimated using an approximate Rayleigh quotient, [2]. At each step i in the dynamic relaxation, a new estimate $(\omega_o)_i$ is computed from (35).

$$(\omega_o) = (q_i^T[K]_{loc}\, q_i/q_i^T[M]q_i)^{\frac{1}{2}} \tag{35}$$

where $[K]_{loc}$ is a diagonal stiffness matrix whose $j^{\underline{th}}$ component is computed from

$$[K]_{loc}|_j = ([T]_i|_j - [T]_{i-1}|_j)/(\Delta\tau p_{i-\frac{1}{2}}|_j) \quad . \tag{36}$$

With each estimate of the fundamental frequency a new value of the damping δ is computed. This has the virtue that the lowest active mode will be found in the event that the fundamental mode is not participating, [2].

7. APPLICATIONS

The first problem while elementary is a relaxation or stress redistribution calculation. A changing stress field provides the maximum feed back in a creep constitutive model, equation (16). The material is naturally occurring bedded halite (NaCl) found some 650 meters below the surface in the southwestern

United States. The properties for use in equation (16) are given in Figure 6 with several of the coefficients combined into a single leading constant for the Norton power law.

The calculation is for the hole closure which takes place as the result of a suddenly created cylindrical hole in a hydrostatically stressed "infinite" body of halite. The surface of the cylindrical hole is traction free. This is a one-dimensional problem in which only a single row of elements proceeding radially outward from the cavity is needed, Figure 6. At time t = 0, the solution is that for an elastic material which gives high deviatoric stresses at the edge of the hole and dying out within 20 to 30 hole radii. Of interest is the hole closure which is influenced by the transient stress redistribution. The goal here is to show the arbitrary size of the load increment in terms of Δt. Based upon a time step selection procedure taken from [26], a critical time step was computed for this problem. The hole closure associated with $\Delta t_{critical}$, 10 x $\Delta t_{critical}$, and 100 x $\Delta t_{critical}$ is given in Figure 7. It is clear from this figure that the choice of Δt influences the accuracy of the calculation but the amount of computational effort associated with each solution is found to be identical. This illustrates the uncoupling of the constitutive model integration from the dynamic relaxation solution algorithm. Both of the smaller Δt solutions show identical hole closure rates at 10 years while the largest Δt has just begun to approach steady creep at 10 years.

The second problem is a two-dimensional calculation concerned with the response around an underground opening. Figure 8 shows the dimensions and boundary conditions. Figure 9 shows the stratigraphy incorporated in the calculation with the mechanical properties given in Table I. Figure 10 shows the finite element mesh used. Zero friction was used on the sliding interfaces representing the clay seams. Again, the opening is assumed to occur instantaneously at t = 0. The calculation is characterized by stress redistribution as a result of the disturbances introduced in the in situ hydrostatic stress field by the drift opening. While there is a great deal of detail in the calculated results which could be displayed, only the floor to ceiling convergence at the room centerline is shown in Figure 11. Shown as well in Figure 11 is the cumulative number of iterations needed to generate the solution as time accumulates. During the early stress redistribution phase, the majority of the calculational effort takes place. Once this period is passed, the problem takes on the character of a creep calculation and very few iterations are needed to obtain the solution.

The third problem considers the necking of the plane strain, steel tensile bar shown in Figure 12. Deformation controlled loading was used with the left end displaced in 100 equal steps to an increase in the axial length of the bar of 78%. The material is modeled as an isotropic hardening, elastic-plastic metal with a bilinear representation for the stress-strain curve. Material property data is shown in Figure 12. To insure that the bar would neck, a geometric imperfection was built into the model. The problem is characterized by large plastic strains both axially and radially as seen in the deformed mesh of Figure 12. The amount of necking, characterized by the non-dimensional width change of the bar, is also plotted in Figure 12. Significant necking begins to occur after the axial strain has exceeded 40%. The solution was found to be sensitive to the convergence tolerance applied to the analysis. Informal comparisons of results with other codes and element choices have shown a convergence tolerance of 0.1% to give acceptable results. Severe distortion of the elements near the neck would normally cause some concern but favorable comparisons with other codes seemed to give no justification for repeating the analysis with a finer mesh.

The last example considers an elastic-perfectly plastic infinite cylinder subjected to an internal pressure. There is an exact solution for this problem [32] so that a comparison can be made with theory. Figure 13 gives the material properties and geometry used in the analysis. The numerical solution matches the theory quite well. The algorithm handles elastic-plastic problems with no difficulty.

TABLE I. MATERIAL PROPERTIES USED IN DRIFT ANALYSIS

	SALT	DEGRADED SALT	ANHYDRITE OR POLYHALITE	ANHYDRITE/ POLYHALITE/SALT
Young's Modulus (Pa)	2.5×10^{10}	2.5×10^{10}	7.3×10^{10}	5.28×10^{10}
Poisson's Ratio	25	.25	.30	.28
Elastic Bulk Modulus (Pa)	1.7×10^{10}	1.7×10^{10}	6.1×10^{10}	$4\ 02 \times 10^{10}$
Elastic Shear Modulus (Pa)	1.0×10^{10}	1.0×10^{10}	2.8×10^{10}	2.06×10^{10}
Creep Constants A ($Pa^{-5}\ s^{-1}$)	$3\ 16 \times 10^{-45}$	1.012×10^{-44}	-------	6.32×10^{-46}
n	5.	5.	-------	5.

Coefficient of Friction $\mu = 0.$

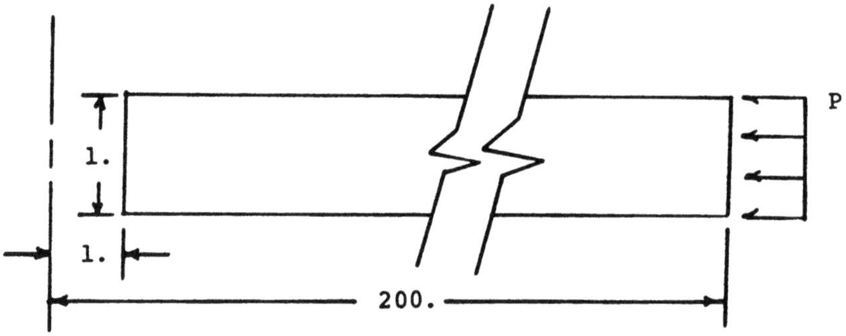

$P = 1.41 \times 10^{7}$ Pa

Material Properties

$G = 9.93 \times 10^{9}$ Pa

$K = 1.65 \times 10^{10}$ Pa

$\dot{\varepsilon}_c = A\bar{\sigma}^n$

$A = 3.47 \times 10^{-44}\ Pa^{-4.9}\ s^{-1}$

$n = 4.9$

Figure 6. Problem Description for the Externally Pressurized Infinite Cylinder.

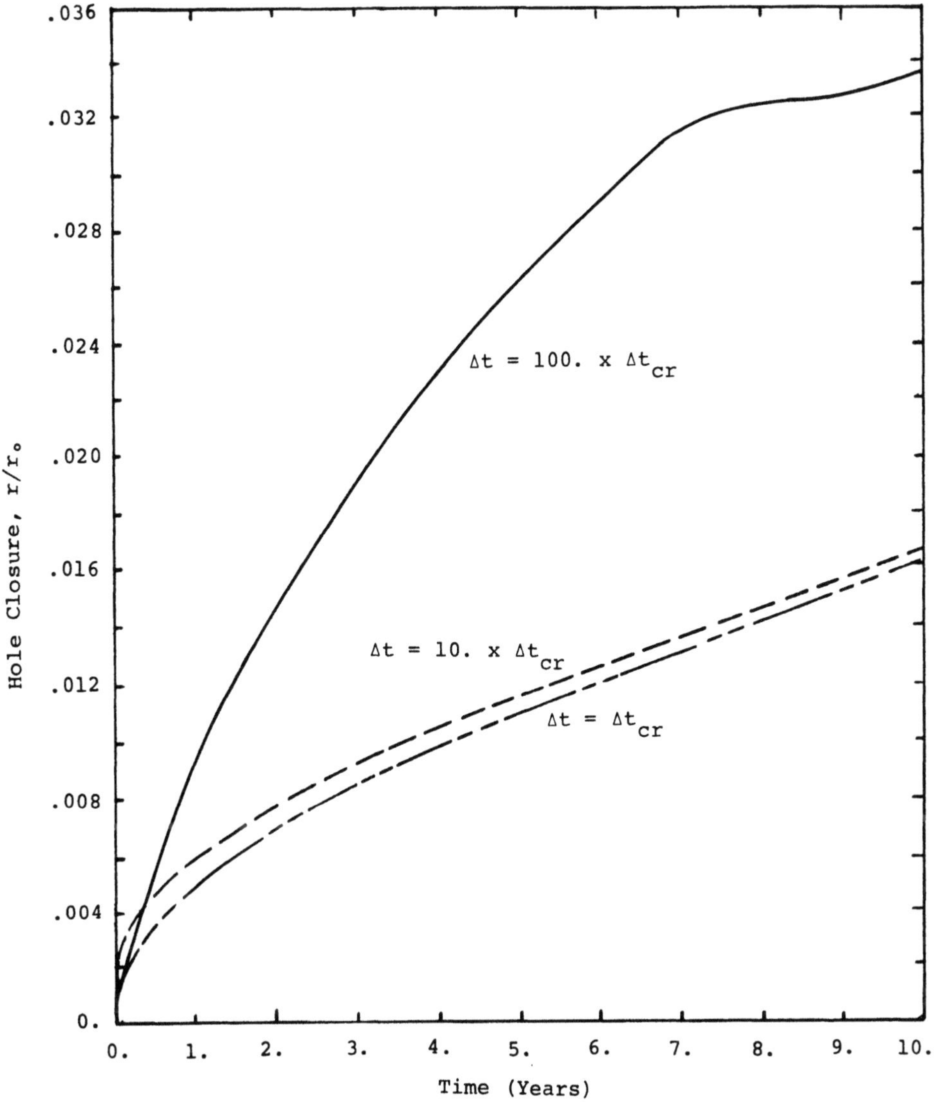

Figure 7. Radial Creep Closure of the Cylindrical Hole for Several Choices of Δt.

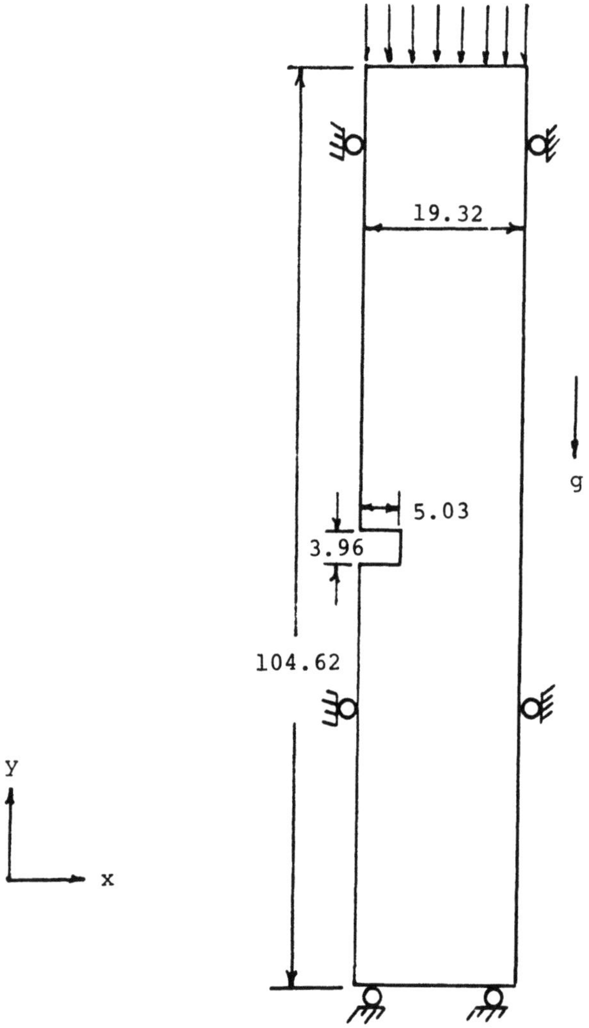

Figure 8. Dimensions and Boundary Conditions for the Two-Dimensional Analysis of an Underground Opening

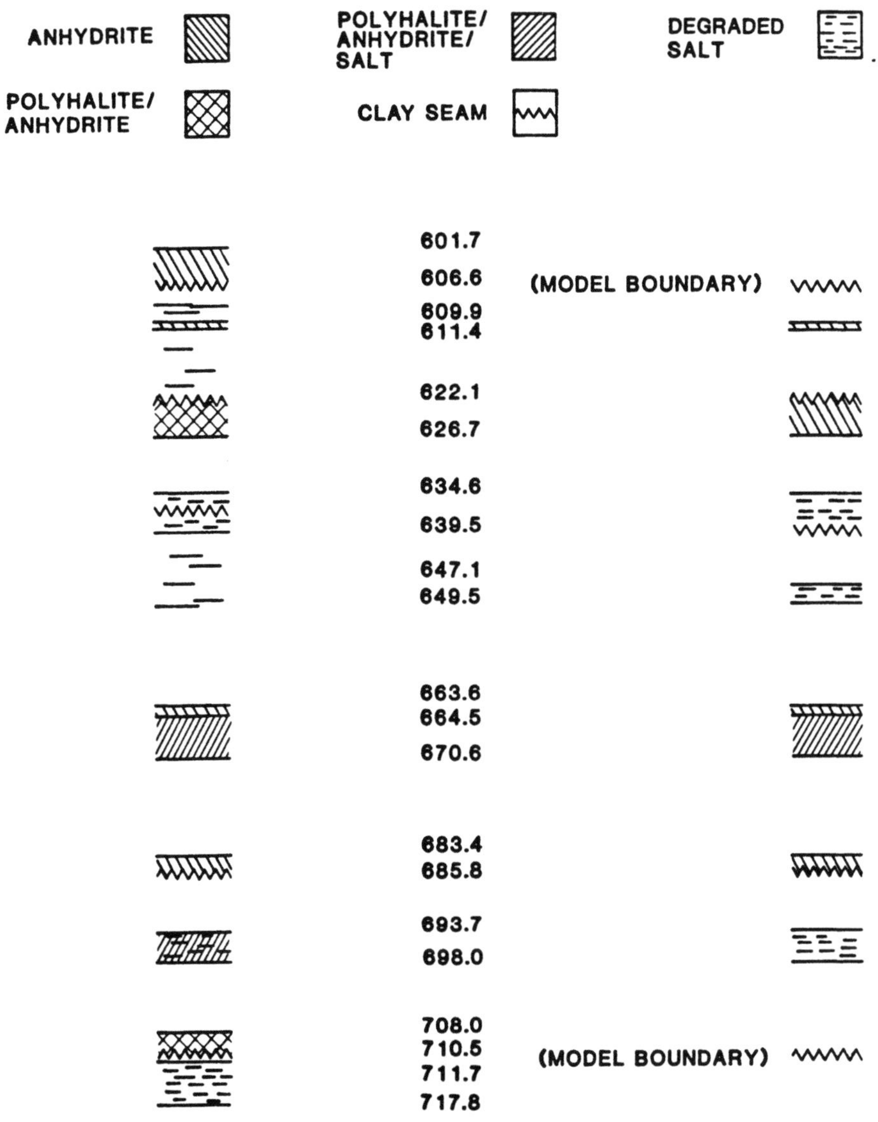

Figure 9. Geologic Profile and Stratigraphy Used in Model of Underground Opening

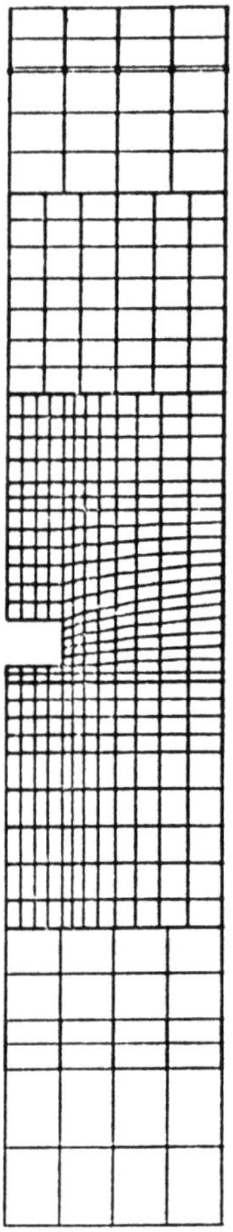

Figure 10. Finite Element Mesh Used for Analysis of Underground Opening

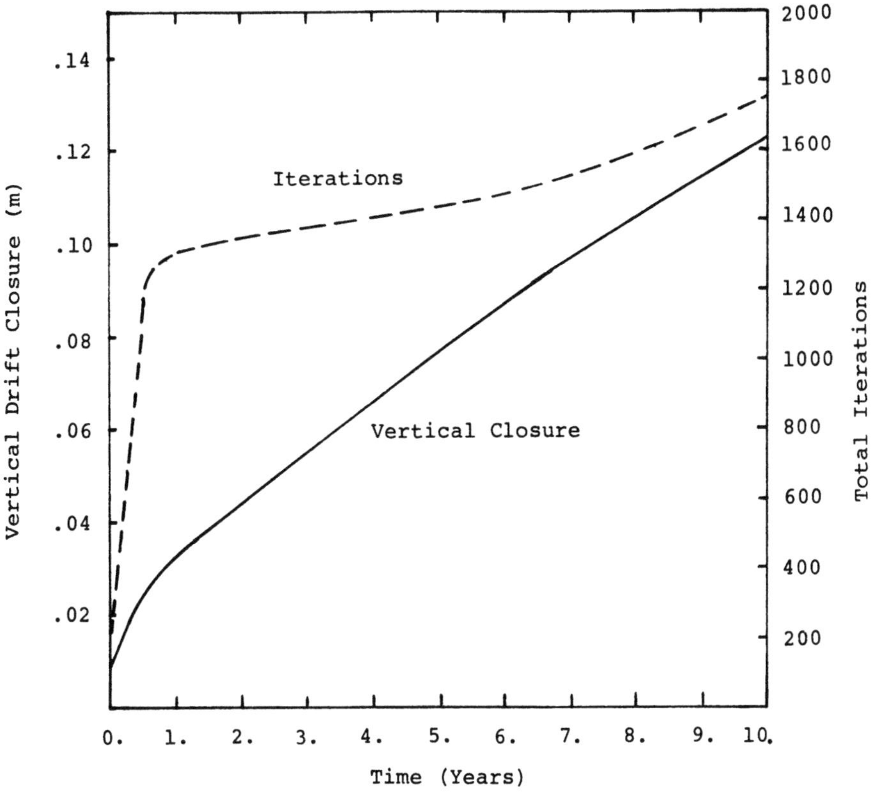

Figure 11. Vertical Drift Closure and Total Number of Solution Iterations Versus Creep Time for a Single Drift in Salt

GEOMETRY

W_A = 3.0 in. (.0762 m)

W_B = 3.0 in. (.0762 m)

W_C = 2.96 in. (.0752 m)

W_D = 2.96 in. (.0752 m)

L = 9.0 in. (.2286 m)

ΔL = 7.0 in. (.1778 m)

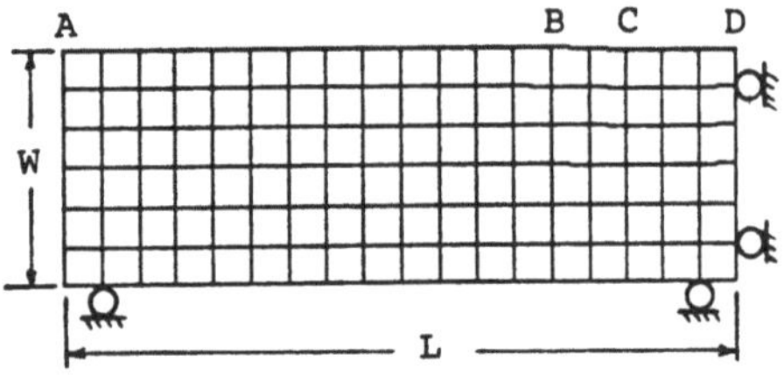

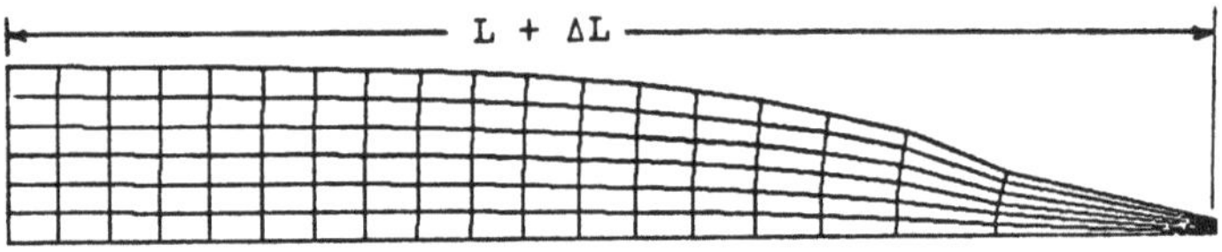

PROPERTIES

$E = 30. \times 10^6$ psi (2.07×10^{11} pa) $H = 5 \times 10^4$ psi (3.447×10^8 pa)

$\sigma_y = 4. \times 10^4$ psi (2.758×10^8 pa) $\nu = .3$

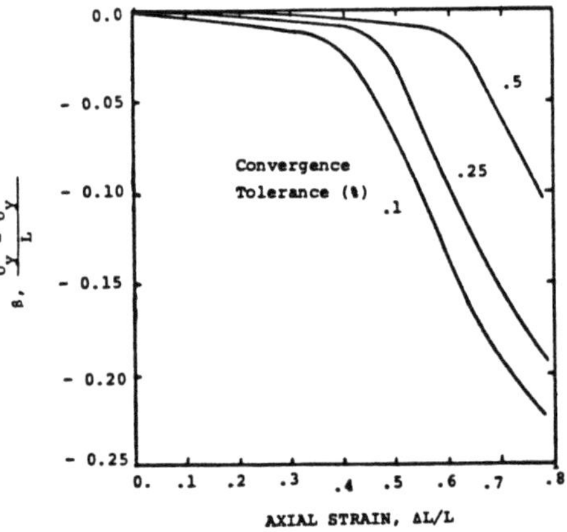

Figure 12. Problem Definition and Results for the Necking of a Plane Strain Steel Tensile Bar.

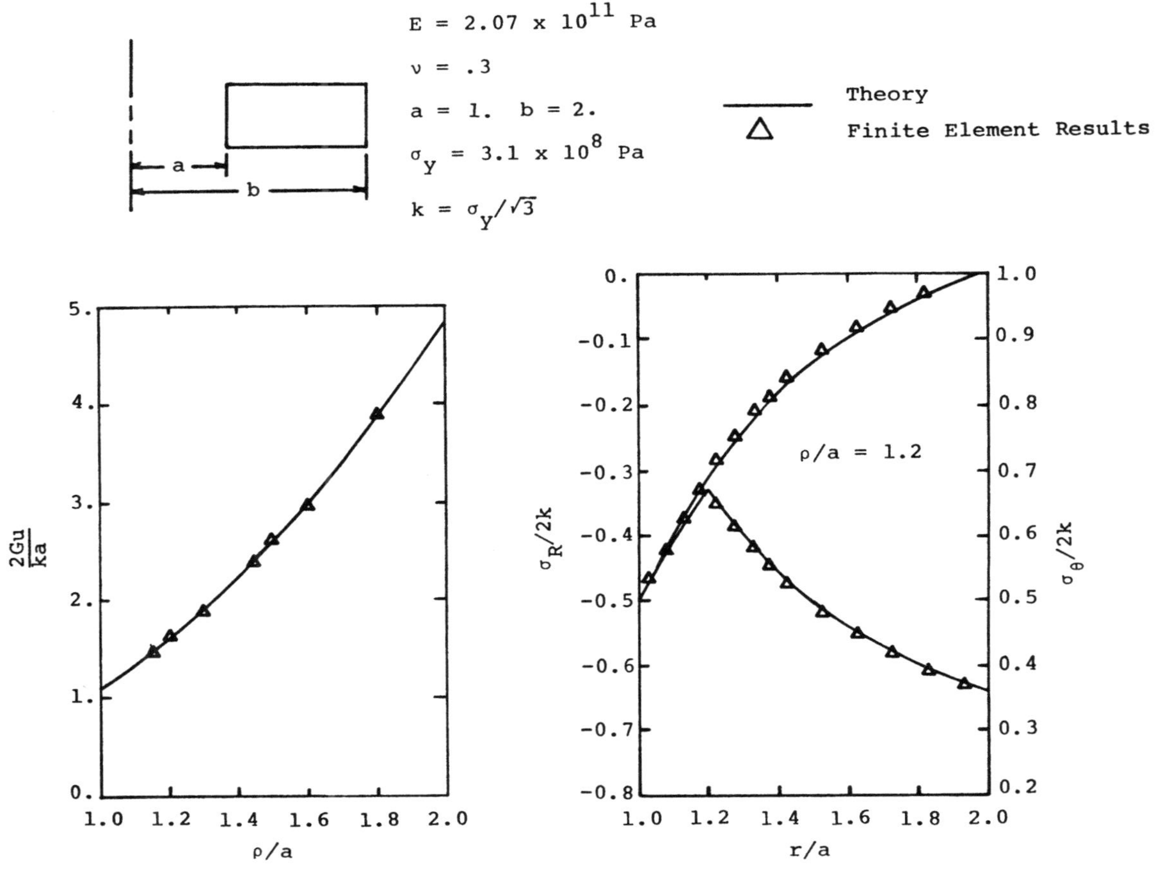

Figure 13. Comparison of Theory and Finite Element Results for an Internally Pressurized Infinite Cylinder

REFERENCES

1. J. R. H. Otter, A. C. Cassell, and R. E. Hobbs, "Dynamic Relaxation," Proc. Institution of Civil Engineers, 35 (1966) pp. 633-656.

2. P. Underwood, "An Adaptive Dynamic Relaxation Technique for Nonlinear Structural Analysis," Lockheed Palo Alto Research Laboratory, Report No. LMSC-D678265, Palo Alto, California, USA, in preparation.

3. R. W. H. Wu and E. A. Witmer, "Finite-Element Analysis of Large Elastic-Plastic Transient Deformations of Simple Structures," AIAA Journal, Vol. 9, No. 9, September 1971.

4. J. F. McNamara and P. V. Marcal, "Incremental Stiffness Method for Finite Element Analysis of the Nonlinear Dynamic Problem," ONR Symposium on Numerical and Computer Methods in Structural Mechanics, Urbana, Illinois, September 1971.

5. M. Hartzman and J. R. Hutchinson, "Nonlinear Dynamics of Solids by the Finite Element Method," J. of Computers and Structures, Vol. 2, pp. 47-77, 1972.

6. J. H. Heifitz and C. J. Costantino, "Dynamic Response of Nonlinear Media at Large Strains," Journal of the Engineering Mechanics Division, ASCE, Vol. 98, No. EM6, December 1972.

7. T. Belytschko and B. J. Hsieh, "Non-linear Transient Finite Element Analysis with Convected Coordinates," IJNME, Vol. 7, 1973, pp. 255-271.

8. J. O. Hallquist, "DYNA2D - An Explicit Finite Element and Finite Difference Code for Axisymmetric and Plane Strain Calculations (User's Guide)," UCRL-52429, Lawrence Livermore Laboratory, University of California, Livermore, California (March 1978).

9. G. Maenchen and S. Sack, "The Tensor Code," Methods in Computational Physics, Alden, Fernback and Rotenberg, eds., Vol. 3, Academic Press, New York, 1964.

10. M. L. Wilkins, "Calculation of Elastic-Plastic Flow," Methods in Computational Physics, Alden, Fernback and Rotenberg, eds., Vol. 3, Academic Press, New York, 1964.

11. L. D. Bertholf and S. E. Benzley, "TOODY II, A Computer Program for Two-Dimensional Wave Propagation," SC-RR-68-41, Sandia Laboratories, Albuquerque, New Mexico, November 1968.

12. D. Bushnell, "A Strategy for the Solution of Problems Involving Large Deflections, Plasticity and Creep," Int. J. Num. Meth's Engr., Vol. 11 (1977) pp. 683-708.

13. D. L. Hicks, "Hydrocode Subcycling Stability," Mathematics of Computation, submitted for publication.

14. C. Truesdell and R. A. Toupin, "The Classical Field Theories," Encyclopedia of Physics, Vol. III/1, Springer-Verlag, Berlin, 1960.

15. P. J. Blatz and W. L. Ko, Application of Finite Elastic Theory to the Deformation of Rubbery Materials," Transactions of the Society of Rheology, Vol. 6, 1962, pp. 223-251.

16. R. P. Goel and L. E. Malvern, "Biaxial Plastic Simple Waves with Combined Kinematic and Isotropic Hardening," JAM, Vol. 37, 1970, pp. 1100-1106.

17. S. W. Key, J. H. Biffle, and R. D. Krieg, "A Study of the Computational and Theoretical Differences of Two Finite Strain Elastic-Plastic Constitutive Models," in Formulas and Computational Algorithms in Finite Element Analysis, edited by Bathe, Oden, and Wunderlich, MIT, Cambridge (1977).

18. R. K. Penny and D. L. Marriott, Design for Creep, McGraw-Hill, London, (1971).

19. J. H. Gittus, Creep, Viscoelasticity and Creep Fracture in Solids, John Wiley and Sons, New York (1975).

20. T. J. R. Hughes and J. Winget, "Finite Rotation Effects in Numerical Integration of Rate Constitutive Equations Arising in Large-Deformation Analysis," Int. J. Num. Meths. Engr., in press.

21. J. O. Hallquist, "NIKE2D: An Implicit, Finite Deformation, Finite-Element Code for Analyzing the Static and Dynamic Response of Two-Dimensional Solids," Lawrence Livermore National Laboratory Report UCRL-52678, University of California, Livermore, March 1979.

22. J. H. Biffle, "JAC: A Two Dimensional Finite Element Computer Program Which Uses a Nonlinear Conjugate Gradient Technique for the Solution of the Nonlinear, Quasistatic Response of Solids," Sandia National Laboratories, Report in preparation, Albuquerque, New Mexico.

23. R. D. Krieg and S. W. Key, "Implementation of a Time Independent Plasticity Theory into Structural Computer Programs," in: J. A. Stricklin and K. J. Saczalski (eds.), Constitutive Equations in Viscoplasticity: Computational and Engineering Aspects, AMD Vol. 20, (The American Society of Mechanical Engineers, New York, 1976), pp. 125-138.

24. R. D. Krieg, "An Efficient Numerical Method for Time Independent Plasticity," SAND77-0943, Sandia Laboratories, Albuquerque, New Mexico (November 1977).

25. R. D. Krieg and D. B. Krieg, "Accuracies of Numerical Solution Methods for the Elastic-Perfectly Plastic Model," J. Press. Vess. Tech. 99 (November 1977) pp. 510-515.

26. J. H. Arqyris, L. E. Vaz and K. J. William, "Improved Solution Methods for Inelastic Rate Problems," Comp. Meth's App. Mech's. and Engr. 16 (1978) pp. 231-277.

27. B. M. Irons, "Engineering Applications of Numerical Integration in Stiffness Methods," AIAA Journal, Vol. 4, No. 11, November 1968.

28. J. C. Nagtegaal, D. M. Parks, and J. R. Rice, "On Numerically Accurate Finite Element Solutions in the Fully Plastic Range," Computer Methods in Applied Mechanics and Engineering, Vol. 4 (1974) pp. 153-177.

29. T. J. R. Hughes, "Equivalence of Finite Elements for Nearly-Incompressible Elasticity," LBL-5237, Lawrence Berkeley Laboratory, University of California, Berkeley, 1976.

30. T. H. H. Pian and S. W. Lee, "Notes on Finite Elements for Nearly-Incompressible Materials," AIAA Journal, Vol. 14, No. 6, June 1976, pp. 824-826.

31. S. W. Key, "A Finite Element Procedure for the Large Deformation Dynamic Response of Axisymmetric Solids," Computer Methods in Applied Mechanics and Engineering, 4 (1974) pp. 195-218.

32. W. Prager and P. G. Hodge, Theory of Perfectly Plastic Solids, John Wiley and Sons, New York (1963).

On the Efficient Solution of Nonlinear Finite Element Systems

H. D MITTELMANN
Universität Dortmund, Germany

Summary

We consider variational problems resp. variational inequalities which correspond to quasilinear elliptic boundary value problems without resp. with linear constraints. Finite element discretization leads to nonlinear algebraic systems resp. linearly constrained nonlinear programming problems. Efficient algorithms are given for the solution of these problems. Convergence is shown and some numerical results are presented. Applications are for the unrestricted case all problems which are given by a minimization problem and for problems with constraints e. g. elastic-plastic torsion or elastic contact problems.

1. Introduction

In the following we will consider optimization problems of the form

$$\begin{aligned} f(x) &= \min_S \\ S &= \{x \in \mathbb{R}^M,\ g(x) = G^T x + g_o \leq 0\} \end{aligned} \tag{1.1}$$

where $f : \mathbb{R}^M \to \mathbb{R}$ is a strictly convex not necessarily quadratic functional and $G \in \mathbb{R}^{M,p}$, $g_o \in \mathbb{R}^p$. We assume that (1.1) is obtained by discretizing an infinite-dimensional variational problem of the form

$$\int_R h(x,\ u,\ \nabla u, \ldots, \nabla^m u)\,dx = \min_K,\ m \geq 1 \tag{1.2}$$

where h is not necessarily quadratic in $\nabla^m u$, $m \geq 1$ and K is a closed convex subset of a suitable function space V of vector functions $u=(u_1,...,u_\ell)$ defined on the domain $R \subseteq \mathbb{R}^n$.

It is therefore natural that (1.1) is not a general linearly constrained nonlinear optimization problem but it has certain special properties. The Hessian matrix $F \in \mathbb{R}^{M,M}$ of f e. g. is a sparse matrix when (1.2) is discretized by finite-difference or finite element methods. Furthermore G is usually a sparse matrix, too. In typical applications from structure mechanics as e. g. the contact problem every column of G has only a few nonzero elements. If the obstacle or the contact problem for membranes is considered p has the same order of magnitude as M but for contact problems of elastic bodies (cf. e. g. | 7 | and the papers cited there) p is considerably smaller than M. Further occurrences of (1.2) in applications are well-known and we need not give a list of them here. It is also not necessary to explain that (1.2) corresponds to a nonlinear variational inequality respectively to a free boundary value problem for its Euler equation.

Problem (1.1) (resp. (1.2)) is often solved by Uzawa's method (cf. e. g. | 4 |). We recall that the dual problem of (1.1) is

$$\begin{aligned} \phi(\lambda^*) &= \max_{\lambda \geq 0} \phi(\lambda), \lambda \in U(\lambda^*) \subseteq \mathbb{R}^p \\ \phi(\lambda) &= \ell_\lambda(x^*) = \min_{x \in U(x^*)} \ell_\lambda(x) \\ \ell_\lambda(x) &= f(x) + \lambda^T g(x) \end{aligned} \tag{1.3}$$

where $\ell_\lambda(x)$ is the Lagrange function of (1.1) and $\phi(\lambda)$ the dual function. Uzawa's method now applies the gradient projection method to (1.3) yielding the algorithm:

Let $\lambda_k \geq 0$, $k \geq 0$, be given. Then compute $\phi(\lambda_k)$, e. g. by solving

$$\nabla_x \ell_{\lambda_k}(x) = \nabla f(x) + \lambda_k^T \nabla g(x) = 0 \tag{1.4}$$

if this is also a sufficient condition for the minimization of $\ell_\lambda(x)$. Denote the solution by x_k. Then improve λ_k by

$$\lambda_{k+1} = \max \{0, \lambda_k + \rho \nabla_\lambda \phi(\lambda_k)\} \tag{1.5}$$

which is not difficult to compute since $\nabla_\lambda \phi(\lambda_k) = g(x_k)$. For the case of quadratic f convergence of the sequence $\{x_k\}$ can be shown for values of ρ in a range

$$0 < \alpha_o \leq \rho \leq \alpha_1$$

(cf. | 4 |, p. 94) with suitable constants α_o, α_1.

In every step of this algorithm an unconstrained nonlinear optimization problem in $\mathbb{R}^M$ respectively a nonlinear system of order M has to be solved.

Recently it was suggested in |13| to use the Augmented Lagrangian method originally developed independently by Hestenes and Powell to problems of the form (1.1) with convex g. This algorithm combines the advantage of reducing the problem via duality to a problem with positivity constraints with the idea of penalization. Under suitable assumptions, e. g. the penalty parameter $r > 0$ need not tend to infinity, Rockafellar proves convergence of this algorithm which becomes arbitrarily fast for increasing r. In every step, however, again an unconstrained problem has to be solved.

Since numerical experience has shown (cf. | 4 |, p. 113) that for many problems of the form (1.1) but with

$$S = \{x \in \mathbb{R}^M, \ c \leq x \leq d\} \tag{1.6}$$

a projected relaxation method is more efficient than Uzawa's method and beyond that solves the restricted problem faster than a similar unrestricted problem (cf. | 4 |), we shall concentrate on relaxation. In its simplest form this is the algorithm:

Let $x^{(k)} \in S = \{x \in \mathbb{R}^M, x \geq 0\}$ be given. Compute $x^{(k+1)}$ according to

$$x_i^{(k+1)} = \max\{0, x_i^{(k)} - \omega_{i,k} \frac{f_i(x_{i,k})}{\delta_{i,k}}\}, \quad i = 1,\ldots,M \quad (1.7)$$

where $\nabla f = (f_i), x_{i,k} = (x_1^{(k+1)},\ldots,x_{i-1}^{(k+1)}, x_i^{(k)},\ldots,x_M^{(k)})$, $\delta_{i,k}$ are suitable positive numbers and $\omega_{i,k}$ relaxation parameters.

The projected point-SOR-N method is obtained for the choice

$$\delta_{i,k} = f_{ii}(x_{i,k}) \quad (1.8)$$

where $F = (f_{ik})$ is the Hessian matrix of f. The convergence properties of (1.7), (1.8) are well known for quadratic f (cf. e. g. | 4 |). In the nonquadratic case global convergence was proved in | 9 | and for a modified relaxation method in | 10|. In computations such methods usually quite rapidly approximate the proper subspace of active constraints corresponding to the (approximation of the) unknown contact region (cf. e. g. | 4 |, p. 135) while most of the time is spent in finding good approximations for the remaining variables.

Thus it is desirable to have an algorithm which is faster than relaxation for unrestricted problems and to use that after a nearly proper subspace has been found. Since one cannot determine when this is the case that algorithm has to be globally convergent for the restricted case, too.
A two-phase algorithm for (1.1), (1.6) in this spirit was given in | 11|. In the following we shall consider this and a completely different two-phase algorithm.

2. A preconditioned conjugate gradient method

In this section we define an algorithm for the approximate solution of

$$f(x) = \min_{S} \qquad (2.1)$$
$$S = \{x\in \mathbb{R}^M,\ G^T x + g_o \geq 0\}$$

where $G = (E_M, -E_M)\in \mathbb{R}^{M,2M}$, E_M the MxM identity matrix and $g_o = (-c^T, d^T)^T$. We denote the columns of G by $g^i, i=1,\ldots,2M$, and define $I(x) = \{i\in\{1,\ldots,2M\},\ g_i^T x + g_{oi} = 0\}$.
Let $G_I = (g_i)_{i\in I}$, $Q_I = E_M - G_I G_I^T$ and for $x = x^{(k)}$ let $I_k = I(x^{(k)})$, $G_k = G_{I_k}$ and Q_k analogously. Q_I is the projector on the subspace of the active constraints. Finally let $r_k = -\nabla f(x^{(k)})$.

Next we assume as in [1o] that the gradient of f may be written as

$$\nabla f(x) = A(x)x - b(x). \qquad (2.2)$$

If (2.1) is obtained by the discretization of a problem corresponding to a quasilinear elliptic equation of second order usually a canonical though not unique decomposition (2.2) exists. The matrix A is given explicitly for two problems in [1o] and compared with the Hessian matrix of f. Usually A has similar properties as symmetry and structure but is easier to compute.
Now we give an algorithmic definition of the projected cg-SMOR-N method and shall comment on in afterwards.

The projected cg-SMOR-N algorithm

Let $x^{(o)}\in S$ and $p_{-1}\in \mathbb{R}^M$ be arbitrary. Set j = k = 0 and $\mu_o = 0$.

Step 1 Set $\lambda_k = 0$. Compute I_k, r_k. Terminate if $G_k^T r_k \leq 0$, $\|Q_k r_k\| <$ eps. Compute $|r_{k\ell}| = \max\{|r_{ki}|, (G_k^T r_k)_i > 0\}$.

Step 2 If $(\|Q_k r_k\| < |r_{k\ell}|$ and $\mu_k = 0$ or $\|Q_k r_k\| <$ eps) then set $I_k = I_k - \{\ell\}$ and $\lambda_k = 1$.

Step 3 Compute z_k as the direction vector taken by one step of the SSOR-algorithm for the linear system $A(x^{(k)})x = b(x^{(k)})$ starting from $x^{(k)}$ and keeping the components corresponding to I_k fixed.

$$\beta_k = \begin{cases} 0, \text{ if } j = 0 \text{ or } \lambda_k = 1 \text{ or } \mu_k = 1 \\ z_k^T r_k / z_{k-1}^T r_{k-1} \text{ otherwise.} \end{cases}$$

$$p_k = z_k + \beta_k p_{k-1}$$

Step 4 Determine $\bar{\alpha}_k$ as the maximal admissible step length in direction p_k. Compute $\tilde{\alpha}_k = z_k^T r_k / p_k^T A(x^{(k)}) p_k$. If $\tilde{\alpha}_k \geq \bar{\alpha}_k$ then set $\alpha_k = \bar{\alpha}_k$, $\mu_{k+1} = 1$ while $\alpha_k = \tilde{\alpha}_k$, $\mu_{k+1} = 0$ otherwise.

Step 5 Set $x^{(k+1)} = x^{(k)} + \alpha_k \tau_k p_k$, for the choice of τ_k see below.

$$j = \begin{cases} 0, \text{ if } j = m, \ m > 0 \\ 1, \text{ if } \lambda_k = 1 \\ j + 1 \text{ otherwise,} \end{cases}$$

$k = k + 1$ and go to Step 1.

This algorithm is in the usual sense a primal feasible direction method and in generalization of Rosen's gradient projection method the direction vectors are the projection of vectors obtained by a back and forth-sweep of the SOR-method for a certain linear system and which are modified by a cg-like rule with a restart after at most m steps. In order to assure global convergence zig-zagging has to be avoided by the strategy used in Step 2.

Although the above algorithm seems to be rather complicated it in fact needs more storage than projected relaxation methods but less operations (cf.[10]). The essential features are the rather weak dependence of the convergence rate on the value of the relaxation parameters used in Step 3 and the fact that the dependence on m is also not very strong with optimal values between m = 1 and m = 5 for many problems.

A thorough study of the choice of the direction taken in each step shows that this is always a feasible one. Especially it is not necessary to insert pure gradient-projection steps. Global convergence is proved in [11] by first showing that the vectors z_k are uniformly related to the projected gradient $Q_k r_k$ and that $p_k^T z_k \geq r_k^T z_k$ for $\tau_k \leq 1/\gamma_k$ where at least bounds for γ_k may be given. The essential fact is that z_k and $Q_k r_k$ are related through

$$z_k = S_k Q_k r_k \tag{2.3}$$

where S_k is a positive definite matrix which in this case is also symmetric. Then the descent properties are stated and the proof is completed by showing that zig-zagging cannot occur.

Since in one step of this algorithm only one constraint may be inactivated and usually only a few may become active there is a certain need for a method which relatively fast approximates the proper subspace and thus yields an acceleration at least in the case when many constraints have to be activated or inactivated during the computation. For this purpose a modified relaxation method namely MOR-N:

$$x_i^{(k+1)} = \max\left(c_i,\ \min\left(d_i, x_i^{(k)} - \omega_{i,k} \frac{f_i(x_{i,k})}{a_{ii}(x_{i,k})}\right)\right) \tag{2.4}$$

$i = 1,\ldots,M$, a_{ii} diagonal element of A in (2.2)

was successfully applied in [11]. For a proper choice of relaxation parameters $\omega_{i,k}$ this algorithm itself was shown to be globally convergent.

The resulting two-phase algorithm solved different nonlinear model problems several times faster than projected relaxation methods alone. Hence, at least for the case of bound-constraints we obtained an efficient method. Generalizations to general linear constraints might be possible, whereas convex nonlinear constraints are more difficult to treat with feasible direction algorithms. Here, however, the augmented Lagrangian method may be applied advantageously.

3. Numerical Results

The cg-SMOR-N and the MOR-N algorithm as described in 2. but without projection were applied in [10] to several unrestricted problems and compared with other related methods as e.g. the classical SOR-N method or the cg-SSOR-N algorithm as suggested in [1]. A considerable improvement was obtained concerning the total amount of work and its dependence on the choice of the parameters involved.

Related bound constrained problems namely the minimal surface over an obstacle, i. e.

$$h(u, \nabla u) = (1 + |\nabla u|^2)^{1/2} \tag{3.1}$$

in (1.2) were solved in [11] with the two-phase algorithm of the last section. The second phase was started when the number of active constraints had the same positive value for two consecutive steps. In order to illustrate the behavior of the method we graphically present typical results for discretization by linear finite elements.

The domain $R = \{(x,y),\ x^2 + y^2 < 2\}$ was triangulated resulting in a triangulation with 475 interior nodes and $g(x,y) \equiv 0$,

$$\phi(x,y) = \begin{cases} a(1-r^2/b^2)^{1/2}, & \text{if } x = 0,\ -b \le y \le b \\ 0 & \text{otherwise} \end{cases}$$

with a = 1.255, b = 1.439 were chosen in (3.1). The starting vector was taken as the points on a sphere of radius 2 with center at the origin. In Fig. 1 the work measured in essential algebraic operations per variable is shown which is needed in phase II to reduce the residual $\| Q_k r_k \|$ to 10^{-6} for varying values of the (constant) relaxation parameter ω. In Fig. 2 the decrease of the residual is shown for the two-phase method in comparison to the projected MOR-N algorithm with the "optimal" value $\omega = 1.9$ given by Fig. 1. Here the amount of work is measured in the equivalent of that necessary for a relaxation step. Finally in Fig. 3 the development of the number of active constraints, it is zero for the starting vector, is shown.

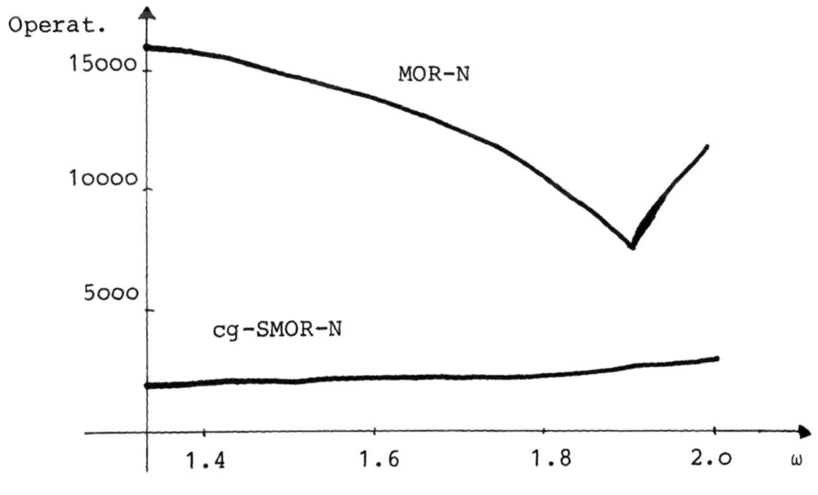

Fig. 1 Number of operations in phase II for different values of ω

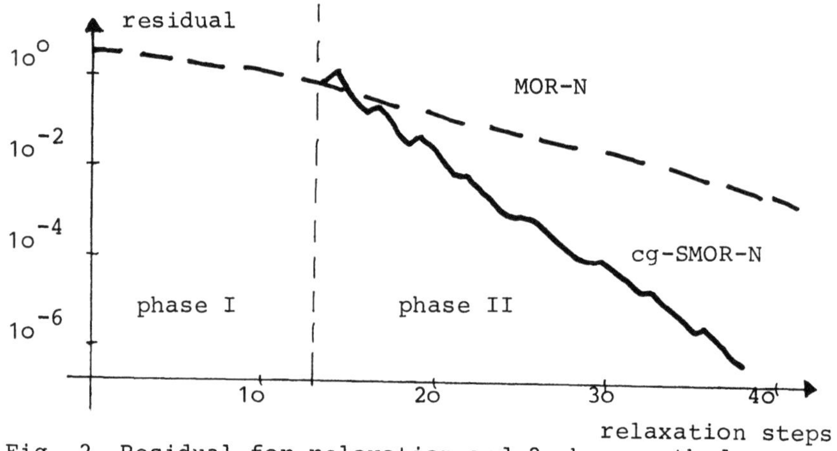

Fig. 2 Residual for relaxation and 2-phase method

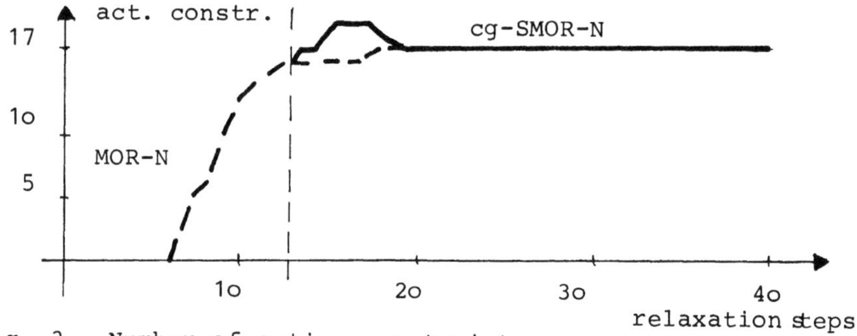

Fig. 3 Number of active constraints

As an example from structure mechanics we consider the indentation problem for the nonlinear membrane. Finite-difference or finite element discretizations of membranes were treated by many authors. We mention here only [12, 16].
In [16] a thin flat unstrained membrane covering the region R in the (x,y)-plane is considered which is loaded by a force acting in the z-direction. If u, v, w denote displacements in the x,y,z-direction, respectively, then under certain simplifying assumptions but with a nonlinear strain-displacement relation the displacement components of the deformed membrane satisfy a coupled system of three nonlinear differential equations. Under further assumptions which are mainly u = v = 0 and after transformation to nondimensional variables the last and most nonlinear of

these reduces to

$$(3w_x^2 + w_y^2)\, w_{xx} + 4w_x w_y w_{xy} + (w_x^2 + 3w_y^2) w_y = -D \quad (3.2)$$

The boundary condition is w = O if the membrane is kept fixed there.

(3.2) is the Euler equation of the functional (1.2) with

$$h(w, \nabla w) = \frac{1}{p} |\nabla w|^p - D, \; p = 4 \quad (3.3)$$

and hence we have to solve the homogeneous Dirichlet problem for a pseudo-Laplace-operator which was also considered e. g. in [3, 14].
The complete solution of the nonlinear membrane problem may be found then as in [16] using the obtained displacement w together with u = v = O as starting solution for the system in u, v, w. It is well-known that for values of p near 1 or considerably greater than 2 a direct treatment of (1.2), (3.3) by classical methods is impossible in practice (cf. [3]). The application of the proposed method in the case p = 4 yielded results comparable to those for the minimal surface problem. The main difference was that the relaxation parameters and the damping factor τ_k in <u>Step 5</u> had to be taken smaller. This is caused by the fact (cf. the inequalities for $\omega_{1,k}$ and τ_k in [11]) that $y^T A(x) y \leq y^T F(x) y$ (the choice of the matrix A is rather canonical) for the membrane problem while the opposite inequality is valid for the first example (cf.[1o]).

In Fig. 4 the z-displacement along the y-axis of a nonlinear membrane over the unit circle is shown which is pressed by the uniform load D=2,6,2o,6o against the obstacle

$$\phi(x,y) = \begin{cases} 1 \text{ for } x^2 + (y-0{,}25)^2 < 0{,}25^2, \\ 1{,}5 \text{ otherwise.} \end{cases}$$

The unit circle is subdivided by a triangulation with average edge length h = 1/12 and 475 interior nodes.

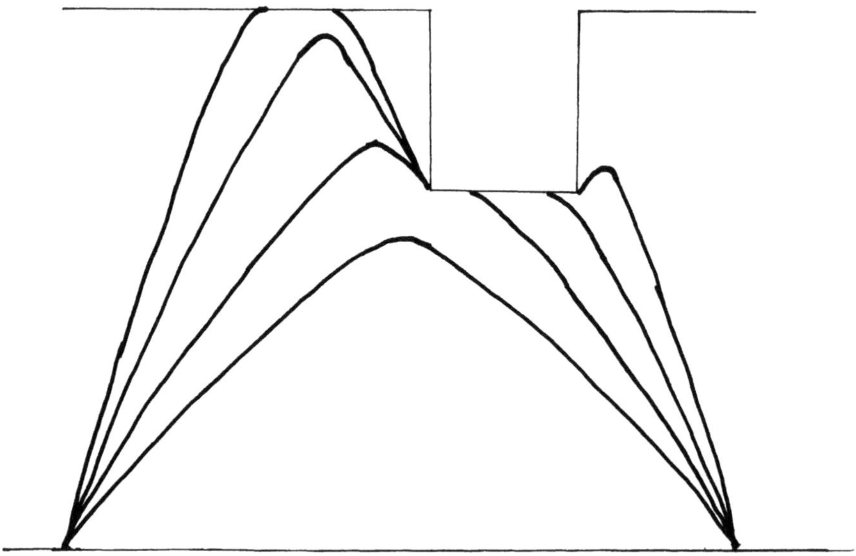

Fig. 4: z-displacement of nonlinear membrane

4. A multigrid method for variational inequalities

One of the methods for the approximate solution of boundary value problems which at least under certain conditions may be quite efficient is the so-called multi-grid method. We shall explain it for an abstract linear boundary value problem given by

$$Au = f. \tag{4.1}$$

For a sequence of stepsizes $\{h_\ell\}$ with

$$h_o > h_1 > h_2 > \dots > h_\ell > \dots > 0$$

(4.1) is to be discretized by a finite-difference or a finite element method yielding the finite-dimensional problems

$$A_\ell u_\ell = f_\ell, \ \ell = 0,1,\dots \tag{4.2}$$

i. e. u_ℓ, f_ℓ are elements of a finite-dimensional vector space V_ℓ, $\ell = 0,1,\ldots$ and $A_\ell : V_\ell \to V_\ell$ is a linear operator. $\Phi_\ell = A_\ell^{-1}$ is assumed to exist and $\tilde{\Phi}_\ell$ is an approximation of Φ_ℓ. Furthermore to connect the grid functions on V_ℓ, $\ell = 0,1,\ldots$ we need a linear and injective prolongation operator

$$p_{\ell,\ell-1}: V_{\ell-1} \to V_\ell, \quad \ell = 1,2,\ldots$$

and a linear and surjective restriction operator

$$r_{\ell-1,\ell}: V_\ell \to V_{\ell-1}, \quad \ell = 1,2,\ldots$$

Finally a smoothing procedure

$$v_\ell \to \mathcal{G}_\ell(v_\ell, f_\ell) = H_\ell(G_\ell v_\ell + f_\ell) \tag{4.3}$$

will be used, where the linear operators G_ℓ, H_ℓ satisfy $A_\ell = H_\ell^{-1} - G_\ell$, I the identity operator. Then in the following quasi-Algol procedure we define a linear MG-algorithm as in [6].

```
prodedure  MGM(ℓ,u,f); integer ℓ; array u,f;
comment    ℓ : actual level number,
           u : input u_ℓ^(i), output: u_ℓ^(i+1),
           f : right-hand side of eq. (4.1);
if ℓ = 0 then u := Φ̃_0*f_0 else
begin integer j; array v,d;
      for j := 1 step 1 until ν do u := 𝒢_ℓ(u,f);
      d := r_ℓ-1,ℓ* (A_ℓ u-f); v := 0;
      MGM(ℓ-1,v,d,);
      u := u - p_ℓ,ℓ-1*v
end MGM;
```

This procedure performs one MG-step starting from an approximation $u_\ell^{(i)}$ to u_ℓ using the levels $\ell, \ell-1,\ldots,0$ and on each of these except the last one at first ν smoothing steps are executed. $\tilde{\Phi}_0$ may, however, itself be an application of a smoothing procedure. This method may be generalized to nonlinear problems (cf.[6]).

In order to apply such methods to problems of the form (1.1), (1.6) one could e. g. use this nonlinear MG-procedure where $\tilde{\Phi}_o$ is the application of ν steps $\mathcal{G}_o^{pr}$ according to relaxation with projection and also $\mathcal{G}_\ell$ in the smoothing steps is substituted by $\mathcal{G}_\ell^{pr}$. We call this method MG1-algorithm. This method was applied to problems similar to those considered in 3. and at least in the computations it did not converge globally ([5]). A global convergent MG-method may however be obtained if in the algorithm defined in 2. Step 3 is changed and the vector p_k is determined as the direction taken by one call of the procedure MGM for the linear problem $A(x^{(k)})x = b(x^{(k)})$ starting from $x^{(k)}$ and keeping the components corresponding to I_k fixed.

Global convergence may then be proved along the lines of the proof in [11] if p_k and $Q_k r_k$ are related as in (2.3) by a matrix S (suppressing the index k) with similar properties, essentially positive-definiteness. For smoothing by relaxation G_ℓ, H_ℓ in (4.3) are given by

$$G_\ell = \frac{1-\omega}{\omega} D_\ell - QL_\ell^T Q, \quad H_\ell = \omega(D_\ell + \omega Q L_\ell Q)^{-1} \tag{4.4}$$

if $A_\ell = D_\ell + L_\ell + L_\ell^T$. Then a simple computation shows that for $\nu=2$ we have on the level ℓ

$$S_\ell = B_\ell + p_{\ell,\ell-1} S_{\ell-1} r_{\ell-1,\ell} G_\ell H_\ell G_\ell H_\ell, \ell = 1,2,... \tag{4.5}$$

where $S_o = \Theta$ and $B_\ell = H_\ell + H_\ell G_\ell H_\ell$.

The matrix S_ℓ is not symmetric. The equally unsymmetric matrix H_ℓ, however, is positive-definite for $0 < \omega < 2$ if A_ℓ has this property. Hence for sufficiently small ω B_ℓ is positive-definite and hence S_ℓ if the second term on the right of (4.5) is multiplied by a damping factor.

If this last method is called MG2-algorithm then we propose as a second two-phase method that obtained by the application of the MG 1-algorithm to approximate in phase I the proper subspace and of the MG2-algorithm as the second phase. In computations performed by W. Hackbusch for problems of the type (1.2), (3.1) discretized by finite-difference methods on a quadratic mesh, R being the unit square, this method also brought a cost-reduction compared to simple projected relaxation which, however, was comparable to though not greater than that for the projected cg-SMOR-N method of 2.

5. Conclusions

We have seen that two-phase methods as proposed in 2. and 4. are efficient algorithms for the solution of variational inequalities, when the feasible set has the form (1.6). Further numerical experience, however, is needed in order to compare it with other algorithms which could be applied to these problems. We mention here especially quasi-Newton methods for problems with a sparse Hessian (cf. e. g.[15, 17]). These methods for unconstrained minimization problems could be generalized to the constrained case. In the convergence proofs for these methods, however, it is assumed that in every step a linear system is solved exactly and one cannot see an easy way how to get rid of this assumption.

References

1 Concus, P., Golub, G. H. and D. P. O'Leary, Numerical solution of nonlinear elliptic partial differential equations by a generalized conjugate gradient method, Computing 19 (197o), 321-339

2 Cottle, R. W., Giannessi, F. and J.-L. Lions (eds.), Variational Inequalities and Complementarity Problems, John Wiley, Chichester 198o

3 Glowinski, R. and A. Marrocco, Sur l'approximation par elements finis d'ordre un, et la reolution par penalisation-dualite, d'une classe de problemes de Dirichlet nonlineaires, Rev. Francaise Automat. Informat. Recherche Operationnelle ser. Rouge Anal. Num. R-2 (1975), 41-76

4 Glowinski, R., J. L. Lions and R. Tremolieres, Analyse Numerique Des Inéquations Variationelles, Dunod, Paris 1976

5 Hackbusch, W., private communication

6 Hackbusch, W., On the convergence of multi-grid iterations, Report 79-4, Universität zu Köln, to appear in Beiträge Num. Math. 9

7 Kikuchi, N. and J. T. Oden, Contact Problems in Elasticity, TICOM Report 79-8, The University of Texas at Austin (1979)

8 O'Leary, D. P., Conjugate gradient algorithms in the solution of optimization problems for nonlinear elliptic partial differential equations, Computing 22(1979), 59-77

9 Mittelmann, H. D., On the approximate solution of nonlinear variational inequalities, Num. Math. 29 (1978), 451 - 462

1o Mittelmann, H. D., On the efficient solution of nonlinear finite element equations I, to appear in Num. Math .

11 Mittelmann, H. D., On the efficient solution of nonlinear finite element equations II, submitted to Num.Math.

12 Oden, J. T. and T. Sato, Finite strains and displacements of elastic membranes by the finite element method, Int. J. Solids Structures 3 (1967), 471 - 488

13 Rockafellar, R. T., Lagrange multipliers and variational inequalities, in [2]

14 Scarpini, F., A numerical approach to the solution of some variational inequalities, in [2]

15 Shanno, D. F., On variable-metric methods for sparse Hessians, Math. Comp. 34 (198o), 499 - 514

16 Shaw, F. S. and N. Perrone, A numerical solution for the nonlinear deflection of membranes, J. Appl. Mech.21 (1954), 117 - 128

17 Toint, Ph., On the superlinear convergence of an algorithm for solving a sparse minimization problem. SIAM J. Num. Anal. 16 (1979), 1o36 - 1o45

The Numerical Calculation of the Contact Problem in the Theory of Elasticity

J. J. KALKER, H. J. C. ALLAERT,
Delft University of Technology, Delft,
The Netherlands

J. DE MUL
SKF-ERC, Nieuwegein,
The Netherlands

Summary

In the present paper, distinction is made between frictional and frictionless contact problems. After their formulation, the epitome of frictional problems, viz. the problem of frictional rolling contact is considered, and the two most recent and successful methods for dealing with it are presented. Then the frictionless problems is considered. Two pairs of methods are presented, each pair consisting of a non-variational method, with its variational counterpart authored by us. The first pair, viz. Paul & Hashemi / Kalker & Allaert, deal with the general half-space frictionless contact problem. The second pair, viz. Reusner / Kalker & De Mul, deal with the contact of bodies of revolution, which may be approximated by half-spaces. The codes implementing the frictionless problems suffer from long operating times. Research is still in progress to speed up the codes.

1. Introduction

Consider two deformable bodies, V^α, $\alpha = 1, 2$, with surfaces ∂V^α, see fig. 1a. They are pressed together with a force N, so that they are brought into contact. In an undeformed state they would intersect, see fig. 1b. The distance h of the bodies at a point of their surfaces is measured along the common normal on the bodies; it is counted negative where the bodies intersect, positive where they do not. In order to counteract the penetration, a displacement u^α (in body α), $\alpha = 1, 2$, comes into being. The contact condition is then that the deformed distance D is non-negative:

$$D \overset{\text{def}}{=} h - u_n^1 - u_n^2 \geq 0 \;;\; 0 : \text{contact}$$

u_n^α : displacement component in the direction of n^α, outer normal on ∂V^α (1)

In addition we will normally have that the normal component of the surface traction in the contact patch is compressive:

$$p_n \leqq 0 \text{ in contact} \quad , \quad = 0 \text{ outside it} \tag{2}$$

Outside contact, the displacement is prescribed as $\tilde{u}^{\alpha}$ in $A_u^{\alpha} \subset \partial V^{\alpha}$, and on the remainder of the surface, the surface load p^{α} is prescribed as $\tilde{p}^{\alpha}$:

$$p^{\alpha} = \tilde{p}^{\alpha} \text{ in } A_p^{\alpha} \quad , \quad u^{\alpha} = \tilde{u}^{\alpha} \text{ in } A_u^{\alpha} \tag{3}$$

In the contact region $K = \{x | \; x \in \partial V^{\alpha} \, , D = 0\}$, frictional forces act. They are commonly modelled in three possible ways:

1) Friction is neglected: the shear traction in the contact area is set zero. This is suitable, e.g. in lubricated contacts in which hydrodynamic effects are small; see also sec. 4.
2) Friction is infinite: the relative tangential velocity of the bodies in the contact patch (the slip) is set zero. This is suitable, e.g. when the friction forces are relatively small while the coefficient of friction (see below) is large.
3) Friction is finite and acts according to Coulomb's law, which reads:

   The tangential traction $|p_T| \leqq - fp_n$, f: coefficient of friction. If the slip $w \neq 0 \Rightarrow w = - \lambda p_T$, $\lambda > 0$, and $|p_T| = - fp_n$
   where the tangential component $p_T \overset{\text{def}}{=} p - p_n n$ (4)

2. Variational formulations in a special case

The above problems may be formulated as variational problems [2, 3, 19].

In the case of one of the bodies being rigid, and the other an elastic half-space $z \geqq 0$, the variational formulation of problem 1) reads:

$$\min_{u,p \text{ together}}! \iint_{z=0} p_n(x,y)\{\tfrac{1}{2}u_n(x,y) - h(x,y)\}dxdy$$

$$\text{sub } p_n(x,y) \leqq 0$$

$$\text{sub } p_T(x,y) = 0$$

$$u \to 0 \text{ as } (x,y,z) \to \infty \qquad (5)$$

By minimizing over "u,p together" we mean that u and p correspond to the same elastostatic field. For instance, in the case of problem 1),

$$u_n(x,y) = \frac{1-\sigma}{2\pi G} \iint_{z=0} p_n(x',y') \frac{dx'dy'}{\sqrt{(x-x')^2 + (y-y')^2}} ;$$

(x,y) : Cartesian coordinates on the surface of the half space (6)

In the case of problems 2) and 3), an incremental argument must be pursued.
Let τ be an increment of time. Thus, problem 2) becomes, when t is a generic instant:

$$\min_{u,p \text{ together}}! \iint_{z=0} p_n(x,y,t)\{\tfrac{1}{2}u_n(x,y,t) -$$

$$- h(x,y,t)\}dxdy + \iint_{K(t-\tau)} p_T(x,y,t)\{\tfrac{1}{2}u_T(x,y,t) -$$

$$- u_T(x,y,t-\tau) - \rho(x,y,t;\tau)\}dxdy \text{ sub } p_n \leqq 0 \qquad (7)$$

where $\rho(x,y,t;\tau)$ is the shift, during the time-interval τ, of the rigid body w.r.t. the half-space at (x,y) at the time t. Note that the contact area $K(t - \tau)$, and the elastic field, are known at the time $(t - \tau)$. These quantities at time t are proveded by the minimization step (7).
In the case of problem 3), the traction bound $g \overset{\text{def}}{=} - fp_n$ must be known beforehand. Again we must work incrementally, and the minimum principle becomes:

$$\min_{u,p \text{ together}}! \iint_{z=0} p_n(x,y,t)\{\tfrac{1}{2}u_n(x,y,t) -$$

$$- h(x,y,t)\}dxdy + \iint_{z=0} p_T(x,y,t)\{\tfrac{1}{2}u_T(x,y,t) -$$

$$- u_T(x,y,t-\tau) - \rho(x,y,t;\tau)\}dxdy$$

$$\text{sub } p_n \leqq 0 \, , \, |p_T|^2 \leqq g^2 \quad (g = - fp_n \, , \text{ known}) \tag{8}$$

We prove the validity of these principles in a heuristic manner; rigorous proofs may be found in 2, 3 .

P r o o f : The contact conditions (1), (2), (4) will be shown to be the Kuhn-Tucker conditions of the above variational principles. To that end we observe:

$$\delta \iint_{z=0} \tfrac{1}{2}p_\beta(x,y,t)u_\beta(x,y,t)dxdy = \iint_{z=0} (\delta p_\beta)u_\beta dxdy \, ,$$

$$\delta \iint_{z=0} p_\beta(x,y,t)g(x,y,t)dxdy = \iint_{z=0} (\delta p_\beta)g(x,y,t)dxdy$$

$\beta = 1, 2$; g: known function; u_β, p_β belong to the same elastostatic field.

Let $D \geqq 0$, $Dp_n = 0$ be the optimal Lagrange multiplier of $p_n \leqq 0$. Let $\lambda \geqq 0$, $\lambda(g^2 - |p_T|^2) = 0$ be the optimal Lagrange multiplier of $g^2 - |p_T|^2 \geqq 0$.

Then:

$$0 = \iint_{z=0} \delta p_n\{u_n(t) - h(t) + D\}dxdy + \iint_{z=0} \delta p_T\{u_T(t) -$$

$$- u_T(t - \tau) - \rho + 2\lambda p_T\}dxdy \; ; \; \forall \delta p_n \, , \, \delta p_T$$

$\Rightarrow$ 1. $h(t) - u_n(t) = D \geqq 0$, deformed distance;

if $p_n < 0 \Rightarrow D = 0 \Rightarrow$ contact,

if $p_n = 0 \Rightarrow D \geqq 0 \Rightarrow$ non-penetration.

$\Rightarrow$ 2. $u_T(t) - u_T(t - \tau) - \rho = - 2\lambda p_T$ (= shift w.r.t. rigid body = slip);

if $|p_T| < g \Rightarrow \lambda = 0$, slip = 0,

if $|p_T| = g \Rightarrow \lambda \geqq 0$, slip = $- 2\lambda p_T$, see (4).

This completes the proof for problem 3).

As to problem 2), it suffices to observe that the "infinite friction" means that g = 0 outside contact, so that λ is, in fact, free and hence slip = $\{u_T(t) - u_T(t - \tau) - \rho\}$ is unrestricted, while inside contact, $|p_T| < g$, so that slip = $\{u_T(t) - u_T(t - \tau) - \rho\} = 0$.

As to problem 1), the extra condition $p_T = 0$ induces the freedom of the slip = $\{u_T(t) - u_T(t - \tau) - \rho\}$, and only the contact formation remains. QED

3. The rolling contact problem

As an example of a frictional problem, we consider the problem of rolling contact [4], viz. "Two bodies roll over each other while friction is present in the interface. Their rigid body speeds are not precisely matched, so that a rigid slip occurs in the interface, which calls up an elastic field. It is required to find the tangential traction p_T in the interface, and in particular its resultant."
This problem is of importance in railway simulations, where the problem needs to be solved very many times, so that speed of caluclation is of the essence.
In the railway application, the contact area and the normal pressure ($-p_n$) carried by it are known, they follow from the theory of Hertz (1881; see e.g. [1]).

3.1. The complete theory

For the sake of solving the equations of elasticity, we approximate the bodies by the half-spaces $\{z \leqq 0\}$, so that the relations of Boussinesq & Cerrutti (see [1]) are valid. In a symbolic form, they read:

$$u_i(x,y) = \iint_K p_j(x',y')H_{ij}(x-x', y-y')dx'dy',$$

$$\text{where, e.g. } H_{33}(x,y) = \frac{1-\sigma}{2\pi G}\sqrt{x^2+y^2}^{\,-1};$$

$$\text{summation convention is observed} \qquad (9)$$

It will be clear, that an implementation [5] of the rolling contact problem on this basis will be very slow. In fact, on an IBM 370/158, calculation times of around 10 s/case seem the best that can be achieved. In this solution method, the tangential traction is discretized as constant over each mesh of a rectangular net in the contact area. The elastic energy due to the tangential traction becomes then:

$$E = \tfrac{1}{2}\iint_K u_i(x,y)p_i(x,y)\,dxdy =$$

$$= \tfrac{1}{2}\sum_{I,J=1}^{N}\ \sum_{i,j=1,2} M_{2I+i,2J+j}\, p_{2I+i}\, p_{2J+j}$$

i,j = 1,2 : indicate components. I,J = 1,...N : indicate meshes, N in total. p_{2I+i} : value of the rectangular traction component p_i (x,y) at the center of the mesh I. (10)

The variational principle (8) becomes:

$$\min_{\{p_i\}}! \left(\sum_{I,J=1}^{N} \sum_{i,j=1,2} (\tfrac{1}{2} M_{2I+i,2J+j}\, p_{2I+i}\, p_{2J+j}) + \right.$$
$$\left. - \sum_{I=1}^{N} \sum_{i=1,2} (u_{2I+i}(t-\tau) + \rho_{2I+i}(t;\tau))\, p_{2I+i}\, Q_I \right)$$

sub $p_{2I+i}^2 + p_{2I+2}^2 = g_I^2$, I = 1,...,N ; Q_I: area of mesh I ; $u_{2I+i}(t - \tau)$: component of displacement at center of mesh I at time (t - τ). p_{2I+i}: analogous (11)

Here the contribution of the (known) normal traction distribution has been omitted. The problem is solved by a mathematical programming algorithm based on Newton's method. Solution of problem (11) is the traction distribution at time t when that at time (t - τ) is known, which latter enters the problem through $u_{2I+i}(t - \tau)$.

In order to find the steady state solution, in which the traction distribution over the contact is independent of time, several elementary time intervals must be traversed. After about 1 - 1.5 contact widths have been traversed by rolling, convergence is normally achieved.

From the description it is readily understood that this is a slow affair, even though:

1. only the total force is required, which reduces N;
2. the matrix $(M_{2I+i,2J+j})$ need be calculated only once;
3. the problem is convex;
4. in Newton's method modified for bounds we possess a fast optimization routine;
5. generally, only one optimization iteration need be performed per time step.

At present, the bottle neck of this method are the "curved"

restrictions $p_{2I+1}^2 + p_{2I+2}^2 \leqq g_I^2$. In the present program, p is represented by the polar representation $(p_{2I+1}, p_{2I+2}) =$ $= (r_I \cos \phi_I , r_I \sin \phi_I)$ when $r_I = g_I$, so that the restriction reduces to the bound $r_I \leqq g_I$.
The rectangular representation is maintained when $r_I < g_I$, because the polar representation becomes singular when $r_I = 0$, and this introduces spurious Kuhn-Tucker points. As a consequence, we shift back and forth between representations, whereby the adaptation of the gradient and Hessian is most consuming.

3.2. The simple theory

It will therefore not cause surprise, that approximate solution methods have been looked for. At present, the most successful and general of these is the so-called "simple theory" [4, 6, 7, 8, 9, 10]. It is based on the replacement of (9) by:

$$u_T(x,y) = Lp_T(x,y) \quad , \; L > 0 \tag{12}$$

where L is constant to be determined. How this is done, is shown in [4, 6]. It may be shown [4] that the steady state rolling contact conditions are:

$$w(x,y) = \text{slip} = \rho(x,y) - L \frac{\partial p_T(x,y)}{\partial x} \quad , \; x\text{: rolling direction} \tag{a}$$

$$|p_T| \leqq g \; ; \; w \neq 0 \Rightarrow w = -\lambda p_T \; , \; \lambda > 0 \; , \; |p_T| = g \tag{13}$$

The following solution method is described in [10]. Let τ be an increment of x, $\tau > 0$; working from high values of x towards low, we rewrite (13a):
$w\tau = \rho\tau - Lp_T(x + \tau;y) + Lp_T(x,y)$.
Call $Lp_H = -\rho\tau + Lp_T(x + \tau;y)$, and set:
$p_T = p_H$ if $|p_H| \leqq g(x,y)$, $p_T = p_H g/|p_H|$ if $|p_H| \geqq g(x,y)$.
Indeed, $|p_T| \leqq g$, and
$w\tau = \rho\tau - Lp_T(x + \tau;y) + Lp_T(x,y) =$
$= - Lp_H + Lp_H = 0$ if $|p_T| < g$, and

$$= L(1 - \frac{|p_H|}{g}) p_T = - \lambda p_T$$

$$\lambda = (\frac{|p_H|}{g}) - 1) L \geqq 0 \text{ if } |p_T| = g$$

which is Coulomb's law.

The algorithm must be provided with a boundary condition, for which reference is made to [10]. It is extremely fast, a reduction factor of 100 in speed of calculation is achieved with respect to the algorithm of sec. 3.1, with errors of an estimated maximum of 15% of the maximum value of the total tangential force. Such high errors are acceptable in rail vehicle dynamics, where owing to pollution of wheels and rails, unpredictable deviations of up to 40% from the theory have been observed.

4. The frictionless contact problem

As was observed in sec. 1, the frictionless contact problem obtains when the contact is lubricated. It also occurs when the contacting bodies may be, for the sake of elasticity, considered as half-spaces with the same elastic constants. Under these circumstances, the problem can be decomposed [11] into a frictionless problem, and a frictional problem of the type considered in sec. 3. This decomposition is of great importance technically; it obtains e.g. in the contact between wheel and rail.

The first contact problem to be solved was a frictionless one (Hertz, 1881, see [1]). In it, the undeformed distance h (see fig. 1 and eq. (1)), is a quadratic function of x and y, Cartesian coordinates on the surface of the half-space. Many problems may be solved with the Hertz theory; we refer to the previous section for an example.

Notable exceptions occur in the railway application when the flange of the wheel comes into contact with the rail, and in bearings, where under high loads the rounded ends of the rollers come into contact with the rings. In both cases, the speed of operation of the algorithms is of the essence, in the case of railway application because of the railway vehicle simulation, in the case of the bearing because there also motion simulations are desired.

Unlike the rolling contact problem, no "simple" truly fast theory has been proposed apart from the Hertz theory.
We will describe four methods of solution in some detail. Two of them, viz. Paul & Hashemi ([12], sec. 4.1) and Reusner ([13], sec. 4.3) are so-called classical methods. Based upon the Boussinesq & Cerrutti integral (6), a number N of displacement quantities are expressed in N independent variables, such as the pressures ($-p_n(x,y)$) and the contact width, under the assumption that contact is established ($D = h - u_n = 0$), which fixes the displacement quantities. This yields N (possibly non linear) equations for the independent variables, which are solved. Then it is verified whether the non-penetration condition $D \geqq 0$ and the compressiveness condition $p_n \leqq 0$ are everywhere satisfied. If they are, the solution has been found; if they are not, the contact area is changed, and a new iteration is performed. The other two methods, viz. Kalker & Allaert and Kalker & De Mul, are variational methods that are described here for the first time. They are not the first variational methods to appear in the literature: priority lies with the method of Fridman & Chernina [14], dating from 1967. Independently, Conry & Seireg [15], published a slightly deviant variational method, and they were the first to introduce the notion of design into contact theory.
The theory was further developed by Páczelt (e.g. [16]), who nowadays is interested in design questions for contact problems. Kalker [17] published a relatively fast method for the calculation of contacts which are slender and fairly sharply pointed. This work is based on line contact theory [17, 18]; in the bearing application, the method can unfortunately be used only when the radius of the rounding at the ends is more than three times larger than the radius of the roller. In practice, these radii may be ten times smaller than the radius of the roller, which renders the line contact method useless for the bearing applications; instead Reusner's method ([13], sec. 4.3) is currently used.

4.1. The method of Paul & Hashemi

The contact area is discretized by rectangles, see fig. 2. The displacement u_I in the centroid of Q_I is expressed in the p_J by Boussinesq & Cerrutti, see (6):

$$u_I = \sum_{J=1}^{N} K_{IJ}\, p_J \;,\; u_I = h_I \tag{14}$$

In accordance with the contact condition $u_I = h_I$, the linear equations (14) are solved for p_J (collocation method). The edge of the contact area is given by $p_n = 0$, while the behaviour of p_n near the edge of the contact is parabolic, see fig. 2b. Based on these new $\{p_J\}$, the points A_i, B_i on the edge of the contact (see fig. 2b) are determined so that the behaviour of fig. 2b is best approximated; between the new A_i and B_i, the same number of elements are conceived as before, and the equations (14) are again set up, etc. The process stops when A_i and B_i have converged, which happens after about four to ten iterations. Note that the coefficients K_{IJ} have to be recomputed for every iteration. As this is a major job, it constitutes the bottleneck for this method.

4.2. The method of Kalker & Allaert

The contact area is discretized by rectangles with sides parallel to the (x,y) axes. First we will consider the case that all rectangles are congruent. Then we calculate the K_{IJ} of (14) as follows:

Let f(x,y) be such that $\frac{\partial^2 F}{dxdy} = f(x,y)$; then

$$\int_{x_1}^{x_2} dx \int_{y_1}^{y_2} f(x,y)\,dy = \left(\left(F(x,y)\right)_{x=x_1}^{x_2}\right)_{y=y_1}^{y_2} \tag{15}$$

Now, if $f(x,y) = \sqrt{x^2 + y^2}^{\,-1} \Rightarrow F(x,y) =$

$= x\,\mathrm{sh}^{-1}(y/|x|) + y\,\mathrm{sh}^{-1}(x/|y|)$; $\mathrm{sh}^{-1} z$ = inverse hyperbolic sine $= \ln(z + \sqrt{z^2 + 1})$ (16)

so that

$$K_{IJ} = \int_{x_J-1}^{x_J+1} dx' \int_{y_J-k}^{y_J+k} dy' \sqrt{(x_I-x')^2 + (y_I-y')^2}^{\,-1} =$$

$$= \left(\left[F(x,y)\right]_{x=x_I-x_J+1}^{x_I-x_J-1}\right)_{y=y_I-y_J+k}^{y_I-y_J-k} \qquad \text{"close"} \qquad (17a)$$

When $(x_I - x_J)^2 + (y_I - y_J)^2 >> l^2 + k^2$ we may simplify this to:

$$K_{IJ} = \frac{4lk}{\sqrt{(x_I-x_J)^2 + (y_I-y_J)^2}} \qquad \text{"far"}$$

if $(x_I - x_J)^2 + (y_I - y_J)^2 = c(l^2 + k^2)$, $c >> 1$, e.g. 4 (17b)

The optimization problem becomes:

$$\min! \sum_{J,I} \tfrac{1}{2}K_{IJ}\, p_K\, p_J - \sum_I p_I j_I \text{ sub } p_J \leqq 0 \qquad (18)$$

This is a simple quadratic programming problem posed long ago (1967) by Fridman & Chernina. As a result, the p_J - representative of the pressure in the center of Q_J - are given, and it is found which Q_J belong to the contact and which do not. In other terms, the boundary of the contact area is given very roughly indeed. The need for improvement of this feature led Paul & Hashemi to their method described in sec. 4.1; in the results, the points A_i, B_i of fig. 2a presumably lie on the contact boundary. We could do the same thing, but we preferred to calculate the boundary of the contact as follows.

In the contact problem with $D(x,y) = h(x,y) - u_n(x,y)$ we shift the bodies over a distance τ w.r.t. each other in the x-direction. The u_n and p_n shift a distance τ too: in fact, the entire contact field shifts over a distance τ. It is then easy to see that $\frac{\partial h_I}{\partial x} = \sum_{I\epsilon \text{contact}, J\epsilon \text{contact}} K_{IJ} \frac{\partial p_J}{\partial x}$, from which $\frac{\partial p_J}{\partial x}$ follows.

Similarly, $\frac{\partial p_J}{\partial y}$ may be found.
These may be used to find a boundary point, see fig. 3; we are of the opinion that this method is as accurate as Paul & Hashemi's to find the boundary. We tend to opine that our method is somwhat better than Paul & Hashemi's on the following grounds:

1. The results are equally accurate.
2. The optimization idea is more general than the P.&H. method, since any combination of rectangles may constitute a contact area. This is not so with P.&H., since their contact must be convex in one direction at least. Very efficient methods are available for the quadratic programming problem, which are equivalent to only a few matrix inversions.
3. The matrix of influence numbers need be calculated only once. When all rectangles are equal, the invariance of the discretization w.r.t. the translation over a mesh implies that the number of influence numbers to be calculated is only $O(N)$ rather than $O(N^2)$, where N is the number of elements. In combination with the simplified element (17b), this renders the calculation of the matrix truly fast.

Whereas in the railway application, congruent rectangles seem appropriate, this is not so in the bearing application, where stress concentrations occur at the transition to the sharp rounding of the roller. The large gradients of the stress necessitate a locally very fine mesh, so fine that the number of equal elements would run into many hundreds or even thousands, which would lead to a very slow algorithm. When the discretization is no longer uniform, the influence numbers of the collocation (14) are no longer symmetric, $K_{IJ} \neq K_{JI}$. This means that there no longer exists a quadratic form $Q(p_J)$ such that $\frac{\partial Q(p_J)}{\partial p_I} = \sum_J K_{IJ} p_J - h_I$, that is, there is no optimization problem fully equivalent to the collocation problem. The most precise way to implement problem 1) is now to solve the following problem:

$$F(x,y) = \tfrac{1}{2}xy^2 \, \mathrm{sh}^{-1}(x/|y|) + \tfrac{1}{2}xy^2 \, \mathrm{sh}^{-1}(y/|x|) +$$

$$- 1/6 \sqrt{x^2 + y^2}^{\,3} ,$$

$$\frac{\partial^4 F(x-x',y-y')}{\partial x \partial x' \partial y \partial y'} = \sqrt{(x - x')^2 + (y - y')^2}^{\,-1} ;$$

$$M_{IJ} = \left[\left[\left[\left[F(x-x',y-y')\right]_{x=x_I-l_I}^{x_I+l_I}\right]_{x'=x_J-l_J}^{x_J+l_J}\right]_{y=y_I-k_I}^{y_I+k_I}\right]_{y'=y_J-k_J}^{y_J+k_J} ,$$

$$H_I = \iint_{Q_I} h(x,y)\,dxdy$$

$$\min!\ \tfrac{1}{2} \sum_{I,J} M_{IJ}\, p_I\, p_J - \sum_I H_I\, p_I \text{ sub } p_I \leqq 0 \quad \forall I ;$$

$$\text{when } (X_I - x_J)^2 + (y_I - y_J)^2 = C\{(l_I - l_J)^2 +$$

$$+ k_I - k_J)^2\} ; \tag{19}$$

with $C >> 1$, e.g. 2:

$$M_{IJ} \cong 16 \sqrt{(x_I - x_J)^2 + (y_I - y_J)^2}^{\,-1}\, l_K l_J k_I k_J \tag{19b}$$

In the present matrix (M_{IJ}) we must calculate 32 logarithms per element which is not covered by (19b), while in the collocation matrix (K_{IJ}) of (17), we need only eight logarithms under the same circumstances. This renders the calculation of (M_{IJ}) roughly 4 times as slow as that of (K_{IJ}). Present research is directed towards diminishing this factor.

4.3. Reusner's method

Reusner's method concerns itself with the calculation of the frictionless contact between two bodies of revolution. Then, the undeformed distance h is given by:

$$h(x,y) = A\,x^2 + B(y) \tag{20}$$

where A is in most cases constant, but may depend on y. Under the assumptions of line contact theory [17, 18], see sec. 4, it may be shown that the normal pressure is semi-elliptical, $p_n = - C \sqrt{a(y)^2 - x^2}$ where a(y) is the half-width of the contact area at y.

Also, the Hertz theory [1] predicts a normal pressure of the

form $p_n = - C \sqrt{1 - (x/a)^2 - (y/b)^2}$, where a and b are the semi-axes of the (elliptical) contact area; i.e. p_n is also semi-elliptical for fixed y.
Reusner then assumes that for fixed y p_n is semi-elliptical whenever (20) holds. His method of calculation assumes a contact area of a form sketched in fig. 4, where each rectangle carries a semi-elliptical pressure distribution. The calculation proceeds roughly as follows.
At the centroid of each rectangle the normal displacement u_n and its second derivative u_{nxx} are calculated. They are equated to B(y) and 2A, respectively (cf. (20)). This yields 2N non-linear equations with the 2N unknowns p_J and a_J (a_J: half-width of contact area at element J), which are solved iteratively. Proper measures are taken to enforce the validity of non-penetration (1) and compressiveness (2) conditions. A correction is made for the finite depth of the body, which is a roller of radius $(\frac{1}{2A})$, rather than a half-space. The speed of calculation on an IBM/370/158 is several minutes.
The advantage over Paul & Hashemi is that with Reusner the number of elements is roughly 1/5th to 1/10th, but the influence numbers are harder to calculate.

4.4. The method of Kalker & De Mul

Just as Kalker & Allaert was designed as an improvement of the method of Paul & Hashemi, the method of Kalker & De Mul, which is at present not yet finalized, is designed to improve upon Reusner. The most important bottleneck of Reusner seemed to be the setting up of the non-linear equations. This seemed to us to be due to the fact that the influence numbers of Reusner are not expressible as elementary functions the way the K_{IJ} (see (17)) and the M_{IJ} (see (19)) are. Consequently, we started to approximate the pressure distribution of fig. 4 by one which is constant over each Reusner mesh. This turned out to be too inaccurate, whereupon we started to work with an element of the shape given in fig. 5. As variables we took the area of the element, 1.5 pa, and the height-width ration (p/a), and we minimized the familiar object function (see (19)):

$$\min!\ C = \tfrac{1}{2} \sum_{I,J} M_{IJ}\, p_I\, p_J - \sum_I H_I\, p_I, \quad \text{sub } (1.5\ pa)_I \leqq 0 \quad I.$$

Note now that the gradient and Hessian of this C consists of M_{IJ} and their first and second derivatives with respect to a_I and a_J, which can, however, easily be expressed in elementary functions. The function C is this no longer quadratic, and the problem must be solved with the methods of non-linear programming, for which we chose Newton's method, modified for bounds. At present the program takes about 15 sec. on an IBM 370/158 computer for a problem which is comparable to the one mentioned at the end of sec. 4.3.
It is hoped that more will be gained when the computation of the Hessian is improved, see sec. 4.2, where it is mentioned that present research is directed towards that end.

5. Conclusions

In the case of frictional rolling contact, it has been shown that variational methods are the only methods available for solving the problem as accurately as one wishes. If a small accuracy is taken into the bargain of an extremely fast operation, recourse may be had to a classical "approximate" method.
In the case of frictionless contact, classical and variational methods are foughly equally fast. That under these circumstances the small edge observed in favor of the variational methods, may very well be due to the "speed-consiousness" of the present authors.

Literature

[1] A.E.H. Love, A treatise on the mathematical theory of elasticity, 4th ed. (1926), Cambridge U.P.

[2] G. Fichera, Problemi elastistatici con vincoli unilaterali: il problema di Signorini con ambigue condizioni al contorno, Mem. Accad. Nat. Lincei, ser. 8, 7 (1964) p. 116-140.

[3] G. Duvaut, J.-L. Lions, Les inéquations en mécanique et en physique, Dunod, Paris (1972).

[4] J.J. Kalker, Survey of wheel-rail rolling contact theory, Veh. Syst. Dyn. 5 (1979) p. 317-358.

[5] J.J. Kalker, The computation of three-dimensional rolling contact with dry friction, Int. J. Num. Meth. Engng, 14 (1979) p. 1293-1307.

[6] J.J. Kalker, Simplified theory of rolling contact, Delft Progr. Rept 1 (1973) p. 1-10.

[7] K. Knothe, D. Moelle, H. Steinborn, ROLCON, ein schnelles, vielseitiges Digitalprogramm zum rollenden Kontakt, ILR Mitt. 55 (1978) (Report Berlin T.U.).

[8] L. Bansagi, Ein analoges Modell und die Simulationsergebnisse zur Nachbildung des rollenden Kontaktes nach der Theorie von Kalker, Report MAN Augsburg (BRD), K 096 556-EDS-08 5.0000.01 (1979).

[9] J.G. Goree, E.H. Law, User's manual for Kalker's simplified non-linear creep theory, Rept US Dept. Transport, FRA/ORD-78/06 (1977).

[10] J.J. Kalker, A fast algorithm for the simplified theory of rolling contact, (1980). Available from the author.

[11] A.D. de Pater, On the reciprocal pressure between two bodies, Proc. Symp. Rolling Contact Phenomena, J.B. Bidwell (Ed.), Elsevier (1962) p. 29-75.

[12] B. Paul, J. Hashemi, An improved numerical method for counterformal contact stress problems, Rept US Dept. Transport, FRA/ORD-78/26,

[13] H. Reusner, Druckflächenbelastung und Oberflächenverschiebung im Wälzkontakt von Rotationskörpern, SKF Schweinfurt (1978).

[14] V.M. Fridman, V.W. Chernina, Iteration methods applied to the solution of contact problems in bodies, (Russian), Mekh. Tverdogo Tela AN SSSR 1 (1967) p. 116-120.

[15] T.F. Conry, A. Seireg, A mathematical programming method for design of elastic bodies in contact, J. Appl. Mech. 38 (1971) p. 387-392.

[16] I. Páczelt, Iteration method applied to the solution of contact problems of elastic systems with elements in unilateral relation, (Russian), Act. Techn. Acad. Sci.

Hung. 76 (1974) 1, 2, p. 217-241.

[17] J.J. Kalker, The surface displacement of an elastic half-space loaded in a slender, bounded, curved surface region with application to the calculation of the contact pressure under a roller, J. Inst. Maths Applics 19 (1977) p. 127-144.

[18] J.J. Kalker, On elastic line contact. J. Appl. Mech. 39 (1972) p. 1125-1132.

[19] J.J. Kalker, Variational principles of contact elastostatics, J. Inst. Maths Applics 20 (1977) p. 100-219.

Captions of figures

Fig. 1 : Two bodies in contact.

Fig. 2a : Discretization according to Paul & Hashemi.

Fig. 2b : The behaviour of p_n near the edge of contact.

Fig. 3 : Finding a boundary point according to Kalker & Allaert.

Fig. 4 : Contact area assumed by Reusner.

Fig. 5 : The element of Kalker & De Mul. The factors ½ may be changed.

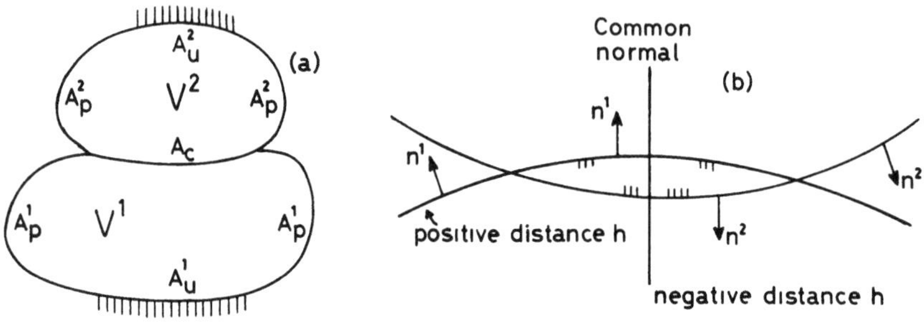

Fig 1

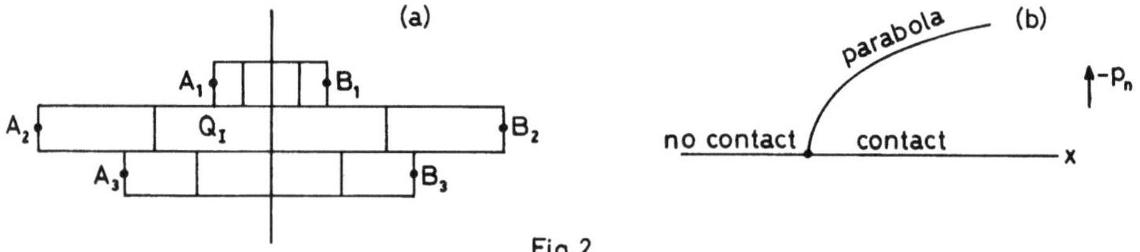

Fig.2

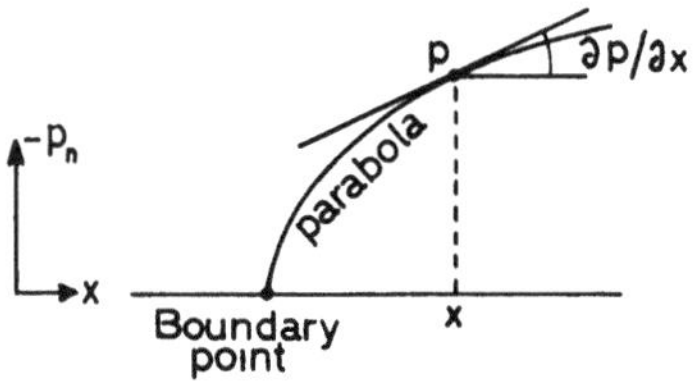

Fig.3

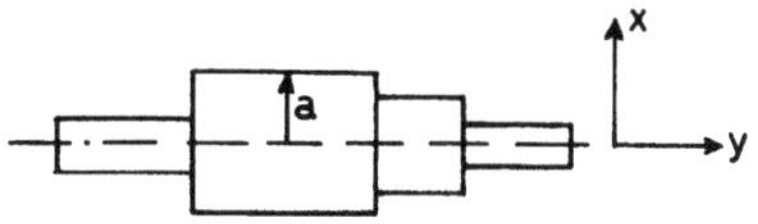

Fig.4

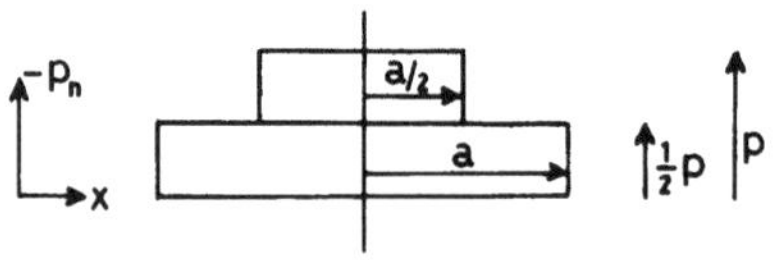

Fig.5

Exterior Penalty Methods for Contact Problems in Elasticity

J. T. ODEN
The University of Texas, Austin, USA

Summary

This paper reviews recent results of Oden, Kikuchi, and Song [4] on the use of exterior penalty methods as a basis for finite element approximations of contact problems in linear elasticity.

Introduction

This note reviews several results recently obtained by the author and his colleagues, N. Kikuchi and Y. J. Song, on the use of exterior penalty methods for unilateral problems in elasticity theory. The physical problem considered is that of a linear elastic body deflected by the application of external forces so that it comes in contact with a rigid frictionless foundation. An appropriate variational statement of this problem is as follows:

$$\left.\begin{aligned}&\text{Find } (\underset{\sim}{u},\sigma) \in V \times M \text{ such that}\\ &B(\underset{\sim}{u},\underset{\sim}{v}) - \langle\sigma,\gamma_n(\underset{\sim}{v})\rangle = f(\underset{\sim}{v}) \qquad \forall\, \underset{\sim}{v} \in V\\ &\langle\tau - \sigma,\gamma_n(\underset{\sim}{u}) - s\rangle \geq 0 \qquad \forall\, \tau \in M\end{aligned}\right\} \tag{1}$$

Here,

$$V = \{\underset{\sim}{v} = (v_1,v_2,v_3) \in (H^1(\Omega))^3 \mid \underset{\sim}{v} = \underset{\sim}{0}$$

$$\text{on } \Gamma_D \subset \Gamma\} = \text{space of admissible displacements.}$$

$\underset{\sim}{u},\underset{\sim}{v} \in V$ = displacement vectors

$M = \{\tau \in H^{-1/2}(\Gamma_C) \mid \tau \leq 0\}$

σ = contact pressure

$$B(\underset{\sim}{u},\underset{\sim}{v}) = 2\mu \int_{\Omega} \varepsilon_{ij}^{D}(\underset{\sim}{u})\varepsilon_{ij}^{D}(\underset{\sim}{v})\, dx \text{ , where}$$

$B : V \times V \to \mathbb{R}$ is the virtual work, μ the shear modulus, ε_{ij}^{D} the deviatoric strains;

$$\varepsilon_{ij}^{D}(\underset{\sim}{u}) = \varepsilon_{ij}(\underset{\sim}{u}) - \frac{1}{3}\varepsilon_{kk}(\underset{\sim}{u}) \;; \quad \varepsilon_{ij}(\underset{\sim}{u}) = \frac{1}{2}(u_{i,j} + u_{j,i})$$

$\langle\cdot,\cdot\rangle$ denotes duality pairing on $W' \times W$, where $W = H^{1/2}(\Gamma_C)$ and Γ_C is the contact boundary:

$$\partial\Omega = \Gamma = \overline{\Gamma}_D \cup \overline{\Gamma}_F \cup \overline{\Gamma}_C \;; \quad \overline{\Gamma}_D \cap \overline{\Gamma}_C = \emptyset ,$$

$\gamma_n : V \to W$ is the normal trace operator;

$\gamma_n(\underset{\sim}{v}) = \underset{\sim}{v} \cdot \underset{\sim}{n}$, $\underset{\sim}{v} \in (C^{0}(\overline{\Omega}))^{3}$, $\underset{\sim}{n}$ being a unit vector normal to the boundary Γ

s = initial gap between the body and the foundation and is given data in W .

$$f \in V' \;; \; f(\underset{\sim}{v}) = \int_{\Omega} \underset{\sim}{f} \cdot \underset{\sim}{v}\, dx + \int_{\Gamma_F} \underset{\sim}{t} \cdot \underset{\sim}{v}\, ds \;;$$

$\underset{\sim}{f}$ the body force and $\underset{\sim}{t}$ the prescribed surface traction, both assumed to be L^2-functions.

The spaces V and W are endowed with norms,

$$\|\underset{\sim}{v}\|_1^2 = \int_{\Omega} v_{i,j}v_{i,j}\, dx \;; \quad \|\phi\|_W$$

$$= \inf \{ \|\underset{\sim}{v}\|_1 \mid \gamma_n(\underset{\sim}{v}) = \phi \}$$

and are Hilbert spaces (meas $\Gamma_D > 0$) . We assume that Ω is a "sufficiently smooth" domain. Additional information on this class of problems and properties of these spaces can be found in the monograph of Kikuchi and Oden [1]. The physical situation is illustrated in Fig. 1.

The notion of exterior penalty methods for constrained minimization problems provides an alternative way to formulate problems of this type. If ε is an arbitrary positive number, we may seek $\underset{\sim}{u}_{\varepsilon} \in V$ such that

$$B(\underset{\sim}{u}_\varepsilon, \underset{\sim}{v}) + \frac{1}{\varepsilon} \langle j(\gamma_n(\underset{\sim}{u}_\varepsilon) - s)_+, \gamma_n(\underset{\sim}{v}) \rangle = f(\underset{\sim}{v}) \qquad \forall\, \underset{\sim}{v} \in V \tag{2}$$

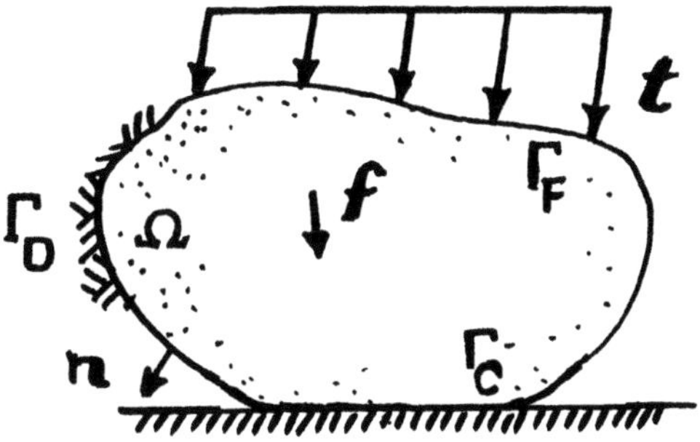

Fig. 1 Elastic body in contact with rigid friction less foundation.

where $j : W \to W'$ is the Riesz map defined on W and $(\phi)_+$ denotes the positive part of ϕ (i.e. $\phi_+ = \sup(\phi, 0)$, "sup" being the supremum relative to the order relation $\leq$ on W).

The fact (which is proved in [4]) that $B(\cdot,\cdot)$ is continuous and V-elliptic makes it possible to show that (2) possesses a unique solution $\underset{\sim}{u}_\varepsilon$ for every $\varepsilon > 0$. Moreover, we also have the condition

$$\left.\begin{array}{l} \text{There exists } \beta > 0 \text{ such that} \\ \beta \|\tau\|_{W'} \leq \sup\limits_{\underset{\sim}{v} \in V} \dfrac{|\langle \tau, \gamma_n(\underset{\sim}{v}) \rangle|}{\|\underset{\sim}{v}\|_1} \quad \forall\, \tau \in W' \end{array}\right\} \tag{3}$$

Moreover, set

$$\sigma_\varepsilon = -\frac{1}{\varepsilon} j(\gamma_n(\underset{\sim}{u}_\varepsilon) - s)_+ \tag{4}$$

Then, under these conditions it is possible to show that

i) the sequence $(\underset{\sim}{u}_\varepsilon, \sigma_\varepsilon)$ converges strongly in $V \times W'$ to the solution $(\underset{\sim}{u}, \sigma)$ of (1) and

ii) there exist constants $C_1, C_2 > 0$, independent of ε, such that

$$\|\underset{\sim}{u} - \underset{\sim}{u}_\varepsilon\|_1 \leq C_1 \frac{\varepsilon}{\beta} \|\sigma\|_{W'}$$

$$\|\sigma - \sigma_\varepsilon\|_{W'} \leq C_2 \|\underset{\sim}{u} - \underset{\sim}{u}_\varepsilon\|_1 \leq C_1 C_2 \frac{\varepsilon}{\beta} \|\sigma\|_{W'}$$

Approximations

The properties of the penalty formulation outlined above make (2) an attractive basis for finite element approximations: 1) instead of two unknown fields as in (1), there is only the unknown regularized displacement field $\underset{\sim}{u}_\varepsilon$. 2) problem (2) is posed on the whole space V rather than in the constraint set $K = \{\underset{\sim}{v} \in V \mid \gamma_n(\underset{\sim}{v}) \leq s\}$. 3) the perturbed solution $\underset{\sim}{u}_\varepsilon$ can be made arbitrarily close to the solution $\underset{\sim}{u}$ by taking ε sufficiently small; and 4) an approximation σ_ε of the contact pressure can be computed using (4) after the problem (2) has been solved, with little extra cost and effort.

It is known, however, from experience with penalty methods for other types of constraints that it may be necessary to integrate the penalty terms in (2) inexactly in order to obtain a satisfactory finite element approximation. In anticipation of similar difficulties here, we prepare for reduced integration schemes by introducing the quadrature rule

$$J(f) = \sum_{e=1}^{E} J_e(f) \ , \quad J_e(f) = \sum_{j=1}^{G} w_j^e \, f(\underset{\sim}{\eta}_j^e) \ ,$$

$$f \in C^0(\Gamma_C) \qquad (5)$$

wherein E is the number of finite elements in a finite element mesh approximating Γ_C , w_j^e are the quadrature weights, and $\underset{\sim}{\eta}_j^e$ are the quadrature points for element $\Gamma_e \subset \Gamma_C$, $1 \leq e \leq E$. We assume here that G can be taken sufficiently large that J( ) defines an exact integration of polynomials f of a given degree: $J(f) \approx \int_{\Gamma_C} f \, ds$.

Next, let us suppose that a family of finite-dimensional subspaces V_h of V has been constructed using regular families of conforming finite elements. Then our finite element approximation of problem (2) (or (1)) assumes the following form:

Find $\underset{\sim}{u}^h_\varepsilon \in V_h$ such that

$$B(\underset{\sim}{u}^h_\varepsilon,\underset{\sim}{v}^h) + \frac{1}{\varepsilon} J[(\gamma_n(\underset{\sim}{u}^h_\varepsilon)-s)_+\gamma_n(\underset{\sim}{v}^h)] = f(\underset{\sim}{v}^h)$$

$$\forall \underset{\sim}{v}^h \in V_h \qquad (6)$$

Behavior of the Approximation

While it may appear that the reduced-integration, exterior penalty, finite element method characterized by (6) involves only the space V_h of approximate displacement fields, there is intrinsic to every such formulation an approximation W_h of W' . To see this, note that if we introduce the perturbed Lagrange functional

$$L_\varepsilon : V_h \times W_h \to \mathbb{R}$$

$$L_\varepsilon(\underset{\sim}{v}^h,\tau^h) = \frac{1}{2} B(\underset{\sim}{v}^h,\underset{\sim}{v}^h) - f(\underset{\sim}{v}^h)$$

$$- J[\tau^h(\gamma_n(\underset{\sim}{u}^h_\varepsilon)-s)] - \frac{\varepsilon}{2} J[(\tau^h)^2] \qquad (7)$$

then its saddle points $(\underset{\sim}{u}^h_\varepsilon,\underset{\sim}{\sigma}^h_\varepsilon)$ are characterized by the system

$$\left.\begin{aligned} &B(\underset{\sim}{u}^h_\varepsilon,\underset{\sim}{v}^h) - J[\sigma^h_\varepsilon\gamma_n(\underset{\sim}{v}^h)] = f(\underset{\sim}{v}^h) \quad \forall \underset{\sim}{v}^h \in V_h \\ &J\{(\tau^h-\sigma^h_\varepsilon)[(\gamma_n(\underset{\sim}{u}^h_\varepsilon)-s) + \varepsilon\sigma^h_\varepsilon]\} \geq 0 \quad \forall \tau^h \in M_h \end{aligned}\right\} \qquad (8)$$

Here W_h is a finite dimensional subspace of W' and M_h is the set of $\tau^h \in W_h$ such that $\tau^h \leq 0$. Thus,

$$J\{(\tau^h-\sigma^h_\varepsilon)[(\gamma_n(\underset{\sim}{u}^h_\varepsilon)-s)_+ + (\gamma_n(\underset{\sim}{u}^h_\varepsilon)-s)_- + \varepsilon\sigma^h_\varepsilon]\} \geq 0$$

from which it follows that we can take

$$\sigma^h_\varepsilon(\underset{\sim}{\eta}^e_j) = -\frac{1}{\varepsilon}(\gamma_n(\underset{\sim}{u}^h_\varepsilon)-s)_+(\underset{\sim}{\eta}^e_j) \qquad (9)$$

where $\underset{\sim}{\eta}_j^e$ are the quadrature points in the definition of $J(\cdot)$. Substituting (9) into $(8)_1$ gives (6).

It follows that the penalty method (6) is equivalent to the perturbed Lagrangian (or perturbed mixed method) (8) and that there is connected with (6) a space W_h defined by each choice of J and V_h. With these observations in mind, we seek to use such spaces W_h which satisfy the following three conditions:

A. The quadrature rule $J(\cdot)$ is such that $J(\tau^h\hat{\tau}^h) = (\tau^h,\hat{\tau}^h) \quad \forall\ \tau^h,\hat{\tau}^h \in W_h$

B. There exists $\sigma_\varepsilon^h \in W_h$ which is uniquely defined by (9) for each solution $\underset{\sim}{u}_\varepsilon^h$ of (2).

C. There exists a constant $\beta_h > 0$ such that

$$\left.\begin{aligned}&\beta_h \|\tau^h\|_0 \leq \sup_{\underset{\sim}{v}^h \in V_h} \frac{|J[\tau^h \gamma_n(\underset{\sim}{v}^h)]|}{\|\underset{\sim}{v}^h\|_1}\\ &\text{for every } \tau^h \in W_h .\end{aligned}\right\} \qquad (10)$$

Under conditions A, B, and C, it is shown in [4] that

i) there exists a unique solution $\underset{\sim}{u}_\varepsilon^h$ to (6) for every $\varepsilon > 0$, (assuming meas $\Gamma_D > 0$, $\Omega = \Omega_h$)

ii) if σ_ε^h is defined by (9), then $(\underset{\sim}{u}_\varepsilon^h,\sigma_\varepsilon^h)$ converges to a mixed finite element approximation $(\underset{\sim}{u}^h,\sigma^h)$ as $\varepsilon \to 0$, where

$$B(\underset{\sim}{u}^h,\underset{\sim}{v}^h) - J[\sigma^h,\gamma_n(\underset{\sim}{v}^h)] = f(\underset{\sim}{v}^h) \qquad \forall\ \underset{\sim}{v}^h \in V_h$$

$$J[(\tau^h-\sigma^h)(\gamma_n(\underset{\sim}{u}^h)-s)] \geq 0 \qquad \forall\ \tau^h \in M_h$$

Here and elsewhere s is assumed to be the trace of a function $\underset{\sim}{s} \in V_h$ on Γ_C.

iii) Under mild additional assumptions on J, there exsits a constant C_1' independent of ε and h, such that

$$\|\underset{\sim}{u} - \underset{\sim}{u}_\varepsilon^h\|_1 \leq C_1' \{ \inf_{\underset{\sim}{v}^h \in V_h} [\ \|\underset{\sim}{u} - \underset{\sim}{v}^h\|_1$$

$$+ \frac{1}{\beta_h} \|\gamma_n(\underset{\sim}{u}-\underset{\sim}{v}^h)\|_{0,\Gamma_C}]$$

$$+ \inf_{\tau^h \in M_h} [\|\sigma-\tau^h\|_{W'} + \quad <\tau^h-\sigma,\gamma_n(\underset{\sim}{u})$$

$$- s>|^{1/2}] + \inf_{\tau \in M} |<\tau-\sigma^h,\gamma_n(\underset{\sim}{u})-s>|^{1/2}$$

$$+ \frac{\varepsilon}{\beta_h} \|\sigma^h\|_{0,\Gamma_C} + E_J^{1/2} \} \qquad (11)$$

where E_J is a measure of quadrature error. A similar estimate of $\|\sigma-\sigma_\varepsilon^h\|_{0,\Gamma_C}$ is given in [4].

Discussion of Results

The key to the success of the reduced-integration penalty method is the use of an integration rule J which will yield a constant β_h independent of h in the discrete stability condition (10). In the following, we list spaces V_h, integration rules J, and computed stability parameters β_h for some typical two-dimensional elements. Complete proofs of these results are given in [5].

We use the notation:

for V_h:
- Q_2 = 9-node biquadratic elements
- Q8 = 8-node isoparametric
- Q_1 = 4-node bilinear
- T_2 = 6-node quadratic triangle
- T_1 = 3-node linear triangle

for J:
- G3 = 3-point Gaussian quadrature
- G2 = 2-point Gaussian quadrature
- G1 = 1-point Gaussian quadrature
- S = Simpson's rule
- T = Trapezoid rule

for W_h:
- P_2^{-1} = discontinuous piecewise quadratic polynomials
- P_1^{-1} = discontinuous piecewise linear polynomials
- P_0^{-1} = discontinuous piecewise constants

P_k^0 = continuous piecewise polynomials of degree $k, (k = 2,1)$

Fifteen examples follow:

	V_h	J	W_h	β_h
1.	Q_2	G3	P_2^{-1}	FAILS
2.	Q_2	G2	P_2^{-1}	FAILS
3.	Q_2	G1	P_0^{-1}	FAILS
4.	Q_2	S	P_2^0	CONSTANT (stable)
5.	Q_2	T	P_1^0	CONSTANT (stable)
6.	T_2	G3	P_2^{-1}	FAILS
7.	T_2	S	P_2^0	CONSTANT (stable)
8.	T_2	T	P_1^0	CONSTANT (stable)
9.	T_1	G2	P_1^{-1}	FAILS
10.	T_1	T	P_1^0	CONSTANT (stable)
11.	Q8	G3	P_2^{-1}	FAILS
12.	Q8	G2	P_1^{-1}	FAILS
13.	Q8	G1	P_0^{-1}	FAILS
14.	Q8	S	P_2^0	CONSTANT (stable)
15.	Q8	T	P_1^0	CONSTANT (stable)

Of course, even though one of the above methods may lead to a constant β_h independent of h and, therefore, a stable approximation of the contact pressures, there may result a loss in the asymptotic rates of convergence of these methods. According to the above table, a conforming approximation is apparently needed to produce a convergent scheme. In the above list, "FAILS" indicates either $\beta_h=0$ or β_h is dependent upon the mesh parameter h (e.g. $\beta_h=0(h)$).

Extensive numerical experiments have been performed using the elements listed above [5]. In all cases, the experiments confirm the predicted instabilities (or stabilities) and, for stable schemes, the rates of convergence.

The problem considered here is the indentation of a rigid cylindrical punch into a rectangular block of isotropic, linearly elastic material with modulus $E = 3000$ and Poisson's ratio $\nu = 0.49999$. Half of the body is modeled with the non-uniform mesh of Q_2-elements shown in Fig. 2. Other dimensions are indicated in the figure. The cylinder is pushed into the body an amount $\delta = 0.8$. Results of a typical numerical experiment are reproduced in Figs. 3 and 4.

Three different integration rules were used to obtain three different solutions of this problem, corresponding to cases 1, 2, and 4 above. Computed distributions of contact pressures for these cases are shown in Fig. 4. As predicted, cases 1 and 2 are unstable; widly oscillating

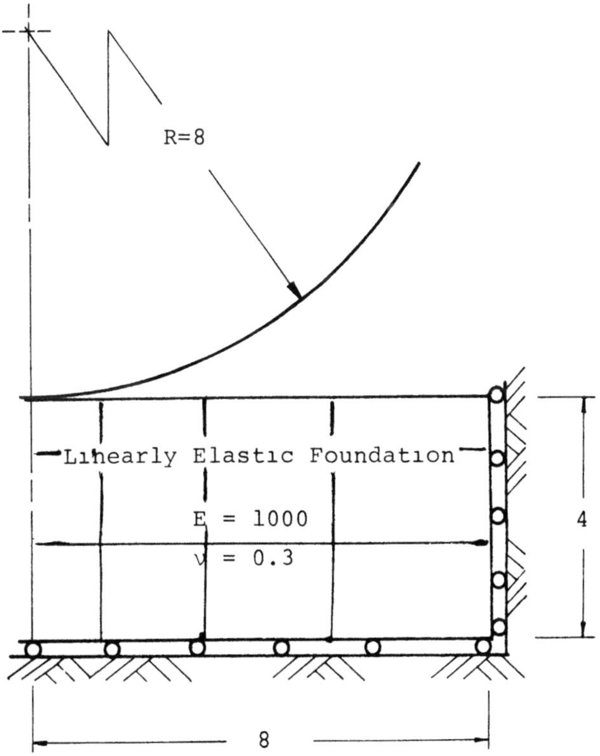

Fig. 2 Model for numerical example of a Signorini problem.

pressure distributions are obtained and further computations indicate that these oscillations become more irratic as the mesh is refined. Use of Simpson's rule, however, leads to a stable scheme and the smooth continuous approximation of the contact pressure indicated in the figure.

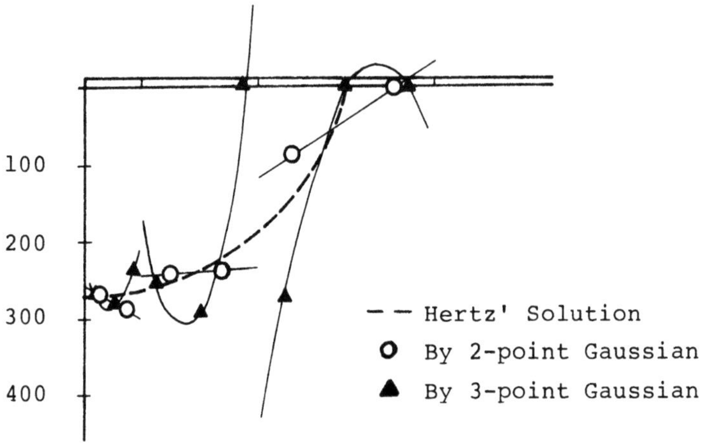

Fig. 3 Contact pressure by Guassian quadrature rules.

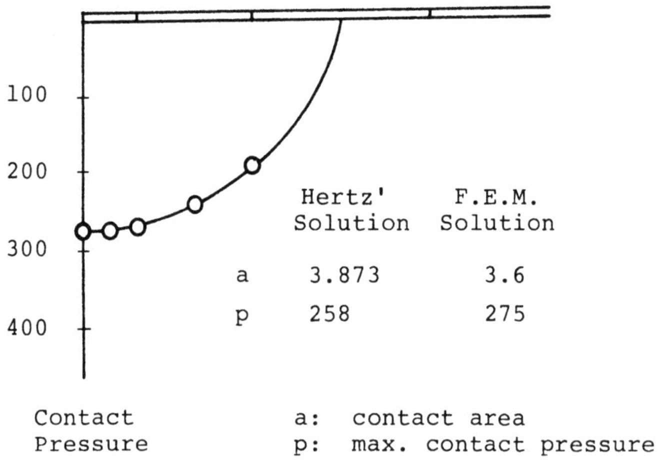

Fig. 4 Contact pressure by Simpson's Rule for numerical integration.

Acknowledgement. The support of the U.S. Air Force Office of Scientific Research under grant F-49620-78-C-0083 is gratefully acknowledged. The work reported here was part of a broad project on finite element methods for constrained problems done in collaboration with Professor N. Kikuchi and Col. Y. J. Song.

References

1. Kikuchi, N.; Oden, J.T.: Contact Problems in Elasticity. SIAM, Philadelphia, (to appear).

2. Oden, J.T.; Kikuchi, N.: Finite Element Methods for Constrained Problems in Elasticity. International Journal for Numerical Methods in Engineering, (to appear).

3. Oden, J.T.; Kikuchi, N.: A Study of Penalty Methods for Problems in Nonlinear Elasticity, IUTAM Symposium Proceedings, Pergamon Press, Oxford, (to appear).

4. Oden, J.T.; Kikuchi, N.; Song, Y.J.: Reduced Integration and Exterior Penalty Methods for Finite Element Approximations of Contact Problems in Incompressible Elasticity, TICOM Report 80-2, Austin, Texas, 1980

5. Song, Y. J.; Oden, J.T.; Kikuchi, N.: R.I.P. Methods for Problems in Elasticity, TICOM Report 80-7, Austin, Texas, 1980.

Approximate Solution of Nonlinear Problems in Incompressible Finite Elasticity

R. GLOWINSKI,
Université Paris VI,
France

P. LE TALLEC,
University of Texas, Austin,
USA

V. RUAS DE BARROS
INRIA, Le Chesnay,
France

1. INTRODUCTION

The main goal of this paper is to describe some numerical methods for solving nonlinear variational problems in incompressible finite elasticity.

In Sec. 2 we discuss a decomposition principle for a large class of variational problems and then derive several iterative methods of solution from this principle. We give in Sec. 3, the formulation of elastostatics two-dimensional and axisymmetric problems for incompressible Mooney-Rivlin materials.

In Sec. 4 we describe the application of the algorithms of Sec. 2 to the iterative solution of the mechanical problems of Sec. 3. In Sec. 5 we give some brief indication on the finite element approximation of the problems of Sec. 3.

Finally some numerical results obtained applying the methods of Sec. 4,5 are presented and discussed in Sec. 6.

2. DECOMPOSITION OF VARIATIONAL PROBLEMS AND ASSOCIATED ALGORITHMS.

2.1. A family of Variational Problems.

Restricting our attention to real Hilbert spaces, we consider two such Hilbert spaces V and H equipped with $\|\cdot\|$ and $|\cdot|$, and inner products $((\cdot,\cdot))$ and $(\cdot,\cdot)$, respectively.

Let $B \in \mathcal{L}(V,H)$ and F,G be two proper, convex, lower semicontinuous functionals from H,V to $\overline{\mathbb{R}}$, respectively, such that

$$\text{(2.1)} \quad \operatorname{dom}(G) \cap \operatorname{dom}(F \circ B) \neq \emptyset \ .$$

We associate with the above V,H,B,F,G the following minimization problem

(P) <u>Find</u> $u \in V$ <u>such that</u> $J(u) \leq J(v) \quad \forall\, v \in V$,

where $J : V \to \overline{\mathbb{R}}$ is defined by

$$J(v) = F(Bv) + G(v). \tag{2.2}$$

It appears from (2.2) that $J(\cdot)$ and therefore (P) have very special structures and taking this into account it is then quite natural to design special methods of solution for (P).

<u>Remark 2.1</u> : Most of the considerations which follow can be applied to variational problems like

$$f \in B^*A_1(Bu) + A_2(u), \tag{2.3}$$

where $f \in V'$ (dual space of V) and where A_1 (resp. A_2) is a <u>monotone operator</u> (possibly multivalued) from H to H' (dual space of H) (resp. V to V') ; the operator $A = B^* \circ A_1 \circ B + A_2$, from $V \to V'$ is not in general the differential (or subdifferential) of a functional J. For many results on these generalizations we refer to P.L. LIONS-B. MERCIER [1] and GABAY [2]. If we suppose that in addition to (2.1) we also have

$$\lim_{\|v\| \to +\infty} J(v) = +\infty \tag{2.4}$$

then (P) has a solution which is unique if J is strictly convex.

<u>Remark 2.2</u> : The applications in finite elasticity that we have in view are actually related to nonconvex minimization problems.

2.2 <u>A decomposition principle</u>.

Let us define $W \subset V \times H$ by

(2.5) $W = \{\{v,q\} \in V \times H,\ Bv-q = 0\}$.

Problem (P) is then clearly equivalent to

$$(\pi) \quad \begin{cases} \underline{\text{Find}}\ \{u,p\} \in W\ \underline{\text{such that}} \\ j(u,p) \le j(v,q) \quad \forall\ \{v,q\} \in W, \end{cases}$$

where

(2.6) $j(v,q) = F(q) + G(v)$.

Remark 2.3 : The new problem (π) has clearly a mixed formulation "flavor" since the linear relation $Bv-q = 0$ suggests the introduction of a Lagrange multiplier.

Remark 2.4 : Problems (P) and (π) are equivalent, but considering (π) we have in some sense simplified the nonlinear structure of (P), at the expense however of the new variable q and of the relation

(2.7) $Bv-q = 0$.

Actually since (2.7) is a linear relation, very efficient techniques may be used to treat it ; in the following this will be done by using, simultaneously, penalty and Lagrange multipliers, via a convenient augmented lagrangian functional.

2.3. An augmented lagrangian functional associated with (π).
Let $r > 0$; define $\mathcal{L}_r : V \times H \times H \to \overline{\mathbb{R}}$ by

(2.8) $\mathcal{L}_r(v,q,\mu) = F(q) + G(v) + \frac{r}{2}\,|Bv-q|^2 + (\mu, Bv-q)$.

It can be proved that if $\{u,p,\lambda\}$ is a saddle-point of $\mathcal{L}_r$ over $V \times H \times H$, i.e., if

$$(2.9) \quad \begin{cases} \{u,p,\lambda\} \in V \times H \times H \text{ and } \forall\ \{v,q,\mu\} \in V \times H \times H \\ \mathcal{L}_r(u,p,\mu) \le \mathcal{L}_r(u,p,\lambda) \le \mathcal{L}_r(v,q,\lambda) \ , \end{cases}$$

then $\{u,p\}$ solves (π), i.e., u solves (P) (with $p = Bu$).

2.4. A first algorithm for solving (P).

To solve (P) and (π) we shall compute the saddle-points of $\mathcal{L}_r$ by an algorithm derived from the duality algorithms considered, for example, in GLOWINSKI-LIONS-TREMOLIERES [3,4], Chap. 2. Such an algorithm applied to the solution of (2.9) is

(2.10) $\lambda^o \in H$, given

then if $n \geq 0$, λ^n given, we compute u^n, p^n, λ^{n+1} by

$$(2.11)\quad \begin{cases} \text{Find } \{u^n,p^n\} \in V\times H \text{ such that} \\ \mathcal{L}_r(u^n,p^n,\lambda^n) \leq \mathcal{L}_r(v,q,\lambda^n) \quad \forall \{v,q\} \in V\times H, \end{cases}$$

$$(2.12)\quad \lambda^{n+1} = \lambda^n + \rho(Bu^n - p^n).$$

Concerning the convergence of algorithm (2.10)-(2.12) it can be proved (see [5, Chap. 3] and [6, Chap. 5]) that under very reasonable assumptions on F,G,B and if

$$(2.13)\quad 0 < \rho < 2r,$$

then we have as $n \to +\infty$

$$(2.14)\quad u^n \to u \text{ strongly in } V,$$

$$(2.15)\quad p^n \to p = Bu \text{ strongly in } H,$$

$$(2.16)\quad \lambda^n \to \lambda \text{ weakly in } H,$$

where u is the solution of (P), and where λ is such that $\{u,p,\lambda\}$ is a saddle-point of $\mathcal{L}_r$ over $V\times H\times H$.

Remark 2.5 : The only nontrivial step in the above algorithm is clearly the solution of the minimization problem (2.11). Actually to solve (2.11), taking into account its special structure, it is very convenient to use a functional block relaxation method (like those discussed in CEA-GLOWINSKI [7];

see [5],[6] for more details on the block relaxation solution of (2.11)). If this relaxation method is used, and if in the calculation of $\{u^n,p^n\}$ we only do <u>one</u> inner relaxation, starting from $\{u^{n-1},p^{n-1}\}$, we obtain the algorithm described in Sec. 2.5.

2.5. <u>A second algorithm for solving</u> (P).
The new algorithm is defined by

(2.17) $\{u^{-1},\lambda^o\} \in V\times H$, <u>given</u>,

<u>then for</u> $n \geq 0$, u^{n-1},λ^n <u>given</u>, <u>we compute</u> p^n,u^n,λ^{n+1} <u>by</u>

(2.18) $$\begin{cases} \underline{\text{Find}}\ p^n \in H\ \underline{\text{such that}} \\ \mathcal{L}_r(u^{n-1},p^n,\lambda^n) \leq \mathcal{L}_r(u^{n-1},q,\lambda^n) \quad \forall q \in H, \end{cases}$$

(2.19) $$\begin{cases} \underline{\text{Find}}\ u^n \in V\ \underline{\text{such that}} \\ \mathcal{L}_r(u^n,p^n,\lambda^n) \leq \mathcal{L}_r(v,p^n,\lambda^n) \quad \forall v \in V, \end{cases}$$

(2.20) $$\lambda^{n+1} = \lambda^n+\rho(Bu^n-p^n).$$

<u>Remark 2.6</u> : Several variants of (2.17)-(2.20) are available. We can for example

(i) Exchange the role of q and v (see also Remark 2.7)

(ii) Update also λ^n between the steps (2.18),(2.19) ; doing so we obtain the following variant (due to GABAY [2]) of (2.17)-(2.20) :

(2.21) $\{u^{-1},\lambda^o\}\in V\times H$, <u>given</u>,

<u>then for</u> $n \geq 0$, u^{n-1},λ^n <u>given we compute</u> $p^n,\lambda^{n+1/2},u^n,\lambda^{n+1}$ <u>by</u>

(2.22) $$\begin{cases} \mathcal{L}_r(u^{n-1},p^n,\lambda^n) \leq \mathcal{L}_r(u^{n-1},q,\lambda^n) \quad \forall\ q \in H, \\ p^n \in H, \end{cases}$$

(2.23) $\lambda^{n+1/2} = \lambda^n + \rho(Bu^{n-1} - p^n)$,

(2.24) $\begin{cases} \mathcal{L}_r(u^n, p^n, \lambda^{n+1/2}) \le \mathcal{L}_r(v, p^n, \lambda^{n+1/2}) & \forall v \in V, \\ u^n \in V, \end{cases}$

(2.25) $\lambda^{n+1} = \lambda^{n+1/2} + \rho(Bu^n - p^n)$;

q and v play a more symmetrical role in (2.21)-(2.25) than in (2.17)-(2.20).

Remark 2.7 : If one uses algorithm (2.17)-(2.20), it is recommended to solve, in the second step, the problem which has the best properties of ellipticity (for more details, see [5,6]). If one uses (2.21)-(2.25), the character of the problems ellipticity does not matter since q and v play a symmetrical role. ∎

Concerning the convergence of (2.17)-(2.20) it is proved in [5,6] that under vary reasonable assumptions on F,G,B we still have (2.14)-(2.15) if

(2.26) $0 < \rho < \frac{1+\sqrt{5}}{2}\, r.$

If G is linear it follows from GABAY-MERCIER [8] that we can again take $0 < \rho < 2r$.

2.6. Remarks on the choice of ρ and r.

For a given r, the optimal choice of ρ is very close to $\rho = r$, as shown by the various numerical experiments that we have done on the algorithms of Secs. 2.5 and 2.6. The choice of r is a more delicate matter. Theoretically, the larger r is, the faster is the convergence of (2.10)-(2.12). Actually, for large values of r the problem (2.11) will not be well-conditioned and its accurate solution will be a costly operation ; moreover, for very large values of r, round-off errors play a significant(negative) role. As we can see, we have therefore two contradictory behaviors as r increases. The global effect of these phenomena on (2.10)-(2.12) is to produce an algorithm which is not very sensitive to the choice of r and which is very robust.

2.7. Relations with Alternating Directions Methods.

2.7.1. Relations between algorithms (2.17)-(2.20), (2.21)-(2.25) and A.D.I.

We suppose for simplicity that V=H and B=I. We suppose also that F and G have as differentials (or subdifferentials) A_1 and A_2, respectively. Note that A_1 and A_2 are necessarily monotone (possibly multivalued) operators. Then (P) is equivalent to

$$(2.27)\quad A_1(u)+A_2(u) = 0$$

where the equal sign has to be replaced by $\ni$ if A_1 and/or A_2 are multivalued. Suppose that $\rho=r$; then by elimination of λ^n in (2.17)-(2.20) we obtain

$$(2.28)\quad u^{-1} \text{ given,}$$

then for $n \geq 0$,

$$(2.29)\quad rp^n + A_1(p^n) = ru^{n-1} - A_2(u^{n-1}),$$

$$(2.30)\quad ru^n + A_2(u^n) = ru^{n-1} - A_1(p^n).$$

Setting $u^{n+1/2} = p^{n+1}$, we finally have

$$(2.31)\quad ru^{n+1/2} + A_1(u^{n+1/2}) = ru^n - A_2(u^n),$$

$$(2.32)\quad ru^{n+1} + A_2(u^{n+1}) = ru^n - A_1(u^{n+1/2}).$$

We recognize in (2.31),(2.32) a Douglas-Rachford Alternating Direction Implicit (ADI) method (see [9]). Similarly, by elimination of λ^n and $\lambda^{n+1/2}$ we obtain from (2.21)-(2.25) (still supposing $\rho=r$)

$$(2.33)\quad ru^{n+1/2} + A_1(u^{n+1/2}) = ru^n - A_2(u^n)$$

$$(2.34)\quad ru^{n+1} + A_2(u^{n+1}) = ru^{n+1/2} - A_1(u^{n+1/2})$$

which is (see [10]) a Peaceman-Rachford ADI method. Such ADI methods involving fairly general monotone operators A_1 and A_2 have been studied in [1,2].

2.7.2. <u>An initial value problem interpretation of (2.17)-(2.20) and (2.21)-(2.25)</u>.

It follows from Sec. 2.7.1 that if $\rho=r$ and $V=H$, $B=I$, then (2.17)-(2.20) and (2.21)-(2.25) can be seen as implicit schemes, based on fractional steps, for the discrete time integration of the initial value problem

$$(2.35)\quad \begin{cases} u(0) = u_o \\ \dfrac{du}{dt} + A(u) = 0 \end{cases}$$

where $A = A_1+A_2$.

From that interpretation r appears as the inverse of a time step Δt (i.e. $r=1/\Delta t$). As shown in [11], this interpretation of the avove algorithms, using the initial value problem (2.35), may be very helpful in obtaining insight concerning the behavior of these algorithms ; for example, the larger r is, the safer the algorithms.

The numerical integration of initial value problems by ADI methods is discussed in [1] under fairly general assumptions on A_1 and A_2.

2.7.3. <u>Further comments</u>.

The solution of problems like (P) by decomposition-coordination methods via augmented lagrangians seems to be due to GLOWINSKI-MARROCCO [12,13,14] (see also POLJAK [15]). For more details and various applications see also [1,2,5,6,8,11,16-22]. Ref. [11],[22] in particular describe the application of the above algorithms to large displacement calculations of flexible, inextensible, bending pipe lines.

To the best of our knowledge the relations between the algorithms of Sec. 2.5 and ADI methods have been observed for the first time by CHAN-GLOWINSKI [23]. In Sec. 2 we have basically followed the presentation of FORTIN-GLOWINSKI [5] and GLOWINSKI [6].

3. - FORMULATIONS OF ELASTOSTATIC PROBLEMS FOR INCOMPRISSIBLE MOONEY-RIVLIN MATERIALS.

3.1. Notation and mechanical assumptions.

A fundamental problem in nonlinear elasticity is the calculation of the deformations of a solid body made of an homogeneous, incompressible, hyperelastic material, subjected to volumetric forces $\rho_0 \underset{\sim}{f}$ (ρ_0 : density in the reference configuration) and superficial forces $\underset{\sim}{S}_0$. In a lagrangian formulation, the related energy functional, corresponding to a displacement field $\underset{\sim}{v}$ is

(3.1) $$\pi(\underset{\sim}{v}) = \int_\Omega \rho_0(\sigma(\underset{\sim}{v}) - \underset{\sim}{f}\cdot\underset{\sim}{v})\,d\underset{\sim}{x} - \int_{\partial\Omega_2} S_0\cdot\underset{\sim}{v}\,d\Gamma ,$$

where Ω is a bounded domain of $\mathbb{R}^N$ corresponding to the reference configuration ; $\partial\Omega(=\partial\Omega_1 \cup \partial\Omega_2)$ is the boundary of Ω, the body being fixed on $\partial\Omega_1$. We have denoted by $\sigma(\underset{\sim}{v})$ the stored energy functional (per mass unit). For a Mooney-Rivlin material we have

(3.2) $$\sigma(\underset{\sim}{v}) = E_1(I_1-2) \ \underline{\text{if}}\ N=2 ,$$

(3.3) $$\sigma(\underset{\sim}{v}) = E_1(I_1-3)+E_2(I_2-3) \ \underline{\text{if}}\ N=3$$

with, in (3.2),(3.3), I_i the i-th invariant of the $\underset{\approx}{F}\underset{\approx}{F}^t$ tensor, where

(3.4) $$\underset{\approx}{F} = \underset{\sim}{\nabla}\underset{\sim}{v}+\underset{\approx}{I}$$

and E_1, E_2 are coefficients which are material dependent. The displacement also has to satisfy the incompressibility condition which here has the following form

(3.5) $$\det \underset{\approx}{F}(\underset{\sim}{v}) = 1 \ \underline{\text{a.e. on}}\ \Omega.$$

Remark 3.1 : We have supposed in (3.1) that $\underset{\sim}{S}_o$ and $\underset{\sim}{f}$ are independent of v. This corresponds to the standard simplifying assumption known as the <u>dead loading</u> hypothesis for which $\underset{\sim}{f}$ and $\underset{\sim}{S}_o$ do not vary during the motion ; if we deal with <u>pressure</u> type forces, then $\underset{\sim}{S}_o$ is a function of $\underset{\sim}{u}$ (the actual displacement), given by

(3.6) $\quad \underset{\sim}{S}_o = -q(\underset{\sim}{x}+\underset{\sim}{u}(\underset{\sim}{x}))(\underset{\approx}{I}+\underset{\sim}{\nabla}\underset{\sim}{u})^{-1}\underset{\sim}{n}$,

where $q(\underset{\sim}{x}+\underset{\sim}{u}(\underset{\sim}{x}))$ is the pressure at the point $\underset{\sim}{x}+\underset{\sim}{u}(\underset{\sim}{x})$ of the actual configuration and $\underset{\sim}{n}$ is the <u>unit outward normal</u> at $\partial\Omega$, in the <u>reference configuration</u>.

3.2. Mathematical formulations.

We shall give in this section several possible formulations of the elasto-static problems ; it is still an open problem to prove their equivalence in general (see [24]-[26] for a discussion of these equivalence, and also the comments of Sec. 3.2.4).

3.2.1. Formulation by minimization.

It is reasonable to suppose that those displacements $\underset{\sim}{u}$ corresponding to <u>stable equilibrium position</u> obey

(3.7) $\quad \begin{cases} \underset{\sim}{u} \text{ is a local minimizer over K of the functional} \\ \underset{\sim}{v} \to \pi(\underset{\sim}{v}), \end{cases}$

with for an <u>incompressible Mooney-Rivlin</u> material

(3.8) $\quad \begin{cases} K = \{\underset{\sim}{v} \in (H^1(\Omega))^N, \underset{\sim}{v}=\underset{\sim}{0} \text{ on } \partial\Omega_1, \det \underset{\approx}{F}(\underset{\sim}{v}) = 1 \\ \text{a.e., } (F^{-1}(\underset{\sim}{v}))^t \in (L^2(\Omega))^{N\times N}\}. \end{cases}$

The existence of solutions for (3.7),(3.8) is proved in [27].

3.2.2. Formulation by equilibrium equations.

The equilibrium positions correspond to the solutions of

the following system of nonlinear partial differential equations

$$(3.9)\quad \begin{cases} \underset{\sim}{u} \in K, \\ (D\pi(\underset{\sim}{u}),\underset{\sim}{v}) + \int_\Omega p[\underset{\sim}{u},\underset{\sim}{v}]\,d\underset{\sim}{x} = 0 \quad \forall\, \underset{\sim}{v} \in X, \end{cases}$$

where $D\pi$ is the differential of π (on $(H^1(\Omega))^N$) and where

$$(3.10)\quad [\underset{\sim}{u},\underset{\sim}{v}] = \frac{\partial}{\partial u_{i,j}} (\det \underset{\approx}{F}(\underset{\sim}{u}))v_{i,j}\ ,$$

$$(3.11)\quad X = \{\underset{\sim}{v} \in (H^1(\Omega))^N,\ \underset{\sim}{v}=\underset{\sim}{0} \text{ on } \partial\Omega_1\}$$

(in (3.10) we have used the standard summation and derivative notations). The above function p, which is clearly a Lagrange multiplier associated with the incompressibility condition (3.5), appears as a pressure.

3.2.3. Formulation by augmented lagrangian.

We proceed as in Sec. 2 by "relaxing" the linear relation (3.4) using an augmented lagrangian. We obtain thus the following formulation :

$$(3.12)\quad \begin{cases} \text{Find } \{\underset{\sim}{u},\underset{\approx}{F},\underset{\approx}{\lambda}\} \in W = X\times Y\times (L^2(\Omega))^{N\times N}, \text{ stationary point} \\ \text{over } W \text{ of the augmented lagrangian} \\ \mathcal{L}_R(\underset{\sim}{v},\underset{\approx}{G},\underset{\approx}{\mu}) = \pi(\underset{\sim}{v}) + \frac{R}{2}\|\underset{\sim}{\nabla}\underset{\sim}{v}+\underset{\approx}{I}-\underset{\approx}{G}\|^2_{L^2} - \\ - \int_\Omega \underset{\approx}{\mu}\cdot(\underset{\sim}{\nabla}\underset{\sim}{v}-\underset{\approx}{I}-\underset{\approx}{G})\,d\underset{\sim}{x}\ , \end{cases}$$

where, in (3.12), we have $R > 0$ and

$$Y = \{\underset{\approx}{F} \in (L^2(\Omega))^{N\times N}\ ,\ (\underset{\approx}{F}^{-1})^t \in (L^2(\Omega))^{N\times N}, \det \underset{\approx}{F}=1 \text{ a.e.}\}.$$

3.2.4. Some relations between formulations (3.7),(3.9) and (3.12).

The following results are proved in [24] :

a) Formulations (3.9) and (3.12) are equivalent.
b) Every "smooth" solution of (3.7) is a solution of (3.9), (3.12),

c) If the functional π is convex (case of a Mooney-Rivlin material if N=2), then every solution of (3.12) is such that u is a minimizer of

$$\underset{\sim}{v} \to \mathcal{L}_R(\underset{\sim}{v},\underset{\approx}{F},\underset{\approx}{\lambda}) \ ;$$

similarly, for R sufficiently large every solution of (3.12) is such that $\underset{\approx}{F}$ is a minimizer of $\underset{\approx}{G} \to \mathcal{L}_R(\underset{\approx}{u},\underset{\approx}{G},\underset{\approx}{\lambda})$.

<u>Remark 3.2</u> : If $\{\underset{\sim}{u},\underset{\approx}{F},\underset{\approx}{\lambda}\}$ is a solution of (3.12), then the condition

$$\partial_{\underset{\sim}{u}}\mathcal{L}_R(\underset{\sim}{u},\underset{\approx}{F},\underset{\approx}{\lambda}) = 0$$

implies

$$\text{(3.13)} \qquad -\partial_j\left(\frac{\partial}{\partial u_{i,j}}(\rho_o\sigma)-\lambda_{ij}\right) = \rho_o f_i .$$

From (3.13), $\underset{\approx}{\lambda}$ appears as this part of the first Piola-Kirchoff stress tensor, corresponding to the incompressibility. We observe also that an algorithm solving (3.12) gives the <u>stress field</u> directly.

3.3. Axisymmetric problems.

3.3.1. Formulation of the problems.

We consider the case of an axisymmetric incompressible hyperelastic body subjected to an axisymmetric system of forces. The problem is to find the axisymmetric positions of equilibrium. We denote by $\underset{\sim}{u} = \{u_1,u_2\}$ the <u>displacement</u>, with u_1 the <u>radial displacement</u> and u_2 the <u>axial</u> one. Using an $\{r,z\}$ system of coordinates we have

$$\text{(3.14)} \qquad \underset{\approx}{I}+\underset{\sim}{\nabla}\underset{\sim}{u} = \begin{pmatrix} 1+u_{1,1} & u_{1,2} & 0 \\ u_{2,1} & 1+u_{2,2} & 0 \\ 0 & 0 & e(\underset{\sim}{u}) \end{pmatrix} ,$$

where $e(\underset{\sim}{u}) = 1+\frac{u_1}{r}$ is the <u>extension ratio</u> in the <u>circumferential direction</u>, and where

$$u_{j,1} = \frac{\partial u_j}{\partial r} \ , \ u_{j,2} = \frac{\partial u_j}{\partial z} \ .$$

Using the notation of Sec. 3.1, we have (with $e=e(\underset{\sim}{u})$)

$$I_1(\underset{\sim}{u}) = e^2+(\delta_{ij}+u_{i,j})^2 ,$$

$$I_2(\underset{\sim}{u}) = e^2(\delta_{ij}+u_{i,j})^2+(1+u_{1,1}u_{2,2}-u_{1,2}u_{2,1}+u_{1,1}+u_{2,2})^2,$$

$$\underline{\det}(\underset{\approx}{I}+\nabla\underset{\sim}{u}) = e(1+u_{1,1}u_{2,2}-u_{1,2}u_{2,1}+u_{1,1}+u_{2,2}).$$

For <u>incompressible</u> materials we have

$$\underline{\det}(\underset{\approx}{I}+\nabla\underset{\sim}{u}) = 1 \text{ a.e.}$$

implying in turn

$$I_2(\underset{\sim}{u}) = e^2(\delta_{ij}+u_{i,j})^2+ 1/e^2.$$

3.3.2. <u>Lagrangian formulation</u>.

Let Ω be the <u>half meridian section</u> of the reference domain ; we associate to Ω the following spaces (with $1 \le p < +\infty$)

$$\mathcal{L}^p = \{\phi \mid \int_\Omega |\phi(x)|^p \, r dr\, dz < +\infty\} ,$$

$$\mathcal{H}^1 = \{\phi \mid \phi, \frac{\partial\phi}{\partial r} , \frac{\partial\phi}{\partial z} \in \mathcal{L}^2\} \quad ;$$

we shall use in the sequel the notation $d\underset{\sim}{x} = rdrdz$.

With respect to the general case we do the following modifications :

a) <u>Spaces</u>

$$X = \{\underset{\sim}{v} \in (\mathcal{H}^1)^2 , \underset{\sim}{v}=\underset{\sim}{0} \ \underline{\text{on}} \ \partial\Omega_1\} ,$$

and with $1 \le i,j \le 2$

$$Y = \{\underset{\approx}{G} = \{g_o,g\} \in L^\infty(\Omega) \times (\mathcal{L}^2)^4, \ g_o > 0, \ g_o\underline{\det}\ \underset{\approx}{g}=1 \text{ a.e.}\},$$

$$Z = \{\underset{\approx}{\mu} = \{\mu_o,\mu_{ij}\} \in \mathcal{L}^1 \times (\mathcal{L}^2)^4\} .$$

b) <u>Augmented lagrangian</u>

$$(3.15)\quad \begin{cases} \mathcal{L}_R(\underset{\sim}{v},\underset{\approx}{G},\underset{\approx}{\mu}) = \pi(\underset{\sim}{v}) + \frac{R}{2}\int_\Omega(\delta_{ij}+v_{i,j}-g_{ij})^2 d\underset{\sim}{x} \\ -\int_\Omega \mu_{ij}(\delta_{ij}+v_{ij}-g_{ij})d\underset{\sim}{x} + \frac{R}{2}\int_\Omega(g_o-e(\underset{\sim}{v}))^2 d\underset{\sim}{x} - \end{cases}$$

$$- \int_\Omega \mu_o (e(\underset{\sim}{v}) - g_o) d\underset{\sim}{x}.$$

Remark 3.3 : A possible variant of $\mathcal{L}_R$ in (3.15) can be obtained by replacing $e(\underset{\sim}{v})$ by g_o in $\pi(\underset{\sim}{v})$ (leading to a functional $\pi(\underset{\sim}{v}, g_o)$).

4. - ITERATIVE SOLUTION OF THE EQUILIBRIUM PROBLEMS.

We apply to the solution of the equilibrium problem (3.12), the iterative methods described in Sec. 2, keeping in mind that (3.12) is a nonlinear, nonquadratic, non convex problem.

4.1. A first algorithm for solving (3.12).

We follow Sec. 2.4 ; using the notation of Sec. 3.2.3 the algorithm is :

$$(4.1) \qquad \underset{\approx}{\lambda}^o \text{ given in } (L^2(\Omega))^{N\times N},$$

then for $n \geq 0$, $\underset{\approx}{\lambda}^n$ known we obtain $\underset{\approx}{u}^n, \underset{\approx}{F}^n$ and $\underset{\approx}{\lambda}^{n+1}$ ($\underset{\approx}{\lambda}^n \in (L^2(\Omega))^{N\times N}$) by

$$(4.2) \qquad \begin{cases} \{\underset{\sim}{u}^n, \underset{\approx}{F}^n\} \in X\times Y \text{ and } \forall \ \{\underset{\sim}{v}, \underset{\approx}{G}\} \in X\times Y \text{ we have} \\ \mathcal{L}_R(\underset{\sim}{u}^n, \underset{\approx}{F}^n, \underset{\approx}{\lambda}^n) \leq \mathcal{L}_R(\underset{\sim}{v}, \underset{\approx}{G}, \underset{\approx}{\lambda}^n), \end{cases}$$

$$(4.3) \qquad \underset{\approx}{\lambda}^{n+1} = \underset{\approx}{\lambda}^n - \rho(\underset{\sim}{\nabla} \underset{\sim}{u}^n + \underset{\approx}{I} - \underset{\approx}{F}^n) \quad , \ \rho > 0.$$

Remark 4.1 :Problem (4.2) is equivalent to the nonlinear system

$$(4.4) \qquad \mathcal{L}_R(\underset{\sim}{u}^n, \underset{\approx}{F}^n, \underset{\approx}{\lambda}^n) \leq \mathcal{L}_R(\underset{\sim}{u}^n, \underset{\approx}{G}, \underset{\approx}{\lambda}^n) \quad \forall \underset{\approx}{G} \in Y, \ \underset{\approx}{F}^n \in Y,$$

$$(4.5) \qquad \partial_{\underset{\sim}{v}} \mathcal{L}_R(\underset{\sim}{u}^n, \underset{\approx}{F}^n, \underset{\sim}{\lambda}^n) \cdot \underset{\sim}{v} = 0 \quad \forall \underset{\sim}{v} \in X, \ \underset{\sim}{u}^n \in X,$$

whose block relaxation solution leads to the algorithm described in Sec. 4.2.

4.2. A second algorithm for solving (3.12).

4.2.1. Description of the algorithm.

The second algorithm is given by

(4.6) $\underset{\sim}{u}^{-1}$ given in X, $\underset{\approx}{\lambda}^{o}$ given in $(L^2(\Omega))^{N\times N}$

then for $n \geq 0$, $\underset{\sim}{u}^{n-1}, \underset{\approx}{\lambda}^{n}$ given, we obtain $\underset{\approx}{F}^{n}, \underset{\sim}{u}^{n}, \underset{\approx}{\lambda}^{n+1}$ from

(4.7) $\mathcal{L}_R(\underset{\sim}{u}^{n-1}, \underset{\approx}{F}^{n}, \underset{\approx}{\lambda}^{n}) \leq \mathcal{L}_R(\underset{\sim}{u}^{n-1}, \underset{\approx}{G}, \underset{\approx}{\lambda}^{n}) \quad \forall\, \underset{\approx}{G} \in Y, \underset{\approx}{F}^{n} \in Y,$

(4.8) $\partial_{\underset{\sim}{v}}\mathcal{L}_R(\underset{\sim}{u}^{n}, \underset{\approx}{F}^{n}, \underset{\approx}{\lambda}^{n})\cdot \underset{\sim}{v} = 0 \quad \forall \underset{\sim}{v} \in X,\ \underset{\sim}{u}^{n} \in X,$

(4.9) $\underset{\sim}{\lambda}^{n+1} = \underset{\sim}{\lambda}^{n} - \rho(\underset{\sim}{\nabla}\underset{\sim}{u}^{n} + \underset{\approx}{I} - \underset{\approx}{F}^{n}).$

4.2.2. On the solution of problems (4.7),(4.8).

Problem (4.8) is equivalent to

$$
(4.10)\quad \begin{cases} \text{Find } \underset{\sim}{u}^{n} \in X \text{ such that} \\ \mathcal{L}_R(\underset{\sim}{u}^{n}, \underset{\approx}{F}^{n}, \underset{\approx}{\lambda}^{n}) \leq \mathcal{L}_R(\underset{\sim}{v}, \underset{\approx}{F}^{n}, \underset{\approx}{\lambda}^{n}) \quad \forall \underset{\sim}{v} \in X \end{cases}
$$

which is an unconstrained minimization problem whose solution is quite easy, particularly if R is sufficiently large. If N=2 then the functional to minimize is quadratic, and therefore solving (4.8),(4.10) is equivalent to solving a linear problem related to a second order partial differential operator which is independent of n and whose discrete variants are linear systems associated with positive definite matrices independent of n (we shall use therefore a prefactorization of these matrices). If N=3 or in the axisymmetric case problem (4.8), (4.10) is no longer linear ;it can be, however, efficiently solved by a preconditioned conjugate gradient algorithm.

Problem (4.7) is a more delicate one (apparently, at least) ; if N=2, (4.7) is reduced to (omitting indice n) :

$$
(4.11)\quad \begin{cases} \text{Find } \underset{\approx}{F} \in Y \text{ and minimizing over Y the functional} \\ \underset{\approx}{G} \rightarrow \int_{\Omega} [RG_{ij}^{2} - 2(R(u_{i,j}+\delta_{ij}) - \lambda_{ij})G_{ij}]d\underset{\sim}{x} \end{cases}
$$

with

$$Y = \{\underset{\approx}{G} \in (L^2(\Omega))^4,\ G_{11}G_{22}-G_{12}G_{21} = 1 \ \underline{\text{a.e. on}}\ \Omega\}.$$

Since derivatives of $\underset{\approx}{G}$ (and $\underset{\approx}{F}$) are not involved in (4.11), we can solve this last problem pointwise. We have therefore to solve an infinity (in theory) of four-dimensional problems of the following class

$$(4.13)\quad \begin{cases} \underline{\text{Find}}\ \{F_{ij}\} \in \mathcal{G}\ \underline{\text{and minimizing over}}\ \mathcal{G}\ \underline{\text{the functional}} \\ \{G_{ij}\} \to RG_{ij}^2 - 2a_{ij}G_{ij}, \end{cases}$$

where

$$(4.14)\quad \mathcal{G} = \{\{G_{ij}\} \in \mathbb{R}^4 \ ,\ G_{11}G_{22}-G_{12}G_{21} = 1\} .$$

We diagonalize the above quadratic relation using as new variables

$$(4.15)\quad \begin{cases} b_1 = (F_{11}+F_{22})/\sqrt{2}\ ,\ b_2 = (F_{11}-F_{22})/\sqrt{2}\ , \\ b_3 = (F_{12}+F_{21})/\sqrt{2}\ ,\ b_4 = (F_{12}-F_{21})/\sqrt{2}. \end{cases}$$

Using $\underset{\sim}{b}$ defined by (4.15), problem (4.13),(4.14) becomes

$$(4.16)\quad \begin{cases} \underline{\text{Find}}\ \underset{\sim}{b} \in C\ \underline{\text{and minimizing over}}\ C\ \underline{\text{the functional}} \\ \underset{\sim}{c} \to Rc_i^2 - 2z_ic_i\ , \end{cases}$$

where

$$(4.17)\quad \begin{cases} C = \{\underset{\sim}{c} \mid \underset{\sim}{c} \in \mathbb{R}^4\ ,\ \varepsilon_i c_i^2 = 2\}\ , \\ \underline{\text{with}}\ \varepsilon_1=\varepsilon_4=1,\ \varepsilon_2=\varepsilon_3=-1. \end{cases}$$

The extremizers of (4.16),(4.17) are given by

$$(4.18)\quad \underset{\sim}{b} \in \mathbb{R}^4\ ,\ b_i=z_i/(R+\varepsilon_i p),\ i=1,2,3,4,$$

where the scalar p (Lagrange multiplier of $\varepsilon_i b_i^2 = 2$) is a solution of

$$(4.19) \qquad (z_1^2+z_4^2)/(R+p)^2 = (z_2^2+z_3^2)/(R-p)^2+2 \; .$$

We suppose that $z_1^2+z_4^2 \neq 0$; then it can be shown that (4.19) has a unique solution in $]-R,+R[$; moreover, using the implicit function theorem (see [24], [25] for more details) one can show that this solution p of (4.19) between -R and R is precisely the one associated with the global minimum of $\underset{\sim}{c} \to Rc_i^2-2z_ic_i$ over $\varepsilon_i c_i^2 = 2$, to which, therefore, corresponds a unique global minimizer given from p in (4.16),(4.17) (actually there is no other minimizer (local or global) than the above global minimizer).

Solving (4.19) on $]-R,+R[$ is a trivial problem ; we have then $\underset{\sim}{b}$ from p and (4.18), and then $\underset{\approx}{F}$ from $\underset{\sim}{b}$ and (4.15).

Remark 4.2 : In the actual computer experiments that we have done, we never encountered the situation $z_1^2+z_4^2 = 0$; in fact we have the feeling that for R sufficiently large this cannot happen in the context of problem (3.12) if N=2.■

The solution of problem (4.7), in the axisymmetric case and for genuinely three-dimensional problems, leads to a far more complicated discussion, however the same general ideas still apply . we refer to [24],[25] (resp. [28]) for the axisymmetric (resp. three-dimensional) case.

5. - FINITE ELEMENT APPROXIMATION OF THE EQUILIBRIUM PROBLEMS.

5.1. Synopsis.

A most important step to the computer use of the iterative methods of Sec. 4, for solving the elasticity problems of Sec. 3, is the reduction of these problems to finite dimensional one via convenient approximation methods. As it can be guessed the main difficulty to overcome is the incompressibility condition (3.5), coupled to the other nonlinearities of the problems under consideration ; a guideline to the numerical treatment of the incompressibility condition is given by those results concerning the numerical analysis of incompressible flows governed by Stokes and Navier-Stokes equations (owing to the very abundant litterature concerning this subject we refer to the corresponding bibliographical references in [24], [30] and also to the papers of Malkus and Oden in these proceedings).

In this paper, restricting our attention to two-dimensional problems we shall describe briefly two families of finite element approximations, preserving the decomposition properties of the continuous problems :

a) A fairly classical one, previously used in [24],[25],[29].
b) A very recent one introduced by the third author in [30].

Corresponding numerical results will be shown in Sec. 6.

5.2. Approximation by quadrilateral finite elements.
The numerical solution of the lagrangian problem (3.12) using the algorithms described in Sec. 4 requires the introduction of finite dimensional spaces X_h, Y_h, Z_h approximating X, Y, Z. An approximate gradient operator s_h from X_h to Z_h must also be defined.

In this section 5.2 the spaces X_h, Y_h, Z_h are constructed using quadrilateral finite elements, with the displacements interpolated at the vertices, the pressure and the tensor $\underset{\approx}{F}$ at the center of each element.

5.2.1. Quadrangulation of Ω.
We suppose that our domain Ω is a polygonal open set in $\mathbb{R}^2$ which can be decomposed into quadrilaterals

$$(5.1) \qquad \Omega = \bigcup_{k \in I_h} \Omega_k \; .$$

Here I_h is a finite set of numbers and Ω_k is the image by a mapping $\mathcal{F}_k$ of a reference rectangle $\hat{\Omega}$ such as the one shown in Fig. 5.1 ; this mapping $\mathcal{F}_k$ is defined by

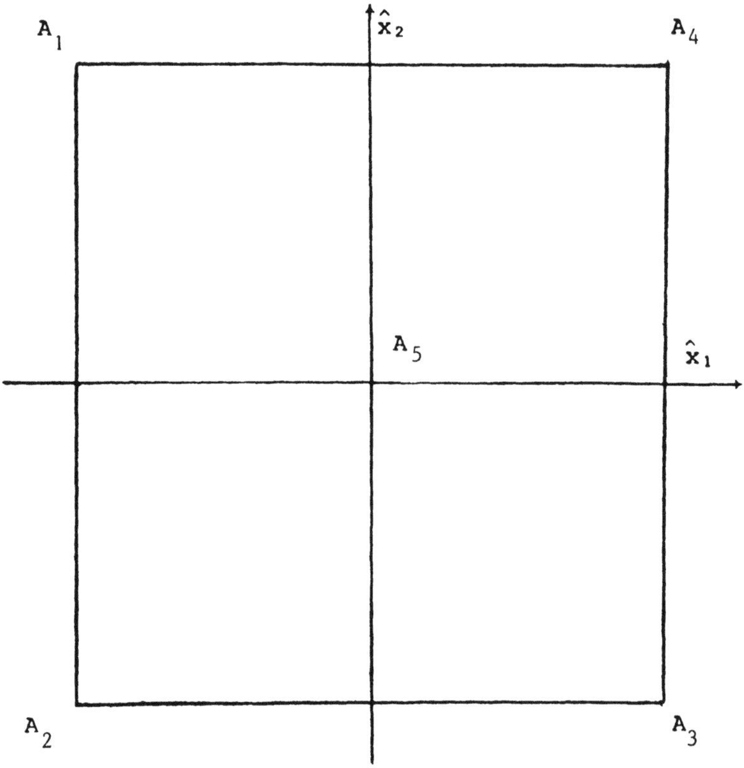

Figure 5.1 : Reference Rectangle

(5.2) $$\underset{\sim}{x} = \mathcal{F}_k(\hat{\underset{\sim}{x}}) = \sum_{\alpha=1}^{4} \underset{\sim}{x}^{\alpha_k} \hat{\phi}(\hat{\underset{\sim}{x}}),$$

where, in (5.2), $\underset{\sim}{x}^{\alpha_k}$ is this vector whose coordinates are those of the node α of the kth element and where $\hat{\phi}^{\alpha}$ is the shape function associated to the node α of the bilinear finite element $\{\hat{\Omega},\{A_\alpha\}_{\alpha=1}^{4}, Q_1(\hat{\Omega})\}$, $Q_1(\hat{\Omega})$ being the space of the bilinear polynomials defined on $\hat{\Omega}$, i.e.

(5.3) $$\hat{Q}_1(\hat{\Omega}) = \{\hat{q}|\hat{q}(\hat{x}) = \hat{a}_{00}+\hat{a}_{10}\hat{x}_1+\hat{a}_{01}\hat{x}_2+\hat{a}_{11}\hat{x}_1\hat{x}_2\}.$$

5.2.2. Discrete spaces and discrete operators.

We can now define the following spaces (with $y_{h|k} = y_{h|\Omega_k}$)

(5.4) $$H_h^1 = \{y_h \in C^0(\bar{\Omega}),\ y_{h|k} = z_h \circ \mathcal{F}_k^{-1},\ z_h \in Q_1(\hat{\Omega})\},$$

(5.5) $P_h = \{q_h | q_h|_K = \underline{\text{const.}}\}$,

(5.6) $X_h = H_h^1 \cap X$,

(5.7) $Z_h = (P_h)^4$,

(5.8) $Y_h = Y \cap Z_h$.

Remark 5.1 : It follows from (5.2),(5.4) that the above finite element approximation is of the isoparametric type. ∎

Concerning the approximate gradient operator s_h, two quite natural choices can be made :

Either

$$(5.9)_1 \quad s_h(\underset{\sim}{u}_h)|_K = \underset{\sim}{\nabla}\underset{\sim}{u}_h(\mathcal{F}_K(A_5)),$$

or

$$(5.9)_2 \quad s_h(\underset{\sim}{u}_h)|_K = L^2\text{-}\underline{\text{projection of}}\ \underset{\sim}{\nabla}\underset{\sim}{u}_h\ \underline{\text{on}}\ Z_h.$$

5.2.3. Formulation of the approximate problems.

We consider only the plane strain problem since the extension to the axisymmetric case is quite obvious ; taking into account the results of Sec. 3.2.3, 3.2.4 the discrete approximation of (3.7),(3.8) that we consider is defined by

$$\underline{\text{Find}}\ \{u_h, F_h, \lambda_h\} \in X_h \times Y_h \times Z_h\ \underline{\text{such that}}$$

$$(5.10)_1 \begin{cases} \partial_{\underset{\sim}{v}}\pi(\underset{\sim}{u}_h)\cdot\underset{\sim}{v}_h + R\displaystyle\int_\Omega(\underset{\approx}{I}+s_h(\underset{\sim}{u}_h)-\underset{\approx}{F}_h)\cdot s_h(\underset{\sim}{v}_h)\,dx\ + \\ -\displaystyle\int_\Omega \underset{\approx}{\lambda}_h\cdot s_h(\underset{\sim}{v}_h)\,d\underset{\sim}{x} = 0 \quad \forall \underset{\sim}{v}_h \in X_h \end{cases}$$

$$(5.10)_2 \begin{cases} \underset{\approx}{F}_h\ \underline{\text{minimizes over}}\ Y_h\ \underline{\text{the functional}} \\ \underset{\approx}{G}_h \to \frac{R}{2}\,||I+s_h(u_h) - \underset{\approx}{G}_h||^2 + \displaystyle\int_\Omega \underset{\approx}{\lambda}_h\cdot\underset{\sim}{G}_h\,d\underset{\sim}{x}, \end{cases}$$

$$(5.10)_3\ \underset{\sim}{I}+s_h(\underset{\sim}{u}_h) = \underset{\sim}{F}_h\ \underline{\text{in}}\ Z_h.$$

It is important to observe that from the construction of Y_h and if we suppose that $\underset{\sim}{u}_h$ and $\underset{\approx}{\lambda}_h$ are known in $(5.10)_2$ then this last minimization problem can be solved elementwise ; therefore the algorithms of Sec. 4 can also be applied to the solution of problem (5.10).

Concerning the convergence of the discrete solutions to the continuous one we refer to LE TALLEC [24].

5.3. Approximation by incomplete quadratic simplicial finite elements.

We shall briefly discuss in this section a new type of finite element approximation, using triangles ; these finite elements have been recently introduced by the third author in [30] (see also [31]) and have been denominated AQL (for Asymmetric Quasi-Linear).

We follow [30] ; this new approximation is defined as follows :

Let $\mathcal{C}_h$ be a triangulation of Ω and let $T \in \mathcal{C}_h$ with vertices $\{\delta_{iT}\}_{i=1}^{3}$; let $P_{4/3}$ be the space of the polynomials in two variables, of degree ≤ 2, defined over T and such that their restriction to two given edges, e_{1T} and e_{2T} for example, are linear functions (e_{iT} is the edge of T, opposite to δ_{1T}) ; we clearly have dim $P_{4/3} = 4$. We introduce now the following set of degrees of freedom $\{a_{iT}\}_{i=1}^{4}$, where :

- a_{iT} is for i=1,2,3 the functional value at δ_{iT},
- a_{4T} is the functional value at the midpoint of edge e_{3T}.

A fundamental property of the AQL element is given by the following

Proposition 5.1 : Let $\underset{\sim}{v} = \{v_1, v_2\}$ be a vector valued function defined over a triangle T and whose components belong to $P_{4/3}$, then

$$\det (\underset{\approx}{I} + \underset{\sim}{\nabla} \underset{\sim}{v}) \text{ is linear over } T.$$

We refer to [30] for the proof of the above proposition. It follows then from Prop. 5.1 that

$$(5.11)\quad \underline{\det}\,(\underset{\approx}{I}+\nabla\underset{\sim}{v})(G_T) = \frac{1}{\text{meas.}(T)}\int_T \underline{\det}(\underset{\approx}{I}+\nabla\underset{\sim}{v}(\underset{\sim}{x}))\,d\underset{\sim}{x}$$

(where G_T = centroid of T), and (5.11) clearly suggests to require the incompressibility condition

$$\det\,(\underset{\approx}{I}+\nabla\underset{\sim}{v}) = 1$$

at G_T only.

Notice that because the nonsymmetric structure of the AQL element we have to make some restriction on the triangulation upon which the space of the approximate displacements is to be defined ; one suggests in [30] the following process that generates nearly as general a mesh as any other finite element triangulation :

We first partition Ω into convex quadrilaterals arbitrarily ; then each quadrilateral is subdivided into two triangles by one of its two diagonals (any of them can be used). The edges over which the restrictions of the second order polynomials of $P_{4/3}$ are allowed to be quadratic are precisely those diagonals.

Using the above AQL finite elements we approximate X,Y,Z by X_h, Y_h, Z_h as follows :

$$(5.12)\quad H_h^1 = \{y_h \in C^0(\bar{\Omega}),\ y_h|_T \in P_{4/3}\quad \forall T \in \mathcal{T}_h\}\ ,$$

$$(5.13)\quad P_h = \{q_h \mid q_h|_T = \underline{\text{const}}.\ \forall T \in \mathcal{T}_h\},$$

$$(5.14)\quad X_h = X \cap H_h^1\ ,$$

$$(5.15)\quad Z_h = (P_h)^4,$$

$$(5.16)\quad Y_h = Y \cap Z_h.$$

Concerning the approximate gradient operator s_h we use

(5.17) $s_h(\underset{\sim}{u}_h)|_T = \underset{\sim\sim}{\nabla u}_h(G_T)$.

Using the above spaces and operator the approximate problem is then defined exactly as in Sec. 5.2.3 by (5.10), and the local properties of $(5.10)_2$ are preserved by the new element.

Remark 5.2 : We refer to [30], [31] for a detailed discussion of the AQL element, and also for various two and three dimensional generalizations.

Remark 5.3 : It will be interesting to use the AQL element (and its generalizations) for the finite element solution of Stokes and Navier-Stokes problems.

6. - NUMERICAL EXPERIMENTS.

All the numerical experiments displayed in this section deal with Mooney-Rivlin materials and either a plane strain or axisymmetric situation ; moreover the numerical results presented here have been obtained using the quadrilateral finite element approximation discussed in Sec. 5.2. Other numerical experiments are done in [24],[25] (including the numerical treatment of plane stress problems) ; numerical experiments using this Ruas' element of Sec. 5.3 (and some of its generalizations are discussed in [30],[31].

6.1. Stretching of a thick cracked rectangular bar.

We consider a thick rectangular slab of Mooney-Rivlin material, with a non-propagating crack in its middle, submitted to vertical stretching forces applied at its extremities. The initial configuration of the lower part of the bar and of the crack is shown on Fig. 6.1 (a). Under the action of the external forces this bar is stretched and its computed equilibrium position (plane strain assumption, with $\sigma(\underset{\sim}{v}) = E_1(I_1(\underset{\sim}{v})-2))$ is shown on Fig. 6.2(b) (the various data concerning this problem are indicated on Fig. 6.2).
Using $\rho=R=8$ we have convergence of algorithm (4.1)-(4.3) in 20 iterations, corresponding to a computational time of 3.2 seconds on CDC 6400. The computed stresses at the boundary

match the applied tension with a 10^{-4} precision.

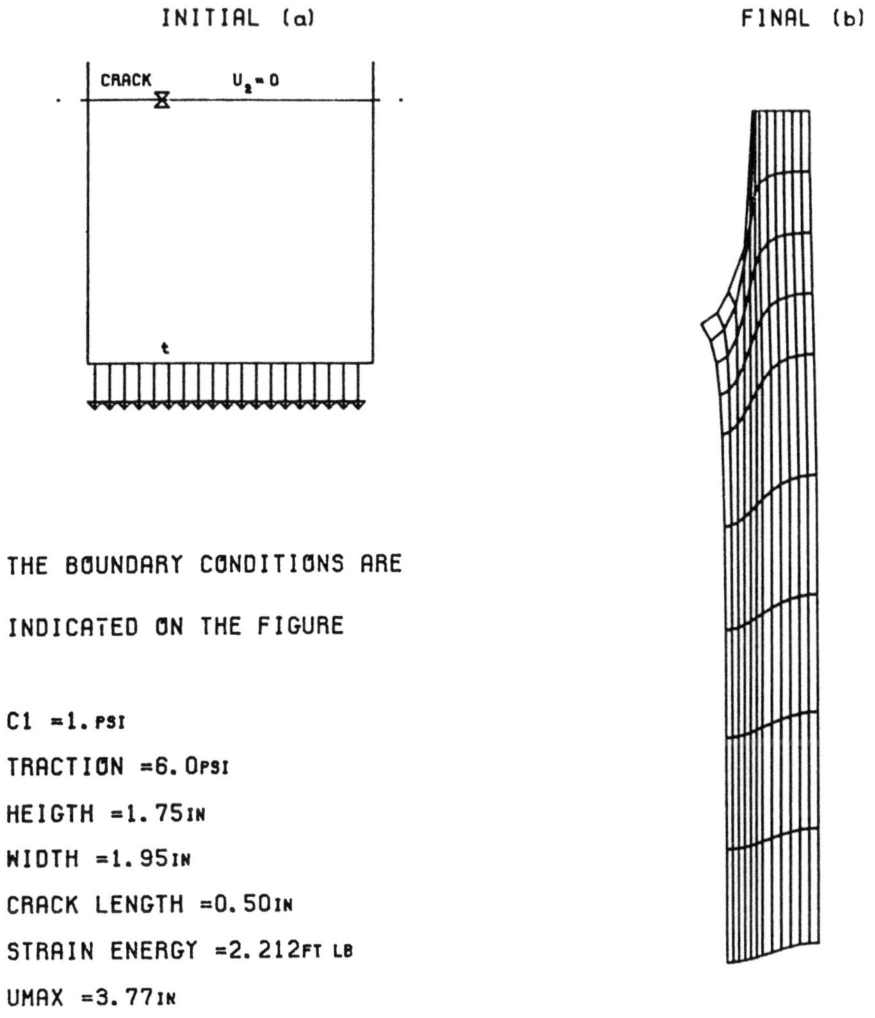

Figure 6.1

6.2. Combined Inflation and Extension of a Circular Cylindrical Tube.

We consider here a circular cylindrical tube, made of an incompressible isotropic elastic material, whose strain function is of Mooney-Rivlin type. This tube is inflated by imposing a fixed radial displacement to the inner surface $\partial\Omega_1$, the outer surface being free of tractions. An analytical solution of this problem is given in CHADWICK-HADDON [32] under the assumption that both extremities are stress free and remain horizontal. We have approximated these conditions by restricting the axial displacement to be zero at the mid cross-section $\partial\Omega_3$ and leaving the upper section traction free. This physical configuration is a close approximation of the case treated in [32] and is described in Fig. 6.2, where we have represented the upper half part of the tube in its initial configuration.

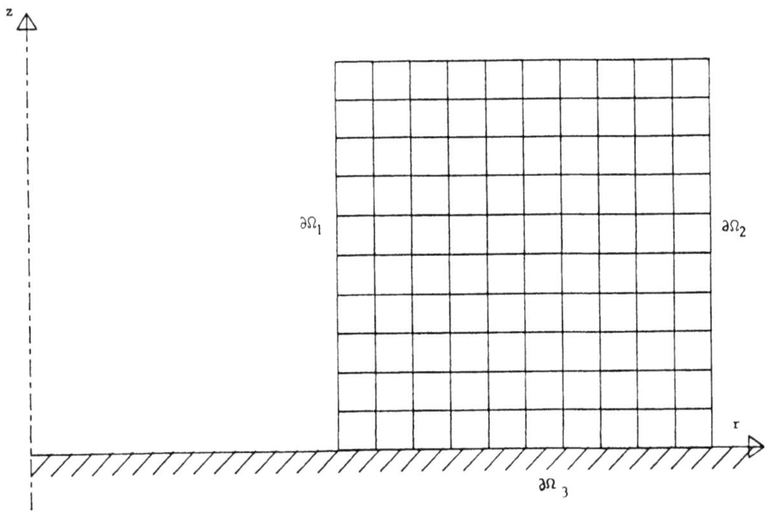

Figure 6.2

Using the notation of [32], the parameters in this problem are

$$2\sigma(\underset{\sim}{v}) = .875(I(\underset{\sim}{v})-3) + .125(I_2(\underset{\sim}{v})-3),$$

N = (outer radius/inner radius) in the reference configuration

Q = final inner radius/initial inner radius.

The numerical values given in Table 6.1, both for the analytical and numerical solution correspond to

EXTV = final height/initial height,

EXTH = final outer radius/initial outer radius.

The computed Cauchy stresses $\sigma_{\theta\theta}$ and σ_{rr} in the mid cross-section are indicated (as functions of r) on Fig. 6.3, for different values of the parameters N and Q ; these computed values (indicated by crosses on Fig. 6.3) are exactly located on the curves obtained analytically in [32] and reproduced on Fig. 6.3 ; this confirms the validity of our computations.

N	1.4	1.4	1.8	1.8	2.2	2.2	2.2
Q	1.2	1.6	1.6	2.0	1.6	2.0	2.2
EXTV analytical	.9460	.8583	.8991	.8432	.9252	.8794	.8578
EXTV computed	.9460	.8582	.8995	.8434	.9261	.8801	.8584
EXTH analytical	1.1191	1.3700	1.2486	1.4334	1.1774	1.3146	1.3879
EXTH computed	1.1192	1.3701	1.2489	1.4339	1.1778	1.3154	1.3882

Table 6.1

Comparison between analytical results (from [32]) and computed results.

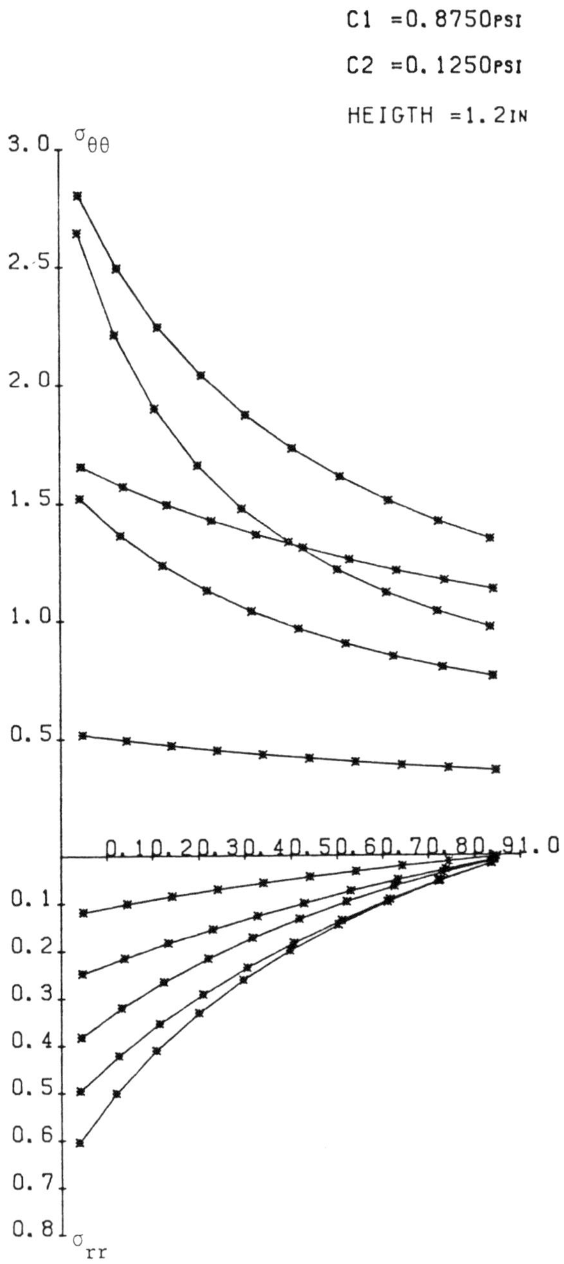
C1 =0.8750PSI
C2 =0.1250PSI
HEIGTH =1.2IN
σθθ
3.0
2.5
2.0
1.5
1.0
0.5
0.1 0.2 0.3 0.4 0.5 0.6 0.7 0.8 0.9 1.0
0.1
0.2
0.3
0.4
0.5
0.6
0.7
0.8
σrr

Figure 6.3

REFERENCES

1 Lions P.L., Mercier B., Splitting algorithms for the sum of two nonlinear operators, SIAM J. Num. Anal., 16, (1979), pp. 964-979.

2 Gabay D., Méthodes Numériques pour l'Optimisation Non Linéaire, Thèse d'Etat, Université Pierre et Marie Curie, Paris, November 1979.

3 Glowinski R., Lions J.L., Trémolières R., Analyse numérique des Inéquations Variationnelles, Vol. 1, Dunod-Bordas, Paris, 1976.

4 Glowinski R., Lions J.L., Tremolières R., Numerical Analysis of Variational Inequalities, North-Holland, Amsterdam (to appear).

5 Fortin M., Glowinski R., Résolution numérique de problèmes aux limites par des méthodes de Lagrangiens augmentés (in preparation).

6 Glowinski R., Numerical Methods for Nonlinear Variational Problems, Lecture Notes, Tata Institute, Bombay and Springer-Verlag, Berlin, 1980.

7 Cea J., Glowinski R., Sur des méthodes d'optimisation par relaxation, Revue Francaise Automatique, Informat., Recherche Opérationnelle, R-3, (1973), pp. 5-32.

8 Gabay D., Mercier B., A dual algorithm for the solution of nonlinear variational problems via finite element approximation, Comp. Math. Appl., Vol. 2, N° 1 (1976), pp. 17-40.

9 Douglas J., Rachford H.H., On the numerical solution of heat conduction problems in two or three space variables Trans. Amer. Math. Soc., 82 (1956), pp. 421-439.

10 Peaceman D.H., Rachford H.H., The numerical solution of parabolic and elliptic differential equations, SIAM J. Appl. Math., 3 (1955), pp. 24-41.

11 Bourgat J.F., Dumay J.M., Glowinski R., Large displacement calculations of flexible pipelines by finite element and nonlinear programming method, SIAM J. Sc. Stat. Comp., 1 (1980), pp. 34-81.

12 Glowinski R., Marrocco A., Sur l'approximation par éléments finis d'ordre un et la résolution par pénalisation dualité d'une classe de problèmes de Dirichlet non lineaire, Compt. Rend. Acad. Sc. Paris, t. 278A (1974), pp. 1649-1652.

13 Glowinski R., Marrocco A., On the solution of a class of nonlinear Dirichlet problems by a penalty-duality method and finite elements of order one, in Optimization Technique : IFIP Technical Conference, G.I. Marchouk, ed. Lecture Notes in Comp. Sciences, Vol. 27, Springer-Verlag, Berlin, 1975, pp. 327-333.

14 Glowinski R., Marrocco A., Sur l'approximation par éléments finis d'ordre un et la résolution par pénalisation-dualité d'une classe de problèmes de Dirichlet non linéaires, Revue Francaise d'Autom. Inf. Rech. Opérationnelle, R-2 (1975), pp. 41-76.

15 Poljak B.T., On the Bertsekas' method for minimization of composite functions, in International Symposium on Systems Optimization and Analysis, A. Bensoussan and J.L. Lions eds., Lecture Notes in Control and Information Sciences, Vol. 14, Springer-Verlag, Berlin, 1979, pp. 179-186.

16 Glowinski R., Marrocco A., Sur l'approximation par éléments finis d'ordre un et la résolution par pénalisation dualité d'une classe de problèmes de Dirichlet non linéaires. Rapport Laboria 115, 1975 (extended version of Ref. 14).

17 Glowinski R., Marrocco A., Numerical solution of two-dimensional magneto-static problems by augmented lagrangian methods, Comp. Meth. Appl. Mech. Eng., 12 (1977), pp. 33-46.

18 Marrocco A., Expériences numériques sur des problèmes non linéaires résolus par éléments finis et lagrangien augmenté, Rapport Laboria 309, 1978.

19 Mercier B., Sur la Théorie et l'Analyse Numérique de problèmes de plasticité, Thèse d'Etat, Université Pierre et Marie Curie, Paris, 1977.

20 Chan T., Glowinski R., Finite element approximation and iterative solution of a class of mildly nonlinear elliptic equations, STAN-CS-78-674, Compupter Sc. Department, Stanford University, 1978.

21 Begis D., Analyse Numérique de l'écoulement d'un fluide visco-plastique de Bingham par une méthode de lagrangien augmenté, Rapport Laboria 355, 1979.

22 Bourgat J.F., Glowinski R., Le Tallec P., Decomposition of Variational Problems and Applications in Finite Elasticity, in Partial Differential Equations in Engineering and Applied Science, R.L. Sternberg, ed., Marcel Dekker, New-York, 1980, pp. 445-480.

23 Chan T., Glowinski R., Numerical methods for a class of mildly nonlinear elliptic equations, Atos do Decime Primeiro Coloquio Brasileiro de Matematica, Vol. I, IMPA, Rio de Janeiro, 1978, pp. 279-318.

24 Le Tallec P., Numerical Analysis of Equilibrium Problems in Incompressible Nonlinear Elasticity, Ph. D. dissertation, The University of Texas at Austin, 1980.

25 Glowinski R., Le Tallec P., Numerical solution of problems in incompressible finite elasticity by augmented lagrangian methods (I) Two-dimensional and axisymmetric problems, submitted to SIAM J. Appl. Math.

26 Le Tallec P., Oden J.T., Existence and characterization of hydrostatic pressure in finite deformation of incompressible elastic bodies, TICOM Report, University of Texas at Austin, 1979.

27 Ball J.M., Convexity conditions and existence theorems in nonlinear elasticity, Arch. Rat. Mech. Anal., 63 (1977), pp. 337-403.

28 Glowinski R., Le Tallec P., Numerical solution of problems in incompressible finite elasticity by augmented lagrangian methods (II) Three-dimensional problems (to appear).

29 Glowinski R., Le Tallec P., Une méthode numérique en élasticité non linéaire incompressible, Comptes Rendus Acad. Sc. Paris, t.290B (1980), pp. 23-26.

30 Ruas V., A class of axymmetric simplicial finite element methods for solving finite incompressible elasticity problems, to appear in Comp. Meth. Appl. Mech. Eng.

31 Ruas V., Sur l'application de quelques méthodes d'éléments finis a la résolution d'un problème d'élasticity incompressible non linéaire, INRIA, Rapport de Recherche 24, May 1980.

32 Chadwick P., Haddon E.W., Inflation-Extension and Eversion of a Tube of Incompressible Isotropic Elastic Material, J. Inst. Maths. Applics., 10, (1972), pp. 258-278.

Incompressible Finite Elements: The LBB Condition and the Discrete Eigenstructure

D. S. MALKUS
Illinois Institute of Technology, Chicago, USA

Summary

The basic stability condition for mixed/Lagrange multiplier variational principles in incompressible media problems is the LBB condition. It plays an essential role in determining whether or not the problem is well-posed and governs the choice of finite elements in the discretization. This paper examines the discrete eigenstructure of a well-known Lagrange multiplier formulation for linear elasticity or Stokes-flow. It shows how the weak incompressibility constraint is reflected in the elementary-divisor structure of the eigenproblem whose solution determines the finite element approximation to the natural modes. In this context the discrete LBB condition can be seen to be a condition determining the limiting disposition, as the mesh parameter decreases, of a matrix pencil with infinite eigenvalues. Pure pressure modes and the load vectors required to transmit them are paired in cyclic subspaces of the infinite eigenspace. This pairing can be related to a well-known heuristic interpretation of the LBB condition. The relationship between the natural mode eigenproblem and the eigenproblems which determine the norm of the inverse of the discrete operator of a static or steady-flow problem is described. Finally, because of the equivalence between classes of mixed and penalty formulations, it is shown that these results apply to penalty/reduced integration finite element methods.

Introduction

This paper discusses certain aspects of the discrete stability condition for finite element models in incompressible elasticity or fluid flow. The term stability is used to denote the uniform invertibility of the discrete operator in the variational statement of a static or steady-flow problem [1-4]. The approach taken here is to view the discrete problem from an algebraic point of view. This approach does have its limitations because, by nature, a stability condition is an analytic rather than algebraic concept. On the other hand, stability conditions often have obvious algebraic consequences, such as the requirement that certain matrices of coefficients in the discrete equations be positive-definite, or that others be of full rank [1-4]. The purpose of this investigation, then, is to frame further analysis of the discrete stability condition for incompressible finite elements in as complete as possible an understanding of its algebraic consequences. As a result, it is hoped that this will contribute to the continuing investigations into determining which incompressible element formulations are computationally reliable and which are not.

<u>Lagrange Multiplier Method</u>

Consider the following familiar variational equation characterizing solutions to problems in incompressible elasticity or steady Stokes-flow:

$$\begin{cases} \text{Find } U \in H \ni \forall\ V \in H \\ B(U,V) = F(V) \end{cases} \tag{1}$$

$F \in H^*$ is a standard load integral. H is the Hilbert space $H^1_o \times H^o$. $U = (\underset{\sim}{u},p) \in H$ is a displacement or velocity/ pressure pair, and the subscript "o" denotes that $\underset{\sim}{u}$ may satisfy appropriate essential boundary conditions. For the purpose at hand, there is no loss of generality in assuming that these conditions are homogeneous, so that H is a linear space. H^* is the dual of H. The form of B(U,V) is assumed to be:

$$B[(\underset{\sim}{u},p),(\underset{\sim}{v},q)] = \int_\Omega [2\mu\, \epsilon_{ij} f_{ij} - \epsilon_{ii} q - f_{ii} p]\, d\Omega \tag{2}$$

where $\epsilon_{ij} = \frac{1}{2}\left(\frac{\partial u_i}{\partial x_j} + \frac{\partial u_j}{\partial x_i}\right)$, $f_{ij} = \frac{1}{2}\left(\frac{\partial v_i}{\partial x_j} + \frac{\partial v_j}{\partial x_i}\right)$ and repeated indices imply tensorial contraction; μ is shear modulus, and $-p/\mu$ is the hydrostatic pressure function for elasticity; μ is a viscosity with p the pressure in Stokes-flow.

It can be shown that B(U,V) satisfies three important conditions which guarantee that the problem of (1) is well-posed, given appropriate conditions on the domain [3] and assuming that the essential boundary conditions are such that the Korn inequality [5] is satisfied. It also may be required that the pressures $p \in H^0$ satisfy some appropriate constraint [3,4]. The three conditions are:

$$\text{i.} \quad |B(U,V)| \le C_1 \|U\|_H \|V\|_H$$

$$\text{ii.} \quad \inf_{\|V\|_H=1} \sup_{\|U\|_H=1} |B(U,V)| \ge C_2 > 0 \quad U,V \in H \tag{3}$$

$$\text{iii.} \quad B(U,V) = B(V,U)$$

$\|\cdot\|_H$ is the usual product norm on H, and C_1 and C_2 are constants. The need for (i-iii) is established in [1-3] and a proof that these conditions are satisfied by (2) is given in [4].

Here consideration is given to the discrete problem arising when (1) is discretized using a finite element trial space $S^h \times T^h \subset H$, with trial functions in S^h satisfying appropriate essential boundary conditions and pressures in T^h satisfying any necessary constraints. The mesh parameter, h, tends to zero as the mesh is refined. Conditions i and iii of (3) carry over to the discrete model and condition ii becomes

$$\text{ii}^*. \quad \inf_{\|V^h\|=1} \sup_{\|U^h\|=1} |B(U^h,V^h)| \ge C_2^h \qquad U^h, V^h \in S^h \times T^h \tag{4}$$

The stability condition of prime importance to the convergence of the discrete model [1-4] is

$$\liminf_{h \to 0} C_2^h \ge C_2^* > 0 \tag{5}$$

which is the uniform invertibility requirement.

The LBB Condition

Under the conditions put forward in the previous section it has been shown [1,2] that the condition (3ii) will be satisfied if and only if

$$\inf_{\|p\|_{H^0}=1} \sup_{\|\underset{\sim}{u}\|_{H^1}=1} |(\epsilon_{ii}, p)| \ge C_3 > 0 \tag{6}$$

$(\cdot\,,\cdot)$ denotes the H^0-inner product on Ω. This has discrete analogue

$$\inf_{\|p^h\|_{H^0}=1} \; \sup_{\|\underset{\sim}{u}^h\|_{H^1}=1} |(\epsilon^h_{ii}, p^h)| \geq C^h_3 \tag{7}$$

Inequality (6) is the continuous LBB condition and inequality (7) together with the requirement

$$\liminf_{h \to 0} C^h_3 \geq C^*_3 > 0 \tag{8}$$

is the discrete LBB condition. That these conditions guarantee the satisfaction of the previously stated conditions in (3ii) and (4-5) follows from the results of [1 and 2] and is implicit in the derivations of [3]. That the discrete LBB condition guarantees uniform invertibility will also follow from the work presented here.

Matrix Equations of the Finite Element Model

$B(U,V) = F(V)$ leads to matrix equations $\underset{\sim}{K}U = F$ which have the following partioned form when the appropriate nodal ordering is chosen:

$$\begin{bmatrix} \underset{\sim}{K}_2 & \bar{\underset{\sim}{K}}_1 \\ \bar{\underset{\sim}{K}}_1^T & 0 \end{bmatrix} \begin{bmatrix} u \\ p \end{bmatrix} = \begin{bmatrix} F \\ 0 \end{bmatrix} \tag{9}$$

Let u,p,etc. denote vectors of nodal values of trial functions $\underset{\sim}{u}^h$, p^h, etc. Then the matrices of (9) are implicitly defined

$$\begin{cases} u^T \underset{\sim}{K}_2\, v = 2\mu \int_\Omega \epsilon^h_{ij} f^h_{ij}\, d\Omega & \forall \underset{\sim}{u}^h, \underset{\sim}{v}^h \in S^h \\ u^T \underset{\sim}{K}_1\, p = -\int_\Omega \epsilon^h_{ii} p^h\, d\Omega & \forall \underset{\sim}{u}^h \in S^h,\ p^h \in T^h \\ v^T F = F(v^h) & \forall \underset{\sim}{v}^h \in S^h \end{cases} \tag{10}$$

The following matrix will also be useful:

$$p^T \underset{\sim}{Q}\, q = \int_\Omega p^h q^h\, d\Omega \qquad \forall p^h, q^h \in T^h \tag{11}$$

These matrices are evaluated with respect to standard f.e.m.

nodal bases of S^h and T^h.

The matrix $\underset{\sim}{K}_2$ plays a special role in this investigation. Because of the assumption that the Korn inequality [5] is satisfied, it means that there exist constants $\sigma > 0$ and $\delta > 0$, independent of h such that

$$\sigma \|\underset{\sim}{u}^h\|_1 \leq [u^T \underset{\sim}{K}_2 u]^{1/2} \leq \delta \|\underset{\sim}{u}^h\|_1 \tag{12}$$

This uniform equivalence between the strain energy and the H^1-norm allows substitution of the strain energy norm for the H^1-norm without loss of generality. Furthermore, the matrix $\underset{\sim}{Q}$ of (11) gives the H^0-norm on T^h. This means that without loss of generality in (3-4 and 6-7), $\|U^h\|_H$ can be replaced by

$$\|U^h\| \equiv [U^T \underset{\sim}{K}_2 \oplus \underset{\sim}{Q}\, U]^{1/2} = [U^T \underset{\sim}{N}\, U]^{1/2} \tag{13}$$

where

$$\underset{\sim}{N} = \underset{\sim}{K}_2 \oplus \underset{\sim}{Q} = \left[\begin{array}{c|c} \underset{\sim}{K}_2 & 0 \\ \hline 0 & \underset{\sim}{Q} \end{array}\right] \tag{14}$$

In what follows the mass matrix $\underset{\sim}{M}$ of S^h will be needed:

$$u^T \underset{\sim}{M}\, v = \int_\Omega \underset{\sim}{u}^h \cdot \underset{\sim}{v}^h \, d\Omega \quad \forall\, \underset{\sim}{u}^h, \underset{\sim}{v}^h \in S^h \tag{15}$$

Three Eigenproblems

A. Natural Modes

Define the weakly constrained Rayleigh quotient

$$R(u,p) \equiv \frac{\int_\Omega [\mu \epsilon_{ij} \epsilon_{ij} - \epsilon_{ii} p] d\Omega}{\int_\Omega u_i u_i \, d\Omega} \tag{16}$$

It is easy to see that this has discrete equations of stationarity

$$\underset{\sim}{K}\, U = \lambda [\underset{\sim}{M} \oplus 0]\, U \tag{17}$$

or

$$\left[\begin{array}{c|c} \underset{\sim}{K}_2 & \bar{\underset{\sim}{K}}_1 \\ \hline \bar{\underset{\sim}{K}}_1^T & 0 \end{array}\right] \left[\begin{array}{c} u \\ \hline p \end{array}\right] = \lambda \left[\begin{array}{c|c} \underset{\sim}{M} & 0 \\ \hline 0 & 0 \end{array}\right] \left[\begin{array}{c} u \\ \hline p \end{array}\right] \tag{18}$$

B. Convergence

Assuming the consistency of the finite element approximation, that is assuming that $S^h \times T^h$ can accurately approximate arbitrary members of H [3], what is needed to establish the convergence of the finite element solutions to (9) towards the exact solution is verification of condition ii* (equation 4). Using the norm given in (13), the satisfaction of ii* can be determined from the following eigenproblem:

$$\underset{\sim}{K}\, U = \lambda\, \underset{\sim}{N}\, U \tag{19}$$

The following fact is surely well-known by most, and is stated here as the first theorem for the sake of completeness. The proof of this result and the proofs of all other theorems stated here may be found in an extended version of this paper [6].

THEOREM 1: If $|\lambda_1| \leq |\lambda_2| \leq \cdots \leq |\lambda_N|$ are the eigenvalues of (19), then

$$\inf_{\|V^h\|=1} \sup_{\|U^h\|=1} |B(U^h,V^h)| = |\lambda_1|$$

C. LBB Eigenproblem

The eigenvalues of the following eigenproblem determine the satisfaction of the discrete LBB condition:

$$\underset{\sim}{K}_1\, u = \lambda\, \underset{\sim}{K}_2\, u \tag{20}$$

where

$$\underset{\sim}{K}_1 = \bar{\underset{\sim}{K}}_1\, \underset{\sim}{Q}^{-1}\, \bar{\underset{\sim}{K}}_1^T \tag{21}$$

which eigenvalue determines the inf sup in (7) is established by

THEOREM 2: Let $k = \dim T^h$ and $n = \dim S^h$. Let $\sigma_1 \leq \sigma_2 \leq \cdots \leq \sigma_k$ be the k largest eigenvalues of (20). The inf sup of (7) (w.r.t. $\|\cdot\|$) is given by $\sqrt{\sigma_1}$ if $k \leq n$, and 0 if $k > n$.

The purpose of the remainder of this paper is to establish the relationship between eigenproblems A, B, and C. It turns out that the natural mode problem A and the convergence eigenproblem B are closely related through the LBB eigenproblem C. In order to see this it is necessary to take into account the form of eigenproblem A, which is rather different from the other two. It is required to draw on knowledge of the canonical form of matrix pencils of arbitrary divisor structure [7-9]. This is because of the singularity of $\underset{\sim}{M} \oplus 0$ and the indefiniteness - and even possible singularity of $\underset{\sim}{K}$. There may fail to be a complete set of eigenvectors. If $C_3^h = 0$ for some h in (7), there may not even be a full set of eigenvalues.

The Eigenstructure of Problem A

If $\sigma_1^h = 0$ for some h, the LBB condition is not satisfied as stated. This can happen for a Dirichlet problem [3,4], but can be easily remedied by applying an appropriate constraint to the pressure unknowns. This reflects the fact that in such a problem, in which there are no stress boundary conditions, the pressure is determined only up to an ambient pressure which is nowhere specified. If it is not already apparent, it will be shown later that the lack of such a constraint will render $\underset{\sim}{K}$ singular by virtue of a rank deficiency in $\bar{\underset{\sim}{K}}_1$ of order one (a constant pressure). For practical purposes in obtaining solutions, one might want to impose the constraint explicitly, but in theory the full statement of the continuous LBB condition can be restricted to range over pressures which are in the quotient space of H^0 by the nullity of the adjoint of the divergence operator [4]. This has the effect of removing the constant pressure mode from consideration in (6-7).

Also $\sigma_1^h = 0$ if $k = \dim T^h > \dim S^h = n$. This can happen for grossly inconsistent (overconstrained) choices of element type and pattern for $\underset{\sim}{u}^h$ and p^h [10]. This can also happen for other element types and patterns - if for very large h - the number of essential boundary conditions "uses up" too many degrees of freedom on $\underset{\sim}{u}^h$. But these last two instances are well-recognized and easy to avoid. Perhaps the most interesting case for which $\sigma_1^h = 0$ is the case in which the discrete LBB condition as stated in (6-7) is purposefully not satisfied. This is done in order to decrease the constraint count by rendering many of the constraints redundant.

The elements of crossed-triangle pattern used by B. Mercier [11] are of this type. The idea is to enlarge the null-space of the discrete divergence operator (represented in weak form by $\bar{\underset{\sim}{K}}_1^T$). In [11] a consistency proof is given for the quadratic member of this family of macro-elements, but the possible effect on the stability of such a choice of elements is not discussed. By enlarging the null-space of $\bar{\underset{\sim}{K}}_1^T$ beyond the size of the null-space predicted by the difference between column and row dimension, one necessarily introduces a rank deficiency of corresponding order to the adjoint of the discrete divergence operator (represented weakly by $\bar{\underset{\sim}{K}}_1$). This rank deficiency is introduced regardless of whether the adjoint of the continuous operator has a null-space, and is potentially of high order (as opposed to the Dirichlet case in which the rank deficiency is of order 1). So assume the following dimensionalities:

$$\begin{cases} n = \dim S^h, & n > 0 \\ k = \dim T^h, & 0 < k \leq n \\ m = \dim \operatorname{Ker} \bar{\underset{\sim}{K}}_1, & 0 \leq m < k \end{cases} \tag{22}$$

From (19) one may easily see that $\sigma_1 = \sigma_2 = \ldots = \sigma_m = 0$.

<u>Definition</u>: $U^h \in S^h \times T^h$ is "mean isochoric" if $U^h = [u \,\vdots\, p^T]^T$ and $\bar{\underset{\sim}{K}}_1^T u = 0$.

Mean isochoric is used to denote those fields for which the $\underset{\sim}{u}^h$ part satisfies the weak form of incompressibility constraint. The word 'mean' indicates that these fields may not be exactly incompressible, though the reverse is true: isochoric implies mean isochoric.

<u>THEOREM 3:</u> There are n-m eigenvectors of problem A. There are n+k-m eigenvalues, n-k+m of which are finite, and 2(k-m) of which are the infinite repeated root. There are n-k+m mean isochoric modes with finite frequency and k pure pressure modes, of these k-m are of infinite frequency. The other m pure pressure modes are ill-disposed and of indeterminate frequency. Problem A is defective of k eigenmodes: There are k-m 2×2 Jordan blocks in the infinite eigenspace, each with a pure pressure mode as an eigenvector and a dilatational mode as cyclic generator. The remaining deficiency in eigenvectors results from the m ill-disposed pure pressure modes, each of which corresponds to a row and column minimal index of zero.

The results of THEOREM 3 can be put into the following

picture of the canonical form of (18) under the equivalence of singular pencils [7]:

COROLLARY 3.1: There exist non-singular matrices $\underset{\sim}{R}$ and $\underset{\sim}{S}$ of real numbers such that $\underset{\sim}{R}(\underset{\sim}{K} - \lambda[\underset{\sim}{M} \oplus 0])\underset{\sim}{S} =$

$$\begin{array}{l} m \left\{ \vphantom{\begin{matrix}0\\ \ddots\\ 0\end{matrix}} \right. \\ 2(k-m) \left\{ \vphantom{\begin{matrix}1\\0\\ \ddots \\ 1\\0\end{matrix}} \right. \\ n-k+m \left\{ \vphantom{\begin{matrix}\lambda\\ \ddots\\ \lambda\end{matrix}} \right. \end{array} \left[\begin{array}{c|c|c} \begin{matrix}0 & & \\ & \ddots & \\ & & 0\end{matrix} & & \\ \hline & \begin{matrix}\begin{bmatrix}1 & \lambda\\ 0 & 1\end{bmatrix} & & \\ & \ddots & \\ & & \begin{bmatrix}1 & \lambda\\ 0 & 1\end{bmatrix}\end{matrix} & \\ \hline & & \begin{matrix}\lambda_{n-k+m}-\lambda & & \\ & \ddots & \\ & & \lambda_1-\lambda\end{matrix} \end{array}\right] \tag{23}$$

This is a block diagonal matrix. Unspecified entries are zero.

It is useful here to introduce an auxilliary eigenproblem:

B^*. <u>Adjoint LBB eigenproblem</u>

$$\bar{\underset{\sim}{K}}_1^T \underset{\sim}{M}^{-1} \bar{\underset{\sim}{K}}_1 p = \lambda \underset{\sim}{Q} p \tag{24}$$

Let $q_1,\ldots,q_n$ be the null eigenvectors of (24) and $p_1,\ldots,p_{k-m}$ be the eigenvectors of (24) with nonzero λ. Let "$\perp$" refer to orthogonality in the norm of T^h (given via $\underset{\sim}{Q}$):

<u>Lemma 3.2.1</u>: The q_i span the kernel of $\bar{\underset{\sim}{K}}_1$, Ker $\bar{\underset{\sim}{K}}_1$, and p_i span the coimage, CoIm $\bar{\underset{\sim}{K}}_1 = \{\text{Ker } \bar{\underset{\sim}{K}}_1\}^{\perp}$. For each p_i, the vector $z_i = \underset{\sim}{M}^{-1} \bar{\underset{\sim}{K}}_1 p_i$ is dilatational, i.e., $\bar{\underset{\sim}{K}}_1^T z_i \neq 0$, and each p_i is the dilatation of the corresponding z_i in that $p_i = \underset{\sim}{Q}^{-1} \bar{\underset{\sim}{K}}_1^T z_i / \lambda_i$.

The lemma provides an orthogonal basis for the pressures in which each well-disposed pressure is obtained by taking the dilatation of a corresponding dilatational mode in similar fashion to the manner in which pressures are calculated in a penalty method [10]. Other choices of basis for the pressures are of course possible.

<u>COROLLARY 3.2</u>: Let the p_i and q_i be as in lemma 3.2.1, then

a) The infinite eigenvectors can be taken to be

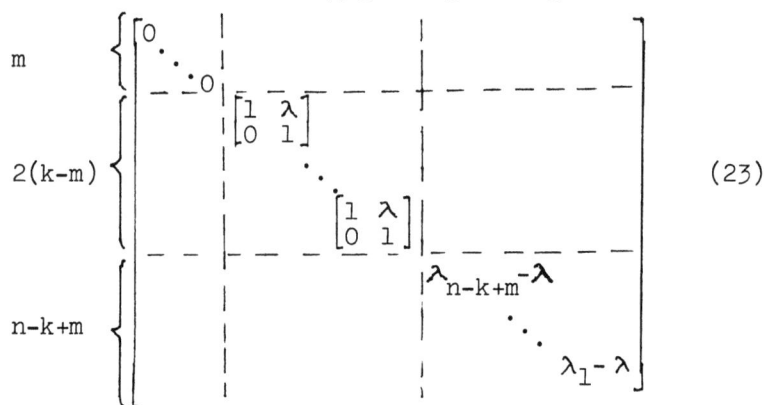

$$x_i = \left[\begin{array}{c} 0 \\ \hline p_i \end{array}\right]$$

b) Each pure pressure mode in (a) is generated by a cyclic vector

$$y_i = \left[\begin{array}{c} \underset{\sim}{M}^{-1} \bar{\underset{\sim}{K}}_1 \, p_i \\ \hline 0 \end{array}\right]$$

in the sense that

$$\underset{\sim}{K} \, x_i = [\underset{\sim}{M} \oplus 0] y_i \tag{25}$$

c) A $\underset{\sim}{Q}$-orthogonal basis of the ill-disposed pressure modes is

$$\left[\begin{array}{c} 0 \\ \hline q_i \end{array}\right]$$

d) Each mean isochoric mode is of the form

$$\left[\begin{array}{c} u_i \\ \hline s_i \end{array}\right]$$

where s_i is some pressure vector.

<u>COROLLARY 3.3</u>: The pure pressure modes satisfy the following static or steady-flow problems

a)
$$\underset{\sim}{K} \left[\begin{array}{c} 0 \\ \hline p_i \end{array}\right] = \left[\begin{array}{c} \bar{\underset{\sim}{K}}_1 \, p_i \\ \hline 0 \end{array}\right] \tag{26}$$

b)
$$\underset{\sim}{K} \left[\begin{array}{c} 0 \\ \hline q_i \end{array}\right] = 0 \tag{27}$$

This implies, in particular, if $m > 0$ $\underset{\sim}{K}$ is singular with rank deficiency m.

<u>Interpretation of THEOREM 3 and COROLLARIES</u>

Each of the k-m independent constraints removes a dilatational mode $\underset{\sim}{M}^{-1} \bar{\underset{\sim}{K}}_1 \, p_i$ from the eigenvector space by "tying it up" in a nonlinear divisor. The fact that for any pure pressure mode $[0 \,\vdots\, p^T]^T$

$$[0 \,\vdots\, p^T][\underset{\sim}{M} \oplus 0] \left[\begin{array}{c} 0 \\ \hline p \end{array}\right] = 0 \tag{28}$$

reflects the fact that such a mode is transmitted without kinetic energy. The infinite frequency of the k-m well-disposed pressure modes represents their infinite acoustic velocity. If $\sigma_i = 0$, the acoustic velocity is indeterminate. If the uniformity condition (7) fails, then there is one or more $\sigma_i \xrightarrow{h} 0$. This implies that, in the limit, the eigenproblem A is tending to a more poorly disposed limiting problem which has proportionately fewer instantaneously propagating pressure modes, and more pressure modes of indeterminate frequency. These are extraneous artifacts of the f.e.m. model.

With each instantaneously propagating pressure mode, besides the dilatational mode it removes from the eigenvector space, there is the closely related $[p_i^T \,\bar{\underset{\sim}{K}}_1^T \vdots\, 0]^T$. This can be considered as a system of distributed loads which generates the corresponding pure pressure mode as a solution to a static or steady-flow problem, according to (26). Then (27) is the negative of the familiar heuristic interpretation of the LBB condition as the assurance that no pressure mode can be transmitted by zero load - except perhaps the ambient constant pressure in a Dirichlet problem. Equation (27) says that the ill-disposed pressures are transmitted with zero load. Again the failure of the uniformity condition (7) has the obvious interpretation for the limiting problem.

The Eigenstructure of Problem B

Eigenproblem B is non-defective because $\underset{\sim}{N}$ is positive definite. In fact, making use of the Cholesky decomposition of $\underset{\sim}{N}$, it is easy to see that there is a complete set of eigenvectors associated with real, finite eigenvalues. In more detail

THEOREM 4: There are n+k eigenvectors of problem B. They are classified as follows

a) $\lambda_1 = \lambda_2 = \dots = \lambda_m = 0$ with eigenvectors

$$\begin{bmatrix} 0 \\ \hline q_i \end{bmatrix}$$

b) $\lambda_{m+1} = \lambda_{m+2} = \dots = \lambda_{n-k+m} = 1$ with eigenvectors

$$\begin{bmatrix} u_i \\ \hline 0 \end{bmatrix}$$

such that $\bar{\underset{\sim}{K}}_1^T u_i = 0$ (mean isochoric).

c) k-m dilatational modes with $\lambda_i^+ > 0$,

$$\begin{bmatrix} u_i \\ \hline \dfrac{\underset{\sim}{Q}^{-1} \bar{\underset{\sim}{K}}_1^T u_i}{\lambda_i^+} \end{bmatrix}$$

with $\lambda = \lambda_i^+ = \dfrac{1 + \sqrt{1 + 4\sigma_i}}{2}$.

d) k-m dilatational modes with $\lambda_i^- < 0$,

$$\begin{bmatrix} u_i \\ \hline \dfrac{\underset{\sim}{Q}^{-1} \bar{\underset{\sim}{K}}_1^T u_i}{\lambda_i^-} \end{bmatrix}$$

with $\lambda = \lambda_i^- = \dfrac{1 - \sqrt{1 + 4\sigma_i}}{2}$.

<u>COROLLARY 4.1:</u> The eigenstructure of problems A and B can be put in 1-1 correspondence as follows:

a) The m ill-disposed eigenvectors of A and the eigenvectors with $\lambda = 0$ in eigenproblem B can be taken to be the same.

b) The $\underset{\sim}{u}^h$ parts of the mean isochoric modes of problem B can be chosen as the $\underset{\sim}{u}^h$ parts of the mean isochoric modes of problem A.

c) The 2×2 Jordan blocks of problem A can be put in correspondence with the invariant subspaces of dimension 2 in problem B generated by the eigenvectors of λ_i^+, λ_i^- pair. Each such invariant subspace of problem B contains a pair of vectors

$$\begin{bmatrix} 0 \\ \hline \underset{\sim}{Q}^{-1} \bar{\underset{\sim}{K}}_1^T u_i \end{bmatrix} \text{ and } \begin{bmatrix} u_i \\ \hline 0 \end{bmatrix}$$

Note that the (c) part of the corollary does not establish the identity of $\underset{\sim}{M}^{-1} \bar{\underset{\sim}{K}}_1 p_i$ in problem A with the u_i here. But these two vectors pay a closely corresponding algebraic role in each eigenproblem: 1) Each dilatational mode is "tied up" in an invariant subspace with a corresponding pressure; 2) each pressure is generated by the dilatation of the corresponding u_i, in the sense of lemma 3.2.1; 3) the well-disposed pressures are H^o-orthogonal to the ill-disposed pressures. Note that in the convergence eigenproblem B, the orthogonality of the ill-disposed to the well-disposed ones arises naturally, rather than by the arbitrary basis choice of problem A.

Interpretation of THEOREM 4 and COROLLARY

For each $\sigma_1 = \sigma_2 = \ldots = \sigma_m = 0$, there is a $\lambda_i = 0$. If $\sigma_i \xrightarrow{h} 0$, this shows up as a negative eigenvalue, λ_i^-, which tends to zero as the mesh is refined. The deterioration in the disposition of eigenproblem A and the corresponding limiting indeterminacy of the frequency of a pressure mode shows up here in the failure of one or more negative eigenvalues of problem B to be bounded away from zero. Since the signature of eigenproblem B is the same as the signature of $\underset{\sim}{K}$, these negative eigenvalues correspond to negative eigenvalues of $\underset{\sim}{K}$.

Penalty Methods

The work of this author and T. J. R. Hughes [10] has shown that the penalty/reduced-selective integration technique produces a static or steady-flow problem

$$(\underset{\sim}{K}_2 + z\,\underset{\sim}{K}_1)u = F \qquad (29)$$

$\underset{\sim}{K}_2$ is the same matrix as in (9), and z is a large penalty parameter. When T^h is constructed based on the integration points of S^h used to evaluate $\underset{\sim}{K}_1$ as described in [10], then relation (21) holds, and (29) produces the same solutions as a slightly perturbed version of (9) [4,10]. More recently Hughes and Malkus [12] have established the correspondence between penalty methods and Lagrange multiplier methods, which allows a much broader choice of pressure trial spaces T^h, yet yields relation (21) relating penalty methods to Lagrange multiplier methods. So in what follows assume that $\underset{\sim}{K}_1$ in the penalty method is given by (21) for $\bar{\underset{\sim}{K}}_1$ and $\underset{\sim}{Q}$ defined for an associated Lagrange multiplier method which also has the same $\underset{\sim}{K}_2$ and F.

The penalty method (29) has the following eigenproblem which determines the norm of the inverse of the discrete operator:

$$(\underset{\sim}{K}_2 + z\underset{\sim}{K}_1)u = \lambda \underset{\sim}{K}_2 u \quad (30)$$

This rearranges to

$$\underset{\sim}{K}_1 u = \sigma \underset{\sim}{K}_2 u \quad (31)$$

where

$$\sigma = \frac{\lambda - 1}{z} \quad (32)$$

Therefore the LBB eigenproblem C is the appropriate discrete stability condition for the penalty method. In more detail:

<u>THEOREM 5</u>: The eigenproblem in (30) determines the uniform invertibility of the operator in (29). If $\sigma_{m+1} \leq \sigma_{m+2} \leq \dots \leq \sigma_k$ are the non-zero eigenvalues of the LBB eigenproblem, the eigenvalues of (30) are

a) $\lambda_1 = \lambda_2 = \dots = \lambda_{n-k+m} = 1$ with u_i such that $\underset{\sim}{K}_1^T u_i = 0$ (mean isochoric).

b) k-m eigenvalues with $\lambda_1 = 1 + z\sigma_i$ and dilatational eigenvectors.

Because pressures in a penalty method are computed from the dilatational part of the solution to (29), the m ill-disposed pressure modes are "invisible" to a penalty method.

<u>COROLLARY 5.1</u>: All pressures computed from a solution to (29) are in the span of the vectors $[0 \vdots p_i^T]^T$ which span CoIm $\bar{\underset{\sim}{K}}_1$.

Furthermore it is clear that the matrix $\underset{\sim}{K}_2 + z\underset{\sim}{K}_1$ is directly invertible for any m.

Interpretation of THEOREM 5 and the COROLLARY

Penalty methods avoid the non-invertibility of $\underset{\sim}{K}$. They avoid the possible contamination of the computed pressures with ill-disposed modes. Therefore the direct violation of the stated LBB condition by having m > 0 is not necessarily fatal to such a method. B. Mercier [11] claims success with his family of crossed triangular elements in a variety of problems. A simple calculation for linear crossed triangles shows that m is equal to the number of circumscribed quadrilaterals, with ill-disposed modes having a variety of oscillating patterns. Mercier uses a penalty method for his calculations. He provides a proof for one member of the family, which shows that in Ker $\bar{\underset{\sim}{K}}_1^T$, a divergenceless field

$\underset{\sim}{u}^h$ can be approximated with optimally consistent accuracy. With other arrangements of linear triangles the dimension of Ker $\bar{\underset{\sim}{K}}_1^T$ would not be large enough.

The work presented in this paper reminds us of two more results which need to be established for such a penalty method. First, the remaining k-m well-disposed pressure modes must be capable of approximating arbitrary pressures in H^0 to appropriate accuracy. Second, the resulting method must still be stable, in the sense that the LBB condition must be satisfied on $\{\text{Ker } \bar{\underset{\sim}{K}}_1\}^{\perp}$, as would be reflected in the uniform boundedness away from zero of $\sigma_{m+1} \leq \sigma_{m+2} \leq \cdots \leq \sigma_k$ with decreasing h. If this condition is satisfied, note that THEOREM 5 implies that the dilatational modes can be given a frequency $\lambda_i = 1 + z\,\sigma_i$ which is arbitrarily large, independent of h. If any of the $\sigma_{m+1}, \ldots, \sigma_k$ tend to zero, then the cardinal sin of penalty methods is committed: For a fixed z, the separation between the two components of the operator spectrum is a function of h.

References

1. Babuška, I.; Oden, J. T.; Lee, J. K.: Mixed-hybrid finite element approximations of second order elliptic boundary-value problems. Comp. Meth. Appl. Mech. Engng. 11 (1977) 175-206.

2. Brezzi, F.: On the existence, uniqueness and approximation of saddle point problems arising from Lagrangian multipliers. R.A.I.R.O. 8 (1974) 129-151.

3. Babuška, I.; Aziz, A. K.: Mathematical Foundations of the Finite Element Method. New York: Academic Press 1972.

4. Bercovier, M.: Perturbation of mixed variational problems. Application to mixed finite element methods. R.A.I.R.O. 12 (1978) 211-236.

5. Hlavaček, I.; Nečas, J.: On inequalities of the Korn type. Arch. Rat. Mech. Anal. 36 (1970) 305-334.

6. Malkus, D. S.: Preprint, available from author.

7. Gantmacher, F. R.: Matrix Theory II. New York, Chelsea 1971.

8. Moler, C.; Stewart, G. W.: An algorithm for generalized matrix eigenvalue problems. S.I.A.M. J. Numer. Anal. 10 (1973) 241-256.

9. Stewart, G. W.: On the sensitivity of the eigenvalue problem $Ax = \lambda Bx$. S.I.A.M. J. Numer. Anal. 9 (1972) 669-686.

10. Malkus, D. S.; Hughes, T. J. R.: Mixed finite elements - reduced and selective integration techniques: a unification of concepts. Comp. Meth. Appl. Mech. Engng. 15 (1978) 63-81.

11. Mercier, B.: A conforming finite element method for two-dimensional incompressible elasticity. Int. J. N. M. E. 14 (1979) 942-945.

12. Hughes, T. J. R. and Malkus, D. S.: Manuscript in preparation.

Part VI: Computational Algorithms

Data Management in Finite Element Analysis

P.J. PAHL
Technische Universität Berlin, Germany

Summary

The global data for finite element analyses are conveniently stored in matrices and table trees. In order to optimize the use of the available storage capacity, the matrix and table tree data structures are implemented on the basis of a common storage structure. The paper describes the properties and functions of storage, matrix and table tree managements for finite element programs.

1. PROPERTIES OF FINITE ELEMENT DATA

1.1 Introduction

During the past years, several technical application programs using different storage and data structures have been developed at the Institut für Allgemeine Bauingenieurmethoden [1-8]. In connection with a proposed program to develop software for nonlinear finite element methods in a cooperative effort at several independent institutions, we have evaluated our experiences with these programs. In addition, we have reviewed publications of comparable programs [9-17]. This paper contains an extract of those aspects of global data management which we consider important for a joint research effort in nonlinear finite element analysis.

1.2 Global Data

The global data of finite element analyses can be categorised as follows:

- Descriptive data in tabular form, eg. nodal coordinates, element topology and loading conditions.

- Computed element properties, eg. element stiffness matrices and element load vectors.
- System equations in matrix form, eg. stiffness matrices, incremental and total displacement and load vectors.
- Computed element results in tabular form, eg. incremental and total strains and stresses.
- Data created by postprocessing, eg. for graphics, in tabular form.

This list shows that tabular and matrix data structures are of equal importance in finite element programs.

1.3 Working Set of the Global Data

The working set of the global data consists of that part of the global data which resides in primary memory at a particular stage of the analysis. In general, the required working set is only a subset of the available working set. For an in-core solution, all required working sets are subsets of the available working set. If storage for an in-core solution is not available, the algorithms should keep the working set as constant as possible to minimize the access to secondary storage.

The following table shows that the required working set changes significantly during the analysis. In particular, the relative volume of tables and matrices in the working set changes. The use of the available primary memory for out-of-core solutions is therefore optimized if the table and matrix data structures are implemented on the basis of a common array storage structure. Space released by matrices can then be used for tables, and vice versa. This concept is the main feature of the data management proposed in this paper.

Working Sets of Global Data for Finite Element Analysis		
Phase of the Analysis	Tables	Matrices
Description of the structure	x	
Checking of the input values	x	
Computation of the element properties	x	x
Assembly of the system equations	x	x
Solution of the system equations		x
Incrementation of the variables	x	x
Postprocessing	x	

2. STORAGE MANAGEMENT

2.1 Storage Structure

The finite element program and its data are stored in a program area on primary memory and in data sets on secondary memory, which are assigned through the operating system.

In primary memory, the global data are separated from the procedures either by defining a COMMON array, or by passing a large FORTRAN array through the argument lists of the subroutines. This array is divided into an administrative zone and a data pool. The administrative zone contains administrative tables of the storage management, the matrix management and the table management. If the operating system is virtual, real memory is subdivided into pages. These are automatically managed by the operating system.

On secondary memory, the global data are stored in one or more data sets. The data sets are subdivided into records, which are accessed with a data management system of the operating system. The storage management of the finite element program divides the data sets into files. Each file is a consecutive set of records containing a specific type of data, eg. a global matrix or table trees. The file is identified by a unique number.

All files are subdivided into arrays of equal size, as shown in Fig. 1. The array size should be a multiple of the record size. The arrays of a file are numbered consecutively. Thus every array of the finite element program is identified by a unique duplet (file number, array number). Arrays are assigned by the storage management upon request by the matrix and table managements, as described in sections 3 and 4 of this paper.

The data pool is subdivided into blocks, whose size equals that of the arrays. For virtual systems, the block size should be a multiple of the page size. The blocks of the pool are numbered consecutively. During execution, blocks are assigned by the storage management upon request of the matrix and table managements. These blocks contain the working set of the arrays. Two special blocks, the scratch pad and the auxiliary array, are reserved for the matrix management.

2.2 Administrative Data of the Storage Management

The administrative data of the storage management are stored in the file table and the pool block table shown in Fig. 2. The file table is indexed with the file number. Each line describes the location, size and occupation of one file. The pool block table is indexed with the block number. Each line describes the status, location and present contents of one block in the pool.

The status parameter of the blocks is used to control the residence of selected arrays. It differentiates between empty, free and fixed blocks. An empty block is assigned without storing its present contents. A free block can only be reassigned after its contents has been saved in a file. A fixed block cannot be reassigned. The matrix management, for instance, controls the pool residence of hypermatrices by storing them in fixed blocks.

2.3 Assignment Algorithms for Arrays

The arrays of a file are assigned sequentially. Individual arrays cannot be deleted or reassigned. The file can, however, be deleted and reused as a whole.

The assignment of blocks in the pool is controlled with priorities. In a first loop, the attempt is made to assign an empty block. The search starts with the block following the last assigned block and ends after all blocks have been considered. If the end of the pool (except the reserved blocks) is reached, the search recommences at the front of the pool.

If the pool does not contain an empty block, the attempt is made in a second loop to assign a block which contains a free array. The sequence of the search is identical with that of the first loop. If a free block is encountered, its present contents is written into its file before the block is reassigned. The transfer to the file is suppressed if the array was previously stored in the file, and has not been changed by writing.

If the pool contains neither empty nor free arrays, the execution is terminated due to lack of storage space.

2.4 Functions of the Storage Management

The storage management is called to execute the following functions:

DEFINE POOL
An empty pool block table is set up.

STORE POOL
All arrays in the blocks of the pool are stored in their files.

DEFINE FILE
The administrative data for a specified file are entered in the file table.

RELEASE FILE
All arrays of the specified file are released in the pool. The file is set to empty.

DELETE FILE
The specified file is released. Its definition in the file table is nullified.

STATUS FILE
The parameters describing the file in the file table are returned.

ASSIGN BLOCK
A block is assigned in the pool by the algorithm described in section 2.3. Its number is returned.

FIX BLOCK
The contents of the specified block is fixed in the pool: the block is not available for reassignment.

FREE BLOCK
The specified block is available for reassignment.

RELEASE BLOCK
The status of the specified block is set to the value empty.

STATUS BLOCK
The parameters describing the block in the pool table are returned.

ASSIGN ARRAY
A new array is assigned in the specified file by the algorithm of section 2.3. Its number is returned.

GET ARRAY
The specified array is transferred from the file to a specified block in the pool.

PUT ARRAY
The array stored in the specified block is transferred to the file registered in the pool block table.

STATUS ARRAY
If the array resides in the pool, the number of the block in which it is stored is returned, otherwise the value zero.

3. MATRIX MANAGEMENT

3.1 Hypermatrix Data Structure

Nonlinear finite element analysis by incremental techniques involves programming and execution of voluminous matrix operations. For out-of-core solvers, the matrix operations are associated with considerable data flow. Sparsity and symmetry properties of matrices must be considered in the algorithms in order to minimize the volume of data and the number of operations. Frequently, this is achieved through the well-known hypermatrix data structure [13].

In the hypermatrix data structure, each matrix is divided into rectangular or square submatrices. Usually, all submatrices of a given matrix are of the same size. The coefficients of a submatrix are stored consecutively (columnwise). Profiles are stored to describe the populated parts of the submatrix. The hypermatrix contains the submatrix numbers. The principle is illustrated in Fig. 3.

It is useful to introduce the hypermatrix scheme as a common data structure for all major matrices of the finite element program. The data flow and the matrix algebra can then be programmed through a standardised set of functions (eg. subroutine calls) which are readily implemented in different environments by replacing minor subroutines of the matrix management with well-defined functions and parameter lists. The program therefore becomes highly portable and adaptable to different computer capacities.

3.2 Storage of Matrices in Arrays

Each matrix of the finite element program is identified by a unique matrix number and is associated with a specific file. As shown in Fig. 4, this file contains only data for the associated matrix. The file can readily be reused if the matrix is deleted, eg. during iteration.

The submatrices of a matrix are identified by consecutive numbers, which are automatically assigned when the submatrices are defined. These submatrix numbers are stored in the hypermatrix in the location corresponding to the hyperindices of the submatrix (see Fig. 3).

Since different matrices of a set of equations may be divided into submatrices of significantly different size (eg. S * S for matrix K and S * L for matrix U in Fig. 3), whereas all arrays are of equal size, a variable number of submatrices is stored per array. A specific submatrix is located by its array number and the displacement d_i shown in Fig. 4.

In order to mark empty rows and columns and to avoid unnecessary operations, the matrix management sets up profiles of hypermatrices and submatrices (Fig. 3), which are stored together with the coefficients (Fig. 4).

3.3 Administrative Data of the Matrix Management

All matrices of a finite element system are recorded in the matrix table shown in Fig. 5. Each line of the table describes the hypermatrix, submatrices and administrative tables of one matrix. During execution, the matrix table resides in the administrative zone of primary memory.

For each defined matrix, the management builds up a hypermatrix containing the submatrix numbers, and a submatrix table containing the submatrix locations. During execution, a block table is built up to record the numbers of the blocks in which the arrays of the matrix are stored in the pool.

The hypermatrix, the submatrix table and the block table are usually stored in fixed blocks of the pool. Between runs, the hypermatrix and the submatrix table are stored in the file associated with the matrix.

3.4 Access Algorithm for Matrices

The submatrix, which is to be accessed, is identified by the matrix number and its hyperindices. These are used to find the submatrix number in the hypermatrix.

If the submatrix number is zero, the submatrix is not defined. A new array is then defined (if required), the next submatrix number is assigned, the displacement in the array is determined and the administrative data are entered in the tables.

If the submatrix is defined, its array number is read in the submatrix table. This is used to read the block number in the block table of the matrix. If the block number is zero, a block is assigned and the array loaded from the file. For positive block numbers, the file and array numbers are checked against the entries in the block table of the storage management. If the block has been reassigned since the last access to the subroutine, they will not coincide. In this case, a new block must be assigned and the array reloaded from the file.

Finally, the pool indices of the first coefficient and of the profiles of the submatrix are returned. Hypermatrices are accessed in a similar manner.

3.5 Functions of the Matrix Management

3.5.1 Conventions

All functions of the matrix management are subject to a common set of conventions:

- Matrices, for which the execution of an operation is requested, must previonsly have been defined.
- All matrix functions (with the exception of SET, ADD and GET GROUP) refer either to complete submatrices or to a complete matrix.
- If a submatrix required to perform a requested function is not available in the pool, it is automatically defined (if applicable) or loaded from the file.
- The hypermatrices, profiles and administrative tables are updated automatically.

3.5.2 General Functions

In addition to the matrix algebra performed on submatrices, the management supports a number of general functions for matrices:

DEFINE MATRIX
The properties of the matrix are entered in the matrix table.

RELEASE MATRIX
All arrays of the matrix are deleted. The file is set to empty.

DELETE MATRIX
The matrix is released. Its definition in the matrix table is nullified.

STORE MATRIX
The arrays of the matrix are stored in the associated file. All blocks assigned to the matrix are released.

PRINT MATRIX
The hypermatrix and administrative tables of the matrix are printed.

STATUS MATRIX
The contents of the associated row in the matrix table is returned.

INDEX MATRIX
The pool indices of the first element of the hypermatrix and its profiles are returned.

DEFINE SUBMATRIX
The specified submatrix is defined and recorded.

SET SUBMATRIX
All coefficients of the submatrix are set to a given value. The profiles are set to zero.

PRINT SUBMATRIX
The coefficients and the profiles of the specified submatrix are printed.

INDEX SUBMATRIX
The pool indices of the first element of the submatrix and of its profiles are returned.

SET GROUP
The coefficients of a specified group are set to given values.

ADD GROUP
The coefficients of a specified group are modified by adding given values.

GET GROUP
The values of the coefficents of a specified group are returned.

The group functions are required primarily to assemble matrices and to read results selectively (eg. displacement components at a node). The group is specified as shown in Fig. 6. It must lie totally within a submatrix.

3.5.3 Matrix Algebra

The algebraic functions of the matrix management are always performed on complete submatrices. Since the solution algorithms are frequenthy programmed for symmetric matrices, all functions can be requested both for the stored submatrices A or R and for their transposes. The dimensions of the submatrices are automatically read in the data structure. The user must make sure that the requested operations are permitted by the rules of matrix algebra.

The proposed set of functions is shown in Fig. 7. It is assumed that all values are real and of the same accuracy. The programmer can store intermediate results in the scratch pad. In addition, he can use one of the operands to store the results. The matrix management automatically recognises this condition and uses the auxiliary array to store temporary results if necessary.

4. TABLE MANAGEMENT

4.1 Table Tree Structure

Tables are rectangular data structures. A specific value is identified by a triplet: table number, line number and column number. It will be assumed that all values in a table are of the same type and length. In addition, it will be assumed that the number of columns is fixed at definition, whereas the number of lines is dynamically adapted to the space requirements.

Typical global values of a finite element program are identified by two or more indices, for example:

Nodal coordinates : node, component
Nodal loads : node, loading condition, component
Element loads : element, loading condition, face, component

Some of the indices assume consecutive values, eg. the component numbers of a vector. Other indices may be sparsely populated, eg. external node and element numbers.

Due to the variable number of indices and the sparcity of some of the indices, the descriptive data and the computational results of a finite element analysis are not conveniently stored in individual tables. Instead, related tables are chained with pointers to form table tree structures. This is illustrated in Fig. 8 for nodal data.

The root of a table tree is always a number table. It is used to compress a sparse external index (eg. node numbers) into a consecutive set of internal index values. All other tables of the tree are data tables containing values and pointers (eg. nodal data such as coordinates, loads and displacements).

4.2 Storage of Trees in Arrays

For iterative nonlinear analyses, the data are so voluminous that secondary storage is frequently required. The tables are therefore divided into subtables containing a multiple of complete lines of the table. Each subtable is stored in an array of its own. In the pool, the table management competes with the matrix management for storage space in the available blocks.

Each table of a finite element program is identified by a unique table number and is associated with a specific file. The table is stored in arrays of this one file only. The file may also contain other tables. The arrays assigned to a table in a file cannot be reused, unless the file as a whole is deleted.

4.3 Administrative Data of the Table Management

The relationship between the arrays of a table tree are described with pointers, which are stored in the subtables and contain two values:

pointer (1) = number of the array containing the dependent subtable

pointer (2) = number of a line within this subtable

In addition to the pointers, the table management uses a table to store administrative data describing the trees. Each line of this table contains the following parameters for a specific table:

Administrative Table for the Table Management	
Column	Contents of one line of the table
1	file number of the table
2	type of the table
3	number of columns in the table
4	number of lines per subtable
5	number of the last assigned array
6	number of the last line assigned in the subtable
7	number of the preceeding table in the table tree
8	column number of pointer in the preceeding table
9-2o	6 pairs of descriptors: array number,block number

Number tables consist of at most 6 subtables. Both the array number and the block number (if resident in the pool) of all subtables are stored in the administrative table.

Data tables consist of an arbitrary number of subtables. At most 6 subtables are stored in the pool at a given time. The numbers of these arrays, as well as the numbers of the blocks in which they are resident, are stored in the administrative table.

4.4 Access Algorithm for Table Trees

The table management accesses one line of a specified table at a time. This line is identified by the table number and one or more indices. The location of the line in the pool is determined by a backward search in the administrative table, followed by a forward search in the subtables. This is illustrated in Fig. 9.

The backward search starts with the specified table. The number of the preceeding table is read in column 7 of the administrative table. The search is continued with this table until a number table is encountered. For all tables, the file number as well as the type and the column number of the pointer in the preceeding array (if applicable) are recorded, as shown in the figure.

The first specified index is used to determine the appropriate number subtable and the local line index K in this subtable. The line contains a single forward pointer (array number and line number). The file and array numbers of the next subtable in the forward search are therefore known. This subtable contains the next pointer in the chain. Depending on the type of the array, the column number of this pointer is either read during the forward search or computed from a specified index. The search ends with the table specified in the access request.

The arrays containing the chained subtables are made available through the storage management. If an empty pointer is encountered during an access for reading in the table, the search is terminated unsuccessfully. If the access is for writing, a new line is assigned in the next table. The array number and line number are recorded in the pointer. The search is continued until the specified values can be entered in the specified line.

4.5 Functions of the table management

All functions of the table management are requested through a common subroutine call:

CALL TABLEj	(function, code, indices, data)
j	identifier for data type and length
function	1/2/3/4 = read/write/status/define
code	completion code of the operation
indices	table number and index values for access
data	FORTRAN-vector, equal in type and dimension with one line of the specified table

The table management consists of definition and access functions:

DEFINE TABLE
The properties of the table are entered in the administrative table.

PUT LINE
The data are written from the FORTRAN-vector into the specified line. If necessary, space for pointers and for the data is automatically assigned in the table tree.

GET LINE
The contents of the specified line is read in the table and stored in the FORTRAN-vector. If the line is defined, the code is set to 1, otherwise to 2.

STATUS LINE
The code is set to the status of the specified line (defined or undefined). The function is used to check the existence of specific data, eg. a node with a particular number.

5. TUNING OF FINITE ELEMENT PROGRAMS

Finite element programs are installed on computers of different capacity and configuration. They are used to solve problems of different type and size. This leads to significant variations in the total required memory, in the available primary memory and in the data management by the operating system. If the solution is to be efficient, the finite element program must be tuned to these conditions.

The tuning parameters describing the individual installation and application all are parameters of the storage management. The total available memory and its assignment to different storage requirements within the program are controlled by the following parameters:

number and length of data sets
number and length of files in a data set

The available primary memory as well as the data flow between primary and secondary memory are controlled by additional parameters:

length of the data pool
length of an array

The storage management is readily implemented in such a way that the user can adjust all of these parameters to the requirements. The management parameters for the data structures, eg. submatrix and subtable dimensions, can be computed internally from the array size and need not be specified by the user. The separation of the data structure from the storage structure thus simplifies the tuning of a finite element program.

Acknowledgement

The concepts described in this paper have been influenced strongly by cooperation and discussions with my colleague at the Institut für Allgemeine Bauingenieurmethoden, Prof. Rudolf Damrath.

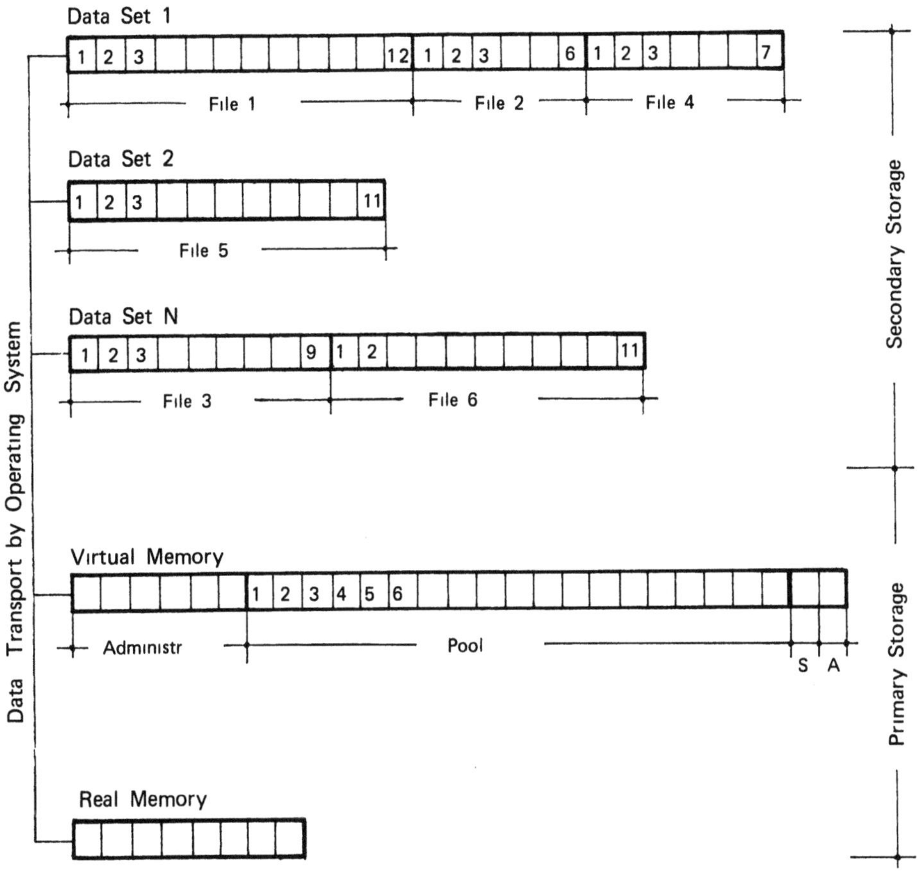

FIGURE 1 · Storage areas of a finite element system

Adress in secondary storage	file number, array number
Adress in primary storage	block number
Special arrays in the pool ·	scratch pad S
	auxiliary array A

File table

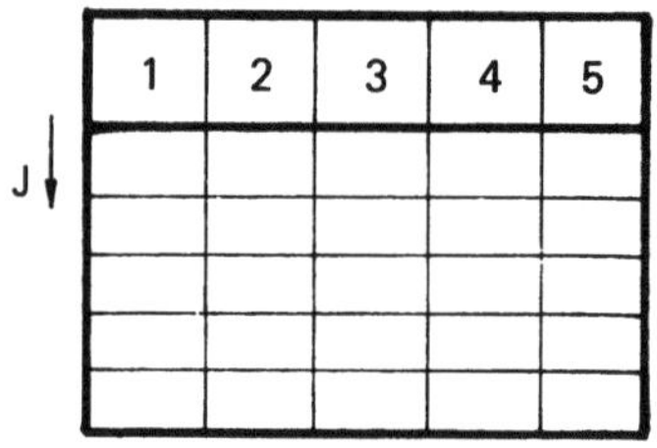

J : file number

Pool block table

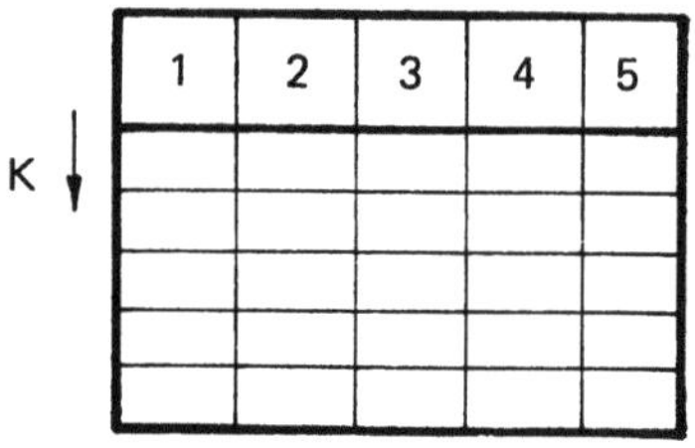

K : block number

Column	Contents of file table
1	number of the data set containing the file
2	number of the first record of the file
3	number of arrays in the file
4	number of records per array
5	number of the last assigned array
Column	**Contents of pool block table**
1	status of the block . empty / free / fixed
2	index of the first word of the block in the pool
3	file number of the stored array
4	array number of the stored array
5	access to the block : reading / writing

FIGURE 2 : Administrative tables of the array management

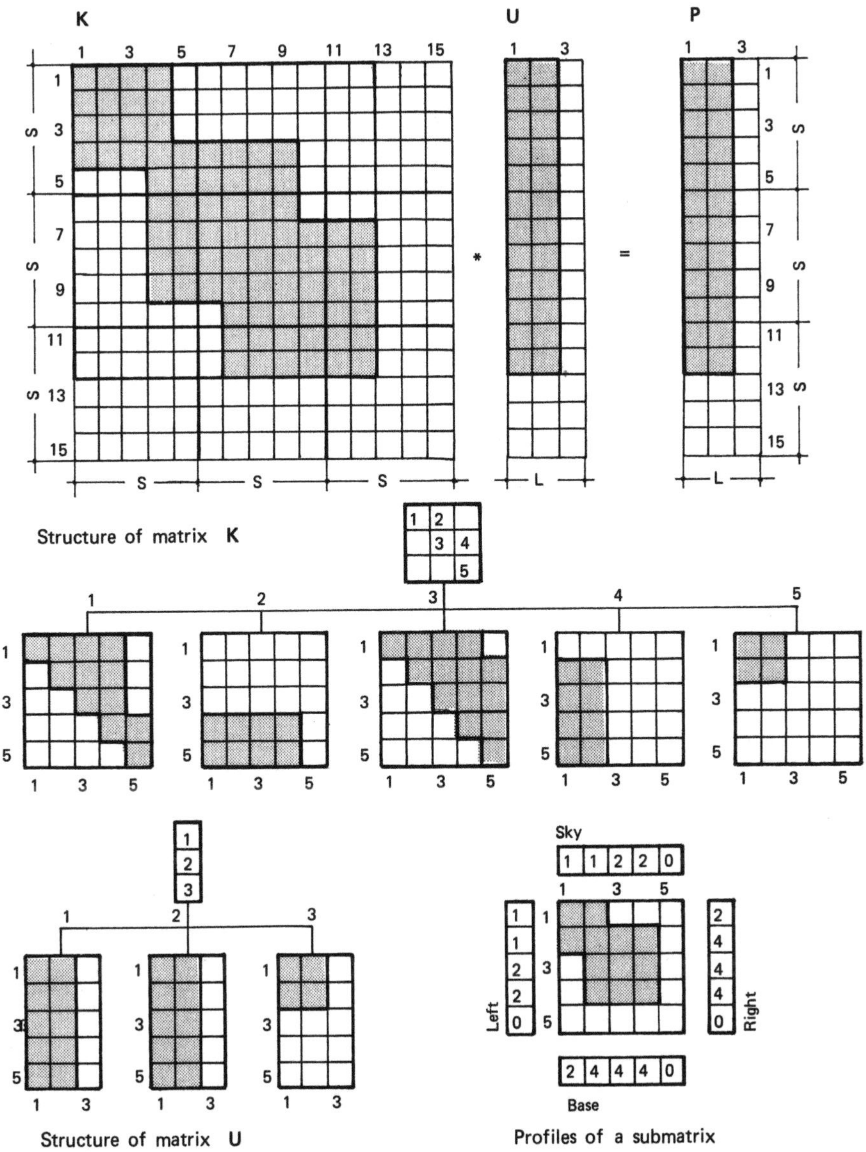

FIGURE 3 Hypermatrix structure for symmetric equations

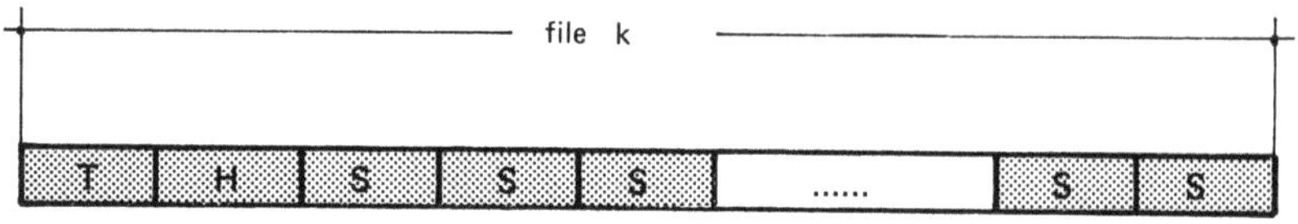

T : array containing submatrix table and block table

H : array containing hypermatrix with profiles

D · array containing submatrices with profiles

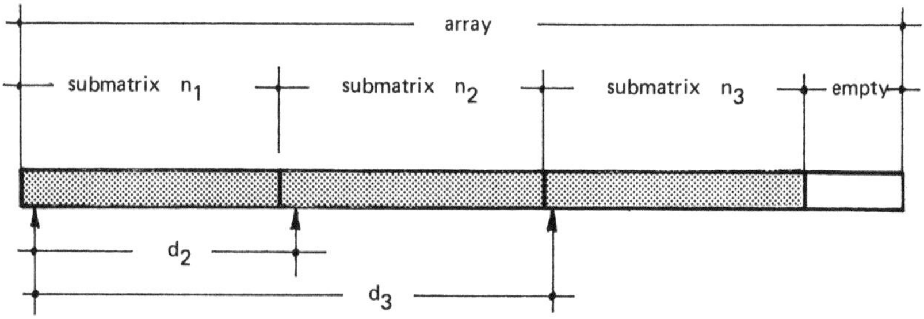

Arrangement of submatrices in an array S

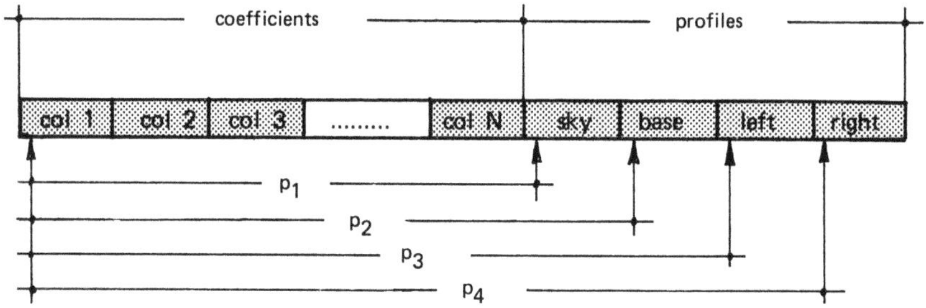

Arrangement of coefficients and profiles of a submatrix

FIGURE 4 : Storage of a matrix in arrays

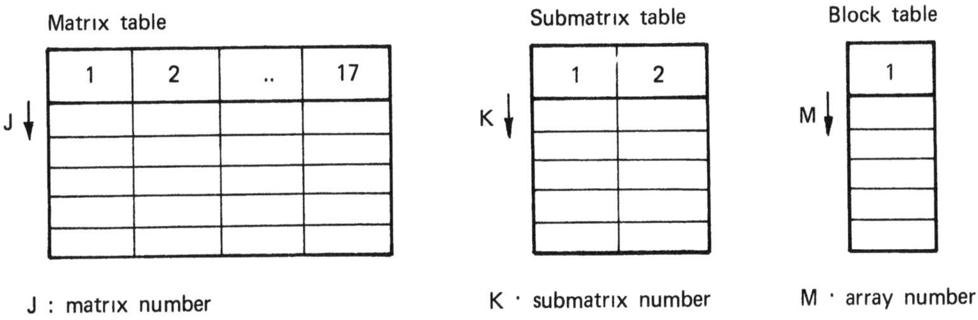

Column	Contents of matrix table
1	status of matrix J . undefined / defined
2	number of the file associated with matrix J
3	array number of submatrix table
4	array number of hypermatrix
5,6	dimensions of hypermatrix
7, .,10	displacements of the hypermatrix profiles
11	displacement of the block table
12,13	dimensions of the submatrices
14, .,17	displacements of the submatrix profiles
Column	**Contents of submatrix table**
1	array number of submatrix K
2	displacement of submatrix K
Column	**Contents of block table**
1	number of block containing array M

FIGURE 5 · Administrative tables of the matrix management

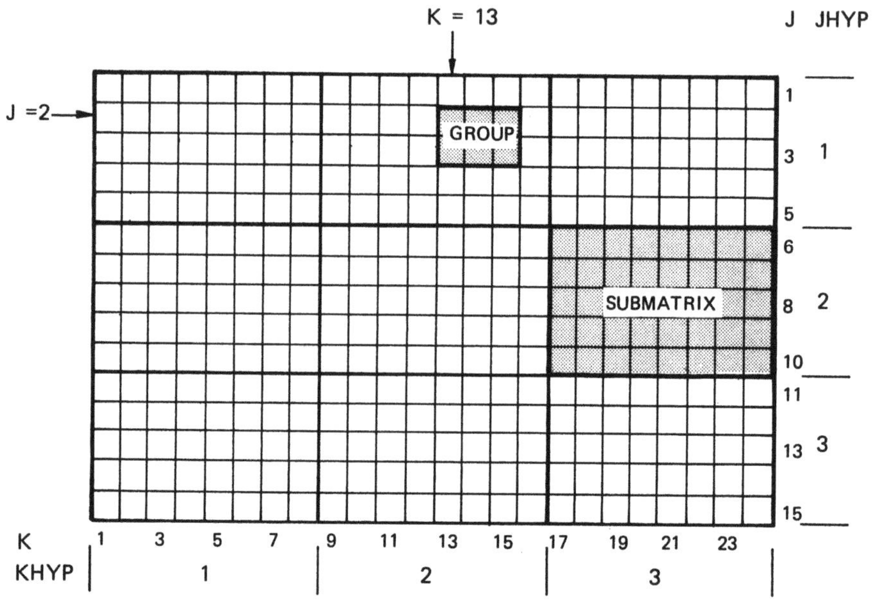

Coefficient mode: specification of a group of coefficients

Number of the matrix M = 9

Indices of apex element (J, K) = (2, 13)

Dimensions of group (JDIM, KDIM) = (2, 3)

Submatrix mode: specification of all coefficients of a submatrix

Number of the matrix . M = 9

Hyperindices of submatrix . (JHYP, KHYP) = (2, 3)

FIGURE 6 . Specification modes for matrix coefficients

Set:	$X = A$	$X = D$
Scale:	$X = c * A$	$X = c * D$
Add.	$X = A_1 \pm A_2$	$X = A \pm D$
Multiply·	$X = A_1 * A_2$	$X = D * A$
		$X = A * D$
Solve	$X = R^{-1} * A$	$X = D^{-1} * A$
	$X = A * R^{-1}$	$X = A * D^{-1}$
Triangularize:	$R^T * R = Q$	determine R
	$L * R = Q$	determine L, R
	$S^T * D * S = Q$	determine D, S

A scratch pad or submatrix

X scratch pad or submatrix, can be identical with A

D diagonal matrix, stored as a vector

L left triangular matrix, coefficients 1.0 on diagonal

R right triangular matrix

S right triangular matrix, coefficients 1.0 on diagonal

Q symmetric submatrix, only diagonal and upper triangle stored

FIGURE 7 Functions for matrix algebra

CALL MATOP (function, op_1, op_2, op_3)

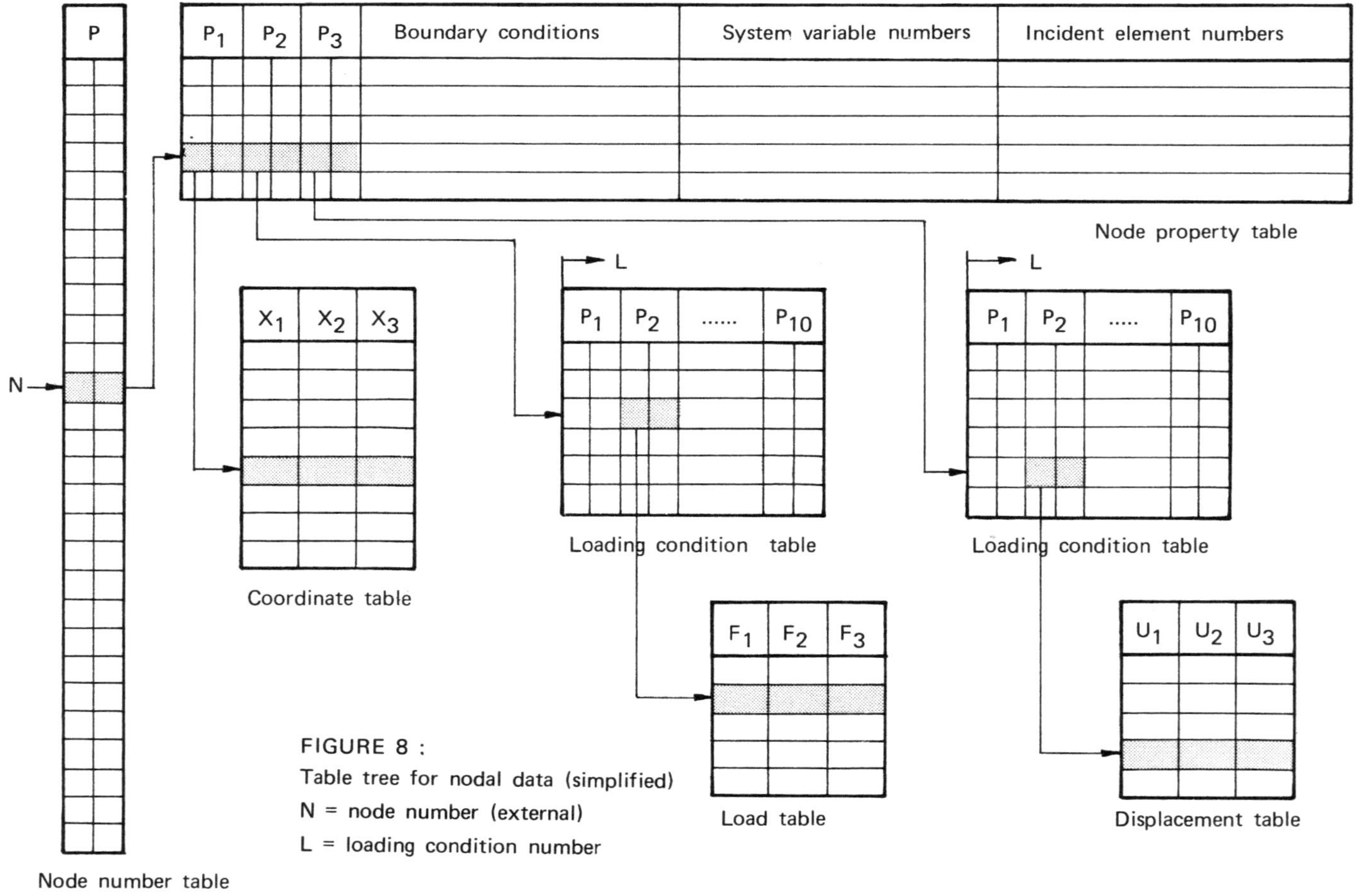

FIGURE 8 :
Table tree for nodal data (simplified)
N = node number (external)
L = loading condition number

Request CALL TABLE (1, NCODE, NPOOL, 65, 815)

Result The line of table 65 assigned to node 815 is supplied in the pool

Administrative data for table trees

	1	2	3	4	5	6	7	8	9	10	11	12	.	20
13 →	4	2	4	256			27	1	42	28				
27 →	4	1	2	512			–	–	17	72	38	75		
65 →	9	4	3	192			13	3	18	33				

Chaining of arrays

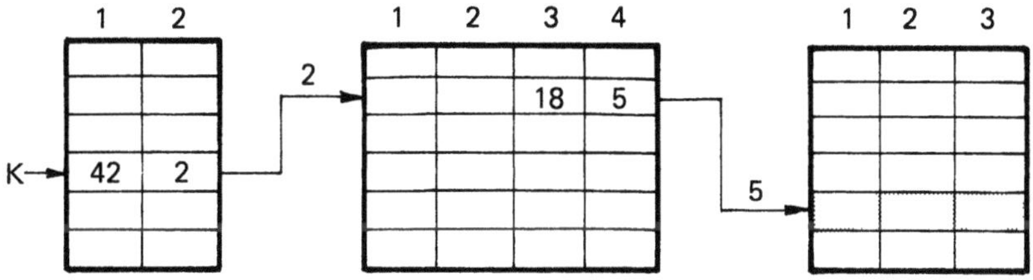

Table	27	13	65
Type·	number	I * 4	R * 8
File	4	4	9
Array	38	42	18
Block	75	28	33
Line.	303	2	5
Column.	1	3	–

FIGURE 9 : Example of a table tree chain for nodal data

Literature

1 Pahl, P.J.: Datenverwaltung des Programmsystems DISKRET. Interner Bericht, Institut für Allgemeine Bauingenieurmethoden, TU Berlin, Mai 1978.

2 Pahl, P.J.; Damrath, R.: Finite Elemente. Vorlesungsskript, Institut für Allgemeine Bauingenieurmethoden, TU Berlin, 1979.

3 Pahl, P.J.; Beilschmidt, L.: Informationssystem Technik, Programmierhandbuch. CAD-Bericht 81, Kernforschungszentrum Karlsruhe, Juli 1978.

4 Beilschmidt, L.: Data and File Management in the Information System Technology. Proceedings, Conception assistee par Ordinateur, Toulouse, 1974, CERT, p. 32-49.

5 Pahl, P.J.; Damrath, R.; Beilschmidt, L.: Prozeduren zur Verwaltung Nummerierter Elemente. Forschungsgruppe Informationssystem Technik, TU Berlin, Bericht 77-07, Aug. 1977.

6 Beilschmidt, L.: Datengruppenhandbuch, IST-Bausteingruppe Hochbau. Forschungsgruppe Informationssystem Technik, TU Berlin, Bericht 77-11, Dezember 1977.

7 Beilschmidt, L.; Lebrecht, F.; Sturm, U.: Datengruppenhandbuch, IST-Bausteingruppe Heizung, Lüftung, Klima, Kälte (HLKK, grafische Darstellung). Forschungsgruppe Informationssystem Technik, TU Berlin, Bericht 79-02, Juni 1979.

8 Damrath, R.; Pahl, P.J.: Data and Storage Structures for a System of Finite Element Programs. U.S.-Germany Symposium on Formulations and Computational Algorithms in Finite Element Analysis, MIT Press, 1977, p. 140-162.

9 Roos, D.: ICES System Design. MIT-Press, Cambridge, 1967.

10 Logcher, R.D.; Connor, J.J.; Nelson, M.F.: ICES STRUDL II, MIT, Reports 68-91, 70-35, 70-77; Internal Logic Manual; 1968-1971.

11 Alcock, Shearing and Partners: GENESYS Reference Manual, The GENESYS Centre, Loughborough, 1971.

12 Enderle, G.; Schlechtendahl, E.G.: The Design of the Integrated CAD-System Regent. Proceedings, Conception assistee par Ordinateur, 1974, CERT, p. 50-79.

13 Schrem, E.: Computer Implementation of the Finite Element Procedure. Numerical and Computer Methods in Structural Mechanics, Academic Press, New York, 1973, 79-121.

14 Schrem, E.: Die Konzipierung eines allgemeinen Rechenprogramms für die Anwendung der Methode der Finiten Elemente. Kolloquium Finite Elemente in der Statik, 302-320. Wilhelm Ernst & Sohn, Berlin, 1973.

15 Brönlund, O.; Kiesbauer, A.; Wilhelmy, V.: TOPAS- ein neues FEM-System, Proceedings, Finite Element Congress, Baden-Baden, 1975, 23-54. IKOSS, Stuttgart.

16 Werner, H.; Axhausen, K.; Katz, C.: Programmaufbau und Datenstrukturen in Entwurfsunterstützenden Programmketten. Finite Elemente in der Baupraxis, Verlag Wilhelm Ernst und Sohn, Berlin, 1978, 328-337.

Performance of Finite Element Algorithms on an Array Processor-Minicomputer Based System

H. A. KAMEL, J. M. TAN
University of Arizona, Tucson, USA

1. Introduction

Finite element analysis is a marriage of classical mechanics, numerical methods and computer technology. As new computer hardware is developed, solution algorithms are affected. In order to be able to take advantage of such advances one has to understand more accurately the behavior of new devices and the interaction between them.

As the technology progresses, issues become more complex. A simple count of the number of floating point operations required to perform certain algorithms may be sufficient for the purpose of comparing alternative core resident algorithms, but is not effective in evaluating the performance of a complex algorithm on a sophisticated system. The overhead imposed by increasingly elaborate operating systems, and the intensive flow of data from one component of a computer system to another, and the contentions and conflicts between such devices and operations make an in-depth study mandatory. Examples of such situations are microcomputer networks [1], array processor/minicomputer combinations [2], as well as mainframe distributed processors [3]. The primary aim of this paper is to attempt an examination of the benefits accured by adding an array processor as a part of a computing system performing finite element analysis and develop a general methodology for the evaluation of algorithm performance on distributed processors.

In order to evaluate the effectiveness of say, the addition of an array processor to an existing system, one may proceed by simply converting programs which have run before on the system to a configuration incorporating such a device. This is indeed one of the goals of this work, and some preliminary results of such an effort are included. However, in order to interpret the results appropriately, and obtain the answers to more important fundamental questions, it is essential that we approach the problem on a more abstract level. The fundamental questions raised here are:

1. How to design an algorithm to run efficiently on a given system configuration.
2. How much advantage is there in installing an array processor?
3. How to configure a computer system for optimum performance, given a particular algorithm.

An approach based on a simplified hardware model, coupled with some average performance measurements is proposed. It is based on simulation, recognizes hardware contention factors, allows for parallel computation and has look-back and ahead features. This program may be used to predict the performance of certain classes of algorithms on specific hardware configurations and design optimum systems for specific algorithms. This simulation program may be used as a "prototype" for an executive for implementing finite element algorithms on such a system.

It would be appropriate here to mention some related efforts. For example, the use of "supercomputer" [4] has been the subject of many other investigations. Noor and Lambiotte [5] have examined the effect of vectorization on finite element analysis of dynamics problems. The use of simulators to predict algorithm performance has been reported by Forslund and Nielson [6], who succeeded in balancing I/O operations with central processor computation, producing a CPU bound algorithm using 2 disk channels and triple buffering. Other examples of simulation may be found in papers by Klein [7], Chlamtac and Franta [8], and Jones et. al. [9].

2. Performance Predictions in a Complex System

A simple model is proposed, in which the main computer appears as a single entity encompassing CPU, floating point processor and memory. Other components of the model are the disk unit(s), including both disk controller(s) and drive(s), the array processor(s), and data busses. Communication between the various components is included in the model as well as the characteristics of each unit. The model allows for device contention. If more than one device communicate over the same bus, a conflict arises which blocks a new data transfer from taking place until an ongoing transfer is ready. The model also allows for parallel computation, which is an important feature of a distributed processor. It is planned that the assumptions underlying this model be tested on a real system in order to validate them.

Once the hardware model has been created, an algorithm is chosen for consideration. The algorithm is then broken down into primitives on a somewhat high level. Examples of these would be matrix multiplications or data block transfers from one device to another. Test programs are then written to measure resources required by each primitive. These experiments have to be selected carefully, and conducted in a controlled environment. Once this is accomplished, one is ready to predict algorithm performance on the given system, estimate changes in performance due to modifications in the hardware configuration, and evaluate alternative algorithms provided they use the same primitives. It is of course possible to define new primitives, as needed, and include them in the model in order to simulate different types of algorithms.

The proposed method of performance prediction in a multiprocessor system is conducted by simulation. Two programs are needed. The "coder" translates the algorithm, written in a higher order language recognizing standard primitives into a set of instructions, indicating interdependence between operations. The inclusion of interdependence is necessary to include parallel computation. The second program, called the "simulator", mimics the execution of these instructions on a given hardware configuration. The simulator must include a definition of hardware characteristics. It has a look-ahead feature, enabling it to utilize system parallelism. In addition, it initiates "execution" of an instruction only if the prerequisites are satisfied and the appropriate device is free. The simulator also serves as a prototype for an "executive" to be developed later for the actual execution of the algorithm.

The first algorithm chosen to test the effectiveness of an array processor-minicomputer combination is that of hypermatrix decomposition. This particular operation was chosen for many reasons. For one thing, It is a representative component of a typical complex computation, whether static analysis, vibrational mode extraction, transient response computation or nonlinear behavior. Secondly, it has distinct primitives, such as matrix multiplication and block data transfers from and to the backing storage. The overall algorithm is, therefore,well suited to simulation as described here, and the submatrix operations are ideal candidates for array processing. The algorithm performance may be influenced by several variable control parameters, including the size of the partitions, and the amount of buffer space allocated within the minicomputer and array processor, in order to minimize input-output operations.

3. Definition of Primitives

The algorithm should be broken down into a number of well-defined primitives, amenable to coding and measurements as independent tasks. Such primitives would include computational as well as data transfer operations. Examples are given in the following sections.

3.1 Disk File Commands

The code generator should allow the declaration of storage areas (files) on various devices, including the disk. These files may be either one- or two-dimensional hypermatrices, where all the submatrices are of identical maximum size. The submatrices may be addressed by their position within the file. It is assumed that the file will incorporate headers which define the sizes of the blocks under consideration. Such hyperarrays may be created by the program at the time of program execution, or attached if they already exist. The following commands may be defined at this point:

DEFINE,device,filename,filetype,hypersize,recordsize

defines a new file on a system component. The "device" is a predefined mnemonic, such as DK (disk), MC (main computer) or AP (array processor). "filetype" may be 1D or 2D. "hypersize" may be 10 for 1D, or 20,30 for 2D. "recordsize" may be 30 (1D) or 40,40 (2D). The command:

ATTACH,device,filename,filetype,hypersize,recordsize

Attaches an already resident file on a device. It should be noted here that no real distinction is made between the various devices. Once a DEFINE or ATTACH command is given, both the device and filename will be declared active until the initialization of the file is complete. Any subsequent commands which address this device or file will have to wait the completion of the initialization process. Here, it may be necessary to engage another device, namely the main computer, for overhead purposes.

3.2 Transfer Commands

A transfer command results in the copying of a record of a particular file on a device to another record on another file and device. For example:

TRANSFER,device1,file1,record1,device2,file2,record2

has the effect of copying the contents of record1, suitably addressed, of file1 on device1 to record2 of file2 on device2. While the transfer is being conducted, both devices and records, as well as the appropriate data busses, are declared active, and subsequent transfers to and from them are suppressed. Both devices may be the same, and both files may be the same.

3.3 Examples of Computational Primitives

The following instruction has the effect of multiplying a sub-matrix, entered as a record on a certain file by a record of another file, and placing the result on a record in a third file. It is assumed that all three files reside on the same device. During the operation it will be assumed that the device and the three records are busy. Depending on the device characteristics, input and output to other files and records on the same device may be allowed. The command format may read:

```
MULT,file1,record1,file2,record2,file3,record3
```

Other examples of computational instructions are the matrix subtaction command, SUBT, the matrix inversion command, MINV, and the matrix transposition command, TPOS. The following samples are self-explanatory:

```
SUBT,array1,record1,array2,record2,array3,record3
MINV,array1,record1,array2,record2
TPOS,array1,record1,array2,record2
```

3.4 Logical Control Statements

In order to handle hypermatrices of arbitrary size, it must be possible to construct loops using arbitrary variables. Standard notations, as used in typical structured languages, such as the WHILE DO construct, are adopted here.

3.5 Measurement of Primitive Parameters

Estimates of resource requirements for the chosen primitives must be obtained before a simulation can be attempted. It was natural to model and measure first the matrix multiplication operation, and compare measured values with predictions.

The results of the comparison between estimates based on knowledge of the hardware characteristics and algorithm demands are quite surprising, showing that the operating system overhead is substantial (see Table 1). As a matter of fact, the amount of time spent doing multiplications and additions is an order of magnitude less than the total CPU time required. The rest is consumed by loop counting, pushing variables on the stack, and other overhead activities. From measurements, it is observed that the computation needs more than eight times the predicted times to run if the computer is lightly loaded, and over nine times the simulation times when the computer is moderately loaded. One observes the large overhead in copying a single doubly-subscripted real variable, whihc is four to five times the time required to multiply two such real variables togenter. It becomes, therefore, a difficult task to model systime performance without a thorough knowledge of the internal organization of the operating system, which is tedious and unreliable, and was abandoned in favor of direct measurements.

Several lessons may be learned from this example. For instance, the actual time measured is an order of magnitude higher than predicted, even though the estimates included an estimate of "direct" overhead. The experiment also shows that data transfers, even from memory to CPU, are a major factor, and should be even more significant if an array processor is used. Also, it is apparent that a recoding of such heavily used portions of the machine language is highly desirable, an already well-known fact.

4. Algorithm Simulation, The Coder and the Simulator

As mentioned before, an algorithm will be chosen, and a coder and simulator will be developed to predict its performance on specific hardware configuration. The following gives this device:

4.1 Example Algorithm

The program, shown in Figure 1, represents the code necessary to triangulate a hypermatrix, using the Gaussian elimination scheme. It uses the language and primitives defined under 3.

4.2. The Coder

The "coder" accepts a program written in a higher-order language, such as the one shown in Figure 1, and produces code such as that shown in Figure 2. It does not need detailed information about the system which will be used, except for the declaration of the device mnemonics, and the assignment of each instruction type to a particular device.

Submatrix Size	Simulation Time (sec.)	Measured Time (sec.)	Ratio	Loop Time (sec.)	Loop Plus Copy (sec.)	Floating Operation Time (mirco sec.)	Copy Time
5 x 5	.00151	.013 .014	8.61 9.27	.002	.012	4	36.4
10 x 10	.01151	.097 .107	8.43 9.30	.011	.083	7	34.3
15 x 15	.03827	.320 .353	8.36 9.22	.035	.272	7.11	34.0
20 x 20	.09002	.749 .824	8.32 9.15	.083	.634	7.19	33.6
25 x 25	.17503	1.449 1.595	8.29 9.11	.161	1.226	7.14	33.4
30 x 30	.30153	2.510 2.744	8.32 9.10	.273	2.104	7.52	33.4
35 x 35	.47779	3.996 4.393	8.36 9.19	.439	3.326	7.81	33.2
40 x 40	.71204	5.941 6.469	8.34 9.09	.650	4.944	7.79	33.1
45 x 45	1.01254	8.405 9.191	8.30 9.08	.926	7.013	7.64	33.0
50 x 50	1.38755	11.553 13.05	8.33 9.41	1.256	9.608	7.78	33.1

Table 1. Simulation and Measurements of Matrix Multiplication.

A few points may be made regarding the generated code. Each instruction is given a sequence number, with which it is identified later. The prerequisites for each instruction are determined by examining all operands involved in the instruction. The last time an operand was referenced by a previous instruction indicates that that instruction is a prerequisite for the instruction under consideration. A "look-back" is, therefore, used in the coder.

4.3 The Simulator

The simulator accepts an instruction stream such as that shown in Figure 2, and simulates its execution on a particular system configuration. The program must be supplied with a

```
ATTACH,DK,STIF,2D,M,M,S,S
DEFINE,MC,WS,1D,5,S
DEFINE,AP,BUF,1D,5,8
D=1
WHILE D<M DO
  TRAN,DK,STIF,D,D,MC,WS,1
  TRAN,MC,WS,1,AP,BUF,1
  MINV,BUF,1,BUF,2
  I=D+1
  WHILE I <=M DO
    TRAN,DK,STIF,D,I,MC,WS,2
    TPOS,WS,2,WS,3
    TRAN,MC,WS,3,AP,BUF,3
    J = I
    WHILE J <= M DO
      TRAN,DK,STIF,I,J,MC,WS,4
      TRAN,MC,WS,4,AP,BUF,4
      TRAN,DK,STIF,D,J,MC,WS,5
      TRAN,MC,WS,5,AP,BUF,5
      MULT,BUF,1,BUF,5,BUF,3
      SUBT,BUF,4,BUF,3,BUF,4
      TRAN,AP,BUF,4,MC,WS,4
      TRAN,MC,WS,4,DK,STIF,I,J
      J = J+1
    ENDWHILE J
    I = I+1
  ENDWHILE I
  D = D+1
ENDWHILE D
```

Figure 1. Program to Triangulate a Hypermatrix

description of the system devices, and their interconnection, as well as the assignment of primitives to specific devices, and estimated execution times for each instruction of the appropriate device.

The model represented here is obviously greatly simplified. For one thing, the logical layout of a modern computer is quite complex. It is assumed here that the system is dedicated to one user, which may not be the case. The current model is deterministic, whereas many events, such as disk access times, are random events. One of the assumptions made is that each instruction is executed on a single device, whereas for example, the CPU will always incur a certain amount of overhead in controlling and setting up of operations executed on other components.

On the other hand,we must state that the approach here may be generalized to include many of the effects described above.

DIJ	##	PRE	OPCODE	DEV	DEV1	FILE	REC	DEV2	FILE	REC	DEV3	FILE	REC
1--	1	0	TRAN	D-M	DK	STIF	1,1	MC	WS	1			
1--	2	1	TRAN	M-A	MC	WS	1	AP	BUF	1			
1--	3	2	INV	AP	AP	BUF	1	AP	BUF	2			
12-	4	0	TRAN	D-M	DK	STIF	1,2	MC	WS	2			
12-	5	4	TPOS	MC	MC	WS	2	MC	WS	3			
12-	6	5	TRAN	M-A	MC	WS	3	AP	BUF	3			
12-	7	6, 3, 2	MULT	AP	AP	BUF	3	AP	BUF	2	AP	BUF	1
122	8	0	TRAN	D-M	DK	STIF	2,2	MC	WS	4			
122	9	8	TRAN	M-A	MC	WS	4	AP	BUF	4			
122	10	0	TRAN	D-M	DK	STIF	1,2	MC	WS	5			
122	11	10	TRAN	M-A	MC	WS	5	AP	BUF	5			
122	12	2,11, 6	MULT	AP	AP	BUF	1	AP	BUF	5	AP	BUF	3
122	13	9, 6,12	SUBT	AP	AP	BUF	4	AP	BUF	3	AP	BUF	4
122	14	13, 8	TRAN	M-A	AP	BUF	4	MC	WS	4			
122	15	14, 8	TRAN	D-M	MC	WS	4	DK	STIF	2,2			
123	16	15	TRAN	D-M	DK	STIF	2,3	MC	WS	4			
123	17	16,14	TRAN	M-A	MC	WS	4	AP	BUF	4			
123	18	11	TRAN	D-M	DK	STIF	1,3	MC	WS	5			
123	19	18,12	TRAN	M-A	MC	WS	5	AP	BUF	5			
123	20	12,19,13	MULT	AP	AP	BUF	1	AP	BUF	5	AP	BUF	3
123	21	17,20	SUBT	AP	AP	BUF	4	AP	BUF	3	AP	BUF	4
123	22	21,17	TRAN	M-A	AP	BUF	4	MC	WS	4			
123	23	22,16	TRAN	D-M	MC	WS	4	DK	STIF	2,3			
133	24	18, 5	TRAN	D-M	DK	STIF	1,3	MC	WS	2			
133	25	24, 6	TPOS	MC	MC	WS	2	MC	WS	3			
133	26	25,20	TRAN	M-A	MC	WS	3	AP	BUF	3			
133	27	26, 7,20	MULT	AP	AP	BUF	3	AP	BUF	2	AP	BUF	1
133	28	23	TRAN	D-M	DK	STIF	3,3	MC	WS	4			
133	29	28,22	TRAN	M-A	MC	WS	4	AP	BUF	4			
133	30	24,19	TRAN	D-M	DK	STIF	1,3	MC	WS	5			

KEY INFORMATION:

DIJ	Current values of loop control parameters
##	Instruction sequence number
PRE	Prerequisites
OPCODE	Operational code TRAN=transfer, INV=invert MULT=multiply, SUBT=subtract, TPOS=transpose
DEV	Device on which instruction is executed, MC=main computer, DK=disk, AP=array processor, D-M=disk/computer bus, M-A=computer/AP bus.
DEV1	Device used with first operand
FILE	File where first operand resides
REC	Record address for first operand
(DEV2,FILE,REC),(DEV3,FILE,REC)	Device, file and record for second and third operands

Figure 2. List of First 30 Instructions Generated by the Code Generator for Simulation Purposes

It is not appropriate at this time, however, to do this, since our main concern is to develop a methodology and obtain answers to some fundamental questions.

During execution or simulation, the executive or simulator, examines the instruction stream constantly for an instruction which has all the necessary prerequisites for execution. For an instruction to be activated, all previous instructions addressing one or more of its operands must have been executed, and the device on which it is to be executed must be free. It is obvious, therefore, that the simulator, and later on the executive, has a look-ahead feature, which may span a large number of instructions, as well as the ability to execute instructions in parallel and out of the original program sequence. It is also obvious that such an executive may not necessarily be as effective on a virtual memory system.

4.4 Choice of Primitive Execution Parameters

Before a simulation may be conducted, it is necessary to estimate, or measure, the performance of specific system components in executing the given primitives. The times shown in Table 2 are for previously defined primitives, assuming 50x50 submatrices, in double precision. Some of these values are based on extrapolated measurements, and some on educated estimates. These are used here for the purpose of illustration

INSTRUCTION	DEVICE	TIME, IN SECS.
TRAN	D-M	.0785
TRAN	M-A	.0785
TPOS	MC	.1
MULT	AP	.188
INV	AP	.184
SUBT	AP	.0025

Table 2. Instruction Times for Typical Primitives

4.5 Simulation Results

Two simulations are attempted here. The first deals with a conventional system, in which three devices are present: a main computer, an array processor (AP) and a disk (DK) -- Configuration 1. Communication between the disk and the main computer (D-M), and between the main computer and the array processor (M-A) both take place via the same bus. The results of the simulation are presented in Figure 3. As expected, the communications bus is the bottleneck. It is fully utilized, whereas the array processor is reasonably busy. The main computer is lightly loaded. However, it can not be used for other activities (in a multiprocessor environment, for example) since input and output to the disk is blocked. In such a configuration, multiprocessing is impractical, although the array processor is well utilized, the main computer is wasted to a great degree.

The second system simulated is one where the array processor and disk may communicate directly with the main computer without conflict -- Configuration 2. In other words, there are two separate communication busses. The busses are assumed to have the same speed. Other system characteristics remain unchanged. The results are presented in Figure 4. In this situation the array processor appears to be fully utilized, whereas the main computer and the two busses, particularly the memory-array processor bus, appear to be reasonably free, thus allowing for other activities. This clearly demonstrates the importance of a proper design of communication paths in such a system.

Other system configurations may also be attempted here. For example, the array processor may be interfaced with the main computer via a faster bus. It is also possible that a disk be attached directly to the array processor, or that a disk may be accessed by both the main computer and array processor independently. Additional primitives may be defined in order to investigate other types of algorithms.

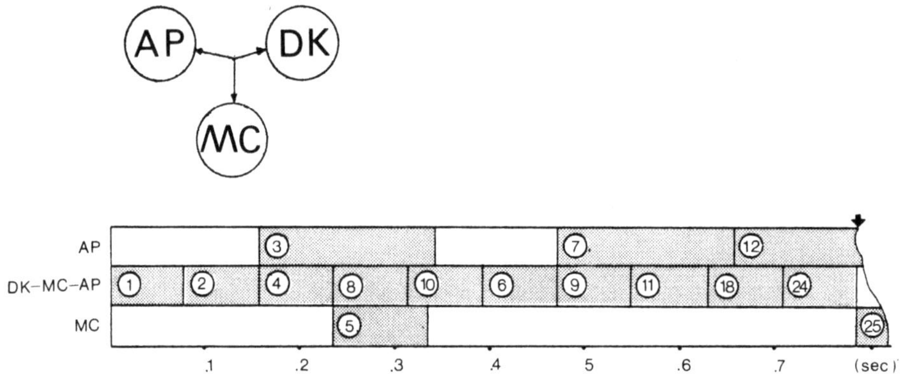

Figure 3. Execution Histogram, Configuration 1

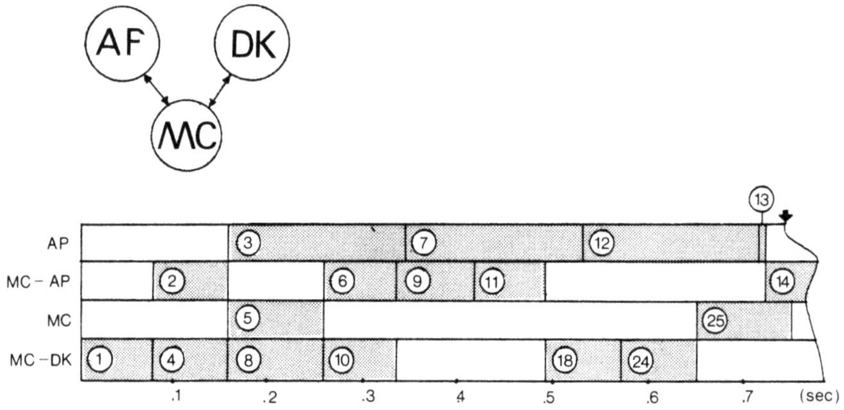

Figure 4. Execution Historgram, Configuration 2

5. Implementation of a Cholesky Matrix Decomposition on a Minicomputer-Array Processor Combination

Finally, the paper presents some measurements undertaken at the laboratory showing the behavior of a hypermatrix decomposition routine, using the Cholesky method, as executed on an ordinary minicomputer, and on the minicomputer with an attached array processor.

The presented measurements were obtained using a 16 bit DGC Eclipse minicomputer, with a CSPI MAP-200 array processor, attached to memory via a data channel, as a user defined device. Transmission of data and instruction packages to control the MAP-200 was done via direct memory access (DMA).

The Eclispe computer operates under the Advanced Operating System (AOS), which is a multi-user timesharing system. This arrangement forced the implementation of the MAP driver as part of the program, instead of imbedding it in the operating system, as is usually possible with a real time operating system. The driver uses a multitasking algorithm in order to perform control functions and communication with the user program. It was necessary to overlay the user program to provide enough space for the code.

The hypermatrix Cholesky decomposition algorithm uses fairly small submatrices (18x18). The computational primitives are matrix multiplication, inversion and transposition. Only the matrix multiplication routine was implemented on the array processor. Instruction overlap (parallel processing) is not fully possible at this time.

5.1 Measurement Results

Two problems were chosen for measurement purposes. They represent shell structures of different sizes. The problem parameters and measured results are shown in Table 3.

Looking over the results, it becomes apparent that the array processor is not well utilized in this particular case. The CPU time is practically unchanged for the small problem, but is reduced by 22 percent for the large problem. The number of blocks transferred from and to disk are approximately the same, which is to be expected. Observing the minicomputer performance, via a special system monitoring program, showed that the CPU was only 40% to 60% active during the computation. One fact which is so far unexplainable is, the elapsed time increased substantially using the array processor over times observed for the minicomputer alone. From the figures measured, it is most probable that the communication with the array processor must be time consuming, pointing out to the need for further optimization of the MAP driver.

	Large Problem		Small Problem	
	Without A.P.	With A.P.	Without A.P.	With A.P.
Elapsed Time (secs)	6,666	25,735	67	97
CPU Time (secs)	7,194	5,607	122	122
No. of Blocks Transferred	138,005	138,033	1102	1074
Problem Parameters				
Number of unknowns	3,840		200	
Number of nodes	688		40	
Number of Elements	575		64	
Half-bandwidth Maximum	318		43	
r.m.s	210		32	

Table 3. Performance Measurements for Cholesky Decomposition on Minicomputer with and without an Array Processor

6. Discussion and Conclusions

The paper consists of two parts and presents a report on work in progress. A simulation methodology has been proposed, which should be helpful in developing finite element algorithms for distributed processors. An important factor here is the proper modeling of the system, and the accurate measurement of primitive execution times.

The second part of the paper describes another effort, so far not directly connected to the simulation approach. It presents results comparing performance of a decomposition algorithm on a minicomputer with and without an array processor. No effort has been as yet devoted to optimizing either the algorithm or the array processor driver. These results are only to be regarded as a starting point for further work. If anything, it does point out that the problem is not straightforward, and must be addressed with care. Particular emphasis is placed on understanding hardware characteristics.

7. Acknowledgements

The authors gratefully acknowledge the support of the Office of Naval Research under contract no. N00014-75-C-0837, and CSP Inc., without which this research would not have been possible. A debt of gratitude is also owed to Ms. Juanita R. Alvarez who typed the manuscript.

8. References

1. Jordan, H.F.; Sawyer, P.L.: A Multiprocessor System for Finite Element Structural Analysis. Computers and Structures, Vol. 10, Nos. 1 & 2, (1979) 21-32.

2. Grosch, C.E.: Poisson Solver on a Large Array Computer. Proc. of the 1978 LASL Workshop on Vector and Parallel Processors. Los Alamos, New Mexico: 1978.

3. Satyanarayanan, M.: Commercial Multiprocessing Systems. Computer, Vol. 13, No. 5, (1980).

4. Kascic, M.J. Jr.: Vector Processing on the Cyber 200. Infotech State of the Art Report "Supercomputers", Infotech Ltd., Maidenhead, U.K., 1979.

5. Noor, A.K.; Lambiotte, J.J. Jr.: Finite Element Dynamic Analysis on CDC STAR-100 Computer. Computers and Structures, Vol. 10, Nos. 1 & 2, (1979) 7-20.

6. Forslund, D.W.; Nielson, C.W.: Vectorized PIC Simulation Codes on the CRAY-1. Proc. of the 1978 LASL Workshop on Vector and Parallel Processors. Los Alamos, New Mexico: 1978.

7. Klein, D.: MMPS -- A Reconfigurable Multi-microprocessor Simulator. Proc. of the National Computer Conference: 1979.

8. Chlamtac, I.; Franta, W.R.: Aids to the Development of Network Simulators. Proc. National Computer Conference: 1979.

9. Jones, A.K.; Chansler, R.J. Jr.; Durham, I.; Feiler, P.H.; Scelza, D.A.; Schwans, K.; Vegdahl, S.R.: Programming Issues Raised by a Multiprocessor. Proc. IEEE Vol. 66, No. 2, 1978.

High-Speed Processors and Implication for Algorithms and Methods

G. F. CAREY
University of Texas, Austin, USA

Summary

In this paper we consider the interrelation of numerical methods and algorithms with new parallel and vector processor capabilities. A brief historical note on the development of computer processors establishes the perspective of this study and leads into an examination of two proposed new processors: (1) an array processor and (2) a vectorized processor with some parallelism. Next we examine typical linear, nonlinear and transient finite element computations, noting where parallelism and vectorization are possible. Substructuring and splitting are taken as special examples, and we summarize some important features concerning direct and iterative methods for linear systems. Finally, the extension to nonlinear systems and the implication for a specific iteration scheme are considered.

Introduction

Development of new computer processors and of methods and algorithms go hand in hand. C. Babbage invented the first "modern" calculating machine in the mid-1800's. His "analytical engine" was capable of tabulating the values of a function and printing the results. Only a part of the machine was constructed due to excessive costs to the British Admiralty, and it was consigned to the Kensington museum. However, the underlying principles were not pursued extensively until the first half of this century. Advanced scientific computing has its beginnings in the 1940's. The first studies of difficult numerical problems are those of Hartree at Manchester. His work represented the first concerted attempt to solve extremely complex physical problems by computational means.

Prior to Hartree's work, the main motivation for developing more advanced automated calculators had been the desire for efficient sustained computation - to do several weeks' work in a day. To some degree this is still valid today as we attempt to solve ever-larger problems. At the same time we should address the important question: do proposed new advances in computer processors open up new fields of numerical and mathematical analysis? This question remains particularly relevant today as we consider the possible impact of parallel and vector processors on scientific computing.

Von Neumann captured the essence of the situation prophetically in 1945 in his introductory memorandum on the Advanced Institute Machine, today termed the "von Neumann Machine." Many of his observations are equally relevant now and we quote segments of the text below:

> A machine as described will certainly revolutionize the purely mathematical approach to the theory of nonlinear differential equations. It will also (practically for the first time) permit extension of . . . compressible, hydro and aerodynamics, as well as the more complicated problem of shock waves . . . . It is likely to make the theories of elasticity and plasticity much more accessible than they have been up to now. It will probably help a great deal in three-dimensional electrodynamical problems . . . . *Apart from these things, the machine, if intelligently used, will completely revolutionize our computing techniques, or to formulate it more broadly, the field of approximation mathematics.* Indeed this device is likely to be, at the very least ten thousand times faster (actually probably more) than the present human computer and multiplying machine methods. However, our present computing methods are developed for these, or for still slower (purely human) procedures. The projected machine will change the possibilities, the difficulties, the emphases, and the whole internal economy of computing so radically, and shift all procedural options and equilibria so completely, that the old methods will be much less efficient than the new ones which have to be developed. These new methods will have to be based on entirely new criteria of what is mathematically simple or complicated, elegant or clumsy.

It is, unequivocally, this development of the high-speed digital computer since the late 1940's that has made *practical numerical* methods such as finite element methods for boundary- and initial-value problems. From this viewpoint we wish to begin to address the implications of evolving computer processors for finite element methods and related numerical algorithms. In particular we shall examine the impact of array and vector processor designs on element calculations and assembly and solution algorithms.

Processor Development and Limitations

An electrical signal (electron) can traverse approximately 15 cm in 1 nanosecond (1 ns = 10^{-9} sec). The maximum "speed" of the central processor is determined by the time required for a gate to change state and by the time required to pass from one gate to another. The worst possible case (path) determines the minimum cycle time interval for the master clock of the computer in question. Most high-speed processors currently in operation have a cycle time of 30 to 50 ns. In the case of the CRAY 1 the maximum signal path has been reduced to a few meters and the cycle time is of the order of 12 ns. In terms of hardware design for increased speed, one goal is a processor with cycle time of 1 ns. This would give the increase in speed required for studying such problems as plasma fusion, reactor failure, weather prediction, Navier Stokes problems, nonlinear three-dimensional solid mechanics problems, fluid-structure interactions, and so on.*

Some exploratory studies on exotic computer hardware with cycle times of a few ns are being conducted using superconductors and Josephson junction circuits [1]. As the junctions must be cooled within a few degrees of absolute zero the device would operate immersed in liquid helium.

* In such applications as weather prediction, war games, and reactor safety, we are really interested in very high speed processing. Most standard engineering computations can be more appropriately treated using small computers with calculations that may take days or even weeks on larger problems. While we can anticipate a growing demand for less expensive dedicated small computers, there are applications of the form cited above when waiting several weeks, days or even hours for a response may not be appropriate.

Practical operating computers of this type necessitate a new technology and are clearly still in the early exploratory stage.

More immediate developments are imminent in the area of array and vector processors. An example is the CRAY 1 mentioned above. Conceptually, the design principle is now to design the computer architecture to exploit parallelism of computations and/or vector manipulations. Both of these ideas are very logical when one considers the nature of approximate numerical solution of boundary and initial-value problems. There is a considerable proportion of parallelism and vector operations in approximate solution using techniques such as finite element analysis. The objective is to exploit these features in designing a new generation of high-speed processors. The design may be highly vectorized or array-oriented or perhaps be intermediate between the extreme forms. This observation has led recently to the idea of developing specific "PDE Machines," "Finite Element Machines" and "Navier Stokes Machines."

There are immediate ramifications for algorithms: in an array processor in which extensive parallelism is to be exploited, certain algorithms will be preferred that may not function well on a highly vectorized machine that has less parallelism. This has particular bearing on the development of linear and nonlinear systems solvers, a point we shall develop further in the second part of this article. At a somewhat subtler level it may have a direct impact on the choice of method itself - e.g. finite difference versus finite element, global weighted residual methods, and so on.

Two Proposed Configurations

The general objective is to obtain numerical solutions to fully three-dimensional unsteady nonlinear problems, using finite difference, finite element or other methods. Reliable and accurate results might be expected in a time frame of 15 minutes on a new scientific computer with a sustained rate of 10^9 operations per second. The simulation may, in fact, be one step in an integrated preliminary design procedure so that several such computations might be required as part of the overall optimization strategy. In [2] it is estimated that most advanced high speed computers in use are approximately a factor of 50 slower than needed for solution of complex three-dimensional nonlinear problems.

There are several other considerations that are pertinent. As processing time for large problems is significant in these calculations, slow-speed access must be avoided so that the demand on memory is increased accordingly. This dictates requirements for parallelism of hardware and the ability to contain the entire problem on-line without having to utilize low-speed mass storage peripherals. It also implies that the process should contain a fast access memory of significant size (8 million words of working storage and 256 million words of secondary storage have been estimated).

Let us briefly examine two processors that have been proposed by Burroughs and CDC, respectively, to meet the above requirements. A more detailed discussion of the designs and performance estimates are given in [2].

The Burroughs device is a highly parallel processing array consisting of a system of 512 computational processors, each with its own local data program memories. These are coordinated by a single control unit and connected by a transportation network to 512 modules of extended memory. An innovative feature of the proposed design concerns the synchronization of processor arrays: between symchronization points the individual processing elements are to operate asynchronously, allowing them a degree of freedom in scheduling instructional sequences. A strategy for elimination of non-productive time resulting from reordering or transposition of data is to be achieved by means of a transposition network.

The CDC device approaches the problem from the standpoint that it will be more reliable to build a super processor with a minimum number of parallel units implemented with the fastest existing or evolving technology. The processor consists of vector units, map unit, scalar unit and a swap unit which can operate concurrently. Eight vector units are utilized in this design. The map unit performs memory access operations for restructuring data and the vector units at the same time are computing on data within the buffers of the vector units.

Finite Element Computations

Let us examine the processing logic for representative finite element analyses of linear, nonlinear and time-dependent problems. We identify, at least qualitatively, those aspects of the computations where parallelism and vectorization can be exploited with relative ease.

(1) Linear Problems

The main processing steps are listed as follows: (1) Input element and nodal data (Parallel); (2) Construct matrix contributions for each element (Parallel); (3) Assemble element contributions (Parallel and Vector); (4) Include boundary data; (5) Linear system solution (Vector or Parallel).

In the input stream we have some opportunities to use parallelism as each element has identified with it specific properties, node numbers, and so on, but this is of lesser significance. The element matrix and vector calculations are almost entirely local. That is, they require only information pertaining to the element in question, and are not serially dependent on information from other elements. In an array processor, such as Processor 1 in the previous section, each small processor in the array can be assigned a subset of elements and compute element contributions in parallel. For example, in a three-dimensional problem with $0(50^3)$ elements, the 512 array processor could treat 0(500) elements in parallel so that each processor would compute 0(250) element contributions. There would be some minor modifications to the data structure to avoid, for instance, duplicate access of data at a common node of adjacent elements. This could be achieved by including pertinent nodal data, such as nodal loads in the element data structure. However, the probability of simultaneous access is obviously small and delay would be minimal even with present data structures. Since the assembly *process simply* requires addition of element contributions into the global matrix and vector locations, each individual processing unit can add the present element contribution directly. Thus the parallel logic can again be exploited, but there will be some slight delay when common global array locations are accessed. Vector operations cannot really be utilized efficiently in assembly, since element variables are stored in a compact form and only related to global variables through a boolean transformation identifying local and global node numbers. Essential and natural boundary data would probably be handled separately and in a sequential manner, although there is some parallelism here.

The final step - the system solver - is most critical as most of the processor effort is in this calculation. Most linear finite element programs employ sparse elimination algorithms and use scalar arithmetic. In large three-dimensional problems and in algorithms where adaptive-mesh refinement, multigrids, and acceleration procedures are used, iterative methods are competitive. When one also introduces the question of array processors and vectorization, the issue becomes even more uncertain. In the latter parts of this article, we consider in greater detail some of these issues. At this point, we shall simply note that: for vector computations with direct solvers, long vectors are desirable and special storage and operation techniques are needed; some iterative methods are eminently suited to array processors - only local information at a node or grid point is utilized and not altered during an iterative sweep.

(2) Nonlinear Problems

The general organization of the program for stationary nonlinear problems does not differ substantially from that of the linear problem. The data structure is essentially unchanged and the major distinction now is the nonlinear solution algorithm, which must be imbedded in the computation. If a standard Newton-Raphson scheme with direct solution of the linear Jacobian system is used, then within each iteration we compute element contributions to the Jacobian, assemble, and solve as before. The parallel and vector processing opportunities may be followed in each Newton iteration. Incremental load approaches, including iteration, can be considered in like manner. Other solution schemes such as SOR-Newton may also be used. We examine some of their properties later.

(3) Transient Problems

Depending now on the choice of explicit, semi-implicit or implicit time integration, quite different solution algorithms are obtained. If the system is lumped and explicit integration used, the solution value at a node for the next time level will depend on the nodal values at the adjacent nodes and current time level. Because of the inherent sparsity of the system, this implies that there is a high degree of parallelism. On the other hand, implicit methods require system solution at each time level so that the level of parallelism is reduced. However, if certain iterative solution algorithms are introduced to solve the resulting linear or nonlinear systems, the proportion of parallelism increases.

Special Techniques

There are a few special techniques used in finite element and finite difference methods that merit particular consideration in the context of parallel and vector processors. Here we limit the treatment to substructuring and splitting strategies.

(1) Substructuring

Substructuring was introduced in the 1960's to solve large scale structural analysis problems, such as aircraft "wing-body" interaction, without heavy dependence on slow out-of-core solution algorithms. The basic idea is to consider the entire structure S as made up of a set of substructures S_i, $i = 1, . . ., n$. For example, the wing and body may be treated as separate structures in the "wing-body" problem. Treating each substructure as a separate domain, the element contributions can be calculated and assembled to the substructure stiffness $\underset{\sim}{K}_i$. Using static condensation (pre-elimination) the substructure stiffness can be reduced in size, retaining only those degrees-of-freedom of interest in the interior and on the interface with the neighboring substructures. Reduced substructures systems are then combined and the final merged system solved.

We see that, to a large extent, each substructure can be handled independently in parallel through much of the calculation. Only a few (less than ten) substructures would typically be encountered in practice. Thus, a design such as the CDC processor with 8 sophisticated vector unit could be used to great advantage. Many of the assembly and reduction calculations are vector oriented (being pre-elimination steps) and could be coded to use the vector capabilities within each of the units. The general principles to be followed are: (1) prescribe the data for each substructure; (2) assign each substructure to a vector unit; (3) element assembly and substructure reduction are vectorized; (4) final reduced total system solved using vector computations.

Remark: Substructuring in actuality corresponds to a specific form of total system partitioning with pre-elimination. Hence, we can equally view the above procedure as a special system solver independent of any physical identification with a structure and substructures. However, the physical structural problem does provide an easy means for explaining the ideas.

(2) Splitting

Splitting techniques have been more extensively utilized in finite difference computations, but they can also be applied in conjunction with finite element methods and other approximation techniques. Splitting can be defined in a general manner to reduce the number of dimensions and/or to separate individual contributions in the governing equation. [3 - 5]. In the former instance, the splitting can be carried out on the differential equation by replacing the differential operator by a sequence of one dimensional differential operators. For example, let the governing differential equation of motion have the general form

$$\frac{\partial u}{\partial t} + \frac{\partial F}{\partial x} + \frac{\partial G}{\partial y} = 0$$

We can split this equation by considering the operators L_x and L_y and advancing the solution in time first through the x-derivative terms alone and then through the y-derivative terms. This yields the sequence

$$\frac{\partial u}{\partial t} + \frac{\partial F}{\partial x} = 0 \quad , \quad \frac{\partial u}{\partial t} + \frac{\partial G}{\partial y} = 0$$

If the numerical solution of these evolution equations is performed using a predictor-corrector strategy, then in the y-split equation

$$u_{ij}^{(p)} = u_{ij} - \frac{\Delta t}{\Delta y} (G_{ij} - G_{ij-1})$$

determines a predicted value of the solution vector. In the corrector we have

$$u_{ij}^{(c)} = \frac{1}{2} \{u_{ij} + u_{ij}^{(p)} - \frac{\Delta t}{\Delta y} (G_{ij+1}^{(p)} - G_{ij}^{(p)})\}$$

In a similar manner, the splitting in the x-direction can be treated by a predictor-corrected integration. The scheme is explicit and combining operations we see that the solution at time level (n+1) is

$$u_{ij}^{(n+1)} = L_x (\frac{\Delta t}{2}) L_y (\frac{\Delta t}{2}) L_x (\frac{\Delta t}{2}) u_{ij}^{(n)}$$

If the problem is such that stability will necessitate too small a timestep, then we may use a split implicit method. Then, at each time step, the splitting produces a sequence of tridiagonal one-dimensional systems for solution. Since the split tridiagonal systems can be treated individually, we can use processors for each of the reduced problems in parallel. Within each processor the solver can be vectorized as indicated later.

In ADI methods the algebraic system is factored to a product of algebraic operators in each of the coordinate directions. Then the algebraic problem is treated as a sequence of lower-dimensional problems through the factorization. Similar strategies to those described above can be applied within each ADI sweep: let $\underset{\sim}{A}\underset{\sim}{u} = \underset{\sim}{b}$ be the linear system to be solved and now split the matrix $\underset{\sim}{A}$ to the sum of two matrices

$$\underset{\sim}{A} = \underset{\sim}{A}_x + \underset{\sim}{A}_y$$

where $\underset{\sim}{A}_x$ is the discrete operator corresponding to the x-derivatives and similarly $\underset{\sim}{A}_y$ corresponds to $L_y u$. For example, in torsion and potential flow we encounter the Laplacian $\Delta \equiv L = L_x + L_y$ where $L_x u = u_{xx}$, $L_y u = u_{yy}$. Thus, the discrete problem for a uniform mesh on a rectangle can be factored accordingly. We solve the one-dimensional systems for

$$(\alpha \underset{\sim}{A}_x + \underset{\sim}{I})\underset{\sim}{u}^{k+1/2} = \underset{\sim}{u}^k - \alpha(\underset{\sim}{A}_y \underset{\sim}{u}^k - \underset{\sim}{b})$$

and

$$(\alpha \underset{\sim}{A}_y + \underset{\sim}{I})\underset{\sim}{u}^{k+1} = \underset{\sim}{u}^{k+1/2} - \alpha(\underset{\sim}{A}_x \underset{\sim}{u}^{k+1/2} - \underset{\sim}{b})$$

where α is a relaxation parameter.

The methods can be generalized to less regular grids of tensor-product elements [6]. These systems are typically tridiagonal for many applications. In the second case, the equations need reordering to achieve the tridiagonal form. This introduces an additional constraint on the processor design if high processing speeds are to be sustained. In the next section we summarize some points related to linear system solution.

Linear System Solution

(1) Direct (Elimination) Methods

Since much of the processing time may be related to repeated system solution, it is essential that the vector capabilities be exploited fully. Ideally, to minimize "startup" time for operations with vectors, we would like to use few vectors of large size. However, this should be obtained without excessive increase in operations or data manipulation.

Let us examine a situation where splitting has reduced the problem to a sequence of many tridiagonal systems. If we elect to vectorize the tridiagonal elimination scheme, the most obvious scheme is to take the system matrix stored as N consecutive vectors by columns. Each vector contains at most $2w + 1 = 3$ non-zero entries where w is the half-bandwidth excluding the diagonal. In a simple matrix * vector product we have N vector operations of length $2w + 1$. If, instead, the system matrix is stored by diagonals, we obtain only 3 vectors and each vector is essentially full. The matrix * vector multiply requires $2w - 1 = 3$ vector operations of length $N-1$.

Crout decomposition of a symmetric global stiffness matrix can be obtained in n steps where r is the number of nodes in the structure. Let d be the number of degrees of freedom at a node so that each nodal entry is a $d \times d$ block and let β be the block half bandwidth. Each decomposition step involves one vector divide, β vector replications, $\beta+1$ vector multiplications and β vector subtractions (fewer in the final β steps). The average length of the vectors in the multiplication and subtraction operations is $\frac{1}{2}\beta d$.

Finally, the solution $\tilde{u}$ is obtained by substitution sweeps using the decomposed system $\underset{\sim}{L}^T \underset{\sim}{D}\, \underset{\sim}{L}\, \underset{\sim}{u} = \underset{\sim}{b}$. Each step of the forward sweep consists of a vector multiplication and a vector subtraction of average length $\frac{1}{2}\beta d$. The back substitution follows with vector multiplications followed by vector subtractions. A complete algorithm and further details are given in [7] for a study of the vector capabilities of the STAR 100.

(2) Iterative Solution

Large sparse systems were initially treated using iterative methods, but since large core storage became available, direct methods have been extensively used. This is particularly the case in two-dimensional finite element programs for linear structural mechanics. For very large two-dimensional problems and particularly for three-dimensional problems, iterative methods again warrant consideration [8].

Rates of convergence depend upon the spectral radius of the iteration matrix concerned and may be very slow in some problems. We can precondition these systems using approximate factorization (again with vectorization) to accelerate convergence. If reasonable bounds on the eigenvalues can be obtained inexpensively, then the relaxation factors of the iterative schemes can be optimized to produce dramatic improvements in rates of convergence. Semi-iterative or Chebyshev acceleration methods need acceleration parameters based on estimates, as does the ADI method.

The Jacobi method is perhaps the simplest linear iteration and provides an apt example of the use of vector and parallel concepts in connection with iterative methods. The Jacobi method is defined by

$$\underset{\sim}{u}^{n+1} = (\underset{\sim}{I} - \underset{\sim}{D}^{-1}\underset{\sim}{A})\underset{\sim}{u}^{n} + \underset{\sim}{k}$$

where $\underset{\sim}{k} = \underset{\sim}{D}^{-1}\underset{\sim}{b}$. The components of the right-hand side can be computed independently in parallel or as blocks in parallel. The calculation involves formation of the rows of the iteration matrix on the right, followed by a matrix × vector product and a vector addition. We can use the vector product form discussed previously for direct solvers. Thus the inefficient scalar Jacobi scheme can be easily implemented as a vectorized parallel scheme in an array processor. However, the Jacobi iteration is inherently so slow that it would require unusual circumstances to make it really practical. SOR, which is superior for scalar processors, is sequential by nature and cannot be so readily adapted to parallel processing. Conjugate gradient and accelerated conjugate gradient techniques appear well-suited to parallel processors since the right-hand sides in the iteration system can be computed simultaneously and some vectorization can be employed. Some Block methods, as noted for Block Jacobi and for substructuring earlier, are evidently well-suited to parallel processors.

In [9] several iterative methods are compared for solution of the Dirichlét problem for Laplace's equation on a unit square using the ASC single pipe machine designed by Texas Instruments. On a uniform mesh with $h = 1/80$ the total time for Jacobi iteration is half the time required using scalar calculations. All points are treated in parallel so the time per Jacobi iteration is less than that of any other method. The importance of parallelism of the data structure is evident when one compares the total time of Jacobi and Gauss-Seidel iterations on the ASC. Let the standard grid numbering scheme be used on the unit square - sequentially left-to-right in rows from bottom to top.

The Jacobi iteration on the pipeline computer takes 85% of the time required by scalar Gauss-Seidel. Although the Gauss-Seidel iteration requires only approximately half as many iterations as Jacobi, the computer time in this test is four times as large for Gauss-Seidel as for Jacobi. This inefficiency is readily identified as due to the data organization. As the most recently computed values are used in the Gauss-Seidel iteration, the arithmetic unit must wait until a grid point value is fully calculated before it can proceed with the next point. This difficulty can be circumvented by restructuring the grid data. For example, we can process the grid diagonally along the lines $x + y = \text{constant}$ (i.e. on $i + j =$ constant for mesh lines $x_i = ih$, $u_j = jh$). This strategy is advocated also for the Gauss-Seidel-Newton nonlinear iterations in the next section, as the basic arguments apply directly to nonlinear point-iterative methods.

Nonlinear Solution

As indicated earlier, straightforward Newton-Raphson iteration leads to a sparse linear Jacobian system to be solved at each iterative step. Both direct and iterative methods can be employed to solve the Jacobian system, and the problem reduces to the linear case discussed above. Thus, we have Newton-elimination and Newton-iteration techniques. For example, if SOR is used to solve the Jacobian system, the scheme is termed Newton-SOR. Alternatively, one can develop a scalar iteration directly for the nonlinear system and solve each nonlinear equation by Newton's method to produce a Jacobi-Newton or SOR-Newton method. A particular iterative method that has proven effective in nonlinear membrane calculations has been classified as belonging to a class of SOR-Newton m_k step iterations [10,11].

Let us write the nonlinear system as

$$\underset{\sim}{F}(\underset{\sim}{u}) = \underset{\sim}{0}$$

and componentwise as

$$F_i(u_1, u_2, \ldots, u_{i-1}, u_i, u_{i+1}, \ldots, u_n) = 0$$

for $i = 1, \ldots, n$.
Motivated by linear iterative schemes, let $u_j^{(k)}$ be the estimate of component u_j at iterate k. Now consider the i'th equation and solve

$$F_i(u_1^{(k)}, \ldots, u_{i-1}^{(k)}, u_i, u_{i+1}^{(k)}, \ldots u_n^{(k)}) = 0$$

which is a nonlinear scalar equation for u_i. We call this the outer (or primary) iteration. The standard Newton iteration for finding the root $u_i = \alpha$ of the scalar equation $F(u) = 0$ is a possible inner (or secondary) iteration. We write

$$u_i^{(k+1,m)} = u_i^{(k+1,m-1)} - F_i(u_i^{(k+1,m-1)}) /$$

$$\{\partial F_i(u_i^{(k+1,m-1)}) / \partial u_i^{(k+1,m-1)}\}$$

where we have chosen as starting iterate $u_i^{(k+1,0)} = u_i^{(k)}$, the current estimate of $u_i = \alpha$. Note that in this formula the dependence on u_j, $j \neq i$ has been suppressed for notational convenience.

In this Jacobi-Newton scheme, each equation of the outer (Jacobi) iteration can be assigned to a processor and inner Newton iterations computed independently. Thus the scheme is very suitable for parallel array processing. If we accelerate the scheme and use most recently calculated values, convergence rates improve but parallelism is lost. In [11] we discuss the convergence and other properties of this family of iterative methods.

The limitations on the data structure noted above for linear systems must be observed for Gauss-Seidel-Newton and SOR-Newton schemes if array or pipeline processors are to be used to advantage. These nonlinear point-iterative methods are employed in [11] for the catenary problem in one dimension and the minimal surface (membrane) problem in two dimensions. In the catenary problem beginning with an initial vector $\underset{\sim}{u}_0 = \underset{\sim}{0}$ at points in the interior, we obtain 2-figure accuracy in 11 outer iterations and 6 figures in 33 outer iterations, using an inner 1-step Newton at each point, so the rate is linear.

As shown in [11] the asymptotic rate is the same for the SOR-Newton m_k step method ($m_k > 1$) as for the SOR-Newton-1 step method. Thus a very effective form of the nonlinear point-iteration scheme is the 1-step Newton method on an array processor in the Jacobi form or with parallel data structure in the Gauss-Seidel or SOR forms. An investigation of the implementation and performance of these nonlinear point-iterative schemes on advanced computer architecture is in progress.

Work estimates give an indication of the relative potential of an array processor. Let FE be the number of function evaluations and C be the number of multiply/divide operations in the scalar Newton step. The work per inner iteration is 2FE + 2C . For k outer iterations, n equations and m_k inner iterations the work estimate W is

$$W \simeq \sum_{k=1}^{k} n(m_k(2FE + 2C))$$

In the 1-step method,

$$W \simeq 2Kn(FE + C)$$

The work estimate for an array processor having p processors is

$$W \quad \overset{k}{\underset{k=1}{}} m_k \frac{n}{p}(2FE + C)$$

and for $m_k = 1$

$$W \simeq 2k \frac{n}{p}(FE + C)$$

The iteration work estimates alone, however, are insufficient to ascertain the effectiveness of advanced scientific computer processors. Reorganization of data and the suitability of a chosen algorithm together with careful and appropriate programming are less tangible factors that must be included in the evaluation.

Concluding Remarks

The difficulties concerning design of advanced computers for scientific computing go significantly beyond the design of processors, data bases and support systems. The numerical methodology as well as algorithms are still evolving and to achieve the required processing speed there must be a strong interdependence between these and the actual processor design. Parallel array processors and vector processors with some parallelism provide a means to achieve the desired improvement in processor speed. This implies that the design must be more profoundly tied to the class of problems of interest and to associated methodology and algorithms. On the other hand, the possibility of computing with vector calculations and parallelism leads one to re-evaluate algorithms from this standpoint and this is the focus of the present study. We examine briefly the structure of finite element programs, including system solution, by direct and iterative methods, and features such as substructuring and splitting.

Acknowledgment

This study has been supported in part by NASA grant NSG-2338 and by ONR contract N00014-78-C-0550. I wish to express my appreciation to David Young and Linda Hayes for their helpful discussions.

References

1 Matisoo, J.: The Superconducting Computer, Scientific American, 242, 5, 50-65, 1980.

2 Petersen, V. L. et al.: Future Computer Requirements for Computational Aerodynamics, Workshop Proceedings, NASA Conference Publication 2032, Feb. 1978.

3 Yanenko, N. N.: The Method of Fractional Steps, (trans M. Holt), Springer, Berlin, 1971.

4 MacCormack, R. W.: An Efficient Numerical Method for Solving the Time-Dependent, Compressible, Navier-Stokes Equations at High Reynolds Number, NASA TN X-73, 1976.

5 Carey, G. F. and R. Krishnan: Navier Stokes Problems: Methods, Solution and Computation, TICOM Report 79-11, Texas Institute for Computational Mechanics, University of Texas at Austin, Austin, Texas, Sept. 1979.

6 Hayes, L. J.: Galerkin Alternating-Direction Methods for Non-Rectangular Regions Using Patch Approximations (to appear).

7 Noor, A. K. and J. J. Lambiotte, Jr.: Finite Element Dynamic Analysis on CDC Star-100 Computer, Computers and Structures, 10, 7-19, 1979.

8 Young, D.: Iterative Solution of Large Linear Systems, Academic Press, New York, 1971.

9 Hayes, L. J.: Timing Analysis of Standard Iterative Methods on a Pipeline Computer, CNA Report 136, CNA, Univ. of Texas, April 1978.

10 Perrone, N. and R. Kao: A General Nonlinear Relaxation Technique for Solving Nonlinear Problems in Mechanics, J. Appl. Mech., 38, 371-376, 1971.

11 Carey, G. F. and R. Krishnan: On a Nonlinear Iterative Method in Applied Mechanics, Part I and II, CMAME, (to appear).

Additional References

Birkhoff, G. and A. George: Elimination by Nested Dissection, Complexity of Sequential and Parallel Numerical Algorithms, J. F. Traub, ed., Academic Press, New York, 1973, pp. 221-269.

Buzbee, B. L.: A Fast Poisson Solver Amenable to Parallel Computation, IEEE Trans. Comput. C-22 (1973), pp. 793-796.

Calahan, D. A.: Complexity of Vectorized Solution of 2-Dimensional Finite Element Grids, Report 91, SEL, Univ. of Michigan, 1975.

Calahan, D. A., W. G. Ames and E. J. Sesek: A Collection of Equation-Solving Codes for the CRAY 1, SEL Report No. 133, University of Michigan, Ann Arbor, Michigan, 1979.

Csanky, L.: Fast Parallel Matrix Inversion Algorithsm, SIAM J. Comput., 5 (1976), pp. 618-623.

George, A., W. G. Poole, Jr. and R. G. Voigt: Analysis of Dissection Algorithms for Vector Computers, ICASE, Hampton, Virginia, 1976.

Graham, W. R.: The Parallel and Pipeline Computers, Datamation, 68-71, April, 1970.

Heller, D.: A Survey of Parallel Algorithms in Numerical Linear Algebra, SIAM Review, 20, 4, 740-777, 1978.

Heller, D., D. K. Stevenson and J. F. Traub: Accelerated Iterative Methods for the Solution of Tridiagonal Linear Systems on Parallel Computers, J. Assoc. Comput. Mach., 23 (1976), pp. 636-654.

Lambiotte, J. J.,Jr. and R. G. Voigt: The Solution of Tridiagonal Linear Systems on the CDC STAR-100 Computer, ACM Trans. Math. Software, 1 (1975), pp. 308-329.

Lazarus, R. B.: Computer Architecture and Very Large Problems, AFIPS Conference Proceedings, 40, 45-49, 1972.

Madsen, N. K., G. H. Rodrigue and J. I. Karush: Matrix Multiplication by Diagonals on a Vector/Parallel Processor, Information Processing Lett., 5 (1976), pp. 41-45.

Miranker, W. L.: A Survey of Parallelism in Numerical Analysis, SIAM Review, 13, 524-545, 1971.

Morice, Ph.: Calcul Parallèle et Décomposition dans la Résolution d'Équations aux Derivées Partielles de Type Elliptique, IRIA, Rocquencourt, France, 1972.

Ortega, J. M. and W. C. Rheinboldt: Iterative Solution of Nonlinear Equations in Several Variables, Academic Press, New York, 1970.

Pease, M. C.: Matrix Inversion Using Parallel Processing, J. Assoc. Comput. Mach., 14 (1967), pp. 757-764.

Poole, W. G., Jr. and R. G. Voigt: Numerical Algorithms for Parallel and Vector Computers: An Annotated Bibliography, CR, 15 (1974), pp. 379-388.

Sameh, A. H.: On Jacobi and Jacobi-like Algorithms for a Parallel Computer, Dept. of Computer Sci., Univ. of Illinois, Urbana, 1975.

Sameh, A. H., S. C. Chen and D. J. Kuck: Parallel Poisson and Biharmonic Solvers, Computing, 17 (1976), pp. 219-230.

Stone, H. S.: Parallel Tridiagonal Equation Solvers, ACM Trans. Math. Software, 1 (1975), pp. 289-307.

Traub, J. F.: Iterative Solution of Tridiagonal Systems on Parallel and Vector Computers, Complexity of Sequential and Parallel Numerical Algorithms, J. F. Traub, ed., Academic Press, New York, 1973, pp. 49-82.

Watson, W. J.: The TI ASC, a Highly Modular and Flexible Super-Computer Architecture, AFIPS Fall 1972, AFIPS Press, Montvale, New Jersy, vol. 41, pt. 1, pp. 221-229.

MIX
Papier aus verantwortungsvollen Quellen
Paper from responsible sources
FSC® C105338

If you have any concerns about our products,
you can contact us on
ProductSafety@springernature.com

In case Publisher is established outside the EU,
the EU authorized representative is:
Springer Nature Customer Service Center GmbH
Europaplatz 3, 69115 Heidelberg, Germany

Printed by Libri Plureos GmbH
in Hamburg, Germany